ANNUAL REVIEW OF ASTRONOMY AND ASTROPHYSICS

EDITORIAL COMMITTEE (1989)

JOHN N. BAHCALL
GEOFFREY BURBIDGE
PETER S. CONTI
HUGH S. HUDSON
DAVID LAYZER
JOHN G. PHILLIPS
MORTON S. ROBERTS
ALLAN SANDAGE

Responsible for the organization of Volume 27
(Editorial Committee, 1987)

JOHN N. BAHCALL
GEOFFREY BURBIDGE
PETER S. CONTI
CARL HEILES
HUGH S. HUDSON
DAVID LAYZER
JOHN G. PHILLIPS
ALLAN SANDAGE
SANDRA M. FABER (Guest)
FRANK H. SHU (Guest)
ROBERT V. WAGONER (Guest)

Production Editor   KEITH DODSON
Subject Indexer   IRENE H. OSTERBROCK

ANNUAL REVIEW OF ASTRONOMY AND ASTROPHYSICS

VOLUME 27, 1989

GEOFFREY BURBIDGE, *Editor*
University of California, San Diego

DAVID LAYZER, *Associate Editor*
Harvard College Observatory

JOHN G. PHILLIPS, *Associate Editor*
University of California, Berkeley

ANNUAL REVIEWS INC   4139 EL CAMINO WAY   P.O. BOX 10139   PALO ALTO, CALIFORNIA 94303-0897

ANNUAL REVIEWS INC.
Palo Alto, California, USA

COPYRIGHT © 1989 BY ANNUAL REVIEWS INC., PALO ALTO, CALIFORNIA, USA. ALL RIGHTS RESERVED. The appearance of the code at the bottom of the first page of an article in this serial indicates the copyright owner's consent that copies of the article may be made for personal or internal use, or for the personal or internal use of specific clients. This consent is given on the conditions, however, that the copier pay the stated per-copy fee of $2.00 per article through the Copyright Clearance Center, Inc. (21 Congress Street, Salem, MA 01970) for copying beyond that permitted by Section 107 or 108 of the US Copyright Law. The per-copy fee of $2.00 per article also applies to the copying, under the stated conditions, of articles published in any *Annual Review* serial before January 1, 1978. Individual readers, and nonprofit libraries acting for them, are permitted to make a single copy of an article without charge for use in research or teaching. This consent does not extend to other kinds of copying, such as copying for general distribution, for advertising or promotional purposes, for creating new collective works, or for resale. For such uses, written permission is required. Write to Permissions Dept., Annual Reviews Inc., 4139 El Camino Way, P.O. Box 10139, Palo Alto, CA 94303-0897 USA.

International Standard Serial Number: 0066-4146
International Standard Book Number: 0-8243-0927-8
Library of Congress Catalog Card Number: 63-8846

Annual Review and publication titles are registered trademarks of Annual Reviews Inc.

∞ The paper used in this publication meets the minimum requirements of American National Standard for Information Sciences—Permanence of Paper for Printed Library Materials, ANSI Z39.48-1984.

Annual Reviews Inc. and the Editors of its publications assume no responsibility for the statements expressed by the contributors to this *Review*.

TYPESET BY AUP TYPESETTERS (GLASGOW) LTD., SCOTLAND
PRINTED AND BOUND IN THE UNITED STATES OF AMERICA

PREFACE

This volume is dedicated to John Phillips, who has been an Associate Editor since 1966. John took up this position early in the development of the series following the resignation of the late Armin Deutsch, and for more than twenty years he has worked very effectively, originally with Leo Goldberg, and since 1973, with me. John has decided to retire, and he will be sorely missed. On behalf of all of us, I would like to thank him for all the work that he has done over the years.

I am very pleased to announce that starting with Volume 28, Allan Sandage has agreed to become an Associate Editor. Allan is well known to everyone as an outstanding scientist, as well as being a prolific contributor of articles to this series.

Volume 27 was planned at a meeting held in Palo Alto on April 25, 1987. Those who attended the meeting included Geoffrey Burbidge (Editor), David Layzer and John Phillips (Associate Editors), Keith Dodson (Production Editor), Peter Conti, Carl Heiles, Hugh Hudson, and Allan Sandage (Committee Members) and guests S. Faber, F. Shu, and R. Wagoner. In the preface to Volume 26 I pointed out that 27 articles were scheduled for this volume. In fact 17 articles are contained here, so that our default rate of about 40% is, unfortunately, holding up. For Volume 28, 32 articles are currently scheduled.

As Editors we try to make the *Annual Review of Astronomy and Astrophysics* representative of worldwide development in the field. How well are we succeeding? Since the beginning, a total of 27 volumes containing 423 articles (excluding prefatory articles) have appeared, with 114 of the articles (27%) written by authors residing outside North America. In the current volume, nearly 40% of the articles come from outside North America. We leave it to our readers to decide whether or not we are providing a fair representation as far as overall research in astrophysics is concerned.

As is probably well known, all readers are encouraged to write to me offering suggestions for articles and the names of potential authors. The final decision as to which topics and authors are chosen is made by the Editor with the assistance of the Associate Editors and the Editorial Committee. Factors that enter into these choices include the timeliness of the subject matter, the proposed reviewer's knowledge of the field, and our judgment (always fallible) as to who can (and will) write a good article and the likelihood that the article will be completed on time. The reviews are not refereed.

THE EDITOR

ERRATUM

Volume 26 (1988)

In "A Morphological Life" by W. W. Morgan, the first two sentences of the third paragraph on p. 8 should read as follows:

The "natural groups" in my spectral classification are not limited to the MK system. Some deal with spectra of very low dispersion, where only a few spectral features may be visible.

SOME RELATED ARTICLES IN OTHER *ANNUAL REVIEWS*

From the *Annual Review of Earth and Planetary Sciences*, Volume 17 (1989):

Achondrites and Igneous Processes on Asteroids, Harry Y. McSween, Jr.

From the *Annual Review of Nuclear and Particle Science*, Volume 39 (1989):

Highest Energy Cosmic Rays, J. Wdowczyk and A. W. Wolfendale

From the *Annual Review of Physical Chemistry*, Volume 40 (1989):

Spectroscopy of the Diatomic 3d Transition Metal Oxides, A. J. Merer

Annual Review of Astronomy and Astrophysics
Volume 27, 1989

CONTENTS

DREAMS, STARS, AND ELECTRONS, *Lyman Spitzer, Jr.*	1
THE STATUS AND PROSPECTS FOR GROUND-BASED OBSERVATORY SITES, *R. H. Garstang*	19
THE ORION MOLECULAR CLOUD AND STAR-FORMING REGION, *Reinhard Genzel and Jürgen Stutzki*	41
X RAYS FROM NORMAL GALAXIES, *G. Fabbiano*	87
POPULATIONS IN LOCAL GROUP GALAXIES, *Paul Hodge*	139
A NEW COMPONENT OF THE INTERSTELLAR MATTER: Small Grains and Large Aromatic Molecules, *J. L. Puget and A. Léger*	161
INTERACTION BETWEEN THE SOLAR WIND AND THE INTERSTELLAR MEDIUM, *Thomas E. Holzer*	199
SURFACE PHOTOMETRY AND THE STRUCTURE OF ELLIPTICAL GALAXIES, *John Kormendy and S. Djorgovski*	235
ABUNDANCE RATIOS AS A FUNCTION OF METALLICITY, *J. Craig Wheeler, Christopher Sneden, and James W. Truran, Jr.*	279
T TAURI STARS: WILD AS DUST, *Claude Bertout*	351
ASTROPHYSICAL CONTRIBUTIONS OF THE INTERNATIONAL ULTRAVIOLET EXPLORER, *Yoji Kondo, Albert Boggess, and Stephen P. Maran*	397
CLASSIFICATION OF SOLAR FLARES, *T. Bai and P. A. Sturrock*	421
DIFFUSE GALACTIC GAMMA-RAY EMISSION, *Hans Bloemen*	469
QUASI-PERIODIC OSCILLATIONS AND NOISE IN LOW-MASS X-RAY BINARIES, *M. van der Klis*	517
KINEMATICS, CHEMISTRY, AND STRUCTURE OF THE GALAXY, *Gerard Gilmore, Rosemary F. G. Wyse, and Konrad Kuijken*	555
SUPERNOVA 1987A, *W. David Arnett, John N. Bahcall, Robert P. Kirshner, and Stanford E. Woosley*	629
CHEMICAL ANALYSES OF COOL STARS, *Bengt Gustafsson*	701
INDEXES	
Subject Index	757
Cumulative Index of Contributing Authors, Volumes 17–27	766
Cumulative Index of Chapter Titles, Volumes 17–27	768

ANNUAL REVIEWS INC. is a nonprofit scientific publisher established to promote the advancement of the sciences. Beginning in 1932 with the *Annual Review of Biochemistry*, the Company has pursued as its principal function the publication of high quality, reasonably priced *Annual Review* volumes. The volumes are organized by Editors and Editorial Committees who invite qualified authors to contribute critical articles reviewing significant developments within each major discipline. The Editor-in-Chief invites those interested in serving as future Editorial Committee members to communicate directly with him. Annual Reviews Inc. is administered by a Board of Directors, whose members serve without compensation.

1989 Board of Directors, Annual Reviews Inc.

Dr. J. Murray Luck, Founder and Director Emeritus of Annual Reviews Inc.
  Professor Emeritus of Chemistry, Stanford University
Dr. Joshua Lederberg, President of Annual Reviews Inc.
  President, The Rockefeller University
Dr. James E. Howell, Vice President of Annual Reviews Inc.
  Professor of Economics, Stanford University
Dr. Winslow R. Briggs, *Director, Carnegie Institution of Washington, Stanford*
Dr. Sidney D. Drell, *Deputy Director, Stanford Linear Accelerator Center*
Dr. Sandra M. Faber, *Professor of Astronomy, University of California, Santa Cruz*
Dr. Eugene Garfield, *President, Institute for Scientific Information*
Mr. William Kaufmann, *President, William Kaufmann, Inc.*
Dr. D. E. Koshland, Jr., *Professor of Biochemistry, University of California, Berkeley*
Dr. Gardner Lindzey, *Director, Center for Advanced Study in the Behavioral Sciences, Stanford*
Dr. William F. Miller, *President, SRI International*
Dr. Charles Yanofsky, *Professor of Biological Sciences, Stanford University*
Dr. Richard N. Zare, *Professor of Physical Chemistry, Stanford University*
Dr. Harriet A. Zuckerman, *Professor of Sociology, Columbia University*

Management of Annual Reviews Inc.

John S. McNeil, Publisher and Secretary-Treasurer
William Kaufmann, Editor-in-Chief
Mickey G. Hamilton, Promotion Manager
Donald S. Svedeman, Business Manager
Ann B. McGuire, Production Manager

ANNUAL REVIEWS OF
Anthropology
Astronomy and Astrophysics
Biochemistry
Biophysics and Biophysical Chemistry
Cell Biology
Computer Science
Earth and Planetary Sciences
Ecology and Systematics
Energy
Entomology
Fluid Mechanics
Genetics
Immunology
Materials Science
Medicine
Microbiology
Neuroscience
Nuclear and Particle Science
Nutrition
Pharmacology and Toxicology
Physical Chemistry
Physiology
Phytopathology
Plant Physiology and
  Plant Molecular Biology
Psychology
Public Health
Sociology

SPECIAL PUBLICATIONS
Excitement and Fascination
  of Science, Vols. 1, 2, and 3

Intelligence and Affectivity,
  by Jean Piaget

A detachable order form/envelope is bound into the back of this volume.

DREAMS, STARS, AND ELECTRONS

Lyman Spitzer, Jr.

Princeton University Observatory, Peyton Hall, Princeton, New Jersey 08544

"What were you doing today, Daddy?" my children often asked me. I would reply, "I was thinking what I would do if I were an electron." Much of my professional work has been based on an ability to visualize a physical system and how it operates and thus to predict its behavior without benefit of mathematics—though mathematics provides a much needed check. This particular facility has led me to work on a variety of physical processes important in astrophysics.

Another characteristic that has greatly affected my research is an attraction to spectacular and difficult projects that strike me as important. Three major such projects, which provided long-term goals for my career, concerned the theory of star formation, a large general-purpose space telescope, and generation of useful power from fusion energy. The objective of the first was to understand the sequence of complex processes occurring; those of the last two involved engineering as well as scientific problems.

This paper deals with my various research activities, especially those that now seem relevant to further work in astronomy. The personal characteristics described above are mentioned because they provide the unifying strands in what might otherwise seem a scattered, somewhat miscellaneous professional career. Some personal items relevant to my scientific life are included also in the subsequent text. In the words of Pooh-Bah (1), perhaps these will give verisimilitude to an otherwise bald and unconvincing narrative.

Early Years

As a boy I enjoyed much of my school work, but a serious interest in science developed only after entering Phillips Academy, Andover, in 1929, at age 15, for two precollege years. There I was introduced to physics by Frederick Boyce, who had a unique capability for making this topic both clear and exciting. It was a wonderful experience to find that by understanding a few simple principles one could master an entire area of physics. In response to a request from several of us, Mr. Boyce gave an astronomy course, supplementing the books by Jeans and Eddington that I had been reading and leaving me permanently fascinated with astrophysics.

At the end of my stay at Andover I became involved in my mind with the first of the big projects that have appealed to me. This concerned the development of a global transportation system, based on the electromagnetic propulsion of levitated cars in evacuated tubes. Others have suggested such ideas. My concept was to integrate intercity travel with local transportation, so that one could get into a car on a high floor of an office building, dial the local or long distance number desired, and get out some minutes later at the desired floor in one's own city or in some remote center. I spent so much time considering various details of such a system that my parents, seriously concerned for my sanity, extracted a promise from me to stop thinking about this enterprise!

From Andover I went on to Yale, where astronomy (as well as the global transportation system) was put aside while majoring in physics. Under Leigh Page I took a general survey course in theoretical physics and found a keen aesthetic pleasure from his elegantly organized mathematical presentations. A limitation of my own in such areas became painfully evident during my final year at Yale, when I took a final oral examination on a physics thesis I had written for some independent work. Said one of my examiners, "Your second equation, Mr. Spitzer, states that the power P equals ir^2. Could you derive this for us, please?" Swiftly going through the elementary steps, I suddenly realized that I had inadvertently written ir^2 instead of the familiar i^2r, and that all the subsequent equations in my thesis were incorrect! My mind tends to make inversions of this sort, and while I have learned to ferret out most of them, they still dog my work from time to time.

In graduate school, first at Cambridge, England, (as a Henry Fellow) in 1935–36 and then at Princeton, I returned to astrophysics, partly because I was so enchanted by the beautiful theories that S. Chandrasekhar had presented at a series of informal evening seminars in his Trinity College rooms, and partly because I was so impressed by the physical lucidity of several contemporary papers by Bengt Strömgren. Studying under Henry Norris Russell was also a unique experience; his enormous knowledge, physical insight, and unfailing enthusiasm made him a very stimulating mentor. The lectures by Edward Condon were particularly helpful in extending my own insight into various aspects of the quantum theory.

My thesis topic, suggested by Russell, was an analysis of high-dispersion spectra that Walter Adams had obtained for three cool supergiant stars. I proposed a fountain model to explain the observed dependence of radial velocity on excitation potential. While later work by others disproved this model, the research was educational for me.

During my postdoctoral year at Harvard (as a National Research Council Fellow) in 1938–39, I found tremendously stimulating the scientific discussions among the active scientists whom Harlow Shapley, Donald Menzel, and Bart Bok had assembled at the Observatory. In particular, the free-ranging exchange of views at Shapley's "hollow square" discussion groups and at the evening seminars organized by Bok at his home gave all of us many exciting, important problems to work on. As pointed out below, my subsequent research in stellar dynamics and in star formation and interstellar matter grew out of these discussions.

In this active ferment of astronomical ideas Martin Schwarzschild was an imposing figure. He greatly impressed me by his insight, originality, and forcefulness. I felt strongly that he would be an ideal colleague if events should make it possible for us to be in the same institution.

Yale and Princeton

A brief discussion of my years at these two institutions will provide a backdrop for a discussion of my research work there. At Yale, where I spent two and a half years before shifting to war research in New York and Washington, and another year and a half afterward, I was alone much of the time as far as astrophysics was concerned. This did not worry me, since the impetus I had received at Princeton and Harvard would suffice for some years, and informal get-togethers of astronomers in the region from Washington to Cambridge were frequent.

My initial appointment in 1939 was as an instructor in the physics department, which gave me a welcome chance to broaden my background in this basic field, as well as to continue some research I had begun at Harvard. Under the influence of Henry Margenau I published a few papers on the collisional broadening of spectral lines, giving relatively exact solutions in certain idealized cases. When I returned to Yale after four years of underwater sound research, I had a joint appointment in physics and astronomy and concentrated my research on problems related to interstellar matter.

In 1947 I was offered a professorship at Princeton and also the directorship of the Observatory, succeeding Russell. I soon accepted; ever since my graduate student days I had felt this would be an ideal position. For this offer I am greatly indebted to three of this country's greatest astronomers—to Russell for his support, to Chandrasekhar for not accepting the earlier offer to him of Russell's professorship, and to Shapley for informing Princeton of my long-standing interest in this particular post and then serving as a helpful intermediary between me and the Princeton administration. In addition to his widespread other activities, Shapley ran sort of an informal employment bureau for young astronomers. Following the tradition established by Russell, I held the directorship (together with the departmental chairmanship) for some three decades; my brilliant colleague Jerry Ostriker succeeded me in 1979, three years before my retirement from the faculty (compulsory then at age 68).

One of my most important achievements at Princeton was to persuade Martin Schwarzschild to join me there. We worked closely and effectively together, his wisdom in practical affairs supplementing my willingness to get absorbed occasionally with administrative details. Our joint objective, when we both moved to Princeton in 1947, was to build a significant graduate program in theoretical astrophysics, with some emphasis also on observations. In support of this latter purpose the Princeton administration agreed to send Schwarzschild and me to the western US (in practice, usually the Mt. Wilson Observatory) for observing, each of us in alternate years for one semester. This arrangement certainly encouraged observational work by the two of us and our students. After 15 years, the shift of our observational research plans to space telescopes forced us to drop, with great reluctance, these regular observing trips to the West Coast.

Interstellar Matter

On arriving at Harvard in 1938 I found widespread discussion of Bethe's now classic explanation of stellar energy generation (half then in published form, half in press), showing that nuclear fusion provided the basic energy source for stars. This conclusion had far-reaching consequences for the age and evolutionary history of stars. From active discussions during that year a simple picture of stars and galaxies began to take shape in my mind.

A basic element of this picture as I then envisioned it was the obvious fact that the highly luminous O and B stars, which are seen in spiral galaxies, must exhaust their hydrogen fuel in much less time than the age of the Universe. That such stars are not seen in elliptical galaxies or in the globular clusters of our Galaxy suggests that these systems are composed of old or primordial stars, formed early in the life of the Universe. It seemed reasonable to attribute the formation of the young stars in spiral galaxies to the interstellar gas, which is a characteristic feature of spiral systems; there were very few traces of gas and dust in elliptical galaxies or globular clusters. Moreover, the absence of pervasive gas in these latter systems is theoretically not unexpected, since the random velocities of atoms are limited by thermal processes and tend to be much less than the random velocities required to support the stars in the observed elliptical or spherical systems.

This broad though tentative picture, which identified the interstellar clouds in spiral galaxies as the birthplace of massive early-type stars, strongly appealed to me and led me to embark on a program of interstellar matter studies, with the goal of understanding how new stars are formed. The essential first step in this program was to determine the physical conditions in the interstellar gas and dust observed in our Galaxy. Hence during my first year at Yale, I computed such items as the mean free path of electrons, positive ions, neutral atoms, and dust grains in interstellar space; I showed that these grains would tend to be negatively charged because of the higher random velocities of electrons as compared with positive ions. An important inference of these early studies was that under normal conditions, the grains would not drift much through the gas. Another conclusion was that different regions of the gas would normally tend toward pressure equilibrium with each other.

In a draft of a paper presenting these detailed results I included an exposition of the broad picture that is described above and that provided the motivation for the work. Several older astronomers advised me not to publish this speculative discussion, which one of them described as "metaphysical." I followed their advice, though with some reluctance. Over subsequent years this picture has become generally accepted, thanks in large part to the detailed observations of elliptical and spiral systems by Walter Baade (2) and others. Quite possibly my published results (3) on the kinetic properties of the interstellar gas would not have received serious attention if they had appeared as part of a grandiose and controversial package.

On returning to Yale after the war, I resumed the analysis of physical conditions in the interstellar gas. In particular, I began a lengthy calculation of the kinetic temperature of this gas, based on the assumption of radiative equilibrium (i.e. that this temperature adjusts itself so that the gains of kinetic energy, per unit of time and volume, just equal the corresponding losses by radiation). For planetary nebulae many of the relevant reactions had been analyzed earlier by Menzel and his collaborators, but for much of the interstellar gas additional physical processes are important; the density, radiation flux, and state of ionization can be quite different in these two environments. Strömgren had pointed out (4) several years earlier that interstellar H atoms in regions of average density or higher must normally be almost entirely neutral or almost entirely ionized; these two types of regions are now generally designated (3) as H I and H II, respectively.

The actual calculations of temperature were made (5) for somewhat idealized atoms, but they sufficed to show a large difference of temperature between H I and H II regions. In H II zones, a strong heating source is provided by the photoionization of neutral H atoms, at a rate set by the number of proton-electron recombinations per unit time and volume. The electrons ejected in the photoionization process carry off kinetic energy, which goes into heating the gas, maintaining a temperature of some 10^4 K despite rapid radiation by collisionally excited heavy ions such as C^+ and O^+. In H I regions the ionizing photons are absent as a result of absorption by the neutral H atoms. Hence this energy source is missing; other such sources are weaker by several orders of magnitude. However, the radiation rates at a given density and temperature are not so very different between the two regions. As a result, the H I regions are necessarily much colder, with T in the general neighborhood of 100 K. In pressure equilibrium the H I region will have a particle density roughly 100 times greater than that of an H II zone, with its temperature of about 10^4 K.

More recent calculations (6, 7) are, of course, much more detailed, taking into account the individual energy levels of the abundant atoms and allowing for important additional processes, such as radiative transfer of ionizing photons and enhanced photoelectric efficiency of dust grains for short wavelengths and small grain radius. For H I regions of low density ($n_H < 1$ cm^{-3}), these more realistic models give significantly higher temperatures than were obtained earlier. The important conditions for possible instability of all these equilibrium temperatures were thoroughly explored by George Field (8), our former graduate student and colleague.

Another physical quantity of some importance in interstellar space is the electrical conductivity, and I resolved to compute this more accurately for fully ionized gas than had been done previously. This work, which spread over several years, was part of my effort in plasma physics and is discussed in the last section below.

At about this time a new observational result was obtained in collaboration with Paul Routly, based on observations made at the Mt. Wilson Observatory over several years. High-resolution spectra of some 20 O and B stars were obtained (9), showing components of the interstellar Na D lines, which could be compared with similar data on the K and H lines of Ca II previously obtained by Adams (10). Quantitative analysis of these data indicated (9) that the ratio of neutral Na to Ca^+ along lines of sight to various clouds varied strongly with cloud velocity; this ratio decreased by about an order of magnitude when the velocity (measured in the local standard of rest) increased from less than 10 to more than 20 km s^{-1}. A later survey, including some 60 stars, gave very similar results (11). These observations may be understood (12) if one assumes that the Ca atoms tend preferentially to be "locked up" in the dust grains (an effect termed "depletion"), but that in high-velocity clouds the grains have evaporated by some mechanism during the process of cloud acceleration, producing a Na-Ca ratio more in accord with cosmic abundance. This assumption was subsequently confirmed by measures of other elements, especially Si and Fe, with the *Copernicus* satellite (see below); these data also indicated that depletion tends to disappear at the higher cloud velocities (13).

A brief research effort combining observations with quantum mechanical theory grew out of a luncheon discussion with Jesse Greenstein, who pointed out the existence of a blue continuous emission in the spectra of planetary nebulae. The upshot was a joint paper (14) that computed for the first time the precise transition probability and the spectral distribution of two-photon emission by hydrogen atoms in the metastable 2s state; our paper used these theoretical results in a preliminary interpretation of the limited astronomical data.

Two physical effects that may be important for interstellar clouds were discussed in joint papers with visiting astronomers at Princeton. The acceleration of clouds by O-type stars was discussed in a paper with Jan Oort (15), where we pointed out that on the side of the cloud facing the star the ionized, heated gas would stream back toward the star, accelerating the cloud by a rocket effect. The detailed theory agrees (16) with the observed brightness distribution of the luminous rims between dark interstellar clouds and H II regions. This process of cloud acceleration may account for an appreciable fraction of the kinetic energies in the observed H I diffuse clouds.

A paper with Leon Mestel (17) discussed how cloud collapse would be affected by the presence of a magnetic field. An important process in this context is the systematic motion of the plasma, including positive ions, electrons, and magnetic lines of force, through the gas of neutral H and He atoms. This process is known in the laboratory as "ambipolar diffusion" (drift of positive and negative ions at the same rate through neutrals) and is sometimes called "plasma drift." Especially in view of the very large cross section (10^{-14} cm^2) for encounters between positive ions and H atoms at low temperature (18), this mechanism is most likely to be important in the later stage of cloud collapse, when the density and opacity are relatively high and the fractional ionization is consequently very low.

My emphasis on the tendency toward pressure equilibrium (12) led to a speculative suggestion (19) concerning the interstellar gas far from the galactic plane. Observations of the interstellar K line in stars at different distance z from the galactic plane indicated (20) that for some of the clouds absorbing this line, z must be at least 1000 pc. To prevent such clouds from rapidly expanding and disappearing, the pressure of a surrounding medium seems required, and if such a medium consists of hydrogen and extends up to $z \approx 1000$ pc its temperature must be of order 10^6 K. I called this a "coronal gas." While subsequent ultraviolet observations from the *Copernicus* and *IUE* satellites, and observations of soft X-rays also, have confirmed the existence of a hot gas within the galactic disk, whether such a gas extends up to high z is still not quite certain (21, 22).

Over the years I took an active interest in the polarization of starlight by magnetically aligned grains and collaborated with other scientists in a number of analyses. Perhaps the most noteworthy of the papers resulting was one with R. V. Jones, a solid-state physicist at Harvard; we pointed out (23) that alignment by dissipative processes, apparently the most effective mechanism (24), would be much enhanced if the Fe atoms in grains were assumed to be distributed in clumps, leading to ferromagnetic or superparamagnetic grains with a much greater energy dissipation rate for rotation in a magnetic field. The increased rotation rate produced by "spin-up" (25) also increases the rate of rotational energy dissipation and may make grain alignment possible even with conventional values for grain paramagnetism. On the other hand, Fe atoms must be present in grains, and their aggregating in clumps seems not unlikely.

One may ask what happened to the initial objective of this interstellar research. How was this knowledge used in theories of star formation? During a dozen or so years after I moved to Princeton I wrote a number of papers, applying to star formation theories the results that I had obtained on the properties of the interstellar gas. At the time I found some of these results rather exciting, but in retrospect they seem to be rather small steps in a long journey. To analyze the condensation of turbulent gas in a rotating galaxy, in the presence of such complicating factors as magnetic fields, cosmic rays, and a full spectrum of electromagnetic radiation, is a very complex undertaking. My own preliminary efforts in this direction (26) led to interesting idealized models and have been of some use as stepping stones for later researchers.

In my opinion there have been two special consequences from my research on these interstellar problems. First was a book—in two successive versions (7)—on this topic that summarized our knowledge of the relevant physical principles. My own research concerned only a fraction of the material presented, but writing the book provided excellent training for me in many subjects. While the text is selective rather than exhaustive, it seems to provide a useful introduction to the physical laws and mechanisms operating in the interstellar gas.

A second consequence was the interstellar research with the *Copernicus* satellite. This program, which led to a substantial jump in our understanding of the interstellar gas, is discussed below in connection with my space astronomy activities.

Stellar Dynamics

My involvement in the dynamics of star systems began one evening in the fall of 1938 in the home of Bart and Priscilla Bok. The occasion was an informal discussion of various research problems, including in particular the equilibrium of a globular cluster. In such clusters random encounters between stars tend to produce a Maxwellian distribution of velocities in a time much shorter than the cluster age. Thus, one would expect these clusters to approach isothermal spheres. But theory shows that such a sphere must have an infinite mass, which makes it a difficult model for an actual cluster to approach. What happens?

After pondering this problem for a while I finally came up with the correct answer, which now seems rather obvious. As random encounters tend to establish a Maxwellian velocity distribution, stars in the tail of this distribution will acquire a velocity greater than the escape velocity (which on the average is only twice the rms random velocity) and promptly leave the cluster. I called this process "evaporation" and sent off a paper (27) discussing an approximate theory of this effect. Not until several years later did I discover that the great Russian astrophysicist V. A. Ambartsumian had published a similar theory (28). At least our results agreed! These theories were based on the totally unrealistic but tractable model of a uniform cluster. To obtain a better approximation would have involved mathematical complications, and I left this more general problem untouched for some 30 years.

A decade later, Martin Schwarzschild pointed out to me the observed increase of velocity dispersion with average age for stars in the galactic disk. It occurred to us that random encounters between stars and massive interstellar clouds might explain this result. My work on the electrical conductivity of a fully ionized gas, which was under way at that time, gave me a good training in the theory of encounters between particles subject to mutual inverse-square forces. Our analysis of encounters between stars and clouds suggested (29) that these could account for the observed stellar velocities if cloud masses were as great as 10^6 $M_\odot$. Detailed numerical simulations (30) have given recent support to this view. The mass required seems high for clouds in the solar neighborhood (31), but extended cloud complexes, such as spiral arms, can under some conditions produce the same acceleration of stars (29) without any systematic velocity of their own other than differential galactic rotation.

A few years later Oort raised informally the question of why so few open clusters could be found with ages exceeding some 10^8 yr. I considered this problem and computed (32) the rate at which a cluster in the galactic disk could be disrupted by encounters with passing clouds, whose transient tidal force must tend to increase the kinetic energy of the cluster stars. To give a simple result I introduced the "impulsive approximation," in which the random stellar motions during the passage of the cloud were ignored. A separate, more exact computation for an idealized cluster in a parabolic gravitational potential made it possible to estimate the error of the impulsive approximation. The results of all this analysis showed that the "large clouds" in the solar neighborhood, evident from optical extinction data and subsequently identified (31) as "giant molecular clouds," would provide the greatest effect and could give maximum ages consistent with those observed for some open clusters. Since that time, more complete analyses have been made (33); the resultant change in predicted disruption times is comparable with the uncertainty resulting from incomplete information on giant molecular clouds.

The discovery of quasars with their very high luminosities led me to consider the evolution of galactic nuclei. Research in collaboration with W. C. Saslaw led to the development of an idealized model (34) that included evaporation of stars from the nucleus (as from a globular cluster) and also direct collision of stars with each other; analysis of this latter process was made possible by a number of drastic physical simplifications. The model calculations showed that the gas released in collisions falls to the center, releasing energy and forming new stars. While the computed maximum luminosity is some 10^{45} erg s^{-1}, comparable with that of quasars, it is not clear whether nuclei have actually reached the compactness required for this process.

Because of my work on this topic, I was invited to attend a small scientific conference on galactic nuclei, held in 1970 by the Pontifical Academy of Sciences. In preparation for this meeting I devoted some further thought to the evolution of spherical star systems. Analysis of equilibrium in a system containing a small fraction of its mass in stars appreciably heavier than the average led to the discovery of the mass stratification instability (35). The heavier stars will lose energy by equipartition with the lighter ones and will sink toward the center of the system. If the fraction of the cluster mass in the heavier stars exceeds a certain small value (about 3×10^{-2} if each heavy star has a mass three times that of a light star), no equilibrium is possible when the heavy stars concentrate toward the center; instead, according to the virial theorem, the self-attraction of these stars will lead to an increase rather than a decrease of their rms random velocity, and the tendency toward equipartition will lead to a continuing contraction of the heavy-star subsystem.

Since the theory leading to this instability was somewhat idealized, I resolved to study the actual evolution of a spherical system by numerical simulation, taking advantage of the fast electronic computers available at

Princeton University. A Monte Carlo method suitable for this purpose, which had been designed by Michel Hénon (36), computed collisional perturbations of orbital elements for each of 1000 representative stars. It seemed to me that a method making more direct use of the velocity diffusion coefficients, which I had already employed in the plasma conductivity research, would have certain advantages. In collaboration with six Princeton students in succession (M. H. Hart, S. L. Shapiro, T. X. Thuan, R. A. Chevalier, J. M. Shull, and R. D. Mathieu), such a Monte Carlo method was designed and gradually extended to explore a wide variety of problems. The emphasis was on globular clusters, since these are known to have relaxation times shorter than their ages.

The results obtained in this research, together with important work by many other theorists, have been summarized in my recent book (37). In addition to verifying the presence of evaporation and of the mass-stratification instability, and computing the relevant rates, the detailed dynamical study of cluster evolution has indicated the dominant importance of the gravothermal instability (38, 39) in leading to collapse of a cluster core. While the nature of postcollapse evolution is still being explored, the evolution until core collapse now seems rather well understood, and in this sense the cluster research begun in 1938 has now reached a satisfying conclusion.

Space Astronomy

As World War II drew near its end, I was approached by a friend on the staff of the RAND Project, an Air Force "think tank." He told me that his group was carrying out a secret study of a possible large artificial satellite, to circle the Earth a few hundred miles up. "Would you be interested," he asked me, "in writing a chapter on how such a satellite might be useful in astronomy?" With my long and ardent background in science fiction, I found this invitation an exciting one and accepted with great enthusiasm. I spent some time analyzing a number of possible research programs for telescopes of different sizes above the atmosphere and prepared a brief description of these.

While no orbiting astronomical telescope resulted from the RAND study, my work on space astronomy convinced me of the paramount scientific importance that a large general-purpose optical telescope in orbit would have. The wartime success of the large German V-2 rockets made the launching of large instruments into Earth orbit seem not too many years in the future. In my thinking, personal association with the development and operation of such a large orbital telescope gradually became a major professional goal. In 1947, when I was urging Martin Schwarzschild to resign from Columbia University and to join me at Princeton, he asked me how likely I was to stay there. I replied that a leadership position with a large space telescope project was one of the very few attractions that might be able to lure me away from Princeton.

It was several years before my fascination with space telescopes led to any actual research in this field. While still at Yale I tried to organize a small program in solar ultraviolet spectroscopy. My attempts to persuade Leo Goldberg to join me in such an effort turned out to be a helpful contribution to astronomy, since the offer to him of a Yale professorship may have been a factor in his promotion to the directorship at Michigan. In the early 1950s Martin's research in solar convection led him into balloon astronomy, thanks to an informal suggestion from James Van Allen (see below). High-resolution imagery above most of the atmosphere did not require orbital altitudes; at about 80,000 ft atmospheric refraction produces very little widening of a stellar image. Accordingly, Project Stratoscope was begun; a 12-inch balloon-launched solar telescope obtained beautifully sharp images in 1957 and 1959, and a 36-inch achieved similar success with planets and galaxies some dozen years later. Since I was convinced that Stratoscope was an important step toward the large space telescope, I encouraged Martin in this effective program by all the pathways that I could think of; in particular, I helped to muster financial support initially and suggested various effects that the unfamiliar environment might conceivably have on the various items of equipment.

The launch of the USSR sputniks, starting in 1957, gave tremendous impetus to the nascent US space program; the launch into orbit of an intermediate-size telescope began to appear possible. An agency of the Air Force asked me if Princeton might wish to carry out some preliminary work on the problems and possibilities of such an instrument. I gave an affirmative answer, of course. A contract was arranged, and a small group of us explored scientific possibilities for interstellar matter research at wavelengths between 1000 and 3000 Å (40) and also possible engineering components for an astronomical satellite (41). In 1959–60 the newly formed NASA (National Aeronautics and Space Administration) gradually took over the support of this program; after the usual combination of technical difficulties, funding shortages, and spacecraft delays, the *Copernicus* satellite was launched into orbit. For me, perhaps the high point of my career was that day in August 1972 when our ultraviolet spectrometer was turned on from the Goddard Space Flight Center, several days after the launch, and was found to be operating properly. I am sure that similar emotions were experienced by the outstanding group of astronomers who shared this historic enterprise with me—Jerry Drake, Kurt Dressler, Ed Jenkins, Don Morton, Don York, and especially Jack Rogerson, who was with this program almost from the start and who became Co-

Principal Investigator; his keen physical insight, his unfailing accuracy, and his dedication to this work were key ingredients in the program's success.

As expected, the combination of high spectral resolution, accurate spectrophotometry, and sensitivity in the far ultraviolet opened up a new chapter in the study of interstellar gas (42, 43). Interstellar H_2 molecules were observed in the spectra of most early-type stars, confirming the ubiquity of this interstellar component and giving information on the kinetic temperature and density within interstellar clouds. The ratio of D to H was measured, indicating (through the theory of element formation during the Big Bang) a material density too small to close the Universe. Observed absorption features of highly ionized atoms such as O^{+5} showed the presence of a hot interstellar gas with T of order roughly 10^6 K in the galactic disk. The relative abundances of many different elements were measured accurately, indicating how the depletion (presumably resulting from condensation on grains) varies from element to element and from cloud to cloud. In addition to the Princeton group, some 170 astronomers in many different countries made effective use of the *Copernicus* telescope-spectrometer for a wide variety of programs. The instrument operated effectively for almost nine years.

The general progress of the US space program after 1957 aroused interest in a possible general-purpose telescope, with a mirror in the 3-m class, providing a unique capability for diffraction-limited imagery and for observations in the ultraviolet and infrared. Because of its high cost and its dedication primarily to use by guest observers, this great instrument required widespread support by astronomers before it could possibly become a reality. A group of us formed a committee (44) under the auspices of the National Academy of Sciences and organized a series of small discussion meetings on various scientific programs for which the Space Telescope (ST) might be particularly helpful. Astronomers unfamiliar with space research were invited, and their participation helped to persuade them that ST would be a very powerful tool. At the beginning there was widespread doubt as to the practicability of such ideas. One astronomer with whom I was discussing a projected orbital telescope remarked to me, "Lyman, you're young. You'll live to see it fail." After the scientific successes obtained with other, though smaller, instruments, these arguments faded away, and by the 1970s most astronomers were enthusiastic supporters of ST.

Almost as difficult as getting support from astronomers was to get Congressional votes for financial support of ST. Fortunately, success with the former group was very helpful with the latter. John Bahcall and I collaborated in urging astronomers to express their support of ST by contacts with Congress as well as with NASA (44). After several years of such efforts, full Congressional approval for this enterprise was finally obtained in 1977. Since then, the program has moved forward on a broad front, despite several delays for technical reasons—especially the 32-month hiatus in space shuttle flights because of the *Challenger* disaster. The launch of the Hubble Space Telescope (HST), as this great observatory has been appropriately named, is now expected in late 1989 or in 1990.

For a variety of reasons my own personal involvement with HST since the program was given a go-ahead by Congress has been very much less than I had hoped for some 40 years ago. However, as an advisor to the Marshall Space Flight Center and as chairman of the AURA (Association of Universities for Research in Astronomy) oversight committee for the Space Telescope Science Institute, I have kept in reasonable touch with what is going on, and it is good to see the program approaching its goals.

I look forward to participating with Bob O'Dell in the analysis of some HST data, especially spectrophotometric observations of interstellar absorption lines. As compared with *Copernicus*, the high-resolution spectrograph on HST has much greater spectral resolution (10^5 as compared with about 2×10^4) and an enormously more rapid photon-gathering capability (up by some two orders of magnitude for interstellar line observations). With these capabilities one can explore the properties of separate clouds, if these have different radial velocities, and can measure in each the relative abundances of different atomic species. Thus we can hope to explore how cloud properties depend both on velocity and on distance from the galactic plane.

Plasma Physics

An important topic that forms part of both plasma physics and stellar dynamics is the set of effects produced by two-body random encounters between particles whose mutual force varies as the inverse square of their separation. These encounters modify the velocity distribution functions for the different types of particles present, tending to make these functions approach the Maxwellian form and to establish equipartition of kinetic energy between particles of different masses.

I became exposed to this type of process in connection with my work on the evaporation of stars from globular clusters (27). To determine the rate at which two groups of stars approach equipartition of energy with each other, I first computed ΔE_A, the change of kinetic energy of a star of group A (and mass m_A) in a single encounter with a star of group B (mass m_B). Multiplying ΔE_A by the rate of such encounters and averaging over all the orbital parameters then gave $\langle dE_A/dt \rangle$, the mean energy exchange rate. A central feature was the assumption of separate Maxwellian velocity distributions for each group. The calculation was trivial but is mentioned here because the resulting formula for $\langle dE_A/dt \rangle$ is so frequently used. It is curious how sometimes the simplest results, which seem in retrospect transparently obvious, are among the most widely quoted.

Some 10 years later, as part of the program to determine the physical properties of the interstellar gas, I became interested in another problem involving similar encounters, this time among electrons and positive ions. My objective was to determine the electrical conductivity of a fully ionized gas, using methods more appropriate to this particular problem than those developed (45) for collisions with mostly neutral particles. In encounters between charged particles, distant collisions producing relatively small velocity changes are dominant, and large changes in velocity are produced by the cumulative effects of many small random changes, as in diffusion or Brownian motion. A recent review paper by Chandrasekhar (46) had provided a detailed discussion of such processes and had discussed the Fokker-Planck equation, which gives the resulting rate of change of the velocity distribution function; this paper provided a basis for the determination of electrical conductivity.

To derive and justify the complex integro-differential equations to be used constituted a PhD thesis for a Yale physics student (R. S. Cohen). To solve these equations numerically provided a year's research for a Princeton graduate student (Routly). These computations were repeated a few years later, including a term that I had previously assumed would vanish, and were extended to include thermal conductivity also (47). Detailed agreement of these results with those obtained by later researchers indicates that the computations are essentially correct.

The behavior of an ionized gas, or plasma, in the presence of a magnetic field was explored in pioneering research by Hannes Alfvén, who had visited Princeton in late 1948 and whose papers and subsequent book (48) were eagerly read and discussed by Martin and me. We both did our best to become familiar with the new field of magnetohydrodynamics so that we could apply it in our respective fields of interest.

An opportunity for such an application came to me in the spring of 1951, when the Argentines announced that they had successfully achieved a controlled thermonuclear reaction in the laboratory. In view of the nearly unlimited supply of nuclear energy available in the deuterium of the oceans, this announcement (which was subsequently shown to be incorrect) raised intense interest, together with considerable skepticism. My wife and I left for a week's ski trip the day after this announcement, and during the long rides on the Aspen single-chair lift I had ample time to consider how a hot thermonuclear plasma might be confined without contact with any material walls. Thomas Cowling was a Visiting Professor in Princeton during that term, and his beautifully systematic lectures on magnetohydrodynamics provided the training needed to consider magnetic confinement of a hot plasma. Such confinement seemed to be a promising possibility, and back in Princeton I worked out (mostly in bed during the middle of the night) a magnetic configuration that I hoped would be effective. The Atomic Energy Commission (AEC) provided funds for a year of theoretical research by a small group that I assembled; the results seemed to confirm earlier hopes, and we embarked on a small experimental program. The dream of a virtually limitless power supply for mankind became for me a primary driving force.

For reasons that I now find quite understandable, the AEC was unwilling to support a major experimental program in controlled fusion at Princeton unless some distinguished experimental physicist was willing to head the effort at least for a few years. Fortunately I was able to persuade James Van Allen to come in this capacity. In addition to his major contributions to our experimental plasma program, he suggested that Martin make use of balloon-launched instruments (one of Van Allen's specialties) for observing the Sun, and he persuaded me to prepare a simple text on plasma physics; the material I put together was designed for our own plasma research group but became the basis for my published plasma physics tract (49), which fortunately appeared at the right time to satisfy a genuine need.

This is not the place to review the progress and problems of the Princeton Plasma Physics Laboratory, as our controlled fusion program was called after its declassification in 1958; this history has been related elsewhere (50, 51). I withdrew from the Laboratory in 1966, after 15 years of mostly half-time participation in the effort. The large staff (now about 1000 persons) engaged on this program have achieved many successes. The theoretical group have made outstanding contributions not only in basic plasma physics but also in plasma astrophysics. On the experimental side, the heating and confinement of a plasma in a magnetic field have been steadily improved and are approaching the levels required for a full-scale power-producing thermonuclear reactor. Of course, as the program has developed over the years my initial concepts and suggestions have formed only a small part of the growing enterprise, as with the Space Telescope also.

In retrospect, I have indeed been fortunate to be engaged in such varied and exciting research and to have had stimulating contacts with so many brilliant teachers, colleagues, and students. While one's early dreams are rarely fulfilled in detail, working toward these goals and taking an active part in fascinating, worthwhile efforts has brought its own rewards.

Literature Cited

1. Gilbert, W. S. 1921. *The Mikado*, Act II
2. Baade, W. 1944. *Ap. J.* 100: 137
3. Spitzer, L. 1941. *Ap. J.* 93: 369
4. Strömgren, B. 1939. *Ap. J.* 89: 526
5. Spitzer, L., Savedoff, M. P. 1950. *Ap. J.* 111: 593
6. Osterbrock, D. E. 1974. *Astrophysics of Gaseous Nebulae*. San Francisco: Freeman
7. Spitzer, L. 1968. *Diffuse Matter in Space*. New York: Wiley-Interscience; 1978. *Physical Processes in the Interstellar Medium*. New York: Wiley
8. Field, G. B. 1965. *Ap. J.* 142: 531
9. Routly, P. McR., Spitzer, L. 1952. *Ap. J.* 115: 227
10. Adams, W. S. 1949. *Ap. J.* 109: 354
11. Siluk, R. S., Silk, J. 1974. *Ap. J.* 192: 51
12. Spitzer, L. 1954. *Ap. J.* 120: 1
13. Spitzer, L. 1976. *Comments Astrophys.* 6: 177
14. Spitzer, L., Greenstein, J. L. 1951. *Ap. J.* 114: 407
15. Oort, J. H., Spitzer, L. 1955. *Ap. J.* 121: 6
16. Pottasch, S. R. 1958. *Bull. Astron. Inst. Neth.* 14: 29
17. Mestel, L., Spitzer, L. 1956. *MNRAS* 116: 503
18. Osterbrock, D. E. 1961. *Ap. J.* 134: 270
19. Spitzer, L. 1956. *Ap. J.* 124: 20
20. Münch, G., Zirin, H. 1961. *Ap. J.* 133: 11
21. Savage, B. D. 1987. In *Interstellar Processes*, ed. D. J. Hollenbach, H. A. Thronson Jr., p. 123. Dordrecht: Reidel
22. Jenkins, E. B. 1987. In *Exploring the Universe with the IUE Satellite*, ed. Y. Kondo, p. 531. Dordrecht: Reidel
23. Jones, R. V., Spitzer, L. 1967. *Ap. J.* 147: 943
24. Davis, L., Greenstein, J. L. 1951. *Ap. J.* 114: 206
25. Purcell, E. M. 1979. *Ap. J.* 231: 404
26. Spitzer, L. 1968. In *Nebulae and Interstellar Matter (Stars and Stellar Systems*, Vol. 7), ed. B. M. Middlehurst, L. H. Aller, p. 1. Chicago: Univ. Chicago Press
27. Spitzer, L. 1940. *MNRAS* 100: 396
28. Ambartsumian, V. A. 1938. *Ann. Leningrad State Univ. No. 22 (Astron. Ser., Iss. 4)*, p. 19; 1985. In *Dynamics of Star Clusters, IAU Symp. No. 113*, ed. J. Goodman, P. Hut, p. 521. Dordrecht: Reidel
29. Spitzer, L., Schwarzschild, M. 1951. *Ap. J.* 114: 385; 1953. *Ap. J.* 118: 106
30. Villumsen, J. V. 1983. *Ap. J.* 274: 632
31. Blitz, L. 1980. In *Giant Molecular Clouds in the Galaxy*, ed. P. M. Solomon, M. G. Edmunds, p. 1. New York: Pergamon
32. Spitzer, L. 1958. *Ap. J.* 127: 17
33. Wielen, R. 1985. In *Dynamics of Star Clusters, IAU Symp. No. 113*, ed. J. Goodman, P. Hut, p. 449. Dordrecht: Reidel
34. Spitzer, L., Saslaw, W. C. 1966. *Ap. J.* 143: 400
35. Spitzer, L. 1969. *Ap. J. Lett.* 158: L139
36. Hénon, M. 1971. *Astrophys. Space Sci.* 13: 284, 14: 151
37. Spitzer, L. 1987. *Dynamical Evolution of Globular Clusters*. Princeton, NJ: Princeton Univ. Press
38. Lynden-Bell, D., Wood, R. 1968. *MNRAS* 138: 495
39. Lynden-Bell, D., Eggleton, P. P. 1980. *MNRAS* 191: 483
40. Spitzer, L., Zabriskie, F. R. 1959. *Publ. Astron. Soc. Pac.* 71: 412; 1988. *Publ. Astron. Soc. Pac.* 100: 509
41. Spitzer, L. 1960. *Astron. J.* 65: 242
42. Spitzer, L., Jenkins, E. B. 1975. *Annu. Rev. Astron. Astrophys.* 13: 133
43. Spitzer, L. 1982. *Searching Between the Stars*. New Haven, Conn: Yale Univ. Press
44. Spitzer, L. 1979. *Q. J. R. Astron. Soc.* 20: 29
45. Chapman, S., Cowling, T. G. 1939. *The Mathematical Theory of Non-Uniform Gases*. Cambridge: Univ. Press
46. Chandrasekhar, S. 1943. *Rev. Mod. Phys.* 15: 1
47. Spitzer, L., Härm, R. 1953. *Phys. Rev.* 89: 977
48. Alfvén, H. 1950. *Cosmical Electrodynamics*. Oxford: Clarendon
49. Spitzer, L. 1956, 1962. *Physics of Fully Ionized Gases*. New York: Wiley-Interscience. 1st, 2nd ed.
50. Bishop, A. S. 1958. *Project Sherwood*. Reading, Mass: Addison-Wesley
51. Bromberg, J. L. 1982. *Fusion: Science, Politics and the Invention of a New Energy Source*. Cambridge, Mass: MIT Press

THE STATUS AND PROSPECTS FOR GROUND-BASED OBSERVATORY SITES

R. H. Garstang

Joint Institute for Laboratory Astrophysics, University of Colorado and National Bureau of Standards, Boulder, Colorado 80309-0440

1. INTRODUCTION

Light pollution is the illumination of the night sky caused by man-made sources. It has been responsible for the deterioration of observing conditions at a number of major observatories as well as at numerous small observatories. This deterioration has gone on for many years. It may be slowed down somewhat by lighting control ordinances, by the replacement of mercury vapor lamps by sodium lamps, by the development of improved luminaires, and by changes in the economic conditions in the surrounding area, but the relentless growth of population in areas of good climate suggests that light pollution will only get worse over the long term.

In this article I use available information on population growth to predict how light pollution at various observatories and prospective observatory sites will grow as populations increase. I make the assumption initially that there is no change in lighting technology. This assumption is incorporated in our use of a standard model for sky brightness calculations. The effects of possible improvements in technology are discussed later. I use the same model for all observatories, with the only change in the basic numerical parameters being the use of a larger ground reflectivity for snow-covered conditions. I have deliberately refrained from making any attempt to fit this model to a particular observatory, even when I have additional observational information about the night sky brightness at that observatory. It would be possible to force a fit if one chose to do so, but such a forced fit would probably not be unique and might be unphysical. I feel that a fairer comparison between sites is obtained by using a realistic standard model. More detailed studies of individual sites can be undertaken if needed and when additional information is available beyond what was available to me for this general survey.

I have made some computer runs on observatories (San Pedro Martir, Cerro Tololo, Haute Provence, Anglo-Australian) outside the US. However, the information readily available to me at this time is insufficient to make predictions for these observatories. Accordingly, I have regretfully confined this article to US observatories.

2. ESSENTIALS OF THE MODEL

2.1 The Model

I base the projections given in this paper upon the model of light pollution described by Garstang (7, and paper in preparation). A city is treated as a point source when it is at a large distance from the observatory and as a circle of uniform brightness (represented by a seven-point discrete approximation) when it is relatively close to an observatory. I assume that the light emission per head of the population is constant, that some of the light is radiated into the upward hemisphere, and that the remainder of the light is radiated downward: Of the latter, some is absorbed by the ground or by objects, and the rest is reflected upward. I further assume (a) Rayleigh scattering by molecules in an atmosphere whose density decreases exponentially with height above sea level; (b) scattering by an average aerosol, using an idealized scattering function with strong forward scattering; (c) an aerosol density that decreases exponentially with height with a relatively small scale height; and (d) an adjustable total atmospheric aerosol content. I allow for the altitudes of the observatory and of the light-polluting cities above sea level. Curved Earth geometry is used in all cases, extinction along the light paths is included, and calculations are performed for the B and V photometric bands. I use a simplified model of the natural night sky background to predict the night sky background brightness at different sites and zenith distances. The natural sky brightness is normalized to the value at Junipero Serra Peak close to sunspot minimum (see further discussion later in this article). Full details of all these assumptions can be found in the papers quoted. In this article the same standard model is used as in these papers, which give the numerical values of all the parameters in the model. I state for reference here, however, that a light emission of 1000 lumens per head of the population is assumed, equivalent to 3.42×10^{18} photons per second per head in the V band. For the B band I use an emission of 1.4×10^{18} photons per second per head. I assume that 10% of the light escapes upward from poorly shielded light sources, and that the ground reflectivity is 15% in the absence of snow cover.

2.2 Population Projections

I obtained population projections from various agencies that have the prediction of populations as one of their missions. In California, population predictions are made by the Department of Finance (3); in Arizona, by the Department of Economic Security (1); in New Mexico, under contract by the Bureau of Business and Economic Research of the University of New Mexico (20); and in Texas, by the Bureau of State Health Planning and Resource Development (15). [In addition, some predictions for Texas have been made by Woods and Poole Economics, Inc. (8).] These sources provide population predictions for counties. In addition, the Arizona projections provide data for cities. Predictions for cities in the San Diego region are made by the San Diego Association of Governments (12, 13). Data for 1980 were taken from the Rand McNally *Road Atlas* (11) and from Bureau of the Census publications (2). Occasionally I used previous editions of these publications for population data from earlier years.

In my model calculations, only the populations of cities and their contiguous suburbs are used, which I believe to be responsible for the major part of the light pollution, and the contributions of areas of very low population and individual farms and ranches are omitted (except in the work on San Benito Mountain, where I felt it essential to include an estimate of the contribution of the farms in the Central Valley of California). In some cases I consulted highway (11) and other more detailed maps to determine what should be included in the total population of a city and its suburbs. For most Arizona cities, I was able to use the city or county projections as provided. For many of the cities in New Mexico and Texas and for a few cities in Arizona and California, I used census data to examine how the ratio of the population of a particular city to the population of the county in which that city is situated has changed over the years from 1950 to 1980. We extrapolated the ratio and used it to estimate the future population of the city from the projected population of the county. In a great many cases we found that the ratio of the city population to the county population is increasing slowly: Thus the population of the cities is increasing rather more rapidly than the projected increase in the county populations. The situation just described applies to many rural counties in which there is only one major city, the county seat. In a few urban situations I could assume that the population of a city is the same as its county population, in a few cases I had to make estimates of the population of several towns in a county, and in the case of Los

Angeles I had to make city by city estimates in several counties. When using census data from different years to extrapolate populations, one should take note of annexations by cities and changes of county boundaries whenever one has appropriate information; otherwise data from different years will not be strictly comparable. However, the appropriate information is often not readily available, and one must manage without it.

2.3 Calculations

Population estimates were prepared in the way described for all the cities involved in the light pollution for a particular observatory. The contribution of each city to the light pollution as calculated for its 1980 population was scaled (the model is linear in the population of a city) to the estimated future population of that city for a series of future dates. The contributions from the various cities were summed to estimate the total light pollution for each of the future dates. The results are given in a series of tables in the following sections of this article.

2.4 Definition of Brightness Increases

In the tabulated results, I give the brightness increase (Δm) caused by cities, expressed in magnitudes per square arcsecond (and omitting the minus sign that should, strictly speaking, be given) relative to the natural sky brightness, whose predicted value is also given in the tables. If it is desired to use a different value of the natural sky background brightness as a reference, the brightness increase caused by cities, expressed in magnitudes, will be altered (it is the intensities, not the magnitudes, that are additive), and the values of Δm should be recalculated.

The natural sky brightness increases considerably as solar activity increases, and the total sky brightness increases correspondingly. The natural sky background values that are used here are normalized to observations by Walker (16–18), many of which were made in 1966 and 1976 and so refer to periods close to sunspot minimum. Total night sky brightness measurements made at times other than sunspot minimum will contain both the contributions of the cities and the additional contribution of the increased airglow intensity. Accordingly, one must take care when comparing any of our predictions with observed brightness measurements. It has not yet been possible to incorporate the solar cycle into our model, but the work of Walker (19) points to the possibility of doing this. In particular, these predictions for the brightness at future dates do not take into account the phase of the solar cycle. I recommend making all comparisons between observatories and sites using background brightness values at or close to sunspot minimum.

3. APPLICATIONS TO OBSERVATORIES

3.1 *Palomar Observatory*

For the San Diego region I used the predictions published by the San Diego Association of Governments (12, 13), which go to the year 2000. I made judicious extrapolations to enable estimates to be made for the years 2010 and 2020. For the Los Angeles basin I used the city populations published in the *California Statistical Abstract* (4, 5) along with the 1980 census populations to estimate the growth rate of each city and extrapolated the populations city by city. I estimated populations for 46 individual cities in the southeastern part of the Los Angeles basin and treated the central and northern part of Los Angeles as a single city. I compared the latest county data (5) with predicted county data (3) and found that the latest populations are larger than the latter publication would suggest. My predicted city populations are also correspondingly larger than earlier projections would have suggested. Of course, my population projections have considerable uncertainties associated with them, but in any event substantial population growth seems certain.

In Table 1, a substantial future increase in the night sky brightness is predicted. The population of the region is steadily increasing. The rate of increase is particularly large on the northern side of San Diego and the southeastern edge of the Los Angeles basin, and thus the growth is in areas at less and less distance from Palomar Observatory. A major source of light pollution is certain to be the Escondido area, which I predict will contribute about a quarter of the total light pollution. The lighting control efforts in San Diego (change to low-pressure sodium lighting) will help to reduce the increases in sky brightness, especially in the *B* band in which less blue photons will be emitted, but the prospects are for a substantial deterioration of the site.

In private discussions, Dr. R. J. Brucato has indicated that the brightness at Palomar may already be larger than the model predicts. A more elaborate study of the developments near Escondido, and elsewhere, would be

Table 1 Predicted light pollution at Palomar Observatory

Band	Direction	Night sky background (mag arcsec^{-2})	Brightness increase (Δm)					
			1980	1990	1995	2000	2010	2020
V	Zenith	21.99	0.44	0.51	0.55	0.59	0.69	0.80
B	Zenith	22.99	0.32	0.38	0.41	0.44	0.51	0.59

Table 2 Predicted light pollution at Mount Wilson Observatory

Band	Direction	Night sky background (mag arcsec^{-2})	Brightness increase (Δm)					
			1980	1990	1995	2000	2010	2020
V	Zenith	22.02	2.27	2.42	2.47	2.52	2.61	2.70
B	Zenith	23.00	2.22	2.37	2.43	2.48	2.57	2.67

justified when it is possible. I urge continued observational monitoring of the sky brightness.

3.2 Mount Wilson Observatory

The serious problems of light pollution at Mount Wilson are well known. I have used population projections from (3), with minor changes where I could use later information (5). It did not seem worthwhile treating even the large cities separately; thus I took nine cities near Mount Wilson and treated them individually and divided the remainder of the Los Angeles basin into a near zone, which consists of the major part of Los Angeles County, and a far zone, which consists of the populated parts of Riverside, Orange, and San Bernardino Counties. The results (Table 2) indicate that the night sky brightness will continue to increase. The rate of increase is somewhat lessened because much of the population growth is taking place in the region that we have called the far zone, and the brightness increase caused by this population increase is relatively small because of the distance of the area from Mount Wilson.

3.3 Lick Observatory

I included in the calculations 80 cities, counties, or parts of counties ranging from Santa Rosa in the north to Salinas in the south to Fresno in the east, using population data from (2, 3, 11). I predict (Table 3) a continuing increase in the light pollution at Lick due to the continuing growth of the population of California. Although the growth in the city centers and inner

Table 3 Predicted light pollution at Lick Observatory

Band	Direction	Night sky background (mag arcsec^{-2})	Brightness increase (Δm)					
			1980	1990	1995	2000	2010	2020
V	Zenith	22.01	1.35	1.49	1.54	1.58	1.65	1.71
B	Zenith	23.00	1.28	1.43	1.48	1.52	1.59	1.65

suburbs of San Francisco and Oakland is minimal, most of the outer suburban areas continue to expand rapidly.

3.4 Kitt Peak National Observatory

For Tucson we used population projections for Pima County. For Phoenix we used projections for Maricopa County. For Nogales we first examined the ratio of the populations of Nogales, Mexico, and Nogales, Arizona, for the census dates from 1950 to 1980; then we extrapolated this ratio and applied it to the projected population of Nogales, Arizona, to estimate the combined population. For completeness we included in our calculations the cities of Sells, Fort Huachuca, Sierra Vista, Casa Grande, and Coolidge: These five contributions added together are probably smaller than the uncertainty in our predictions for Tucson alone, but it seemed best to include them. Predictions are not available for Sells, but I assumed no change in population, since the contribution of Sells is likely to remain unimportant unless the population increases by an order of magnitude. In Tables 4 and 5, predictions for the zenith and for six azimuths at zenith distance 45° are given to the year 2035, the last year for which the Arizona Department of Economic Security issued population projections. My calculations indicate that at the zenith for B magnitudes Tucson contributes about 85% of the total light pollution, with Phoenix contributing a further 7%. For V magnitudes the corresponding values are 80% and 12%. Tucson and Phoenix are predicted to grow at nearly equal rates, so Tucson will remain the major source of light pollution at Kitt Peak.

Table 4 Predicted light pollution at Kitt Peak National Observatory

	Increase in brightness (Δm) in B magnitudes[a]						
		Zenith distance 45°					
Date	Zenith	AZ 0°	AZ 60°	AZ 120°	AZ 180°	AZ 240°	AZ 300°
1980	0.057	0.081	0.124	0.078	0.063	0.066	0.065
1990	0.076	0.109	0.166	0.105	0.084	0.088	0.087
1995	0.087	0.124	0.188	0.120	0.096	0.100	0.099
2000	0.099	0.140	0.21	0.135	0.109	0.113	0.112
2010	0.124	0.175	0.26	0.170	0.137	0.142	0.141
2020	0.148	0.21	0.31	0.20	0.164	0.169	0.168
2035	0.185	0.26	0.38	0.25	0.20	0.21	0.21

[a] These results are relative to an assumed night sky background brightness $B = 22.99$ mag arcsec^{-2} at the zenith and $B = 22.83$ mag arcsec^{-2} at zenith distance 45°. Abbreviation: AZ, azimuth.

26 GARSTANG

Table 5 Predicted light pollution at Kitt Peak National Observatory

		Increase in brightness (Δm) in V magnitudes[a]					
			Zenith distance 45°				
Date	Zenith	AZ 0°	AZ 60°	AZ 120°	AZ 180°	AZ 240°	AZ 300°
1980	0.085	0.120	0.174	0.115	0.098	0.106	0.101
1990	0.114	0.160	0.23	0.154	0.132	0.141	0.136
1995	0.130	0.183	0.26	0.175	0.151	0.161	0.155
2000	0.147	0.21	0.29	0.198	0.170	0.181	0.175
2010	0.183	0.26	0.36	0.25	0.21	0.23	0.22
2020	0.22	0.30	0.42	0.29	0.25	0.27	0.26
2035	0.27	0.37	0.51	0.36	0.31	0.33	0.32

[a] These results are relative to an assumed night sky background brightness $V = 21.99$ mag arcsec^{-2} at the zenith and $V = 21.791$ mag arcsec^{-2} at zenith distnce 45°. Abbreviation: AZ, azimuth.

3.5 McDonald Observatory

For McDonald Observatory I used the county projections (15) and made city projections for Fort Davis, Marfa, Alpine, Pecos, and El Paso as described above. These projections were given for years up to 2000, from which I extrapolated to the year 2005. There is little economic growth in the area. Predictions are given in Table 6 for the zenith and for a zenith distance 45° in the azimuth of Fort Davis. Fort Davis is and will likely remain the principal source of light pollution, with Alpine the second most important contributor. As mentioned above, I have a second set of population predictions (8). These are for the year 2005 only. This set predicts hardly any population growth for the small cities of interest here, so that if this population scenario should prove to be correct, the light pollution would be even less than the first set of population estimates

Table 6 Predicted light pollution at McDonald Observatory

Band	Direction	Night sky background (mag arcsec^{-2})	Brightness increase (Δm)					
			1980	1990	1995	2000	2005	2005[a]
V	Zenith	22.018	0.008	0.010	0.010	0.011	0.012	0.009
V	45° Fort Davis	21.795	0.021	0.025	0.027	0.029	0.031	0.021
B	Zenith	22.993	0.007	0.009	0.009	0.010	0.011	0.007
B	45° Fort Davis	22.842	0.018	0.022	0.024	0.026	0.028	0.019

[a] See text for these alternative predictions.

would suggest. The site is a very dark one, and there is little likelihood of a significant increase in man-made light.

3.6 Mauna Kea Observatory

I do not have a reliable set of future population projections for the island of Hawaii. Thus, I have made several estimates of future brightness under various assumptions. The first assumption is that the various towns will grow at the same rates as they did during 1960 to 1980. I used census data (2) to estimate these rates for the towns used in the 1980 brightness calculation and then applied the rates to predict the future sky brightness. The 1980 brightness was calculated by including Hilo and 15 small towns on the island of Hawaii, along with Honolulu and the island of Maui (the latter two being barely significant). My estimate does not take into account the clouds that often form over Hilo, so that the sky may be even darker than I predict. The predictions are given in Table 7. Mauna Kea is a very dark site, as is well known from observations (9).

I plan a more detailed study of Mauna Kea but do not have sufficient data at present to pursue this. Instead, I report calculations on two possible scenarios that were suggested to me by Dr. S. M. Faber. In each case I added the predicted brightness due to Hilo and other towns outside the area of the scenario, using my predictions for the year 2000. The calculations were performed only for V magnitudes.

Table 7 Predicted light pollution at Mauna Kea Observatory

Band	Direction	Night sky background (mag arcsec^{-2})	Brightness increase (Δm)			
			1980	1990	1995	2000
Normal predictions						
V	Zenith	21.980	0.023	0.027	0.030	0.032
B	Zenith	22.959	0.022	0.026	0.028	0.030
V	ZD 45° Hilo	21.774	0.043	0.050	0.055	0.059
B	ZD 45° Hilo	22.786	0.043	0.047	0.051	0.056
V	ZD 45° West	21.774	0.031	0.036	0.040	0.043
B	ZD 45° West	22.786	0.027	0.032	0.035	0.038
Intermediate scenario[a]						
V	Zenith	21.980				0.086
V	ZD 45° West	21.774				0.131
Worst-case scenario[a]						
V	Zenith	21.980				0.53
V	ZD 45° West	21.774				0.83

[a] See text for explanation.

3.6.1 INTERMEDIATE SCENARIO   We assume that a coastal strip stretching from 3 km north of Kawaihae to 3 km south of Kailua (Kona) and having a width of 0.8 km is populated to a density of 750 km^{-2}. We assume that four towns (Kailua, Kawaihae, Waimea, and Waikoloa) each have 25,000 people who operate the infrastructure supporting the coastal resorts. I applied the standard model in its point source form to the four towns and to the coastal strip (divided into two pieces) and added the results to obtain the brightness listed in Table 7. My prediction is that the sky would be slightly less bright than Kitt Peak is today.

3.6.2 WORST-CASE SCENARIO   This scenario assumes that the whole area of the west coast of the island of Hawaii, between Kawaihae and Kailua (Kona) and inland as far as Waimea on the northeast side and as far as the 6000-ft contour on the southwest slope of Mauna Kea, is populated to the density of the Los Angeles basin, which we took to be 750 km^{-2}. I divided the total area into four pieces, applied the model to each piece, and added the results. The calculated brightnesses are given in Table 7. My prediction is that the sky would be about as bright as Mount Palomar or Mount Lemmon is today.

3.7 *Sacramento Peak Observatory*

For Sacramento Peak I included Alamogordo, Hollomon Air Force Base, El Paso, La Luz, Tularosa, Las Cruces, Cloudcroft, and Roswell, and for observation to the northeast I also included Ruidoso. Projections for New Mexico cities were made starting with the county population projections (20). I used some data from (10, 11) to examine trends in the city/county population ratios over the years from 1950 to 1980, extrapolated these ratios, and then applied the ratios to the county projections to obtain city projected populations. The same population projections for El Paso were used as for McDonald Observatory. Alamogordo is the most important source of light pollution, with significant contributions from Hollomon, El Paso, and Cloudcroft. Predictions are given in Table 8 for the zenith and for zenith distance 45° in five azimuths. The site is not a good one from the point of view of light pollution, since it is appreciably brighter than the IAU (International Astronomical Union) recommended maximum brightness (14) for a dark site and is predicted to increase steadily.

3.8 *Mount Hopkins*

The cities of Tucson, Green Valley, Nogales, Sierra Vista, Fort Huachuca and Huachuca City, Patagonia, Benson, Bisbee, Cananea, and Douglas and Agua Prieta were included for the Mount Hopkins calculations. The first four of these make major contributions to the light pollution. No official

Table 8 Predicted light pollution at Sacramento Peak Observatory

Magnitude	Direction	Night sky background (mag arcsec^{-2})	Brightness increase (Δm)				
			1980	1990	1995	2000	2005
V	Zenith	21.988	0.106	0.141	0.157	0.172	0.186
V	45° Northeast	21.786	0.126	0.168	0.193	0.20	0.22
V	45° West	21.786	0.163	0.22	0.24	0.26	0.28
V	45° South	21.786	0.127	0.169	0.187	0.20	0.22
V	45° Alamogordo	21.786	0.22	0.27	0.32	0.35	0.38
V	45° El Paso	21.786	0.133	0.175	0.194	0.21	0.23
B	Zenith	22.980	0.104	0.141	0.158	0.172	0.187
B	45° Northeast	22.820	0.118	0.160	0.179	0.196	0.21
B	45° West	22.820	0.158	0.21	0.24	0.26	0.28
B	45° South	22.820	0.119	0.160	0.178	0.195	0.21
B	45° Alamogordo	22.820	0.21	0.29	0.32	0.35	0.38
B	45° El Paso	22.820	0.121	0.162	0.180	0.197	0.21

projections of the population of Green Valley were available, so rough estimates were made as best as I could, and these introduce additional uncertainty into the predicted sky brightnesses. My calculations do make it clear that if the Green Valley area continues to grow at anything like its present rate, then it will become a major polluting source from the year 2000 onward, and one must pay special attention to lighting control in the area, as well as in Tucson and Nogales. The results (Table 9) suggest that

Table 9 Predicted light pollution at Mount Hopkins

	Increase in brightness (Δm) in V and B magnitudes[a]			
	Zenith		Zenith distance 45° toward Tucson	
Date	V	B	V	B
1980	0.137	0.112	0.25	0.198
1990	0.184	0.151	0.33	0.27
1995	0.21	0.175	0.38	0.31
2000	0.24	0.20	0.43	0.35
2010	0.32	0.27	0.55	0.45
2020	0.41	0.35	0.68	0.57
2035	0.56	0.49	0.91	0.80

[a] These results are relative to an assumed night sky background brightness $V = 21.986$ mag arcsec^{-2} at the zenith, $V = 21.787$ mag arcsec^{-2}, at zenith distance 45°, $B = 22.979$ mag arcsec^{-2} at the zenith, and $B = 22.819$ mag arcsec^{-2} at zenith distance 45°.

the sky brightness at Mount Hopkins is growing more rapidly than the brightness at Miller Peak (see below) and may exceed the latter after about the year 2010.

3.9 Mount Lemmon

Tucson is the major contributor of light pollution for Mount Lemmon. I also considered the cities of Phoenix, San Manuel, Oracle, and Nogales and included their contributions, even though they are barely significant. Mount Lemmon is already a rather bright site, and my prediction (Table 10) is that it will continue to get significantly brighter.

3.10 Lowell Observatory

I have made predictions (Table 11) for the Mars Hill site and for the Anderson Mesa site, in both cases assuming no snow cover. The projections (1) of the populations of Flagstaff and Phoenix were used, with the Flagstaff population divided into Flagstaff and East Flagstaff in the ratio 10:18, which is the present ratio. I have assumed that the mean distance of the "center of light" of the cities from the sites does not change, which means that I assume that much of the future development takes place in a southerly direction. If future developments take place at significantly larger distances from Mars Hill, then the rate of increase of the brightness at Mars Hill would be rather less than is predicted. If the developments mostly took place in a southeasterly direction, the brightness at Anderson Mesa might increase rather more rapidly than is predicted.

Table 10 Predicted light pollution at Mount Lemmon

	Increase in brightness (Δm) in V and B magnitudes[a]			
	Zenith		Zenith distance 45° toward Tucson	
Date	V	B	V	B
1980	0.45	0.45	0.85	0.85
1990	0.58	0.58	1.05	1.04
1995	0.64	0.64	1.14	1.14
2000	0.71	0.71	1.23	1.23
2010	0.84	0.84	1.42	1.41
2020	0.95	0.95	1.57	1.56
2035	1.10	1.10	1.76	1.75

[a] These results are relative to an assumed night sky background brightness $V = 21.986$ mag arcsec^{-2} at the zenith, $V = 21.784$ mag arcsec^{-2} at zenith distance 45°, $B = 22.976$ mag arcsec^{-2} at the zenith, and $B = 22.815$ mag arcsec^{-2} at zenith distance 45°.

Table 11 Predicted light pollution at Lowell Observatory

	Increase in brightness (Δm) in V and B magnitudes[a]			
	Mars Hill zenith		Anderson Mesa zenith	
Date	V	B	V	B
1980	1.47	1.59	0.16	0.16
1990	1.87	1.99	0.24	0.25
1995	1.98	2.11	0.27	0.28
2000	2.08	2.21	0.30	0.30
2010	2.30	2.46	0.37	0.38
2020	2.51	2.64	0.44	0.44
2035	2.73	2.87	0.52	0.53

[a] These results are relative to an assumed night sky background brightness $V = 21.996$ mag arcsec^{-2} and $B = 22.998$ mag arcsec^{-2} at the zenith.

It should be noted that my calculations do not take into account the known darkening of the sky after 11 P.M., when many lights are turned off.

3.11 US Naval Observatory, Flagstaff

I included Flagstaff and Phoenix for the US Naval Observatory calculations and used the same population projections as for Lowell Observatory. The results are given in Table 12. I used a ground reflectivity of 15% for Flagstaff in summer and for Phoenix, and 60% for Flagstaff when there is snow cover. It is difficult to predict exactly where future growth

Table 12 Predicted light pollution at US Naval Observatory, Flagstaff

	Increase in brightness (Δm) in V and B magnitudes[a]			
	No snow at Flagstaff		Snow cover at Flagstaff	
Date	V	B	V	B
1980	0.24	0.25	0.44	0.49
1990	0.35	0.38	0.64	0.70
1995	0.39	0.42	0.70	0.76
2000	0.43	0.46	0.76	0.82
2010	0.53	0.57	0.91	0.98
2020	0.61	0.65	1.03	1.11
2035	0.73	0.77	1.18	1.27

[a] These results are relative to an assumed night sky background brightness $V = 22.00$ mag arcsec^{-2} and $B = 23.00$ mag arcsec^{-2} at the zenith when there is no snow cover at Flagstaff, and $V = 21.97$ mag arcsec^{-2} and $B = 22.94$ mag arcsec^{-2} at the zenith when there is snow cover at Flagstaff.

of Flagstaff will occur. I have assumed that the "center of light" will remain at the same distance from the Observatory as it is at present. Large developments in southwesterly or westerly directions will lead to light pollution greater than is predicted. Large growths at the east end of Flagstaff will be less harmful. More detailed predictions of light pollution at this Observatory will be useful when additional information is available on development prospects for Flagstaff.

4. APPLICATION TO PROSPECTIVE OBSERVATORY SITES

4.1 *Mount Graham*

In the Mount Graham calculations I used projections for Tucson, Phoenix, and Nogales made in the way described above for Kitt Peak. I also included in the calculations the cities of Globe, Safford, Thatcher, Pima, Clifton, Willcox, Sierra Vista, and Fort Huachuca. I increased the projected populations for Globe and Clifton to allow for neighboring areas not included in the projections; the effect of these adjustments was very small. My calculations (Tables 13, 14) indicate that in 1990, for V magnitudes at the zenith, Tucson is the largest single source of light pollution, that the total contributions of Safford, Thatcher, and Pima are about equal to that of Tucson, and that Phoenix is also a significant contributor. Tucson and Phoenix are predicted to grow at nearly equal rates, while Safford, Thatcher, and Pima are predicted to grow very slowly. Thus, in later years Tucson will be the major source of light pollution at the Mount Graham zenith.

Table 13   Predicted light pollution at Mount Graham

	Increase in brightness (Δm) in V magnitudes[a]		
Date	Zenith	Zenith distance 45° toward Tucson	Zenith distance 45° toward Safford
1980	0.045	0.069	0.078
1990	0.055	0.087	0.093
1995	0.060	0.095	0.099
2000	0.065	0.105	0.106
2010	0.076	0.124	0.120
2020	0.086	0.143	0.133
2035	0.102	0.171	0.153

[a] These results are relative to an assumed night sky background brightness $V = 21.984$ mag arcsec^{-2} at the zenith and $V = 21.780$ mag arcsec^{-2} at zenith distance 45°.

Table 14 Predicted light pollution at Mount Graham

	Increase in brightness (Δm) in B magnitudes[a]		
Date	Zenith	Zenith distance 45° toward Tucson	Zenith distance 45° toward Safford
1980	0.037	0.055	0.070
1990	0.044	0.066	0.080
1995	0.046	0.071	0.083
2000	0.050	0.077	0.087
2010	0.056	0.089	0.094
2020	0.061	0.100	0.101
2035	0.070	0.117	0.112

[a] These results are relative to an assumed night sky background brightness $B = 22.970$ mag arcsec^{-2} at the zenith and $B = 22.805$ mag arcsec^{-2} at zenith distance 45°.

One point of concern is that while the superb all-sky photograph (21) taken from Mount Graham shows the light domes above Tucson and Phoenix, there is little evidence of such a light dome above Safford, Thatcher, and Pima. One possible explanation is that the light emission from these small cities is spread over about five times the range of azimuth as seen from Mount Graham as is the light from Tucson and Phoenix, and so is less apparent at large zenith distances. Another possible explanation is that the light emission per head in these smaller cities is less than the model assumes. Should this be the case, the model would somewhat overestimate the Mount Graham sky brightness.

It may also be noted that my calculations take no account of the shielding of the lights of Tucson by intervening mountains. Such shielding would lower the Mount Graham sky brightness somewhat. In a study of the profiles of the mountains between Tucson and Mount Graham, I made a preliminary estimate of the screening, and estimated that screening might reduce the scattering of light from Tucson by 6%. This would reduce the estimated total brightness at the Mount Graham zenith by 0.001 mag arcsec^{-2}.

4.2 Junipero Serra Peak

For these calculations I included all major cities, some smaller cities, and several heavily populated counties stretching from Marin County in the north to San Luis Obispo in the south to Fresno in the east, using population data from (2, 3, 11). Monterey, Salinas, King City, Greenfield, and Soledad are important contributors to the light pollution, as is San Jose, and small contributions come from many other cities. Nevertheless, Junipero Serra Peak is a good site (Table 15). It will deteriorate somewhat less

Table 15 Predicted light pollution at Junipero Serra Peak

Band	Direction	Night sky background (mag arcsec^{-2})	Brightness increase (Δm)					
			1980	1990	1995	2000	2010	2020
V	Zenith	21.990	0.092	0.112	0.122	0.131	0.148	0.165
B	Zenith	22.990	0.059	0.075	0.081	0.088	0.100	0.114

rapidly than Kitt Peak. But it is a poorer site than Mount Graham from the point of view of light pollution because the population of California continues to grow at an astonishing rate.

4.3 Miller Peak

I included in the Miller Peak calculations the cities of Sierra Vista, Fort Huachuca, Tucson, Nogales, Douglas and Agua Prieta, Huachuca City, Bisbee, Benson, Green Valley, and Cananea. The first four of these provide major contributions to the light pollution: From the early 1990s onward, Sierra Vista will probably be the largest source of light. Miller Peak has a rather bright sky (Table 16) at the present time and would appear to be a poor choice of site for a future observatory. However, my calculations suggest that the sky brightness will not grow as rapidly as the brightness at Mount Hopkins.

Table 16 Predicted light pollution at Miller Peak

	Increase in brightness (Δm) in V and B magnitudes[a]			
	Zenith		Zenith distance 45° toward Sierra Vista	
Date	V	B	V	B
1980	0.186	0.193	0.36	0.38
1990	0.22	0.23	0.43	0.44
1995	0.25	0.25	0.47	0.48
2000	0.27	0.27	0.50	0.51
2010	0.31	0.31	0.58	0.58
2020	0.36	0.35	0.65	0.65
2035	0.43	0.42	0.75	0.74

[a] These results are relative to an assumed night sky background brightness $V = 21.976$ mag arcsec^{-2} at the zenith, $V = 21.784$ mag arcsec^{-2} at zenith distance 45°, $B = 22.977$ mag arcsec^{-2} at the zenith, and $B = 22.815$ mag arcsec^{-2} at zenith distance 45°.

4.4 South Baldy

This site was originally included in my calculations at the request of Dr. K. Davidson. I included Socorro, Magdalena, Albuquerque, and the Belen area. Population projections were obtained by starting with the county projections (20) and estimating city populations in the same way as described above for Sacramento Peak Observatory. The results (Table 17) show that the site is a very dark one, and that the brightness is likely to increase rather slowly.

4.5 San Benito Mountain

I have made predictions for San Benito Mountain, which has been used in recent years by Walker (19) for night sky brightness studies. I used the same population estimates for many cities that were used in the predictions for Lick Observatory and Junipero Serra Peak. In addition, I included many cities in the Central Valley of California, including places as far away as Bakersfield, in making estimates of the populations (3, 5). I also felt it necessary to include an admittedly rather rough estimate for the contribution of the farms by using the number of farms in the Valley and assuming a single poorly shielded mercury light in every farmyard. This contribution is appreciable for this site: In most parts of the southwestern US the ranches are too sparse to make a significant contribution. The results (Table 18) indicate a brightness and brightness trend similar to those at Junipero Serra Peak.

5. EFFECT OF OTHER FACTORS

I reiterate that these projections are mainly intended to demonstrate the probable effects of normal population growth alone. There are other factors that will affect the future growth of light pollution.

Table 17 Predicted light pollution at South Baldy

Magnitude	Direction	Night sky background (mag arcsec^{-2})	Brightness increase (Δm)				
			1980	1990	1995	2000	2005
V	Zenith	21.985	0.019	0.024	0.027	0.029	0.031
V	45° Socorro	21.782	0.034	0.045	0.050	0.054	0.058
B	Zenith	22.972	0.016	0.020	0.022	0.024	0.026
B	45° Socorro	22.807	0.029	0.038	0.043	0.046	0.050

Table 18 Predicted light pollution at San Benito Mountain

Band	Direction	Night sky background (mag arcsec^{-2})	Brightness increase (Δm)					
			1980	1990	1995	2000	2010	2020
V	Zenith	21.991	0.104	0.125	0.135	0.143	0.162	0.181
B	Zenith	22.994	0.060	0.074	0.080	0.087	0.098	0.111

5.1 Changes in Light Sources

Many cities are changing from mercury vapor street lighting to high- or low-pressure sodium lighting. Some mercury lamps may remain for some time into the future along with fluorescent lamps, incandescent lamps, and an increasing number of metal halide lamps. It is also possible that new types of light sources will be developed. Predictions are difficult, but in view of the preponderance of street lighting in outdoor illumination it seems likely that the number of photons per second per head of the population emitted in the blue (B) photometric band will be reduced. This reduction will be particularly significant if large numbers of low-pressure sodium lamps are installed. The parameter in my model for the emission of blue photons per head of the population was normalized to reproduce sky brightnesses at several observatories in the years from 1966 to 1980. The phasing out of most mercury vapor lighting will lead to increases (Δm) in sky brightness in the B band somewhat smaller than my predictions indicate, and the brightnesses will grow at a slower rate. I do not yet have adequate data to estimate the reduction in the B and V photon outputs caused by the change to sodium lighting.

The change to sodium lamps, and especially to low-pressure sodium lamps, leads to lower power consumption for a given luminous output and thus to corresponding financial economies (after the initial changeover costs have been recovered). In some cases the luminous output may be reduced modestly without serious loss of night-time street illumination. For example, in Boulder, Colorado, 3685 mercury vapor lamps, each consuming 175 W and emitting about 8000 lumens, were replaced by 70-W high-pressure sodium lamps, each emitting about 5800 lumens (6).

5.2 Changes in Luminaires

In addition to new light sources, there is a gradual improvement in the design of luminaires. In my model, I allowed for light that escapes upward and assumed in numerical calculations that this upward-escaping light was 10% of the total effective lumens emitted by the source. This upward light is a major contributor to the light scattered in the high atmosphere above an observatory, particularly when the light escapes at relatively small angles above the horizontal, as is the case for many light sources. Modern luminaire design can reduce the upward-escaping light to zero. Such luminaires could in time be fitted to all street lights. Some retrofitting would be needed, as well as new fittings of improved design. Model calculations for several cases show that if the upward-escaping light could be reduced to zero, the light pollution would be reduced to one third of its present amount. That goal is unattainable for a whole city, though it may be nearly attainable in smaller developments near an observatory. But a significant reduction in light pollution would be attained if upward-escaping light could be reduced by one half, a goal that is realistic in most cities. This is the least goal that should be sought in many communities. Combined with the change away from mercury vapor lighting, one might hope to reduce the brightness and the rate of growth of brightness in the B band substantially and effect a moderate reduction in the V band. Dr. C. Pilachowski has informed me of recent night sky brightness measurements at Kitt Peak, made before solar activity started to increase, that are somewhat fainter than the model predicts, and that could conceivably be an indication that the lighting control efforts in Tucson have had some effect. An alternative explanation is that the solar minimum night sky brightness used in the model is a little too bright: Such a possibility is suggested by the measurements of Walker (19). It is clear that such matters can be clarified only by a long series of observations that cover both time of night and phase of the solar cycle.

5.3 *Unforeseen Developments*

There may be unforeseen developments that affect the population growth in an area. World War II dramatically affected the growth of many southwestern states. The oil crisis and glut has affected several states. Major tourist developments, such as may give rise to the scenarios that we discussed in the section on Mauna Kea Observatory, can have effects far beyond the normal population growth effects. Local problems, such as the construction of sports facilities close to an observatory, may be hard to control and impossible to prevent without causing unreasonable alienation of the public. The ease of automobile transport and the power of modern mechanical equipment may lead to housing developments in what had been thought would remain virgin areas. The effect of all such unforeseen developments cannot be predicted in a general way. Astronomers must be aware of the possibilities in their areas and try to minimize the potential damage of such developments as they occur.

6. CONCLUSIONS

By way of a summary, I have shown in two diagrams my predictions of the increase in the night sky brightness caused by man-made light pollution for V magnitudes for the zenith of all the observatories and sites discussed in this paper. The more heavily polluted sites are shown in Figure 1, and the less heavily polluted sites are shown in Figure 2. The differences between various sites are clearly shown.

In the absence of major unforeseen developments, the change in light sources and the use of improved luminaires together should effect a reduction in the absolute amount of light pollution and in its rate of growth, but unless the growth of population moderates considerably, the

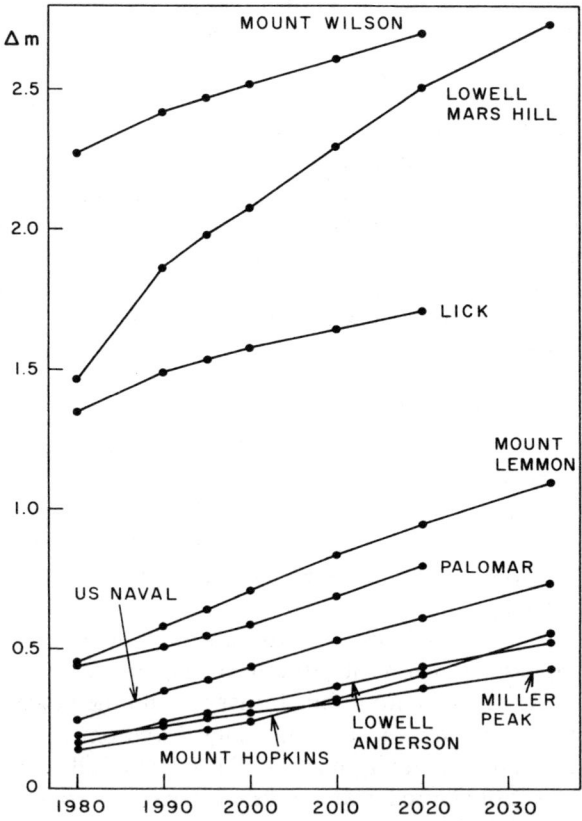

Figure 1 Predicted growth in man-made contributions to the night sky brightness. Δm is in V magnitudes per square arcsecond relative to the natural night sky background brightness at sunspot minimum.

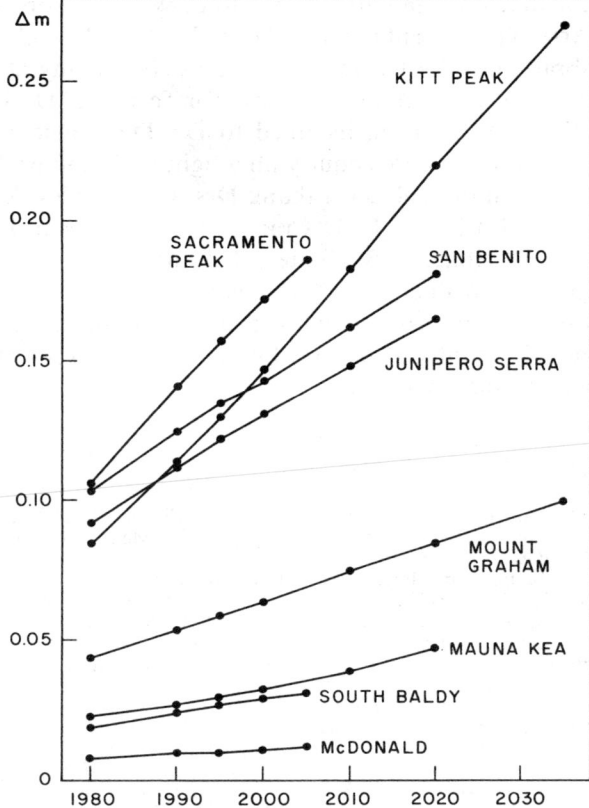

Figure 2 Predicted growth in man-made contributions to the night sky brightness. Δm is in V magnitudes per square arcsecond relative to the natural night sky background brightness at sunspot minimum. The special scenarios for Mauna Kea are not shown on the figure.

prospects in most cases are for long-term deterioration of astronomical sites.

ACKNOWLEDGMENTS

I am indebted to the Government Documents Librarian at Pomona College, who helped me find the population projections for California during a visit I made to the College for the Astronomical Society of the Pacific meeting in July 1987. I am also indebted to the Science and Technology Librarian at the San Diego Public Library for help with the populations of San Diego County and its cities. I thank the staff at the New Mexico State Library in Santa Fe, who showed me Lynn Wombold's projections for that state and allowed me to copy them. Population projections for Arizona were sent to me by Dr. P. Hintzen. I thank both Tami Andrews, Librarian of the University of Texas Astronomy Department, for searching out the population projections for Texas and Dr. H. J. Smith for sending them to me. I am indebted to Dr. Don Henderson of the National Park Service, whose enquiry on a light pollution problem stimulated my interest in this subject. I thank Drs. H. D. Ables, R. F. Berry, R. J. Brucato, F. H. Chaffee, D. L. Crawford, K. Davidson, J. A. Dawe, S. M. Faber, A. R. Hoag, T. B. Hunter, R. L. Millis, C. A. Pilachowski, A. R. Upgren, M. F. Walker, and D. Westpfahl, among many others, with whom I have enjoyed discussions and correspondence on problems of light pollution, and whose interest in modeling and predictions were a major encouragement during my work.

Literature Cited

1. Arizona Department of Economic Security. 1986. *Arizona Population Projections—Counties and Places 1985–2035*, Table 1, prepared in March 1986 by the Population Statistics Unit
2. Bureau of the Census. *1980 Census of Population, Volume 1, Characteristics of the Population, Chapter A, Number of Inhabitants: Part 4, Arizona, 1982; Part 6, California, 1982; Part 13, Hawaii, 1981; Part 33, New Mexico, 1982; Part 45, Texas, 1982*
3. California Department of Finance. 1977. *Population Projections for California Counties 1975–2000 With Age/Sex Detail to 2000. Series E-150. Rep. 77-P-3*, Sacramento, Calif.
4. *California Statistical Abstract*. 1977. Sacramento, Calif., Table B-6
5. *California Statistical Abstract*. 1987. Sacramento, Calif., Table B-4
6. Cornett, L. 1985. *Sunday Camera*. Boulder, Colo., March 3, pp. 1A, 4A
7. Garstang, R. H. 1986. *Publ. Astron. Soc. Pac.* 98: 364–75
8. Holdrich, M. K. 1986. *Texas State Profile*, Woods and Poole Economics, Inc., Washington, DC, Table 31
9. Morrison, D., Murphy, R. E., Cruikshank, D. P., Sinton, W. M., Martin, T. Z. 1973. *Publ. Astron. Soc. Pac.* 85: 255–67
10. *New Mexico Statistical Abstract*. 1979– 1980 edition. Bur. Bus. Econ. Res., Univ. N. Mex., Albuquerque, pp. 88–95
11. Rand McNally *Road Atlas*. 1984 edition
12. San Diego Association of Governments. 1984. *Comprehensive Plan for the San Diego Region, Volume 10; Series 6 Regional Growth Forecasts*. San Diego, Calif., pp. 27, 28, 31, 53
13. San Diego Association of Governments. 1986. *Economic Development Guide and Extract: Volume I: San Diego Regional Economic Development Guide*. San Diego, Calif., Table A-17
14. Smith, F. G. 1979. *Trans IAU* 17A (Part 1): 220
15. Texas Department of Health. 1986. *Total Population Summary Tables—Texas Counties 1970–2000*, Bur. State Health Plan. Resour. Dev., Austin, Tex.
16. Walker, M. F. 1970. *Publ. Astron. Soc. Pac.* 82: 672–98
17. Walker, M. F. 1973. *Publ. Astron. Soc. Pac.* 85: 508–19
18. Walker, M. F. 1977. *Publ. Astron. Soc. Pac.* 89: 405–9
19. Walker, M. F. 1988. *Publ. Astron. Soc. Pac.* 100: 496–505
20. Wombold, L. B. 1985. *Projections of the Population of New Mexico by County 1980–2005*, Bur. Bus. Econ. Res., Univ. N. Mex., Albuquerque, Table 4
21. Woolf, N. 1985. *Sky Telesc.* 70: 424–25

THE ORION MOLECULAR CLOUD AND STAR-FORMING REGION

Reinhard Genzel and Jürgen Stutzki

Institut für Extraterrestrische Physik, Max-Planck-Institut für Physik und Astrophysik, D-8046 Garching, Federal Republic of Germany

1. INTRODUCTION

The Orion giant molecular cloud, at a distance of about 450 pc, is the nearest region showing recent OB star formation. Orion is unique for studying interstellar matter and star formation. The first candidate protostars were discovered here (17, 151), as were most interstellar molecules (32, 37, 38, 139, 163, 164, 265, 285, 296, 322). Its proximity and brightness have made possible many pioneering investigations, resulting in important discoveries and a wealth of new information in the past decade.

This review emphasizes the interstellar matter and star formation in the Orion region, as well as the interaction of the young stars with their parental cloud. It has four main parts. Section 2 covers the large-scale molecular cloud and its structure, origin, and kinematics. A discussion of the stellar component in and around the cloud follows in Section 3. Section 4 then deals with the interaction of the massive stars in Orion A with their surroundings, followed by a discussion in Section 5 of the BN-KL infrared cluster and its environment (the most recent region of massive star formation). Section 6 contains conclusions. This article is certainly not complete, and for additional information we refer the reader to earlier reviews (336, 347), the summary by Goudis (94) of observational material, and the proceedings of the Henry Draper Symposium (88).

2. THE GIANT MOLECULAR CLOUD

2.1 Large-Scale Distribution of Gas

Large-scale mapping of the Orion Molecular Cloud in the 2.6-mm $J = 1 \to 0$ rotational transition of ^{12}CO (295) revealed very extended molecular emission associated with the dark cloud L1630 (188). It stretches over nearly 4°, roughly north-south from the Horsehead Nebula and NGC 2024 (Orion B) to the reflection nebulae NGC 2068 and NGC 2071. Kutner et al. (166) found an even larger molecular cloud associated with the dark clouds L1640, L1641, and L1647 that extends about 6° south from the Orion Nebula (NGC 1976, Orion A) and is elongated parallel to the Galactic plane at $-19.4°$ Galactic latitude. The extensive CO survey by Maddalena et al. (189; see Figure 1a) shows that these two clouds, named Orion A and B after the most prominent H II regions in them, are physically connected by a bridge of low-level emission. They cover 29 and 19 $\deg^2$ on the sky, respectively, with gas masses of each of about 10^5 $M_\odot$ (189). Masses derived by virial theorem arguments, by estimates of CO optical depth and abundance, and by empirical conversion factors between integrated ^{12}CO $1 \to 0$ flux and H_2 column density all agree within about 50%. The molecular clouds are embedded in lower density H I gas of total mass $\approx 7 \times 10^4$ $M_\odot$, which is apparent as an enhanced emission plateau on the wide H I emission associated with the Galactic plane (93).

The large-scale, 12- to 120-μm infrared image obtained with the *Infrared Astronomical Satellite* (*IRAS*) (22, 246; Figure 1b is a gray-scale version) shows striking similarity with the large-scale CO distribution. Molecular and dust emission show a ridge with a sharp drop-off along its western edge. Peaks of CO emission along this ridge near the H II regions Orion A, NGC 1977, NGC 2023, NGC 2024, NGC 2068, and NGC 2071 (all known to be cores with ongoing, massive star formation) are also prominent on the *IRAS* picture. The more distributed continuum emission detected with *IRAS* traces dust of temperature ~ 50 K. Except for the warm cloud cores near star formation regions, the infrared flux distribution is a measure of dust column density.

About 8° east of the Orion A cloud is the Mon R2 cloud (distance 850 pc; see Figure 1). It is presently unclear whether it is related to the Orion clouds. The NGC 2149 cloud may be a filament connecting the Orion A and Monoceros clouds at their southern end (189), although their respective velocities do not match very well. The NGC 2149 cloud is part of a chain of filaments (named the "southern filament" and the "crossbones") extending over 20° on the sky and inclined by about 50° to the Galactic plane. This chain connects the southern end of the Orion A cloud with the Galactic plane. Parallel to this chain runs the "northern filament," connecting the northern tip of the Orion B cloud to the Galactic plane.

2.2 Origin of the Cloud Complex

An interesting question is whether the entire cloud complex has a common origin. Franco et al. (70) propose that the progenitor of the Orion/

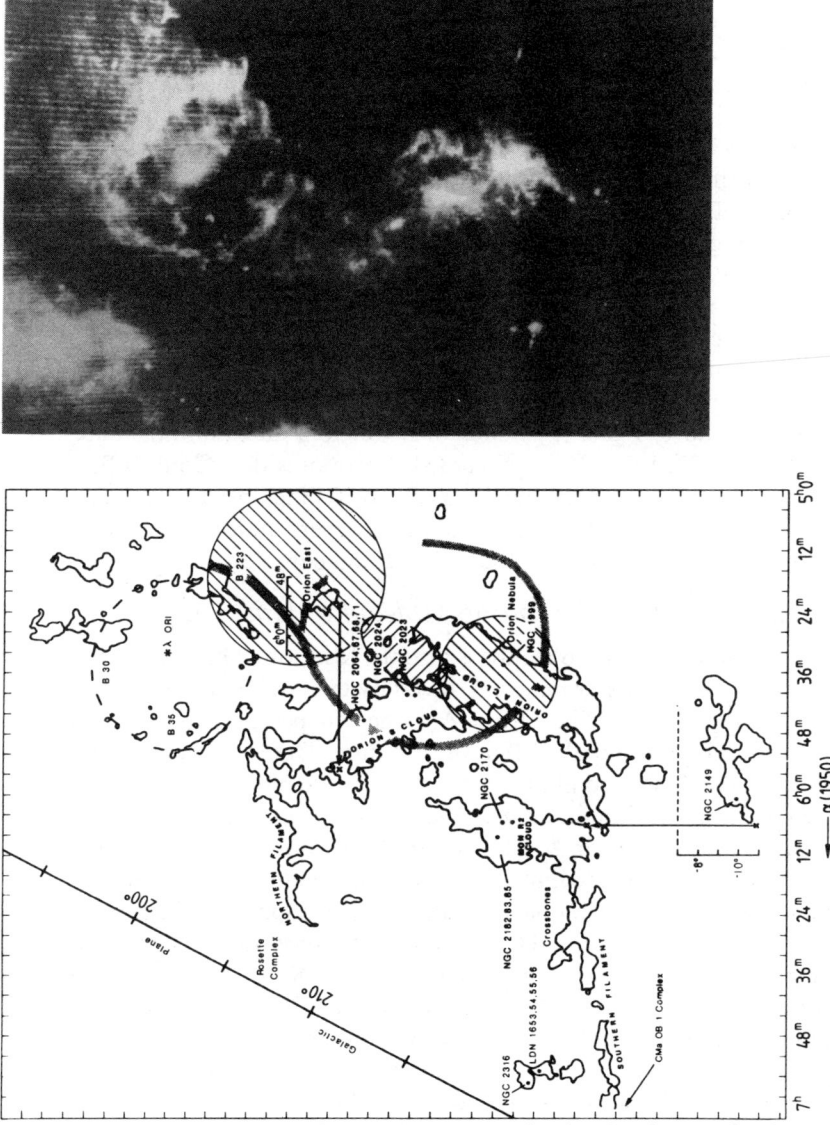

Figure 1 Large-scale distribution of clouds and star formation in Orion. *Left (a)*: Outlines of the molecular cloud complexes in Orion and Monoceros [from the CO mapping by Maddalena et al. (189)] along with the standard designations of clouds and individual star-forming regions (adapted from 189). The hatched circles are the Orion Ia, Ib, and Ic OB associations (from northwest to southeast). The Orion Id OB association around θ^1 C is located at the position of the Orion Nebula. The shaded arc is Barnard's loop. *Right (b)*: Composite 12- to 120-μm *IRAS* image of the Monoceros, Orion A, Orion B and λ Orion *(top)* regions. Clearly recognizable are the bright OB star-forming regions (adapted from 22, 246).

Monoceros complex formed by the collision of a high-velocity cloud falling from the southern Galactic hemisphere with the Galactic disk about 6×10^7 yr ago. The cloud fragmented, owing to tidal forces, into the Orion and Monoceros complexes while oscillating once through the Galactic plane and is now about 150 pc below the plane. This model has some very attractive features. It accounts for the 150-pc displacement from the Galactic plane with the same mechanism that created the molecular cloud complex. It also explains why the first generation of massive stars (see Section 3.1) formed on the side of the cloud complex away from the Galactic plane. This was the side where the high-velocity cloud hit and caused the strongest compression. However, the model does not explain the origin of the parallel northern and southern filaments. They could consist of cloud material dragged along out of the Galactic disk at the edge of the impact area.

Gould's Belt, an expanding, irregular disklike structure tilted about 20° with respect to the Galactic plane, is apparent in the distribution of local OB supergiants and H I gas. It extends only out to a maximum distance of about 450 pc (177, 192, 225). Weaver (312) proposes that Gould's Belt may also have formed by the impact of gas clouds falling onto the Galactic plane from high latitudes. Its small extent, however, indicates that it is probably not physically related to the Orion cloud complex.

2.3 Small-Scale Distribution of Molecular Material

The ridge of the Orion A and B clouds is also the location of higher density $[n(H_2) \geq 10^4$ cm$^{-3}]$ molecular material traced by the large-scale surveys in OH (14) and H_2CO (44). The Orion A ridge (40 pc × 2 pc) has been mapped in detail by Bally et al. (12) in the 2.7-mm ^{13}CO $1 \to 0$ line (Figure 2). It consists of an S-shaped filament of about 1° (8.6 pc) length, approximately centered on the Orion Nebula, and a V-shaped structure of length 4° with the tip near the Orion Nebula and pointing northwest toward the center of the I Orion OB association. The whole cloud has a very clumpy or filamentary structure in the spatial and velocity domains (Figure 2). Bally et al. identify more than 100 individual condensations, which typically cover areas of 30 to 100 arcmin^2 (length scale 0.8 to 1.5 pc) and have masses of a few tens to several hundred $M_\odot$ of gas. All of the individual sources in the Orion A and B clouds have very interesting features worth discussing in some detail. In the following we concentrate on the region near the Orion Nebula, which is the best-studied region.

This prominent molecular emission region (referred to henceforth as OMC 1) extends several parsecs north-south in high-density tracing molecules like H_2CO (66, 106, 163–165, 285), HCN (42), and CS (184). There are local peaks near the infrared source OMC 2 (73, 128) and a sharp

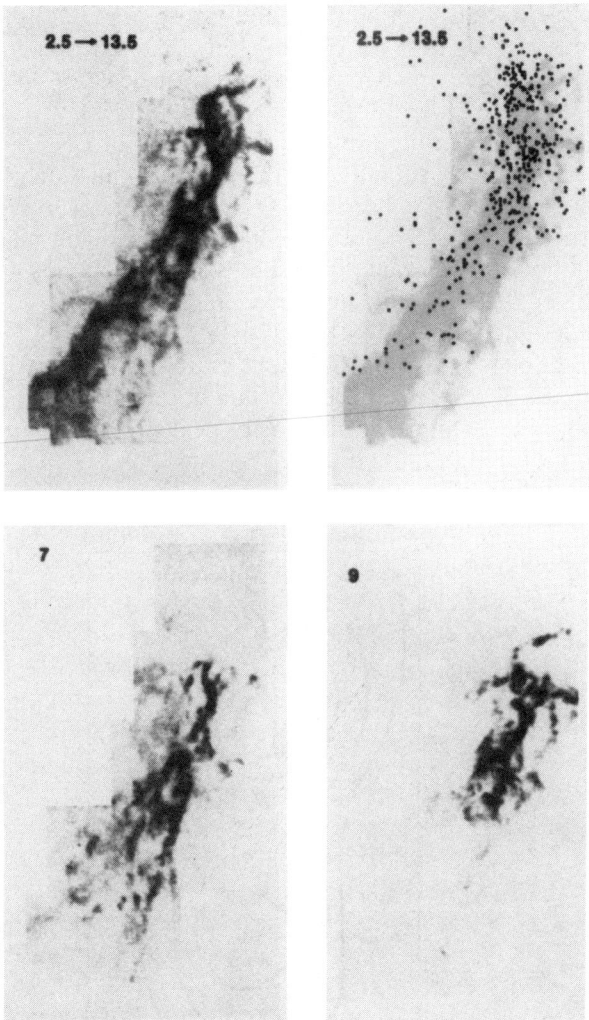

Figure 2 Large-scale structure and star-forming activity in the Orion A region. *Top left:* Velocity-integrated ^{13}CO $1 \to 0$ map of Orion A at 90″ resolution (from 12). The northern boundary of the map is at declination $-4°$, the southern boundary at $-9°$. *Top right:* The same, overlaid with the distribution of Hα emission-line stars (227; adapted from 146). *Bottom left and right:* Typical ^{13}CO velocity channel maps at $v_{LSR} = 6$ km s^{-1} and 9 km s^{-1}, respectively, showing the fine structure and overall velocity gradient in the cloud.

interface toward the H II region NGC 1977. This ridge is the central part of the S-shaped filament seen in ^{13}CO (12).

Observations by Ziurys et al. (340) and Batrla et al. (13) of the 1.3-cm inversion lines of NH$_3$ were the first to show that this ridge consists of a chain of individual clumps barely resolved on a scale of 40″ (0.1 pc) or smaller (103). More recently, the general clumpiness of the ridge has become very obvious in high-resolution mapping of CS $2 \to 1$ (90, 108, 213, 214) and millimeter/submillimeter continuum emission (40, 148, 213, 248). Figure 3 is a 22″ resolution C^{18}O $J = 1 \to 0$ map by Wilson et al. (324), superposed on a 35″ resolution 400-μm continuum map by Keene et al. (148). Interferometric mapping shows that the clumps have cores of typical sizes ≈ 0.05 pc, molecular hydrogen densities of $\geq 10^5$ cm^{-3}, and masses of several ten $M_\odot$ (103, 213, 327).

Several of the clumps [e.g. OMC 2, peak 2 (BN-KL) and peak 4 (OMC 1-South) in Figure 3] contain embedded stars observed in the near-infrared (73, 143, 199), are far-infrared luminosity peaks (137), or are associated

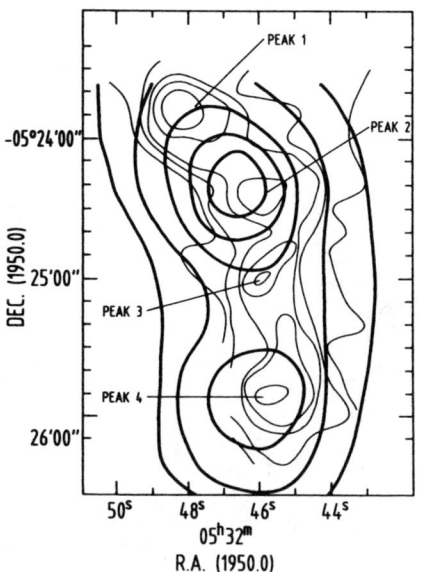

Figure 3 Submillimeter continuum map [heavy contours: 400 μm (148), 35″ beam] and C^{18}O $J = 1 \to 0$ velocity integrated map [thin contours, 2.6 mm (324), 22″ beam] of the central ridge near BN-KL (adapted from 324). Peak 4 is identical to OMC 1-South [using the nomenclature of Keene et al. (148)]. Peak 2 is identical with the hot core and the dense cloud core near BN-KL (see also Figure 6).

with H_2O masers and mass outflows (12, 78, 331). For example, OMC 1-South contains a molecular outflow source [as indicated by high-velocity wings in the SiO $v = 0\ 1 \rightarrow 0$ (342) and CO $7 \rightarrow 6$ (253) transitions] and at least three embedded, compact, near-infrared sources (199). It is, however, less luminous ($L \approx 10^4\ L_\odot$) and has a lower mass ($\approx 50\ M_\odot$) and lower dust temperature (45 K; 56, 137) than the BN-KL nebula ($10^5\ L_\odot$, 150 $M_\odot$, 70 K), which indicates that it contains less luminous stars or is in an earlier stage of evolution.

Other clumps are cold and may be prime candidates for collapsing clouds (103, 213, 327). Mundy et al. (214) note that the OMC 1 ridge of clumps may have been compressed externally by the expanding Orion A H II region. The condensations are not unique to OMC 1. Mezger et al. (206), for example, find a number of compact (10″) dust clumps in NGC 2024 from 1-mm continuum mapping with the IRAM 30-m telescope. Lada (170) finds a series of condensations of dense gas in the Orion B cloud from a CS $2 \rightarrow 1$ survey with the Bell Labs antenna.

These observations clearly show that the dense molecular clumps are the locations of recent and ongoing star formation. The submillimeter/millimeter dust continuum peaks may be the best candidates yet for locations of protostars or stars in very early evolutionary stages.

2.4 *Kinematics and Energetics of the Gas*

There is a large-scale velocity change along the Orion A cloud (12 km s^{-1} LSR in the northwest to 5 km s^{-1} in the southeast) that has been interpreted as a rotation of the entire cloud along an east-west axis (166). The average gradient is 0.135 km s^{-1} pc^{-1}, implying a rotation period of 4×10^7 yr. The sense of rotation would be opposite to the differential Galactic rotation. Simple overall rotation, however, is not sufficient for an adequate description of the measured velocity field. Velocities are about constant in condensations of about 1° size (44, 166). The central S-shaped filament shows a rather complicated velocity structure, including a sharp change in velocity near the BN-KL nebula (Section 5.2). In addition, the velocity field may be disturbed by interaction with ionization and shock fronts from the H II region around the Trapezium OB stars (186, 250, 278, 320). As in other clouds, the supersonic line widths (1 to 5 km s^{-1}) suggest the presence of local turbulent motions.

Outflows from embedded young stars represent one possible source of kinetic energy that could drive these turbulent motions. Plausible arguments have been presented both in favor of (190, 218) and against (11) outflows as a major source of turbulent energy. Much depends on the time scale of dissipation of the turbulence. Scalo & Pumphrey (249) estimate that the dissipation time scale is about one order of magnitude larger than the free-fall time scale, or 5×10^6 yr. Other important factors are the number and size of the outflows. Fukui et al. (71) find a total of seven outflows from an unbiased survey in Orion A (see also 10). They estimate that these outflows may deposit about 2 $L_\odot$ of mechanical luminosity into the cloud, excluding the energetic outflow(s) from BN-KL (200 $L_\odot$; 266). If the present number of outflows is representative of the average number in the cloud at any time, then these flows can supply the turbulent energy in the cloud of $\approx 10^{48}$ ergs, if Scalo & Pumphrey's estimate of the dissipation time is applicable. The flows may be able to affect a significant fraction of the cloud, as they are distributed rather uniformly throughout Orion A. Further evidence for the effect of outflows on the overall structure of the cloud comes from the ^{13}CO map of Orion A (12), which shows many bubbles or shell-like cavities around outflow sources and Herbig-Haro objects (243). However, especially the larger bubbles may result from other effects, such as the expansion induced by the thermal pressure in photodissociation regions around embedded B stars (J. Bally, private communication).

A second possibility that needs to be considered is that magnetic fields support the turbulence (see discussion of observations of the magnetic field in Section 2.5). A magnetic field of a few 10 μG is required for equipartition between magnetic, gravitational, and kinetic (turbulent) energy (214). At this field strength the linewidth is about equal to the Alfvén speed. The energy dissipation in clump-clump collisions in the cloud is substantially reduced by the presence of the magnetic field, which could support the turbulent linewidths over the estimated lifetime of the cloud. Puget (242) and Pudritz (241) have discussed scenarios of how, under such conditions, the clumpy structure of the molecular cloud could be maintained or possibly created, following the original suggestion by Arons & Max (5).

2.5 The Magnetic Field

Optical polarization measurements of magnetically aligned dust grains provide information on the orientation of the magnetic field component in the plane of the sky (50). Following initial work by Appenzeller (3, 4), Vrba et al. (302) have presented polarization measurements of stars selected to be in the vicinity of the L1641 dark cloud. The distribution of polarization position angles of 71 stars with polarization greater than 1% peaks near a position angle 110° east of north, with a FWHM spread of 33°. About 40 of the 71 stars are highly reddened and thus are probably embedded within the cloud. The average magnetic field direction is within 30° of the long axis of the Orion A molecular cloud. The estimated angle between the magnetic field and the line of sight of about 70° (3) is also consistent with the magnetic field lines running along the major axis of the molecular cloud. Hence, the Orion A cloud cannot have contracted to its present elongated or flattened shape along magnetic field lines. Rather, the magnetic field may support the elongated structure of the Orion A cloud.

Heiles (114, 116) finds that the line of sight magnetic field reverses its direction across the molecular cloud, pointing toward the Sun on the side of the Orion A cloud toward the Galactic plane and away from the Sun at lower Galactic latitudes. The reversal of the field across the cloud matches the observed reversal of the line-of-sight component of the magnetic field below and above the Galactic plane (209). This finding supports the idea that the Orion clouds may be fragments being pushed out of the Galactic plane by the impact of a high Galactic latitude cloud.

Appenzeller (4) favored a model in which a magnetic pocket supports the Orion complex at low Galactic latitudes. However, with the new information about the large-scale distribution of the molecular material in Orion now available, a close inspection of his polarization map [Figure 1 in (4)] actually shows very good agreement with the interpretation discussed above, i.e. that the magnetic field lines are trapped in and oriented along the long axis of the clouds.

As indicated by the large scatter of polarization position angles, the orientation of the magnetic field on small scales does not necessarily trace the large-scale average orientation discussed above. Rather, the small-scale magnetic field orientation is correlated with the local distribution of the cloud in the case of NGC 2024 (308, 338). Similarly, the polarization angles of the embedded objects in the KL nebula and OMC 1-South derived from near-infrared absorption measurements (58, 59, 152) and far-infrared emission studies (49, 52, 125) give a magnetic field direction aligned very well along the axis of the local outflow. On the other hand, the directions of several optical outflows are aligned with the large-scale magnetic field (274), which suggests that the magnetic field plays an important role in the presently ongoing star formation in the clouds, even if it may not have dominated the formation of the complex as a whole.

Line-of-sight magnetic field strengths for several positions in Orion A and B derived from H I 21-cm and OH 18-cm Zeeman measurements range from 50 to 125 μG (30, 48, 113, 115, 147, 293, 298). The field strength inferred for the OH maser sources in BN-KL is a few microgauss (34, 100, 219), and 40 mG in the BN-KL H_2O maser spots (67), all pointing toward the observer. The total field strengths are likely more than twice as large as the line-of-sight fields, as the field in Orion is probably mainly in the plane of the sky. Total field strengths thus derived cover gas clouds with hydrogen densities ranging from several times 10^2 cm^{-3} (H I) to about 10^{10} cm^{-3} (H_2O). They are consistent with theoretical models predicting that the magnetic field is about 5 μG up to densities of about 10^3 cm^{-3} and then increases proportional to (density)$^{1/2}$ (147, 211, 214, 292). The models are based on an assumed equipartition between magnetic, turbulent, and gravitational energy and include the process of ambipolar diffusion.

In summary then, the measurements provide compelling evidence that magnetic fields are dynamically important for support of the molecular cloud against gravitational collapse and critically control the star-forming process in the densest cores.

3. THE STELLAR COMPONENT

3.1 *The I Orion OB Association*

At visible wavelengths the outstanding feature in the Orion region is the H II region M42 (Orion A), which is excited by the θ^1 OB stars (the "Trapezium"). The θ^1 OB stars are the brightest members of the youngest subgroup of the I Orion OB association. The I Orion OB association contains 56 O6 through B2 stars (23). The four subgroups Ia–Id are progressively younger, with ages of 12, 8, 6 and 4 Myr, respectively. The older groups have larger diameters. Relating size and age gives expansion velocities near 2 km s^{-1}, in good agreement with the velocity dispersion of 2.3 km s^{-1} measured in the youngest cluster (Section 3.4; 146). The older subgroups are at a larger (projected) distance from the molecular cloud than are the younger ones (Figure 1). A similar trend is observed for H I gas, for which the H I column densities are correlated inversely with age of the subgroups (93). These facts strongly support the concept of externally triggered star formation (e.g. 62, 63). Ionization/shock fronts from supernova explosions in the first generation of high-mass stars initiate the formation of the next generation of stars out of the same giant molecular cloud complex. Then the original molecular cloud gets successively disrupted by the ongoing high-mass star formation. The Orion complex is one possible example of this process, with others being M17, Centaurus-Scorpio, W3, or CMa R (64, 168).

Distance estimates for the stars in the OB association range between 400 and 500 pc (28, 142, 178, 183, 226, 258, 259, 273, 304, 307) and are in good agreement with the direct distance determination to the BN-KL star-forming core in OMC 1 by H$_2$O proper motion measurements [470±70 pc; (79) and R. Genzel, M. J. Reid & J. M. Moran, private communication]. There may be a distance gradient in the OB subgroups, with the older subgroups being closer than the younger ones (28, 46, 102, 273, 307). The average distance between subgroups is about 50 pc. This distance and the age difference between the subgroups transform into a propagation velocity of the star-forming activity of around 25 pc Myr^{-1}, or 25 km s^{-1}. Such a high velocity is easily explained if the sequence of star formation events were propagated by the combined supernova shocks of the earlier generation of stars.

3.2 Supernova Explosions, High-Velocity Gas, and the Structure of the Cloud

Optical diffuse emission lines (244) and UV absorption-line measurements (45) show that the I Orion OB association is surrounded by rapidly expanding material. Cowie et al. (45) interpret high-velocity gas with LSR velocities of about -100 km s^{-1} as a spherically expanding shock in its radiative phase, triggered by the last supernova event about 3×10^5 yr ago. Lower velocity, denser gas (15–25 km s^{-1}) may be material swept up and accelerated by the subsequent series of several tens of supernovae that have occurred since the oldest subgroup in the association formed. The inner edge of this material has been ionized, and part of this ionized shell forms Barnard's loop (Figure 1; 45, 220). Barnard's loop may interact with the Orion A and Orion B clouds, as there is a larger velocity dispersion in those areas where Barnard's loop intersects the molecular clouds.

Bally et al. (12) suggest that the structure of the Orion A cloud is influenced by this high-velocity material. In this scenario the narrow V-shaped feature pointing towards the northwest in the direction of the center of the oldest OB association is a bow shock resulting from the refraction of the rapidly expanding material at the high-density material in the core of the cloud. A similar interaction is indicated by filaments or streamers in the Orion B cloud and by large cometary globules located about 9° west of Orion A, seen in ^{13}CO (J. Bally, private communication), pointing toward the oldest OB subgroup.

Bally et al.'s (12) interpretation also gives a clue to the spatial configuration of the Orion region. The alignment of the V-shaped feature toward the center of the OB association, taken together with the different distance of the oldest subgroup and the Orion Nebula, implies a tilt of the Orion A cloud of about 45° with respect to the plane of the sky, with the southern region being farther away. In this geometry the velocity change along the Orion A cloud may actually result from the interaction with the expanding material, which has already accelerated the northern part of the cloud now moving away and thus appearing redshifted.

3.3 Distribution of Low-Mass Stars

Surveys of Hα emission-line stars (118, 150, 227) as tracers of low-mass star formation show a distribution nicely tracking the molecular ridge in the Orion A cloud [Figure 2 (*top right*)]. The distribution of young low-mass stars peaks in the area of the "Orion Nebula Cluster," but there is also a larger, lower density distribution of stars following the molecular emission region. The Orion Nebula Cluster contains at least 1000 stars (146). It extends about 2.5 pc east-west and 4.5 pc north-south along the central S-shaped molecular ridge. The near-infrared survey of the L1641 dark cloud by Nakajima et al. (217) shows the detected pre-main-sequence stars to be closely associated with the dark cloud in the southern part of the Orion A cloud. Cohen & Kuhi (43) find that the T Tauri stars in Orion are younger than a few times 10^6 yr, similar to the age of the high-mass stars that formed in the Orion Nebula Cluster. Also indicative of the youth of a number of the pre-main-sequence stars is their unusual X-ray activity (159).

3.4 The Trapezium Cluster: Current Low-Mass Star Formation in Orion A

Trumpler (294) and Baade & Minkowski (7) first noted a concentration of faint visible stars near the θ^1 OB association, now generally referred to as the "Trapezium Cluster" (TC; 119). New information has come in the past few years from narrowband charge-coupled detector (CCD) imaging and optical spectroscopy, from proper motion studies, and from near-infrared imaging, now greatly facilitated by the advent of two-dimensional detector arrays (e.g. 68). As an example, Figure 4 shows a 2-μm (K-band) image of a 5.3' × 4.7' area centered on θ^1 C, created by mosaicing 126 individual images with the new 1- to 5-μm Infrared Camera IRCAM on the United Kingdom Infrared Telescope (198, 201). This image reveals 477 near-infrared stars brighter than $K = 15$.

Herbig & Terndrup (120) determined a color-magnitude diagram for 68 stars of the Trapezium Cluster within 0.4 pc projected distance to θ^1 C, combined with spectroscopic information for the brighter members. They conclude that most of the TC stars are young ($<$ a few times 10^6 yr), with little indication of a population of stars older than 10^7 yr. Most of the stars have masses between 0.5 and 2 $M_\odot$. It is thus improbable that the visible Trapezium Cluster has formed throughout the lifetime of the molecular cloud ($>$ a few times 10^7 yr). The cluster's stellar density is 2×10^3 stars pc^{-3} or larger, corresponding to an average density of at least 10^5 hydrogen nuclei cm^{-3}, typical of the densest cores of the molecular cloud. The stars, therefore, must have formed at high efficiency ($>10\%$) from a very dense condensation in OMC 1.

One important question is the spatial extent of the cluster. Do the young low-mass stars penetrate a significant part of the molecular cloud behind, or are they a local concentration near the OB association and at the surface of the cloud? Hyland et al. (133) surveyed a 1' × 4' region at K and find 88 point sources brighter than $K = 13$. They conclude that there are in fact reddened pre-main-sequence stars embedded in the molecular cloud,

Figure 4 Composite mosaic of 126 2-μm (K-band) continuum images of a 5.3' × 4.7' area from the work of McCaughrean (198) and McLean et al. (201). The transfer function chosen is proportional to the cube root of intensity in order to emphasize the brightest stars (θ^1 C and θ^2 A in the center of image, BN toward the northwest of θ^1 C) as well as the weakest stars and nebulous emission. The faintest stars are approximately $K = 15$, and the faintest nebulosity is about $K = 17$ arcsec^{-2}. A cluster of faint stars clearly can be seen to be concentrated on θ^1 C. This is the "Trapezium Cluster" discussed in Section 3.4.

similar in intrinsic color to the T Tauri stars sampled by Cohen & Kuhi (43). Herbig & Terndrup (120) compared these 2-μm observations with the predicted K-magnitude distribution extrapolated from the density of visible stars at the surface for different values of the total 2-μm extinction A_K through OMC 1. Estimates for A_K range from 5 to 15, as derived from submillimeter or near-millimeter dust continuum emission (85, 148, 252) or $C^{18}O$ column densities (252, 324). For $A_K = 10$, the number of stars at $K \geq 10$ (133) is smaller than the predicted number by about a factor of 2. Herbig & Terndrup, therefore, conclude that the very high stellar density of the Trapezium Cluster is local to the near side of the molecular cloud. A local enhancement of young low-mass stars near θ^1 C is also strongly supported by the spatial distribution of the visible and near-infrared stars (146, 198).

The situation may still be more complicated, however. On the one hand, the more recent 2-μm IRCAM image (199; shown in Figure 4) has a substantially larger area density of stars fainter than $K = 10$ compared with the work by Hyland et al. (133). Bally et al.'s (12) ^{13}CO mapping of OMC 1 shows that the molecular cloud is highly filamentary, indicating that the line-of-sight depth is 0.5–2 pc, significantly smaller than the 4 pc assumed by Herbig & Terndrup (120). McCaughrean et al. (199) have reanalyzed the comparison between visible and infrared star counts, taking into account these new aspects. They find that the observed K distribution can be well explained by a cluster penetrating OMC 1, as long as the depth of the cloud is ≈ 2.5 pc and its K extinction is ≈ 4. For the smaller depths and larger extinctions indicated by the millimeter measurements, there are even too many near-infrared stars. On the other hand, McCaughrean et al. do not find strong evidence for a correlation between K-magnitude and near-infrared color. Such a correlation would be expected if the fainter stars are located deeper in the cloud, as in Herbig & Terndrup's model used for the above analysis.

Perhaps the most likely explanation is that the low-mass cluster around θ^1 C in fact only penetrates the small and dense molecular core near the OB stars. A conclusion of this general type is further supported by the very interesting 2-μm imaging of dense molecular clumps in Orion B by Lada (170). She finds that four out of the five most massive concentrations of molecular gas ($M_{gas} > 500$ $M_\odot$) show local clusters of faint embedded stars with typically > 50 members brighter than $K = 14$. Each one of them (NGC 2023, NGC 2024, NGC2068, NGC 2071) also contains at least one OB star. Moreover, the only massive clump that does not show embedded faint stars also does not contain OB stars.

The obvious interpretation of all these findings is that the triggering of star formation by external events is important. Either external compression waves have induced coeval formation of both low- and high-mass stars relatively recently ($<$ a few times 10^6 yr ago), or else the high-mass stars form first and then trigger low-mass star formation in their vicinity.

Another important question is the dynamical state of the Trapezium Cluster. Recent studies of radial and proper motions of the stars in the

Orion Nebula Cluster find velocity dispersions of 2.3 ± 0.1 km s^{-1} (for 900 members) and 1.5 ± 0.2 km s^{-1} (for the 49 brightest members), respectively (146, 297). These velocities are too high to keep the Orion Nebula or Trapezium Clusters bound or in virial equilibrium (146). The combined system of molecular cloud, young low-mass stars, and OB association, however, probably does satisfy the virial criterion.

In summary then, the most likely scenario is that a very dense cluster of young pre-main-sequence stars has been forming near θ^1 C for a few times 10^6 yr, probably in an unusual concentration of dense gas. Its density at present is still more than two orders of magnitude larger than that of ordinary young clusters (120). The cluster will most likely disperse in the future as the OB stars in the Trapezium and in BN-KL remove the gas that is necessary to bind the overall system. This scenario is consistent with Herbig's original proposal (117) that low-mass star formation activity in a cloud ceases as soon as massive stars are formed that disperse the cloud (see also 169).

4. INTERACTION BETWEEN THE ORION A OB STARS AND THE CLOUD

4.1 *The Orion A H II Region*

Zuckerman (345) was the first to propose a model for Orion A in which the compact H II region (diameter 0.5 pc) is in front of and eating into the dense Orion molecular cloud [the "blister" model; see also (9, 347)]. In this picture the BN-KL infrared nebula is 0.1–0.2 pc behind θ^1 C near the front surface of OMC 1 (164, 344, 345). The blister model is still a good description at the present time (229) and can also be understood theoretically in terms of the so-called champagne phase of H II regions (284). In this model, the line and radio continuum emission originate predominantly from gas near the photoionized front surface of the cloud. There are several compact thermal radio knots embedded in the Orion A H II region (41, 72) that may be dense, externally ionized fragments remaining from the dense molecular cloud that is now being disrupted by radiation and winds from the OB stars (172). The evaporating ionized gas streams away from the cloud surface into the H II region and toward the observer, thereby explaining the ~ 14 km s^{-1} blueshift of the hydrogen recombination lines relative to the neutral gas. Optical, radio, and infrared spectroscopic measurements of the ionized gas have given detailed information on the H II region's average electron density ($n_e = 5000$ cm^{-3}; 180), electron temperature ($T_e = 8500$ K; 323) and ionic abundances (179, 229, 230). Simpson et al. (262) have presented a detailed model of the optical and infrared line emission of the central nebula. Adopting the line-blanketed, LTE stellar atmospheres of Kurucz (162), they derive a best-fit, mean effective temperature of the UV radiation field of ≤37,000 K. The observationally derived effective temperature of θ^1 C is ~39,000 K (O6p; 174). As an explanation of this discrepancy, Simpson et al. propose either that dust inside the nebula softens the UV field or, more likely, that the theoretical atmospheres are not of sufficient accuracy.

4.2 The H II Region/Molecular Cloud Interface

UV photons longward of the Lyman edge of atomic hydrogen can penetrate from the ionized region quite deep into the surrounding molecular cloud. There the UV radiation can photodissociate molecules and photoionize atoms (such as carbon and silicon) with ionization potentials < 13.6 eV. This "photo-dissociation region" (PDR) associated with the molecular cloud ($v_{LSR} = 8$ km s^{-1}, $\Delta v = 5$ km s^{-1}) was first apparent in optical C I line emission (126, 212) and in studies of C II radio recombination lines (8, 136, 160, 346).

Measurements during the past few years now clearly show that dense photodissociation regions (PDR) exposed to intense UV fields are copious emitters of atomic and molecular, infrared, and microwave line emission. Table 1 contains a summary of the relevant observations and the derived physical parameters for the Orion A interface region. The PDR signposts include intense emission lines in the far-infrared/submillimeter fine-structure transitions of [C II] (247), [O I] (202, 268), and [C I] (138, 231). Bright molecular line emission from the UV-heated interface has been observed in the H$_2$ 2-μm ro-vibrational, quadrupole lines (112), as well as in millimeter, submillimeter, and far-infrared CO rotational lines (27, 156, 250, 253, 270). The observations indicate hydrogen densities of at least 2×10^5 cm^{-3} and gas kinetic temperatures of 100–500 K for the atomic and molecular gas, well in excess of the dust temperature (47, 269, 270). Hydrogen column densities in the PDR exceed 10^{22} cm^{-2} ($A_v = 5$; Table 1).

Figure 5 shows maps of the 158-μm [C II] (270) and 372-μm CO $J = 7 \to 6$ lines (253), superimposed on the distribution of 2.6-mm CO $1 \to 0$ peak temperature (250). The right panel of Figure 5 gives a comparison of the [C II] (27) and CO $7 \to 6$ (270) line profiles toward θ^1 C. The similarity in profiles, spatial extents, and distributions of the warm molecular and atomic gas definitely show that the interface region is physically associated with the quiescent molecular cloud. It is also clear that the ^{12}CO millimeter and submillimeter "spike" emission in Orion A is dominated by the interface region (e.g. 224). Note that the Hα and radio continuum distributions are also very similar to that of the [C II] line, although the kinematics of the ionized gas are quite different, as discussed above. [C II] and warm

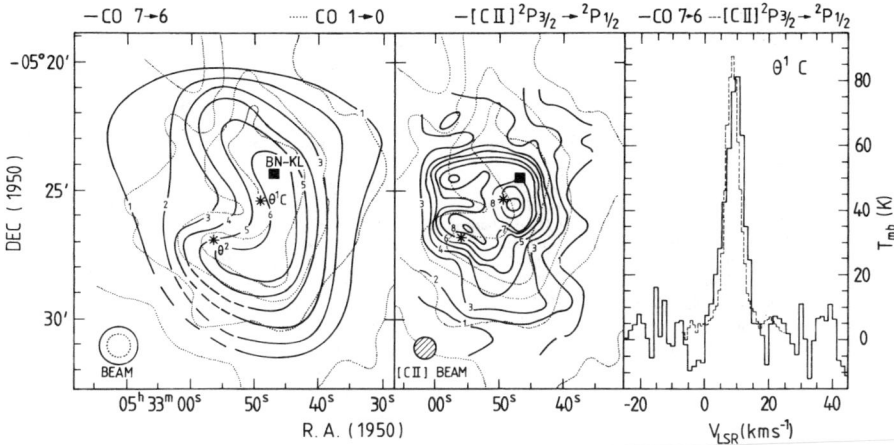

Figure 5 The Orion A photodissociation region. *Left:* 372-μm CO $J = 7 \to 6$ velocity-integrated map [heavy contours, 90″ beam (253)], superimposed on a map of the CO $1 \to 0$ brightness temperature [peak 100 K, 50″ beam (251)]. The $7 \to 6$ data were taken with the Max-Planck-Institut für Radioastronomie heterodyne spectrometer on the *Kuiper Airborne Observatory* (*KAO*) on approximately full-beam sampling. θ^1 C and θ^2 A are indicated as asterisks, and the position of BN-KL is a filled square. *Middle:* Map of the 158-μm [C II] $^2P_{3/2} \to {}^2P_{1/2}$ fine-structure transition [heavy contours (from 270)], superimposed on the CO $1 \to 0$ brightness temperature map. The [C II] data, taken with the University of California, Berkeley tandem Fabry-Perot on the *KAO*, are fully sampled with 55″ resolution. The bar (near θ^2 A) and dark bay (just north of it) can clearly be seen. *Right:* Comparison of 158-μm [C II] [45″ beam (27)] and CO $7 \to 6$ [26″ beam (270)] spectra toward θ^1 C. Both lines are centered near 9 km s^{-1} and have a width of 5 km s^{-1}, indicative of gas associated with the molecular cloud.

CO emission probably come from the neutral back side of the H II blister. The prominent bar, visible in infrared line and continuum emission (e.g. 18), may be the edge of the blister tangential to the line of sight (287).

The bar is also prominent in the 3.3-, 6.2-, 7.7-, and 11.3-μm dust emission features (74). These features are now commonly believed to come from very small (10–50 Å) dust grains or from very large molecules, such as polycyclic aromatic carbohydrates (PAHs; 176, 290). A small dust grain is transiently heated to a very high temperature when it absorbs a UV photon (257). The UV-heated, small dust grains and the warm photodissociated gas are intimately connected.

Theoretical models of neutral clouds illuminated by UV radiation predict a well-defined C II/C I/CO transition region (171, 305, 313). Tielens & Hollenbach (288, 289) and Sternberg & Dalgarno (271) have calculated detailed models of such dense, warm interface regions for cases of high UV energy densities [$\varepsilon_{UV} \approx 10^5 \times \varepsilon$ (interstellar radiation field)], as appro-

Table 1 Components of molecular and atomic gas in Orion-KL

	Quiescent cloud material			
	Core of extended ridge	Compact ridge	Hot core	Photodissociated gas at surface of cloud/H II
Tracers	Molecular lines ^{13}CO, C^{18}O (250, 324) CS (108, 214) NH$_3$ (13, 127, 340) HCN (301) HCO$^+$ (300) Millimeter/submillimeter dust continuum (85, 148, 263, 318)	Molecular lines H$_2$CO (144) CH$_3$OH thermal + masers (195, 196, 205, 238, 326) HDO (237)	Molecular lines NH$_3$ (81, 124, 210, 228, 291, 306) HC$_3$N (194) HDO (237), CH$_3$CN (187, 279b) Dust: mm/submm continuum (75, 192, 334) Vibrationally excited gas pumped by IR: CH$_3$CN, HC$_3$N (89, 91) HCN (341), NH$_3$ (197)	Far-IR/submm atomic [C II]/[O I]/[Si II] (26, 97, 269, 270) [C I] (233, 343) IR/submm molecules CO* (27, 253, 270) H$_3^+$ (74, 112) mm molecules CO (111, 250) [C II] recombination lines (135) visible C I (126)
Source size (FWHM)	>40″ along position angle 30° with embedded knots	15″, centered on IRc4/5	5–10″, centered 3″ south of IRc2; clumps ≤1″	5′ along ridge
v_{LSR} (centroid) [km s^{-1}] Δv_{FWHM}	8–10 3–5	7–8 3–4	5–6 5–15 turbulent knots	8 5
T_{kin} [K]	70 gas ≈ dust	100–150 (24, 215)	200 range from 150 (dust, NH$_3$) to ≥250 (^{13}CO, CH$_3$CN)	100 (CO, mm) ≥400 (CO, submm) 150–500 (C II/O I)
n(H$_2$) [cm^{-3}]	10^5	10^6?	1–3 × 10^7 cm^{-3} (NH$_3$, ^{13}CO)	10^5–10^6
N(H$_2$) [cm^{-2}]	3 × 10^{23} average, approaching 10^{24} in knot 30″ NE of IRc2	≥3 × 10^{23}	10^{24} (^{13}CO)–a few 10^{24} (mm/submm continuum)	1–3 × 10^{22}
M [$M_\odot$]	100 to 150 within 0.1 pc of IRc2	3	≥10	10 within 0.1 pc of BN-IRc2
Chemical characteristics	Standard ion-molecule gas phase chemistry; carbon-species much enhanced (C I?, H$_2$CO, CCH) (25, 141, 221)	High abundance of complex oxygen-rich molecules (CH$_3$OH, OCS, HCOOCH$_3$); interaction of ridge with low-velocity flow? (25, 141, 221)	Recent evaporation of fully hydrogenated molecules from dust grains [NH$_3$, H$_2$O(HDO), HCN, C$_2$H$_5$CN] (25, 31, 141, 221, 306)	Photodissociation of dense molecular cloud (271, 288, 289)
Remarks	Rotation of ridge (108, 301, 324) around IRc2/KL, or two separate clouds? Enhancement of C I due to foreground photodissociated gas (270)?		Submm dust continuum optically thick? $Q_d \approx \lambda^{-1.3}$ between 300 μm and 3 mm? (334) Contribution of lines to continuum at 1 mm ≥50% (279a)	

	Outflowing and shocked gas			
Low-velocity flow + shocked gas	High-velocity flow (plateau)	High-velocity shock-excited gas		
Maser flows H_2O (79, 208), OH (101, 219) SiO (222, 333) Thermal flow SiO $v = 0$ (332) NH_3 (81, 228) Shocked gas in flow SO, SO_2, HCN (234–236) HCN $9 \rightarrow 8$ (104, 275)	CO wings (65, 110, 155, 161, 167, 193, 223, 264, 325) Millimeter lines HCO^+ (300) SO_2 (235, 251) H_2O masers (79) SiO $v = 0$ wings (332) H_2CO (330) Far-IR/submm HCN $9 \rightarrow 8$ (104, 275) OH far-IR (203, 204, 299)	Near-IR rotation and ro-vibrational lines of H_2^* (16, 19, 21, 76, 216, 255) Far-IR/submm CO lines (27, 47, 156, 239, 254, 267, 272, 299, 303, 311) far-IR ^{13}CO (84) far-IR OH (203, 204, 299) far-IR [O I] [Si II] (97, 317)	Tracers
0.1″ (SiO masers on IRc2) 5–15″ (SiO $v = 0$, SO, SO_2) 30″ H_2O masers	10″ (SiO $v = 0$) 20–40″ (CO)	40″ (submm/far-IR CO) 60″ (H_2^*, [O I])	Source size (FWHM)
5 for flow near IRc2 8–9 for shocked gas FWZP: 35 km s^{-1} bipolar (?), flow along position angle 40–60°	7–9 FWZP: 150–200 bipolar along position angle $-45°$	0 ± 10 25–40 FWZP: 150	v_{LSR} (centroid) [km s^{-1}] Δv_{FWHM}
100–500	120 (SO_2) and higher (CO)	200–2000 (CO) 1000–3000 (H_2^*)	T_{kin} [K]
10^5 average $\geq 10^7$ (HCN $9 \rightarrow 8$) $> 10^9$ (H_2O and SiO masers) } clumps	10^5–10^6	10^6 (CO) 10^7–10^8 (HCN $9 \rightarrow 8$, OH)	$n(H_2)$ [cm^{-3}]
10^{23}	5×10^{22}	3×10^{21} (CO) 3×10^{19} (H_2^*)	$N(H_2)$ [cm^{-2}]
5	5	0.5 (500–1000 K) 0.02 (2000 K)	M [$M_\odot$]
Si/H_2O-rich near star; shock chemistry (sulfur-bearing molecules abundant) (SO, SO_2) (25, 141, 221)	Shock chemistry (25, 141, 221)		Chemical characteristics
Low-velocity flow bipolar?		Clump-clump collisions for explanation of high-velocity wings? C and J-type shocks?	Remarks

priate for Orion A. The models can account for most observed line intensities and the high observed gas temperatures by the combined effects of photoelectric heating and far-UV pumping of H_2. A major shortcoming of the models at present is their failure to fully account for the observed high intensities of the submillimeter and far-infrared CO lines, which require temperatures of 100–500 K in the CO emission region (270; see Table 1).

Photodissociation regions may also be the areas in which the extended 610-μm [C I] fine-structure line emission in Orion originates, discovered by Phillips & Huggins (233). They find that the submillimeter atomic carbon emission extends at least 20' (3 pc) east and west of the Orion A H II region, very similar in spatial extent to that of the CO $1 \to 0$ line. Phillips & Huggins conclude that the [C I] submillimeter emission must originate throughout molecular clouds, possibly as a result of the young chemical age of the cloud ($t < 10^6$ yr), the production of atomic carbon from CO photodissociated by cosmic-ray-induced UV photons (282), or a C/O abundance ratio >1 that leaves atomic carbon in the gas phase (149). Recently, Stacey et al. (270) have found strong 158-μm [C II] emission spread over about the same region as [C I]. Unlike [C I], [C II] emission of the observed brightness requires the presence of relatively intense (stellar) UV radiation, which can then also account for the [C I] emission. The most likely origin for the extended photodissociation region in Orion is a clumpy structure of the molecular cloud that allows UV photons from Orion A or from embedded B stars to penetrate several parsecs into the cloud (e.g. 276).

5. THE BN-IRc2 STAR FORMATION REGION

5.1 *The BN-IRc2 Infrared Cluster*

Following the discovery of the Becklin-Neugebauer object (17) and of the cooler, extended Kleinmann-Low nebula (151) 10" south of it, Rieke et al. (245) reported that at higher resolution the BN-KL infrared nebula splits up into a number of compact sources of different color temperatures. Figure 6a shows a contour map of a recent 12-μm camera image at 0.9" resolution by Gezari (86). This image and other high-resolution, 2- to 30-μm maps (51, 96, 175, 337) show at least 10 compact "sources" embedded in extended emission, with standard designations marked on Figure 6a. The KL nebula is composed of the sources IRc2-5 and surroundings.

Two important questions come to mind. First, are these sources self-luminous? Second, what is their luminosity and their evolutionary state? Until a few years ago, the common interpretation has been that the infrared nebula is a cluster of newly formed stars in a very early stage of evolution

(335)—that is, protostars in Larson's (173) or Wynn-Williams' (336) sense. In the following, we discuss how measurements of the last few years have changed this picture considerably.

The total mid- and far-infrared luminosity of the complex is between 6×10^4 and 1.2×10^5 $L_\odot$ (56, 137, 287, 314), which indicates that the young stars in the cluster are luminous and massive. Only about 10% of that luminosity emerges at $\lambda \leq 30$ μm, where large ground-based telescopes allow the separation into individual sources. This fact also makes clear that radiative transport effects at near- and mid-infrared wavelengths are very important. Hence, until very recently, the assignment of luminosities to individual sources in the BN-KL region has been a major issue of debate.

Absorption features in the infrared continua due to, for example, water-ice at 3.1 μm or silicates at 9.7 μm give detailed information on the composition and column density of the dust in the nebula (87, 153). Rieke et al. (245) first noted that the integrated depth of the 9.7-μm silicate absorption feature is stronger in the sources of the KL nebula than in BN itself. Measurements at higher spatial and spectral resolution show that the silicate absorption depth (and shape) varies greatly from source to source, with IRc2, 3, and 4 having the deepest features (1, 51, 96). If the silicate absorption is due to cooler dust in front of the compact sources, correction for the overlying absorption then leads to the conclusion that the sources in the KL nebula, although much less conspicuous than BN at most near/mid-infrared wavelengths, are intrinsically at least as luminous as BN. Downes et al. (51) proposed that IRc2, then already known to be intimately connected to the massive outflows (see Section 5.2), is responsible for most of the region's luminosity of 10^5 $L_\odot$, whereas Aitken et al. (1) concluded that BN, IRc2, and IRc4 all have a luminosity of only about 5000 $L_\odot$.

For the present interpretation of the infrared nebula, the 3.8-μm polarization observations by Werner et al. (316) and the high-resolution 2–30 μm mapping by Lee et al. (175) and Wynn-Williams et al. (337) are particularly relevant. Briefly, these observations show the following:

1. The nebula's radiation between 2 and 4 μm is highly polarized. The high degree of polarization (up to 50%), its wavelength dependence, and, in particular, its large-scale systematic spatial pattern unambiguously demonstrate that the polarized radiation is due to scattered light from a few central sources (132, 315). BN and IRc2 are the main sources illuminating this reflection nebula (Figure 7). Dyck et al. (57) first found the 10-μm emission to be polarized as well. Aitken et al. (2) inferred from the large-scale pattern of the 8–13 μm polarization that

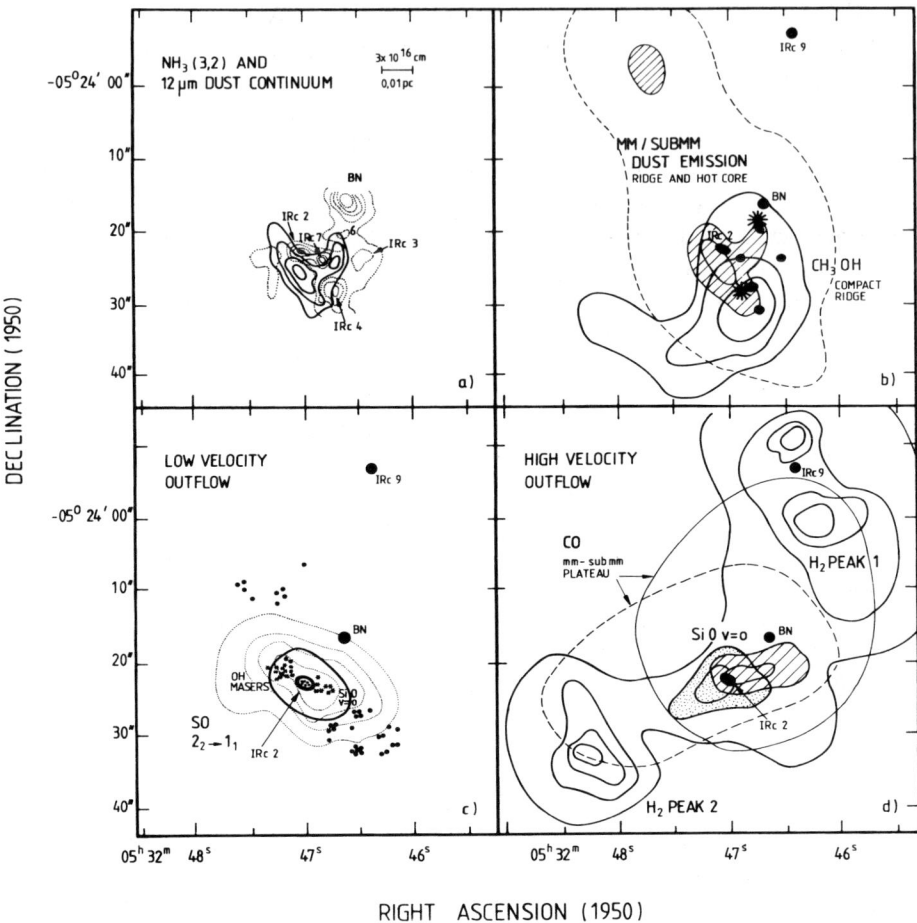

Figure 6 Composite of spatial distributions of different gas components at the core of BN-KL. *Top left* (*a*): Comparison of a 0.9″ resolution 12-μm infrared continuum image [dotted contours (86)] with a 1.2″ velocity-integrated Very Large Array (VLA) map of the 1.2-cm $(J, K) = (3, 2)$ inversion transition of NH_3 [heavier contours (81)], along with the standard designations of the compact infrared sources. The 12-μm image was taken with the Gezari/Goddard Space Flight Center array camera. The comparison, at the highest spatial resolution available in either wavelength range, clearly shows an anticorrelation between 12-μm continuum and line emission. *Top right* (*b*): Quiescent gas/dust maps. Shaded regions mark the 3-mm dust emission from the Owens Valley interferometer maps (194), along with the lower brightness millimeter and submillimeter dust emission detected with single dishes (dashed contours). The dust emission traces the hot core (near IRc2, like NH_3 on left side) and the extended ridge. The heavy contours mark the compact ridge, as represented by the 3-mm CH_3OH 8_0-7_1 transition (238), after subtraction of the CH_3OH masers (whose positions are

the grains are aligned by the intrinsic angular momentum of photons originating from IRc2.
2. The most prominent 8–12.5 and 12.5–20 μm color temperature peaks are BN and IRc2, followed by IRc4, which suggests that they contain the warmest dust and are close to heating sources. Elsewhere in the region, the color temperature is relatively constant.
3. The zones of strong scattering at 2.2 and 3.8 μm (e.g. IRc3, 4, 5, 7) coincide with the most intense 20- and 30-μm peaks (Figure 7), which indicates that their emission at the longer wavelengths is largely thermal reradiation of photons from BN and IRc2 absorbed at 2–20 μm.
4. Column density maxima in molecular gas coincide with peaks of 20–30 μm dust opacity and intensity minima at 8–30 μm, which clearly demonstrates the presence of patchy extinction. The anticorrelation of infrared and molecular emission is apparent on the overlay in Figure 6a of the 12-μm emission (86) on a 1.3″ resolution VLA map of the 1.2-cm $(J, K) = (3, 2)$ inversion line of NH_3 [which emphasizes the dense, warm gas in the "hot core"; see Section 5.2 and (81, 228)].

Werner et al. (316) and Wynn-Williams et al. (337) have proposed a self-consistent model of the central 30″ of the nebula in which the geometry of the BN-KL region is that of a clumpy cavity (diameter 10^{17} cm), rather than that of a number of isolated objects. The low average infrared opacity and gas/dust column densities through the cavity [$\tau(3.8$ μm$) \leq 1$ or $N(H_2) \leq 10^{23}$ cm^{-2}] probably result from the violent mass outflow from IRc2. The cavity is centered on or near IRc2, which is the source of most of the luminosity of the region ($L_{IRc2} = 2$–$10 \times 10^4 \ L_\odot$). IRc2 is intrinsically a powerful 2–30 μm source; however, it is heavily obscured by the edge of an optically thick gas and dust ridge centered about 2″

---

indicated by asterisks). The positions of the compact IR sources are indicated by filled circles and an ellipse (IRc2). *Bottom left (c)*: The low-velocity outflow, traced in 1.3-cm H_2O masers [dots (79)], 18-cm OH masers [asterisks (101, 219)], 3-mm SiO $v = 0 \ J = 2 \rightarrow 1$ emission [heavy contour at the FWHM contour, from Hat Creek interferometry by Wright et al. (332)], and in contours of SO 2_2-1_1, 3-mm line emission [dotted, Hat Creek interferometer (234)]. The positions of BN, IRc9 (filled circles), and IRc2 (open ellipse) are indicated. *Bottom right (d)*: High-velocity flow and shocked gas. The blueshifted (shaded) and redshifted (stippled) high-velocity gas close to IRc2 is traced in the 3-mm $v = 0 \ J = 2 \rightarrow 1$ transition of SiO [Hat Creek (332)]. The extended high-velocity flow (plateau) is traced in the millimeter transitions of CO and HCO^+. The extent (FWHM) of the blueshifted gas is indicated by a continuous curve, and that of the redshifted gas by a dashed curve. Outside of the plateau are the contours (6″ resolution) of the 2-μm $H_2 \ v = 1 \rightarrow 0 \ S(1)$ line from the work by Beckwith et al. (19).

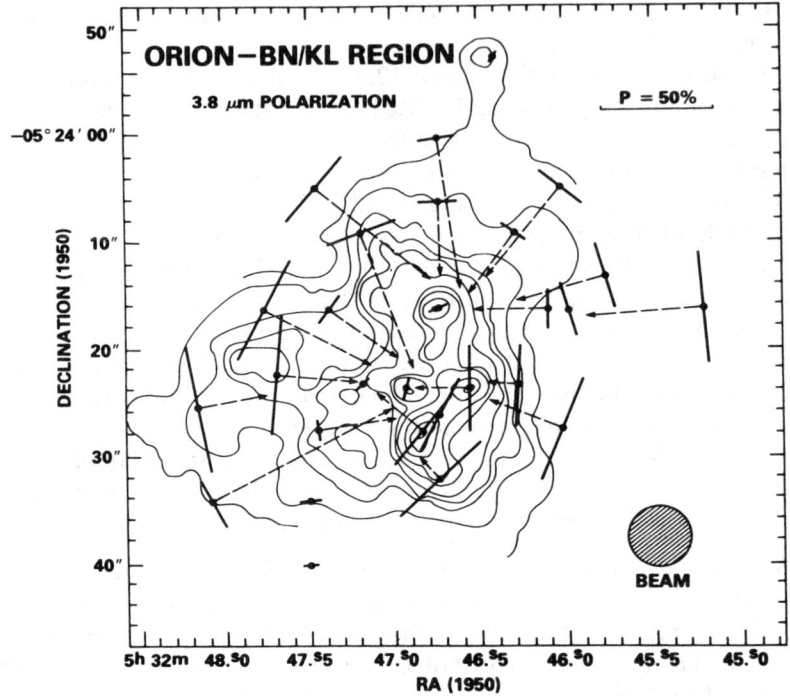

Figure 7 Distribution of electric vector of 3.8-μm polarization (316) superposed on the 2″ resolution 20-μm map by Downes et al. (51). In the interpretation of Werner et al., the polarized 3.8-μm emission is due to scattering, and the dashed arrows indicate the possible origins of the scattered radiation [adapted from Werner et al. (316)].

south of the source [Figure 6*a*; $N(H_2) = 10^{24}$ cm^{-2} or $\tau(10\text{–}30\ \mu m) = 10\text{–}20$]. BN ($L = 10^4\ L_\odot$) and IRc9 ($L > 200\ L_\odot$) are the other two sources in the region that clearly are self-luminous.

The model that best fits all the infrared and radio data (337) implies that most of the other peaks in the Kleinmann-Low nebula (IRc3, 5, 6, 7, etc.) are irregularities in the material within, at the edge of, and surrounding the cavity, rather than major self-luminous sources. IRc4 is a prominent peak of gas temperature in the "compact ridge" [see Section 5.2 (228, 238, 326)] and may contain an intrinsic energy source as luminous as $10^4\ L_\odot$; this limit is set mainly by confusion with the processed radiation from IRc2 and BN.

Another interesting but still open question is whether there is also a concentration of low-mass, pre-main-sequence stars in the vicinity of BN-IRc2. Such a concentration was first proposed by Lonsdale et al. (185), on the basis of spatially chopped 2-μm observations, but was later challenged by Hyland et al. (133). The recent IRCAM data confirm several of the Lonsdale et al. sources, leaving the question of a low-mass cluster near BN-IRc2 open (199).

In summary, the structure of the infrared emission in BN-KL is almost certainly determined by radiative transport in a very dense, clumpy molecular core. Most of the radiation originates from only a few major luminosity sources, of which IRc2 and BN are the most important. Based on their luminosity, these young stars must have masses between 15 (BN) and 25 (IRc2) $M_\odot$. The mass function of the cluster (339) cannot be tested by current observations, as the number of confidently identified young stars is too small.

5.2 The Orion-KL Cloud Core

Multiwavelength infrared and microwave spectroscopy, high-resolution line mapping (especially with the VLA and the Hat Creek and Owens Valley interferometers) and detailed measurements of molecular abundances now give a fairly coherent physical picture of the molecular gas in terms of spatially, kinematically, and chemically distinct regions. The measurements demonstrate that the violent mass outflows from IRc2/BN have substantially affected the structure and physical conditions of the cloud core. Shocks and grain mantle evaporation in the warm circumstellar environment have substantially altered the chemistry. A summary of the observations and derived physical parameters of the molecular and atomic gas and dust, together with a (necessarily incomplete) list of references, is given in Table 1. Figure 6 shows in partly schematic form the spatial distribution of these different components and their relationship to the infrared sources. Figure 8 gives an impression of the rich chemistry of the molecular gas, as derived from an extensive spectral scan at the Owens Valley millimeter telescope containing more than 800 lines in the 1.3-mm atmospheric window (24, 25, 280). We also refer to the other spectral surveys of Orion A (140, 141, 221, 319, 320). The following conclusions emerge.

5.2.1 RIDGE OF QUIESCENT GAS The core of the quiescent molecular cloud ridge contains several clumps or condensations (Figures 3, 6a,b) with a total gas mass of about 150 $M_\odot$ within 0.1 pc of IRc2. Typical parameters are a temperature of 70 K, a molecular hydrogen volume density of 10^5 cm^{-3}, and a column density of 3×10^{23} cm^{-2}. Chemically this region is characterized by standard gas phase, ion-molecule chemistry (121), with an abundance of carbon-rich species (e.g. CS, CN, CCH), and a lack of oxygen-rich molecules (107). The observed molecular abundances

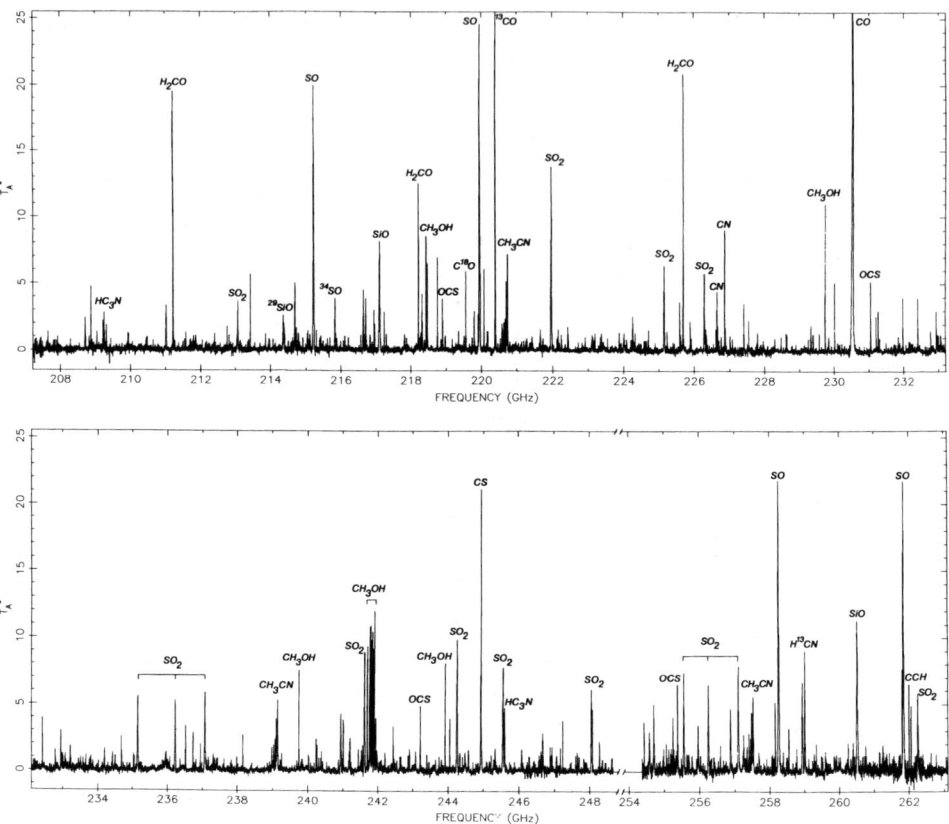

Figure 8 The Owens Valley 208–262 GHz spectral line scan of Orion, showing about 800 lines (24, 280). Most of these lines are identified and give detailed information on the chemistry of the region [adapted from Blake et al. (25)].

are reasonably well fit by recent ion-chemistry models (122, 240). Additional oxygen may be in the form of H_2O and O_2, which cannot be easily observed (e.g. 92, 135). Apart from the 1.3-cm masers, H_2O or $H_2^{18}O$ has been observed in near-millimeter rotational transitions (135, 232, 309; M. A. Frerking, private communication) and possibly in a 2.6-μm rovibrational transition (154). The derived fractional H_2O abundance ranges between 5×10^{-7} and 10^{-5}, not large enough to account for the oxygen underabundance. Several authors propose that oxygen-bearing molecules are depleted onto dust grains at low temperature (cf. 99, 329).

5.2.2 COMPACT RIDGE   Generally considered separate from the extended ridge is a 15″ size condensation of quiescent gas at $v_{LSR} = 8$ km s^{-1}, located at the southwestern edge of the KL nebula (Figure 6b). This "compact ridge," first noted spectrally in the 3-mm Onsala survey (141), is warmer ($T = 100–150$ K) than the rest of the ridge and has a significantly different chemistry. Complex oxygen-rich molecules, such as CH_3OH, $HCOOCH_3$, or CH_3OCH_3, are highly abundant. The compact ridge also contains several CH_3OH masers (195, 205, 238). Blake et al. (25) have proposed that mixing of the outflows with ridge gas in this region releases oxygen into the gas phase, which is then incorporated into the large oxygen-rich molecules by radiative association reactions (121, 182). The proposed interaction is also made plausible by the spatial coincidence between the compact ridge and the low-velocity H_2O masers (Figure 6b,c). The masers presumably are the locations of clump-cloud shocks (83).

There is a velocity difference of about 2 km s^{-1} between the ridge gas just southwest and northeast of BN-KL. Ho & Barrett (127) propose that two clouds are colliding, whereas others (108, 301) interpret the velocity gradient as evidence for rotation of the inner ridge. The ridge velocity appears to diverge approaching IRc2 from either side, a signature consistent with Keplerian rotation about a central object. Vogel et al. (301) derive a central mass of 25 $M_\odot$ for an assumed inclination of the ridge of 40° and conclude that the enclosed mass is dominated by the central star within 25″ of IRc2. From the constant ridge velocity farther out, Hasegawa et al. (108) estimate a mass of about 100 $M_\odot$ within 0.1 pc of IRc2. While these values are reasonably consistent with the sum of the stellar and gaseous masses estimated by other means, it is necessary to carefully check the evidence for rotation of the ridge by higher resolution measurements (324).

5.2.3 HOT CORE The "hot core" is apparent in Figure 6a,b as a 10″ diameter, elongated gas and dust ridge centered 2″ south of IRc2. Its hydrogen column density is at least 10^{24} cm^{-2}, as estimated from millimeter and far-infrared ^{13}CO measurements and from millimeter and submillimeter dust continuum observations (84, 148, 191). The total gas mass is 10 $M_\odot$ or more. As a result of the large column densities, the dust emission from the hot core has been detected even at 3 mm (192, 334). Furthermore, the hot core is optically thick in the dust radiation at mid- and far-infrared wavelengths. In comparing their 3-mm continuum fluxes with the 400-μm data from Keene et al. (148), Wright & Vogel (334) conclude that the dust emissivity scales like $\lambda^{-1.3}$ between these wavelengths, in marked contrast to the theoretically expected λ^{-2} dependence for bulk grains (55). The presence of 0.03 $M_\odot$ of millimeter-sized or larger dust grains, or of two-dimensional, layered amorphous carbon or silicates (290, 328), could explain such a shallow submillimeter/millimeter emissivity law. Alternatively, if the λ^{-2} dependence of the emissivity is correct, even the 400-μm radiation may be optically thick, resulting in 2 to 4 times larger dust column densities than estimated above (213; P. G. Mezger, private communication).

Large optical depths are also characteristic of the millimeter, submillimeter, and far-infrared transitions of the most abundant molecules. Hence, radiative trapping of the line radiation is often very important in the hot core, in that it lowers the hydrogen densities required in the optically thin limit in some cases by two orders of magnitude. Current best estimates of the molecular hydrogen density range between 10^7 and 3×10^7 cm^{-3}, as derived (for example) from NH$_3$ and ^{13}CO observations (84, 124, 210). The dust temperature is about 150 K, and the gas kinetic temperature is near 200 K. Molecular line excitation temperatures range between 120 and 260 K and may reflect actual physical variations in gas temperature. Pumping by far-infrared continuum radiation is important for a number of molecules for excitation of rotational levels in the ground vibrational state (e.g. NH$_3$, OH, H$_2$O; 124, 203, 204, 291). Mid-infrared pumping of the vibrational states has been found to account for millimeter rotational transitions of molecules in vibrationally excited states, such as HC$_3$N, CH$_3$CN, HCN, and NH$_3$ (89, 197, 341). Radiation from IRc2 almost certainly is responsible for the pumping. The source-averaged excitation temperature of the vibrationally excited HC$_3$N is about 150 K, in very good agreement with the dust and radiation temperature near the hot core (91). Furthermore, Plambeck (236) finds that the excitation temperature of vibrationally excited HC$_3$N peaks on IRc2, whereas its emission centroid is displaced from IRc2.

The hot core ridge appears to consist of a turbulent ensemble of clumps with sizes of 1" (7.2 $\times$ 10^{15} cm) or smaller on VLA maps of 1.2-cm NH$_3$ inversion transitions (Figure 6a; 81, 123, 228). Its velocity centroid (v_{LSR} = 5 km s^{-1}) is identical to the centroid of the SiO masers on IRc2, which probably is a reliable indicator of the stellar velocity. Hence, IRc2 and the hot core are probably physically related. Figures 6a and 6b also demonstrate that the clumpy hot core appears to stretch around and in front of IRc2, thus determining the near- and mid-infrared appearance of the BN-KL nebula (337; Section 5.1). The hot core may be the remnant of the dense parental cloud out of which IRc2 has formed and that now interacts with its outflow and radiation (194).

The chemistry of the hot core is characterized by unusually high abundances of hydrogen-saturated molecules, such as NH$_3$ or H$_2$O (observed in its isotopic form HDO), or CH$_3$CN and CH$_2$CH$_3$CN. A likely interpretation is that, analogous to comets, the gas phase is enriched with fully hydrogenated species by evaporation of grain mantles that have been heated by radiation from IRc2 (25, 31, 281, 288).

A very interesting puzzle is that for several molecules the ratio of relative abundances in deuterated to hydrogenated species is at least two orders of magnitude greater than the interstellar D/H ratio (e.g. 237, 306). A high fractional abundance of deuterated species is usually interpreted as a clear signature of fractionation in cold gas and cannot easily be understood for the high-temperature environment of the hot core if the chemistry is in equilibrium. Walmsley et al. (306), Brown et al. (31), and Plambeck & Wright (237) propose that the hot core has been heated only relatively recently ($t \leq 10^4$ yr), and thus the chemistry still reflects the original grain composition.

The hot core is not a massive disk rotating about IRc2. None of the high-resolution interferometer maps show evidence for rotation. It is also unlikely that stars form presently in the hot core region. The Jeans mass at 150 K and $n(H_2) = 3 \times 10^7$ cm^{-3} is 10 $M_\odot$. An individual clump in the hot core has a mass of 0.1–1 $M_\odot$, a value clearly not large enough for collapse. The velocity dispersion of the entire ensemble (≈ 4.5 km s^{-1}) is probably sufficient to stabilize the hot core as a whole against gravitational collapse. On the other hand, it is not clear whether the entire hot core region can be heated by IRc2 alone, or whether there must be additional internal heating sources (P. G. Mezger, private communication).

Another question of interest is the relationship between compact ridge and hot core. The two zones are clearly morphologically and kinematically related (228, 237, 337). Mauersberger et al. (196) have recently found a similarly high abundance of deuterated species in the compact ridge as well. It is conceivable that the two components represent two parts of the same region, but with substantially higher column densities, volume densities, and temperatures in the hot core (238).

5.2.4 OUTFLOWS The parameters of the dynamically active gas are separated in Table 1 into those of the "low-velocity" (or "18 km s^{-1}" or "expanding doughnut") flow, the "high velocity" flow (or "plateau"), and the "high-velocity shock-excited gas" because of obvious differences in kinematics, spatial distribution (Figure 6c,d), and excitation. The flows represent gas streaming away from the center of the BN-IRc2 cluster. This has been demonstrated by the measurement of proper motions of H_2O masers (79) and of Herbig-Haro objects (6, 145). Furthermore, spectroscopic observations at near-infrared (98, 256) and far-infrared wavelengths (204) show the blueshifted, high-velocity gas in absorption against the dust continuum, again suggesting an overall outflow of the high-velocity gas. The center of the outflows is within a few arcseconds of IRc2.

Their dynamical age is a few times 10^3 yr. The inferred mass loss rate, as derived from that age and the total mass involved, is about $10^{-3\pm0.3}$ $M_\odot$ yr^{-1} (80).

The low-velocity flow, with a fairly well-defined maximum velocity spread of about 35 km s^{-1}, is best recognized in the SiO, OH, and low-velocity H$_2$O masers, as well as in thermal millimeter and submillimeter rotational transitions of several heavy top molecules (SiO, HCN, SO$_2$, SO). The gas flow extends northeast-southwest, more or less along the dense quiescent ridge (Figure 6c). It can be traced from within 0.2" of IRc2 (see Figure 10) to about 20" from the star, with about a constant velocity range. The double-peaked appearance of the H$_2$O and SiO "shell" masers strongly suggests a spherically symmetric outflow, where the greatest maser gain is along the front and back sides of the expanding shell (82). The flowing gas probably is clumped, as strong emission lines from molecular transitions are seen that require very high densities for their excitation (275). A significant change in velocity centroid appears between gas within a few arcseconds of IRc2 ($v_{LSR} = 5$ km s^{-1} for shell H$_2$O and SiO masers and SiO $v = 0$ $J = 1 \to 0$) and the OH and low-velocity H$_2$O masers at 5" or more from IRc2 ($v_{LSR} = 9$ km s^{-1}). Furthermore, the gas farther away from IRc2 also has a very high abundance of sulfur-rich molecules (SO, SO$_2$) compared with the silicon-rich flow in the circumstellar environment. All these facts may be accounted for in a model whereby the initially free outflow from the star plunges into the dense molecular ridge at $R = 10^{17}$ cm. In the shocks at this interface, the low-velocity H$_2$O masers are created, as is a thin, but very dense, zone of molecular species that require shock chemistry for their formation. This is the "expanding doughnut" of Plambeck et al. (234), although it is clear now that this feature has about 10 times less H$_2$ column density than was proposed originally by those authors. Hydrogen densities in this shocked shell, or in shocked clumps within the flow, are a few times 10^7 cm^{-3} or more (275). The low-velocity outflow also appears to be the prominent spectral feature in several newly detected, high-excitation molecular transitions in the 800-GHz range (277). The inner, relatively low-density, free outflow zone is identical with the "infrared cavity" of Section 5.1. The observational data are in good agreement with model calculations of the chemistry and the formation of masers in such shocks (25, 61, 105, 134, 283).

A second flow of larger, but less well-defined, velocity range ($\Delta v \leq 250$ km s^{-1}) stretches predominantly in the northwest-southeast direction—that is, approximately perpendicular to the low-velocity flow and the ridge. This high-velocity flow (Figure 6d) has been investigated in much detail in the millimeter rotational transitions of CO (see references in Table 1). It has a weakly bipolar velocity structure (65), similar to those found in many other regions of star formation (11). As in the case of the low-velocity flow, the SiO molecule traces the high-velocity gas closest to the center, which is within a few arcseconds of IRc2 (332; Figure 6d). The centroid of the high-velocity CO emission is 10″ north of IRc2, most likely because the millimeter CO transitions sample relatively low-density gas $[n(H_2) \geq 10^4 \text{ cm}^{-3}]$ located mostly north of IRc2 and because the dense gas in the KL region impedes the flow south of IRc2 (192). Vogel et al. (300) find that the intensity of the 3-mm HCO^+ $1 \to 0$ line in the high-velocity flow is at least an order of magnitude greater than in the low-velocity flow. This fact could be a consequence either of the lower density and column density in the higher velocity gas or of a greater formation rate of HCO^+ in high-velocity shocks.

The fact that the low-velocity flow is mostly along the dense ridge, whereas the higher velocity gas appears to stream mainly perpendicular to it, has been interpreted as the channeling of one single, high-velocity flow by the surrounding cloud (33, 157). However, the two flows are distinct to within about 2″ (10^{16} cm) of IRc2, far inside the outflow cavity. Furthermore, the mean gas density in the outflow cavity is not sufficient for dynamic channeling of the high-velocity flow. Hence, either the two flows emerge from different stars, or else the outflowing gas has to be channeled in the immediate environment of the star, possibly by a circumstellar disk. Unfortunately, there is currently no plausible quantitative model for the latter scenario.

In addition to the main outflow source of size $\leq 60″$, Hasegawa (111) and White et al. (321) also find straight and very elongated ($\geq 3′$) streamers emanating from BN-IRc2 that are possibly connected with the outflow phenomena and the Orion A Herbig-Haro objects (6, 145).

5.2.5 HIGH-VELOCITY SHOCKED GAS   The high-velocity flow plunges into the surrounding cloud about 30″ from the dynamical center, creating high-velocity H_2O masers and a zone of shocked molecular gas cooling in infrared and submillimeter lines. The shocked zone has been studied in detail in the near-infrared lines of excited molecular hydrogen (20, 260) and in the far-infrared lines of rotationally excited CO and OH (299, 310, 311).

The molecular hydrogen is in a thermalized distribution with excitation temperatures ranging from about 1000 K for the rotationally excited states (15, 21) to 3000 K for rotational states in the $v \geq 2$ vibrational level (21, 29). It is not clear whether the clumpy 2-μm H_2 distribution (19; Figure 6d) is due to local variations in extinction (255) or to actual column density variations in the hot molecular hydrogen, located behind a screen of spatially constant extinction with $A_v = 20$ mag (16). A related issue is the appearance of the H_2 line profiles, which have more blueshifted than redshifted emission, especially near the brightest emission region (peak 1; see Figure 6d). Nadeau et al. (216) and Scoville et al. (255) interpret the prevalence of blueshifted gas as the result of differential extinction within the expanding source, which suppresses the redshifted gas at the back side of the source more than the blueshifted gas in the foreground. More recently, Geballe et al. (76), investigating line profiles of 2- and 4-μm H_2 lines, have concluded that the asymmetry near peak 1 is intrinsic to the kinematics of the source. In Geballe et al.'s interpretation, the wide angle bipolar outflow strikes the cloud northwest of BN-IRc2 mostly toward the observer, thus leading to a prevalence of blueshifted shocked gas. This interpretation is also supported by line profiles of the far-infrared CO rotational lines, which also show more blueshifted than redshifted emission (27, 47, 239).

The high-velocity, far-infrared CO emission predominantly comes from lower temperature gas (500–1000 K) having two orders of magnitude higher column density than inferred from the 2-μm H_2 lines (272, 311). The large column density of this medium-temperature gas, having only moderately high densities [$n(H_2) = 10^6$ cm^{-3}], cannot be produced in non-dissociative, purely gas-dynamic J-shocks (for "jump" transition). Draine (53) first proposed that a "softer" excitation of gas can be obtained in a magnetohydrodynamic shock where magnetosonic waves propagate faster than the shock front. This magnetic precursor sweeps up ions, which in turn heat the neutral gas by collisions. Such C-type shocks (for "continuous" transition), with magnetic fields of about 1 mG and shock speeds near 40 km s^{-1}, account successfully for the absolute and relative intensities of most CO and H_2 infrared lines (36, 54, 130).

Not all of the observed characteristics of the hot, high-velocity gas in Orion can be accounted for by the standard C-shock models. First, there is atomic and ionized gas associated with the shocked region. Werner et al. (317) find high-velocity neutral atomic oxygen with about the same distribution as the 2-μm H_2 emission. There is [Si II] emission from the shocked region (131). Hasegawa & Akabane (109) report a possible detection of high-velocity H51α line emission associated with peak 1. McKee & Hollenbach (200) propose that additional, fast gas-dynamic shocks in the outflow explain the [O I], [Si II], and H51α observations, although the [O I] emission by itself may also come from a C-shock if the preshock atomic oxygen abundance is high (54). Furthermore, Geballe & Garden (77) find unexpectedly strong near-infrared emission from vibrationally excited CO. Either much larger densities than inferred from the rotationally excited CO or much larger column densities of $T \geq 2000$ K gas than inferred from the 2-μm H_2 data are required to account for the vibrationally excited CO. Third, Brand et al. (29) find significant deviations from the standard C-type models based on measurements of H_2 lines over a wide range of excitations. They conclude that their data can be better explained by a simple cooling zone of initially very high temperature. Finally, it has been clear for some time that the high-velocity wings in the H_2, CO, and OH infrared lines cannot be accounted for by the relatively slow, C- or J-type cloud shocks. Clearly, more refined theoretical models, probably including collisions of dense cloudlets in the flow (39) and a combination of C- and J-type shocks, are required to account for the more recent data.

5.3 *A Close-Up View of BN and IRc2*

Grasdalen's (95) discovery of H I Brα emission from BN was the first strong evidence that the infrared source (size <0.1"; 69) contains a central star of temperature $>10^4$ K. Radio observations demonstrate that the circumstellar H II region is very small (<20 AU), with electron densities of $\geq 10^7$ cm^{-3} (207). The electron densities in the infrared recombination-line zone must be even higher ($>2 \times 10^8$ cm^{-3}), as there is no detectable 12.8-μm [Ne II] emission from BN (E. Serabyn & R. Genzel, unpublished). More recently, weak radio emission (2 mJy at 14 GHz) has also been detected from IRc2, which supports the view that this source contains a hot central star as well (41, 72).

High-resolution near-infrared spectroscopy, mainly with the Kitt Peak Fourier Transform Spectrometer, has given detailed information on the circumstellar environment of BN (98, 256). Scoville et al. (256) identify four distinct regimes in the circumstellar and line-of-sight gas. First, the ultradense H II region discussed above contains an ionized gas mass of $3 \times 10^{-6} M_\odot$. High-velocity wings in the infrared H I recombination lines ($\Delta v = 200$ km s^{-1}) may indicate supersonic ionized gas flows. Second, 2.3-μm high-*J* band-head emission of CO indicates the presence of a highly confined region of very hot ($T \sim 3500$ K) and very dense [$n(H_2) = 10^{12}$ cm^{-3}] molecular gas that is probably directly adjacent to the ionized region. Third, emission in the 2.3- and 4.6-μm ro-vibrational CO lines shows the presence of a molecular circumstellar envelope with $n(H_2) = 10^7$ cm^{-3} and temperature 600 K. This envelope is probably identical with the infrared "dust photosphere." Fourth, there is lower excitation gas at $T \approx 150$ K from the molecular ridge and plateau in front of BN, apparent in CO absorption. The latter three components can be recognized in Figure 9, which gives composite 2.3- and 4.6-μm spectra from Scoville et al.'s (256) work.

In the model of Scoville and coworkers, BN is a hot zero-age main-sequence star with a UV emission rate equivalent to a B0.5 star ($T_{\text{eff}} = 26,000$ K, $L = 10^4 L_\odot$). The star is surrounded by ionized gas

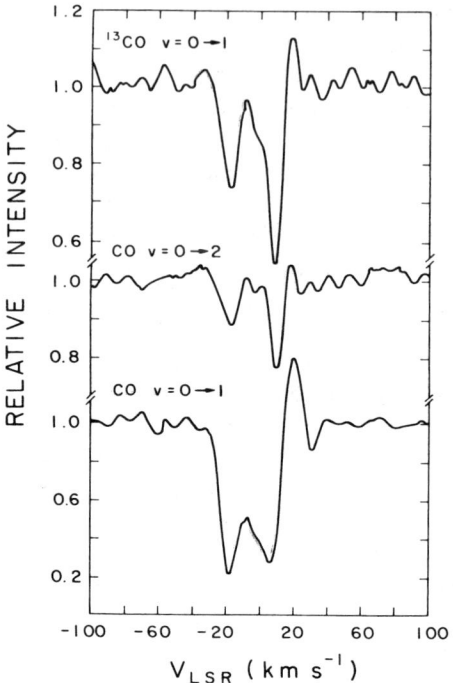

Figure 9 Composite 2.3- and 4.6-μm CO spectra toward the BN continuum [adapted from Scoville et al. (256)]. The data were taken at 7 km s^{-1} resolution with the Fourier Transform Spectrometer on the 4-m Kitt Peak telescope (256). Several components are seen in absorption or emission, with different strengths in ground-vibrational, overtone, and isotopic transitions. The measurements indicate different physical gas components in the circumstellar environment of BN.

flowing out at ≥ 100 km s^{-1}. The inferred mass outflow rate is 4×10^{-7} $M_\odot$ yr^{-1}. The compact H II region in turn is surrounded by a dense circumstellar envelope or disk falling in at a velocity of 10 km s^{-1} toward the star. The CO band-head emission comes from the inner rim (thickness a few astronomical units) of this neutral envelope at a distance of about 20 AU from the star. A very interesting and puzzling result of the spectroscopic observations is the fact that BN's systemic velocity is $v_{LSR} = 21$ km s^{-1}, which implies a motion of 12 km s^{-1} relative to the molecular cloud. If BN is not in a binary system, it will leave the cloud core on a time scale of only a few thousand years. A similar conclusion also applies to IRc2, whose systemic velocity ($v_{LSR} = 5$ km s^{-1}) differs by 4 km s^{-1} from that of the cloud core.

A clear-cut determination of the stellar and circumstellar characteristics of BN is not yet possible because of the difficulties of a complete theoretical modeling. Krolik & Smith (158) point out the importance of collisional excitation of the infrared recombination lines at the high electron densities near the star. Simon et al. (261) and Thompson (286) show that at these high densities, the $n = 2$ level of hydrogen can be collisionally populated and then Balmer photons can contribute to the ionization. In a recent paper, Höflich & Wehrse (129) conclude that the circumstellar H II regions can only be quantitatively understood if non-LTE atmospheres, photoionization by Balmer photons, bound-bound collisions, large optical depths in lines, and high photon densities in the continua are fully taken into account in model calculations. In the Höflich & Wehrse model, BN has an effective temperature of 28,000 K and a stellar radius of 5.2 $R_\odot$ ($L = 10^4 L_\odot$). The outer radius of the ionized zone is at 7 AU. The derived electron density at the base of the assumed R^{-2} envelope is 3×10^{12} cm^{-3}.

The inference of mass outflow from the broad line wings is also not straightforward. These wings could be the result of Stark broadening and Thomson scattering if the base density of the ionized envelope is $>10^{12}$ cm^{-3}, as in the models of Höflich & Wehrse (129) and P. Höflich (private communication).

Even less is known about the nature of IRc2, other than that it has a luminosity approaching $10^5 L_\odot$ and a circumstellar H II region similar to that of BN. The main reason is that, owing to the combined effects of line-of-sight extinction (through the hot core) and the large size of the infrared photosphere ($\sim 2''$; 35, 181), no detailed circumstellar infrared spectroscopy has been possible up to now.

From modeling of the SiO maser associated with IRc2, Elitzur (60) has derived a scenario for the circumstellar environment. In his model, the outflow starts at a radius of a few times 10^{14} cm from the star. The SiO masers occur at $R = 5 \times 10^{14}$ cm, where the outflow (driven by radiation pressure on grains) has fully developed. In order to pump the masers, the molecular hydrogen density must be just under 10^{12} cm^{-3}. Strong turbulent motions heat the gas above the temperature of the dust, a necessary condition for maser pumping. Elitzur derives a mass outflow rate of about $10^{-3} M_\odot$ yr^{-1}, consistent with the other estimates given above.

Figure 10 shows the distribution of infrared emission and masers in the $2''$ zone around the star. The 12-μm intensity distribution (86) is elongated and double peaked, strikingly similar to that of the H$_2$O shell masers (208). The resemblance of the maser and infrared continuum distributions suggests that the dust emission traces the outflowing gas. Blueshifted and redshifted H$_2$O masers occur close to both 12-μm peaks, making it unlikely that they just represent the front and back sides of the expanding shell, or that there are two stars. A single source of the outflow is also supported

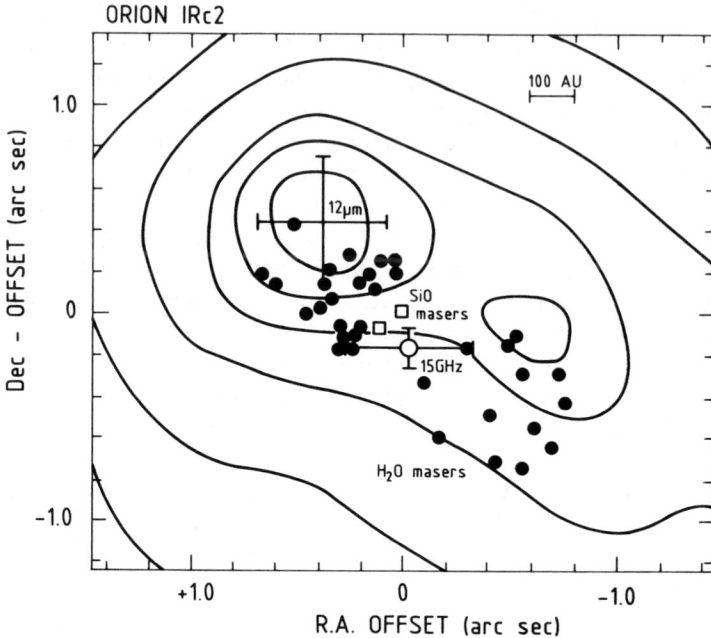

Figure 10 The circumstellar region around IRc2 (from 236). Contours represent the 12-μm infrared continuum emission [0.9" resolution image by Gezari (86)]. Heavy black dots are individual "shell" H_2O masers (208). The two open squares give the positions of the SiO shell masers (333; R. L. Plambeck, private communication), and the open circle is the position of the supercompact 15-GHz radio continuum source found by Garay et al. (72) and Churchwell et al. (41). The (0, 0) base position is at R.A. = $05^h 32^m 47.023^s$, Dec. = $-05°24'23.83"$ (1950). The 1σ positional error bars are $\pm 0.1"$, if not otherwise indicated in the figure. Note that the best astrometric infrared position of IRc2 (as indicated) places the infrared emission of BN right on top of BN's radio continuum counterpart (198). Because of the uncertainty in absolute astrometry, the infrared and radio tracers can be shifted by $\pm 0.5"$ with respect to each other.

by the fact that the SiO masers and the supercompact H II region are located approximately in between the two hot spots. The elongation strongly suggests that the low-velocity outflow is not spherically symmetric but rather disklike or bipolar. A bipolar outflow may also be favored by the structure and kinematics of the low-velocity flow on larger scales. The general alignment along p.a. 45–60° (Figure 6c) and the strong velocity asymmetries of the OH masers (101, 219) strengthen the impression that either the low-velocity flow or the quiescent gas interacting with it are distributed anisotropically about IRc2. There is a small offset between blueshifted and redshifted SiO and H_2O maser features (208, 333). A similar offset is seen in the distribution of the low-velocity flow at larger scales (e.g. 234). This offset could indicate rotation of the gas around the outflow axis or, perhaps more likely, an inclination of approximately 60° for the flattened or bipolar distribution with respect to the line of sight (234).

The exact location of the hot core and its physical association to IRc2 remain somewhat uncertain in this qualitative picture. Clearly it is in front of IRc2 (Section 5.1) and probably in front of the inner part of the outflow cavity. Harris et al. (104) find a sharp self-absorption notch at $v_{LSR} = 5$ km s^{-1} against the emission from the low-velocity flow in the HCN $9 \rightarrow 8$ transition, an observation in support of the latter conclusion. Therefore, IRc2 is probably detached (distance $1-3 \times 10^{16}$ cm) from the hot core, as described in Section 5.2. Cartoons of the possible geometry of the outflow zone around IRc2 can be found in various publications (82, 194, 337).

6. CONCLUSIONS AND OUTLOOK

The Orion Molecular Cloud clearly has a complex and intricate structure, with dense filaments, sheets, and clumps interspersed with low-density bubbles and cavities. Clumping of the interstellar matter in the cloud can now be traced from scales of several parsecs to several times 10^{-3} parsec. The interplay of pressure waves from supernovae, stellar winds, and outflows, together with the effects of radiation and magnetic fields, leads to an ever-shifting, highly dynamic evolution of the cloud on time scales of 10^6 yr or less.

The end stages of this evolutionary process are the OB associations and groupings of lower mass stars that have dispersed most of their parental cloud. The earliest stages of the evolution are the very dense condensations of gas and dust that the new millimeter and submillimeter observations have now begun to uncover and that may be the birthplaces of new stars. We can be optimistic that high-resolution, submillimeter continuum and line work will give much new information on these regions in the near future.

Triggering of stellar formation by the last generation of stars appears to be important for the formation of very dense clusters of low- as well as high-mass stars in these condensations. From near-infrared spectroscopic measurements and their theoretical analysis, we also slowly have acquired the probes for investigating the characteristics of the deeply embedded stars themselves. More detailed work with the new infrared cameras, sampling large areas at high sensitivity, is now possible and will be essential for the important goal of finally establishing a reliable initial mass function in a region of active star formation.

The BN-IRc2 region is our best laboratory for studying early evolution of massive stars. In this region we can directly observe how the violent mass outflows from the young stars are disrupting their environment and are changing the chemical composition of the interstellar gas. We have begun to understand quantitatively the fireworks of infrared and microwave line emission, created by shock waves or by ultraviolet radiation impinging on the cloud's surfaces. These tools can now be applied to other regions.

ACKNOWLEDGMENTS

We thank the many members of the "Orion Community" who have sent preprints, have given us helpful comments, and have let us use data prior to publication. We are especially grateful to J. Bally, G. Blake, D. Downes, J. Franco, T. Geballe, D. Gezari, A. Harris, T. Hasegawa, C. Henkel, G. Herbig, J. Jackson, E. Lada, T. Lee, R. Maddalena, M. McCaughrean, I. McLean, P. Mezger, L. Mundy, R. Plambeck, J. Schmid-Burgk, G. Stacey, A. Stark, A. Sternberg, E. Sutton, P. Thaddeus, M. Walmsley, T. Wilson, and H. Zinnecker.

Literature Cited

1. Aitken, D. K., Roche, P. F., Spenser, P. M., Jones, B. 1981. *MNRAS* 195: 921
2. Aitken, D. K., Bailey, J. A., Roche, P. F., Hough, J. M. 1985. *MNRAS* 215: 815
3. Appenzeller, I. 1966. *Z. Astrophys.* 64: 296
4. Appenzeller, I. 1974. *Astron. Astrophys.* 36: 99
5. Arons, J., Max, C. E. 1975. *Ap. J. Lett.* 196: L77
6. Axon, K., Taylor, K. 1983. *MNRAS* 207: 4
7. Baade, W., Minkowski, R. 1937. *Ap. J.* 86: 119
8. Balick, B., Gammon, R. H., Doherty, L. H. 1974. *Ap. J.* 188: 45
9. Balick, B., Gammon, R. H., Hjellming, R. M. 1974. *Publ. Astron. Soc. Pac.* 86: 616
10. Bally, J. 1982. See Ref. 88, p. 191
11. Bally, J., Lada, C. J. 1983. *Ap. J.* 265: 824
12. Bally, J., Langer, W. D., Stark, A. A., Wilson, R. W. 1987. *Ap. J. Lett.* 312: L45
13. Batrla, W., Wilson, T. L., Bastien, P., Ruf, K. 1983. *Astron. Astrophys.* 128: 129
14. Baud, B., Wouterloot, J. G. 1980. *Astron. Astrophys.* 90: 297
15. Beck, S. C., Bloemhof, E. E., Serabyn, E., Townes, C. H., Tokunaga, A. T., et al. 1982. *Ap. J. Lett.* 253: L83
16. Beck, S. C., Beckwith, S. 1983. *Ap. J.* 271: 175
17. Becklin, E. E., Neugebauer, G. 1967. *Ap. J.* 147: 799
18. Becklin, E. E., Beckwith, S., Gatley, I., Matthews, K., Neugebauer, G., et al. 1976. *Ap. J.* 207: 770
19. Beckwith, S., Persson, S. E., Neugebauer, G., Becklin, E. E. 1978. *Ap. J.* 223: 464
20. Beckwith, S. 1982. See Ref. 88, p. 118
21. Beckwith, S., Evans, N. J., Gatley, I., Gull, G., Russell, R. W. 1983. *Ap. J.* 223: 464
22. Beichman, C. A. 1988. *Astrophys. Lett. Commun.* 27: 67
23. Blaauw, A. 1964. *Annu. Rev. Astron. Astrophys.* 2: 213
24. Blake, G. A., Sutton, E. C., Masson, C. R., Phillips, T. G. 1986. *Ap. J. Suppl.* 60: 357
25. Blake, G. A., Sutton, E. C., Masson, C. R., Phillips, T. G. 1987. *Ap. J.* 315: 621

26. Boreiko, R. T., Betz, A. L., Zmuidzinas, J. 1988. *Ap. J. Lett.* 325: L47
27. Boreiko, R. T., Betz, A. L., Zmuidzinas, J. 1989. Submitted for publication
28. Borgman, J., Blaauw, A. 1964. *Bull. Astron. Inst. Neth.* 17: 358
29. Brand, P. W. J. L., Moorhouse, A., Burton, M. G., Geballe, T. R., Bird, M., Wade, R. 1988. *Ap. J. Lett.* 334: L103
30. Brooks, J. W., Murray, J. D., Radhakrishnan, V. 1971. *Astrophys. Lett.* 8: 121
31. Brown, P. D., Charnley, S. B., Millar, T. J. 1988. *MNRAS* 231: 409
32. Buhl, D., Snyder, L. E. 1970. *Nature* 228: 267
33. Canto, J., Rodriguez, L. F., Barral, J. F., Carral, P. 1981. *Ap. J.* 244: 102
34. Chaisson, E. J., Beichman, C. A. 1975. *Ap. J. Lett.* 199: L39
35. Chelli, A., Perrier, C., Lena, P. 1984. *Ap. J.* 280: 163
36. Chernoff, D. F., Hollenbach, D., McKee, C. F. 1982. *Ap. J. Lett.* 259: L97
37. Cheung, A. C., Rank, D. M., Townes, C. H., Thornton, D. D., Welch, W. J. 1968. *Phys. Rev. Lett.* 21: 1701
38. Cheung, A. C., Rank, D. M., Townes, C. H., Thornton, D. D., Welch, W. J. 1969. *Nature* 221: 626
39. Chevalier, R. A. 1980. *Astrophys. Lett.* 21: 57
40. Chini, R. 1989. In *Submillimetre and Millimetre Wave Astronomy*, ed. A. S. Webster. Dordrecht: Kluwer. In press
41. Churchwell, E., Felli, M., Wood, D. O. S., Massi, M. 1987. *Ap. J.* 321: 516
42. Clark, F. O., Buhl, D., Snyder, L. E. 1974. *Ap. J.* 190: 545
43. Cohen, M., Kuhi, L. V. 1979. *Ap. J. Suppl.* 41: 743
44. Cohen, R. J., Matthews, N., Few, R. W., Booth, R. S. 1983. *MNRAS* 203: 1123
45. Cowie, L. L., Songaila, A., York, D. G. 1979. *Ap. J.* 230: 469
46. Crawford, D. L., Barnes, J. V. 1966. *Astron. J.* 71: 610
47. Crawford, M. K., Lugten, J. B., Fitelson, W., Genzel, R., Melnick, G. 1986. *Ap. J. Lett.* 303: L57
48. Crutcher, R. M., Kazes, I. 1983. *Astron. Astrophys.* 125: L23
49. Cudlip, W., Furniss, I., King, K. J., Jennings, R. E. 1982. *MNRAS* 200: 1169
50. Davis, L., Greenstein, J. L. 1951. *Ap. J.* 114: 206
51. Downes, D., Genzel, R., Becklin, E. E., Wynn-Williams, C. G. 1981. *Ap. J.* 244: 869
52. Dragovan, M. 1986. *Ap. J.* 308: 270
53. Draine, B. T. 1980. *Ap. J.* 241: 1021
54. Draine, B. T., Roberge, W. G. 1982. *Ap. J. Lett.* 259: L91
55. Draine, B. T., Lee, H. M. 1984. *Ap. J.* 285: 89
56. Drapatz, S., Haser, L., Hofmann, R., Oda, N., Iyengar, K. V. 1983. *Astron. Astrophys.* 128: 207
57. Dyck, H. M., Capps, R. W., Forrest, W. J., Gillett, F. C. 1973. *Ap. J. Lett.* 183: L99
58. Dyck, H. M., Beichman, C. A. 1974. *Ap. J.* 194: 57
59. Dyck, H. M., Lonsdale, C. J. 1979. *Astron. J.* 84: 1339
60. Elitzur, M. 1982. *Ap. J.* 262: 189
61. Elitzur, M., Hollenbach, D. J., McKee, C. F. 1988. *Ap. J. Lett.* In press
62. Elmegreen, B. G., Lada, C. J. 1976. *Astron. J.* 81: 1089
63. Elmegreen, B. G., Lada, C. J. 1977. *Ap. J.* 214: 725
64. Elmegreen, B. G. 1980. In *Giant Molecular Clouds in the Galaxy*, ed. P. Solomon, M. Edmunds, p. 19. Oxford: Pergamon
65. Erickson, N. R., Goldsmith, P. F., Snell, R. L., Berson, R. L., Huguenin, G. R., et al. 1982. *Ap. J. Lett.* 261: L103
66. Evans, N. J., Morris, G., Sato, T., Zuckerman, B. 1975. *Ap. J.* 199: 383
67. Fiebig, D., Güsten, R. 1988. Submitted for publication
68. Forrest, W. J., Moneti, A., Woodward, C. H., Pipher, J. L., Hoffman, A. 1985. *Publ. Astron. Soc. Pac.* 97: 183
69. Foy, R., Chelli, A., Léna, P., Sibille, F. 1979. *Astron. Astrophys.* 79: L5
70. Franco, J., Tenorio-Tagle, G., Bodenheimer, P., Rozyczka, M., Mirabel, I. F. 1988. *Ap. J.* 333: 826
71. Fukui, Y., Sugitani, K., Takaba, H., Iwata, T., Mizuno, A., et al. 1986. *Ap. J. Lett.* 311: L85
72. Garay, G., Moran, J. M., Reid, M. J. 1987. *Ap. J.* 314: 535
73. Gatley, I., Becklin, E. E., Matthews, K., Neugebauer, G., Penston, M. V., Scoville, N. Z. 1974. *Ap. J. Lett.* 191: L121
74. Gatley, I., Kaifu, N. 1987. In *Astrochemistry*, ed. M. S. Vardya, S. P. Tarafdar, p. 153. Dordrecht: Reidel
75. Gear, W., Robson, I., Sandell, G., Hughes, R., Duncan, W. 1988. *Protostar* 5: 2
76. Geballe, T. R., Persson, S. E., Simon, T., Lonsdale, C. J., McGregor, P. J. 1986. *Ap. J.* 302: 500

77. Geballe, T. R., Garden, R. 1987. *Ap. J. Lett.* 317: L107
78. Genzel, R., Downes, D. 1977. *Astron. Astrophys. Suppl.* 30: 145
79. Genzel, R., Reid, M. J., Moran, J. M., Downes, D. 1981. *Ap. J.* 244: 884
80. Genzel, R., Downes, D. 1982. In *Regions of Recent Star Formation*, ed. R. S. Roger, P. E. Dewdney, p. 251. Dordrecht: Reidel
81. Genzel, R., Downes, D., Ho, P. T. P., Bieging, J. H. 1982. *Ap. J. Lett.* 259: L103
82. Genzel, R., Downes, D. 1983. In *Highlights of Astronomy*, ed. R. West, p. 689. Dordrecht: Reidel
83. Genzel, R. 1986. In *Masers, Molecules and Mass Outflows in Star Forming Regions*, ed. A. D. Haschick, p. 233
84. Genzel, R., Poglitsch, A., Stacey, G. J. 1988. *Ap. J. Lett.* 333: L59
85. Gezari, D. Y., Joyce, R. R., Righini, G., Simon, M. 1974. *Ap. J. Lett.* 191: L11
86. Gezari, D. Y. 1989. Submitted for publication
87. Gillett, F. C., Forrest, W. J. 1973. *Ap. J.* 179: 483
88. Glassgold, A. E., Huggins, P. J., Schucking, E. L., eds. 1982. *Symposium on the Orion Nebula to Honor Henry Draper. Ann. NY Acad. Sci.*, Vol. 395
89. Goldsmith, P. F., Krotkov, R., Snell, R. L., Brown, R. D., Godfrey, P. 1983. *Ap. J.* 274: 184
90. Goldsmith, P. F. 1984. In *Galactic and Extragalactic Infrared Spectroscopy*, ed. M. F. Kessler, J. P. Phillips, p. 233. Dordrecht: Reidel
91. Goldsmith, P. F., Krotkov, R., Snell, R. L. 1985. *Ap. J.* 299: 405
92. Goldsmith, P. F. 1985. *Proc. Int. Symp. Millimeter and Submillimeter Wave Radio Astron.*, p. 256. Granada: URSI
93. Gordon, C. P. 1970. *Astron. J.* 75: 914
94. Goudis, C. 1982. *The Orion Complex: A Case Study of Interstellar Matter. Astrophys. Space Sci. Libr.*, Vol. 90. Dordrecht: Reidel
95. Grasdalen, G. L. 1976. *Ap. J. Lett.* 205: L83
96. Grasdalen, G. L., Gehrz, R. D., Hackwell, J. A. 1981. In *Infrared Astronomy*, ed. C. G. Wynn-Williams, D. P. Cruikshank, p. 179. Dordrecht: Reidel
97. Haas, M. R., Hollenbach, D. J., Erickson, E. F. 1986. *Ap. J. Lett.* 301: L57
98. Hall, D. N. B., Kleinmann, S. G., Ridgway, S. T., Gillett, F. C. 1978. *Ap. J. Lett.* 223: L47
99. Hall, D. N. B. 1984. In *Galactic and Extragalactic Infrared Spectroscopy*, ed. M. F. Kessler, J. P. Phillips, p. 269. Dordrecht: Reidel
100. Hansen, S. S., Moran, J. M., Reid, M. J., Johnston, K. J., Spencer, J. H., Walker, R. C. 1977. *Ap. J. Lett.* 218: L65
101. Hansen, S. S., Johnston, K. J. 1983. *Ap. J.* 267: 635
102. Hardie, R. H., Heiser, A. M., Tolbert, C. R. 1964. *Ap. J.* 140: 1472
103. Harris, A. I., Townes, C. H., Matsakis, D. N., Palmer, P. 1983. *Ap. J. Lett.* 265: L63
104. Harris, A. I., Graf, U., Jaffe, D. T., Stutzki, J., Genzel, R. 1989. In preparation
105. Hartquist, T. W., Oppenheimer, M., Dalgarno, A. 1980. *Ap. J.* 236: 182
106. Harvey, P. M., Gatley, I., Werner, M. W., Elias, J. H., Evans, N. J., et al. 1974. *Ap. J. Lett.* 189: L87
107. Harwit, M. 1982. See Ref. 88, p. 56
108. Hasegawa, T., Kaifu, N., Inatani, J., Morimoto, M., Chikada, Y., et al. 1984. *Ap. J.* 283: 117
109. Hasegawa, T., Akabane, K. 1984. *Ap. J. Lett.* 287: L91
110. Hasegawa, T. 1986. *Astrophys. Space Sci.* 118: 421
111. Hasegawa, T. 1987. In *Star Forming Regions*, ed. M. Peimbert, J. Jugaku, p. 123. Dordrecht: Reidel
112. Hayashi, M., Hasegawa, T., Gatley, I., Garden, R., Kaifu, N. 1985. *MNRAS* 215: 31P
113. Heiles, C., Troland, T. H. 1982. *Ap. J. Lett.* 260: L23
114. Heiles, C. 1985. *Bull. Am. Astron. Soc.* 17: 570
115. Heiles, C., Stevens, M. 1986. *Ap. J.* 301: 331
116. Heiles, C. 1987. In *Interstellar Processes*, ed. D. J. Hollenbach, H. A. Thronson, p. 171. Dordrecht: Reidel
117. Herbig, G. H. 1962. *Adv. Astron. Astrophys.* 1: 47
118. Herbig, G. H., Rao, N. K. 1972. *Ap. J.* 174: 401
119. Herbig, G. H. 1982. See Ref. 88, p. 64
120. Herbig, G. H., Terndrup, D. M. 1986. *Ap. J.* 307: 609
121. Herbst, E., Klemperer, W. 1973. *Ap. J.* 185: 505
122. Herbst, E., Leung, C. M. 1986. *Ap. J.* 310: 378
123. Hermsen, W., Wilson, T. L., Bieging, J. H. 1988. *Astron. Astrophys.* 201: 276
124. Hermsen, W., Wilson, T. L., Walmsley, C. M., Henkel, C. 1988. *Astron. Astrophys.* 201: 285
125. Hildebrand, R. H., Dragovan, M., Novak, G. 1984. *Ap. J. Lett.* 284: L51

126. Hippelein, H., Münch, G. 1978. *Astron. Astrophys.* 68: 17
127. Ho, P. T. P., Barrett, A. H. 1978. *Ap. J. Lett.* 224: L23
128. Ho, P. T. P., Barrett, A. H., Myers, P. C., Matsakis, D. N., Cheung, A. C., et al. 1979. *Ap. J.* 234: 912
129. Höflich, P., Wehrse, R. 1987. *Astron. Astrophys.* 185: 107
130. Hollenbach, D., McKee, C. F. 1988. Submitted for publication
131. Hollenbach, D., Erickson, E. F., Haas, M. 1989. In preparation
132. Hough, J. H., Axon, D. J., Burton, M. G., Gatley, I., Sato, S., et al. 1986. *MNRAS* 222: 629
133. Hyland, A. R., Allen, D. A., Barnes, P. J., Ward, M. J. 1984. *MNRAS* 209: 465
134. Iglesias, E. R., Silk, J. 1978. *Ap. J.* 226: 851
135. Jacq, T., Jewell, P. R., Henkel, C., Walmsley, C. M., Baudry, A. 1988. *Astron. Astrophys.* 199: L5
136. Jaffe, D. T., Pankonin, V. 1978. *Ap. J.* 226: 869
137. Jaffe, D. T., Davidson, J. A., Dragovan, M., Hildebrand, R. H. 1984. *Ap. J.* 284: 637
138. Jaffe, D. T., Harris, A. I., Silber, M., Genzel, R., Betz, A. L. 1985. *Ap. J. Lett.* 290: L59
139. Jefferts, K. B., Penzias, A. A., Wilson, R. W. 1970. *Ap. J. Lett.* 161: L87
140. Jewell, P. 1989. In *Submillimetre and Millimetre Wave Astronomy*, ed. A. S. Webster. Dordrecht: Kluwer. In press
141. Johansson, L. E. B., Andersson, C., Elldér, J., Friberg, P., Hjalmarson, A., et al. 1984. *Astron. Astrophys.* 130: 227
142. Johnson, H. L., Hiltner, W. A. 1956. *Ap. J.* 123: 267
143. Johnson, J. J., Gehrz, R. D., Jones, T. J., Smith, J. R., Hackwell, J. A., Grasdalen, G. L. 1989. *Ap. J.* In press
144. Johnston, K. J., Palmer, P., Wilson, T. L., Bieging, J. H. 1983. *Ap. J. Lett.* 271: L89
145. Jones, B. F., Walker, M. F. 1985. *Astron. J.* 90: 1320
146. Jones, B. F., Walker, M. F. 1988. *Astron. J.* 95: 1755
147. Kazes, I., Crutcher, R. M. 1986. *Astron. Astrophys.* 164: 328
148. Keene, J., Hildebrand, R. H., Whitcomb, S. E. 1982. *Ap. J. Lett.* 252: L11
149. Keene, J., Blake, G. A., Phillips, T. G., Huggins, P. J., Beichman, C. A. 1985. *Ap. J.* 299: 967
150. Kholopov, P. N. 1959. *Astron. Zh.* 36: 434 (Engl. transl., *Sov. Astron. AJ* 3: 425)
151. Kleinmann, D. E., Low, F. J. 1967. *Ap. J. Lett.* 149: L1
152. Knacke, R. F., Capps, R. W. 1979. *Astron. J.* 84: 1705
153. Knacke, R. F., McCorkle, S. M. 1987. *Astron. J.* 94: 972
154. Knacke, R. F., Larson, H. P., Knoll, K. S. 1988. *Ap. J. Lett.* 335: L27
155. Knapp, G. R., Phillips, T. G., Huggins, P. J., Redman, R. O. 1981. *Ap. J.* 250: 201
156. Koepf, G. A., Buhl, D., Chin, G., Peck, D. D., Fetterman, H. R., et al. 1982. *Ap. J.* 260: 584
157. Königl, A. 1982. *Ap. J.* 261: 115
158. Krolik, J. H., Smith, H. A. 1981. *Ap. J.* 219: 141
159. Ku, W. H. M., Righini-Cohen, G., Simon, M. 1982. *Science* 215: 61
160. Kuiper, T. B. H., Evans, N. J. 1978. *Ap. J.* 219: 141
161. Kuiper, T. B. H., Zuckerman, B., Rodriguez-Kuiper, E. N. 1981. *Ap. J.* 251: 201
162. Kurucz, R. L. 1979. *Ap. J. Suppl.* 40: 1
163. Kutner, M., Thaddeus, P. 1971. *Ap. J. Lett.* 168: L67
164. Kutner, M., Thaddeus, P., Jefferts, K. B., Penzias, A. A., Wilson, R. W. 1971. *Ap. J. Lett.* 164: L49
165. Kutner, M. L., Evans, N. J., Tucker, K. D. 1976. *Ap. J.* 209: 452
166. Kutner, M. L., Tucker, K. D., Chin, G., Thaddeus, P. 1977. *Ap. J.* 215: 521
167. Kwan, J., Scoville, N. Z. 1976. *Ap. J. Lett.* 210: L39
168. Lada, C. J. 1980. In *Giant Molecular Clouds in the Galaxy*, ed. P. Solomon, M. Edmunds, p. 239. Oxford: Pergamon
169. Lada, C. J. 1987. In *Star Forming Regions*, ed. M. Peimbert, J. Jugaku, p. 1. Dordrecht: Reidel
170. Lada, E. 1989. PhD thesis. Univ. Tex., Austin
171. Langer, W. D. 1976. *Ap. J.* 206: 699
172. Laques, P., Vidal, J. L. 1979. *Astron. Astrophys.* 73: 97
173. Larson, R. B. 1969. *MNRAS* 145: 297
174. Lee, T. A. 1968. *Ap. J.* 152: 913
175. Lee, T. J., Beattie, D. H., Geballe, T. R., Pickup, D. A. 1983. *Astron. Astrophys.* 127: 417
176. Leger, A., Puget, J. L. 1984. *Astron. Astrophys.* 137: L5
177. Lesh, J. R. 1968. *Ap. J. Suppl.* 16: 371
178. Lesh, J. R. 1968. *Ap. J.* 152: 905
179. Lester, D. F., Dinerstein, H. L., Rank, D. M. 1979. *Ap. J.* 232: 139
180. Lester, D. F., Dinerstein, H. L., Werner, M. W., Watson, D. M., Genzel, R. 1983. *Ap. J.* 271: 618
181. Lester, D. F., Becklin, E. E., Genzel,

R., Wynn-Williams, C. G. 1985. *Astron. J.* 90: 2331
182. Leung, C. M., Herbst, E., Huebner, W. F. 1984. *Ap. J. Suppl.* 56: 231
183. Lindblad, P. O. 1967. *Bull. Astron. Inst. Neth.* 19: 34
184. Liszt, H. S., Wilson, R. W., Penzias, A. A., Jefferts, K. B., Wannier, P. G., Solomon, P. M. 1974. *Ap. J.* 190: 557
185. Lonsdale, C. J., Becklin, E. E., Lee, T. J., Stewart, J. M. 1982. *Astron. J.* 87: 1819
186. Loren, R. B. 1979. *Ap. J. Lett.* 234: L207
187. Loren, R. B., Mundy, L. G. 1984. *Ap. J.* 286: 232
188. Lynds, B. T. 1962. *Ap. J. Suppl.* 7: 1
189. Maddalena, R. J., Morris, M., Moscowitz, J., Thaddeus, P. 1986. *Ap. J.* 303: 375
190. Margulis, M., Lada, C. J., Snell, R. L. 1988. *Ap. J.* 333: 316
191. Masson, C. R., Berge, G. L., Claussen, M. J., Heiligman, G. M., Leighton, R. B., et al. 1984. *Ap. J. Lett.* 283: L37
192. Masson, C. R., Claussen, M. J., Lo, K. Y., Moffett, A. T., Phillips, T. G., et al. 1985. *Ap. J. Lett.* 295: L47
193. Masson, C. R., Lo, K. Y., Phillips, T. G., Sargent, A. I., Scoville, N. Z., Woody, D. P. 1987. *Ap. J.* 319: 446
194. Masson, C. R., Mundy, L. G. 1988. *Ap. J. Lett.* In press
195. Matsakis, D. N., Cheung, A. C., Wright, M. C. H., Askne, J. I. H., Townes, C. H., Welch, W. J. 1980. *Ap. J.* 236: 481
196. Mauersberger, R., Henkel, C., Jacq, T., Walmsley, C. M. 1988. *Astron. Astrophys.* 194: L1
197. Mauersberger, R., Henkel, C., Wilson, T. L. 1989. *Astron. Astrophys.* In press
198. McCaughrean, M. J. 1988. PhD thesis. Univ. Edinburgh, Scotl.
199. McCaughrean, M. J., McLean, I. S., Rayner, J. T., Aspin, C. 1989. In preparation
200. McKee, C. F., Hollenbach, D. J. 1987. *Ap. J.* 322: 275
201. McLean, I. S., McCaughrean, M. J., Rayner, J. T., Aspin, C. 1988. In preparation
202. Melnick, G., Gull, G. E., Harwit, M. 1979. *Ap. J. Lett.* 227: L29
203. Melnick, G., Genzel, R., Lugten, J. B. 1987. *Ap. J.* 321: 530
204. Melnick, G. J., Lugten, J. B., Poglitsch, A., Stacey, G. J., Genzel, R. 1989. In preparation
205. Menten, K. M., Walmsley, C. M., Henkel, C., Wilson, T. L. 1988. *Astron. Astrophys.* 198: 253
206. Mezger, P. G., Chini, R., Kreysa, E., Wink, J. E., Salter, C. J. 1988. *Astron. Astrophys.* 191: 44
207. Moran, J. M., Garay, G., Reid, M. J., Genzel, R., Wright, M. C. H., Plambeck, R. L. 1983. *Ap. J. Lett.* 271: L31
208. Moran, J. M., Silber, M., Reid, M. J., Genzel, R. 1989. In preparation
209. Morris, D., Berge, G. L. 1964. *Ap. J.* 139: 1389
210. Morris, M., Palmer, P., Zuckerman, B. 1980. *Ap. J.* 237: 1
211. Mouschovias, T. C. 1987. In *Physical Processes in Interstellar Clouds*, ed. G. E. Morfill, M. Scholer, p. 453. Dordrecht: Reidel
212. Muench, G., Hippelein, H. 1982. See Ref. 88, p. 170
213. Mundy, L. G., Scoville, N. Z., Baath, L. B., Masson, C. R., Woody, D. P. 1986. *Ap. J. Lett.* 304: L51
214. Mundy, L. G., Cornwell, T. J., Masson, C. R., Scoville, N. Z., Baath, L. B., Johansson, L. E. B. 1988. *Ap. J.* 325: 382
215. Myers, P. C., Goodman, A. A. 1988. *Ap. J. Lett.* 326: L27
216. Nadeau, D., Geballe, T. R., Neugebauer, G. 1982. *Ap. J.* 253: 154
217. Nakajima, T., Nagata, T., Nishida, M., Sato, S., Kawara, K. 1986. *MNRAS* 221: 483
218. Norman, G., Silk, J. 1981. *Ap. J.* 238: 158
219. Norris, R. P. 1984. *MNRAS* 207: 127
220. O'Dell, C. R., York, D. G., Henize, K. G. 1967. *Ap. J.* 150: 835
221. Ohishi, M., Kaifu, N., Suzuki, H. 1988. *Publ. Astron. Soc. Jpn.* In press
222. Olofsson, H., Hjalmarson, A., Rydbeck, O. E. H. 1981. *Astron. Astrophys.* 100: L30
223. Olofsson, H., Elldér, J., Hjalmarson, A., Rydbeck, G. 1982. *Astron. Astrophys.* 113: L18
224. Padman, R., Scott, P. F., Vizard, D. R., Webster, A. S. 1985. *MNRAS* 214: 251
225. Palous, J. 1987. *Proc. Eur. Reg. Meet. IAU, 10th*, ed. J. Palous, p. 209
226. Parenago, P. P. 1954. *Publ. Sternberg Astron. Inst. No. 25*
227. Parsamian, E. S., Chevira, E. 1982. *Bol. Inst. Tonantzintla* 3: 69
228. Pauls, T. A., Wilson, T. L., Bieging, J. H., Martin, R. N. 1983. *Astron. Astrophys.* 124: 23
229. Peimbert, M. 1982. See Ref. 88, p. 24
230. Peimbert, M. 1987. In *Star Forming Regions*, ed. M. Peimbert, J. Jugaku, p. 111. Dordrecht: Reidel
231. Phillips, T. G., Huggins, P. J., Kuiper, T. B. H., Miller, R. E. 1980. *Ap. J. Lett.* 238: L103

232. Phillips, T. G., Kwan, J., Huggins, P. J. 1980. In *Interstellar Molecules*, ed. B. Andrews, p. 21. Dordrecht: Reidel
233. Phillips, T. G., Huggins, P. J. 1981. *Ap. J.* 251: 533
234. Plambeck, R. L., Wright, M. C. H., Welch, W. J., Bieging, J. H., Baud, B., et al. 1982. *Ap. J.* 259: 617
235. Plambeck, R. L. 1987. In *Galactic and Extragalactic Star Formation*, ed. R. Pudritz, M. Fich, p. 253. Dordrecht: Kluwer
236. Plambeck, R. L. 1988. Private communication
237. Plambeck, R. L., Wright, M. C. H. 1987. *Ap. J. Lett.* 317: L101
238. Plambeck, R. L., Wright, M. C. H. 1988. *Ap. J. Lett.* 330: L61
239. Poglitsch, A., Geis, N., Genzel, R., Stacey, G. J. 1988. In preparation
240. Prasad, S. S., Huntress, W. T. Jr. 1982. *Ap. J.* 260: 590
241. Pudritz, R. 1989. In *Submillimetre and Millimetre Wave Astronomy*, ed. A. S. Webster. Dordrecht: Kluwer. In press
242. Puget, J. L. 1989. In *Submillimetre and Millimetre Wave Astronomy*, ed. A. S. Webster. Dordrecht: Kluwer. In press
243. Reipurth, B., Graham, J. 1988. *Astron. Astrophys.* 202: 219
244. Reynolds, R. J., Ogden, P. M. 1979. *Ap. J.* 229: 942
245. Rieke, G. H., Low, F. J., Kleinmann, D. E. 1973. *Ap. J. Lett.* 186: L7
246. Robinson, L. J. 1984. *Sky Telesc.* 67: 4
247. Russell, R. W., Melnick, G., Gull, G., Harwit, M. 1980. *Ap. J. Lett.* 240: L99
248. Sandell, G. 1989. In *Submillimetre and Millimetre Wave Astronomy*, ed. A. S. Webster. Dordrecht: Kluwer. In press
249. Scalo, J. M., Pumphrey, W. A. 1982. *Ap. J. Lett.* 258: L29
250. Schloerb, F. P., Loren, R. B. 1982. See Ref. 88, p. 32
251. Schloerb, F. P., Friberg, P., Hjalmarson, A., Höglund, B., Irvine, W. M. 1983. *Ap. J.* 264: 161
252. Schloerb, F. P., Snell, R. L., Schwartz, P. R. 1987. *Ap. J.* 319: 426
253. Schmid-Burgk, J., Densing, R., Krügel, E., Nett, H., Röser, H. P., et al. 1989. Submitted for publication
254. Schultz, G. V., Durwen, E. J., Röser, H. P., Sherwood, W. A., Wattenbach, R. 1985. *Ap. J. Lett.* 291: L59
255. Scoville, N. Z., Hall, D. N. B., Kleinmann, S. G., Ridgway, S. T. 1982. *Ap. J.* 253: 136
256. Scoville, N., Kleinmann, S. G., Hall, D. N. B., Ridgway, S. T. 1983. *Ap. J.* 275: 201
257. Sellgren, K., Werner, M. W., Dinerstein, H. L. 1983. *Ap. J. Lett.* 271: L13
258. Sharpless, S. 1952. *Ap. J.* 116: 251
259. Sharpless, S. 1962. *Ap. J.* 136: 767
260. Shull, J. M., Beckwith, S. 1982. *Annu. Rev. Astron. Astrophys.* 20: 163
261. Simon, M., Felli, M., Cassar, L., Fischer, J., Massi, M. 1983. *Ap. J.* 266: 623
262. Simpson, J. P., Rubin, R. H., Erickson, E. F., Haas, M. R. 1986. *Ap. J.* 311: 395
263. Smith, J., Lynch, D. K., Cudaback, D., Werner, M. W. 1979. *Ap. J.* 234: 902
264. Snell, R. L., Scoville, N. Z., Sanders, D. B., Erickson, N. R. 1984. *Ap. J.* 284: 176
265. Solomon, P. M., Jefferts, K. B., Penzias, A. A., Wilson, R. W. 1971. *Ap. J. Lett.* 168: L107
266. Solomon, P. M., Huguenin, G. R., Scoville, N. Z. 1981. *Ap. J. Lett.* 245: L19
267. Stacey, G. J., Kurtz, N. T., Smyers, S. D., Harwit, M., Russell, R. W., Melnick, G. 1982. *Ap. J. Lett.* 257: L37
268. Stacey, G. J., Smyers, S. D., Kurtz, N. T., Harwit, M. 1983. *Ap. J. Lett.* 265: L7
269. Stacey, G. J. 1985. PhD thesis. Cornell Univ., Ithaca, N.Y.
270. Stacey, G. J., Geis, N., Genzel, R., Poglitsch, A., Graf, U., et al. 1989. In preparation
271. Sternberg, A., Dalgarno, A. 1989. *Ap. J.* In press
272. Storey, J. W. V., Watson, D. M., Townes, C. H., Haller, E. E., Hansen, W. L. 1981. *Ap. J.* 247: 136
273. Strand, K. Aa. 1958. *Ap. J.* 128: 14
274. Strom, K. M., Strom, S. E., Wolff, S. C., Morgan, J., Wenz, M. 1986. *Ap. J. Suppl.* 62: 39
275. Stutzki, J., Genzel, R., Harris, A. I., Herman, J., Jaffe, D. T. 1988. *Ap. J. Lett.* 330: L125
276. Stutzki, J., Stacey, G. J., Genzel, R., Harris, A. I., Jaffe, D. T., Lugten, J. B. 1988. *Ap. J.* 332: 379
277. Stutzki, J., Genzel, R., Graf, U. U., Harris, A. I., Jaffe, D. T. 1989. Submitted for publication
278. Sugitani, K., Fukui, Y., Ogawa, H., Kawabata, K. 1986. *Ap. J.* 303: 667
279a. Sutton, E. C., Blake, G. A., Masson, C. R., Phillips, T. G. 1984. *Ap. J. Lett.* 283: L41
279b. Sutton, E. C., Blake, G. A., Genzel, R., Masson, C. R., Phillips, T. G. 1986. *Ap. J.* 311: 921
280. Sutton, E. C., Blake, G. A., Masson, C. R., Phillips, T. G. 1985. *Ap. J. Suppl.* 58: 341
281. Sweitzer, J. S. 1978. *Ap. J.* 225: 116
282. Tarafdar, S. P., Prasad, S. S., Huntress, W. T., Villere, K. R., Black, D. C. 1985. *Ap. J.* 289: 220

283. Tarter, J. C., Welch, W. J. 1986. *Ap. J.* 305: 467
284. Tenorio-Tagle, G. 1979. *Astron. Astrophys.* 71: 59
285. Thaddeus, P., Wilson, R. W., Kutner, M., Penzias, A. A., Jefferts, K. B. 1971. *Ap. J. Lett.* 168: L59
286. Thompson, R. I. 1984. *Ap. J.* 283: 165
287. Thronson, H. A., Harper, D. A., Bally, J., Dragovan, M., Mozurkewich, D., et al. 1986. *Astron. J.* 91: 1350
288. Tielens, A. G. G. M., Hollenbach, D. 1985. *Ap. J.* 291: 722
289. Tielens, A. G. G. M., Hollenbach, D. 1985. *Ap. J.* 291: 747
290. Tielens, A. G. G. M., Allamandola, L. J. 1987. In *Interstellar Processes*, ed. D. J. Hollenbach, H. A. Thronson, p. 397. Dordrecht: Reidel
291. Townes, C. H., Genzel, R., Watson, D. M., Storey, J. W. V. 1983. *Ap. J. Lett.* 269: L11
292. Troland, T. H., Heiles, C. 1986. *Ap. J.* 301: 339
293. Troland, T. H., Crutcher, R. M., Kazes, I. 1986. *Ap. J. Lett.* 304: L57
294. Trumpler, R. J. 1931. *Publ. Astron. Soc. Pac.* 43: 255
295. Tucker, K. D., Kutner, M. L., Thaddeus, P. 1973. *Ap. J. Lett.* 186: L13
296. Turner, B. E., Zuckerman, B. 1974. *Ap. J. Lett.* 187: L59
297. van Altena, W. F., Lee, J. T., Lee, J. F., Lu, P. K., Upgren, A. R. 1988. *Astron. J.* 95: 1744
298. Verschuur, G. L. 1969. *Nature* 223: 140
299. Viscuso, P. J. 1986. PhD thesis. Cornell Univ., Ithaca, N.Y.
300. Vogel, S. N., Wright, M. C. H., Plambeck, R. L., Welch, W. J. 1984. *Ap. J.* 283: 655
301. Vogel, S. N., Bieging, J. H., Plambeck, R. L., Welch, W. J., Wright, M. C. H. 1985. *Ap. J.* 296: 600
302. Vrba, F. J., Strom, S. E., Strom, K. M. 1988. *Astron. J.* 96: 680
303. Walker, C. K., Schulz, A., Krügel, E., Gillespie, A. R. 1988. *Astron. Astrophys.* 205: 243
304. Walker, M. F. 1969. *Ap. J.* 155: 447
305. Walmsley, C. M. 1975. In *H II Regions and Related Topics*, ed. T. L. Wilson, D. Downes, p. 17. Berlin: Springer-Verlag
306. Walmsley, C. M., Hermsen, W., Henkel, C., Mauersberger, R., Wilson, T. L. 1987. *Astron. Astrophys.* 172: 311
307. Warren, W. H., Hesser, J. E. 1978. *Ap. J. Suppl.* 36: 497
308. Warren-Smith, R. F., Gledhill, T. M., Scarrott, S. M. 1985. *MNRAS* 215: 75P
309. Waters, J. W., Gustincic, J. J., Kakar, R. K., Kuiper, T. B. H., Roscoe, H. K., et al. 1980. *Ap. J.* 235: 57
310. Watson, D. M. 1982. PhD thesis. Univ. Calif., Berkeley
311. Watson, D. M., Genzel, R., Townes, C. H., Storey, J. W. V. 1985. *Ap. J.* 298: 316
312. Weaver, H. 1974. In *Highlights of Astronomy*, ed. G. Contopoulos, 3: 423. Dordrecht: Reidel
313. Werner, M. W. 1970. *Astrophys. Lett.* 6: 81
314. Werner, M. W., Gatley, I., Harper, D. A., Becklin, E. E., Loewenstein, R. F., et al. 1976. *Ap. J.* 204: 420
315. Werner, M. W. 1982. See Ref. 88, p. 79
316. Werner, M. W., Dinerstein, H. L., Capps, R. W. 1983. *Ap. J. Lett.* 265: L13
317. Werner, M. W., Crawford, M. K., Genzel, R., Hollenbach, D. J., Townes, C. H., Watson, D. M. 1984. *Ap. J. Lett.* 281: L81
318. Westbrook, W. E., Werner, M. W., Elias, J. H., Gezari, D. Y., Hauser, M. G., et al. 1976. *Ap. J.* 209: 94
319. White, G. J., Monteiro, T. S., Richardson, K. J., Griffin, M. J., Rainey, R. 1986. *Astron. Astrophys.* 162: 253
320. White, G. J., Phillips, J. P. 1988. *Astron. Astrophys.* 197: 253
321. White, G. J. 1989. In *Submillimetre and Millimetre Wave Astronomy*, ed. A. S. Webster. Dordrecht: Kluwer. In press
322. Wilson, R. W., Jefferts, K. B., Penzias, A. A. 1970. *Ap. J.* 161: L43
323. Wilson, T. L., Pauls, T. 1985. *Astron. Astrophys.* 138: 225
324. Wilson, T. L., Serabyn, E., Henkel, C., Walmsley, C. M. 1986. *Astron. Astrophys.* 158: L1
325. Wilson, T. L., Serabyn, E., Henkel, C. 1986. *Astron. Astrophys.* 167: L17
326. Wilson, T. L., Johnston, K. J., Henkel, C., Menten, K. M. 1989. *Astron. Astrophys.* In press
327. Wilson, T. L., Johnston, K. J. 1989. *Ap. J.* In press
328. Woody, D. P., Scott, S. L., Scoville, N. Z., Mundy, L. G., Sargent, A. I., et al. 1989. Preprint
329. Wootten, A., Evans, N. J., Snell, R., Vanden Bout, P. 1978. *Ap. J. Lett.* 225: L143
330. Wootten, A., Loren, R. B., Bally, J. 1984. *Ap. J.* 277: 189
331. Wouterloot, J. G. A., Walmsley, C. M. 1986. *Astron. Astrophys.* 168: 237
332. Wright, M. C. H., Plambeck, R. L., Vogel, S. N., Ho, P. T. P., Welch, W. J. 1983. *Ap. J. Lett.* 267: L41
333. Wright, M. C. H., Plambeck, R. L. 1983. *Ap. J. Lett.* 267: L115
334. Wright, M. C. H., Vogel, S. N. 1985. *Ap. J. Lett.* 297: L11

335. Wynn-Williams, C. G., Becklin, E. E. 1974. *Publ. Astron. Soc. Pac.* 86: 5
336. Wynn-Williams, C. G. 1982. *Annu. Rev. Astron. Astrophys.* 20: 587
337. Wynn-Williams, C. G., Genzel, R., Becklin, E. E., Downes, D. 1984. *Ap. J.* 281: 271
338. Zaritzky, D., Shaya, E. J., Scoville, N. Z., Sargent, A. I., Tytler, D. 1987. *Astron. J.* 93: 1514
339. Zinnecker, H. 1982. See Ref. 88, p. 226
340. Ziurys, L. M., Marin, R. N., Pauls, T. A., Wilson, T. L. 1981. *Astron. Astrophys.* 104: 288
341. Ziurys, L. M., Turner, B. E. 1986. *Ap. J. Lett.* 300: L19
342. Ziurys, L. M., Friberg, P. 1987. *Ap. J. Lett.* 314: L49
343. Zmuidzinas, J., Betz, A. L., Boreiko, R. T., Goldhaber, D. M. 1986. *Ap. J. Lett.* 307: L75
344. Zuckerman, B., Buhl, D., Palmer, P., Snyder, L. E. 1970. *Ap. J.* 160: 485
345. Zuckerman, B. 1973. *Ap. J.* 183: 863
346. Zuckerman, B., Ball, J. A. 1974. *Ap. J.* 190: 35
347. Zuckerman, B., Palmer, P. 1974. *Annu. Rev. Astron. Astrophys.* 12: 279

X RAYS FROM NORMAL GALAXIES

G. Fabbiano

Harvard-Smithsonian Center for Astrophysics, 60 Garden Street, Cambridge, Massachusetts 02138

1. INTRODUCTION

The study of the X-ray properties of normal galaxies as a class was made possible by the launch of the *Einstein Observatory* in November 1978 (Giacconi et al. 1979). Before then, with the exclusion of the bright X-ray sources associated with Seyfert nuclei (Elvis et al. 1978, Tananbaum et al. 1978), only four galaxies had been detected in X rays: the Milky Way, M31, and the Magellanic Clouds (see Helfand 1984a, and references therein). The *Einstein* X-ray observations of well over 100 galaxies have been reported in the literature to date, and data on a similar number can still be found in the *Einstein* data bank. Some galaxies were detected with enough detail to allow a study of their X-ray morphology, spectra, and individual sources and to make comparisons with optical, infrared, and radio data. For all the galaxies, values of the X-ray flux, or even upper limits to this flux in the case of nondetections, can be used to explore average sample properties. These observations have shown that normal galaxies of all morphological types are spatially extended sources of X-ray emission with luminosities in the range of $\sim 10^{38}$ erg s^{-1} to 10^{42} erg s^{-1}. Although this is only a small fraction of the total energy output of a normal galaxy, X-ray observations are uniquely suited to study phenomena that are otherwise elusive. These include the end products of stellar evolution (supernova remnants and compact remnants) and a hot phase of the interstellar medium, discovered in bright early-type galaxies. This review gives an up to date and hopefully complete account of these *Einstein* results and of the published results from the European *EXOSAT* and Japanese *Ginga* satellites. Spiral and elliptical galaxies are reviewed separately, since their X-ray properties differ and suggest different origins for their X-ray emission.

2. THE X-RAY EMISSION OF SPIRAL GALAXIES

2.1 *Sources of X Rays*

Spiral galaxies are extended and complex X-ray sources with total luminosities in the *Einstein* band (~ 0.2–3.5 keV) of $\sim 10^{38}$ erg s^{-1} to a few 10^{41} erg s^{-1} (Fabian 1981, Long & Van Speybroeck 1983, Fabbiano 1984, 1986a). Observations of the Milky Way and of the Local Group galaxies (e.g. see Fabian 1981, Helfand 1984a,b) suggest that a good fraction of this X-ray emission is due to a collection of individual bright sources, such as close accreting binaries with a compact companion and supernova remnants, with luminosities ranging from $\sim 10^{35}$ erg s^{-1} up to a few 10^{38} erg s^{-1}. Stars also emit coronal X rays with luminosities of 10^{28} erg s^{-1} to 10^{33} erg s^{-1} (Vaiana et al. 1981). However, except perhaps at the lowest energies and in some starburst regions, stars do not contribute significantly to the total X-ray emission, since the X-ray to optical ratios measured in spiral galaxies (Long & Van Speybroeck 1983, Fabbiano & Trinchieri 1985) are larger than those expected from a normal stellar population (Topka et al. 1982, Helfand & Caillault 1982), and since the average X-ray spectrum of spiral galaxies appears harder than that of the stellar emission [$kT > 2$ keV in galaxies (Fabbiano & Trinchieri 1987); $kT \sim 0.5$–1 keV in stars (Helfand & Caillault 1982)]. Nuclear sources, either connected with star formation activity or with nonthermal Seyfert-like activity, can also be present and contribute various amounts to the X-ray emission. In this paper, however, we do not discuss "classical" Seyfert-type galaxies, where the emission is totally dominated by the nucleus (e.g. Elvis et al. 1978). Neither diffuse X-ray emission from inverse Compton scattering of the radio electrons off the optical-infrared photons nor synchrotron emission is likely to contribute significantly to the total X-ray emission (Fabbiano et al. 1982). A more likely source is diffuse thermal emission from a hot phase of the interstellar medium, heated by supernovae, and this is discussed later in this review.

2.2 *X-Ray Observations of the Local Group*

It is not surprising that the most detailed work on individual X-ray sources in galaxies and their identifications has been done as a result of the *Einstein* observations of Local Group galaxies (distance ≤ 1 Mpc). This work has been reviewed comprehensively by Helfand (1984a). Here I summarize the main results and review the more recent work not included in Helfand's paper. Table 1, adapted from Helfand (1984a), summarizes the results on the X-ray sources detected and identified with objects belonging or likely to belong to these galaxies. Although this summary offers a way to intercompare the different galaxies in the Local Group, the reader should take care in using it because of the different limiting sensitivities and completenesses of the different surveys. In particular, the Magellanic Clouds have been surveyed down to limiting luminosities for point-source detection of 10^{34}–10^{35} erg s^{-1}, whereas the other galaxies have limiting luminosities closer to 10^{37} erg s^{-1}.

In the Large Magellanic Cloud (LMC) most of the detected and identified sources are supernova remnants (see also Long & Helfand 1979, Helfand & Long 1980, Helfand 1982, 1984b, Tuohy et al. 1982, Mathewson et al. 1983, 1984, 1985, Cowley et al. 1984), of which three are Crab Nebula–like and one definitely contains a pulsar (Clark et al. 1982, Seward et al. 1984, Chanan et al. 1984). Studies of this sample of remnants has provided new insight on the supernova rate, the pulsar birthrate, and the evolution of the remnants (see Helfand 1984a, and references therein). One fourth of the sources found in the Small Magellanic Cloud (SMC) are likely to be supernova remnants. The X-ray population of the Clouds also includes a number of massive young Population I close accreting binaries, to which belong the brightest X-ray binaries known in the pre-*Einstein* era (Clark et al. 1978), and a number of still unidentified fainter sources with $L_X \sim 10^{34}$–10^{36} erg s^{-1} (e.g. Long et al. 1981a, Bruhweiler et al. 1987). The latter authors speculate that these sources in the SMC may be analogous to the wide Be-neutron star binaries that are found in the Galaxy (White et al. 1982, Tuohy et al. 1988), or that they could be normal O and B stars whose X-ray luminosity might have been enhanced by a factor of $\sim 10^4$ because of the low metallicity of the SMC.

Supernova remnants instead constitute a less important component of the X-ray sources detected in the two spiral systems M31 (Figure 1) and M33, which appear to be dominated by binary X-ray source candidates. One should remember that the luminosity threshold for these galaxies is higher than in the Clouds (see above), and that at least in the edge-on M31 the soft X-ray emission of supernova remnants might be affected by interstellar extinction, so it is not possible to draw a strong conclusion from this result. However, seven remnants with $L_X > 10^{37}$ erg s^{-1} are found in the LMC, while only two identifications with supernova remnants have been reported for M31 in this luminosity range, and possibly one for M33 (Long & Van Speybroeck 1983, Long et al. 1981b). The pointlike X-ray sources detected in M33 appear to be associated with young Population I indicators, with the exception of a strong nuclear source that is discussed later (Long et al. 1981b, Markert & Rallis 1983, Trinchieri et al. 1988).

90 FABBIANO

Table 1 X-ray sources in Local Group galaxies

Galaxy	Type	Total L_X (erg s^{-1})	No. of sources	Interlopers[a]	SNR	Binaries (young Pop I)	Binaries (older population)	Unidentified in galaxies	References[f]
LMC[b]	Ir I	6.6×10^{38}	102(52)	~46(~19)	32(21)	7(6)	2(2)	~15(~4)	1, 4, 5, 6
SMC	Ir I	6.1×10^{37}	57	~14	~12	1	0	~30	1, 5, 7, 8, 9
M31	Sb	3.6×10^{39}	117	~6	2	~26	~23GC + ~60[c]	—	1, 5, 10, 11, 12
M32	E2	5.4×10^{37}	1	—	—	—	1	—	2, 5, 12
M33	Sc	$1.1 \times 10^{39\text{d}}$	17	~3	1	~12	0	—	1, 5, 13, 14, 15
IC 1613	Ir I	—	0	—	—	—	—	—	1, 5
NGC 6822	Ir	1×10^{37}	2	~1	~1	—	—	—	1
NGC 205	E5	$<9 \times 10^{36}$	0	—	—	—	—	—	1
U Mi	E	$\leq 3.2 \times 10^{35}$	3	—	—	—	—	≤3	1
Maffei 1	E?	1×10^{39}	≤3[e]	—	—	—	—	extended	1
Milky Way	~Sc	$~3 \times 10^{39}$	~125[e]	—	~10	~40	8GC + ~67[c]	—	3, 5, 16

[a] These are either background or foreground sources not belonging to the galaxies.
[b] Numbers in parentheses are from the complete X-ray-flux-limited sample.
[c] GC are globular cluster sources.
[d] This includes the bright nuclear source.
[e] Approximate number of galactic X-ray sources with $L_X > 10^{35}$ erg s^{-1}.
[f] References: (1) Markert & Donahue 1985, and references therein; (2) G. Trinchieri, private communication, 1988; (3) Fabian 1981; (4) Long et al. 1981a; (5) Helfand 1984a; (6) Cowley et al. 1984; (7) Seward & Mitchell 1981; (8) Inoue et al. 1983; (9) Bruhweiler et al. 1987; (10) Van Speybroeck et al. 1979; (11) Long & Van Speybroeck 1983; (12) Crampton et al. 1984; (13) Long et al. 1981b; (14) Markert & Rallis 1983; (15) Trinchieri et al. 1988; (16) Helfand 1985.

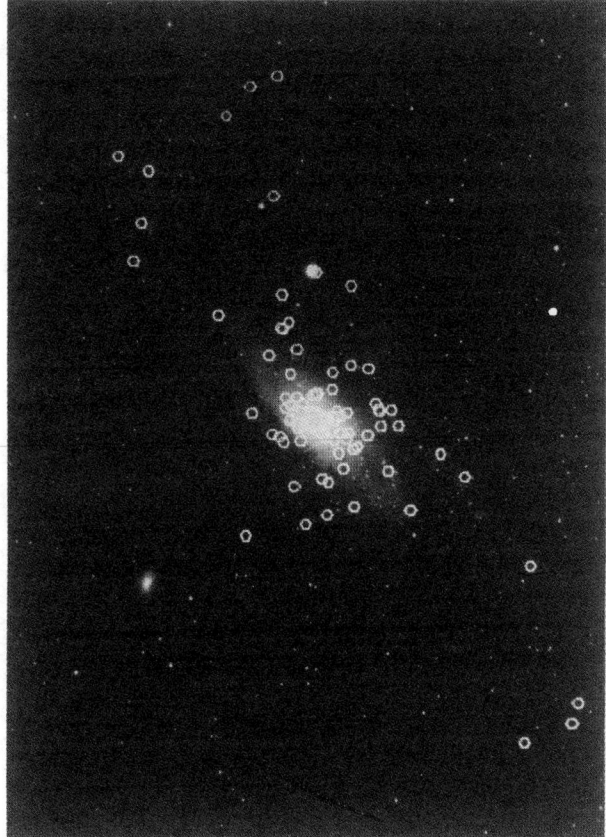

Figure 1 The circles show the positions of the X-ray sources of M31, superimposed onto an optical photograph (courtesy of L. Van Speybroeck). Notice the clustering of sources in the bulge.

One of these sources (M33 X-7) has a variable light curve that is consistent with a 1.8-day eclipsing binary period, similar to those of some massive galactic X-ray binaries (Peres et al. 1989). In M31 (Van Speybroeck et al. 1979, Van Speybroeck & Bechtold 1981, Long & Van Speybroeck 1983; see Helfand 1984a) most of the X-ray sources are instead likely to belong to an older stellar population. The luminosity of these sources is in the range of that of Galactic low-mass binaries, and some variability has been reported (Van Speybroeck & Bechtold 1981, McKechnie et al. 1984), consistent with the hypothesis of their being powered by accretion onto a compact object. Of these X-ray sources in M31, 19 have been identified with globular clusters tabulated by Sargent et al. (1977) and Battistini et al. (1980), and 4 more sources could also be associated with globular clusters (Crampton et al. 1984); ~30 sources are associated with the disk and the spiral arms; 19 pointlike sources plus a number of confused sources are detected in the inner bulge, within 2′ (400 pc) of the nucleus; and 22 additional sources lie in the outer bulge. The average X-ray spectrum of the bulge sources is similar to those of low-mass X-ray binaries in the Galaxy (Fabbiano et al. 1987b, Makishima et al. 1989). Only a few sources have been detected in the less massive members of the Local Group, but this is not surprising if the X-ray source formation rate is somehow linked to the mass or to the stellar content of a galaxy (see later).

Table 1 also gives an approximate estimate of the X-ray source content of the Galaxy (from Helfand 1984a; see also Fabian 1981). These estimates undoubtedly suffer from the unavoidable biases of all Galactic observation—namely, the difficulties in estimating distances and membership in a given stellar population or galactic structure, and the obscuration of softer X-ray sources (among which are the supernova remnants) by the interstellar medium in the Galactic plane. Even if we keep in mind all the necessary cautions, however, a comparison of the different entries of Table 1 suggests differences in the X-ray source composition in galaxies of different morphology. In particular, it appears that there is a shift from X-ray sources belonging to the young Population I to X-ray sources belonging to an older stellar population, going from the later-type galaxies (LMC, SMC, and M33, which are dominated by the spiral arm and disk stellar component) to the earlier-type ones (the Milky Way and definitely M31, with prominent bulges and a large number of globular clusters). Helfand (1985) shows that the percentage per unit mass of sources belonging to the spiral arm population increases with the morphological type in Local Group galaxies, going from bulge-dominated to disk-dominated galaxies; conversely, the percentage of bulge-type sources and globular cluster sources decreases with morphological type. Moreover, in M31 and the Galaxy the brightest X-ray sources belong to the older stellar population, while the opposite is observed in the Clouds and in M33. Another effect, first pointed out by Clark et al. (1978) (who ascribed it to the lower metallicity of the accreting gas), is the higher luminosity of the X-ray binary sources in the Clouds (see also Long & Van Speybroeck 1983, Crampton et al. 1984, Helfand 1984a). This effect could be responsible for the enhancement of the detection of young Population I sources in later-type galaxies. Alternatively, the reason could lie in an intrinsic higher mass for the accreting compact object: In particular, two of the four best black hole candidates are found in the LMC [LMC X-1 and LMC X-3 (Hutchings 1984; see Helfand 1985)].

Detailed comparisons have been made between the different X-ray source populations of the two most massive members of the Local Group: the Galaxy and M31 (Van Speybroeck & Bechtold 1981, Long & Van Speybroeck 1983, Battistini et al. 1982, Crampton et al. 1984). In particular, Long & Van Speybroeck (1983) remarked that M31 appears to have both a larger population of bulge sources and a larger and more luminous population of globular cluster sources than does the Galaxy. These results should be explainable in terms of the differences between these two galaxies: M31 has a larger population of globular clusters [~ 600 (Crampton et al. 1985) versus 180 in the Galaxy (Harris & Racine 1979)]. Crampton et al. (1984) remark that the fractions of X-ray-emitting globular clusters in the two galaxies are similar, and that the higher X-ray luminosities seen in the M31 globular clusters may simply reflect the fact that M31 contains more globular clusters and that the X-ray-bright ones, which are also the optically brightest and more condensed, lie on the high-luminosity tail of an overall distribution that is similar to the one in the Galaxy. Other authors (Long & Van Speybroeck 1983, Huchra et al. 1982, Battistini et al. 1982) have debated whether these M31 globular clusters are peculiar, and whether a metallicity–X-ray luminosity effect is possible. The bulge of M31 has a central concentration of ~ 20 sources within ~ 400 pc that appear more luminous on average than a similar number of sources in the outer bulge (Van Speybroeck et al. 1979). If these sources are distributed spherically, they would constitute a structure unlike any seen in the Galaxy, where only a few sources are seen in a similar volume (Long & Van Speybroeck 1983). These authors, however, suggest the alternative that the bulge X-ray sources in both galaxies might form similar, flattened barlike structures (see also Van Speybroeck & Bechtold 1981). The coincidence of this enhanced source distribution in M31 with a reported hole in the distribution of optical novae led Vader et al. (1982) to suggest that these sources could be the result of the evolution of a dead cataclysmic variable population that had long since undergone its nova phase. More recent observations, however, have dispelled the notion of a nova hole in M31 (Ciardullo et al. 1987). These authors suggest that the enhanced X-ray source content and the high specific nova rate that they find in the bulge of M31 could both be connected with the disruption of globular clusters in the bulge, which would then release the binaries that had formed in their cores. Although controversial (van Paradijs & Lewin 1985, Vader et al. 1982), a similar mechanism has been suggested by Grindlay (1984, 1985) for the formation of low-mass binaries in the Galaxy.

The results discussed so far are concerned with the detectable individual source content of the Local Group galaxies. However, below the single source threshold, it is still possible to detect the integrated emission of the fainter, individually undetectable sources and of any truly diffuse component, such as a hot phase of the interstellar medium. This type of emission has been reported in the Galaxy, where it accounts for <10% of the integrated emission of the resolved sources in the 2–10 keV range [the "Galactic ridge" (e.g. Worrall et al. 1982, Warwick et al. 1985, Koyama et al. 1986)]. A soft diffuse X-ray background ($E < 0.284$ keV), which is likely to be of Galactic origin, has also been observed (e.g. McCammon et al. 1983, Marshall & Clark 1984). A soft (0.25 keV) X-ray survey has revealed the presence of diffuse emission in the LMC (Singh et al. 1987). This emission has also been seen in the *Einstein* survey and has a total X-ray luminosity of 3×10^{38} erg s^{-1} (D. Helfand, private communication, 1988). The only other galaxy of the Local Group for which diffuse emission has been reported to date is M33 (Trinchieri et al. 1988), where the diffuse component accounts for $\sim 1/3$ of the total nonnuclear emission (which is not surprising given the much higher source detection threshold of these observations) and can be separated into a spectrally hard ($kT > 2$ keV) and a soft ($kT < 1$ keV) component. The hard component could be due to the integrated contribution of several lower luminosity compact accreting systems and young supernova remnants; the soft component is most likely due to the integrated emission of stellar coronae, with a possible contribution from a hot phase of the interstellar medium.

2.3 *Detailed Observations of Bright Spiral Galaxies*

Only a few very bright X-ray sources can be detected in the *Einstein* images of more distant galaxies, which typically appear as extended X-ray emission regions (e.g. Fabbiano & Trinchieri 1987). However, there is reason to believe that most of the X-ray emission of these galaxies is due to sources akin to those detected in the Local Group. A comparison of the fraction of the X-ray emission resolved in individual bright sources versus that which appears diffuse is consistent with the bulk of the emission originating from individual sources below threshold: More distant galaxies, with higher point-source thresholds, have a relatively larger "diffuse" emission than do less distant galaxies (Fabbiano 1988a). Moreover, the X-ray spectra of these galaxies, although ill defined, are consistent with the hard spectra expected from binary X-ray sources (Fabbiano & Trinchieri 1987, Trinchieri et al. 1988, Fabbiano 1988b).

Only a very few of the sources detected in spiral galaxies can be chance superpositions of background or foreground objects, given the statistics of the serendipitous *Einstein* source detections (Gioia et al. 1984; see Fabbiano & Trinchieri 1987). Most of these sources are therefore in the galaxies, often in the spiral arms, and their X-ray luminosities can be very high indeed. They are typically well above the Eddington limit for accretion onto a 1-$M_\odot$ compact object, which is $\sim 1.3 \times 10^{38}$ erg s^{-1}, and can be as bright as a few $\times$ 10^{39} erg s^{-1}. These sources have been detected in a number of spirals, including M83, M51, NGC 253, NGC 4631, NGC 6946, M101, and M81 (Long & Van Speybroeck 1983, Fabbiano & Trinchieri 1984, 1987, Trinchieri et al. 1985, Palumbo et al. 1985, Fabbiano 1988a), and eight bright sources have also been reported in the central starburst region of M82 (Watson et al. 1984). We exclude bright nuclear regions from the present discussion and concentrate instead on the more puzzling galactic sources. One of the most extreme cases is that of a source reported in M100 with an X-ray luminosity of $\sim 10^{40}$ erg s^{-1} (Palumbo et al. 1981). However, a very recent analysis of the same data by this reviewer, and of a subsequent longer exposure with the same high resolution, suggests that this source might be spurious: It is not detected in the longer exposure, and it is only marginally detected as an *extended* feature in the first observation. A similarly luminous source, however, is detected in NGC 4631 (Fabbiano & Trinchieri 1987). About 36 sources more luminous than $\sim 2 \times 10^{38}$ erg s^{-1} have been reported to date, 16 of which are more luminous than 10^{39} erg s^{-1}.

What are these sources? In one case the answer is simple: One of these sources is SN 1980K, which was detected in NGC 6946 $\sim$35 days after maximum light (Canizares et al. 1982b). The variability reported for some bright sources in M101 suggests pointlike objects [possibly bright accretion binaries (Long & Van Speybroeck 1983)]. If these sources are mostly complex emission regions, we would be faced with several bright sources (e.g. 10^{37} erg s^{-1}) in volumes with typical dimensions of a few hundred parsecs to a kiloparsec (Fabbiano & Trinchieri 1987). These sources are not typically in bulges, where such crowding could be expected [e.g. M31 (see earlier discussion)]. If these sources are truly single objects, they could indicate the presence of massive black holes in these galaxies. It is possible, however, that the distances of some galaxies might have been overestimated, which would make these sources appear more luminous than they are in reality. For instance, estimates of the distance of NGC 4631 range from 12 Mpc (Sandage & Tammann 1981) to 3 Mpc (Duric et al. 1982); using the lower estimate, the luminosity of the source reported by Fabbiano & Trinchieri (1987) as $\sim 1.4 \times 10^{40}$ erg s^{-1} would become $\sim 9 \times 10^{38}$ erg s^{-1}. However, this still exceeds the Eddington luminosity of an accreting neutron star.

The observations of the Local Group suggest differences in the X-ray sources in galaxies of different morphological type. However, differences also seem to occur in galaxies of similar morphology. An example is given by a comparison of the X-ray properties of the two Sb galaxies M31 and M81 (Fabbiano 1988a). The latter shows a number of individual X-ray sources, all more luminous than the most luminous sources of M31, and it is highly unlikely that this is due to an overestimate of the distance of M81. With the present data it is impossible to discriminate between an intrinsically more luminous X-ray source population in M81 or a more numerous population with the same luminosity function as that of M31. An overall comparison between these two galaxies also shows that M81 is overluminous in both X-ray and radio continuum emission, and to a lesser extent in far-infrared emission, relative to their optical luminosities. This suggests differences in the star formation history of these two galaxies, which resulted in a more efficient production of X-ray, cosmic ray, and far-infrared sources in M81. All these results are very tantalizing and show the importance of future sensitive X-ray observations in furthering our understanding of the global properties of spiral galaxies and of the detailed physical properties of their stellar remnants.

The presence of bulge (and globular cluster) X-ray sources on the one hand, and of often very bright X-ray sources associated with the spiral arms and H II regions on the other, is immediately demonstrated by the X-ray images of the Local Group and of other relatively nearby galaxies. The close resemblance between the radial profile of the X-ray surface brightness of a few face-on spirals and that of the optical light of their exponential disk suggests the presence of a third component of the X-ray emission, one associated with the stellar population of the disks (see Fabbiano 1986a). This effect was first seen in M83 (Figure 2; Trinchieri et al. 1985) and then in M51 (Palumbo et al. 1985) and possibly NGC 6946 (Fabbiano & Trinchieri 1987) and M81 [Fabbiano 1988a; see also M33 (Trinchieri et al. 1988)]. In particular, the X-ray profile in M51 is significantly different from the Hα profile (where the arms are very prominent) and follows the exponential disk distribution, as do the radio continuum and the CO profiles (Figure 3). These observations open interesting possibilities for our understanding of the origin of low-mass X-ray binaries. The nature of these sources, which constitute $\sim 60\%$ of the Galactic sources and which are also called "galactic bulge sources" and "Population II" sources in the X-ray literature (e.g. van den Heuvel 1980, Helfand 1984a), is one of the open problems of "classical" X-ray astronomy. Suggestions on their origin have included capture of neutron stars in the Galactic bulge (van den Heuvel 1980), remnants of disrupted globular clusters (Grindlay 1984, 1985), or the evolved remnants of low-mass binary systems in the Galactic disk (e.g. Gursky 1976, Rappaport et al. 1982, Nomoto 1984, van den Heuvel 1984, and references therein). The observations of external galaxies morphologically similar to the Milky Way suggest that at least a good fraction of these sources may originate from the evolution of binary systems belonging to the disk stellar population, rather than from dynamical evolution (Fabbiano 1985a, 1986a, Trinchieri

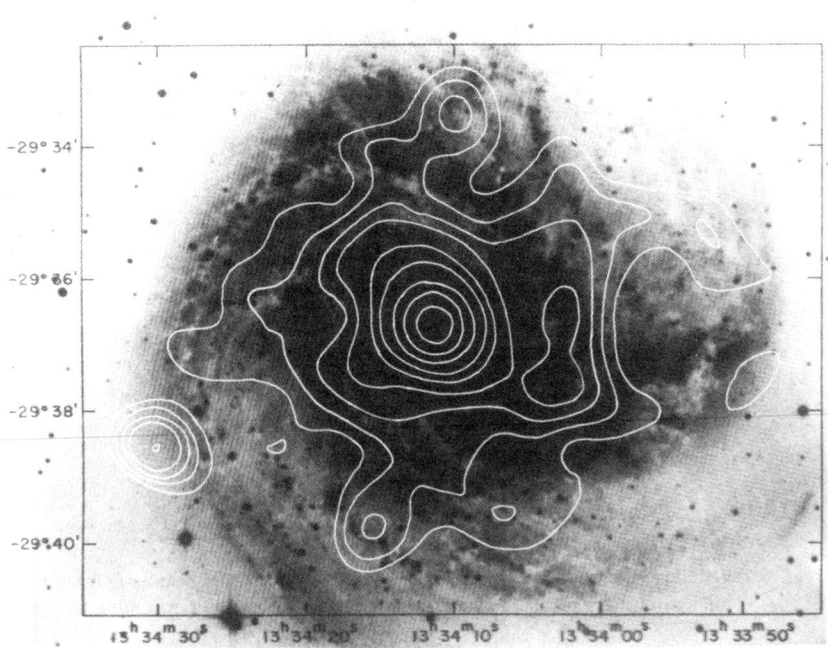

Figure 2 The *Einstein* HRI (high-resolution imager) X-ray map of M83 (Trinchieri et al. 1985). X-ray emission can be seen from most of the optical disk. A bright source is visible on a spiral arm.

et al. 1985). However (Fabbiano 1985a, 1986a), there is a relative excess of X-ray emission over the disk emission in the innermost disk region seen in both M83 and M51 (Trinchieri et al. 1985, Palumbo et al. 1985), which, if we scale by distance and dimensions, roughly coincides with what has been called the X-ray Galactic bulge. This excess emission could either (*a*) indicate an intrinsically brighter population of X-ray sources, analogous to the bright "bulge" Galactic sources, (*b*) point to an enhanced past episode of star formation in the inner disk, in contrast perhaps with steady star formation in the disk as a whole (e.g. Vader et al. 1982), or (*c*) suggest the presence of an additional component of the X-ray binary source population, which could be related to the disruption of globular clusters (Grindlay 1984, 1985).

However, there is one case in which the X-ray surface brightness profile clearly does not follow the optical light but instead follows quite well the radio continuum profile: NGC 253 (Fabbiano 1988b). Enhanced radio and X-ray emission is observed in the inner disk of NGC 253, and I discuss

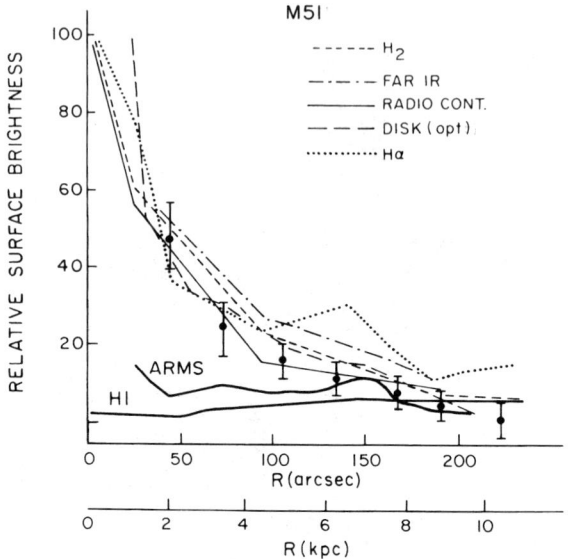

Figure 3 Radial profile of the X-ray surface brightness of M51 [the points (Palumbo et al. 1985)], together with radial profiles at other wavelengths from Scoville & Young (1983) and Klein et al. (1984).

in the next section the implications of this and other observations for a connection between X-ray sources and radio continuum emission.

Another source of X-ray emission, besides the contributions of individual sources, has been predicted in spiral galaxies. This is the thermal emission of the interstellar medium, heated by supernovae, which release $\sim 10^{42}$ erg s^{-1} in the galaxy. It has been suggested that hot gaseous coronae, or galactic fountains, could be produced and should be visible in soft X rays in the *Einstein* range (e.g. Spitzer 1956, Cox & Smith 1974, Bregman 1980a,b, Corbelli & Salpeter 1988). There is evidence of soft thermal diffuse emission both in the Galactic plane and in the LMC (e.g. McCammon et al. 1983, Marshall & Clark 1984, Singh et al. 1987), and perhaps in M33 (Trinchieri et al. 1988). A search for this type of emission in more distant galaxies has been made by Bregman & Glassgold (1982) and McCammon & Sanders (1984). The former searched two edge-on galaxies, NGC 3628 and NGC 4244, for coronal emission and set limits of 10^{39} and 2×10^{38} erg s^{-1}, respectively, on the X-ray luminosity of a gaseous halo; the latter instead analyzed a large face-on galaxy, M101, and concluded that the limits on the diffuse X-ray emission require that the temperature of any hot gas that is radiating $\geq 10\%$ of the average supernova power be less than $10^{5.7}$ K, and that hot bubbles occupy at most 25% of the region between 10 and 20 kpc from the galactic center. Subsequently, Cox & McCammon (1986) have used these measurements to constrain the density of the interstellar medium of M101 and the characteristics of the population of supernova remnants evolving in the disk. The lack of intense diffuse soft X-ray emission could imply that most of the supernova energy is radiated in the unobservable far-ultraviolet (Cox 1983). The only reported instance of this type of soft X-ray emission in a spiral galaxy is in the edge-on NGC 4631, where this component could have an X-ray luminosity of 5×10^{39} erg s^{-1}, which represents $\sim 13\%$ of the total emission in the *Einstein* band (Fabbiano & Trinchieri 1987).

Shostak et al. (1982) interpret an arc-shaped structure aligned with a ridge of radio continuum emission in NGC 1961 as evidence of a shock-heated interstellar medium that is being stripped by a hot intergalactic medium. However, without clear spectral confirmation, this interpretation is not unique: Enhanced star formation resulting from the interaction could also be responsible for the excess emission (e.g. Fabbiano et al. 1982; see next section). Similarly, a plume of extended emission emanating from NGC 4438, a spiral galaxy in the Virgo cluster, is reported by Kotanyi et al. (1983) as evidence of ram pressure sweeping of the interstellar gas from the disk of the galaxy. This plume, however, seems to emanate from the bright nucleus of this galaxy and could perhaps be similar to the gaseous plumes detected near the starburst nuclei of NGC 253 (Fabbiano & Trinchieri 1984) and M82 (Watson et al. 1984). I discuss these later in this review.

2.4 Average Sample Properties and Correlations With Other Wavebands

Although the *Einstein* images of some galaxies allow us to study them in detail and so reach some understanding of their X-ray emission components, most of the data are not of this high quality. Some 50 spiral galaxies were surveyed as part of the original *Einstein* observing program, and the results of these observations have been used to study the average properties of the sample and to explore correlations with the emission at other wavebands. This sample is not statistically complete; however, it can be regarded as representative of "normal" spiral galaxies of different morphologies and absolute magnitudes (Fabbiano & Trinchieri 1985). These galaxies have X-ray luminosities ranging between 10^{38} erg s^{-1} and 10^{41} erg s^{-1}, which are linearly correlated with their emission in the optical B band. This correlation is similarly tight for early-type bulge-dominated spirals and for late-type disk/arm-dominated galaxies: For all of them the ratios of monochromatic (2 keV) X-ray to optical (B) flux densities cluster around 10^{-7}. This result suggests that the X-ray emission is mostly due to sources constituting a constant fraction of the stellar population, in agreement with the conclusion of the detailed X-ray observations discussed above, which show that the X-ray-emitting population is likely to be dominated by binary X-ray sources (Long & Van Speybroeck 1983, Fabbiano et al. 1984b, Fabbiano 1984, Fabbiano & Trinchieri 1985). Even Sa galaxies follow this correlation, which suggests that their X-ray emission is due to the same type of sources responsible for the general emission of spiral galaxies and does not require an additional large gaseous emission component, such as is seen in bright elliptical galaxies (see later in this review).

Correlations have also been found between the X-ray emission and other variables, including the radio continuum, the near-infrared H band, and the far-infrared *IRAS* emission (Fabbiano & Trinchieri 1985, Fabbiano et al. 1988). These correlations are all very tight in late-type galaxies. In the subsample of bulge-dominated galaxies, however, the correlations between radio continuum and/or far-infrared luminosities with any of the other emission bands show a considerable amount of scatter and sometimes also a shift in zero point. In particular, for a given X-ray luminosity, there is a clear deficiency of radio continuum emission when bulge-dominated spirals are compared with disk/arm-dominated galaxies, whereas no differences are seen in the X-ray and optical (B) correlations. Since most of the optical and near-infrared emission of early-type spirals is dominated by the emission of the bulge (Kent 1985), these differences suggest that the radio continuum and the far-infrared are mainly related to the stellar population of the disk, whereas the X rays originate in both the disk and the bulge components. The latter conclusion is in agreement with the results of the X-ray observations of M31, which were discussed earlier.

Fabbiano & Trinchieri (1985; see also Fabbiano et al. 1984b) find that the correlation between X-ray and B-band emission is stronger than those between X-ray and either H-band or the $B-H$ color, which are both indicators of older stellar content and/or galaxy mass (e.g. Whitmore 1984). This result suggests that the X-ray sources belong predominantly to the blue-emitting stellar Population I. A link with the youngest Population I, to which the massive binary X-ray sources belong, is suggested in particular by a comparison between spiral galaxies with "normal" average colors and galaxies with blue peculiar colors, indicative of extensive and recent star formation activity (see Larson & Tinsley 1978). For a given optical luminosity, the X-ray luminosity is enhanced in relatively bluer galaxies (Fabbiano et al. 1982, 1984b, Fabbiano & Panagia 1983). These results also have implications for the nature and evolution of the low-mass X-ray binaries, which represent a very large component of the X-ray emission of the Milky Way: In particular, they suggest that most of these sources are likely to belong to the old Population I and possibly originate from the evolution of native binary systems, rather than having a dynamical origin. These conclusions are supported by the presence of an exponential disk in the X-ray emission of face-on spiral galaxies, as discussed earlier (see Fabbiano 1985a). Of all the correlations studied in late-type spirals, only two—the X-ray/B and the radio/far-infrared—imply strict proportionality between the emission at the two different wavelengths; the others follow power laws with exponents significantly different from unity. Trying to understand which correlations are intrinsically stronger, and therefore more directly connected to the underlying phenomena we wish to discover, and why not all the correlations scale simply with luminosity can give us new insight on the stellar components and evolution of spiral galaxies. Implications for the initial mass function (IMF), the average dust content of the disks, and the presence of compact star-forming regions are reviewed and discussed by Fabbiano et al. (1988). In particular, these correlations suggest the preferential occurrence of obscured starburst components in the more luminous galaxies.

An interesting possibility, raised both by single-galaxy studies and by statistical comparisons, is that of a connection between X-ray and radio continuum emission in spiral and irregular galaxies (Fabbiano et al. 1984b, Fabbiano & Trinchieri 1985, 1987, Palumbo et al. 1985, Fabbiano 1988b). This connection could be through recent star formation, but it is possible that there could be a more direct link between X-ray sources and cosmic-ray production (Fabbiano & Trinchieri 1985, Fabbiano et al. 1988). In particular, there is strong evidence of particle acceleration in X-ray binaries: Relativistic jets have been detected in the massive X-ray binary SS 433 and have been suggested to explain the radio morphology of the low-mass binary Sco X-1 (Geldzahler et al. 1981, Hjellming & Johnston 1981, Watson et al. 1983); and gamma-ray emission has been reported from the X-ray binaries Cyg X-3, Her X-1, and Vela X-1, suggesting intense cosmic-ray production (Samorski & Stamm 1983, Dowthwaite et al. 1984, Protheroe et al. 1984). Although strong, the X-ray/radio correlation has a power law exponent different from unity. Fabbiano & Trinchieri (1985), assuming proportionality between the sources of cosmic rays and the X-ray-emitting population, suggested that this result could imply a luminosity dependence of the intensity of the magnetic field of spiral galaxies. However, the proportionality between radio and far-infrared emission suggests that the sources of cosmic-ray electrons are a constant fraction of the stars responsible for heating the dust to far-infrared temperatures, and therefore that the nonlinear X-ray/radio correlation could be the result of these obscured regions contributing relatively less to the X-ray emission than to the radio and far-infrared emission (Fabbiano et al. 1988).

Although limited to relatively small samples, these comparisons have shown the potential of a multifrequency approach to the study of global galaxy properties. Their extension to larger and better defined samples, through systematic searches of the *Einstein* data bank (e.g. D. Burstein et al., in preparation, 1989) and through future X-ray observations, will be needed to confirm some of the present results and to gain a more general understanding of the structure and evolution of spiral galaxies.

3. STARBURST ACTIVITY AND LOW-ACTIVITY NUCLEI

3.1 *Widespread Starburst Activity in Peculiar Galaxies*

It was mentioned in the previous section that bluer "starburst," often-interacting galaxies tend to have enhanced X-ray emission compared with galaxies having redder, more normal colors (Fabbiano et al. 1982, 1984b, Stewart et al. 1982). The X-ray emission of these galaxies tends to originate from spatially extended regions, excluding a purely nonthermal nuclear origin, and their X-ray spectra exclude on average very soft emission, which suggests that the X-ray emission is not dominated by the thermal emission of a gaseous halo (Fabbiano et al. 1982). There are, however, exceptions to this second statement: The interacting pair NGC 4038/9 (the Antennae) possibly has a softer component of the X-ray emission that is of gaseous origin (Fabbiano et al. 1982, Fabbiano & Trinchieri 1983), and a similar component has been suggested in Arp 220 (Eales & Arnaud 1988); gaseous emission has been suggested to occur in ring galaxies (Ghigo et al. 1983), although there is no proof of its existence; and such emission has been reported in spiral-rich compact groups (Bahcall et al. 1984). Detailed observations of nearby galaxies with starburst nuclei show extended gaseous components emanating from the nuclear regions (e.g. Watson et al. 1984, Fabbiano & Trinchieri 1984, Fabbiano 1988b).

The bulk of the X-ray emission of these galaxies can be understood in terms of a number of young supernova remnants and massive X-ray binaries (with X-ray luminosity possibly enhanced by the low metallicity of the accreting gas) similar to those observed in the Magellanic Clouds (Fabbiano et al. 1982, Stewart et al. 1982). Although this explanation is not applicable in general (Fabbiano et al. 1982), the integrated coronal emission of the young stellar population [see Vaiana et al. (1981) for typical values] may dominate in very young starbursts, where the ultraviolet *IUE* spectra suggest the presence of a very large number of OB stars (Moorwood & Glass 1982, Fabbiano & Panagia 1983). Recently, Ward (1988) has reported a correlation between the Brackett-γ line emission from a sample of starburst nuclei and their X-ray emission, which is interpreted in terms of a relationship between the number of ionizing photons produced by the OB stars and the associated X-ray binary population, as estimated by Fabbiano et al. (1982). These authors, by comparing the observed radio and X-ray luminosities of their sample of peculiar galaxies with the expected output of a population of supernova remnants and X-ray binaries, also point out that the nonthermal radio emission from supernova remnants is not likely to account for the entire observed radio power, and thus that a different radio emission mechanism is required, as had already been observed by Biermann & Friecke (1977) in their study of Markarian galaxies.

3.2 Starburst Nuclei

There are galaxies in which the starburst activity is confined to the nuclear regions. The first reported instance of X-ray emission from this type of nucleus is that of NGC 7714 by Weedman et al. (1981), who associate this emission with the type of activity discussed above, as opposed to Seyfert-type nonthermal activity. A number of starburst nuclei, often embedded in an otherwise normal spiral galaxy, have been studied in X rays. They include the Galactic center region (Watson et al. 1981), M82 (Van Speybroeck & Bechtold 1981, Watson et al. 1984, Biermann 1984, Kronberg et al. 1985, Schaaf et al. 1989, Fabbiano 1988b), NGC 253 (Van Speybroeck & Bechtold 1981, Fabbiano & Trinchieri 1984, Fabbiano 1988b), M83 (Trinchieri et al. 1985), M51 (this nucleus also contains a Seyfert component, which does not dominate the X-ray emission; Palumbo et al. 1985), NGC 6946 and IC 342 (Fabbiano & Trinchieri 1987), and NGC 3628 (G. Fabbiano et al., in preparation, 1989). A common characteristic of the emission spectra of these nuclei is the intense far-infrared component (see Figure 9 of Fabbiano & Trinchieri 1987, and references therein), indicative of dusty nuclear regions heated by newly formed early-type stars. The X-ray-emitting regions are seen to be extended whenever observed with high-enough spatial resolution, and in M82 there is evidence of a population of bright individual sources (Watson et al. 1984). Typical X-ray luminosities of these nuclear regions are in the 10^{39} erg s^{-1} range, except for the Galactic center region, which is ~ 1000 times less luminous. To explain this emission requires, in different cases, different amounts of evolved sources (supernova remnants and X-ray binaries) superimposed on the integrated stellar emission from a young stellar population. The X-ray spectra of two of these nuclei, NGC 253 and M82 (Fabbiano 1988b, Schaaf et al. 1989), are intrinsically absorbed, consistent with the presence of a dusty emission region. In M82, in particular, it is possible that two different spectral components are present: a softer one, possibly dominated by the newly formed stars and by the interstellar medium shock heated by the frequent supernovae; and a harder one, which could be due to either binary X-ray sources with large intrinsic absorption cutoff or inverse Compton emission resulting from the interaction of the infrared photons in the nucleus with the relativistic electrons responsible for the radio emission. The latter mechanism was invoked by Kruegel et al. (1983) to explain the apparent excess X-ray luminosity of M82 when compared with NGC 253, but it had been dismissed by Watson et al. (1984; see also Fabbiano & Trinchieri 1984) on the basis of a spatial comparison of the X-ray, radio, and far-infrared emission. Schaaf et al. (1989), however, argue that it cannot be excluded that inverse Compton effects are responsible for a sizable fraction of the X-ray emission. It is clear that future observations at higher spectral resolution will be essential for distinguishing thermal and nonthermal emission components.

Line-of-sight extinction affects differently the optical B-band and the X-ray emission in the *Einstein* band, depending on the "hardness" of the X-ray spectrum. Using this effect, Trinchieri et al. (1985) constrained the maximum allowable extinction in the nuclear region of M83 by comparing the observed X-ray to optical flux ratio with the expected ratio for early-type stars. Any other assumption on the source of X-ray emission would lead to a smaller upper bound on A_V. The limits on the extinction thus calculated for M83 and for IC 342 (Fabbiano & Trinchieri 1987) are of order $A_V < 13$ mag, typically smaller than the mid-infrared estimates [35 mag in M83 (Lebofsky & Rieke 1979) and 15 mag in IC 342 (Becklin et al. 1980)]. This result suggests a nonhomogeneous distribution of the dust in the nuclear region, in agreement with the picture of Becklin et al. (1980), based on a similar discrepancy between the 10-μm infrared extinction and that estimated to the stars seen in the near-infrared.

Perhaps the most unexpected result from the *Einstein* observations of these nuclei has been the discovery of extended emission components, suggestive of gaseous outflows from the nuclear regions, in the edge-on galaxies M82, NGC 253 (Figure 4), and (more recently) NGC 3628. In M82, a correspondence between the region of extended X-ray emission ($\sim 90''$ radially from the nucleus) seen in the *Einstein* high-resolution imager (HRI) and the Hα filaments was first noticed by Van Speybroeck & Bechtold (1981). Watson et al. (1984) then suggested that this X-ray "halo" is likely to be thermal emission of shock-heated gas escaping the nuclear region. They point out that only 2% of the energy released by supernovae in the nucleus, exploding at a rate of 0.2 yr^{-1} over a time scale of 10^7 yr, is needed to heat the gas to X-ray temperatures. In NGC 253 the presence of an extended source, positionally coincident with a region of noncircular motions (Demoulin & Burbidge 1970), was also first reported by Van Speybroeck & Bechtold (1981). Fabbiano & Trinchieri

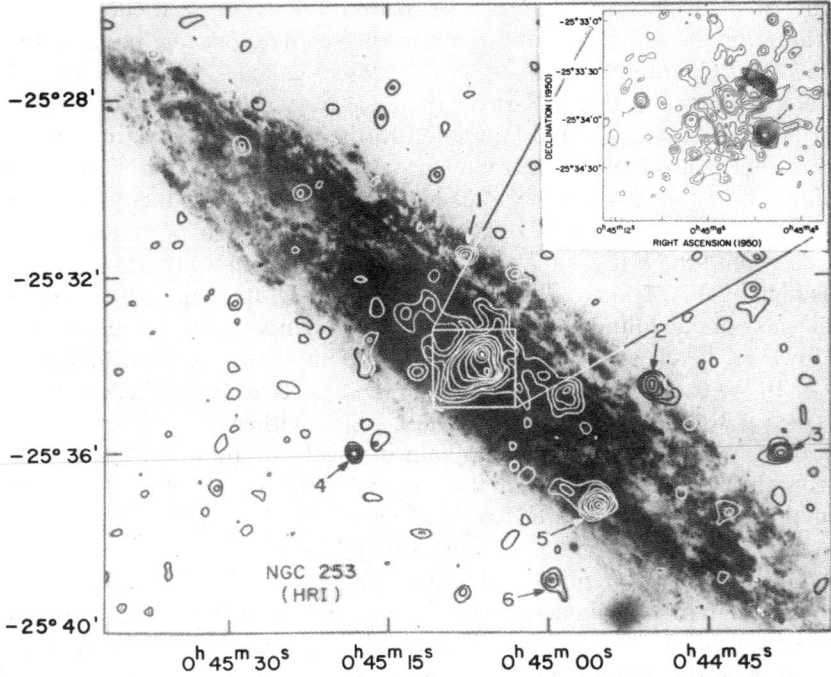

Figure 4 The *Einstein* HRI contour map of NGC 253, smoothed with a 7″ Gaussian. The insert shows a higher resolution map of the nuclear region. Here the shaded oval represents the nuclear starburst region seen in the radio and infrared. A plume of X-ray emission can be seen extending along the southern minor axis, and this was interpreted as evidence of hot outflowing gas (see Fabbiano & Trinchieri 1984, and references therein).

(1984) studied this nuclear region and identified both an emission region associated with the starburst nucleus proper and a "jet-like" feature, or "plume" of emission, extending for $\sim 60''$ along the southern minor axis. They suggested that the latter feature could be due to a bipolar nuclear outflow, similar to the one seen in M82, collimated by the galaxy disk and seen in projection along the minor axis. Fabbiano & Trinchieri further proposed that the northern side of the outflow would not be visible because the soft X-ray photons would be absorbed by the interstellar medium in the disk of NGC 253. The subsequent report of an OH line emission plume from the dusty northern side (Turner 1985) confirms this picture. Optical work on the emission line gas velocity fields in these two galaxies is also in agreement with the proposed gaseous outflows (McCarthy et al. 1987, Bland & Tully 1988). Theoretical models of this phenomenon have been offered by Chevalier & Clegg (1985) and Tomisaka & Ikeuchi (1988). Analyzing the lower resolution, but more sensitive, *Einstein* Imaging Proportional Counter (IPC) images of these two galaxies, Fabbiano (1988b) found evidence of diffuse X-ray emission at large radii in the northern side of NGC 253, which could be related to the nuclear outflow; in M82, on the other hand, there is clear evidence of an X-ray halo that is elongated along the minor axis and extends as far as ~ 9 kpc from the nucleus (see also Kronberg et al. 1985). This halo is not likely to be bound to the galaxy, and the hot gas may be leaving the system at a rate that could be as high as 0.7 $M_\odot$ yr^{-1}. These estimates are now uncertain, since neither the gas volume filling factor nor its emission temperature is really known. However, taken at face value, they would imply a maximum lifetime of 7×10^8 yr for the starburst, since the mass present in the nuclear region is $\sim 5 \times 10^8$ $M_\odot$ (Rieke et al. 1980), unless either the outflowing gas, cooling at large radii, flows in again in a galactic fountain or fresh gas from the intergalactic medium flows in to fuel the nucleus. Relatively short lifetimes ($< 2 \times 10^9$ yr) are also suggested by the OH data for the starburst at the nucleus of NGC 253 (Turner 1985).

Fabbiano (1984) remarked that these gaseous outflows should be visible in X rays in many galaxies, since starburst or low-activity nuclei are quite common (Keel 1983). With the present data it is impossible to distinguish them from the underlying disk emission in face-on galaxies such as M83 and M51 (Trinchieri et al. 1985, Palumbo et al. 1985); they should, however, be obvious in edge-on galaxies. Very recently, a reanalysis of the *Einstein* data of the edge-on galaxy NGC 3628 in different energy bands has shown an elongated soft emission region associated with the nucleus, suggestive of this phenomenon. The presence of a gaseous plume has been confirmed by subsequent optical observations (G. Fabbiano et al., in preparation, 1989). Different authors have pointed out that these outflows, if generally associated with violent star formation activity, could be responsible for the formation and enrichment of a large part of the gaseous intracluster medium (Heckman et al. 1987, Fabbiano 1988b). In particular, if a relatively small galaxy like M82 can expel of the order of 1 $M_\odot$ yr^{-1}, a primordial large elliptical system could expel 1000 times this amount. Therefore, some 1000 such systems in a cluster, undergoing violent star formation over a period of 10^8 yr, could produce the $\sim 10^{14}$ $M_\odot$ of gas that are now found in clusters of galaxies (Jones & Forman 1984). This type of scenario has been modeled by Mathews (1988a).

3.3 *Low-Activity Nuclei*

The sample of "normal" spiral galaxies observed with the *Einstein* satellite was selected to exclude known Seyfert nuclei, but some of these galaxies host nuclei with low-level activity not directly related to star formation. These galaxies include M51, whose nuclear region is, however, extended in X rays and therefore dominated by a starburst or by a hot gaseous component (Palumbo et al. 1985); M81, which hosts a small Seyfert nucleus (Peimbert & Torres-Peimbert 1981, Shuder & Osterbrock 1981) detected in X rays as a pointlike source (Elvis & Van Speybroeck 1982, Fabbiano 1988a); and two more galaxies, M33 and NGC 1313, for which the evidence of nuclear activity rests only on the X-ray data (Long et al. 1981b, Markert & Rallis 1983, Gottwald et al. 1987, Fabbiano & Trinchieri 1987, Trinchieri et al. 1988, Peres et al. 1989). A bright nuclear region ($L_X \sim 1.5 \times 10^{40}$ erg s^{-1}) was also observed in M100; however, this nucleus appears extended or complex (Palumbo et al. 1981). Other galaxies with relatively low-luminosity nuclei, although more luminous than the ones just mentioned, were surveyed to study their nuclear emission (e.g. Maccacaro & Perola 1981, Maccacaro et al. 1982). In at least one of these, the X-ray emission is extended and therefore not dominated by the nucleus [NGC 1365 (Maccacaro et al. 1982)].

The nucleus of M81 appears as an unresolved source in the *Einstein* images, with a luminosity $L_X \sim 1.6 \times 10^{40}$ erg s^{-1} (Elvis & Van Speybroeck 1982, Fabbiano 1988b). Variability of this source in the 2-10 keV range over a 5-month time scale has been reported by Schaaf et al. (1989). Fabbiano (1988b) reports that the *Einstein* IPC spectrum appears to be soft and intrinsically absorbed. This X-ray spectrum is reminiscent of the soft spectral components reported in QSO and bright active nuclei (e.g. Elvis et al. 1985, Arnaud et al. 1985, Pounds et al. 1986, Wilkes & Elvis 1987); an extrapolation of this spectrum to the UV suggests that there is enough photoionizing continuum to excite the optical emission lines, whereas this continuum would be missing for a more conventional ($\alpha_E \sim 0.7$) spectral power law (Bruzual et al. 1982). By using an accretion disk model (Bechtold et al. 1987, Czerny & Elvis 1987) to interpret this emission, one can constrain the mass of the central black hole to the range $\leq 10^{4-5} M_\odot$ and the accretion rates to values $\leq 10^{-(4-5)} M_\odot$ yr^{-1} (Fabbiano 1988b). This nucleus could therefore be just a low-luminosity specimen of a normal active galactic nucleus. The nucleus of M33 and the unresolved source in NGC 1313 are a factor of 10 less luminous than the nucleus of M81 and have similar spectral characteristics, within the fitting uncertainties (Trinchieri et al. 1988, but see also Gottwald et al. 1987, Fabbiano & Trinchieri 1987). The nucleus of M33 is also variable in X rays (Markert & Rallis 1983), most dramatically in the soft X-ray band (Peres et al. 1989). However, very little or no indication of activity is seen in optical and radio observations of these last two nuclei (von Kapp-her et al. 1978, Glass 1981, Gallagher et al. 1982, O'Connell 1983, Rubin & Ford 1986, J. Gallagher, private communication, 1987). One could speculate (Fabbiano 1988b) that these nuclei might represent a new type of source, possibly the radio-quiet counterpart of sources like the nucleus of 3C 264, which similarly shows no sign of optical activity (Elvis et al. 1981, Fabbiano et al. 1984a). If this is so, it is quite appropriate for these sources to be found in spiral galaxies, in analogy with what is observed in radio-loud and radio-quiet optically active nuclei (Miller 1985).

4. ELLIPTICAL AND S0 GALAXIES

4.1 Discovery and Properties of the Hot Interstellar Medium

The absence of tracers of a cold interstellar medium (ISM) in most early-type galaxies (e.g. Sandage 1957, Gallagher et al. 1975, Faber & Gallagher 1976) has prompted the construction of models to explain the removal of the gas shed from stars during their evolution (Mathews & Baker 1971, Bregman 1978, White & Chevalier 1983), which at the present rates would amount to $10^{9-10} M_\odot$ in a Hubble time in galaxies with an optical luminosity of $10^{10-11} L_\odot$ (e.g. Faber & Gallagher 1976). Now, X-ray observations have revealed this long-sought interstellar medium.

A hot gaseous halo had been known for some time to be associated with the Virgo cluster and M87 (Kellogg et al. 1975, Mitchell et al. 1976, Malina et al. 1976, Serlemitsos et al. 1977, Gorenstein et al. 1977, Fabricant et al. 1978, Lea et al. 1979, Canizares et al. 1979), but the presence of a hot gaseous medium in more normal ellipticals was revealed for the first time by the *Einstein* survey of the Virgo cluster (Forman et al. 1979). Five early-type galaxies in Virgo were detected in X rays in this first paper, with X-ray luminosities of $5-70 \times 10^{39}$ erg s^{-1}, over a factor of ~ 100 less luminous than M87. The X-ray image of one of them, M86, which has a large velocity relative to the Virgo cluster mean, was particularly important for establishing the gaseous nature of the emission. The X-ray isophotes of this galaxy appear significantly asymmetric with respect to the optical image (see Figure 7 of the *Annual Reviews* paper of Forman & Jones 1982), which suggests the presence of a gaseous halo of $\sim 6 \times 10^9 M_\odot$ that experiences ram pressure stripping in its approach to the cluster core (Forman et al. 1979; see also Fabian et al. 1980, Takeda et al. 1984, Forman et al. 1984b). Another galaxy in Virgo, NGC 4472 (Figure 5), was later found to have a similar X-ray morphology (Forman et al. 1985, Trinchieri et al. 1986). Forman et al. (1979) also concluded that the hot gas detected in the Virgo galaxies had to be indigenous, since the cooling time of the intracluster medium is too long to make accretion onto the galaxies possible.

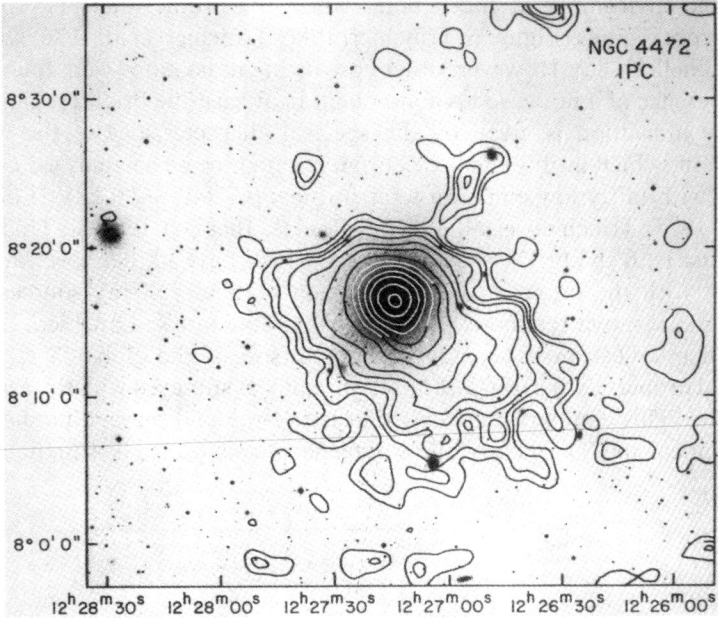

Figure 5 The *Einstein* IPC X-ray map of NGC 4472. Notice the asymmetric halo.

Subsequent X-ray observations and data analysis led to the detection of many more early-type galaxies in Virgo, in poor clusters and groups, and in the field, showing the ubiquity of their X-ray emission [Kriss et al. 1980, 1983, Biermann et al. 1982, Biermann & Kronberg 1983, Nulsen et al. 1984, Dressel & Wilson 1985, Forman et al. 1985, Trinchieri & Fabbiano 1985, Mason & Rosen 1985 (who report *EXOSAT* observations), Killeen et al. 1986, Trinchieri et al. 1986, Canizares et al. 1986, 1987, Killeen & Bicknell 1988a]. Typical X-ray luminosities in the *Einstein* band range from 10^{39} to 10^{42} erg s^{-1}, except for group or cluster-dominant galaxies, which tend to be more luminous (in the 10^{43} erg s^{-1} range). The sources detected with better signal to noise are clearly extended, with radii of ~ 40–70 kpc, generally comparable with those of the stellar distribution. This spatial extent is not in itself proof of the gaseous nature of the emission, because none of these galaxies is close enough to allow the spatial resolution of single X-ray sources with the *Einstein* instruments, as in the case of nearby spirals. Moreover, the radial distributions of the X-ray and optical surface brightnesses in those galaxies for which both profiles are available tend to follow each other, at least if one excludes the outermost radii, where the uncertainties in the field background subtraction and possible environmental effects could affect the profiles and perhaps the inner core regions (Figure 6; Trinchieri 1986, Trinchieri et al. 1986, Killeen & Bicknell 1988a). However, other convincing indications were found for the presence of a hot gaseous component in at least the brightest galaxies.

One indication is given by the spectral characteristics of the X-ray emission, which in those galaxies bright enough to be so analyzed can be fitted with fairly low emission temperatures [$kT \sim 0.5$–2.0 keV (Forman et al. 1985, Trinchieri et al. 1986, Killeen & Bicknell 1988a)]. This is in contrast with the bright spiral galaxies, which have harder spectra, consistent with the presence of a population of binary X-ray sources and young supernova remnants [$kT > 2$ keV (Fabbiano & Trinchieri 1987)]. Another indication is given by a comparison of the global X-ray and optical properties of early- and late-type galaxies surveyed with the *Einstein* satellite. This comparison shows that the X-ray and optical luminosities of elliptical and S0 galaxies follow a steeper correlation ($\alpha \sim 1.6$) than does

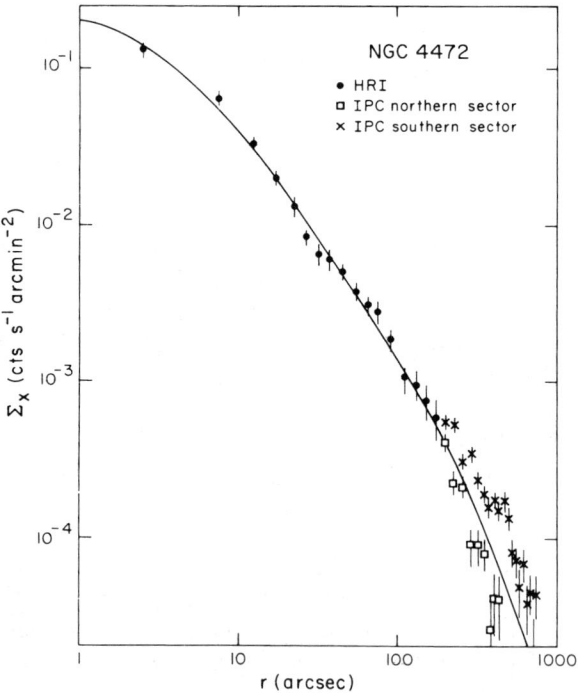

Figure 6 Radial profile of the X-ray surface brightness of NGC 4472 [the points (Trinchieri et al. 1986)] together with the optical profile from King (1978)]. Note the north-south asymmetry at large radii.

the linear one observed in spirals, which suggests that a different emission mechanism or at least an additional emission component is present in early-type galaxies (Fabbiano 1984, Forman et al. 1985, Trinchieri & Fabbiano 1985, Canizares et al. 1987). Those galaxies more luminous in X rays for which we have some spectral information are also those generally presenting excess emission relative to the relationship of the spirals (Figure 7). Therefore, this "excess" X-ray emission can be attributed to a hot gaseous component.

Once the presence of a hot gaseous component in bright early-type galaxies has been established, we would like to know how much hot gas is present in galaxies of different optical and X-ray luminosities.

How much of the X-ray emission of early-type galaxies is due to the hot

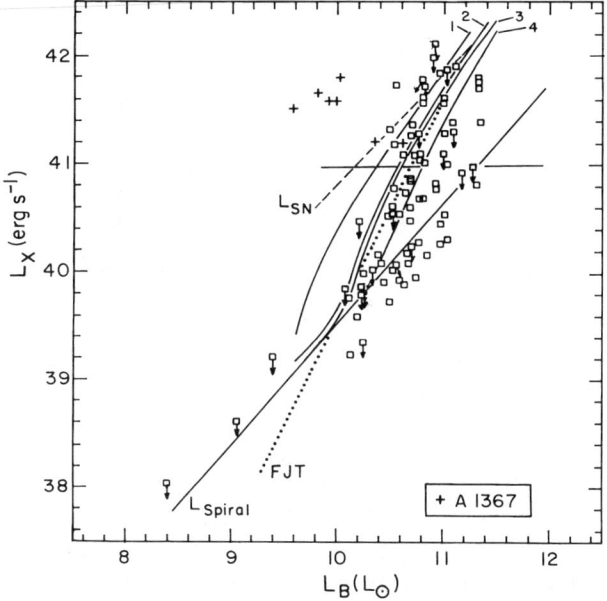

Figure 7 X-ray (in erg s^{-1}) and optical (in solar units) luminosities of E and S0 galaxies observed in X rays (from Canizares et al. 1987). The crosses are the galaxies in A1367 discussed by Bechtold et al. (1983). The solid line (L_{Spiral}) is the best-fit line of early-type spirals from Fabbiano et al. (1988); the horizontal solid line delimits the maximum X-ray luminosity of spiral galaxies. The dashed line (L_{SN}) is the estimate by Canizares et al. (1987) of supernova heating of the halos; the dotted line (FJT) is the estimate by Forman et al. (1985) (see Section 4.2); and the curves labeled 1, 2, 3, and 4 are the models of Sarazin & White (1988) for massive halos and supernova heating, supernova heating without massive halos, massive halos but no supernova heating, and no massive halos and no supernova heating, respectively.

gaseous component and how much can instead be ascribed to a population of evolved galactic sources, similar to those found in late-type galaxies, is still a matter of debate. Forman et al. (1985; see also Jones 1987, Forman & Jones 1988) conclude that all of the emission of early-type galaxies with absolute blue magnitude $M_B \leq -19$ is due to gaseous halos. Trinchieri & Fabbiano (1985), Canizares et al. (1987), and Fabbiano et al. (1989) note, however, that the data require the presence of a gaseous component only in the more X-ray-luminous galaxies, whereas the emission of galaxies less luminous than $\sim 10^{40.5-41}$ erg s^{-1} (these galaxies can be optically very bright, with $M_B \sim -22$) can be easily explained with a population of accreting low-mass binary sources. These different conclusions are partly due to the different choices of benchmarks for the expected contributions of discrete X-ray sources, and they illustrate the uncertainties inherent in the present data. Forman et al. (1985) used the *Einstein* X-ray observations of Cen A [NGC 5128 (Feigelson et al. 1981)] as a benchmark. However, the complexity of the X-ray emission of Cen A makes the evaluation of its "binary" component uncertain. Trinchieri & Fabbiano (1985) and Canizares et al. (1987) based their estimate on an extrapolation of the X-ray properties of the bulge of M31, which is dominated by pointlike sources, and of its globular cluster system (Van Speybroeck et al. 1979, Van Speybroeck & Bechtold 1981, Long & Van Speybroeck 1983). This choice has been criticized by Forman & Jones (1988), who remark that the bulge of M31 could be peculiarly bright in X rays. However, the stellar content of this bulge is indistinguishable from that of well-studied elliptical galaxies such as NGC 4472 (Oke et al. 1981, Faber 1983, Bohlin et al. 1985), and therefore it represents our best nearby and well-studied example of such a system (Canizares et al. 1987). More recently, Fabbiano et al. (1989) have performed a direct comparison of the global X-ray and optical luminosities of ellipticals and S0s with those of early-type spirals and found that only in galaxies with X-ray luminosities greater than 10^{41} erg s^{-1} is there a departure from a linear relationship between X-ray and optical luminosities, such as that observed in spirals, which then requires an additional, nonstellar emission component (see Figure 7).

An uncontroversial way to resolve this issue would be to perform a spectral analysis of early-type galaxies of different luminosities, since the spectral signatures of binary X-ray sources and gaseous halos differ. In particular, the X-ray spectrum of the bulge of M31 is typical of low-mass binary X-ray sources (Fabbiano et al. 1987b, Makishima et al. 1989) and differs from those of bright early-type galaxies. This analysis was attempted by Canizares et al. (1987) but proved inconclusive because of the poor statistics of the data. The resolution of this issue will have to wait for future, more sensitive X-ray observations.

There is a consensus, however, that the hot gaseous component dominates the X-ray emission of the more luminous galaxies. The X-ray data can then be used to derive the physical properties of this gas, such as central density, cooling time, and total mass. For the brightest galaxies, radial profiles of density and cooling time can also be derived. Two approaches have been followed (e.g. Fabian et al. 1981, Canizares et al. 1983, 1987, Fabricant & Gorenstein 1983, Stewart et al. 1984a, Forman et al. 1985, Thomas et al. 1986, Trinchieri et al. 1986, Killeen & Bicknell 1988a). One is to use an "onion skin" technique to deproject the X-ray surface brightness and to infer the gas density distribution directly from the data. The other is to assume a parameterization of the form

$$S_X(r) = S_X(0) \times [1+(r/a_X)^2]^{-3\beta+1/2} \qquad 1.$$

for the X-ray surface brightness profile (where r is the radial distance from the centroid of the X-ray surface brightness distribution, and a_X is an X-ray core radius) and then from here derive the deprojected electron density profile $n_e(r)$ [and therefore the gas density $\rho_{gas}(r)$] under the assumption that the gas is isothermal, i.e.

$$n_e(r) = n_e(0) \times [1+(r/a_X)^2]^{-3\beta/2}. \qquad 2.$$

Fits to the X-ray data generally give $\beta \sim 0.4$–0.6 [i.e. a radial dependence of the X-ray surface brightness of the type $S_X(r) \propto r^{-(1.4-2.6)}$ and of the electron density $n_e(r) \propto r^{-(1.2-1.8)}$], although departures from simple power laws and from spherical symmetry are common (see Figure 6). The parameter a_X is less well defined but could be of the order of a few kiloparsecs or less.

Estimates of the gas parameters made by Canizares et al. (1987; see also Forman et al. 1985, Trinchieri et al. 1986, Thomas et al. 1986) give typical central densities of ~ 0.1 cm^{-3}, central cooling times ranging between 10^6 and 10^8 yr, and gas masses ranging between 10^{11} and $10^8 M_\odot$ for galaxies detected in X rays. The amount of hot gas detected in bright galaxies is therefore consistent with the amount expected assuming the present gas injection rates over a Hubble time (Forman et al. 1985), although this could be a coincidence, since the mass shed by the first generations of stars might have been larger than the present rate (Renzini & Buzzoni 1986, Jones 1987, Loewenstein & Mathews 1987a). However, upper limits on the gas masses can be as low as $10^6 M_\odot$, showing a deficiency of hot interstellar medium in the less luminous galaxies. It should also be noted that the density and mass estimates given above are likely to be upper limits for these galaxies, since the X-ray emission of these galaxies could contain a sizable contribution from the "discrete source" component.

4.2 Physical Status of the Halos

Many authors have tried to understand the present status and the evolution of hot gaseous halos by comparing the X-ray observations with different theoretical scenarios. The temperature of the halos suggests the need for heating in addition to that supplied to a static halo by the same gravitational field experienced by the stars (e.g. Fabbiano 1986a). The high X-ray luminosities and corresponding large gas densities of some galaxies rule out the existence of galactic winds (Mathews & Baker 1971, Faber & Gallagher 1976, MacDonald & Bailey 1981, White & Chevalier 1983) because the mass supply from normal stellar processes falls several orders of magnitude short of that required to replenish such a wind (Nulsen et al. 1984, Forman et al. 1985, Sarazin 1986, Loewenstein & Mathews 1987a, Sarazin & White 1988). Moreover, the radial dependence of the gas density ($\rho_{\rm gas} \propto r^{-1.5}$ on average) is flatter than that expected of a wind [$\rho_{\rm gas}({\rm wind}) \propto r^{-2}$ (Forman et al. 1985)]. All of this points to the need for a confining mechanism.

There are two indications that the halos are "hotter" than the stars. One is given by a comparison between the temperature inferred from the X-ray spectral fits [$T \sim (5.8-23) \times 10^6$ K, with best fits of $\sim 1.2 \times 10^7$ K] and that predicted under the assumption of thermal equilibrium with the stellar component. For typical line-of-sight velocity dispersions of $\sigma \sim 150-300$ km s^{-1} (e.g. Whitmore et al. 1985), and under the assumption of isotropy, the latter would be given by $T = \mu m_{\rm p}\sigma^2/k \sim 1.4-5.4 \times 10^6$ K, which is less than the best-fit temperature and probably near or below the lowest allowed by the data (Mathews & Baker 1971; see also Fabbiano 1986a, Killeen & Bicknell 1988a). The other indication is given by the radial dependence of the gas density as compared with that of the stellar light. Since the X-ray and optical surface brightness profiles tend to follow each other, and since the X-ray bremsstrahlung emission measure is a function of $n_{\rm e}^2$, it is easy to see that $\rho_{\rm gas}^2 \sim \rho_{\rm stars}$. If the gas and the stars are two isothermal spheres experiencing the same gravitational potential, this implies that $T_{\rm gas} \sim 2T_{\rm stars}$ (Cavaliere & Fusco-Femiano 1976; see Killeen & Bicknell 1988a). However, other heating mechanisms in addition to the motion of the mass-losing stars are readily available. They include supernova heating and gravitational heating. The latter would result from the dense gas cooling radiatively and then falling in the potential well in accreting "cooling flows" (Mathews & Baker 1971, MacDonald & Bailey 1981, Nulsen et al. 1984, White & Chevalier 1984, Sarazin 1986, 1987a,b, Thomas et al. 1986, Mathews & Loewenstein 1986, Canizares 1987, Fabian et al. 1987a, Canizares et al. 1987, Loewenstein & Mathews 1987a, Sarazin & White 1987, 1988, Umemura & Ikeuchi 1987, Vedder et al. 1988).

Nonthermal heating by the relativistic electrons in radio sources (e.g. M87) and heating by conduction from the surrounding intracluster medium have also been suggested (Tucker & Rosner 1983, Bertschinger & Meiksin 1986, Rosner & Tucker 1989). Once heated, the gaseous halos must be confined within their emitting volumes. This can be achieved by the pressure of an external medium (e.g. Fabian et al. 1980, Binney & Cowie 1981, Forman et al. 1984b, White & Chevalier 1984, Vedder et al. 1988), by radiative cooling (e.g. MacDonald & Bailey 1981, White & Chevalier 1984), and by gravity (e.g. Bahcall & Sarazin 1977, Mathews 1978, Nulsen et al. 1984, Forman et al. 1985; see Canizares et al. 1987).

Figure 7 shows different model predictions compared with the X-ray/optical scatter diagram of early-type galaxies. In first approximation the functional dependences of the models are easy to understand: In the supernova-dominated model the X-ray and optical luminosities would be linearly related because the supernova energy input is proportional to the stellar luminosity for a constant supernova rate (Tammann 1974); in the gravitational-cooling flow model the X-ray and optical luminosities would be related as $L_X \propto L_B^{1.5-1.7}$ because the X-ray luminosity will be proportional to the stellar mass loss rate and to the square of the stellar velocity dispersion ($L_X \propto \dot{m}\sigma^2$) and $\dot{m} \propto L_B$, while the Faber-Jackson relation (1976; also Tonry 1981) gives $L_B \propto \sigma^{3-4}$ (White & Chevalier 1984, Nulsen et al. 1984, Sarazin 1986, 1987a, Canizares et al. 1987). Sarazin & White (1988), however, find that models can deviate somewhat from these simple approximations, because in parts of the models the temperature can be fairly low, so that the X-ray luminosity is not as simply related to cooling. If supernova heating is the dominant source of energy, the gas could be confined by a combination of radiative cooling and external pressure; if gravitational heating prevails, gravity would confine the gaseous halos (Canizares et al. 1987). In Figure 7 is also plotted the $L_X \propto L_B^2$ line of Forman et al. (1985) for hydrostatic halos accumulated in the galaxy potential over a Hubble time. This model, however, does not take into account cooling and heating mechanisms, or at least assumes that they balance closely (Sarazin 1987a). Moreover, it assumes the same emitting volume for all the galaxies, since the volume also figures in the emission measure (Fabbiano 1986a).

One common characteristic of these predictions is that they can easily explain only the more X-ray-luminous galaxies. The supernova-dominated models, in particular, predict X-ray luminosities in agreement with the upper envelope of the scatter diagram. Therefore, for many galaxies either the supernova rate is significantly lower than Tammann's (1982) estimate or the bulk of supernova energy is dissipated or radiated outside the X-ray band. Gravitational-heating models also predict large X-ray luminosities.

Smaller luminosities could be obtained if the mean stellar mass loss rate were smaller than that used in the calculations, or if a considerable fraction of $\dot{m}$ were not incorporated into the accretion flow or were removed from the flow by thermal instabilities before the flow had reached the center of the galaxy, although in the latter case the reduction is not dramatic (Thomas 1986, Thomas et al. 1986, Fabian et al. 1987a, Canizares et al. 1987, Sarazin & White 1988, Vedder et al. 1988, C. L. Sarazin & G. A. Ashe, submitted for publication, 1988). Moreover, the models cannot account for the observed scatter in the data. This scatter and the possibility that hot halos do not dominate the X-ray emission of the less X-ray-luminous galaxies (see the previous section) might be related to the effect of ram pressure stripping (Forman et al. 1984a, Sarazin & White 1988, Vedder et al. 1988) or to the presence of winds or partial winds (MacDonald & Bailey 1981, White & Chevalier 1984, Sarazin & White 1987; see also Loewenstein & Mathews 1987a, Umemura & Ikeuchi 1987) in some of the galaxies. A. Renzini (preprint, 1988) argues against cooling flows and in favor of winds for the low-X-ray-luminosity galaxies because the rate of stellar mass return is too high to reconcile the cooling flow model with the observed X-ray luminosities.

Pressure confinement has been invoked (Forman et al. 1985) to explain the presence of the very bright coronae in Abell 1367 reported by Bechtold et al. (1983), which (as shown by Figure 7) would be over two orders of magnitude more luminous than early-type galaxies of comparable optical luminosity in the field and in Virgo. Canizares et al. (1987) disagreed with this explanation on the basis that the density of the cluster medium of A1367 is similar to that of Virgo (Forman et al. 1979, Bechtold et al. 1983). More recently, Vedder et al. (1988) developed steady-state cooling flow models for early-type galaxies and concluded that similar galaxies with similar gas content should have similar X-ray luminosities regardless of their location, since the external pressure around a galaxy would not affect the luminosity of the gas within the galaxy, but rather its temperature, and it would only affect the X-ray surface brightness near the outside of the galaxy. However, if conduction is important, heat could flow from the cluster to the galaxy and increase the X-ray luminosity (C. Sarazin, private communication, 1988).

The X-ray/optical correlation does not allow one to discriminate between the different models. However, there are two other observational constraints that need to be satisfied: the radial distribution of the halos, and their emission temperature. Comparison with models of the observed surface brightness profiles have been attempted by various authors, usually by using "average" profiles, which disregard the asymmetries and departures from power laws that are sometimes observed at large radii (see

Trinchieri et al. 1986). Since these events could be due to interaction with the environment, this approach might appropriately describe an undisturbed early-type galaxy. Sarazin (1986; see also Sarazin 1987a,b, Loewenstein & Mathews 1987a, Sarazin & White 1988) pointed out that if heating and cooling balance locally in a halo, and if the heating of the gas per unit mass is independent of position, then $\rho_{\text{gas}}^2 \propto \rho_{\text{stars}}$, and thus the X-ray and optical surface brightness profiles would then follow each other, as observed (Trinchieri et al. 1986, Killeen & Bicknell 1988a). Cooling flow calculations, however, in the absence of conduction heating and under the simplest assumption that the gas is a one-phase medium, generate models that typically have total luminosities that are too large (see above) and have a central peak of the X-ray emission that is not observed (Thomas 1986, Thomas et al. 1986, Loewenstein & Mathews 1987a, Sarazin & White 1988, Vedder et al. 1988). Modifying the models by allowing "sinks" of mass from the flow to prevent accretion at the cores, or even cooling outflows, produces profiles in closer agreement with the observations (Thomas 1986, Thomas et al. 1986, White & Sarazin 1987a,b,c, Vedder et al. 1988, C. L. Sarazin & G. A. Ashe, submitted for publication, 1988). It is also possible that many different phases are present in the cooling gas (Nulsen 1986, Thomas et al. 1987). At present the data do not allow us to discriminate readily between different models with different inputs of supernova energy, dark massive halos, and external pressure. Different models, however, produce different temperature profiles (e.g. Sarazin & White 1987, Vedder et al. 1988), and future X-ray observations will be able to give us an answer. I discuss in the next sections the constraints on the mass of the galaxies imposed by this type of model fitting.

4.3 Cooling Flows

As discussed above, with the exclusion of the static halo scenario of Forman et al. (1985), most models for the gaseous halos involve cooling flows to the galaxy cores. The presence and the amount of these flows have been objects of debate, chiefly because observational evidence at wavelengths other than the X ray is not clear-cut. Although a full discussion of this subject is beyond the scope of this review, I summarize here the aspects that are more relevant for the halos of elliptical galaxies. These flows have been suggested because the cooling times implied by the X-ray data for the central regions of most observed galaxies, and in some cases for most of the galaxy volume, are considerably shorter than the Hubble time (Nulsen et al. 1984, Trinchieri et al. 1986, Sarazin 1987a, Canizares et al. 1987). In the absence of other sources of heat, therefore, the gas in a volume element will cool and accrete subsonically in pressure-driven flows toward the center of the galaxy potential (e.g. Silk 1976, Fabian &

Nulsen 1977, Cowie & Binney 1977, Mathews & Bregman 1978). Derived mass accretion rates range from a few to $\sim 1000\ M_\odot\ \mathrm{yr}^{-1}$ for dominant cluster galaxies (Fabian et al. 1981, Mushotzky et al. 1981, Canizares et al. 1982a, Lea et al. 1982, Stewart et al. 1984a,b, Arnaud et al. 1984, Matilsky et al. 1985, Canizares et al. 1988) to $\sim 1\ M_\odot\ \mathrm{yr}^{-1}$ for normal ellipticals (Nulsen et al. 1984, Trinchieri et al. 1986, Canizares et al. 1987).

Although there is evidence for cooler gas in optical filaments and emission-line regions associated with cluster cooling flows and with the central regions of some early-type galaxies [e.g. Fabian et al. 1984b (and references therein), 1985, 1987b, Demoulin-Ulrich et al. 1984, Hu et al. 1985, Phillips et al. 1986, Crawford et al. 1987], the main problem of this scenario has been with finding incontrovertible evidence of star formation, which should be both substantial (given the mass cooling rates) and happening at all radii (Stewart et al. 1984a, Nulsen et al. 1984, Fabian et al. 1984b, 1987a, Nulsen & Carter 1987, White & Sarazin 1987b,c, 1988). Star formation with a normal IMF might be taking place in some galaxies (Silk et al. 1986, Romanishin 1987, Johnstone et al. 1987, Burstein et al. 1988, Bertola 1988); however, in most cases the galaxies' colors and 2.3-μm CO absorption index suggest that only low, or very low, mass stars may be forming (Fabian et al. 1982, Sarazin & O'Connell 1983, Fabian et al. 1984a, Arnaud & Gilmore 1986, Johnstone et al. 1987, O'Connell & McNamara 1988). This unusual IMF has been justified on the grounds that the large pressure in the flows will decrease the Jeans mass (Jura 1977, Fabian et al. 1982, Sarazin & O'Connell 1983), but the inconsistency of this explanation with the formation of massive stars in dense molecular clouds has led some authors to invoke additional sources of heat in the halos (e.g. Silk et al. 1986). Heating by conduction, in particular, has been suggested for galaxies in clusters (Tucker & Rosner 1983, Bertschinger & Meiksin 1986, Rosner & Tucker 1989), but it might only apply for a very limited range of physical parameters (e.g. Bregman & David 1988). This is very much an open field of investigation, and there is no doubt that it will continue to be pursued from both an observational and a theoretical point of view.

Another way of testing some of the cooling flow theories is to try to find evidence of a colder interstellar medium in early-type galaxies and then compare these observations with the X-ray data. Canizares et al. (1987) searched for possible correlations between the X-ray emission and either the optical emission line (e.g. Phillips et al. 1986, Véron-Cetty & Véron 1986; see also Sadler 1988) or the H I emission (e.g. Knapp et al. 1985, Wardle & Knapp 1986), that could suggest a direct connection between the hot and the cooler interstellar medium, but they found none. However, the subsets of galaxies considered in each case are rather small and, at least for the H I, are dominated by nondetections. This lack of correlation would be consistent with the suggestion that the H I could have been accreted as a result of close encounters between galaxies (e.g. Knapp et al. 1985). However, Bregman et al. (1988) report the discovery of 1.5×10^8 $M_\odot$ of neutral hydrogen in NGC 4406 that they believe might be created within a cooling flow, and a similar conclusion has been reached by Huchtmeier et al. (1988) from their detection of CO emission from NGC 4472. The neutral hydrogen in NGC 4406 is centrally peaked, and there is no evidence for rotational support, as is often found in elliptical galaxies, where the H I tends to lie in rotationally supported extensive rings that are thought to be of external origin (e.g. van Gorkom et al. 1986). For completeness, it should be mentioned that Nulsen et al. (1984) have suggested that rotationally supported H I rings, not necessarily aligned with the galaxy's rotation axis, could be excreted from a cooling flow as a way of dissipating angular momentum.

Work on the *IRAS* data has shown that cool dust is present in early-type galaxies (Jura 1986, Wrobel et al. 1986, Thronson & Bally 1987). A possible, although not strong, correlation was reported between the far-infrared and the X-ray emission (Jura 1986, Knapp 1988, Kim 1988) that suggested a link between the hot and the cold interstellar medium in these galaxies, possibly through a common origin for the dust and the hot gas from stellar ejection. In particular, Jura (1986) concluded that the cold far-infrared-emitting dust should exist in pockets separated from the hot X-ray-emitting gas; otherwise, the dust would be destroyed by thermal sputtering in a short time. This might be evidence for the existence of a multiphase interstellar medium, which has been suggested in the framework of cooling flow modeling of the X-ray halos (e.g. Thomas et al. 1987).

In a recent series of papers, Mathews (1988a,b,c) has explored the constraints provided by the observational mass-to-light ratios to the mass of optically dark stars, originating from cooling flows in the cores of elliptical galaxies, and from these to the IMF and the history of winds and cooling inflows in the interstellar medium. He concludes that a component of dark stars in the cores is unlikely to be more massive than 30 times the core mass of luminous stars (Mathews 1988c) and suggests that the IMF of stars in elliptical galaxies is significantly flatter than the Salpeter IMF. In this scenario strong winds have expelled from the galaxies the mass shed by stars in early times, creating the intracluster medium, while present-day ellipticals are likely to be closed systems (Mathews 1988a). A. Renzini (preprint, 1988) also points out the presence of winds in young galaxies, given the enhanced supernova rate, and suggests that these winds might still be prevalent in the less X-ray-luminous ellipticals.

4.4 The Mass of Early-Type Galaxies

X-ray observations have been used to measure the mass of early-type galaxies (previous reviews on this subject include Canizares 1987, Sarazin 1987b, Fabian & Thomas 1987, Fall 1987, Trimble 1987). These measurements are based on the assumption that the X-ray emission is thermal bremsstrahlung from hot gaseous halos, and that the gas traces the potential but does not contribute to it significantly. This is a reasonable assumption (Forman et al. 1985), since $M_{gas}/M_{stars} < 7\%$ if one assumes that $M_{stars}/L_{stars} = 6$ (Faber & Gallagher 1979). Three approaches have been used: The first has been to measure the binding mass within a certain radius, under the assumptions of hydrostatic equilibrium and spherical symmetry. This method was first applied to M87 and has more recently been used to estimate the mass of less X-ray-luminous galaxies. The second approach has made use of more detailed modeling to match theoretical predictions to the observed X-ray surface brightness profiles. Finally, Fabian et al. (1986a) have devised a method to estimate lower limits to the *total* mass of a galaxy. In the following I discuss the results of these three types of analyses, with an eye to examining the assumptions made and the limitations of the current data.

The discovery of the extended thermal X-ray source in the Virgo cluster centered on M87 prompted the first attempts at estimating the binding mass of an early-type galaxy within radii well away from the optical core, to which optical measurements have been restricted. The first estimates, based on the assumption of the gas being in hydrostatic equilibrium in the galaxy potential at a temperature $T \sim 3 \times 10^7$ K, suggested very large binding masses, between 10^{13} and $\geq 10^{14}$ $M_\odot$ (Bahcall & Sarazin 1977, Mathews 1978). However, Binney & Cowie (1981) showed that the X-ray data then available could also be fitted with a model of pressure-confined cooling atmosphere surrounding a low-mass galaxy. The observations of M87 with the *Einstein* IPC, which produced an image and spatially resolved spectral information, have led to more accurate measurements of the binding mass and have confirmed the earlier suggestions of the presence of a massive dark halo (Fabricant et al. 1980, Fabricant & Gorenstein 1983). In particular, Fabricant & Gorenstein (1983) could exclude a radial dependence of the temperature of the gaseous halo consistent with Binney & Cowie's model and found that the integral mass-to-light ratio (M/L) of M87 in solar units must increase from 5–15 at a radius of 1' [~ 4.4 kpc (from the optical data of Sargent et al. 1978)] to over 180 at 20' (~ 87 kpc), which implies that a dark massive halo is present. They derived a total mass $M \sim 3$–6×10^{13} $M_\odot$ within 60' (~ 260 kpc) from the core. Similar parameters were found for other central galaxies in clusters, including

NGC 1275 in Perseus (Fabian et al. 1981, Branduardi-Raymont et al. 1981) and NGC 4696 (Matilsky et al. 1985), the dominant galaxy of the Centaurus cluster. An apparently isolated galaxy, identified as the counterpart of a serendipitous X-ray source [1E0116.3−0116 (Maccagni et al. 1987)], and central galaxies in poor clusters (Kriss et al. 1983, Canizares et al. 1983, Biermann & Kronberg 1983, Malumuth & Kriss 1986) have also been reported to have similar X-ray luminosities (a few 10^{43} erg s^{-1}) and M/L ratios.

The equation used by Fabricant et al. (1980) to estimate the binding mass of M87 was derived from the equation of hydrostatic equilibrium in combination with the ideal gas law, under the assumption of spherical symmetry, and is given below:

$$M(<r) = -(r_{gas}kT_{gas}/G\mu m_H)(d\log\rho_{gas}/d\log r + d\log T_{gas}/d\log r), \quad 3.$$

where G is the gravitational constant, and μm_H is the mean gas particle mass (μ is taken to be 0.6, and m_H is the mass of the hydrogen atom).

Equation 3 shows that the measurement of the binding mass depends on four variables: the radius within which the mass is estimated, the gas temperature at this radius, and the logarithmic gradients of both temperature and gas density at the same radius. Each of these variables is a potential source of uncertainty. While the observations of M87 yielded enough signal to noise to allow a relatively well-constrained mass estimate, the same unfortunately is not true when this method is applied to early-type galaxies not at the center of clusters. These galaxies typically are ~100 times less luminous than and at least as distant as M87, and therefore their radial profiles and especially temperature information are not as well constrained. Forman et al. (1985) pointed out that the assumption of hydrostatic equilibrium for these galaxies is reasonable even in the presence of cooling flows because these flows would be largely subsonic. They derived M/L ratios ranging from ~10 to 90 for a sample of 13 galaxies by assuming that the X-ray emission is due to an isothermal gas at $T = 1.2 \times 10^7$ K, adopting as r_{gas} the outermost radius at which X-ray emission could be detected, and using an average fitted value for the density gradient. Taken at face value, this result would suggest that dark extended halos are a common feature of early-type galaxies. A supporting argument given by these authors for the widespread presence of massive dark halos is the suppression of galactic winds (see the previous section). However, radiative wind suppression in the inner regions and an external pressure from a surrounding intracluster or intergalactic medium (e.g. Fabian et al. 1980, White & Chevalier 1984) could also be responsible for the retention of the hot interstellar medium (Canizares 1987, Canizares et al. 1987). The temperature in this case should increase at the outer radii. Temperature profiles are needed to exclude this possibility (Sarazin & White 1987, Vedder et al. 1988). Forman et al. (1985) performed a spectral fit of the X-ray image of NGC 4472 in three radial annuli and concluded that the data are consistent with isothermality. However, a sharp increase of the temperature at the outer radii, as required by the pressure-confined models, is also possible, given the large uncertainties of the fit. Moreover, Trinchieri et al. (1986) suggest a complex spectrum for this X-ray source, which would thus increase the uncertainties on the fitted temperature.

Subsequent work (Trinchieri et al. 1986; see also Fabbiano 1985b, 1986a,b, Canizares 1987) has shown that the uncertainties of this type of mass measurement can be large. In particular, detailed analysis of the *Einstein* data of six elliptical and S0 galaxies, which are among those detected with higher signal to noise in the sample of Forman et al. (1985), shows both departures from spherical symmetry, which can be ascribed either to interlopers in the field or to the interaction with the surrounding medium, and peculiar variations of the gas density gradient at the outermost radii. Even ignoring some of these effects, the uncertainties of the gas temperature and of its gradient are such that mass estimates differing by up to a factor of 10 are possible for a single galaxy (Figure 8). Jones (1987) and Forman & Jones (1988) have pointed out that the assumption of isothermality is justified in galaxies with a gravitationally bound hot halo (e.g. Norman & Silk 1979; however, see the discussion above about nongravitational confinement). The uncertainties in the IPC spectral fits, however, are such that the range of possible temperatures is still very large, even if one were to assume isothermality (Trinchieri et al. 1986; see Figure 8). Within these uncertainties, the X-ray measurements are generally consistent with the mass estimates from optical velocity dispersion. Therefore, large dark halos are allowed by these data but not required by this analysis. The use of less luminous galaxies or of Sa galaxies will introduce an additional uncertainty that cannot be resolved with the present data. This is the possibility that most of the X-ray emission is due to the evolved stellar component and not to a hot gaseous medium (Fabbiano & Trinchieri 1985, Trinchieri & Fabbiano 1985, Canizares et al. 1987, Fabbiano et al. 1989), thus invalidating the whole approach. In this regard, as remarked by Knapp (1987; see also Fall 1987), it may be significant that the masses of the Sombrero galaxy (an Sa) and of NGC 5128 (Cen A), estimated from H I rotation curves (Bajaja et al. 1984, van Gorkom 1987), are significantly smaller than the estimates of Forman et al. (1985). Although the H I measurements stop at radii smaller than those used for the X-ray measurements, the discrepancy would imply in both cases that the rotation curves, which are flat in the H I measurements, would then rise steeply between the H I and X-ray radius. This effect might be real [a radial increase of the

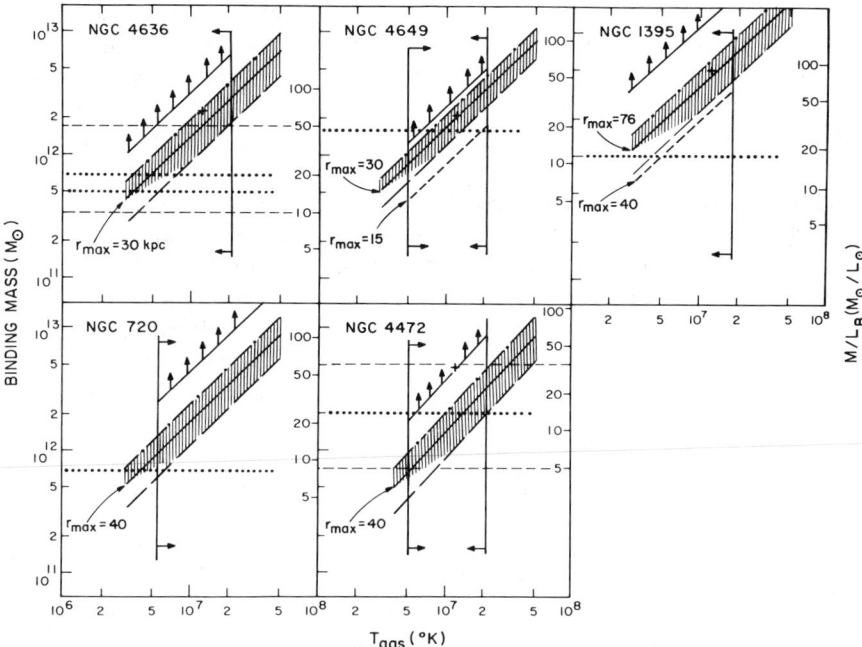

Figure 8 Mass estimates and related uncertainties for five early-type galaxies (adapted from Trinchieri et al. 1986). The shaded areas are the allowed regions from Equation 3, as in the above paper; the central solid line is the estimate for an isothermal halo; and the dashed and dot-dashed lines are the estimates for halos with moderate radial dependences of the temperature ($T \sim r^{0.5}$ and $T \sim r^{-0.5}$, respectively). The vertical lines restrict the temperature ranges to the values obtained by the spectral fits of the X-ray data (at the 90% confidence level). The diagonal short-dashed lines are isothermal estimates of the binding mass for more conservative choices of the halo outermost radii, which exclude regions where the radial surface brightness profiles depart from a power law (see Trinchieri et al. 1986). The horizontal dashed and dotted lines are mass-to-light ratio determinations from optical velocity dispersions (see Trinchieri et al. 1986, and references therein). The diagonal lines with upward-pointing arrows are the lower limits to the total binding mass (and mass-to-light ratios) calculated following Fabian et al. (1986a, Equation 4 herein) for the less conservative estimates of the outer radii (larger r_{max}); if one uses the smaller r_{max} values, these estimates will be displaced downward.

circular velocity occurs in M87 if the halo is gravitationally confined (Binney & Cowie 1981, Sarazin 1987b)], but it could also indicate some problem with the X-ray estimate, since rotation curves generally tend to flatten out at large radii (Knapp 1987).

A different approach to mass measurements relies on a comparison of the observed radial profile of the X-ray surface brightness with model predictions. The method generally followed is to assume a potential that describes the stellar radial distribution and the observed stellar velocity dispersion and then to generate, under various assumptions, the X-ray (gas) surface brightness profile (and/or temperature), to be then compared with the data. This approach was used on M87 and poor clusters (e.g. Mathews 1978, Binney & Cowie 1981, Stewart et al. 1984a, Canizares et al. 1983). In particular, Stewart et al., using the *Einstein* spectral and spatial data of M87, chose a family of models, which describe the radial mass distribution, in agreement with the optical data and with the X-ray data of Fabricant & Gorenstein (1983). Model fitting of less X-ray-luminous, normal early-type galaxies gives less clear results. Using ad hoc models, Trinchieri et al. (1986) showed that the observed gas distributions do not necessarily require the presence of dark halos; Canizares et al. (1987) concluded similarly that the observed correlation between X-ray and optical luminosities does not put strong constraints on the existence of massive halos; and Vedder et al. (1988) pointed out that the only visible effect of massive halos would be in the temperature profiles, which cannot be measured with the *Einstein* data. Sarazin & White (1987, 1988) instead conclude that cooling flow models, possibly without supernova heating but with massive halos, better reproduce the data, although they stress the large uncertainties of their result. A similar conclusion had been reached by Thomas (1986), who applied a cooling flow model to NGC 4472. However, the X-ray isophotes of this galaxy are not circular at large radii, which suggests interaction with the intracluster medium. The binding mass within 40 kpc, which is the radius up to which the "undisturbed" gas distribution should extend in this asymmetric halo (see Trinchieri et al. 1986), is in very good agreement both with the estimate of Trinchieri et al. (1986) and with optical estimates. Killeen & Bicknell (1988a) have estimated the binding mass of NGC 1399, using both model comparison and Equation 3, and conclude that the determination is rather uncertain and does not exclude M/L values in agreement with optical measurements: Thus the presence of a large massive halo, although allowed, is not required. A different approach has been followed by Mathews & Loewenstein (1986), Loewenstein & Mathews (1987a; see also Loewenstein & Mathews 1987b), and Umemura & Ikeuchi (1987), who have studied the time history of gaseous halos in early-type galaxies and find luminosities in agreement with the data only in the presence of dominant dark halos.

Instead of relying on a measure of the binding mass within a certain observed radius, Fabian et al. (1986a) devised a method for estimating a lower limit to the *total* binding mass. Their method is based upon three assumptions: (*a*) that the gas within the observed radius is confined by a hydrostatic outer atmosphere, (*b*) that the halo is convectively stable, and

(c) that the pressure gradient is always negative. With these assumptions they obtain the equation

$$M(r_\infty) \geq 5r_0 kT_0[1-(P_\infty/P_0)^{2/5}]/2G\mu m_H(1-r_0/r_\infty), \qquad 4.$$

where r_0 is the outermost detected radius of the halo, P_0 and T_0 are the gas pressure and temperature at this radius, and r_∞ and P_∞ are the outer radius of the halo and the gas pressure at this radius, respectively. From this equation, and using the outermost radii (r_0) and best-fit values of temperatures from the literature, they derive mass-to-light ratios for a sample of early-type galaxies, which lead them to conclude that there is overwhelming evidence for large massive halos in these galaxies. However, the same uncertainties apply here as in the calculation of the binding mass from Equation 3. Using the five galaxies for which the binding mass had been measured by Trinchieri et al. (1986) and temperature estimates that take into account the uncertainties of the spectral fits, together with the values of r_0 derived in that paper, I have calculated lower limits to the mass with Equation 4. They are shown in Figure 8. For two of the galaxies (NGC 4649 and NGC 4472) these limits do not require particularly large M/L ratios in excess of those inferred from optical measurements. For the other three galaxies (NGC 4636, NGC 1395, and NGC 720), however, values of $M/L > 20$ are clearly required. Therefore, there is evidence for large dark halos in some normal elliptical galaxies at least, unless the three basic assumptions of Fabian et al. do not apply. This might be the case for galaxies experiencing considerable external forces, as, for example, is the case for M86, which could be ram pressure confined (Fabian et al. 1980; see also Fabian et al. 1986a). Also, as remarked by the latter authors, if there is a significant confining pressure due to intracluster or intragroup gas, the lower mass limit may not apply to the galaxy but to the group as a whole.

Considering these results, I believe that, purely on *observational* grounds, there is still insufficient evidence to prove that dark massive halos are a general feature of early-type galaxies, although there is evidence of their existence in cluster dominant galaxies and perhaps in some of the most X-ray-luminous normal ellipticals. The use of X-ray data to measure the mass of these galaxies is very promising, but future more sensitive X-ray observations are needed to get a more definite answer.

Assuming that dark halos are a common feature of early-type galaxies, one may ask the question if all galaxies have similar dark halos. Forman et al. (1985) suggested that this may be the case, and moreover that the ratio between total and luminous mass in early-type galaxies could be the same as those of clusters and groups (Blumenthal et al. 1984). This would indicate that the dark matter of clusters might just be the superposition of single galaxy halos. It is also possible that normal ellipticals might be less massive than group or cluster dominant galaxies. In particular, Mathews (1978) remarked that the presence of an extended luminous X-ray halo in M87 and the absence of comparable X-ray emission from NGC 4472, which is also in the Virgo cluster and is optically more luminous than M87, argued for an exceptional massive component in the latter (see also Forman et al. 1984a).

One can investigate further this issue, insofar as the data will allow it, by plotting the mass-to-light ratios of different galaxies (including dominant galaxies in clusters and groups, and normal early-type galaxies) as a function of the ratio between their X-ray and optical luminosities, which is a measure of their gaseous content. This is shown in Figure 9a. There is a suggestion of a possible correlation in this plot (if one disregards the large uncertainties on each point) in the sense that cD galaxies, which retain a larger amount of hot gas, may be more massive than normal ellipticals, and therefore different galaxies might have different amounts of dark matter relative to their luminous matter. It is interesting that there is no correlation of the central stellar velocity dispersions with the X-ray to optical ratio, in the same range of ratios (Figure 9b), consistent with the dark matter being in extended halos so as not to disturb the stellar component. However, galaxies with X-ray to optical ratios consistent with those of spirals ($L_X/L_B < 10^{30}$ erg s^{-1} $L_\odot^{-1}$; this corresponds to a monochromatic flux ratio of $10^{-6.5}$), which therefore might not be able to retain gaseous halos, tend to have smaller central velocity dispersions. Given the paucity of points in Figure 9a, however, and the large error bars, the possibility that all early-type galaxies have similar mass-to-light ratios cannot be discounted either. Also, these ratios depend on the measured extent of the X-ray emission and therefore are technically lower limits (see Forman et al. 1985). What has to be explained in this case is why there is a systematic difference between the X-ray luminosities of different types of galaxies, some of which are at the same distance (in Virgo) and were observed with comparable sensitivity. In particular, why can dominant galaxies in clusters retain far larger gaseous halos than those in poor groups, which are still more successful in this endeavor than general field or nondominant cluster galaxies? And, even in the nondominant cluster galaxies, why can galaxies with similar optical luminosity or central velocity dispersion have such a large range of X-ray luminosities (Canizares et al. 1987, Sarazin & White 1988)? Although other interpretations, invoking dynamical stripping of galaxies orbiting the cluster, cannibalism, or past enhanced accretion, are certainly possible (e.g. Fabian et al. 1981, Takeda et al. 1984, Stewart et al. 1984b), it is possible that different amounts of dark matter in otherwise similar galaxies might ab initio be responsible for these results.

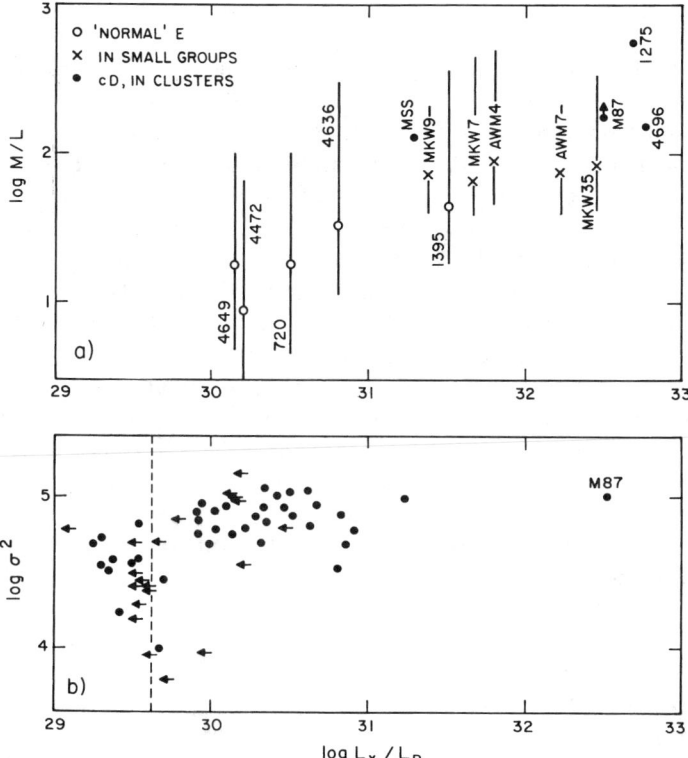

Figure 9 (*a*) Mass-to-light ratio (in solar units) versus the ratio of the X-ray (in erg s^{-1}) to the optical (in $L_\odot$) luminosity for "normal" elliptical galaxies and for group and cluster dominant galaxies (see text for references). MSS is the galaxy found by Maccagni et al. (1987) in the serendipitous *Einstein* survey. (*b*) Square of the central stellar velocity dispersion versus the X-ray to optical ratio. The vertical dashed line represents the average ratio for spiral galaxies. The M87 velocity dispersion is from Sargent et al. (1978), and the other velocity dispersions are from Whitmore et al. (1985). X-ray data are from the compilations of Canizares et al. (1987) and Matilsky et al. (1985).

And finally, getting onto even more speculative ground, we can ask, what would the nature of this dark matter be? One possibility is that of exotic particles, as in the cold dark matter scenario (e.g. Blumenthal et al. 1984; see also Forman et al. 1985). Another possibility is that the dark matter is composed of ordinary stellar remnants or low-mass concentrations, as suggested by Fabian et al. (1986b, 1987a) in the framework of the cooling flow scenario [but see White & Sarazin (1987b) as a dissenting voice: the distribution of the matter deposition from cooling flows may be more similar to the light distribution than to extended massive halos].

4.5 Radio Sources and Gaseous Halos

The importance of hot gaseous halos for both fueling nuclear radio sources through accreting cooling flows and confining extended radio lobes has long been recognized (e.g. Cowie & Binney 1977, MacDonald & Bailey 1981, Norman & Silk 1979). A correlation found between radio power and X-ray emission of central cluster galaxies gave empirical evidence of the first phenomenon; the effectiveness of the intracluster gas in confining the radio sources was demonstrated by detailed comparisons of radio and X-ray data (Burns et al. 1981, Forman & Jones 1982, Valentijn & Bijleveld 1983, Jones & Forman 1984, Harris et al. 1984, Feretti et al. 1984a, Morganti et al. 1988, Valentijn 1988). Similar comparisons show that the hot interstellar medium of more isolated early-type galaxies is also effective in confining the radio sources and is likely to play a role in their origin (Biermann & Kronberg 1983, Stanger et al. 1984, Dressel & Wilson 1985, Stanger & Warwick 1986, Thomas et al. 1986, Fabbiano et al. 1987a, 1989, Killeen et al. 1986, 1988). The extended hot gaseous halos can also be responsible for the depolarization observed at long radio wavelengths (e.g. 49 cm) in the bridges of double-lobed powerful radio galaxies (Strom & Jägers 1988).

If one excludes the more "active" specimens (e.g. 3CR galaxies, Cen A, For A), early-type galaxies tend to have relatively faint radio emission confined within their cores. These radio sources in some cases have clearly the same morphology as the more powerful radio galaxies and could therefore be considered physically similar to them and related to nuclear activity (e.g. Ekers & Ekers 1973, Bieging & Biermann 1977, Condon & Dressel 1978, Ekers & Kotanyi 1978, Hummel et al. 1983, Birkinshaw & Davies 1985, Fabbiano et al. 1987a, 1989). Correlations between the radio power and the X-ray luminosity of small samples of relatively radio-faint early-type galaxies suggested a possible link between the hot interstellar medium and the nuclear activity, since the X-ray emission of these galaxies is typically dominated by the extended thermal component (Dressel & Wilson 1985, Fabbiano et al. 1987a; see also Trinchieri 1988). More recently, Fabbiano et al. (1989) have compared the X-ray and radio continuum properties of the larger sample of early-type galaxies studied by Canizares et al. (1987) and report a correlation between the radio "core" power and the X-ray to optical ratio. The latter is a measure of the excess X-ray emission over the linear correlation observed in spiral systems and thus can then be considered as a direct indicator of the gaseous component (see previous discussion). This correlation therefore reinforces the connection between the hot interstellar medium and the nuclear radio sources and points to accreting cooling flows as the fuel.

The hot interstellar medium can also play an important role in the formation of extended radio structures. Fabbiano et al. (1989) notice that in their sample of early-type galaxies, very extended radio lobes, such as those of Cen A (NGC 5128) and For A (NGC 1316), tend to be associated with galaxies with relatively small X-ray to optical ratios. By contrast NGC 1399, which has a comparable core component, and therefore a comparably powerful central engine, but a larger X-ray to optical ratio, has the radio lobes well contained within its optical body. This result suggests that the gaseous halos also play a fundamental role in disrupting the radio jets and confining the extended radio structures. This conclusion had been previously reached by de Ruiter & Parma (1984), who observed significant distortions of the less extended sources in maps of radio galaxies of low to moderate power, suggesting interaction and bending of the jets by the interstellar medium. It has also been addressed in theoretical papers by Soker & Sarazin (1988) and Norman et al. (1988; see also Killeen & Bicknell 1988b). Fabbiano et al. (1989), in particular, demonstrate that the equation of Soker & Sarazin (1988) for the critical luminosity of the radio source, below which the jet is likely to be disrupted by shocks at the sonic radius of a galaxy cooling flow, correctly predicts the typical radio power of a few 10^{29} erg s^{-1} Hz^{-1} at 5 GHz below which extended lobes are not found (Colla et al. 1975, Jenkins 1982, Feretti et al. 1984b, Fabbiano et al. 1989). Once the jets have been disrupted, the thermal pressure of the hot gas is typically effective in confining the less powerful radio lobes well within the galaxies (see also Stanger & Warwick 1986, Killeen et al. 1988).

5. GALAXIES AND THE X-RAY BACKGROUND

The extragalactic X-ray background was discovered in 1962 in the same rocket flight that led to the discovery of the first extrasolar source of X-rays, Sco X-1 (Giacconi et al. 1962). Since then a great deal of effort has been spent trying to understand if this radiation is due to the integrated contributions of different classes of discrete sources or if diffuse processes are responsible for it (see Boldt 1987, Giacconi & Zamorani 1987). Based on the four galaxies then known to emit X rays (Milky Way, M31, and the Magellanic Clouds), Silk (1973) estimated that normal galaxies would contribute $\sim 10\%$ to the X-ray background. This estimate was then revised downward to less than $\sim 1\%$ (Rowan-Robinson & Fabian 1975, van Paradijs 1978, Worrall et al. 1979; see Fabian 1981) for a variety of reasons. One is the assumption that the X-ray luminosity of normal galaxies, which is dominated by the low-mass binary sources, would not be correlated with the optical luminosity but with some function of the galaxy mass (Rowan-Robinson & Fabian 1975); however, we now know that this is not so, as discussed earlier in this review. Other reasons are the effect of the redshift on the spectrum of the galactic sources (van Paradijs 1978), and the failure to detect a volume-limited sample of galaxies with the *HEAO-1* satellite (Worrall et al. 1979). However, this survey observed galaxies that are all of very low optical luminosity and as such cannot be considered representative of normal galaxies as a whole (Fabian 1981, Elvis et al. 1984).

Based on the X-ray to optical ratios of normal galaxies observed with the *Einstein* Observatory (Fabbiano & Trinchieri 1985, Trinchieri & Fabbiano 1985), Giacconi & Zamorani (1987) estimated instead that the integrated emission of normal galaxies contributes $\sim 13\%$ of the 2-keV extragalactic X-ray background (see also Setti 1985). If one includes in this estimate the contribution of low-activity nuclei present in a fraction of these galaxies and the effect of starburst activity, it could be significantly larger: Elvis et al. (1984) estimate that the low-activity nuclei could contribute $\sim 20\%$ of the X-ray background in the absence of evolution; and Weedman (1987), using X-ray to optical ratios measured in starburst galaxies (Fabbiano et al. 1982) and the 60-μm luminosity function derived from the *IRAS* survey, finds that in the absence of evolution these galaxies could account for $\sim 13\%$ of the 2-keV background (see also Giacconi & Zamorani 1987). If one assumes that starburst activity was much more pronounced in the past (Bookbinder et al. 1980, Stewart et al. 1982), starburst galaxies could be responsible for the bulk of the background (Giacconi & Zamorani 1987, Weedman 1987). Although this possibility is rejected by Giacconi & Zamorani because it would predict a surface density of 21–23 mag galaxies inconsistent with optical searches, the presence of a large amount of dust in these systems, suggested by the *IRAS* data, leaves this a still viable option (Weedman 1987).

Even if galaxies contribute substantially to the 2-keV X-ray background, their contribution to the X-ray background in a harder spectral range is uncertain and rests upon the spectral characteristics of their X-ray emission (van Paradijs 1978, Giacconi & Zamorani 1987, Weedman 1987). The spectra of spiral galaxies are consistent with a relatively hard X-ray emission ($kT > 2$ keV; Fabbiano & Trinchieri 1987), and X-ray binaries can have hard X-ray spectra (with $kT \sim 20$ keV). Moreover, a hard spectral component may be present in the starburst galaxy M82 (Fabbiano 1988b, Schaaf et al. 1989). However, it is unlikely that the galactic contribution would be substantial above 10 keV, especially considering that the spectra of faraway galaxies would be redshifted. The soft X-ray spectrum of the nucleus of M81 (Fabbiano 1988a) introduces an additional source of uncertainty to the estimate by Elvis et al. (1984) of the contribution of low-activity nuclei to the 2–10 keV X-ray background, since most of the X-ray luminosity of these sources could be emitted in a softer energy range.

On the other hand, we do not know how common and how bright the optically quiet X-ray-active nuclei, such as those of M33 and NGC 1313, are (see earlier in this paper). Their inclusion could raise the estimate of the galactic contribution to the X-ray background, although their X-ray spectra (Fabbiano & Trinchieri 1987, Trinchieri et al. 1988) suggest that even this type of source should contribute mainly in the soft energy range. Future X-ray observations will help to constrain the spectral range in which galaxies may contribute to the extragalactic X-ray background and will give us a better estimate of the contribution of low-activity nuclei.

6. THE FUTURE

What we have learned so far about the X-ray properties of normal galaxies has been the result of limited exploratory observations. Future X-ray satellites, with increased sensitivity and higher spatial and spectral resolution, will be essential for answering the many open questions resulting from the present work and for expanding and deepening our knowledge of these systems. The German X-ray satellite *ROSAT*, which is scheduled to be launched in 1990, will increase the number of galaxies mapped in X rays with a good sensitivity to low-surface-brightness features, and the Japanese *ASTRO-D* (to be launched in 1993) will allow the study of galactic spectral properties with a tenfold increased spectral resolution (but only 2' spatial resolution). The next major US X-ray astronomy endeavor, *AXAF*, with its subarcsecond spatial resolution and good spectral resolution, will allow the study of the luminosity function of X-ray sources in nearby spiral galaxies down to limiting luminosities at least 100 times smaller than present ones, and it will be able to detect single early-type stars in the Magellanic Clouds. Spectral parameters or X-ray colors will be measured for these sources and should help in establishing their nature (e.g. black hole candidates vs. X-ray pulsars; see White & Marshall 1984). With *AXAF* and the European *XMM*, with its larger collective area and sensitivity to low-surface-brightness features, the astronomical community will be able to address some of the outstanding questions on the X-ray properties of elliptical galaxies. These include firmly establishing the nature of the X-ray emission in the less X-ray-luminous galaxies, i.e. if it comes from a collection of binary X-ray sources or from a hot gaseous halo; measuring temperatures, temperature gradients, and metallicities of these halos; studying their interaction with the surrounding medium; and, finally, measuring with good accuracy the mass of these galaxies. It will also be possible to establish the luminosity functions and the spectral characteristics of individual X-ray sources in different spiral galaxies and thus to investigate the nature and evolution of these sources in different environments. With the plenitude of X-ray data to come from these missions, this may well be the last review that can cover the entire topic of X rays from normal galaxies.

ACKNOWLEDGMENTS

I thank P. Biermann, C. Canizares, M. Elvis, J. Grindlay, D. Helfand, W. Mathews, A. Renzini, C. Sarazin, D. Schwartz, F. Seward, P. Slane, H. Tananbaum, and G. Trinchieri for sending me preprints of their work and for discussions and comments on the manuscript. This work was supported by NASA contract NAS8-30751.

Literature Cited

Arnaud, K. A., Branduardi, G., Culhane, J. L., Fabian, A. C., Hazard, C., et al. 1985. *MNRAS* 217: 105
Arnaud, K. A., Fabian, A. C., Eales, S. A., Jones, C., Forman, W. 1984. *MNRAS* 211: 981
Arnaud, K. A., Gilmore, G. 1986. *MNRAS* 220: 759
Bahcall, J. N., Sarazin, C. L. 1977. *Ap. J. Lett.* 213: L99
Bahcall, N. A., Harris, D. E., Rood, H. J. 1984. *Ap. J. Lett.* 284: L29
Bajaja, E., van der Burg, G., Faber, S. M., Gallagher, J. S., Knapp, G. R., Shane, W. W. 1984. *Astron. Astrophys.* 141: 309
Battistini, P., Bonoli, F., Braccesi, A., Fusi Pecci, F., Malagnini, M. L., Marano, B. 1980. *Astron. Astrophys. Suppl.* 42: 357
Battistini, P., Bonoli, F., Buonanno, R., Corsi, C. E., Fusi Pecci, F. 1982. *Astron. Astrophys.* 113: 39
Bechtold, J., Czerny, B., Elvis, M., Fabiano, G., Green, R. F. 1987. *Ap. J.* 314: 699
Bechtold, J., Forman, W., Giacconi, R., Jones, C., Schwarz, J., et al. 1983. *Ap. J.* 265: 26
Becklin, E. E., Gatley, I., Matthews, K., Neugebauer, G., Sellgren, K., et al. 1980. *Ap. J.* 236: 441
Bertola, F. 1988. In *Cooling Flows in Clusters and Galaxies, Proc. NATO Adv. Res. Workshop*, ed. A. C. Fabian, p. 127. Dordrecht: Kluwer
Bertschinger, E., Meiksin, A. 1986. *Ap. J. Lett.* 306: L1
Bieging, J. H., Biermann, P. 1977. *Astron. Astrophys.* 60: 361
Biermann, P. 1984. In *Frontiers of Astronomy and Astrophysics, Eur. Reg. Astron. Meet., 7th*, ed. R. Pallavicini, p. 191. Florence: Ital. Astron. Soc.

Biermann, P., Friecke, K. 1977. *Astron. Astrophys.* 54: 461
Biermann, P., Kronberg, P. P. 1983. *Ap. J. Lett.* 268: L69
Biermann, P., Kronberg, P. P., Madore, B. F. 1982. *Ap. J. Lett.* 256: L37
Binney, J., Cowie, L. L. 1981. *Ap. J.* 247: 464
Birkinshaw, M., Davies, R. L. 1985. *Ap. J.* 291: 32
Bland, J., Tully, R. B. 1988. *Nature* 334: 43
Blumenthal, G. R., Faber, S. M., Primack, J. R., Rees, M. J. 1984. *Nature* 311: 517
Bohlin, R. C., Cornett, R. H., Hill, J. K., Hill, R. S., O'Connell, R. W., Stecher, T. P. 1985. *Ap. J. Lett.* 298: L37
Boldt, E. 1987. *Phys. Rep.* 146: 215
Bookbinder, J., Cowie, L. L., Krolik, J. H., Ostriker, J. P., Rees, M. 1980. *Ap. J.* 237: 647
Branduardi-Raymont, G., Fabricant, D., Feigelson, E., Gorenstein, P., Grindlay, J., et al. 1981. *Ap. J.* 248: 55
Bregman, J. N. 1978. *Ap. J.* 224: 768
Bregman, J. N. 1980a. *Ap. J.* 236: 577
Bregman, J. N. 1980b. *Ap. J.* 237: 681
Bregman, J. N., David, L. P. 1988. *Ap. J.* 326: 639
Bregman, J. N., Glassgold, A. E. 1982. *Ap. J.* 263: 564
Bregman, J. N., Roberts, M. S., Giovanelli, R. 1988. *Ap. J. Lett.* 330: L93
Bruhweiler, F. C., Klinglesmith, D. A. III, Gull, T. R., Sofia, S. 1987. *Ap. J.* 317: 152
Bruzual, G. A., Peimbert, M., Torres-Peimbert, S. 1982. *Ap. J.* 260: 495
Burns, J. O., Gregory, S. A., Holman, G. D. 1981. *Ap. J.* 250: 450
Burstein, D., Bertola, F., Buson, L. M., Faber, S. M., Lauer, T. R. 1988. *Ap. J.* 328: 440
Canizares, C. R. 1987. In *Dark Matter in*

the Universe, IAU Symp. No. 117, ed. J. Kormendy, G. R. Knapp, p. 165. Dordrecht: Reidel
Canizares, C. R., Clark, G. W., Jernigan, J. G., Markert, T. H. 1982a. Ap. J. 262: 33
Canizares, C. R., Clark, G. W., Markert, T. H., Berg, C., Smedira, M., et al. 1979. Ap. J. Lett. 234: L33
Canizares, C. R., Donahue, M., Trinchieri, G., Stewart, G., McGlynn, T. 1986. Ap. J. 304: 312
Canizares, C. R., Fabbiano, G., Trinchieri, G. 1987. Ap. J. 312: 503
Canizares, C. R., Kriss, G. A., Feigelson, E. D. 1982b. Ap. J. Lett. 253: L17
Canizares, C. R., Markert, T. H., Donahue, M. E. 1988. In Cooling Flows in Clusters and Galaxies, Proc. NATO Adv. Res. Workshop, ed. A. C. Fabian, p. 63. Dordrecht: Kluwer
Canizares, C. R., Stewart, G. C., Fabian, A. C. 1983. Ap. J. 272: 449
Cavaliere, A., Fusco-Femiano, R. 1976. Astron. Astrophys. 49: 137
Chanan, G. A., Helfand, D. J., Reynolds, S. P. 1984. Ap. J. Lett. 287: L23
Chevalier, R. A., Clegg, A. W. 1985. Nature 317: 44
Ciardullo, R., Ford, H. C., Neill, J. D., Jacoby, G. H., Shafter, A. W. 1987. Ap. J. 318: 520
Clark, D. H., Tuohy, I. R., Long, K. S., Szymkowiak, A. E., Dopita, M. A., et al. 1982. Ap. J. 255: 440
Clark, G., Doxsey, R., Li, F., Jernigan, J. G., van Paradijs, J. 1978. Ap. J. Lett. 221: L37
Colla, G., Fanti, C., Fanti, R., Gioia, I., Lari, C. 1975. Astron. Astrophys. 38: 209
Condon, J. J., Dressel, L. L. 1978. Ap. J. 221: 456
Corbelli, E., Salpeter, E. E. 1988. Ap. J. 326: 551
Cowie, L. L., Binney, J. 1977. Ap. J. 215: 723
Cowley, A. P., Crampton, D., Hutchings, J. B., Helfand, D. J., Hamilton, T. T., et al. 1984. Ap. J. 286: 196
Cox, D. P. 1983. In Supernova Remnants and Their X-Ray Emission, IAU Symp. No. 101, ed. J. Danziger, P. Gorenstein, p. 385. Dordrecht: Reidel
Cox, D. P., McCammon, D. 1986. Ap. J. 304: 657
Cox, D. P., Smith, B. W. 1974. Ap. J. Lett. 189: L105
Crampton, D., Cowley, A. P., Hutchings, J. B., Schade, D. J., Van Speybroeck, L. P. 1984. Ap. J. 284: 663
Crampton, D., Cowley, A. P., Schade, D., Chayer, P. 1985. Ap. J. 288: 494
Crawford, C. S., Crehan, D. A., Fabian, A. C., Johnstone, R. M. 1987. MNRAS 224: 1007
Czerny, B., Elvis, M. 1987. Ap. J. 321: 305
Demoulin, M.-H., Burbidge, E. M. 1970. Ap. J. 159: 799
Demoulin-Ulrich, M.-H., Butcher, H. R., Boksenberg, A. 1984. Ap. J. 285: 527
de Ruiter, H. R., Parma, P. 1984. Astron. Astrophys. 141: 189
Dowthwaite, J. C., Harrison, A. B., Kirkman, I. W., Macrae, H. J., Orford, K. J., et al. 1984. Nature 309: 691
Dressel, L., Wilson, A. 1985. Ap. J. 291: 668
Duric, N., Crane, P. C., Seaquist, E. R. 1982. Astron. J. 87: 1671
Eales, S. A., Arnaud, K. A. 1988. Ap. J. 324: 193
Ekers, R. D., Ekers, J. A. 1973. Astron. Astrophys. 24: 247
Ekers, R. D., Kotanyi, C. G. 1978. Astron. Astrophys. 67: 47
Elvis, M., Maccacaro, T., Wilson, A. S., Ward, M. J., Penston, M. V., et al. 1978. MNRAS 183: 129
Elvis, M., Schreier, E. J., Tonry, J., Davis, M., Huchra, J. P. 1981. Ap. J. 246: 20
Elvis, M., Soltan, A., Keel, W. C. 1984. Ap. J. 283: 479
Elvis, M., Van Speybroeck, L. 1982. Ap. J. Lett. 257: L51
Elvis, M., Wilkes, B. J., Tananbaum, H. 1985. Ap. J. 292: 357
Fabbiano, G. 1984. In X-Ray Astronomy '84, ed. M. Oda, R. Giacconi, p. 333. Tokyo: Inst. Space Astronaut. Sci.
Fabbiano, G. 1985a. Jpn.-US Sem. Galactic and Extragalactic Compact X-ray Sources, ed. Y. Tanaka, W. H. G. Lewin, p. 233. Tokyo: Inst. Space Astronaut. Sci.
Fabbiano, G. 1985b. Proc. ESA Workshop Cosmic X-Ray Spectroscopy Mission, ESA SP-239, p. 33
Fabbiano, G. 1986a. Publ. Astron. Soc. Pac. 98: 525
Fabbiano, G. 1986b. In Gaseous Halos of Galaxies, ed. J. N. Bregman, F. J. Lockman, p. 203. Charlottesville, Va: NRAO/AUI
Fabbiano, G. 1988a. Ap. J. 325: 544
Fabbiano, G. 1988b. Ap. J. 330: 672
Fabbiano, G., Feigelson, E., Zamorani, G. 1982. Ap. J. 256: 397
Fabbiano, G., Gioia, I. M., Trinchieri, G. 1988. Ap. J. 324: 749
Fabbiano, G., Gioia, I. M., Trinchieri, G. 1989. Submitted for publication
Fabbiano, G., Klein, U., Trinchieri, G., Wielebinski, R. 1987a. Ap. J. 312: 111
Fabbiano, G., Miller, L., Trinchieri, G., Longair, M., Elvis, M. 1984a. Ap. J. 277: 115
Fabbiano, G., Panagia, N. 1983. Ap. J. 266: 568

Fabbiano, G., Trinchieri, G. 1983. *Ap. J. Lett.* 266: L5
Fabbiano, G., Trinchieri, G. 1984. *Ap. J.* 286: 491
Fabbiano, G., Trinchieri, G. 1985. *Ap. J.* 296: 430
Fabbiano, G., Trinchieri, G. 1987. *Ap. J.* 315: 46
Fabbiano, G., Trinchieri, G., Macdonald, A. 1984b. *Ap. J.* 284: 65
Fabbiano, G., Trinchieri, G., Van Speybroeck, L. S. 1987b. *Ap. J.* 316: 127
Faber, S. 1983. In *Highlights of Astronomy*, ed. R. M. West, 6: 165. Dordrecht: Reidel
Faber, S., Gallagher, J. 1976. *Ap. J.* 204: 365
Faber, S. M., Gallagher, J. S. 1979. *Annu. Rev. Astron. Astrophys.* 17: 135
Faber, S., Jackson, R. E. 1976. *Ap. J.* 204: 668
Fabian, A. C. 1981. In *The Structure and Evolution of Normal Galaxies*, ed. S. M. Fall, D. Lynden-Bell, p. 181. Cambridge: Cambridge Univ. Press
Fabian, A. C., Arnaud, K. A., Nulsen, P. E. J., Mushotzky, R. F. 1986b. *Ap. J.* 305: 9
Fabian, A. C., Arnaud, K. A., Thomas, P. A. 1987a. In *Dark Matter in the Universe, IAU Symp. No. 117*, ed. J. Kormendy, G. R. Knapp, p. 201. Dordrecht: Reidel
Fabian, A. C., Crawford, C. S., Johnstone, R. M., Thomas, P. A. 1987b. *MNRAS* 228: 963
Fabian, A. C., Hu, E. M., Cowie, L. L., Grindlay, J. 1981. *Ap. J.* 248: 47
Fabian, A. C., Ku, W. H.-M., Malin, D. F., Mushotzky, R. F., Nulsen, P. E. J., Stewart, G. C. 1985. *MNRAS* 196: 35P
Fabian, A. C., Nulsen, P. E. J. 1977. *MNRAS* 180: 479
Fabian, A. C., Nulsen, P. E. J., Arnaud, K. A. 1984a. *MNRAS* 208: 179
Fabian, A. C., Nulsen, P. E. J., Canizares, C. R. 1982. *MNRAS* 201: 933
Fabian, A. C., Nulsen, P. E. J., Canizares, C. R. 1984b. *Nature* 310: 733
Fabian, A. C., Schwarz, J., Forman, W. 1980. *MNRAS* 192: 135
Fabian, A. C., Thomas, P. A. 1987. In *Structure and Dynamics of Elliptical Galaxies, IAU Symp. No. 127*, ed. T. deZeeuw, p. 155. Dordrecht: Reidel
Fabian, A. C., Thomas, P. A., Fall, S. M., White, R. E. III. 1986a. *MNRAS* 221: 1049
Fabricant, D., Gorenstein, P. 1983. *Ap. J.* 267: 535
Fabricant, D., Lecar, M., Gorenstein, P. 1980. *Ap. J.* 241: 552
Fabricant, D., Topka, K., Harnden, F. R. Jr., Gorenstein, P. 1978. *Ap. J. Lett.* 226: L107
Fall, S. M. 1987. In *Nearly Normal Galaxies*, ed. S. M. Faber, p. 326. New York: Springer-Verlag
Feigelson, E., Schreier, E., Delvaille, J., Giacconi, R., Grindlay, J., Lightman, A. 1981. *Ap. J.* 251: 31
Feretti, L., Gioia, I. M., Giovannini, G., Gregorini, L., Padrielli, L. 1984a. *Astron. Astrophys.* 139: 50
Feretti, L., Giovannini, G., Gregorini, L., Parma, P., Zamorani, G. 1984b. *Astron. Astrophys.* 139: 55
Forman, W., Jones, C. 1982. *Annu. Rev. Astron. Astrophys.* 20: 547
Forman, W., Jones, C. 1988. *Proc. Symp. in Honor of Bill Liller*. Cambridge: Cambridge Univ. Press. In press
Forman, W., Jones, C., DeFaccio, M. 1984a. *Proc. ESO Workshop Virgo Cluster, ESO Conf. Proc. 20*, ed. O. G. Richter, B. Binggeli, p. 323
Forman, W., Jones, C., Tucker, W. 1984b. In *Clusters and Groups of Galaxies*, ed. F. Mardirossian, G. Giuricin, M. Mezzetti, p. 297. Dordrecht: Reidel
Forman, W., Jones, C., Tucker, W. 1985. *Ap. J.* 293: 102
Forman, W., Schwarz, J., Jones, C., Liller, W., Fabian, A. 1979. *Ap. J. Lett.* 234: L27
Gallagher, J. S., Faber, S. M., Balick, B. 1975. *Ap. J.* 202: 7
Gallagher, J. S., Goad, J. W., Mould, J. R. 1982. *Ap. J.* 263: 101
Geldzahler, B. J., Fomalont, E. B., Hilldrup, K., Corey, B. E. 1981. *Astron. J.* 86: 1036
Ghigo, F. D., Wardle, J. F. C., Cohen, N. L. 1983. *Astron. J.* 88: 1587
Giacconi, R., Branduardi, G., Briel, U., Epstein, A., Fabricant, D., et al. 1979. *Ap. J.* 230: 540
Giacconi, R., Gursky, H., Paolini, F., Rossi, B. 1962. *Phys. Rev. Lett.* 9: 439
Giacconi, R., Zamorani, G. 1987. *Ap. J.* 313: 20
Gioia, I. M., Maccacaro, T., Schild, R. E., Stocke, J. T., Liebert, J. W., et al. 1984. *Ap. J.* 283: 495
Glass, I. S. 1981. *MNRAS* 197: 1067
Gorenstein, P., Fabricant, D., Topka, K., Tucker, W., Harnden, F. R. Jr. 1977. *Ap. J. Lett.* 216: L95
Gottwald, M., Pietsch, W., Hasinger, G. 1987. *Astron. Astrophys.* 175: 45
Grindlay, J. E. 1984. *Adv. Space Res.* 3: 19
Grindlay, J. E. 1985. *Jpn.-US Sem. Galactic and Extragalactic Compact X-Ray Sources*, ed. Y. Tanaka, W. H. G. Lewin, p. 215. Tokyo: Inst. Space Astronaut. Sci.
Gursky, H. 1976. In *Structure and Evolution of Close Binary Systems, IAU Symp. No. 73*, ed. P. Eggleton, S. Mitton, J. Whelan, p. 19. Dordrecht: Reidel

Harris, D. E., Costain, C. H., Dewdney, P. E. 1984. *Ap. J.* 280: 532
Harris, W. E., Racine, R. 1979. *Annu. Rev. Astron. Astrophys.* 17: 241
Heckman, T. M., Armus, L., Miley, G. 1987. *Astron. J.* 93: 276
Helfand, D. J. 1982. In *Supernovae: A Survey of Current Research*, ed. M. J. Rees, R. J. Stoneham, p. 529. Dordrecht: Reidel
Helfand, D. J. 1984a. *Publ. Astron. Soc. Pac.* 96: 913
Helfand, D. J. 1984b. In *Structure and Evolution of the Magellanic Clouds*, ed. S. van den Bergh, K. deBoer, p. 293. Dordrecht: Reidel
Helfand, D. J. 1985. *Jpn.-US Sem. Galactic and Extragalactic Compact X-Ray Sources*, ed. Y. Tanaka, W. H. G. Lewin, p. 207. Tokyo: Inst. Space Aeronaut. Sci.
Helfand, D. J., Caillault, J.-P. 1982. *Ap. J.* 253: 760
Helfand, D. J., Long, K. S. 1980. In *X-Ray Astronomy*, ed. R. Giacconi, G. Setti, p. 47. Dordrecht: Reidel
Hjellming, R. M., Johnston, K. J. 1981. *Ap. J. Lett.* 246: L141
Hu, E. M., Cowie, L. L., Wang, Z. 1985. *Ap. J. Suppl.* 59: 447
Huchra, J., Stauffer, J., Van Speybroeck, L. 1982. *Ap. J. Lett.* 259: L57
Huchtmeier, W. K., Bregman, J. N., Hogg, D. E., Roberts, M. S. 1988. *Astron. Astrophys.* 198: L17
Hummel, E., Kotanyi, C. G., Ekers, R. D. 1983. *Astron. Astrophys.* 127: 205
Hutchings, J. B. 1984. In *Structure and Evolution of the Magellanic Clouds*, ed. S. van den Bergh, K. deBoer, p. 305. Dordrecht: Reidel
Inoue, H., Koyama, K., Tanaka, Y. 1983. In *Supernova Remnants and Their X-ray Emission*, IAU Symp. No. 101, ed. J. Danziger, P. Gorenstein, p. 535. Dordrecht: Reidel
Jenkins, C. R. 1982. *MNRAS* 200: 705
Johnstone, R. M., Fabian, A. C., Nulsen, P. E. J. 1987. *MNRAS* 224: 75
Jones, C. 1987. In *Nearly Normal Galaxies*, ed. S. M. Faber, p. 109. New York: Springer-Verlag
Jones, C., Forman, W. 1984. *Ap. J.* 276: 38
Jura, M. 1977. *Ap. J.* 212: 634
Jura, M. 1986. *Ap. J.* 306: 483
Keel, W. C. 1983. *Ap. J.* 269: 466
Kellogg, E., Baldwin, J. R., Koch, D. 1975. *Ap. J.* 199: 299
Kent, S. M. 1985. *Ap. J. Suppl.* 59: 115
Killeen, N. E. B., Bicknell, G. V. 1988a. *Ap. J.* 325: 165
Killeen, N. E. B., Bicknell, G. V. 1988b. *Ap. J.* 324: 198
Killeen, N. E. B., Bicknell, G. V., Carter, D. 1986. *Ap. J.* 309: 45
Killeen, N. E. B., Bicknell, G. V., Ekers, R. D. 1988. *Ap. J.* 325: 180
Kim, D.-W. 1988. PhD thesis. Univ. Calif., Los Angeles
King, I. R. 1978. *Ap. J.* 222: 1
Klein, U., Wielebinski, R., Beck, R. 1984. *Astron. Astrophys.* 135: 213
Knapp, G. R. 1987. In *Structure and Dynamics of Elliptical Galaxies*, IAU Symp. No. 127, ed. T. deZeeuw, p. 145. Dordrecht: Reidel
Knapp, G. R. 1988. In *Cooling Flows in Clusters and Galaxies*, Proc. NATO Adv. Res. Workshop, ed. A. C. Fabian, p. 93. Dordrecht: Kluwer
Knapp, G. R., Turner, E. L., Cunniffe, P. E. 1985. *Astron. J.* 90: 454
Kotanyi, C., van Gorkom, J. H., Ekers, R. D. 1983. *Ap. J. Lett.* 273: L7
Koyama, K., Makishima, K., Tanaka, Y. 1986. *Publ. Astron. Soc. Jpn.* 38: 121
Kriss, G. A., Canizares, C. R., McClintock, J. E., Feigelson, E. D. 1980. *Ap. J. Lett.* 235: L61
Kriss, G. A., Cioffi, D. F., Canizares, C. R. 1983. *Ap. J.* 272: 439
Kronberg, P. P., Biermann, P., Schwab, F. R. 1985. *Ap. J.* 246: 751
Kruegel, E., Tutukov, A., Loose, H. 1983. *Astron. Astrophys.* 124: 89
Larson, R. B., Tinsley, B. 1978. *Ap. J.* 219: 46
Lea, S. M., Mason, K. O., Reichert, G., Charles, P. A., Riegler, G. 1979. *Ap. J. Lett.* 227: L67
Lea, S. M., Mushotzky, R., Holt, S. 1982. *Ap. J.* 262: 24
Lebofsky, M. J., Rieke, G. H. 1979. *Ap. J.* 229: 111
Loewenstein, M., Mathews, W. G. 1987a. *Ap. J.* 319: 614
Loewenstein, M., Mathews, W. G. 1987b. In *Nearly Normal Galaxies*, ed. S. M. Faber, p. 96. New York: Springer-Verlag
Long, K. S., D'Odorico, S., Charles, P. A., Dopita, M. A. 1981b. *Ap. J. Lett.* 246: L61
Long, K. S., Helfand, D. J. 1979. *Ap. J. Lett.* 234: L77
Long, K. S., Helfand, D. J., Grabelsky, D. A. 1981a. *Ap. J.* 248: 925
Long, K. S., Van Speybroeck, L. P. 1983. In *Accretion-Driven X-Ray Sources*, ed. W. Lewin, E. P. J. van den Heuvel, p. 117. Cambridge: Cambridge Univ. Press
Maccacaro, T., Perola, G. C. 1981. *Ap. J. Lett.* 246: L11
Maccacaro, T., Perola, G. C., Elvis, M. 1982. *Ap. J.* 257: 47
Maccagni, D., Gioia, I. M., Maccacaro, T., Schild, R. E., Stocke, J. T. 1987. *Ap. J.* 316: 132

MacDonald, J., Bailey, M. E. 1981. *MNRAS* 197: 995
Makishima, K., et al. 1989. *Publ. Astron. Soc. Jpn.* In press
Malina, R., Lampton, M., Bowyer, S. 1976. *Ap. J.* 209: 678
Malumuth, E. M., Kriss, G. A. 1986. *Ap. J.* 308: 10
Markert, T. H., Donahue, M. E. 1985. *Ap. J.* 297: 564
Markert, T. H., Rallis, A. D. 1983. *Ap. J.* 275: 571
Marshall, F. J., Clark, G. W. 1984. *Ap. J.* 287: 633
Mason, K. O., Rosen, S. R. 1985. *Space Sci. Rev.* 40: 675
Mathews, W. G. 1978. *Ap. J.* 219: 413
Mathews, W. G. 1988a. Preprint
Mathews, W. G. 1988b. *Astron. J.* 95: 1047
Mathews, W. G. 1988c. In *Cooling Flows in Clusters and Galaxies*, *Proc. NATO Adv. Res. Workshop*, ed. A. C. Fabian, p. 279. Dordrecht: Kluwer
Mathews, W., Baker, J. 1971. *Ap. J.* 170: 241
Mathews, W. G., Bregman, J. N. 1978. *Ap. J.* 224: 308
Mathews, W. G., Loewenstein, M. 1986. *Ap. J. Lett.* 306: L7
Mathewson, D. S., Ford, V. L., Dopita, M. A., Tuohy, I. R., Long, K. S., Helfand, D. J. 1983. *Ap. J. Suppl.* 51: 345
Mathewson, D. S., Ford, V. L., Dopita, M. A., Tuohy, I. R., Mills, B. Y., Turtle, A. J. 1984. *Ap. J. Suppl.* 55: 189
Mathewson, D. S., Ford, V. L., Tuohy, I. R., Mills, B. Y., Turtle, A. J., Helfand, D. J. 1985. *Ap. J. Suppl.* 58: 197
Matilsky, T., Jones, C., Forman, W. 1985. *Ap. J.* 291: 621
McCammon, D., Burrows, D. N., Sanders, W. T., Kraushaar, W. L. 1983. *Ap. J.* 269: 107
McCammon, D., Sanders, W. T. 1984. *Ap. J.* 287: 167
McCarthy, P. J., Heckman, T., van Breugel, W. 1987. *Astron. J.* 93: 264
McKechnie, S. P., Jansen, F. A., deKorte, P. A. J., Hulscher, F. W. H., van der Klis, M., et al. 1984. In *X-Ray Astronomy '84*, ed. M. Oda, R. Giacconi, p. 373. Tokyo: Inst. Space Aeronaut. Sci.
Miller, J. S. 1985. In *Astrophysics of Active Galaxies and Quasi-Stellar Objects*, ed. J. S. Miller, p. 367. Santa Cruz, Calif: Univ. Sci. Books
Mitchell, R. J., Culhane, J. L., Davison, P. N., Ives, J. C. 1976. *MNRAS* 175: 29P
Moorwood, A. F. M., Glass, I. S. 1982. *Astron. Astrophys.* 115: 84
Morganti, R., Fanti, R., Gioia, I. M., Harris, D. E., Parma, P., deRuiter, H. 1988. *Astron. Astrophys.* 189: 11
Mushotzky, R. F., Holt, S. S., Smith, B. W., Boldt, E. A., Serlemitsos, P. J. 1981. *Ap. J. Lett.* 244: L47
Nomoto, K. 1984. In *Problems of Collapse and Numerical Relativity*, ed. D. Bancel, M. Signore, p. 89. Dordrecht: Reidel
Norman, C., Silk, J. 1979. *Ap. J. Lett.* 233: L1
Norman, M. L., Burns, J. O., Sulkanen, M. 1988. Preprint
Nulsen, P. E. J. 1986. *MNRAS* 221: 377
Nulsen, P. E. J., Carter, D. 1987. *MNRAS* 225: 939
Nulsen, P. E. J., Stewart, G. C., Fabian, A. C. 1984. *MNRAS* 208: 185
O'Connell, R. W. 1983. *Ap. J.* 267: 80
O'Connell, R. W., McNamara, B. R. 1988. In *Cooling Flows in Clusters and Galaxies*, *Proc. NATO Adv. Res. Workshop*, ed. A. C. Fabian, p. 103. Dordrecht: Kluwer
Oke, J. B., Bertola, F., Capaccioli, M. 1981. *Ap. J.* 243: 453
Palumbo, G. G. C., Fabbiano, G., Fransson, C., Trinchieri, G. 1985. *Ap. J.* 298: 259
Palumbo, G. G. C., Maccacaro, T., Panagia, N., Vettolani, G., Zamorani, G. 1981. *Ap. J.* 247: 484
Peimbert, M., Torres-Peimbert, S. 1981. *Ap. J.* 245: 845
Peres, G., Reale, F., Collura, A., Fabbiano, G. 1989. *Ap. J.* 336: 140
Phillips, M. M., Jenkins, C. R., Dopita, M. A., Sadler, E. M., Binette, L. 1986. *Astron. J.* 91: 1062
Pounds, K. A., Stanger, V. J., Turner, T. J., King, A. R., Czerny, B. 1986. *MNRAS* 224: 443
Protheroe, R. J., Clay, R. W., Gerhardy, P. R. 1984. *Ap. J. Lett.* 280: L47
Rappaport, S., Joss, P. C., Webbink, R. F. 1982. *Ap. J.* 254: 616
Renzini, A., Buzzoni, A. 1986. In *Spectral Evolution of Galaxies*, ed. C. Chiosi, A. Renzini, p. 195. Dordrecht: Reidel
Rieke, G. H., Lebofsky, M. J., Thompson, R. I., Low, F. J., Tokunaga, A. T. 1980. *Ap. J.* 238: 24
Romanishin, W. 1987. *Ap. J. Lett.* 323: L113
Rosner, R., Tucker, W. H. 1989. *Ap. J.* 338: 761
Rowan-Robinson, M., Fabian, A. C. 1975. *MNRAS* 170: 199
Rubin, V. C., Ford, W. K. Jr. 1986. *Ap. J. Lett.* 305: L35
Sadler, E. M. 1988. In *Cooling Flows in Clusters and Galaxies*, *Proc. NATO Adv. Res. Workshop*, ed. A. C. Fabian, p. 263. Dordrecht: Kluwer
Samorski, M., Stamm, W. 1983. *Ap. J. Lett.* 268: L17
Sandage, A. 1957. *Ap. J.* 125: 422
Sandage, A., Tammann, G. A. 1981. *A Revised Shapley-Ames Catalog of Bright Galaxies*. Washington, DC: Carnegie Inst. Washington

Sarazin, C. L. 1986. In *Gaseous Halos of Galaxies*, ed. J. N. Bregman, F. J. Lockman, p. 223. Charlottesville, Va: NRAO/AUI

Sarazin, C. L. 1987a. In *Dark Matter in the Universe, IAU Symp. No. 117*, ed. J. Kormendy, G. R. Knapp, p. 183. Dordrecht: Reidel

Sarazin, C. L. 1987b. In *Structure and Dynamics of Elliptical Galaxies, IAU Symp. No. 127*, ed. T. deZeeuw, p. 179. Dordrecht: Reidel

Sarazin, C. L., O'Connell, R. W. 1983. *Ap. J.* 268: 552

Sarazin, C. L., White, R. E. III. 1987. *Ap. J.* 320: 32

Sarazin, C. L., White, R. E. III. 1988. *Ap. J.* 331: 102

Sargent, W., Kowal, C., Hartwick, F. D. A., van den Bergh, S. 1977. *Astron. J.* 82: 947

Sargent, W. L. W., Young, P. J., Boksenberg, A., Shortridge, K., Lynds, C. R., Hartwick, F. D. A. 1978. *Ap. J.* 221: 731

Schaaf, R., Pietsch, W., Biermann, P. L., Kronberg, P. P., Schmutzler, T. 1989. *Ap. J.* 336: 722

Scoville, N., Young, J. S. 1983. *Ap. J.* 265: 148

Serlemitsos, P. J., Smith, B. W., Boldt, E. A., Holt, S. S., Swank, J. H. 1977. *Ap. J. Lett.* 211: L63

Setti, G. 1985. In *Nonthermal and Very High Temperature Phenomena in X-Ray Astronomy*, ed. G. C. Perola, M. Salvati, p. 159. Rome: Ist. Astron., Univ. "La Sapienza"

Seward, F. D., Harnden, F. R. Jr., Helfand, D. J. 1984. *Ap. J. Lett.* 287: L19

Seward, F. D., Mitchell, M. 1981. *Ap. J.* 243: 736

Shostak, G. S., Hummel, E., Shaver, P. A., van der Hulst, J. M., van der Kruit, P. C. 1982. *Astron. Astrophys.* 115: 293

Shuder, J. M., Osterbrock, D. E. 1981. *Ap. J.* 250: 55

Silk, J. 1973. *Annu. Rev. Astron. Astrophys.* 11: 269

Silk, J. 1976. *Ap. J.* 208: 646

Silk, J., Djorgovski, S., Wyse, R. F. G., Bruzual, G. 1986. *Ap. J.* 307: 415

Singh, K. P., Nousek, J. A., Burrows, D. N., Garmire, G. P. 1987. *Ap. J.* 313: 185

Soker, N., Sarazin, C. L. 1988. *Ap. J.* 327: 66

Spitzer, L. 1956. *Ap. J.* 124: 20

Stanger, V. J., Warwick, R. S. 1986. *MNRAS* 220: 363

Stanger, V. J., Warwick, R. S., Schwarz, J. 1984. In *X-Ray Astronomy '84*, ed. M. Oda, R. Giacconi, p. 377. Tokyo: Inst. Space Astronaut. Sci.

Stewart, G. C., Canizares, C. R., Fabian, A. C., Nulsen, P. E. J. 1984a. *Ap. J.* 278: 536

Stewart, G. C., Fabian, A. C., Jones, C., Forman, W. 1984b. *Ap. J.* 285: 1

Stewart, G. C., Fabian, A. C., Terlevich, R. J., Hazard, C. 1982. *MNRAS* 200: 61P

Strom, R. G., Jägers, W. J. 1988. *Astron. Astrophys.* 194: 79

Takeda, H., Nulsen, P. E. J., Fabian, A. C. 1984. *MNRAS* 208: 261

Tammann, G. 1974. In *Supernovae and Supernovae Remnants*, ed. C. B. Cosmovici, p. 155. Dordrecht: Reidel

Tammann, G. 1982. In *Supernovae: A Survey of Current Research*, ed. M. Rees, R. J. Stoneham, p. 371. Dordrecht: Reidel

Tananbaum, H., Peters, G., Forman, W., Giacconi, R., Jones, C., Avni, Y. 1978. *Ap. J.* 223: 74

Thomas, P. A. 1986. *MNRAS* 220: 949

Thomas, P. A., Fabian, A. C., Arnaud, K. A., Forman, W., Jones, C. 1986. *MNRAS* 222: 655

Thomas, P. A., Fabian, A. C., Nulsen, P. E. J. 1987. *MNRAS* 228: 973

Thronson, H. A., Bally, J. 1987. *Ap. J. Lett.* 319: L63

Tomisaka, K., Ikeuchi, S. 1988. *Ap. J.* 330: 695

Tonry, J. 1981. *Ap. J. Lett.* 251: L1

Topka, K., Avni, Y., Golub, L., Gorenstein, P., Harnden, F. R. Jr., et al. 1982. *Ap. J.* 259: 677

Trimble, V. 1987. *Annu. Rev. Astron. Astrophys.* 25: 425

Trinchieri, G. 1986. In *Gaseous Halos of Galaxies*, ed. J. N. Bregman, F. J. Lockman, p. 215. Charlottesville, Va: NRAO/AUI

Trinchieri, G. 1988. In *Cooling Flows in Clusters and Galaxies, Proc. NATO Adv. Res. Workshop*, ed. A. C. Fabian, p. 273. Dordrecht: Kluwer

Trinchieri, G., Fabbiano, G. 1985. *Ap. J.* 296: 447

Trinchieri, G., Fabbiano, G., Canizares, C. R. 1986. *Ap. J.* 310: 637

Trinchieri, G., Fabbiano, G., Palumbo, G. G. C. 1985. *Ap. J.* 290: 96

Trinchieri, G., Fabbiano, G., Peres, G. 1988. *Ap. J.* 325: 531

Tucker, W. H., Rosner, R. 1983. *Ap. J.* 267: 547

Tuohy, I. R., Buckley, D. A. H., Remillard, R. A., Bradt, H. V., Schwartz, D. A. 1988. Preprint

Tuohy, I. R., Dopita, M. A., Mathewson, D. S., Long, K. S., Helfand, D. J. 1982. *Ap. J.* 261: 473

Turner, B. E. 1985. *Ap. J.* 299: 312

Umemura, M., Ikeuchi, S. 1987. *Ap. J.* 319: 601

Vader, J. P., van den Heuvel, E. P. J., Lewin, W. H. G., Takens, R. J. 1982. *Astron. Astrophys.* 113: 328

Vaiana, G., Cassinelli, J. P., Fabbiano, G., Giacconi, R., Golub, L., et al. 1981. *Ap. J.* 245: 163

Valentijn, E. A. 1988. In *Cooling Flows in Clusters and Galaxies*, Proc. *NATO Adv. Res. Workshop*, ed. A. C. Fabian, p. 189. Dordrecht: Kluwer

Valentijn, E. A., Bijleveld, W. 1983. *Astron. Astrophys.* 125: 223

van den Heuvel, E. P. J. 1980. In *X-Ray Astronomy*, ed. R. Giacconi, G. Setti, p. 119. Dordrecht: Reidel

van den Heuvel, E. P. J. 1984. In *Frontiers of Astronomy and Astrophysics*, *Eur. Reg. Astron. Meet.*, *7th*, ed. R. Pallavicini, p. 167. Florence: Ital. Astron. Soc.

van Gorkom, J. H. 1987. In *Structure and Dynamics of Elliptical Galaxies*, *IAU Symp. No. 127*, ed. T. deZeeuw, p. 421. Dordrecht: Reidel

van Gorkom, J. H., Knapp, G. R., Raimond, E., Faber, S. M., Gallagher, J. S. 1986. *Astron. J.* 91: 791

van Paradijs, J. 1978. *Ap. J.* 226: 586

van Paradijs, J., Lewin, W. H. G. 1985. *Astron. Astrophys.* 142: 361

Van Speybroeck, L., Bechtold, J. 1981. In *X-Ray Astronomy With the Einstein Satellite*, ed. R. Giacconi, p. 153. Dordrecht: Reidel

Van Speybroeck, L., Epstein, A., Forman, W., Giacconi, R., Jones, C., et al. 1979. *Ap. J. Lett.* 234: L45

Vedder, P. W., Trester, J. J., Canizares, C. R. 1988. *Ap. J.* 332: 725

Véron-Cetty, M.-P., Véron, P. 1986. *Astron. Astrophys. Suppl.* 66: 335

von Kapp-her, A., Berkhuijsen, E. M., Wielebinski, R. 1978. *Astron. Astrophys.* 62: 51

Ward, M. J. 1988. *MNRAS* 231: 1P

Wardle, M., Knapp, G. R. 1986. *Astron. J.* 91: 23

Warwick, R. S., Turner, M. J. L., Watson, M. G., Willingale, R. 1985. *Nature* 317: 218

Watson, M. G., Stanger, V., Griffiths, R. E. 1984. *Ap. J.* 286: 144

Watson, M. G., Willingale, R., Grindlay, J. E., Hertz, P. 1981. *Ap. J.* 250: 142

Watson, M. G., Willingale, R., Grindlay, J. E., Seward, F. D. 1983. *Ap. J.* 273: 688

Weedman, D. W. 1987. In *Star Formation in Galaxies*, ed. C. J. Lonsdale Persson, p. 351. Washington, DC: US Govt. Print. Off.

Weedman, D. W., Feldman, F. R., Balzano, V. A., Ramsey, L. W., Sramek, R. A., Wu, C.-C. 1981. *Ap. J.* 248: 105

White, N. E., Marshall, F. E. 1984. *Ap. J. Lett.* 281: 354

White, N. E., Swank, J. H., Holt, S. S., Parmar, A. N. 1982. *Ap. J.* 263: 277

White, R. E. III, Chevalier, R. A. 1983. *Ap. J.* 275: 69

White, R. E. III, Chevalier, R. A. 1984. *Ap. J.* 280: 561

White, R. E. III, Sarazin, C. L. 1987a. *Ap. J.* 318: 612

White, R. E. III, Sarazin, C. L. 1987b. *Ap. J.* 318: 621

White, R. E. III, Sarazin, C. L. 1987c. *Ap. J.* 318: 629

White, R. E. III, Sarazin, C. L. 1988. *Ap. J.* 335: 688

Whitmore, B. C. 1984. *Ap. J.* 278: 61

Whitmore, B. C., McElroy, D. B., Tonry, J. L. 1985. *Ap. J. Suppl.* 59: 1

Wilkes, B. J., Elvis, M. 1987. *Ap. J.* 323: 243

Worrall, D. M., Marshall, F. E., Boldt, E. A. 1979. *Nature* 281: 127

Worrall, D. M., Marshall, F. E., Boldt, E. A., Swank, J. H. 1982. *Ap. J.* 255: 111

Wrobel, J. M., Neugebauer, G., Miley, G. K. 1986. *Ap. J. Lett.* 310: L11

POPULATIONS IN LOCAL GROUP GALAXIES

Paul Hodge

Astronomy Department, University of Washington, Seattle, Washington 98195

1. STELLAR POPULATIONS

The concept of stellar populations, developed primarily by W. Baade in the 1950s, was inspired by the apparent dichotomy of galaxy content. Some galaxies (spirals and irregulars) seemed to be made up largely of bright, blue stars, with an admixture of interstellar material, whereas others (ellipticals) seemed to be made up exclusively of faint red stars (13). Baade called the two kinds of stars Population I and Population II, respectively. One of Baade's important insights was the relation of these two types of galaxy content with the properties of the stars in the Milky Way Galaxy (hereafter MWG), which provided him with kinematical information that eventually led to an astrophysical understanding of the differences in terms of the evolution of the MWG.

Much of Baade's original inspiration was derived from his detailed study of Local Group galaxies, and this particular sample of examples has continued to play an important part in the further development of our understanding of stellar populations. The history of this development has been the subject of a number of papers, reviews, and books, most recently that edited by Norman et al. (94). This review, therefore, limits itself primarily to a discussion of the most recent events, with a demonstration of both the failings and the strengths of Baade's arguments, especially in terms of the Local Group galaxies.

The Two-Point Diagram

Baade's original criteria for distinguishing Population I from Population II included four parameters: location, color, association with interstellar material, and kinematics. However, he found that these features are inter-

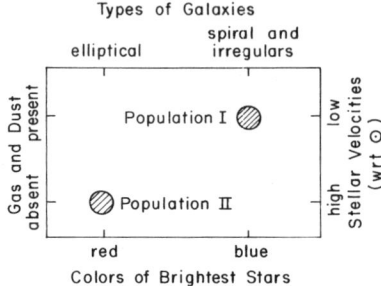

Figure 1 The traditional two-point population diagram.

related in such a way that all stars seemed to be classifiable by only two parameters. Figure 1 illustrates this division into types, represented by two points in a plane.

The Population Line

As a better understanding of stellar evolution was developed, and as ideas about galaxy evolution progressed, these criteria were supplanted by two more fundamental ones: stellar ages and stellar chemical abundances. It appeared to be the case, at least for the MWG, that all of the other observed differences could be understood in terms of these two variables, and therefore that stellar populations could be represented by two segments of a line that described the chemical enrichment of a stellar group due to stellar evolution (Figure 2). As implied by this representation, there can be a variety of different ages among Population I or II stars, with a corresponding range in heavy element abundances. Intermediate populations were introduced, especially by those astronomers trying to understand the kinematics and spatial distribution of stars in the MWG, and the simple two-point system broke down, to be supplanted by a linear progression of types.

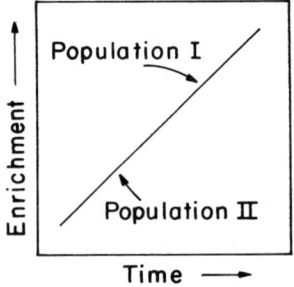

Figure 2 The two populations as evolutionary line segments.

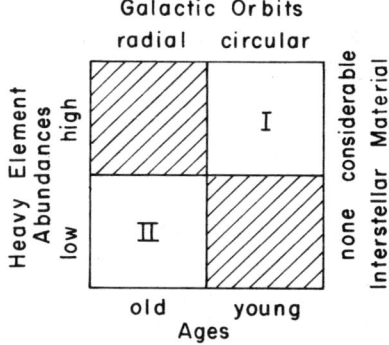

Figure 3   Morphology of the evolutionary plane.

The Population Plane

Figure 3 represents a more morphological approach, including as it does possibilities of old metal-rich stars and young metal-poor stars. In the 40 years since the introduction of the two populations, increasing evidence has suggested that such stars do exist in certain environments, especially in other galaxies, leading to a further breakdown in the simple two-population classification scheme.

Figure 4 illustrates how this situation arises, using the MWG and the Small Magellanic Cloud (SMC) as examples. Probably because of its lower mass, and thus of a consequent different star formation rate history, the SMC has experienced a slower enrichment of heavy elements. Therefore, it contains some stars that are relatively young but still metal poor. We do not, on the other hand, find or expect to find well-mixed galaxies or segments of galaxies that have old metal-rich stars together with young

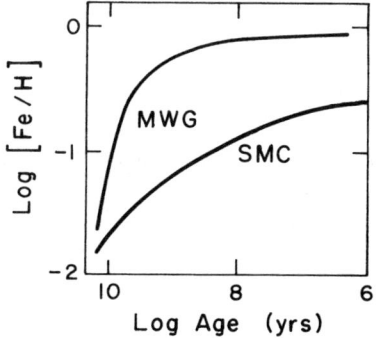

Figure 4   Schematic enrichment lines for the Milky Way Galaxy (MWG) and the Small Magellanic Cloud (SMC).

metal-poor ones, as enrichment, of course, only proceeds in one direction. Thus, the entire area of Figure 3 is not inhabitable by a given star group, though a composite galaxy, such as the MWG, can have a spread in these parameters.

2. THE POPULATION BOX

Recent detailed studies of galaxies have shown that even the fairly elaborate representation of Figure 3 is not adequate to address the newfound differences among galaxies' stellar populations. In particular, it is found that the rate of star formation as a function of time is not simple, nor is it common to all galaxies, even those of the same Hubble type. For example, the Sculptor and Fornax galaxies, both classified as dwarf ellipticals (sometimes the term "spheroidal" is used for what are now irrelevant historical reasons), cannot be characterized as having similar populations of stars. Sculptor seems to consist entirely of stars of great age (about 15 Gyr) but with a range in heavy-element abundances (29). Fornax, on the other hand, while also showing a range in heavy-element abundances, seems to show a somewhat greater mean enrichment and has a significant population of intermediate-age stars (20, 39). Thus, there seem to be three important variables, somewhat interconnected, that determine the stellar population of a galaxy: time, abundances, and the star formation rate.

To illustrate population differences for different galaxies, I have chosen to use a three-dimensional representation, as defined in Figure 5, with the three axes being the ages of the stars, their heavy-element abundances, and the star formation rates. For the schematic population box diagrams that are given in the next section, I plot the logarithms of these parameters, with the range in values dictated by observation, as given in Figure 5.

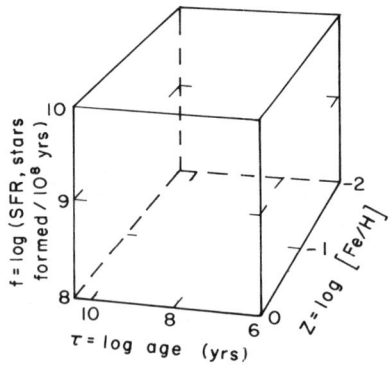

Figure 5 The population box.

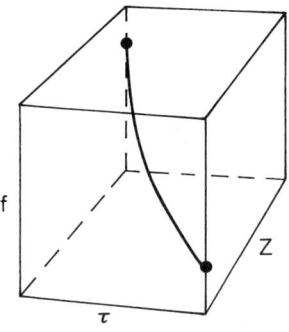

Figure 6 A simple galactic history plotted in the population box.

Figure 6 gives an example of what a very simple galactic history would produce. In this case, the line in the box shows that the galaxy began star formation abruptly about 15 Gyr ago and then formed stars at a steadily decreasing rate as the interstellar raw materials were used up. As time progressed, the chemical abundances of the material in the stars that formed gradually changed as stellar evolution produced increasing amounts of heavy elements. Of course, this is an unrealistic example, because a real galaxy would not be able to maintain a perfectly uniform enrichment history, nor would star formation necessarily proceed perfectly smoothly.

Another simple example is given in Figure 7, which represents the classical view of a Population II galaxy. Most of the star formation took place in an initial burst, and very little heavy-element enrichment had time to occur in this galaxy.

A further simple example of a population box is Figure 8, which represents a somewhat unphysical pure Population I galaxy, which has had a uniform rate of star formation since it was formed, and whose abundances are and have been approximately solar.

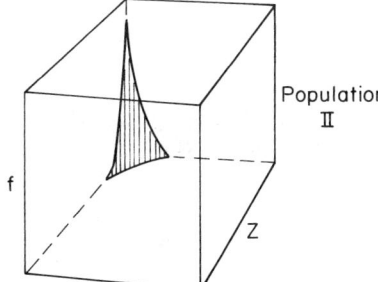

Figure 7 The volume occupied in the population box by a traditional Population II galaxy.

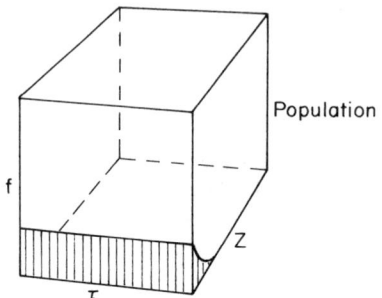

Figure 8 The volume occupied in the population box by a (nonphysical) pure Population I galaxy.

A final example is given in Figure 9, which shows a highly schematic picture of the population characteristics of the MWG. The surface in the box indicates that stars formed in an initial burst at a high rate at low-Z values, and then that star formation continued at a slowly decelerating rate, with Z gradually but not uniformly increasing to the present value. This very uncertain and schematic treatment of the Galaxy is the only inclusion of it in this review; in the following, only the other Local Group members are discussed in any detail.

3. DIAGNOSES FOR LOCAL GROUP GALAXIES

Galaxies of the Local Group are all near enough that rather reliable methods can be used to diagnose what kinds of stellar populations each contains. There are still many uncertainties in detail, but experience has shown that the problem is at least tractable for the local sample, whereas it becomes extremely difficult and uncertain for more distant galaxies, where only integrated properties can be measured [see the excellent dis-

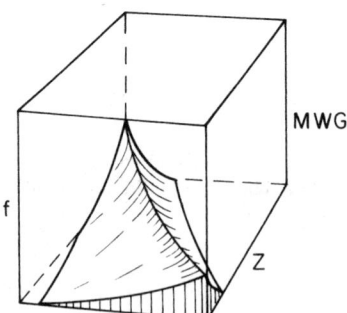

Figure 9 A schematic version of the MWG's volume in the population box.

cussions in Norman et al. (94)]. There is usually no difficulty in ascertaining whether or not a galaxy has any very young stars or uncondensed interstellar material, though even this question becomes tricky when very small amounts are being sought. The problem becomes more difficult as older components are looked for because of the faintness of older stars and because of the ambiguity in age dating them. (It is difficult to distinguish a low-mass old star from a low-mass young star.) Even in the Magellanic Clouds, where we can measure stars down to luminosities fainter than the Sun's, one cannot be certain about how to interpret a given field star color-magnitude diagram (CMD) because of the unknown and probably variable star formation rate (SFR), and, possibly, a variable initial-mass function (IMF).

One method that at least provides fairly reliable age data is the use of star clusters as probes of the galaxy. From good charge-coupled device color-magnitude diagrams it is possible to determine the ages and chemical compositions of star clusters in the Magellanic Clouds and to obtain less accurate data for clusters in the more distant Local Group members, such as M31 and M33, where integrated colors and spectra can be used (with calibration from the MWG and the Magellanic Clouds). But there is no good way to compensate for the fact that star clusters disintegrate in time at a rate that probably depends on both the properties of the clusters and those of the host galaxies. Thus, though it is relatively simple to determine the present star formation rate for stars in clusters in a galaxy, it is not possible to trace the rate back in history farther than a time that is about the mean lifetime of the clusters, which is about 10^8 yr for the MWG (131). Attempts to do this have been made for the Magellanic Clouds (38, 62) and for the dwarfs NGC 6822 and IC 1613 (65). These tracings go back only a fraction of the lifetimes of the galaxies, however, and so we are forced to reconstruct the oldest times from other evidence.

4. CRITERIA FOR THE PRESENCE OF 15-GYR-OLD STARS

There are three reliable criteria that can be used to establish the presence of very old stars in Local Group galaxies. Here I call the oldest stars "15-Gyr-old stars" on the basis of the recent results of globular cluster main sequence fitting, which indicate that most of these stars have ages of 14–17 Gyr (53).

Main Sequence Fitting

So far, main sequence fitting for 15-Gyr-old *clusters* is only possible for the MWG and the Magellanic Clouds. In the latter case, only recently has it become possible to measure the main sequences of such clusters to as faint as two magnitudes or so below the turnoff. For the nearest dwarf elliptical galaxies, main sequence fitting is possible for the system as a whole.

RR Lyrae Stars

From both theoretical arguments and observational evidence (they occur only in truly old clusters in the MWG and the Magellanic Clouds), RR Lyrae stars should be found only among 15-Gyr-old stars. This criterion is sufficient but not necessary; there are bona fide old Galactic globular clusters without RR Lyrae stars, and, of course, very old metal-rich stars may not become normal RR Lyrae stars.

Globular Clusters

The presence of true globular clusters can be used as an indicator that a galaxy has at least some 15-Gyr-old stars. The problem, however, is to identify the clusters as true globular clusters, as a glance at the checkered history of our knowledge of the globular cluster population of the Magellanic Clouds will quickly show. For unresolved clusters, we must rely on colors, structure, and spectra. Use of *UBV* colors alone is not sufficient, as the age relations for this color system are double valued at certain colors. Because of the effect of the horizontal branch, which almost mimics the effect of a hotter main sequence turnoff, it is difficult to tell a 15-Gyr-old cluster from a 0.8-Gyr-old one in the Magellanic Clouds. The addition of another color, such as *R*, can help.

Table 1 gives the results from the application of these criteria to Local Group galaxies. Only 16 have been demonstrated so far to have a population of 15-Gyr-old stars. It is likely, on the basis of less stringent criteria such as the presence of an envelope of faint red stars, that the other members also contain at least a few very old members. A more detailed discussion of this problem can be found in Hodge (68).

5. POPULATIONS IN LOCAL GROUP GALAXIES

In this section, each member of the Local Group is discussed in turn, following the order in Table 1, which is approximately in decreasing order of luminosity. (It is approximate because in some cases the luminosities are still not well determined.) Not included are a few objects for which there is very little information or whose membership in the Local Group is still doubtful.

Table 1 The presence of 15-Gyr-old stars[a]

Galaxy	Main sequence fitting	Sample reference	RR Lyrae stars	Sample reference	Globular clusters	Sample reference
M31			×	101	×	17
MWG	×		×		×	
M33					(×)	23
LMC	×	118	×	93	×	118
SMC	×	119	×	51	×	119
NGC 205					×	62
M32						
NGC 6822					×	68
NGC 185					×	57
IC 1613					(×)	68
NGC 147			×	106	×	60
Fornax			(×)	86	×	20
GR8						
Sculptor	×	29	×	124		
Leo I			(×)	73		
Leo II			×	121		
Ursa Minor			×	123		
Draco			×	15		
Carina			×	100		

[a] ×, 15-Gyr-old stars present, based on listed criterion. Parentheses indicate uncertain identification. Adapted from Hodge (68).

M31

The oldest stars of M31 are detected in its abundant globular clusters (17, 28, 111) and in its halo RR Lyrae variables (101). Mould & Kristian (92) have detected halo giant stars, which are also presumably 15 Gyr old. There is, furthermore, the presumption that the central bulge of M31 [which makes up a large fraction of its luminosity (36)] also contains some old stars, judging from its colors (113) and from various kinds of spectrophotometry (96, 116), but no accurate measurement of the ages of the bulge stars is yet available. As a first approximation, these data indicate that M31 may have formed fractionally more stars in its initial period than did the MWG.

The old stars of the bulge of M31 do show a gradient in color (109, 113) and in spectral line intensities (18, 25), which may result from a range in mean age, in the mean abundances, or in the IMF. In any case, there is plenty of evidence that the bulge contains more than just old, metal-poor stars, as was once believed. This evidence includes the spectral synthesis studies cited above, as well as abundant evidence of young stars (from the

UV excess) (32), gas (24, 78) and dust (49, 64, 115). Deharveng et al. (32) estimate the SFR in the bulge presently to be less than about 7×10^{-5} solar masses per year.

The M31 disk has been studied recently with increasingly better data. Since Baade's (14) pioneering work it has been known that the disk is rich in young stars. Attempts to reproduce its populations have been based on narrowband photometry (122) and on CMDs of variously located disk fields (16, 71, 72, 76, 87). These studies seem to indicate the presence of an old population, but they are unable to distinguish between a 2-Gyr-old population and a 15-Gyr-old one, for instance, and therefore there is uncertainty about the early history of the disk. On the other hand, these and other studies do show that there is considerable star formation going on at present in the disk. Spectra of individual stars (76) and the ratio of the number of carbon stars to M giants (102) both indicate that the chemical abundances in the disk are similar to those in the solar neighborhood.

All of these studies indicate that the populations of stars in M31 are similar to those in the MWG, but that there may be some minor but interesting differences. In particular, there may have been a longer or more intensive period of star formation in M31's earliest history, and there may now be a less vigorous mean rate of star formation than in the MWG. These tentative differences are shown schematically in Figure 10, when compared with Figure 9.

M33

Because relatively little has been published about any but its brightest individual stars, most information about populations in M33 has to come from studies of integrated light and of star clusters. It is obvious from photographs, which show large amounts of high-luminosity stars, H II regions, and dust (77), that M33 is currently experiencing a vigorous rate

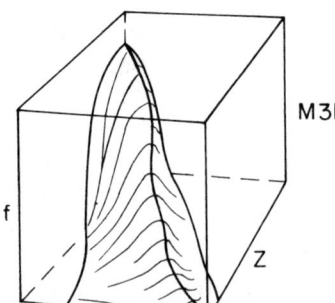

Figure 10 The population history of M31.

of star formation, but the lack of a clear bulge and halo leaves the distant past's star formation rate an open question. Until RR Lyrae stars are discovered and clear examples of old giants are identified, we are limited to what can be learned of old clusters. Recent work on 128 of the 250 known cluster candidates (23) indicates that there is a wide range in the ages of the clusters, from extremely young objects to those that are probably about 15 Gyr old. The age-abundance relation shows a mean that lies close to that of the Large Magellanic Cloud (LMC), intermediate between the MWG and the SMC. There is a spread in their derived values that, if real, would indicate about an order-of-magnitude spread in abundances for a given age. Although part of this spread is probably due to observational errors, some of it may also be intrinsic, which would thicken the figure for M33 in its population box. I have not plotted a figure for M33, however, because of the lack of sufficient information on the variation of the star formation rate with time.

The LMC

There is still a lack of agreement concerning some of the details about the history of the LMC, but there is no doubt about the gross features of the development of its population. It has large numbers of RR Lyrae variables (93, and many other references) and at least a few globular clusters of great age (118, and many other references), so it is clear that there was a significant period of star formation about 15 Gyr ago involving the formation of low heavy-element abundance stars. Subsequent star formation has surely gone on, but different studies find different patterns in the deduced rate of star formation. The views can be roughly separated into two extremes: Either the subsequent star formation rate has been approximately constant, with a gradually increasing heavy-element abundance (26, 66), or else there have been two (or perhaps more) episodes of vigorous star formation, dated at approximately 10^8 yr ago and 4×10^9 yr ago (45, 89, and many others). A review of the situation has been given by Lequeux (85) and is contained in the papers edited by van den Bergh & de Boer (128). Very recent discussions of this issue (37, 38, 43, 79, 105) continue to reach disparate conclusions about both the variation in the star formation rate and the abundance history for the LMC. Therefore, for this galaxy I show in Figure 11 a somewhat muted version of a nonuniform star formation history, in which, after a modest start, the LMC became particularly active a few billion years ago [possibly because of tidal effects, as discussed by Fujimoto & Murai (48)] and is again especially active now. It probably shows an unusually large dispersion in its present heavy-element abundances (88) because of its relatively relaxed dynamics and the consequent slow global mixing of its interstellar material.

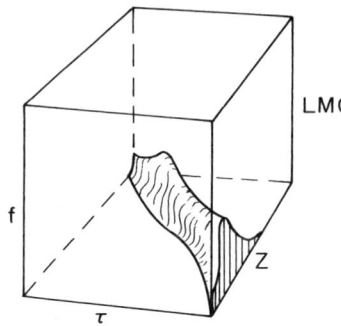

Figure 11 The population history of the Large Magellanic Cloud (LMC).

The SMC

Many of the papers mentioned in the previous paragraph also discuss the evolutionary history of the SMC, which is connected to the LMC by a bridge of H I and probably also by a common set of intergalactic experiences. Most studies indicate a somewhat more homogeneous and more uniform rate of star and heavy-element formation compared with the LMC, but this may be at least partly the result of there being significantly less information for the SMC, which has many fewer old clusters and fewer well-studied young clusters. There is general agreement, however, that the present mean heavy-element abundances in the SMC are smaller than those in the LMC by about a factor of two (98, 105, and many other references). The rate of cluster formation in the two Clouds differs by a factor of 8 (69), which is roughly the ratio of their masses, suggesting that the star formation rates of the two are very similar when normalized by mass.

NGC 205

Although usually classified as an elliptical galaxy, NGC 205, a close companion to M31, has long been known to be anomalous (13). More or less centered on the nucleus is a clear young population of stars, including about 100 resolved O and B stars and several dust clouds (62) and some H I (80). Ultraviolet *IUE* spectra reveal the presence of this young population also (132), as does visual spectrophotometry (104). Underlying it and spread out more widely in space is an old, metal-poor population. Figure 12 schematically shows the probable population history of this galaxy; a large fraction of its mass went into stars in an initial burst of star formation, followed by a small amount of continuing star formation up to the present with an attendant increase in heavy-element abundance. It should be pointed out, however, that much of the detail in Figure 12 is

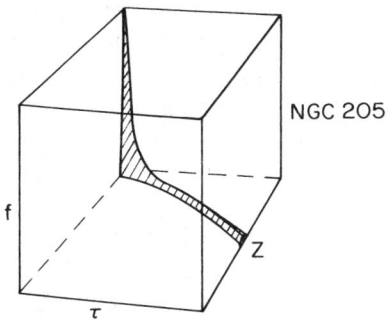

Figure 12 The probable population history of NGC 205.

based more on guess than evidence. Present data do not clearly distinguish between a continuous, slow star formation rate or a sporadic, episodic pattern. Furthermore, the percentage of mass involved in the old population, its abundances, and the duration of its formation are also uncertain.

M32

Because of its brightness and its proximity, a great deal of work has been done on the elliptical galaxy M32, also a companion to M31, going back to the early work of Stebbins & Whitford (117) and even before. It has recently been studied by Faber (41), Pritchet (100), Frogel et al. (47), Wu et al. (133), O'Connell (97), Gunn et al. (54), Burstein et al. (21), Rose (104), and others. All agree that M32 is not just an overluminous globular cluster in its stellar population. Both from spectrophotometry and from integrated colors, it clearly differs from Galactic globular clusters, either metal poor or metal rich. Although at long wavelengths M32 matches 10–15-Gyr-old metal-rich globular clusters, at short wavelengths it is distinctly different, both in its excess blue light and in its line strengths. It is not easy to distinguish between the effects of (a) a small number of young stars, (b) an anomalously large population of blue stragglers, (c) a shorter global age, (d) a strong metal-poor component, or (e) a significant population of intermediate-age stars. The extensive analyses of O'Connell (97) and Rose (104) both argue convincingly for the last of these. Rose (104), for example, finds that most of the light at 4000 Å comes from slightly metal-poor dwarf stars of intermediate age. Thus, although there is a metal-poor old population (contributing perhaps 10% of the light at blue wavelengths), many stars were forming as recently as 10 Gyr ago, though very little star formation has occurred more recently.

NGC 6822

The dwarf irregular galaxy NGC 6822 has a conspicuous population of recently formed and forming stars [see references in Hodge et al. (70) and

Azzopardi et al. (11)]. Color-magnitude diagrams (63, 74, 81), neutral hydrogen measurements (50), H II region studies (70, 75, 82, 99), and the existence of Wolf-Rayet stars (7, 11, 129) all have contributed to our knowledge of the young population of NGC 6822. Most results on the abundances in NGC 6822 suggest that it is just a little more metal poor than the LMC, with $Z = 0.0045$, a value that conforms to its low Wolf-Rayet star population (11).

It is less clear just what kind of old population this galaxy has. There is at least an intermediate population, as shown by the presence of fairly luminous giants, but little has been deduced about the relative numbers of intermediate-age stars or the numbers of very old stars. The oldest clusters (68) suggest that the initial burst of star formation, if there was one, was relatively weak; at the least, it failed to produce more than one or two rather small globularlike clusters.

NGC 185

A distant companion of M31, NGC 185 is quite similar to NGC 205 in almost all respects. Though a little less luminous, it has all of the composite qualities that made NGC 205 so puzzling for so long: a basic elliptical galaxy content of very old stars, plus some O and B stars and dust clouds (13, 60), and a small amount of H I (80). Its star formation history probably looks something like Figure 12.

IC 1613

The low-surface-brightness irregular galaxy IC 1613 is somewhat similar to NGC 6822, though somewhat less luminous. Its stellar content (44, 64, 110), its H I content (84), its H II regions (108), and its enigmatic star clusters (63a, 68, 126) all show that it has an important young population and that star formation is still going on. Its present metals abundance is very similar to that of the SMC ($Z = 0.002$), and the small number of Wolf-Rayet stars (one) is consistent with such a low value (11). Sandage (108), from an examination of deep Palomar 5-m plates, found evidence for an underlying component of very old stars, though it was not possible, of course, from this alone to gauge their actual age. A small number of very faint, old clusters have been reported (63a), but these objects have not been examined further, nor have the clusters in this remarkably cluster-free galaxy been fully confirmed independently (44). Figure 13 shows the probable history of stellar populations in IC 1613, with a nearly uniform, slow rate of star formation and a gradual enrichment of heavy elements.

NGC 147

This close companion of NGC 185 is apparently a "pure" elliptical galaxy. It contains no detected young stars, no dust, and no gas. The brightest

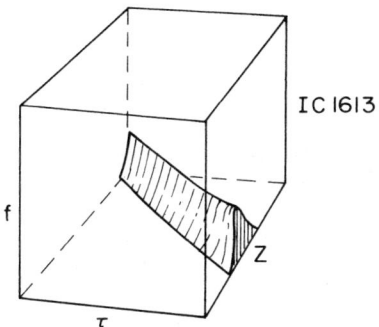

Figure 13 The population history of IC 1613.

stars are red giants (12), and there are large numbers of RR Lyrae stars (106). It appears to have had a burst of star formation about 15 Gyr ago and to have been quiescent ever since.

Fornax

Only a little less luminous than NGC 147, Fornax is the brightest of the Local Group's extreme dwarf ellipticals. It is sufficiently massive that it has its own family of six globular clusters (58), and it marks the boundary between the ellipticals that do (all of those brighter than Fornax) and the ellipticals that do not (all of those fainter). Information on the stellar populations in these galaxies comes from color-magnitude diagrams, variable star studies, and measurements of the characteristics of their carbon stars; see reviews by Aaronson et al. (6), Zinn (136), DaCosta (30), Aaronson (1), and Aaronson & Olszewski (5).

For Fornax, there is evidence of an overall dominance of very old stars [as shown by CMDs published by Demers et al. (35), Buonanno et al. (20), and Gratton et al. (52) and confirmed by the luminosity function derived by Eskridge (39)]. The globular clusters (20, 31, 55, 58, 125, 137) have integrated colors and CMDs that are very similar to those of normal old globular clusters in the Galaxy, which also indicates that a 15-Gyr-old population exists in the Fornax system.

However, it is clear that there are also younger stars in Fornax, as is shown by the width of the giant branch (35), the presence of faint, blue, probable main sequence stars (20), and the existence of large numbers of luminous carbon stars (4, 34, 46, 103, 130). Therefore, the star formation history of Fornax (Figure 14) looks something like that of NGC 205, except that star formation stopped about 3 Gyr ago. The figure shows a widening of the volume, indicative of the spread in metal abundances suggested by the CMDs.

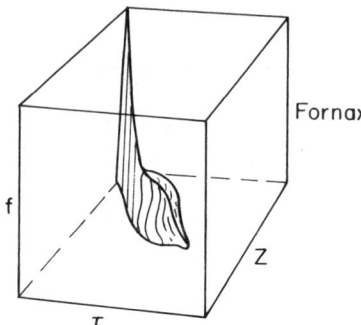

Figure 14 The population history of the Fornax dwarf elliptical.

Sculptor

Smaller and less luminous than Fornax, Sculptor is also somewhat simpler. It contains no globular clusters, but also none of the various indicators of recent star formation (H II regions, O and B stars). Its CMD (29, 59, 83, 95) shows a well-populated giant branch with a main sequence turnoff that DaCosta (29) detects and interprets as indicating either a low helium abundance or a slightly younger age than that of Galactic globular clusters. In addition, there are a few stars on the main sequence above the turnoff; these are either "blue stragglers" like those found in some Galactic globular clusters, or else they are young stars that represent a population similar to but much smaller than the intermediate-age stars of Fornax. Eskridge (40), from a global luminosity function for Sculptor, similarly concluded that there are blue strugglers, and he detected a range in the brightness of the turnoff suggesting that star formation occupied an interval of a few billion years before it stopped.

An additional matter of interest is the chemical inhomogeneity of the Sculptor giants. The spectra of Norris & Bessell (95) showed that either [Fe/H] or [Ca/H] varies, or both. Subsequent panoramic photometry by Smith & Dopita (114) confirmed the inhomogeneity and suggested that CN and Ca variations were correlated, similar to the case for ω Centauri.

Leo I

The two dwarf elliptical galaxies in Leo are both about three times as distant as Sculptor, and consequently they have not been as thoroughly studied. Leo I has had the further disadvantage of lying so close in the sky to the star Regulus that it is difficult to work on without concern for scattered light, or even without saturating whatever detector is being used. A color-magnitude diagram has been published for its giants by Fox & Pritchet (42), who showed that its dominant population is of old, red stars.

Hodge & Wright (73) found a large population of anomalous Cepheids, which they interpreted to indicate a significant fraction of stars more massive than the turnoff. A carbon star was reported by Aaronson et al. (6), and subsequent searches by Azzopardi et al. (9, 10) have turned up an additional 18. As discussed by Richer & Westerlund (103), Aaronson et al. (6), and others, the relative number of carbon stars is correlated with the heavy-element abundances in a galaxy. Leo I's large number of carbon stars goes along with its low abundances. [Suntzeff et al. (120) derive $[Fe/H] = -1.5 \pm 0.25$.]

Leo II

Swope (121) derived an instrumental (uncalibrated) CMD for the galaxy Leo II. A more recent result was published by Demers & Harris (33), who found a moderately wide giant branch that resembles that of Fornax, including an extension of the giant branch to unusually red values ($B-V = 2.2$). Five of the reddest stars were subsequently identified as carbon stars (6, 9). Known variable stars in Leo II are primarily RR Lyrae stars (121), but there are also a few anomalous Cepheids (136). The mean metals abundance is low, with $[Fe/H] = -1.9 \pm 0.25$ (120). Leo II may have had most, if not all, of its star formation occur in a single, early period.

Ursa Minor

The variable stars of the Ursa Minor dwarf elliptical are mainly RR Lyrae stars plus a few anomalous Cepheids (123). Its CMD, first examined in uncalibrated form by van Agt (123), has been investigated by Schommer et al. (112), who found (after eliminating foreground stars by means of proper-motion measurements), that the stars fit a metal-poor globular cluster CMD quite well; the giant branch is blue and steep. However, there is a minor component (approximately 25% of the giants) that scatters to the redward side of the giant branch. Among these is a relatively metal-rich ($[Fe/H] = -1.4$) star for which spectra were obtained by Zinn (135). The giants on the main part of the giant branch showed a lower metals abundance ($[Fe/H] = -2.5$, according to Zinn). The peculiar property of Ursa Minor's CMD is its very blue horizontal branch; otherwise, it is very much like Draco (see below). While there is no evidence that an extended period of star formation went on after most of Ursa Minor's stars first formed, the apparent spread in its abundances gives us reason to believe that its history has not been entirely uneventful since that time. Olszewski's discovery of obviously recently determined structure in the system provides further intriguing evidence of this.

Ursa Minor contains at least one carbon star (6).

Draco

Draco is probably the most thoroughly studied, as well as the most uninteresting, of the Local Group dwarf galaxies. It was the first for which a CMD was obtained [by Baade & Swope (15), who also studied its RR Lyrae variables]. Spectrophotometry by Zinn (134) showed that its abundances were very low, a result that confirmed earlier photometry (8, 22, 27, 56). Aaronson et al. (3) found three carbon stars, and Azzopardi et al. (10) found a fourth candidate. Like Ursa Minor, there seems to be a range of metals abundance in Draco but no evidence that there was an extended period of latter-day star formation.

Carina

The Carina dwarf galaxy is a very different matter. Although there are RR Lyrae variables (107) and carbon stars (91) present, and thus a normal population (including a basic old population) is probably present, most of the galaxy's stars seem to be of intermediate age (90). The age determined from the turnoff is 7.5 ± 1.5 Gyr, and Mould & Aaronson find that the luminosity function does not permit a significant number of very old stars. Thus, Carina seems to be a unique object in the Local Group, an elliptical galaxy that did not form until after the Universe was old, but that only formed stars for a relatively short time. Figure 15 illustrates schematically its probable star formation history: a very slow start, a peak at about half the age of the Universe, and then a rapid decrease to quiescence.

CONCLUSION

The Local Group has a wide variety of population types among its members. Although the early picture of two population types proposed by Baade can very roughly divide the members into two lists, we now

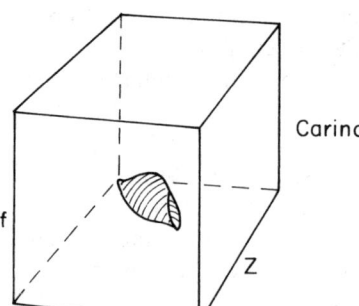

Figure 15 The population history of the Carina dwarf elliptical.

know that there are more than just two variables involved in determining a galaxy's population. The variety provided us by nature, even within our Local Group, will keep us working for some time before all the factors are separated out and understood.

Literature Cited

1. Aaronson, M. 1987. In *Nearly Normal Galaxies*, ed. S. Faber, p. 57. New York: Springer-Verlag
2. Deleted in proof
3. Aaronson, M., Liebert, J., Stocke, J. 1982. *Ap. J.* 254: 507
4. Aaronson, M., Mould, J. R. 1985. *Ap. J.* 290: 191
5. Aaronson, M., Olszewski, E. 1989. In *Evolution of Large-Scale Structures in the Universe, IAU Symp. No. 130*. In press
6. Aaronson, M., Olszewski, E. W., Hodge, P. W. 1983. *Ap. J.* 267: 271
7. Armandroff, T. E., Massey, P. 1985. *Ap. J.* 291: 685
8. Arp, H. 1961. *Science* 134: 810
9. Azzopardi, M., Lequeux, J., Westerlund, B. E. 1985. *Astron. Astrophys.* 144: 388
10. Azzopardi, M., Lequeux, J., Westerlund, B. E. 1986. *Astron. Astrophys.* 161: 232
11. Azzopardi, M., Lequeux, J., Maeder, A. 1988. *Astron. Astrophys.* 189: 34
12. Baade, W. 1949. *Ap. J.* 100: 147
13. Baade, W. 1951. *Publ. Univ. Mich. Obs.* 10: 7
14. Baade, W. 1963. *Evolution of Stars and Galaxies*. Cambridge, Mass: Harvard Univ. Press
15. Baade, W., Swope, H. H. 1961. *Astron. J.* 66: 300
16. Baade, W., Swope, H. H. 1963. *Astron. J.* 68: 435
17. Battistini, P., Bonoli, F., Braccesi, A., Federici, L., Fusi Pecci, F., et al. 1987. *Astron. Astrophys. Suppl.* 67: 447
18. Blair, W. P., Kirshner, R., Chevalier, R. 1982. *Ap. J.* 254: 50
19. Deleted in proof
20. Buonanno, R., Corsi, C. E., Fusi Pecci, F., Hardy, E., Zinn, R 1985. *Astron. Astrophys.* 152: 65
21. Burstein, D., Faber, S., Gaskell, C. M., Krumm, N. 1984. *Ap. J.* 287: 586
22. Canterna, R., Schommer, R. 1978. *Ap. J. Lett.* 219: L119
23. Christian, C. A., Schommer, R. A. 1988. *Astron. J.* 95: 704
24. Ciardullo, R., Rubin, V. C., Ford, H. C., Ford, W. K. 1988. *Astron. J.* 95: 438
25. Cohen, J. 1979. *Ap. J.* 228: 405
26. Cohen, J. 1982. *Ap. J.* 258: 143
27. Cowley, A. P., Hartwick, F. D. A., Sargent, W. L. W. 1978. *Ap. J.* 220: 453
28. Crampton, D., Cowley, A. P., Schade, D., Chayer, P. 1985. *Ap. J.* 288: 494
29. DaCosta, G. S. 1984. *Ap. J.* 285: 483
30. DaCosta, G. S. 1988. See Ref. 53, p. 217
31. Danzinger, I. J. 1973. *Ap. J.* 181: 641
32. Deharveng, J. M., Joubert, M., Monnet, G., Donas, J. 1982. *Astron. Astrophys.* 106: 16
33. Demers, S., Harris, W. 1983. *Astron. J.* 88: 329
34. Demers, S., Kunkel, W. E. 1980. *Publ. Astron. Soc. Pac.* 91: 761
35. Demers, S., Kunkel, W. E., Hardy, E. 1979. *Ap. J.* 232: 84
36. de Vaucouleurs, G. 1958. *Ap. J.* 128: 465
37. Elson, R. W., Fall, M. 1985. *Ap. J.* 299: 211
38. Elson, R. W., Fall, M. 1988. *Astron. J.* 96: 1383
39. Eskridge, P. B. 1987. *Astron. J.* 94: 1564
40. Eskridge, P. B. 1988. *Astron. J.* 95: 445
41. Faber, S. M. 1973. *Ap. J.* 179: 731
42. Fox, M. F., Pritchet, C. J. 1987. *Astron. J.* 93: 1381
43. Frantsman, J. L. 1988. *Astrophys. Space Sci.* 145: 251
44. Freedman, W. 1988. *Astron. J.* 96: 1248
45. Frogel, J. A. 1984. *Publ. Astron. Soc. Pac.* 96: 856
46. Frogel, J. A., Blanco, V. M., McCarthy, M. F., Cohen, J. G. 1982. *Ap. J.* 252: 133
47. Frogel, J. A., Persson, S. E., Aaronson, M., Matthews, K. 1978. *Ap. J.* 220: 75
48. Fujimoto, M., Murai, T. 1984. See Ref. 128, p. 115
49. Gallagher, J. S., Hunter, D. A. 1981. *Astron. J.* 86: 1312
50. Gottesman, S., Weliachew, L. 1978. *Astron. Astrophys.* 61: 523
51. Graham, J. A. 1984. See Ref. 128, p. 37
52. Gratton, R. G., Ortolani, S., Richter, O.-G. 1986. *Mem. Astron. Soc. Ital.* 57: 561
53. Grindlay, J., Philip, A. G. D., eds. 1988. *The Harlow Shapley Symposium*

on Globular Cluster Systems in Galaxies, *IAU Symposium No. 126.* Dordrecht: Reidel
54. Gunn, J., Stryker, L., Tinsley, B. 1981. *Ap. J.* 249: 48
55. Harris, H. C., Canterna, R. 1977. *Astron. J.* 82: 798
56. Hartwick, F. D. A., McClure R. D. 1974. *Ap. J.* 193: 321
57. Hodge, P. W. 1969. *Astron. J.* 68: 691
58. Hodge, P. W. 1965. *Ap. J.* 141: 806
59. Hodge, P. W. 1965. *Ap. J.* 142: 1390
60. Hodge, P.W. 1971. *Annu. Rev. Astron. Astrophys.* 9: 35
61. Deleted in proof
62. Hodge, P. W. 1973. *Ap. J.* 182: 671
63. Hodge, P. W. 1977. *Ap. J. Suppl.* 33: 69
63a. Hodge, P. W. 1978. *Ap. J. Suppl.* 37: 145
64. Hodge, P. W. 1980. *Astron. J.* 85: 376
65. Hodge, P. W. 1980. *Ap. J.* 241: 125
66. Hodge, P. W. 1981. In *Astrophysical Parameters for Globular Clusters, IAU Colloq. No. 68,* ed. A. G. D. Philip, D. S. Hayes, p. 205. Schenectady, NY: L. Davis Press
67. Deleted in proof
68. Hodge, P. W. 1986. *Mem. Soc. Astron. Ital.* 57: 553
69. Hodge, P. W. 1988. *Publ. Astron. Soc. Pac.* 100: 1051
70. Hodge, P. W., Kennicutt, R. C. Jr., Lee, M. G. 1988. *Publ. Astron. Soc. Pac.* 100: 917
71. Hodge, P. W., Lee, M. G. 1988. *Ap. J.* 329: 651
72. Hodge, P. W., Lee, M. G., Mateo, M. 1988. *Ap. J.* 324: 172
73. Hodge, P. W., Wright, F. W. 1978. *Astron. J.* 83: 228
74. Hoessel, J. G., Anderson, N. 1986. *Ap. J. Suppl.* 60: 507
75. Hubble, E. P. 1925. *Ap. J.* 62: 409
76. Humphreys, R. M. 1979. *Ap. J.* 234: 854
77. Humphreys, R. M., Sandage, A. 1980. *Ap. J. Suppl.* 44: 319
78. Jacoby, G., Ford, H., Ciardullo, R. 1985. *Ap. J.* 290: 136
79. Jensen, J., Mould, J. R., Reid, N. 1988. *Ap. J. Suppl.* 67: 77
80. Johnson, D. W., Gottesmant, S. 1983. *Ap. J.* 275: 549
81. Kayser, S. 1967. *Astron. J.* 72: 134
82. Killen, R. M., Dufour, R. 1982. *Publ. Astron. Soc. Pac.* 94: 444
83. Kunkel, W. E., Demers, S. 1977. *Ap. J.* 214: 21
84. Lake, G., Skillman, E. 1988. Preprint
85. Lequeux, J. 1984. See Ref. 128, p. 405
86. Light, R. M., Armandroff, T. E., Zinn, R. 1986. *Astron. J.* 92: 43
87. Massey, P., Armandroff, T. E., Conti, P. S. 1986. *Astron. J.* 92: 1303
88. Mateo, M. 1987. PhD thesis. Univ. Wash., Seattle
89. Mould, J. R., Aaronson, M. 1982. *Ap. J.* 263: 629
90. Mould, J. R., Aaronson, M. 1983. *Ap. J.* 273: 530
91. Mould, J. R., Cannon, R. D., Aaronson, M., Frogel, J. A. 1982. *Ap. J.* 254: 500
92. Mould, J. R., Kristian, J. 1986. *Ap. J.* 305: 591
93. Nemec, J. M., Hesser, J. E., Ugarte, P. 1985. *Ap. J. Suppl.* 57: 287
94. Norman, C. A., Renzini, A., Tosi, M., eds. 1986. *Stellar Populations.* New York: Cambridge Univ. Press
95. Norris, J., Bessell, M. S. 1978. *Ap. J. Lett.* 225: L49
96. O'Connell, R. W. 1976. *Ap. J.* 206: 370
97. O'Connell, R. W. 1980. *Ap. J.* 236: 430
98. Pagel, B. E. J., Edmunds, M. G., Fosbury, R. A. E., Webster, B. L. 1978. *MNRAS* 184: 569
99. Pagel, B. E. J., Edmunds, M. G., Smith, G. 1980. *MNRAS* 193: 219
100. Pritchet, C. J. 1977. *Ap. J. Suppl.* 35: 397
101. Pritchet, C. J., van den Bergh, S. 1987. *Ap. J.* 316: 517
102. Richer, H. B., Crabtree, D. R. 1985. *Ap. J. Lett.* 298: L13
103. Richer, H. B., Westerlund, B. E. 1983. *Ap. J.* 264: 114
104. Rose, J. A. 1985. *Astron. J.* 90: 1927
105. Russell, J., Bessell, M. S. 1988. Preprint
106. Saha, A., Hoessel, J. G. 1987. *Astron. J.* 94: 1556
107. Saha, A., Monet, D. G., Seitzer, P. 1986. *Astron. J.* 92: 302
108. Sandage, A. R. 1971. *Ap. J.* 166: 13
109. Sandage, A. R., Becklin, E. E., Neugebauer, G. 1969. *Ap. J.* 157: 55
110. Sandage, A. R., Katem, B. 1976. *Astron. J.* 81: 743
111. Sargent, W. L. W., Kowal, C. T., Hartwick, F. D. A., van den Bergh, S. 1977. *Astron. J.* 82: 947
112. Schommer, R., Olszewski, E. W., Cudwork, K. M. 1982. In *Astrophysical Parameters for Globular Clusters, IAU Colloq. No. 68,* ed. A. G. D. Philip, D. S. Hayes, p. 453. Schenectady, NY: L. Davis Press
113. Sharov, A. S., Lyutyi, V. M. 1983. *Astron. Zh.* 57: 449
114. Smith, G. H., Dopita, M. A. 1983. *Ap. J.* 271: 113
115. Soifer, B. T., Rice, W. L., Mould, J. R., Gillett, F. C., Rowan-Robinson, M., Habing, H. J. 1986. *Ap. J.* 304: 651

116. Spinrad, H., Taylor, B. J. 1971. *Ap. J. Suppl.* 22: 445
117. Stebbins, J., Whitford, A. 1948. *Ap. J.* 108: 413
118. Stryker, L. L. 1983. *Ap. J.* 266: 86
119. Stryker, L. L., DaCosta, G. S., Mould, J. R. 1985. *Ap. J.* 298: 544
120. Suntzeff, N. B., Aaronson, M., Olszewski, E., Cook, K. H. 1986. *Astron. J.* 91: 1091
121. Swope, H. H. 1967. *Publ. Astron. Soc. Pac.* 79: 439
122. Tinsley, B., Spinrad, H. 1971. *Astrophys. Space Sci.* 12: 118
123. van Agt, S. L. T. J. 1967. *Bull. Astron. Inst. Neth.* 19: 275
124. van Agt, S. L. T. J. 1978. *Publ. David Dunlap Obs.* 3: 205
125. van den Bergh, S. 1969. *Ap. J. Suppl.* 19: 145
126. van den Bergh, S. 1979. *Ap. J.* 230: 95
127. Deleted in proof
128. van den Bergh, S., de Boer, K., eds. 1984. *Structure and Evolution of the Magellanic Clouds, IAU Symp. No. 108.* Dordrecht: Reidel. 425 pp.
129. Westerlund, B. E., Azzopardi, M., Breysacher, J., Lequeux, J. 1983. *Astron. Astrophys.* 123: 159
130. Westerlund, B. E., Edvardsson, B., Lundgren, K. 1987. *Astron. Astrophys.* 178: 41
131. Wielen, R. 1988. See Ref. 53, p. 393
132. Wilcots, E., Böhm-Vitense, E., Hodge, P. W., Eskridge, P. B. 1988. *Bull. Am. Astron. Soc.* 20: 1039
133. Wu, C. C., Faber, S. M., Gallagher, J. S., Peck, M., Tinsley, B. M. 1980. *Ap. J.* 237: 290
134. Zinn, R. 1978. *Ap. J.* 225: 790
135. Zinn, R. 1981. *Ap. J.* 251: 52
136. Zinn, R. 1985. *Mem. Soc. Astron. Ital.* 56: 223
137. Zinn, R., Persson, S. E. 1981. *Ap. J.* 247: 849

A NEW COMPONENT OF THE INTERSTELLAR MATTER: Small Grains and Large Aromatic Molecules

J. L. Puget

Laboratoire de Physique de l'Ecole Normale Superiéure, 24 rue Lhomond, F-75005 Paris, France, and DEMIRM, Observatoire de Meudon, F-92195 Meudon, France

A. Léger

Groupe de Physique des Solides de l'ENS, Université Paris VII, Tour 23, 4 Place Jussieu, F-75251 Paris, France

1. OBSERVATIONAL EVIDENCE FOR VERY SMALL GRAINS

Dust models have been constructed to account for the interstellar extinction curve. Unfortunately, when emission data became available, they were in conflict with the predictions of these models. The aim of this review is to show that the introduction of small grains and large aromatic molecules as a new component of the interstellar matter can solve this conflict.

1.1 Predictions of Standard Dust Models for the Reemission of Absorbed Energy

We call "standard dust models" those models consisting of a set of solid materials with a typical size or size distribution that account well for the average extinction, absorption, and polarization of starlight by dust in the diffuse interstellar medium. Although many such models have been advocated over the years, two models can be considered as representative. The first one is based on pure silicate and pure graphite grains, with a differential size distribution $N(a)$ extending in radius a from 0.01 μm to 0.25 μm with a power-law distribution, $N(a) = a^{-3.5}$. These particles are believed to be produced mostly in the atmospheres of cool stars (Mathis et al. 1977, Draine & Lee 1984). The second model invokes three families of grains with different characteristic sizes. This model gives an important role to a dust component generated in the dense cold interstellar clouds: the organic refractory residue remaining after processing of the molecular mantles that are thought to condense on interstellar grains in dense molecular clouds (Greenberg 1985). For a review on interstellar dust, see Tielens & Allamandola (1987).

The expected infrared (IR) emission of interstellar grains has been computed for both of these models (Mezger et al. 1982, Mathis et al. 1983, Greenberg 1985). In these models all grains are in thermal equilibrium with the interstellar radiation field in which they are embedded. Their heat capacity is large enough so that cooling between the absorption of two photons is negligible. The smallest grains considered in these models have a typical radius of 0.01 μm.

The total energy reradiated in the IR must be equal to the total energy absorbed, and this quantity can be computed directly from the empirically known quantities u_v and $\sigma_H(v)$, where u_v is the spectrum of the interstellar radiation field, and $\sigma_H(v)$ is the absorption cross section of interstellar matter normalized per H atom. We find for this quantity the following:

$$L_{ir,H} = c \int \sigma_H(v) u_v \, dv = 5.7 \times 10^{-31} \text{ W (H atom)}^{-1}. \qquad 1.$$

Mathis et al. (1983) give a slightly lower value [4.2×10^{-31} W (H atom)$^{-1}$]. The difference comes from the uncertainties in the far-ultraviolet (UV) absorption properties of diffuse interstellar dust and in the interstellar radiation field in the same wavelength range. The observed value measured by Boulanger & Pérault (1987) for high-latitude dust is $L_{ir,H} = 6.1 \times 10^{-31}$ W (H atom)$^{-1}$. Here the main uncertainty is in the parts of the spectrum that are not covered at all by the *IRAS* photometric bands: $\lambda < 7.5$ μm, and $\lambda > 120$ μm. The good agreement (within the uncertainties) between the observed value and the predicted ones confirms that the empirical quantities involved are, as expected, rather well known.

What is model dependent, however, is the spectral distribution of the reradiated energy, which depends critically on the temperature distribution of the grains. This temperature distribution in turn depends on the optical properties in the visible and UV parts of the spectrum (where the energy is absorbed) and in the IR (where the energy is reradiated), as well as on the size of the particles. The absorption cross section is proportional to the mass of the particle when the wavelength is much larger than the size of the particle (which is always the case for the reradiated energy but not for the absorbed radiation). For standard dust models, Draine & Lee (1984) find that temperatures range from 17 to 20 K for the graphite particles, and from 15 to 18 K for the silicate particles, over the Mathis et al. (1977) size distribution (MRN size distribution).

The predicted reemitted spectrum by Mathis et al. (1983) for the diffuse Galactic emission is shown in Figure 1. Draine & Anderson (1985) predicted the color ratio of intensities I_v in the *IRAS* photometric bands for optically thin interstellar H I clouds (the so-called cirrus clouds) with an MRN size distribution:

$$\frac{I_v(60 \ \mu m)}{I_v(100 \ \mu m)} = 0.11, \qquad \frac{I_v(25 \ \mu m)}{I_v(100 \ \mu m)} = 2 \times 10^{-6},$$

$$\frac{I_v(12 \ \mu m)}{I_v(100 \ \mu m)} = 1 \times 10^{-13}. \qquad\qquad 2.$$

Obviously, these predictions mean that any significant emission from the interstellar medium at $\lambda < 60$ μm far from sources indicates a fundamental problem with the standard model.

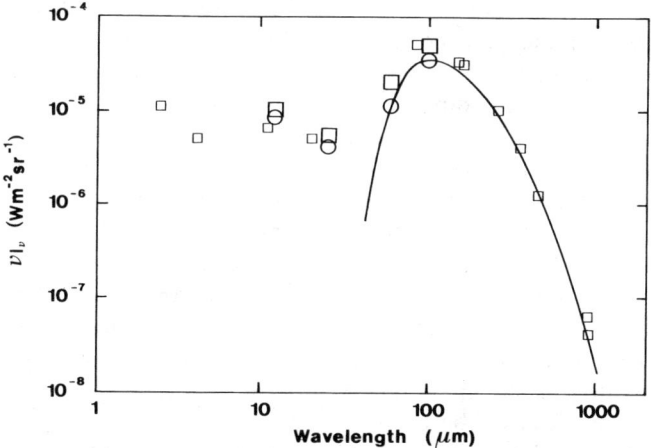

Figure 1 Energy distribution of the Galatic emission for $10° < l < 35°$, taken from Pérault et al. (1988). The squares are the photometric data (see references in Pérault et al.), and the circles indicate the diffuse emission (excluding strong extended sources). The curve is the Mathis et al. (1983) prediction.

1.2 Photometric Data on the IR Emission of the Interstellar Medium: Contradictions with the Predictions of the Standard Models

The first evidence for a significant discrepancy between the spectrum of dust emission and the predictions of the standard dust model was noticed by Andriesse (1978) in the photometric data for the H II region M17. He suggested that the presence of very small particles could account for the data.

Price (1981) mapped the diffuse Galactic emission at 11 and 20 μm. These fluxes, combined with the measurements from balloons and from the ground at shorter and longer wavelengths, lead to the spectrum shown in Figure 1. The *IRAS* data give a more reliable but basically identical overall spectrum. The diffuse Galactic emission at 11 and 20 μm is larger by several orders of magnitude than the predictions of Mathis et al. (1983) for dust emission (Figure 1). Although Mezger et al. (1982) argued that diffuse emission could be due to the integrated emission of cool stars, Caux et al. (1985) noticed that for large H II regions/molecular cloud complexes, the 11- and 20-μm emission correlated very well with the 150-μm emission on scales larger than 100 pc in regions where no hot dust was expected. Since cool stars have no reason to follow the same distribution as the gas, they cannot account for this emission. Pajot et al. (1986) derived the temperature distribution of the dust reradiating the diffuse Galactic IR radiation; in addition to the expected cool component, they found a strong, hot component that could not be explained by dust in the immediate vicinity of stars that had been evaluated by de Muizon & Rouan (1985). The huge discrepancy between the photometric observations of the diffuse Galactic emission and the predictions of the standard dust models around 10 μm became overwhelming with the *IRAS* data.

A parallel development at shorter wavelengths led to a solution to this dilemma. Photometric studies of reflection nebulae at shorter wavelengths (2-5 μm) by Sellgren (1984) not only showed an excess emission at these wavelengths, but also a characteristic independence of the color temperature with distance to the star. This was interpreted by Sellgren (1984) as evidence for emission by grains heated to 1000 K following single-photon absorption. Such emission depends only on the photon energy and the properties of the particles and not on the intensity of the radiation field as long as the particle can radiate most of the energy of the absorbed photon before the next photon is absorbed.

From this idea, Puget et al. (1985) derived a model for the emission of the interstellar matter that predicted a strong excess around 10 μm for the cirrus clouds. Such an excess is actually observed (Boulanger et al. 1985, Weiland et al. 1986, Puget 1988a), as shown in Figure 2.

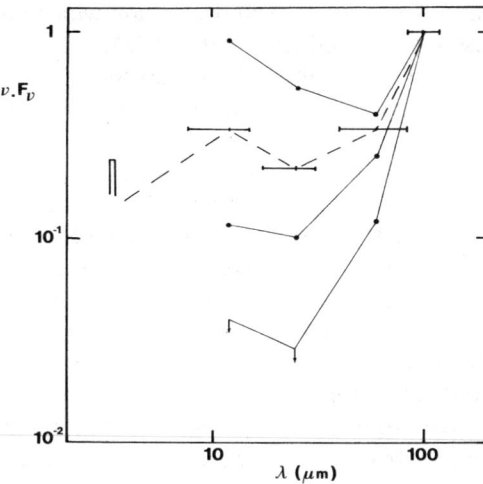

Figure 2 Energy distribution of the cirrus emission, normalized to the 100-μm value. The flux for each of the *IRAS* bands is given (together with the typical width of the filter over which it is measured). The intensity of the feature at 3.3 μm is deduced from the data of Giard (1988) and Giard et al. (1988a,b) for cirrus emission. The energy distributions given in the form of dots connected by thin lines are taken from the work of Boulanger et al. (1988b) on molecular clouds and illustrate the variations of the energy distribution in molecular clouds.

The comparison of the predictions of the standard dust model with the observed $I_v(60\ \mu m)/I_v(100\ \mu m)$ ratio also led Draine & Anderson (1985) to conclude that the dust size spectrum should be extended down to a few angstroms.

In molecular clouds, similar excesses have been found.

The crucial point is that the need for *nonstationary behavior* of the thermal dust emission in the interstellar medium leads to a major revision of the dust models.

1.3 *IR Spectroscopy of Dust Features: Emission and Absorption*

In absorption, many solid-state features are observed in dense regions, and some of them [9.7 and 18 μm (attributed to silicates) and 3.4 μm (attributed to saturated hydrocarbons)] are also seen in the diffuse medium (Willner 1984). However, except for the immediate vicinity of stars, where dust is hot, most of these features have not been observed in emission.

There is only one family of diffuse features observed in many astronomical sources, including the integrated spectrum of galaxies. It is composed of bands at 3.3, 6.2, 7.7, 8.6, and 11.3 μm, which have been referred to since their discovery by Gillett et al. (1973) as the "unidentified" IR features. Duley & Williams (1981) noted that the 3.3- and 11.3-μm bands were characteristic of vibrational transitions of C–H bonds on aromatic carbon atoms.

The fact that these features were also found in reflection nebulae, which showed evidence of transient heating of very small particles (Sellgren 1984), led Léger & Puget (1984) to identify these very small particles with polyciclic aromatic molecules (PAHs). Because alternative explanations (e.g. hydrogenated amorphous carbon material) also attribute them to aromatic species, we shall refer to these bands in the following as the aromatic IR (AIR) bands. A detailed discussion of the comparison between the interstellar and laboratory spectra is given in Section 2.6.

As we have shown above, most of the observational evidence for very small particles comes from photometric data. Nevertheless, the chemical identification rests mostly on the diffuse features. It is thus critical to demonstrate the close link between the IR emission measured photometrically and the AIR bands. This has been done in three cases. Ryter et al. (1987) have shown that a substantial fraction of the flux measured in *IRAS* band 1 (12 μm) in reflection nebulae is accounted for by the emission in the AIR bands and their associated broad pedestals. Similarly, in the starburst galaxy M82, the spectrum obtained by Willner et al. (1977) shows that the nonstellar energy radiated in the 3–15 μm region of the spectrum is not from a continuum but from the AIR bands. The most spectacular demonstration that the particles emitting the features are present everywhere in the interstellar medium, including the cirrus clouds, has been given by Giard (1988) and Giard et al. (1988a,b), who measured the 3.3-μm diffuse Galactic emission with a spectrophotometric balloon-borne experiment that compared the flux collected in a narrow filter centered on the feature with that collected in a broad filter centered at the same wavelength but excluding the feature. They have shown that the 3.3-μm feature is present in the Galactic emission up to a latitude of 6° and has a distribution very similar to the *IRAS* 12-μm emission.

These features have also been observed in planetary nebulae by Gillett et al. (1973). The strong correlation found by Willner (1984) and Cohen et al. (1986) between the fraction of the energy radiated in the 7.7-μm band and the carbon-to-oxygen ratio confirms the basic identification with hydrocarbon compounds.

Table 1 compares the energy in the 3.3-μm band, all the AIR bands, and the total broadband photometric emission, including the far-IR emission for the various sources for which these data are available.

Many near- and mid-IR features have been observed in the spectra of

Table 1 Energy in various sources

	$\dfrac{L_{\text{PAH}}{}^{b}}{L_{\text{total}}}$ (%)	$\dfrac{L_{3.3\,\mu m}}{L_{\text{total}}}$ (%)	$\dfrac{I_\nu(60\,\mu m)}{I_\nu(100\,\mu m)}$	U_ν (eV cm^{-3})
Cirrus^d (5 kpc)	19	0.65	0.27	1.0
Galactic disk^d	13	0.20^a	0.25	—
Cirrus^c (10 kpc)	20	—	0.21	0.12
NGC 2023^e	11.5	1.03	1.5	50
NGC 7023^e	5	0.20	2	100
Orion^f (30′)	—	0.6–1	0.65	1
Orion^g (30″)	4	0.12	2.4	200
M82^h	3.8	0.16^a	1.4	50
NGC 7027^e	6.1	0.20	3.7	700

^{a} Not corrected for absorption.
^{b} Estimated as the radiation at $\lambda < 15\,\mu m$.
^{c} Values deduced from Boulanger & Pérault (1987).
^{d} Values deduced from Pérault et al. (1988).
^{e} Values deduced from Sellgren et al. (1985).
^{f} Values deduced from Giard (1988).
^{g} Values deduced from Sellgren (1981).
^{h} Values deduced from Willner et al. (1977).

sources deeply embedded in molecular clouds. They give evidence for silicate grains, for the formation of mantles through condensation of molecules like CO and H$_2$O, and for the processing of these mantles by cosmic rays and ultraviolet radiation modifying the chemical nature of the condensed material. In the Greenberg (1985) dust model, this mechanism is the basis for the production of the organic refractory material.

The striking fact is *the complete absence of overlap between the emission features and those seen in absorption in the same wavelength range.* For example, the absorption at 3.4 μm is attributed to aliphatic C–H bonds in the organic part of the grain mantles, and according to Tielens & Allamandola (1987) these mantles contain 24% of all interstellar carbon. Nevertheless this feature, when seen in emission, is much weaker than the 3.3-μm emission. On the other hand, the aromatic 3.3-μm feature, which has been shown by Giard (1988) and Giard et al. (1988a,b) to be ubiquitous in the interstellar medium emission, is not seen in absorption. Thus, *there must be a fundamental difference between the physical states of these two closely related chemicals to explain this difference.* The existence of large grains covered with mantles containing aliphatic hydrocarbons on the one hand, and of very small isolated particles chemically dominated by aromatic compounds on the other, accounts in a simple way for this difference. The aliphatic bonds are never seen in emission at short wavelengths because the grains in which they are included are always too cold. The aromatic bonds, on the other hand, are seen in emission because the particles containing them are small enough to be heated to high temperatures by single photons. We see this as a strong argument in favor of a component of *free* aromatic molecules in the interstellar medium, in contrast with models requiring aromatic molecules within hydrogenated amorphous carbon (HAC) grains (Duley & Williams 1988).

2. PHYSICS OF IR EMISSION BY THERMAL FLUCTUATIONS

2.1 *Emission During Temperature Fluctuations*

The idea of temperature fluctuations for small particles was suggested many years ago (Greenberg 1968, Duley 1973, Allen & Robinson 1975, Purcell 1976). However, Andriesse (1978) and Sellgren (1984) were the first to discuss the process in the extreme case of temperature excursions much larger than the average temperature, and they proposed this process as a major source for the near-IR emission of the interstellar matter.

The pioneering paper by Andriesse (1978) explained the 10–100 μm emission of M17 by temperature fluctuations of 10-Å grains up to 150 K. Sellgren (1984) concluded that for reflection nebulae, grains as small as 4 Å containing no more than 50 atoms were emitting with a color temperature of 1000 K in a region where the grain equilibrium temperature could not be over 60 K.

This model could explain a key feature of the observed emission—the nondependence of the color temperature on the distance from the exciting star—but it did not explain the presence of the prominent unidentified IR bands in the spectrum.

2.2 *Molecular Nature of the Very Small Particles: the PAH Model*

Léger & Puget (1984) discussed the nature of these very small particles. Given the necessary resistance of these particles against sublimation when heated, Léger & Puget were able to eliminate ices and silicates as candidates but retained graphitic materials. Since the binding energy between carbon atoms in graphite is very anisotropic, a 50-atom cluster is more likely planar than spherical. The presence of dangling bonds at the periphery of these clusters and of hydrogen atoms in the surrounding gas led Léger & Puget to propose hydrogenated graphite platelets as candidates for the very small particles. Such structures are known in organic chemistry as polycyclic aromatic hydrocarbon molecules (PAHs).

Since a 50-atom cluster lies on the border of the class of objects described by solid-state physics and by molecular physics, either description should be used only after careful consideration of the validity of the approximations involved.

The suggestion that PAHs could be present in the interstellar medium was first made by Platt (1956) and Donn (1968). The molecules were proposed to account for parts of the extinction curve, but the absence of clear spectroscopical identification prevented further development of this hypothesis for many years.

Taking advantage of existing IR absorption data on aromatic molecules, Léger & Puget (1984) calculated the expected emission using the largest measured PAH at that time: coronene ($C_{24}H_{12}$). They found a spectrum with strong bands that impressively resembled the unidentified IR features. This led them to propose the PAH model—i.e. "a *mixture of free* PAH molecules is a major and ubiquitous component of the interstellar matter."

Allamandola et al. (1985) reached the same conclusion by considering the spectra of other PAHs and pointed out that these molecules should be ionized in strongly irradiated regions (e.g. reflection nebulae).

Although PAHs are not identified in the same way as molecules containing only a few atoms, i.e. by using rotational transitions in the radio domain, this model is well enough defined that laboratory measurements and calculations can be performed, leading to predictions that can be observationally tested.

The main alternative explanation for the interstellar near-IR emission is the hydrogenated amorphous carbon (HAC or QCC) grain models proposed by Duley & Williams (1981, 1988) and Sakata et al. (1984). The main difference between these models and the PAH model is not so much the nature of the species (both are aromatic) but their size. In the HAC or QCC model, the grains have the classical size (200–2000 Å). The crucial problem is whether the energy of an incident photon can remain localized in a molecule weakly bound to the rest of the grain during the few seconds required for the IR emission. This would be in conflict with all the failed attempts made in physical chemistry to localize energy in molecules for times longer than 10^{-11} s! (e.g. Hutchinson et al. 1983). This point remains a very serious difficulty with the HAC model. In this review we concentrate on the free PAHs and very small grains (VSGs) hypothesis and do not further discuss the model of large temperature inhomogeneities within amorphous grains. The VSGs considered further (Section 3.2) are defined as particles small enough to fluctuate in temperature when absorbing a single photon but larger than PAHs.

2.3 Energy Redistribution and Statistical Description for a Molecule After One Photon Absorption

Before comparing the PAH model with observations, let us consider the physics of emission by large molecules.

The case of ions is examined because PAHs are expected to be ionized in the regions that are presently best observed (Allamandola et al. 1985, Omont 1986), but the case of neutral species is similar.

The electronic and vibrational levels of a large molecular ion are schematically shown in Figure 3 (Leach 1987). The absorption of a UV photon induces a transition to an upper electronic state. The system then makes rapid (10^{-10} s) isoenergetic transitions to the ground electronic state D_0 with high vibrational energy, where it remains during the cooling, apart from short excursions to the D_1 state.

This evolution toward a vibrationally hot position occurs as a consequence of the ergodic principle because the ground electronic state D_0 has a much higher density of states. In the solid-state physics description, it corresponds to the transfer of the photon energy to the system lattice. The ion is in a state of internal thermodynamical equilibrium because its internal coupling is much stronger than the external one (photoemission).

The hot ion cools subsequently by IR emission [also called IR fluorescence (Allamandola et al. 1987)] and visual fluorescence, including Poincaré fluorescence from the D_1 level (Léger et al. 1988c).

For an isolated system, a specific statistical treatment (microcanonical) should be applied (Allamandola et al. 1985). However, Léger et al. (1988a) have shown that a thermal (canonical) model, based on the concept of *vibrational temperature*, is valid when the energy localized in one mode is small compared with the total energy of the system. For IR emission, it implies that the internal energy of the ion is much higher (typically a factor of 10 or larger) than the energy of emitted IR photons, which is the case during

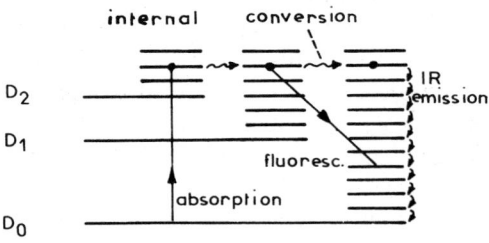

Figure 3 Electronic and vibrational levels of a large molecular ion (X^+ or X^-). Vibrational levels are schematically represented: In reality, they become more and more closely spaced as the vibrational energy increases.

most of the cooling process when the absorbed photon is a UV photon. This condition is not fulfilled for visual fluorescence or dissociation processes.

When valid, the concept of vibrational temperature provides a simple and convenient physical concept and is used in this paper. The relation between the internal energy per vibrational mode of the molecule and its temperature is given in Figure 7 of Léger et al. (1988b).

2.4 Visual and IR Fluorescence From a Hot Molecule

When an isolated ion has absorbed a UV photon, it can emit by visual fluorescence, either 10^{-7} s after the absorption (ordinary fluorescence) or during its recurrent excursions to the D_1 state (Poincaré fluorescence). The latter process is specific to isolated species and can be efficient for small species ($N = 10$–100 atoms). Many reflection nebulae and high-latitude clouds have a red emission excess (Chlewicki & Laureijs 1987, Witt & Schild 1988), which has been interpreted as HAC fluorescence (Duley 1985) or PAH fluorescence (d'Hendecourt et al. 1986, Ryter & d'Hendecourt 1988). This subject is recent, and much work remains to be done. In the case of interstellar PAHs the relative spectral powers observed to be emitted in the visual and the IR indicate that the fluorescence channel is less important ($\leq 10\%$) than the IR one for the energy relaxation. We now derive the emission of hot molecules in the IR regime.

Cherchneff & Barker (1988) have shown direct experimental evidence of IR emission from UV-excited molecules. They have measured the emission profile of the 3.3-μm band in azulene and found it in agreement with expectations. However, absorption measurements are usually easier in the laboratory. Within the thermal model, Léger et al. (1988b) have shown that the emitted spectral power P_λ can be calculated from the absorption cross section σ_λ as

$$P_\lambda = 4\pi B_\lambda(T)\sigma_\lambda(T), \qquad 3.$$

where $B_\lambda(T)$ is the Planck function. Measurements of $\sigma_\lambda(T)$ in conditions similar to the interstellar ones are not yet available because the molecules should be at high temperature, isolated, ionized, and partially dehydrogenated (see below), but most laboratory measurements have been performed at room temperature in condensed phases (molecular solids) for neutral and fully hydrogenated PAHs.[1] This may be a source of discrepancy for the position of the bands and their intensity. However, some measurements have recently been performed on hot and isolated samples

---

[1] By hydrogenated PAHs, we mean aromatic molecules with their normal H coverage and *not* species chemically attacked by H to saturate their carbon double bonds, as is usually meant in chemistry.

(Bernard et al. 1988, Blanco et al. 1988, Léger et al. 1988b), and they do not indicate too drastic a change. Cross sections for a model PAH deduced from laboratory measurements are reported in Table 2 and can be used for astronomical calculations.

The vibrational temperature decreases as the radiation is emitted according to

$$P(T)\,dt = -C(T)\,dT, \qquad 4.$$

where $P(T) = \int P_\lambda(T)\,d\lambda$, and $C(T)$ is the molecular specific heat. $C(T)$ can be calculated in the harmonic approximation when the different vibrational modes of the molecule are known. Explicit functions are given by Omont (1986) and Léger et al. (1988a).

The resulting emission of a PAH molecule in its different IR bands during the cooling process is given in Figure 4 per temperature interval. The cooling time from 1000 to 500 K is 9.6 s.

The initial temperature, or peak temperature T_p, is related to the energy of the absorbed UV photon by $U(T_p) \sim h\nu_{uv}$, where h is the Planck constant, ν_{uv} is the frequency of the absorbed photon, and $U(T)$ is the internal energy of the hot ion, given by

$$U(T_p) = (3N-6)\int_0^{T_p} c(T)\,dT. \qquad 5.$$

Here $c(T)$ is the specific heat per mode of the molecule containing N atoms, so that U is proportional to the number of vibrational modes $3N-6$. This assumes that the internal energy prior to the absorption is negligible. For 10 eV absorbed in a 60-atom PAH, the peak temperature

Table 2 Absorption cross sections of a model PAH (Léger et al. 1988b)[a]

	λ_i (μm)					
	3.3	6.2	7.7	8.6	11.3	>14
σ_i (10^{-21} cm^2)	$35N_H$	$4.1N_C$	$2.9N_C$	$3.0N_H$	$47N_H$	$3.3 \times 10^{-3}(\lambda/\text{cm})^{-1}N_C$
$\Delta\lambda_i$ (μm)	0.04	0.17	0.7	0.4	0.3	—

[a] The positions and integrated cross sections ($\sigma_i\Delta\lambda_i$) are equal to the mean values of the laboratory measurements on several PAHs. The bandwidths ($\Delta\lambda_i$) are equal to the astronomically observed ones and are plausible for a mixture of PAHs. The cross sections σ_i result. N_H and N_C are the number of H and C atoms in the molecule, respectively. For $\lambda > 14$ μm the positions of the bands vary from species to species, and a continuous mean value is adopted.

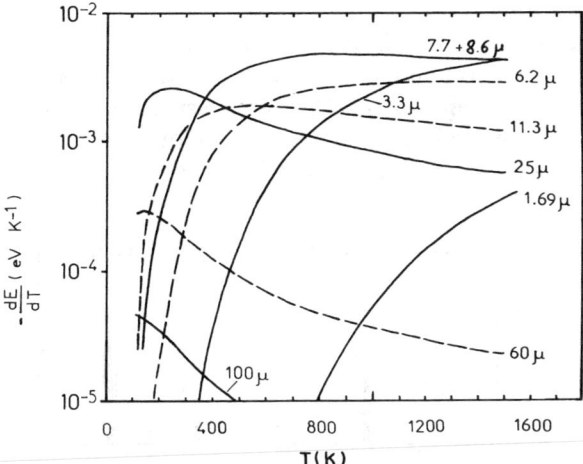

Figure 4 Energy emitted at different wavelengths, per interval of temperature, during the cooling of a 60-atom PAH with a hydrogen coverage factor $x_H = 30\%$. The wavelength indications correspond to the different emission bands of the model PAH (Table 2). The 1.69-μm band is the harmonic of the 3.3-μm one. The 25-, 60-, and 100-μm bands represent the long-wavelength emission of the molecule in the *IRAS* bands. The emitted energy is calculated from Equations 3 and 4, using the values of Léger et al. (1988b). The emission from another size of molecule is obtained by simply scaling by $(3N-6)/174$. The peak temperature T_p from which the cooling starts is obtained from the absorbed photon energy and Equation 5, and it depends on the molecular size.

is 1150 K. Per heat impulse, the total IR emission in a band is the integral of the corresponding curve in Figure 4 from T_p to low temperature.

The emission of a PAH in the interstellar medium is obtained by summing over the photons of the radiation field, taking into account the UV absorption of the molecule. This last quantity has recently been measured in the laboratory (Léger et al. 1988d), yielding a mean energy of the absorbed photons of 9 eV for PAHs in the interstellar radiation field.

However, when the next photon absorption occurs before the system has cooled to a low enough temperature ($T_{initial} \ll T_{peak}$), a multiphoton treatment is required (Désert et al. 1986). This formalism should be used for large species, which are the ones radiating at long wavelengths (e.g. in the 25–60 μm and 100-μm *IRAS* bands) or in high-intensity fields. The limiting case is the quasi-stationary temperature of big grains.

2.5 *Color Temperature and Mean Molecule Size*

An interesting indication of the temperature and size of the emitting PAHs can be obtained from the study of the 11.3- and 3.3-μm bands. These bands are attributed to two modes of the same radical (νCH and γCH), and since they occur at significantly different wavelengths, their intensities will then depend sensitively on the peak temperature, as can be seen in Figure 5. If $E_{3.3, 11.3}$ and $(\sigma\Delta\lambda)_{3.3, 11.3}$ are, respectively, the band-emitted energies and integrated cross sections (proportional to their oscillator strength), a mean color temperature T_c is defined as a solution of the equation

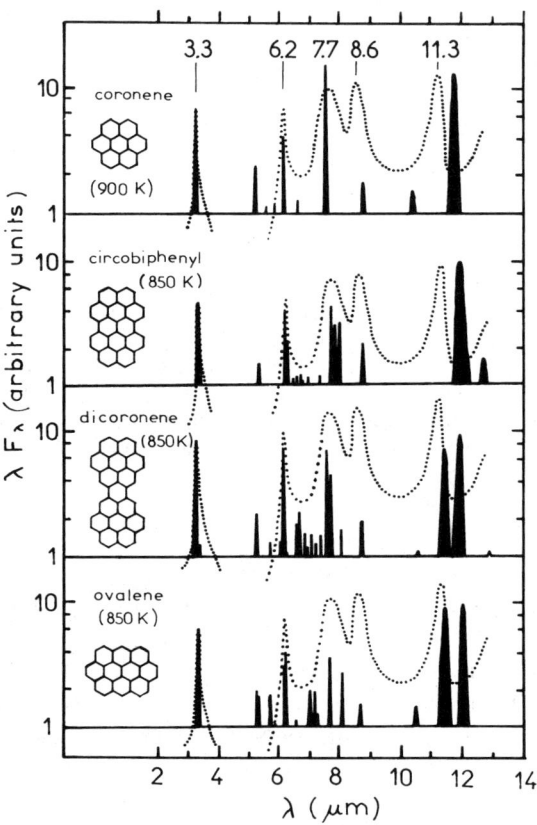

Figure 5 Emission spectra of several compact PAHs calculated from their laboratory-measured absorption spectra (in solid phase) using Equation 3 (from Léger et al. 1988b). The spectrum of the reflection nebula NGC 2023 is reported for comparison (dotted line). It is deduced from observations by Sellgren et al. (1985) except for the 6.2-μm band, which was not resolved and whose shape is assumed to be similar to that of M82 (Willner et al. 1977). The only free parameter in the calculation is the emission temperature, which is adjusted to reproduce the observed ratio of the CH bands at 3.3 and 11.3 μm (adopted T_c is in parentheses). Note the accurate agreement of bands at 3.3 and 6.2 μm between the observations and the compact PAH spectra. The poor wavelength match at 11–14 μm is discussed in Section 2.6.

$$\frac{E_{11.3}}{E_{3.3}} = \frac{B_{11.3}(T_c)}{B_{3.3}(T_c)} \frac{(\sigma\Delta\lambda)_{11.3}}{(\sigma\Delta\lambda)_{3.3}}.$$ 6.

The temperature T_c can be inferred from the observed spectra if the laboratory cross sections are reliably known. This last condition is the weak point of the derivation, as the IR oscillator strengths of the modes are variable and dependent on the surroundings of the radical. They could be different for the interstellar species and their laboratory analogues. The best present estimates are based on the absorption of a PAH (coronene) when isolated in a rare gas matrix. They give values of T_c in the range 500–700 K, as shown further.

A color temperature implies a mean molecule size. For a single molecule size, Figure 4 shows that one T_p corresponds to a value of T_c, giving an energy per mode of $U/(3N-6)$ from Equation 5. On the other hand, the initial energy U is provided by the absorbed UV photon. The number of atoms N in the molecule results. This reasoning is basically the same that led Sellgren (1984) to her estimate of 50 atoms for the VSGs. If one uses $\langle hv_{uv} \rangle = 9$ eV and $T_c = 600$ K in reflection nebulae, a number of 90 atoms per molecule is found. These are really big molecules when compared with the individually identified interstellar molecules. It must be emphasized that this is a mean value for species emitting in the 3–11 μm range, but that *a size distribution is expected*, and that a mean size is meaningless if the whole spectrum is considered.

2.6 *Confrontation With Observations*

In principle, the IR emission resulting from the PAH model can be tested by using spectroscopic data on laboratory analogues of interstellar PAHs, calculating their emission (Section 2.4), and comparing it with the observations. In practice, there are intrinsic difficulties in making precise identifications:

1. On Earth, the natural mixtures of PAHs contain a great variety of species (e.g. Peaden et al. 1980), which suggests an analogous situation in space. So we cannot expect a one-to-one identification of the observed lines, such as is obtained between the rotational lines of CH_3OH and the corresponding features in radio spectra. Fortunately, as opposed to radio data, IR spectra have bands that depend mainly on local atomic groups (e.g. C–C, C=C, C–H, C=O, . . .) and their immediate surroundings (like the other bonds of the carbon atom, for example) and that are common to a whole family of species. This allows one to not only identify the bonds (like C–H, for example) but also to say that the hydrocarbons involved have aromatic carbon bonds and not aliphatic ones, as was pointed out

by Duley & Williams (1981). *If radio spectroscopy were the only tool available to identify molecules, this class of very abundant species (see further) would probably have never been detected* because of their extremely complex rotation spectra and the confusion limit.

2. Present laboratory analogues of interstellar PAHs are only approximate. The situation can be schematically summarized as follows:

INTERSTELLAR	LABORATORY
• $\langle N_C \rangle \sim 80$ atoms	• $N_C < 40$ atoms
• Isolated species	• Molecular solids (most data); isolated (few data—gas phase and rare gas matrix)
• Ionized (expected in most regions where spectroscopic data are presently available)	• Neutral
• Partially hydrogenated ($\alpha_H \sim 30\%$)	• Fully hydrogenated ($\alpha_H = 100\%$)
• Temperatures: 600–1000 K	• 300 K (mostly); 4–70 K (matrices); 500 K (gas phase)
• Emission	• Absorption + calculation; emission (one experiment: Cherchneff & Barker 1988)

Léger et al. (1988b) have recently published absorption spectra of large ($N_C = 24$–38 atoms) PAHs and compared their calculated emission with observations. They distinguish compact, noncompact, and heteroatom-containing subfamilies. Clearly, a better fit is obtained with the compact PAHs than with the other groups (Figures 5, 6). In agreement with other authors (Allamandola et al. 1987), Léger et al. (1988b) identify the observed bands with various C–C and C–H modes.

The similarity between the spectra strongly suggests that the unidentified IR emission bands can be explained by a mixture of compact PAHs. (An exception to this similarity, the 11–14 μm region, is discussed below.) Léger et al. (1988b) find that the differences between the interstellar species and their present laboratory analogues are within the changes expected on the basis of the set of differences listed above. From the point of view of analytic IR spectroscopy, the observations of the *whole set* of the 3.3-, 6.2-, and 11.3-μm bands are *highly characteristic of aromatic molecules.* These observations are much more specific than just indicating the presence of very small particles containing carbon atoms. For instance, Figure 6 shows that the presence, in significant amounts, of oxygen atoms in alde-

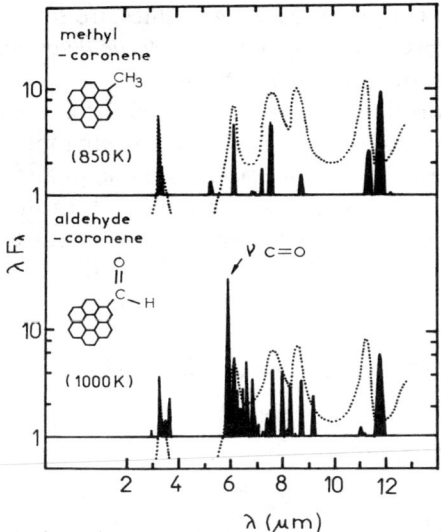

Figure 6 Emission spectra of PAHs with attached radical. The presence of methyl groups in the interstellar mixture appears possible, but that of abundant aldehyde groups is excluded.

hyde groups is excluded because it would give a strong band at 5.9 μm, which is *not* observed.

The 11–14 μm region deserves special attention. It corresponds to out-of-plane C–H bending modes, but the exact position of the band depends upon the number of adjacent CH bonds that are present on the rings: solo (11.0–11.6 μm), duo (11.6–12.5 μm), or trio (12.4–13.3 μm) (Bellamy 1966). The species shown in Figure 5 have mostly duo H, as can be seen in both their spectra and chemical formulae. Interstellar spectra have a main peak at 11.3 μm, although a broader structure at up to 13.5 μm has also been detected (Cohen et al. 1985, de Muizon et al. 1986), indicating *the presence of different hydrogen sites with a predominance of solo*.

This is an argument in favor of dehydrogenation. A simple way to have peripheral hydrogen atoms with no immediate neighbors is to withdraw a large fraction of them at the periphery of a compact PAH.

Examination of Figure 5 indicates that C–H modes are less intense relative to C=C modes in the spectrum of the nebula NGC 2023 than in the laboratory PAH spectra. This also supports the idea of dehydrogenation of PAHs in space.

Quantitatively, the hydrogen coverage x_H [= (H present)/(sites for H)] can be deduced from the intensities of H and C bands if one assumes that dehydrogenation does not upset too much the oscillator strengths of the remaining modes. Hydrogen coverage values are reported in Table 3 for astronomical objects, and they point to a *large dehydrogenation of PAHs* ($x_H = 10$–30%) *in strongly irradiated regions*.

The far-IR absorption has been measured for a few PAHs by Léger et al. (1988b). The spectral features in the 15–150 μm range seem to be specific to each molecule and vary in position from one to another. In the future, satellites such as the *IR Space Observatory* (ISO) will make this spectral range accessible. However, if the interstellar mixture does not contain species with abundance larger than 1%, observable spectral bands are not to be expected.

3. AN INTERSTELLAR DUST MODEL INCLUDING VERY SMALL PARTICLES

3.1 *The Size Spectrum*

Since the Planck function $B_\lambda(T)$ has a rather narrow distribution in λ, there is a simple relationship between the temperature of the grain and the wavelength at which the thermal energy is radiated when the grain cools by $\delta \ln T = 1$ around T.

This energy is of the order $TC(T)$, where $C(T)$ is the heat capacity defined in Section 2. The wavelength λ where the function $\lambda' B_{\lambda'}(T)$ has its maximum is related to T by

Table 3 Color temperature (T_c) and hydrogen coverage (x_H) deduced from laboratory IR cross sections $\sigma\Delta\lambda$ (Table 2, using a H/C ratio of 0.5) and observed emitted energies E_i

		λ_i (μm)			
		3.3	6.2	7.7	11.3
$(\sigma\Delta\lambda)_i/(\sigma\Delta\lambda)_{3.3}$		1	1.0	2.9	10.0
$E_i/E_{3.3}$	(NGC 2023)[a]	1	4.6	9.8	4.2
	(Red Rectangle)[b]	1	4.3	8.1	2.0
	(M82)[c]	1	4.1	17.6	1.8
T_c	(NGC 2023)				600
(K)	(Red Rectangle)				710
	(M82)				730
x_H[d]	(NGC 2023)		28	28	
(%)	(Red Rectangle)		18	19	
	(M82)		18	8	

[a] Values deduced from data in Sellgren et al. (1985) and Cohen et al. (1986).
[b] Values deduced from data in Cohen et al. (1986).
[c] Values deduced from data in Willner et al. (1977).
[d] The two sets of x_H are obtained by fitting the 6.2- and 7.7-μm bands, respectively.

$$\lambda T = 4000 \; \mu\text{m K}. \qquad 7.$$

As a first approximation, the spectrum radiated by a very small particle cooling from the temperature T_p it reaches just after absorbing one photon is given by

$$\lambda f_\lambda \approx T(\lambda) C(T(\lambda)), \qquad 8.$$

for all wavelengths such that $\lambda > \lambda_{\min} = 4000/T_p$, and where $T(\lambda)$ is given by Equation 7. If the Planck function were infinitely narrow, Equation 8 would be exact. Otherwise, it is a good approximation as long as $C(T)$ is not too steep. At $T < 1000$ K, $C(T)$ can be fitted by

$$C(T) = (2.3 \times 10^{-12} T^4 - 6.4 \times 10^{-9} T^3 + 5 \times 10^{-6} T^2)(3N-6)k, \qquad 9.$$

where the temperature T is expressed in Kelvins, and k is the Boltzmann constant. For illustration purposes, we concentrate on cases where the temperature stays below 350 K and $C(T)$ can be fitted by a parabola in T. The spectrum emitted by grains of a given size falls off as λ^{-3} for wavelengths larger than $\lambda_{\min}$. The integral under each of these spectra is proportional to the energy absorbed and thus to the amount of carbon in the particles of each size. The part of the average interstellar spectrum attributed to temperature fluctuations, shown in Figure 1, slowly rises from 3.3 to 60 μm (with a maximum of energy around 10 μm). We now show that the only way to get an overall spectrum λF_λ (after summation over the size distribution) roughly independent of wavelength is to have a comparable amount of carbon in particles of various sizes.

Taking the average energy of the absorbed photons for $U(T_p)$, the maximum temperature T_p is a solution of Equation 5. In the same approximation as above, $U(T_p)$ is given by

$$U(T_p) = 3.3 \times 10^{-22} T_p^3 (a/\text{Å})^3 \; \text{erg}. \qquad 10.$$

We thus get $T_p(a) = 1700(hv)^{1/3}(a/\text{Å})^{-1}$. A more refined calculation leads to the relationship shown in Figure 7. The wavelength at which most of the energy will be radiated by a grain of radius a is given by

$$(\lambda/\mu\text{m}) = 1.13(a/\text{Å}) \quad \text{for} \quad \lambda \gtrsim 10 \; \mu\text{m}. \qquad 11.$$

The energy spectrum from PAHs is then approximately given by

$$\lambda F_\lambda = \frac{d\tilde{N}_\text{C}}{d \ln a} c \bar{v} u_{\bar{v}} \sigma_{\text{PAH,C}}(\bar{v}),$$

where $\bar{v}$ is the average frequency of the absorbed UV photons, $a(\lambda)$ is given by Equation 11, and $\tilde{N}_\text{C}$ is the distribution function of the number of carbon atoms per H atom and per size, or

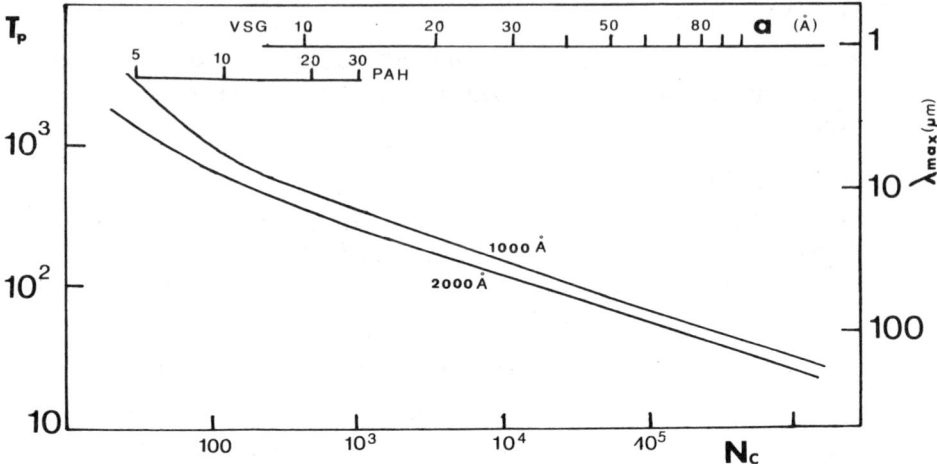

Figure 7 Maximum temperature reached by a very small grain as a function of mass (expressed in carbon atom numbers) after absorption of a single photon. The wavelength at which most of the energy is radiated is also given. Two wavelengths of the incident photon are considered: 1000 Å and 2000 Å.

$$\lambda F_\lambda = 3.4 \times 10^{-27} \frac{d\tilde{N}_C}{d\ln a} \text{ W (H atom)}^{-1}. \qquad 12.$$

The observed spectral distribution of the IR energy is directly related to the size distribution by Equation 12. An example of the size distribution needed to explain the cirrus emission is given in Figure 8.

3.2 Chemical Composition of Very Small Particles

The relationship between the typical grain size and the wavelength at which the grains radiate most of their energy shows that the PAHs, which have been proposed as the family of particles radiating the IR bands between 3 and 12 μm, contain less than $\sim 10^3$ carbon atoms (with maximum radii of 15 Å if they are spherical particles, 30 Å if they are planar molecules). For the very small particles filling the gap between the PAHs and the classical interstellar grains (spherical particles with typical radii of 15–100 Å), we have, so far, no spectral signature in the wavelength region where they radiate: 15–60 μm. We refer to these particles as very small grains (VSGs). They do not necessarily have the same chemical nature as the PAHs.

The absence of silicate features in emission in external galaxies and in reflection nebulae has been used by Désert et al. (1986) to conclude that the VSGs are not silicate grains. A rather natural extrapolation of the

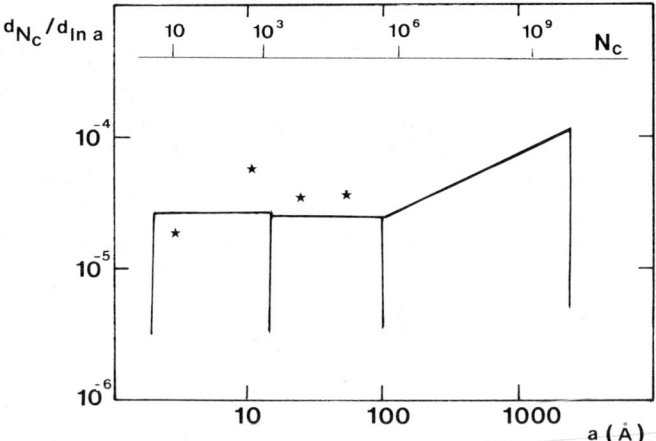

Figure 8 Interstellar carbon particle (from left to right, PAHs, carbonaceous VSGs, and grains) size distribution for the three-component model (discussed in text) as a function of the typical radius a (in Å) for spherical particles; the corresponding number of carbon atoms is also given. The stars give the size distribution from Equation 12 for the cirrus energy distribution.

PAH identification leads to the assumption that the VSGs are partly graphitic grains, but other types of VSGs cannot be excluded. Léger & Puget (1984) argued that graphite particles could best explain the 2–5 μm emission observed by Sellgren (1984) in reflection nebulae because these particles are the most refractory of the chemicals likely to be abundant in the interstellar medium. This argument *does not apply* to the VSGs with radii larger than 15 Å, which never get hotter than 500 K. Metallic or metal oxide particles have been suggested in the past and cannot be excluded. The Magellanic Clouds are an interesting case in this respect, in that the IR spectrum, the UV extinction curve, and the chemical composition of the Clouds are very different from their Galactic equivalents. The very low 12- and 25-μm emission in the Small Magellanic Cloud might be related to the very low carbon abundance in its interstellar medium, in which case the rather strong 60-μm emission should be due to noncarbonaceous VSGs.

For the Galaxy we develop a simple three-component model consisting of (*a*) PAH molecules (planar structure 3–25 Å in radius), (*b*) VSGs [spherical graphite (or HAC) grains with radii from 15 to 100 Å], and (*c*) classical interstellar grains (with an MRN size distribution, but with a reduced amount of large graphite grains to account for carbon locked up in smaller particles).

3.3 Modeling the Extinction Curve

The introduction of PAHs and VSGs to the model of interstellar matter is not just a minor modification aimed at explaining the IR emission features. From the data discussed in Section 1, one finds that 28% of the energy is radiated between 2 and 25 μm, and another 16% is radiated through temperature fluctuations between 25 and 70 μm. Altogether, the classical interstellar grains (particles with radii > 100 Å) radiate (and thus absorb) only 56% of the energy absorbed in the interstellar medium. Thus the standard dust models, in which the big particles account for all the absorption, have to be revised substantially. To account for the energy radiated, almost half of the absorption must be due to PAHs and VSGs.

The absorption properties of various mixtures of PAHs have been measured by Léger et al. (1988b). Even though these are not the interstellar mixture (for example, their average size is too small), such measurements give us a way to estimate the typical absorption cross section of interstellar PAHs.

For PAHs irradiated by the interstellar radiation field (u_ν), we get

$$L_{\text{PAH}} = c \int u_\nu \sigma_{\text{PAH}}(\nu)\, d\nu = 2.24 \times 10^{-27} \text{ W (C atom)}^{-1}, \qquad 13.$$

where $\sigma_{\text{PAH}}(\nu)$ is the absorption cross section normalized per carbon atom.

A similar estimate for the graphitic VSGs can be done using the absorption cross section computed by Draine & Lee (1984) for 30-Å graphite grains, when size effects do not upset the optical constants [i.e. for $\lambda < 800a$ or $\lambda < 2.4$ μm (Puget et al. 1985)]. We then get

$$L_{\text{VSG}} = c \int u_\nu \sigma_{\text{VSG}}(\nu)\, d\nu = 1.5 \times 10^{-27} \text{ W (C atom)}^{-1}. \qquad 14.$$

The energy attributed to PAHs (2–15 μm) can be related to the 12-μm IRAS flux by $L_{\text{PAH}} = 1.1 \nu F_\nu(12\ \mu\text{m})$ using the reflection nebulae spectrum (Sellgren et al. 1985). For the average cirrus (Boulanger & Pérault 1987), we find 1.2×10^{-31} W (H atom)$^{-1}$. This amounts to 20% of the total IR output evaluated as $L_{\text{total}} \sim \Sigma_i \nu_i F(\nu_i)$ on the four IRAS bands.

To account for this energy, we need a carbon abundance in PAHs relative to hydrogen of

$$N_{\text{C in PAH}}/N_{\text{H}} = 5.4 \times 10^{-5},$$

or 15% of cosmic carbon, using C/H $= 3.6 \times 10^{-4}$ (Lang 1980).

If the VSGs were carbonaceous, the fraction of carbon needed to account for the emission through temperature fluctuations between 15 and 70 μm

(23% of P_{total}) would be in conflict with the strength of the 2200-Å feature. We adopt

$$N_{\text{C in VSG}}/N_{\text{H}} = 4.6 \times 10^{-5},$$

or 13% of cosmic carbon, which accounts for about half of this energy.

These results call for two comments: (a) The boundary set at 15 μm between the emission of PAHs and VSGs is not precisely defined, but the total amount of carbon needed to account for the 2–70 μm emission does not depend much on its exact value, as the absorption efficiency is not very different for PAHs and VSGs. (b) There must be VSGs with different chemical compositions to account for about half of the emission between 15 and 70 μm.

The size distribution implied by the abundances of the three-component model is shown in Figure 8. The PAHs and VSGs have been displayed assuming within each size range a size spectrum such that $d\tilde{N}_{\text{C}}/d\ln a$ is constant. The overall distribution is in good agreement with the distribution obtained with Equation 12.

3.4 The Extinction Curve

We can now see how the three components of this model contribute to the Galactic extinction curve. First we compute

$$\sigma_{\text{H,PAH}} = N_{\text{C in PAH}}/N_{\text{H}}\sigma_{\text{PAH}}(v) \qquad 15.$$

and

$$\sigma_{\text{H,VSG}} = N_{\text{C in PAH}}/N_{\text{H}}\sigma_{\text{VSG}}(v). \qquad 16.$$

As the largest of these particles are such that $2\pi a < \lambda$, even at the shortest wavelength of interest (912 Å), we neglect scattering by the very small grains. Both $\sigma_{\text{H,PAH}}(v)$ and $\sigma_{\text{H,VSG}}(v)$ can be subtracted from the extinction cross section of diffuse interstellar matter per hydrogen atom, $\sigma_{\text{H}}(v)$. The remaining part is attributed to the classical interstellar grains having radii larger than about 100 Å. The various contributions are displayed in Figure 9. It is interesting to note the weak frequency dependence of this extinction cross section due to large interstellar grains in the UV part of the spectrum. This is what would be expected from absorbing particles for which $2\pi a > \lambda$.

The other notable feature of this model is that the far-UV part of the extinction curve is due to the smallest particles (PAHs). This is basically because the cross section measured by Léger et al. (1988b) for coronene at 1000 Å is very large: 2×10^{-17} cm^{-2} per carbon atom. (A similar value is found for benzene.) Incoming radiation falling on an infinite graphitic plane from all directions has a probability of 0.2 of being absorbed if this cross section applies. The number of layers in a 30-Å-thick piece of graphi-

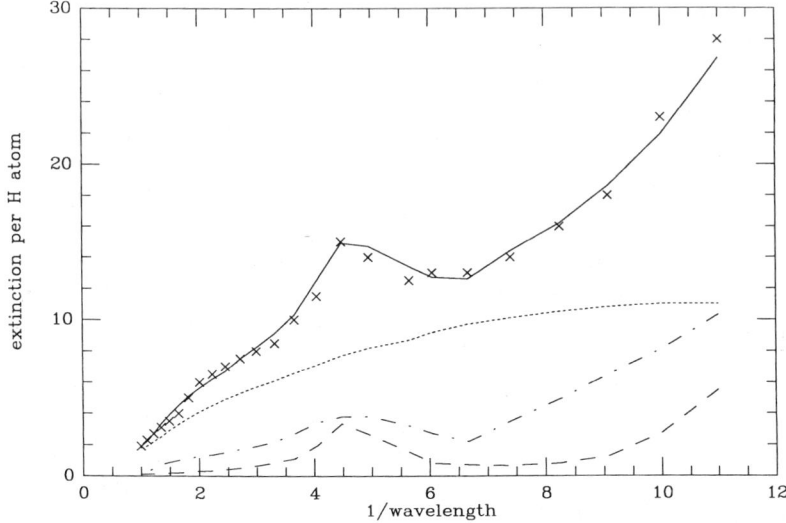

Figure 9 Measured average extinction curve (crosses), expressed as σ_H, as a function of the inverse wavelength (in microns). The dotted line shows the standard grain contribution, the dashed line the carbonaceous VSG contribution, and the dot-dash line the PAH contribution. The sum is shown as the solid line. The unit for the cross section in 10^{-22} cm^{-2}.

tic material is 10, so even a VSG of that size is not optically thin in the far-UV. We thus understand why very small graphitic grains with radii between 15 and 100 Å are less efficient than planar PAH molecules (for the same amount of carbon) in absorbing UV radiation.

It has been shown that the *IRAS* color ratios that are characteristic of the relative amounts of PAHs, VSGs, and large grains change by large factors in molecular clouds and filaments (Beichman et al. 1988, Boulanger et al. 1988b, Puget 1988). These variations cannot be due only to variations of the UV radiation field. They imply that the size spectrum of the very small particles is highly variable in the interstellar medium (Boulanger et al. 1988b).

The three-component model leads to a close connection between the UV extinction curve and the spectrum of the IR radiation emitted. The difficulty in establishing the existence of this correlation is obvious in view of the strong color gradients found in molecular clouds (see Section 1). The UV extinction that is observed in absorption in the spectrum of hot stars is measured only on a limited number of lines of sight with vanishing solid angle. On the other hand, the spectral distribution of the IR emission is derived from the *IRAS* data, which are very sensitive but have a large beam (typically 5′ in the cross-scan direction). Furthermore, the gas close to the star "sees" a stronger radiation field and might dominate the emission, although it does not necessarily dominate the absorption.

Cox & Leene (1987) looked for, but failed to find, a correlation between the shape of the UV extinction curve (both 2200-Å bump strength and far-UV rise) and the IR colors as seen by *IRAS* in the same direction. Leene & Cox (1987) studied the case of another sample of stars from Fitzpatrick & Massa (1986) for which the IR emission and extinction are dominated by matter close to the star. In this case, they find an anticorrelation between the strength of the bump and the 60-μm/100-μm ratio, which they interpret as evidence for destruction of the small particles responsible for the bump by the strong UV radiation field. They did not find any effect on the 12-μm/100-μm ratio. This last result is in contradiction with detailed studies of χ Per (Boulanger et al. 1988a) and σ Sco (Ryter et al. 1987), which show that the 12-μm/100-μm ratio drops in the vicinity of the star (see Section 4.1). The case of σ Sco is especially interesting, because in this case the drop of the 12-μm/100-μm ratio goes together with an absence of steep far-UV rise in the extinction curve, in agreement with the predictions of the model.

In conclusion, the correlation between the spectral distribution of the IR emission and the shape of the UV extinction curve predicted by the three-component model faces conflicting evidence. The difficulties mentioned above and those associated with the subtraction of the zodiacal emission or the Galactic background are such that no answer can be obtained without a detailed study of each line of sight.

3.5 *Model Predictions of the IR Emission of Interstellar Clouds*

Since the heating of very small particles is increasingly dominated by far-UV radiation as the size of the particles becomes smaller, one expects the spectral distribution of IR radiation to depend strongly on the incident radiation field and thus on the depth in the cloud at which the IR radiation is emitted. This expectation was stressed by Beichman et al. (1988) in connection with the observation of the B5 molecular cloud. It is only on the basis of such predictions that one can discuss quantitatively possible abundance variations of PAHs and VSGs in the interstellar medium.

We consider two cases: a molecular cloud with total visual extinction of 5 to the center of the cloud, and an H I cloud with a total visual extinction of 0.22 to the center. Figure 10 shows the expected brightness distribution at 100 μm, as well as the $I_\nu(12\ \mu m)/I_\nu(100\ \mu m)$ ratio for two different incident radiation fields: the interstellar radiation field, and a UV field typical of the heating by a nearby B star [approximated by $vu_\nu = 1$ eV cm^{-3} in the far-UV and UV, dropping in the visible (0.5 eV cm^{-3}), and

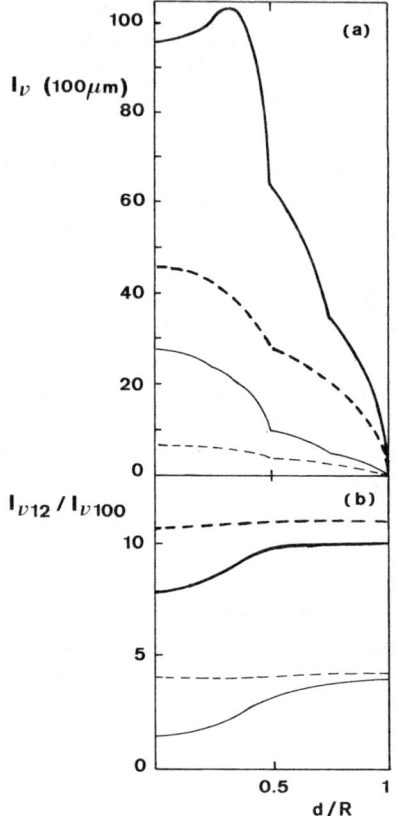

Figure 10 Prediction of the model discussed in the text for the IR emission of interstellar clouds as a function of impact parameter (d) of the line of sight to the center of the cloud. Values expressed in fraction of the cloud radius. The top frame shows the brightness $I_\nu(100\ \mu\mathrm{m})$. The dashed lines refer to a cirrus-type cloud with a visual extinction of 0.22 to the center, and the solid lines refer to a molecular cloud with a visual extinction of 2.1 to the center. The thick lines are for the case of UV heating, and the thin lines are for heating by the interstellar radiation field. The bottom frame shows the ratio $I_\nu(12\ \mu\mathrm{m})/I_\nu(100\ \mu\mathrm{m})$ for the same cases as on the top.

falling to zero in the near-IR]. *The abundances of PAHs and VSGs are assumed constant* and adjusted to fit the average cirrus emission, as discussed in Section 3.1.

The main results apparent in Figure 10 are the following: Heating by the UV radiation leads to a $I_\nu(12\ \mu\mathrm{m})/I_\nu(100\ \mu\mathrm{m})$ ratio twice as large as that due to heating by the interstellar radiation field. In the molecular cloud case, a limb-brightening effect is predicted, similar to what was predicted by Beichman et al. (1988). The $I_\nu(12\ \mu\mathrm{m})/I_\nu(100\ \mu\mathrm{m})$ gradient is stronger for the case of the interstellar radiation field. It decreases monotonically from edge to center, and a minimum is predicted for the line of sight going through the center of the cloud.

Three cases of observations of molecular clouds are reported by Puget (1988). Although the qualitative behavior of the color ratio is in good agreement with the model, as also shown for the B5 cloud by Beichman et al. (1988), there are interesting deviations from the model that prompted Boulanger et al. (1988b,c) to conclude that abundance variations exist for PAHs in molecular complexes.

4. DESTRUCTION AND FORMATION OF PAHs AND VSGs

4.1 *Evidence for Destruction by a Strong UV Radiation Field*

Ryter et al. (1987) and Boulanger et al. (1988a) have shown that the ratio of 12-μm to total IR radiation drops for lines of sight as they approach an O or B star. Figure 3 in Boulanger et al. (1988a) gives a most striking illustration of this phenomenon. Gosh et al. (1986) have shown that the same correlation holds for Galactic H II regions and starburst galaxies.

Using the model presented in Section 3, these color ratios can be converted to PAH abundances in increasing UV radiation fields; this is done in Table 4. There is very strong evidence for the destruction of PAHs radiating in the 12-um *IRAS* band by a strong UV radiation field.

A similar effect can be seen in the anticorrelation between the $I_v(12\ \mu m)/I_v(25\ \mu m)$ and $I_v(60\ \mu m)/I_v(100\ \mu m)$ color ratio, first noticed by Pajot et al. (1986) for galaxies. Helou (1986) confirms this anticorrelation for a much larger sample of galaxies.

Finally as already mentioned in Section 3.4, Leene & Cox (1987) have shown that the strength of the 2200-Å bump anticorrelates with the $I_v(60\ \mu m)/I_v(100\ \mu m)$ ratio. They interpret this as evidence for destruction by strong UV radiation fields of the small particles giving rise to this bump.

We conclude that there is convincing evidence, mostly from the near- and mid-IR emission but also from the UV extinction curve, that very small particles are destroyed by strong UV radiation, and that this destruction affects primarily the smallest particles (the PAHs). There is one exception to this behavior, namely the color ratio observed in carbon-rich planetary nebulae (see Table 1). This case is discussed in the context of the formation of the very small particles in Section 4.3.

Table 4 PAH abundance as a function of UV energy density

UV (eV cm^{-3})	0.15	0.8	1.25	2.2	5	20
$\dfrac{[PAH]}{[PAH]_{cirrus}}$	1	0.90	0.70	0.35	0.20	0.06

4.2 Destruction Mechanisms for Very Small Particles

The main destruction mechanism discussed for standard interstellar grains (see the review by Seab 1987)—namely, erosion by collision with the gas in shocks—is also obviously important for very small particles. The efficiency of this mechanism is reduced by the presence of interstellar magnetic fields, which soften the interstellar shocks. The largest PAHs will be destroyed by shocks on a time scale of a few times 10^8 yr (Omont 1986). Several other destruction mechanisms have also been discussed in the literature. Duley & Williams (1986) and Duley (1985) have considered the chemical interaction of powdered graphite with hydrogen and oxygen to produce CH_4, CO, and CO_2. They argue that photooxidation of PAHs would lead to a rapid destruction of the aromatic structure. Their basic argument is that the thermal spiking at temperatures around 10^3 K allows tunneling through the activation barrier for formation of CO and CO_2. They find that the lifetime of PAHs in diffuse clouds can be as short as 3000 yr and argue that PAHs could be found only very close to where they are formed. However, there is competition between this mechanism and the growth of PAHs by C^+ accretion, as discussed in Section 5. Here we concentrate on destruction by UV radiation, for which there is convincing observational evidence.

The first mechanism to be considered by Puget et al. (1985) and Omont (1986) was photothermodissociation, which leads to a minimum PAH size dependent on the UV radiation field intensity and on the hydrogen density. This mechanism has been reconsidered more recently by Léger et al. (1988a), who point out that (*a*) an appropriate statistical physics treatment must be applied to the isolated system constituted by the hot PAH molecule because it is a regime where the thermal approximation fails completely (Section 2.3), and (*b*) the Poincaré fluorescence can stabilize the smallest particles (see Section 5). A good estimate of the importance of this process is difficult, since there are no reliable values for the minimum sizes allowed by photothermodissociation in various UV environments.

Leach (1986) has suggested another photodestruction mechanism that could be efficient. He shows that the double ionization of a PAH leads with high probability to a Coulomb explosion, where the doubly ionized PAH splits into two singly ionized fragments. Double ionization in two steps can be achieved even with photons of energy less than 13.6 eV for most PAHs containing more than 20 carbon atoms. An interesting suggestion is thus that the combination of photothermodissociation (for the smaller PAHs) and double ionization (for the larger ones) could account for the destruction of PAHs by UV radiation, which seems to be shown by observations. In addition, the Coulomb explosion is expected to be very efficient within H II regions, where hard photons are available. This is in agreement with observational evidence (Aitken & Roche 1983).

4.3 Formation of PAHs and VSGs

Omont (1986) and Tielens et al. (1987) suggest the splitting of graphite grains in shocks as a possible origin for small planar graphitic structures. This is an interesting idea, but it requires observational evidence to be settled. Désert et al. (1986) have proposed a similar mechanism to explain why the 10-μm-emitting particles are graphitic and not silicates. These authors suggest that the difference in behavior of the two types of solids in shocks (easy splitting/three-dimensional cohesion) could be the reason for this selection.

The formation of PAH molecules in the atmosphere of carbon-rich cool stars has been studied in detail by Keller (1987a,b) and Gail & Seldmayr (1987), and their studies show that cyclic molecules can be formed via acetylene polymerization. This theory is in agreement with the low condensation temperature observed in circumstellar shells around late-type carbon stars. Frenklach & Feigelson (1988) have constructed a chemical reaction mechanism for PAH formation. Their results differ from those of Keller (1987a,b), who concludes that PAH growth in circumstellar envelopes proceeds rapidly until extremely large molecules are formed. Frenklach & Feigelson find a PAH mean size of about 40 carbon atoms per molecule, and they claim that carbon-rich circumstellar envelopes could be the source of interstellar PAHs if their acetylene-to-hydrogen ratio and residence time in the 900–1000 K window are sufficiently high.

Direct evidence of PAH condensation has been obtained for Nova Cen 1986, in which the 3.3-μm feature (together with a broader feature at 3.4–3.5 μm) has been seen (Hyland & McGregor 1988). Other direct evidence for PAH condensation comes from the work of Willner (1984) and Cohen et al. (1986) on planetary nebulae. The large ratio of AIR to total IR emission obtained in carbon-rich planetary nebulae (see Table 1) and the correlation with the carbon-to-oxygen ratio give a strong indication that PAHs can condense out as soon as carbon becomes significantly more abundant than oxygen.

The evidence for destruction of PAHs by UV radiation implies that an efficient mechanism exists to replenish the interstellar medium with them. This mechanism cannot be condensation in carbon-rich circumstellar envelopes, because matter is not cycled often enough in such stars. PAH formation could take place in molecular clouds either in the gas phase or in grain mantles. So far, the observational evidence from IR absorption is of little help because the oscillator strength of the 3.3-μm aromatic C–H stretch is smaller by about an order of magnitude than the 3.4-μm aliphatic

C–H stretch. Thus, there is no useful upper limit on the amount of aromatic carbons in the organic mantles formed on grains in molecular clouds. If a significant amount of aromatic carbon is formed, together with the aliphatic carbon that is observed, one would expect that the recycling of interstellar matter from regions shielded from the UV to more diffuse regions where only the most refractory elements survive (including PAHs) might build up the typical abundances of PAHs seen in the interstellar medium. Such a mechanism could help us in understanding the large variations in PAH abundance that seem to be present in molecular clouds.

In diffuse regions the reconstruction of PAHs by C^+ ions probably plays a major role (see Section 5.6).

5. IMPLICATIONS AND OPEN QUESTIONS FOR THE INTERSTELLAR PHYSICS AND CHEMISTRY

If the PAH model is correct, these species are the most abundant free organic molecules. They involve $\sim 15\%$ of the cosmic carbon and process $\sim 20\%$ of the energy absorbed by the interstellar medium. It can be expected that the presence of such a component has major implications in several domains of interstellar physics and chemistry. Some of these are unavoidable consequences of the PAH model (e.g. the contribution to the UV extinction curve), but others are much more uncertain or speculative. In the latter case, PAHs are only candidate solutions for explaining an enigma (e.g. the diffuse interstellar band carriers).

5.1 *Contribution to the Extinction Curve: the Absence of UV Features*

This question is connected with the discussion in Section 3. The absorption of neutral PAHs has been measured in the visible–UV (Léger et al. 1988d) and is strong for $\lambda < 3300$ Å. In the diffuse H I medium, PAHs are expected to be mostly neutral (Omont 1986), so the energy they absorb is the direct product of this absorption by the interstellar radiation field. To obtain 20% of the total energy input, *PAHs must contribute a very substantial part of the UV extinction (e.g. most of the far-UV rise)*. This conclusion was considered to be a problem for the PAH model because of the possible conflict between the sharp bands exhibited in the UV by individual PAHs (Donn 1968) and their absence in the extinction curve. The measurements of PAH mixtures from coal pitches show that the individual bands merge into a rather smooth continuum when numerous species are present; this solves the apparent contradiction and supports the idea of a mixture of many different PAHs in space.

5.2 The Diffuse Interstellar Bands (DIBs)

The DIB part of the extinction curve makes a minor contribution to the energy absorption, but it carries considerable spectroscopical information (~ 50 bands). It is frustrating that these bands have not brought any insight on the interstellar medium more than a half-century after their discovery.

Since the observed bandwidths favor molecules rather than impurities in solids, the appearance of a new class of large interstellar molecules makes them potential candidates for the DIB carriers. This was proposed almost simultaneously by several authors less than one year after Léger & Puget's (1984) paper (van der Zwet & Allamandola 1985, Léger & d'Hendecourt 1985, Crawford et al. 1985).

There are different arguments in favor of PAHs, as reviewed by these authors (see also van der Zwet 1987): (a) They are large molecules, so that the rotational structure of their transitions and the broadening by internal conversion (Douglas 1977) make their line shapes compatible with the observations. (b) They are very stable and can survive the UV radiation field, which gives them a serious advantage over the initial proposition of large organic molecules made by Douglas (1977), who considered long carbon chains. Such chains would be much more sensitive to photodissociation. (c) Their abundance, as derived from the IR, is high enough to account for the observed DIB equivalent widths: If an oscillator strength $f = 1$ is assumed for a 90-atom PAH, the species responsible for the largest DIB (800 mÅ mag^{-1} at 4430 Å) would require only 0.3% of the carbon involved in the total PAH mixture. This abundance is plausible when compared with individual abundances in terrestrial natural mixtures (up to 3% in coal pitches) and is also compatible with the requirements outlined in Section 5.1. (d) PAH ions and radicals (e.g. dehydrogenated species) have strong transitions in the visual (Khan 1988). In the diffuse medium, PAHs are mostly neutral but a sufficient fraction ($\sim 10\%$) is expected to be singly ionized to account for the DIBs. Unfortunately, the presently available laboratory data include an experimental broadening and do not indicate whether the bandwidths in the visual are narrow enough to fit the observations (down to 0.7 Å). On the other hand, the widths of the UV transitions are much larger because the internal conversion broadening increases rapidly with the energy of the transition, and this is in agreement with the absence of sharp structures in the UV extinction.

A difficulty with this proposal is the selection process required to select the few PAH species that could be responsible for the observed bands among the huge variety of these molecules. In principle, it can be imagined that the physics and chemistry of the diffuse medium would give such a selection, but in practice there is a serious difficulty because our poor understanding of the formation and evolution of the PAHs does not permit us to choose the proper species to study in the laboratory. Presently, a spectroscopical fit is clearly the necessary test of any proposition concerning the DIBs.

5.3 Heating of the Interstellar Gas

The heating of the H I interstellar gas used to be a difficult problem. On the one hand, the temperature of the gas and, in a few cases, its cooling mechanism lines (C II 158 μm) are well established (Pottasch et al. 1979), but on the other, none of the proposed heating mechanisms was able to provide enough power. The most attractive mechanism was the photoelectric effect on grains, but to reach the required heating, an unrealistic yield of $\sim 50\%$ was required.

PAHs are attractive for two reasons. First, if their abundance as derived in Section 3 is correct, they are responsible for a major part of the far-UV extinction, which is the most favorable energy range for the photoelectric effect. Second, the *yield for PAHs is naturally higher*. For solids, it is intrinsically low because the absorption length of a photon is ~ 100 Å and the mean free path of electrons of a few eV is ~ 10 Å (Sébenne et al. 1975). Even if *each* photon would produce *one* electron inside the solid, the overall yield would be about 5% because most of the electrons could not escape. Still, much lower values are obtained experimentally. Since PAHs are two-dimensional objects, there is no such problem. Each photoelectron does go away, and the global yield can be high. In fact, the photoelectric yield of PAHs is a measurable quantity in the laboratory. This basic idea was already suggested by Jura (1976) when he proposed that very small grains (~ 50 Å) could have a higher photoelectric yield and supply the required energy.

The suggestion that PAHs could be the major source of heating of the H I gas was first made by d'Hendecourt & Léger (1987) on the basis of cross sections for PAHs estimated from benzene and graphite data. Recent measurements of ionization cross sections performed on large PAHs (pyrene and coronene) by Verstraete et al. (1988) confirm the process. There are still some uncertainties—for instance, on the fraction of the energy that remains in the hot ion and that which goes into the gas as kinetic energy of the outgoing electron. However, the PAH abundance that Verstraete et al. estimated necessary to account for the observed heating corresponds to $\leq 15\%$ of the cosmic carbon, which is compatible with the independent evaluation of Section 3.

An interesting implication and future test of this proposal is the expected correlation between the energy deposited in the gas (as measured by the C II 158-μm line emission) and the energy absorbed by PAHs (as measured by their IR emission).

5.4 Visible Fluorescence

Poincaré fluorescence (Section 2.3) seems to be a very general process, depending almost exclusively on the size of the molecule and the position of its first excited state. For middle-size molecules or ions ($N < 30$ atoms) with a transition at 2 eV, it is quite important and has several consequences. First, it may be the origin of the red luminescence, as already mentioned. Second, the observed ratio of emission in the visual to that in the IR gives a lower limit on the size of abundance species. Third, as it is an efficient channel for relaxing the energy of small excited species, Poincaré fluorescence can reduce the probability of photodissociation and stabilize these molecules.

Laboratory studies are needed of this fluorescence, which is specific to isolated molecules and has important consequences in astronomy.

5.5 Radio Emission

The rotational temperature of PAHs in reflection nebulae is probably in the range of 100 K. The resulting spectrum for a single species is very complicated (rotational level $J \sim 10^2$) and for a mixture of many PAHs is probably confused. However, the integrated emission could be an important source of radiation in the centimetric range (Harwit 1988, Rouan et al. 1988).

5.6 Interstellar Chemistry

In any model of the interstellar matter that includes PAH molecules, the geometric area of this component is larger than that of big grains. Therefore, it is likely to have an important role in the chemistry.

Omont (1986) gave the first detailed review of this question. He pointed out that PAHs, on the basis of their large number of atoms, share several basic properties with large grains, but with radical differences—namely, the clear role of (better-defined) peripheral sites, and the easy desorption of formed molecules due to the large thermal fluctuations.

An initial question is the probability that impinging species will "stick." Because of the rapidly increasing density of states with the vibrational energy of PAHs and the consequent reduction of the "bouncing" probability, Omont suggests no sticking for interaction energy $E_{int} \leq 0.1$ eV (physisorption of close shell species on neutral PAHs) and good sticking for $E_{int} > 1$ eV [e.g. H/PAH$^+$, X^+/PAH^0 (X being any atom), e$^-$/PAH^0, chemisorption].

The role of PAHs in the formation of H_2 and hydrides (CH, OH, etc.) is uncertain because it depends upon whether the atomic species H, C, O can chemisorb on the carbon lattice of PAHs. If they do, the corresponding route becomes dominant for the formation of H_2.

Omont (1986) also studied the formation of radical PAHs by dehydrogenation. He concludes that dehydrogenation is high in strongly irradiated regions, in agreement with indications from presently available IR spectroscopy (Table 3), and is low in diffuse H I regions. This finding suggests interesting predictions for the spectroscopy of the diffuse H I regions when space IR observatories make such spectroscopy possible.

The accretion of O and C^+ on active sites at the molecule periphery is found to be efficient. Oxygen can produce heteroatom species or molecular erosion (combustion) by ejection of CO. There is no spectroscopic evidence of the former (i.e. Figure 6), but the latter was proposed by Duley & Williams (1986) to be such an efficient destruction mechanism that it proved to be a difficulty for the PAH model. However, the accretion of C^+ ions is also efficient and leads to a continuous reconstruction of the molecules. The reaction rate of the latter mechanism is expected to be even higher than that for the former because the cross section for ions (C^+) is larger than for neutrals (O). The net result is in agreement with the ubiquitous evidence for the presence of PAHs in the interstellar medium.

In molecular clouds, PAHs (if not depleted on grains) are expected to bear most of the negative charges because they present a large cross section to electrons. They are also the main sink for metal ions, which, according to Omont (1986), makes irrelevant the "high metal abundance" model discussed by Graedel et al. (1982). Introducing PAHs in chemical reaction routes causes profound changes in the resulting abundances. It favors the production of C-bearing species [e.g. C I and C_3H_2 are enhanced by two orders of magnitude according to Lepp & Dalgarno (1988)].

5.7 Testable Predictions of the PAH Model

The PAH model is specific and leads to predictions that can be used as tests of its validity (as should be the case for any model!). We summarize these predictions here, the first three of which have been verified at the time of this writing.

1. The ubiquitous and surprising 12-μm cirrus emission detected by *IRAS* (Boulanger et al. 1985) was predicted by Puget et al. (1985).
2. These authors also predicted that the 3.3-μm feature should be present everywhere in the diffuse Galactic emission. It has been found by the AROME balloon-borne experiment (Giard 1988, Giard et al. 1988a,b).
3. The study of several PAHs in the laboratory showed the systematic

presence of a band at 5.3 μm (Léger & d'Hendecourt 1987), which has just been detected (Allamandola 1988).
4. The anharmonicity of CH modes is high, and one can expect to observe their harmonics in the astronomical spectra (Barker et al. 1987). Assuming a color temperature of 700 K and a single molecule size, Léger et al. (1988b) have predicted a band at 1.68 ± 0.01 μm with an integrated intensity relative to the 3.3-μm band of 10^{-2} before reddening.
5. With the exception of the observations by AROME, the presently available spectra are of strongly irradiated regions. They often show a dominating 11.3-μm band (solo H). When spectra of regions with low irradiation (cirrus, edges of molecular clouds) are taken, more hydrogen atoms are expected in duo and trio positions with the corresponding 12- and 13-μm structures (Sections 2.6, 5.6). The plateau in emission seen in several H II regions should be even stronger in this case.
6. A correlation is expected between the C II 158-μm line when it dominates the cooling of the interstellar gas and the heating by PAHs as traced by their IR emission (Section 5.3). However, it may be necessary to model this heating in order to account for additional heat sources, as well as the color of the radiation field.

6. CONCLUSIONS

The existence of IR emission through temperature fluctuations following single-photon absorption is well established. It is observed in a variety of regions of the interstellar medium: H II regions, reflection nebulae, planetary nebulae, cirrus clouds, and edges of molecular clouds. The dependence of the distribution of this emission upon UV sources shows with little doubt that UV photons dominate this mechanism.

For most of the interstellar medium, the evidence relies mainly on photometric data. Nevertheless, it is well demonstrated that the set of IR emission features known for many years as the unidentified IR features is part of this emission.

Chemical identification with a family of molecules—compact PAHs—has resulted from a comparison of the features with laboratory measurements made on such molecules. This identification is different from that of individual small molecules by rotational lines. The size of the PAH molecules and their likely large variety make such an identification probably impossible in the radio domain, although they are expected to be the most abundant free organic molecules in the Universe ($\sim 17\%$ of the cosmic carbon). Their identification is based on analytic IR spectroscopy and is specific to this class of species.

Two different models for the IR emission have been advocated based on this (so far unique) chemical identification: (a) PAHs are parts of amorphous carbon grains (but the required localization of the absorbed energy over periods of a few seconds is a very serious difficulty), or (b) PAHs and VSGs are free in the interstellar medium (for which there is no such a difficulty). The second hypothesis leads to a rather smooth continuation of the MRN grain-size distribution toward the small species. This modification of the classical dust model is not minor, as it completely changes the reemitted radiation spectrum.

The size distribution for the smallest particles is variable both in regions of high UV radiation field and in molecular clouds.

Mechanisms able to explain the photodestruction have been studied. Evidence of formation of these particles in planetary nebulae is clear, but whether this is sufficient is still an open question.

Finally, the existence of these particles, which absorb $\sim 40\%$ of the stellar energy reradiated in the IR, has many implications for the physics and chemistry of the interstellar medium.

Literature Cited

Aitken, D. K., Roche, P. F. 1983. *MNRAS* 202: 1233

Allamandola, L. J. 1988. See Allamandola & Tielens 1988. In press

Allamandola, L. J., Tielens, A. G. G. M., eds. 1988. *Interstellar Dust, Proc. IAU Symp. No. 135.* Dordrecht. In press

Allamandola, L. J., Tielens, A. G. G. M., Barker, J. R. 1985. *Ap. J. Lett.* 290: L25

Allamandola, L. J., Tielens, A. G. G. M., Barker, J. R. 1987. In *Interstellar Processes*, ed. D. J. Hollenbach, H. A. Thronson, p. 471. Dordrecht: Reidel

Allen, M., Robinson, G. W. 1975. *Ap. J.* 195: 81

Andriesse, C. D. 1978. *Astron. Astrophys.* 66: 169

Barker, J. R., Allamandola, L. J., Tielens, A. G. G. M. 1987. *Ap. J. Lett.* 315: L61

Beichman, C. A., Wilson, R. W., Langer, W. D., Goldsmith, P. F. 1988. *Ap. J.* press

Bellamy, L. J. 1966. *Infrared Spectra of Complex Molecules.* New York: Wiley

Bernard, J. P., d'Hendecourt, L., Léger, A. 1988. Submitted for publication

Blanco, A., Borghesi, A., Fonti, S., Orofino, V., Bussoletti, E., Colangeli, L. 1988. In *Dust in the Universe*, ed. Bailey, D. A. Williams. Cambridge: Univ. Press

Boulanger, F., Baud, B., van Albada, G. D. 1985. *Astron. Astrophys.* 144: L9

Boulanger, F., Beichman, C., Désert, F. X., Helou, G., Pérault, M., Ryter, C. 1988a. *Ap. J.* In press

Boulanger, F., Falgarone, E., Helou, G., Puget, J. L. 1988b. In preparation

Boulanger, F., Falgarone, E., Helou, G., Puget, J. L. 1988c. See Allamandola & Tielens 1988. In press

Boulanger, F., Pérault, M. 1987. *Ap. J.* 330: 964

Caux, E., Puget, J. L., Serra, G., Gispert, R., Ryter, C. 1985. *Astron. Astrophys.* 144: 37

Cherchneff, I., Barker, J. R. 1988. See Allamandola & Tielens 1988. In press

Chlewicki, G., Laureijs, R. 1987. See Léger et al. 1987, p. 235

Cohen, M., Allamandola, L. J., Tielens, A. G. G. M., Bregman, J., Simpson, J. P., et al. 1986. *Ap. J.* 302: 737

Cohen, M., Tielens, A. G. G. M., Allamandola, L. J. 1985. *Ap. J. Lett.* 299: L93

Cox, P., Leene, A. 1987. *Astron. Astrophys.* 174: 203

Crawford, M. K., Tielens, A. G. G. M., Allamandola, L. J. 1985. *Ap. J. Lett.* 293: L4

de Muizon, M., Geballe, T. R., d'Hendecourt, L. B., Baas, F. 1986. *Ap. J. Lett.* 306: L105

de Muizon, M., Rouan, D. 1985. *Astron. Astrophys.* 143: 160

Désert, F. X., Boulanger, F., Shore, S. 1986. *Astron. Astrophys.* 160: 295

d'Hendecourt, L. B., Léger, A. 1987. *Astron. Astrophys.* 180: L9
d'Hendecourt, L. B., Léger, A., Olofsson, G., Schmidt, W. 1986. *Astron. Astrophys.* 170: 91
Donn, B. 1968. *Ap. J. Lett.* 152: L129
Douglas, A. E. 1977. *Nature* 269: 130
Draine, B. T., Anderson, N. 1985. *Ap. J.* 292: 494
Draine, B. T., Lee, H. M. 1984. *Ap. J.* 285: 89
Duley, W. W. 1973. *Astrophys. Space Sci.* 23: 43
Duley, W. W. 1985. *MNRAS* 215: 259
Duley, W. W., Williams, D. A. 1981. *MNRAS* 196: 269
Duley, W. W., Williams, D. A. 1986. *MNRAS* 219: 859
Duley, W. W., Williams, D. A. 1988. *MNRAS* 231: 969
Fitzpatrick, E. L., Massa, D. 1986. *Ap. J.* 292: 494
Frenklach, M., Feigelson, E. D. 1988. Submitted for publication
Gail, M. P., Seldmayr, E. 1987. In *Physical Processes in Interstellar Clouds*, ed. G. E. Morfill, M. Scholer, p. 275. Dordrecht: Reidel
Giard, M. 1988. Thesis. Univ. Paul Sabatier, Toulouse, Fr.
Giard, M., Pajot, F., Lamarre, J. M., Serra, G., Caux, E., et al. 1988a. *Astron. Astrophys.* 201: L1
Giard, M., Pajot, F., Lamarre, J. M., Serra, G. 1988b. Submitted for publication
Gillett, F. C., Forrest, W. J., Merrill, K. M. 1973. *Ap. J.* 183: 87
Gosh, S. K., Drapatz, S., Peppel, U. C. 1986. *Astron. Astrophys.* 167: 341
Graedel, T. E., Langer, W. D., Frerking, M. A. 1982. *Ap. J. Suppl.* 48: 321
Greenberg, J. M. 1968. In *Nebulae and Interstellar Matter* (*Stars and Stellar Systems*, Vol. 7), ed. B. M. Middlehurst, L. H. Aller, p. 221. Chicago: Univ. Chicago Press
Greenberg, J. M. 1985. In *Birth and Infancy of Stars*, ed. R. Lucas, A. Omont, p. 139. Amsterdam: North-Holland
Harwit, M. 1988. In *Comets to Cosmology*, ed. A. Lawrence, p. 385. Berlin: Springer-Verlag
Helou, G. 1986. *Ap. J. Lett.* 311: L33
Hutchinson, J. S., Reinhardt, W. P., Hynes, J. T. 1983. *J. Chem. Phys.* 79: 4247
Hyland, A. R., McGregor, P. J. 1988. See Allamandola & Tielens 1988. In press
Jura, M. 1976. *Ap. J.* 204: 12
Keller, R. 1987a. Thesis. Tech. Univ. Berlin, Germ.
Keller, R. 1987b. See Léger et al. 1987, p. 387
Khan, Z. H. 1988. *Spectrochim. Acta* 44A: 313

Lang, K. R. 1980. *Astrophysical Formulae.* Berlin: Springer-Verlag
Leach, S. 1986. *J. Electron Spectrosc.* 41: 427
Leach, S. 1987. See Léger et al. 1987, p. 99
Leene, A., Cox, P. 1987. *Astron. Astrophys.* 174: L1
Léger, A., Boissel, P., Désert, F. X., d'Hendecourt, L. 1988a. *Astron. Astrophys.* In press
Léger, A., Boissel, P., d'Hendecourt, L. 1988c. *Phys. Rev. Lett.* 60: 92
Léger, A., d'Hendecourt, L. B. 1985. *Astron. Astrophys.* 146: 81
Léger, A., d'Hendecourt, L. B. 1987. See Léger et al. 1987, p. 223
Léger, A., d'Hendecourt, L. B., Boccara, N., eds. 1987. *Polycyclic Aromatic Hydrocarbons and Astrophysics.* Dordrecht: Reidel
Léger, A., d'Hendecourt, L. B., Défourneau, D. 1988b. *Astron. Astrophys.* In press
Léger, A., Puget, J. L. 1984. *Astron. Astrophys.* 137: L5
Léger, A., Verstraete, L., Dutuit, O., d'Hendecourt, L., Schmidt, W., et al. 1988d. See Allamandola & Tielens 1988. In press
Lepp, S., Dalgarno, A. 1988. *Ap. J.* 324: 553
Mathis, J. S., Mezger, P. G., Panagia, N. 1983. *Astron. Astrophys.* 128: 212
Mathis, J. S., Rumpl, W., Nordsieck, K. H. 1977. *Ap. J.* 217: 425
Mezger, P. G., Mathis, J. S., Panagia, N. 1982. *Astron. Astrophys.* 105: 372
Omont, A. 1986. *Astron. Astrophys.* 164: 159
Pajot, F., Boissé, P., Gispert, R., Lamarre, J. M., Puget, J. L., Serra, G. 1986. *Astron. Astrophys.* 157: 393
Peadon, P. A., Lee, M. L., Hirata, Y., Novotny, M. 1980. *Anal. Chem.* 52: 226
Pérault, M., Boulanger, F., Puget, J. L., Falgarone, E. 1988. Submitted for publication
Platt, J. R. 1956. *Ap. J.* 123: 486
Pottasch, S. R., Wesselius, P. R., van Duinen, R. J. 1979. *Astron. Astrophys.* 74: L15
Price, S. D. 1981. *Astron. J.* 86: 193
Puget, J. L. 1988. See Allamandola & Tielens 1988. In press
Puget, J. L., Léger, A., Boulanger, F. 1985. *Astron. Astrophys.* 142: L19
Purcell, E. M. 1976. *Ap. J.* 206: 685
Rouan, D., Léger, A., Omont, A., Giard, M. 1988. See Allamandola & Tielens 1988. In press
Ryter, C., d'Hendecourt, L. 1988. Submitted for publication
Ryter, C., Puget, J. L., Pérault, M. 1987. *Astron. Astrophys.* 186: 312
Sakata, A., Wada, S., Tanabé, T., Onaka, T. 1984. *Ap. J. Lett.* 287: L51

Seab, C. G. 1987. In *Interstellar Processes*, ed. D. J. Hollenbach, H. A. Thronson, p. 491. Dordrecht: Reidel

Sébenne, C., Bolmont, D., Guichar, G., Balkanski, M. 1975. *Phys. Rev. B* 12: 3280

Sellgren, K. 1981. *Ap. J.* 245: 138

Sellgren, K. 1984. *Ap. J.* 277: 623

Sellgren, K., Allamandola, L. J., Bregman, J. D., Werner, M. W., Wooden, D. H. 1985. *Ap. J.* 299: 416

Tielens, A. G. G. M., Allamandola, L. J. 1987. In *Interstellar Processes*, ed. D. J. Hollenbach, H. A. Thronson, p. 397. Dordrecht: Reidel

Tielens, A. G. G. M., Seab, C. G., Hollenbach, D. J., McKeen, C. F. 1987. *Ap. J. Lett.* 319: L109

van der Zwet, G. P. 1987. See Léger et al. 1987, p. 351

van der Zwet, G. P., Allamandola, L. J. 1985. *Astron. Astrophys.* 146: 76

Verstraete, L., Léger, A., Dutuit, O., d'Hendecourt, L., Défourneau, D. 1988. See Allamandola & Tielens 1988. In press

Weiland, J. L., Blitz, L., Dwek, E., Hauser, M. G., Magnani, L., Rickard, L. J 1986. *Ap. J. Lett.* 306: L101

Willner, S. P. 1984. In *Galactic and Extragalactic Infrared Spectroscopy*, ed. M. F. Kessler, J. P. Phillips, p. 37. Dordrecht: Reidel

Willner, S. P., Soifer, B. T., Russell, R. W., Joyce, R. R., Gillett, F. C. 1977. *Ap. J. Lett.* 217: L121

Witt, A. N., Schild, R. E. 1988. *Ap. J.* 325: 837

INTERACTION BETWEEN THE SOLAR WIND AND THE INTERSTELLAR MEDIUM

Thomas E. Holzer

High Altitude Observatory, National Center for Atmospheric Research, Boulder, Colorado 80307

1. INTRODUCTORY REMARKS

Studies of the interaction between the solar wind and the interstellar medium have been carried out with varying degrees of intensity over more than three decades (e.g. early work by Davis 1955, 1962, Parker 1961, 1963, Axford et al. 1963; reviews by Axford 1972, 1973, Fahr 1974, Holzer 1977, Thomas 1978, Lee 1988). The intensity of activity in this field has been determined primarily by the availability of relevant observations, and a recent surge of activity has been generated by the suggestion of a possible indirect detection of the shock transition terminating the supersonic solar wind flow (Kurth et al. 1984, 1987, Lee 1988), by the consequent realization of the possibility that a deep space probe may soon cross this terminal shock, and by recently increased activity in the observational study of the local interstellar medium (e.g. Frisch 1986, Cox & Reynolds 1987). It thus seems an appropriate time for a critical review of the observational and theoretical work on which our current understanding of this subject is based.

The basic dynamical interaction between the solar wind and the interstellar medium involves the relaxation toward pressure equilibrium between the solar and interstellar magnetized plasmas. This interaction leads to the formation of a cavity in the interstellar medium carved out by the solar plasma, which we refer to as the heliospheric cavity, or simply the heliosphere. It is not difficult to determine what heliospheric and interstellar parameters are likely to be important in the interaction between the solar wind and the interstellar medium, and in Section 2 we provide an overview of the observationally inferred values of these parameters, including the uncertainties in the inferences. Then, in Section 3 we examine from a theoretical point of view the basic physical processes that are likely to be important in this interaction, and we conclude in Section 3.8 by combining this theoretical information with the observational information of Section 2 in an effort to develop the currently most likely overall picture of the heliosphere that is shaped by the local interstellar medium.

2. OBSERVATIONAL OVERVIEW

We begin with an overview of currently available observational inferences concerning (*a*) the relevant properties of the solar wind in its asymptotic flow regime, (*b*) the nature of the very local interstellar medium (VLISM), and (*c*) the various modes of interaction between the solar wind and the VLISM. The degree of uncertainty associated with these observational inferences varies greatly, with the inferences concerning the in-ecliptic solar wind being the least uncertain and those concerning the VLISM, particularly the interstellar magnetic field, being the most uncertain. In the present section we simply note the degrees of uncertainty, and in the following sections we discuss the implications of these uncertainties for our understanding of the interaction between the solar wind and the interstellar medium. Some observations that are relevant to the problem at hand are not considered in this section, but these are discussed at an appropriate point in Section 3.

2.1 *The Distant Solar Wind*

As noted in Section 1 and discussed in detail in Section 3, the basic dynamical interaction between the solar wind and the VLISM involves the relaxation toward pressure equilibrium between the solar and interstellar magnetized plasmas. The principal solar wind parameter controlling this interaction is the ram pressure of the supersonic flow, ρu^2, where ρ is the mass density, and u the flow speed of the solar wind. In contrast, the penetration of interstellar hydrogen atoms into the heliosphere and their effects on the solar wind in the regions of supersonic and subsonic flow are largely controlled by the solar wind proton flux density $n_p u$ (where n_p is the proton density), the solar wind flow speed u in the supersonic region, and the solar wind temperature T in the subsonic region. Thus, the three directly observable solar wind parameters in which we are most interested are $n_p u$, u, and ρu^2 in the supersonic solar wind.

The region of supersonic solar wind flow is generally organized (except near the maximum of the 11-yr solar activity cycle) by the dipole component of the solar magnetic field, with relatively low-speed wind flowing near the dipole equator and relatively high-speed wind flowing at higher magnetic latitudes (e.g. Hundhausen 1977, Hundhausen et al. 1981). Throughout much of the 11-yr cycle (namely, the minimum and postminimum phases of the cycle) the dipole axis is nearly coincident with the solar rotation axis, but during the declining (postmaximum) phase of the cycle the dipole axis is tilted significantly (some 30°) with respect to the solar rotation axis. During this latter period the nonalignment of the rotational and magnetic axes gives rise to a strong interaction between high-speed and low-speed solar wind at low and middle solar latitudes (e.g. Hundhausen 1977, Pizzo 1986, Burlaga 1988). Finally, near the maximum of the solar activity cycle, the organizing dipole structure largely disappears, and relatively slow wind flows over most solar latitudes and longitudes (e.g. Sime 1983, Kojima & Kakinuma 1987).

In situ solar wind observations (e.g. Feldman et al. 1977, Schwenn 1983) indicate that the wind parameters of particular interest to us are well organized by solar wind speed, so we should be able to infer reasonable average values for these parameters ($n_p u$, u, and ρu^2) for all solar latitudes and all phases of the solar activity cycle. (The complicating effects of solar wind stream interactions at low and middle solar latitudes during the declining phase of the solar cycle are discussed in Section 3.) The major uncertainty we confront in determining these values arises from the uncertainty in the absolute measurement of proton density (Feldman et al. 1977). Although the relative accuracy of density measurements by a given instrument is quite good, the absolute (systematic) error may be more than 30% (Feldman et al. 1977), as is evidenced by the intercalibration of the HELIOS and IMP8 plasma instruments (Schwenn 1983). With this difficulty in mind, we summarize the average values of our three solar wind parameters (cf. Table 1) appropriate to low-speed and high-speed wind, and we indicate the smallest and largest values of the two density-dependent parameters ($n_p u$, ρu^2) that seem consistent with the expected uncertainties in density determination. The largest values are taken from Feldman et al. (1977), and the smallest are 20% smaller (at both low and high speeds) than those given by Schwenn (1983) and correspond to the indirect inferences of solar wind proton flux density drawn from Ly-α backscatter observations (e.g. Lallement et al. 1985). Note that the values in Table 1 apply to a heliocentric distance of $r = r_E = 1$ AU; in the supersonic flow beyond 1 AU, u can be assumed constant, and both $n_p u$ and ρu^2 can be assumed to vary as r_E^2/r^2.

2.2 *The Very Local Interstellar Medium (VLISM)*

Our knowledge of the interstellar medium very near the solar system derives to some extent from observations of the interstellar medium (ISM)

Table 1 Average values of solar wind parameters in high-speed and low-speed flows at $r = r_E = 1$ AU

Wind parameter	Low-speed flows	High-speed flows
u (km s^{-1})	330	700
$n_p u$ (cm^{-2} s^{-1})[a]	3.9×10^8	2.7×10^8
$n_p u$ (cm^{-2} s^{-1})[b]	2.6×10^8	1.6×10^8
ρu^2 (dyn cm^{-2})[a]	2.1×10^{-8}	3.2×10^{-8}
ρu^2 (dyn cm^{-2})[b]	1.7×10^{-8}	1.9×10^{-8}

[a] Largest expected average values (from Feldman et al. 1977).
[b] Smallest expected average values: 20% below those given by Schwenn (1983), as inferred by Lallement et al. (1985); note the small variation of ρu^2 (Steinitz & Eyni 1980). Uncertainties associated with electrostatic analyzer observations (Feldman et al. 1977, Schwenn 1983) may soon be substantially reduced through the analysis of Faraday cup observations during the same time period (Lazarus & Belcher 1988; A. J. Lazarus, private communication, 1989).

over very long lines of sight (namely, over distances of much more than 100 pc), but more useful information is gained from observations restricted to the local interstellar medium (LISM), which we arbitrarily define to lie within 100 pc of the Sun. Of course, the most useful information for our purposes would come from observations of the VLISM, in which the heliosphere is immersed, and which we take to lie within 0.01 pc of the Sun. Unfortunately, our observations of the VLISM are confined to the interstellar neutral gas and Galactic cosmic rays, both of which can penetrate deeply into the heliosphere, but both of whose characteristics may be significantly modified during this penetration.

The general picture of the LISM that has emerged over the last several years is thoughtfully discussed by Cox & Reynolds (1987), and it is recommended that their review be read in conjunction with the present one. We can summarize this picture of the LISM by referring to Figure 1, which is adapted from Figure 2 of Cox & Reynolds (1987). It appears that the Sun is currently located in a volume called the Local Bubble, whose dimensions range from some 70 pc across in the Galactic plane to some 300 pc across perpendicular to the Galactic plane. This volume is filled predominantly by a hot, low-density, X-ray-emitting plasma characterized by an electron density of about 5×10^{-3} cm^{-3} and a temperature of about 10^6 K. The Sun, however, is immersed in a small-scale (a few parsecs or less) feature, called the Local Fluff, that has a much higher density, ($\sim 10^{-1}$ cm^{-3}) and a much lower temperature ($\sim 10^4$ K) than the Local Bubble. It is not known whether the Local Fluff is an equilibrium or nonequilibrium structure characteristic of the Local Bubble or of an expanding Loop I Bubble (cf. Figure 1), which is thought to be the remnant of a supernova explosion.

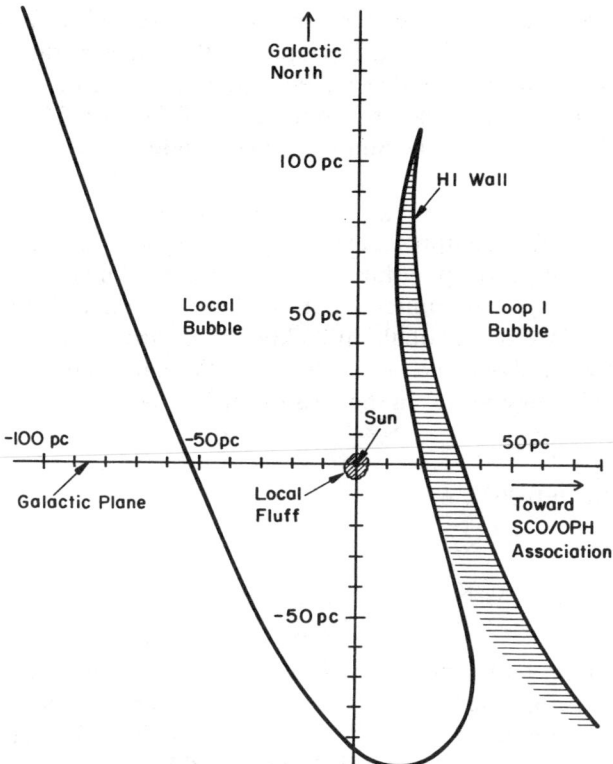

Figure 1 A cross section of the Local Bubble in the Galactic meridional plane, the right half of which corresponds to a Galactic longitude between 330° and 350° (based on Figure 2 of Cox & Reynolds 1987). The shaded region around the Sun is the Local Fluff, in which the heliosphere is thought to be immersed. The other shaded region represents a wall of neutral hydrogen separating the Local Bubble from the Loop I Bubble.

We can, however, place significant constraints on the parameters characterizing the various components of the VLISM, and perhaps of the Local Fluff as a whole.

2.2.1 NEUTRAL ATOMIC COMPONENT OF THE VLISM   Neutral atoms in the interstellar medium penetrate relatively freely into the heliosphere (e.g. Patterson et al. 1963, Hundhausen 1968, Blum & Fahr 1970, Holzer & Axford 1971), and solar UV radiation resonantly scattered by these atoms has been observed from space over the past two decades (e.g. early papers by Bertaux & Blamont 1971, Thomas & Krassa 1971, Paresce & Bowyer 1973; reviews by Tinsley 1971, Fahr 1974, Thomas 1978; recent papers by Chassefière et al. 1986, 1988a,b, Ajello et al. 1987, and references therein).

Extensive analyses of backscattered radiation observed in the resonance lines of H I $\lambda 1216$ and He I $\lambda 584$, coupled with models of the penetration of interstellar neutrals into interplanetary space, have led to the following inferred properties of the neutral component of the VLISM. The velocity of the VLISM relative to the Sun has a magnitude

$$u_I = 23 \pm 2 \text{ km s}^{-1}$$

and is directed toward approximately 74.5° ecliptic longitude and $-7.5°$ ecliptic latitude (corresponding to 175° Galactic longitude and $-21°$ Galactic latitude). This implies a motion of the neutral VLISM relative to the local standard of rest of about 20 km s^{-1} toward a Galactic longitude of about 120° and very nearly in the Galactic plane. The H I and He I densities and temperatures in the VLISM are

$$n_H = 0.10 \pm 0.03 \text{ cm}^{-3},$$

$$n_{He} = 0.010 \pm 0.005 \text{ cm}^{-3},$$

$$T_H = (9 \pm 2) \times 10^3 \text{ K},$$

$$T_{He} = (9 \pm 2) \times 10^3 \text{ K}.$$

The relatively large uncertainties in the hydrogen and helium densities reflect conflicts among different observational inferences, rather than the uncertainty estimates attached to any of the individual inferences (which are generally considerably smaller). Evidently, the ratio n_{He}/n_H is consistent with the cosmic value of [He]:[H] = 0.1 for a rather large range of ionization fractions (cf. Blum et al. 1980, Meier 1980, Weller & Meier 1981), so inference of the ionization fraction of the VLISM from UV backscatter observations must await reduction of the density uncertainties. Values of $u_I = 25$ km s^{-1}, $n_H = 0.12$ cm^{-3}, and $T_H = 1.15 \times 10^4$ K inferred for the Local Fluff (Frisch 1986) lie at the high end of the ranges given above for the VLISM, and at present these two sets of observations cannot be considered inconsistent. Of course, there is no reason to believe that average values for the Local Fluff should correspond exactly to VLISM values, because the VLISM makes up only a tiny fraction of the Local Fluff volume.

2.2.2 IONIZED, THERMAL COMPONENT OF THE LOCAL FLUFF Since we cannot presently draw useful inferences concerning the VLISM ionization state from UV backscatter observations, we must rely on line-of-sight average values of the electron density deduced for the Local Fluff from observations of nearby stars. The results of such observations are summarized by Frisch et al. (1987) as a prelude to their discussion of a specific study of the Local Fluff through observation of Mg I $\lambda 2852$ in the direction of α Oph. If it is assumed that in a warm gas the Mg ionization balance between the first two stages is determined by photoionization of Mg I and dielectronic recombination of Mg II, then it is possible to estimate the absorption in Mg I $\lambda 2852$ that would be produced by the Local Fluff with a given electron density. Frisch et al. (1987) conclude that an appropriate upper limit on the electron density in the Local Fluff is $n_e \lesssim 3 \times 10^{-3}$ cm^{-3}, which is a factor of 5 smaller than the theoretical predictions of McKee & Ostriker (1977) when applied to a gas with temperature 10^4 K and density 0.1 cm^{-3}. Of course, the VLISM is not necessarily in the same ionization state as the Local Fluff as a whole, but it seems that the results of Frisch et al. (1987) provide a strong indication of a low VLISM electron density. For the purpose of the calculations in Section 3, we take

$$n_e < 0.1 n_H.$$

2.2.3 THE VLISM MAGNETIC FIELD   The magnetic field is, perhaps, the most difficult of the interstellar parameters to determine. A variety of methods have been used in attempts at this determination, including the analysis of pulsar rotation and dispersion measures, the detection of the Zeeman effect, and the comparison of Galactic synchrotron emission with the Galactic cosmic-ray electron spectrum at high energies (e.g. review by Heiles 1976; more recent papers by Thompson & Nelson 1980, Brown & Chang 1983, Troland & Heiles 1986, and references therein). The most attractive result of such studies is the correlation of magnetic field with gas density suggested by Brown & Chang (1983), but unfortunately the method used in obtaining this correlation seems to be statistically invalid (Troland & Heiles 1986). Troland & Heiles (1986), using the data of Thompson & Nelson (1980) and employing reasonable assumptions about the "background" and "fluctuating" components of the Galactic magnetic field, conclude that a field of about 5 μG is most likely to characterize a warm, low-density gas like the Local Fluff. The uncertainties in this estimate, however, are rather large, and we shall take the VLISM magnetic field to be

$$B_I = 5 \pm 3 \ \mu G.$$

The parameter in which we are most interested is, of course, the magnetic pressure (cf. Section 3), and it is evidently highly uncertain, with its lower and upper bounds differing by more than an order of magnitude.

2.2.4 GALACTIC COSMIC RAYS   Galactic cosmic rays make a significant contribution to the energy density of the ISM and thus might be expected to play an important role in the interaction between the solar wind and the

VLISM. We can observe Galactic cosmic rays directly with interplanetary spacecraft-borne detectors (e.g. Webber 1987), but the observed cosmic-ray energy spectrum is significantly modified at low energies through interaction with the expanding solar wind, so there is some uncertainty concerning the contribution to the total Galactic cosmic-ray pressure by the low-energy particles (<300 Mev nucleon^{-1}). Although it has been suggested that the low-energy cosmic rays could provide a pressure as large as 6×10^{-12} dyn cm^{-2} (Suess & Dessler 1985), a rather thorough analysis of the problem by Axford & Ip (1986; see also Axford 1976, Ip & Axford 1985) has demonstrated that this is likely to be a substantial overestimate. Axford & Ip (1986) have considered the possibilities that we are either far from or near to (or immersed in) a supernova remnant (namely, that associated with the Loop I Bubble discussed above), and they have concluded that it is highly probable that low-energy cosmic rays make only a small contribution to the total cosmic-ray pressure. Consequently, we assume the following for our calculations in Section 3:

$$p_{cr}(\text{total}) \approx (1.3 \pm 0.2) \times 10^{-12} \text{ dyn cm}^{-2},$$

$$p_{cr}(<300 \text{ Mev nucleon}^{-1}) \approx (3 \pm 2) \times 10^{-13} \text{ dyn cm}^{-2}.$$

2.2.5 INTERSTELLAR DUST We assume (e.g. Greenberg 1978) that dust in the VLISM has a typical grain radius of 0.05 μm, which is appropriate for metallic grains, and that the dust-to-gas ratio by mass is 0.01, leading to a mass density for dust of

$$\rho_d \approx 2 \times 10^{-27} \text{ g cm}^{-3}.$$

2.3 Interaction Between the Solar Wind and the VLISM

The only relatively direct observational evidence we have of the interaction between the solar wind and the VLISM involves the penetration of interstellar neutral atoms and Galactic cosmic rays into interplanetary space, which we discussed above (cf. Sections 2.2.1, 2.2.4). The observation most relevant to the basic dynamical wind/VLISM interaction would be detection of the shock front expected to characterize the transition from supersonic to subsonic solar wind flow. As yet, there has been no certain detection of this shock, but there are three lines of observational evidence that are currently thought to point to a location of the shock just outside the orbit of Pluto.

The first of these involves the detection, beginning in late 1983, of radio frequency signals near 2 and 3 kHz by the *Voyager 1* and *2* plasma wave instruments (Kurth et al. 1984, 1987, Kurth 1988). At the time of first detection, the ambient solar wind plasma frequency was below the observed frequency range, so the signals could be interpreted as resulting from freely propagating radio waves in the outer heliosphere. Before the separate spectral peaks at 2 and 3 kHz were resolved, Kurth et al. (1984) suggested the possible interpretation that these radio waves were generated by the shock terminating supersonic solar wind flow at twice the downstream plasma frequency. A plasma frequency between 1 and 1.5 kHz corresponds to a downstream electron density between 0.012 and 0.028 cm^{-3} and (for a strong shock) to an upstream density between 0.003 and 0.007 cm^{-3}. Extrapolating this density range back to 1 AU, assuming a constant solar wind flow speed, we see from Table 1 that it corresponds to a range of shock distances 34 AU $< R_s <$ 63 AU for low-speed flows and to a range 18 AU $< R_s <$ 36 AU for high-speed flows.

The existence of two separate spectral peaks with a frequency ratio of 1.6 (eliminating the possibility of harmonic emission) complicates the interpretation. One suggested resolution (Lee 1988) is that the two frequencies correspond to twice the plasma frequencies upstream and downstream of a shock [modified by cosmic-ray pressure (Lee & Axford 1988, Drury 1988)], with a density jump of 2.56 rather than 4. This interpretation leads to shock distances of 25 AU $< R_s <$ 31 AU for low-speed flows and 14 AU $< R_s <$ 18 AU for high-speed flows. These values are a bit too small, since a shock in these distance ranges would already have been crossed by deep space probes. Another possible interpretation of the two spectral peaks (Kurth et al. 1987) involves a modification of the first (Kurth et al. 1984) interpretation through the assumption of an asymmetric termination shock, with the two different frequency bands originating from two different regions of the shock; we return to a discussion of this possibility in Section 3. Finally, McNutt (1988) has recently suggested that the radio emissions are triggered by two anomalous high-speed solar wind streams, and he places the shock distance in the range 70 AU $< R_s <$ 140 AU. The detailed arguments underlying this hypothesis will have to be worked out before its viability can be evaluated.

A second line of evidence indicating a not-too-distant termination shock is based on a combination of observations of the cosmic-ray anomalous component (e.g. Cummings & Stone 1988) and theoretical models of the acceleration of the anomalous component (Pesses et al. 1981, Fisk 1986, Jokipii 1986; L. A. Fisk, unpublished work, 1982) through the injection and first-order Fermi acceleration of interstellar pickup ions (cf. Section 3.6) at the termination shock (see the review by Lee 1988). The cosmic-ray anomalous component is thought to arise from the ionization of interstellar neutral atoms, which implies that it consists of energetic singly charged atoms, in contrast to the highly charged atoms making up most of the cosmic-ray spectrum. If the anomalous component does originate from interstellar neutral atoms that are ionized in the solar wind, then hydrogen (which is not detected as part of the anomalous component, presumably because of obscuration by Galactic cosmic-ray protons) should dominate the anomalous component pressure. Observations, together with the inferred presence of hydrogen, indicate that the anomalous component pressure is increasing sufficiently rapidly out to 20 AU that it is likely to be comparable to the solar wind ram pressure inside 50 AU if the anomalous component is accelerated at the termination shock [Cummings & Stone 1988 (these authors do, however, present evidence that the anomalous component pressure may not increase as rapidly outside 20 AU as it does nearer the Sun)]. Evidently, if this is the case, the shock location must be $R_s < 50$ AU.

Another inference of a termination shock inside 50 AU comes from observations of H I $\lambda 1216$ radiation scattered by interstellar hydrogen atoms that have penetrated into interplanetary space. The *Pioneer 10* Ly-α data apply to the downstream direction (with respect to the interstellar wind) in which the spacecraft is traveling, and it shows an anomalously rapid decrease in backscattered Ly-α beyond about 39 AU (Wu et al. 1988). This decrease is interpreted as a possible signature of a nearby termination shock, beyond which the relatively high-density, high-temperature subsonic solar wind flow produces a more rapid ionization of interstellar H atoms. It is not clear, however, how the relatively sharp drop in Ly-α intensity beyond 39 AU can be consistent with a shock-related decline in hydrogen density, which we would expect to be relatively gradual. At present, therefore, we resist the temptation to take this UV observation as an indication of a termination shock inside 50 AU.

3. MODELS OF THE INTERACTION BETWEEN THE SOLAR WIND AND THE INTERSTELLAR MEDIUM

The dynamical interaction between the solar wind and the interstellar medium involves the relaxation toward pressure equilibrium between the solar and interstellar magnetized plasmas. This relaxation is characterized principally by the transition from supersonic to subsonic solar wind flow (presumably a shock transition, which we refer to as the terminal shock) and by the turning of the subsonic solar wind flow to achieve compatibility with the local interstellar structure. As mentioned in Section 1, the location of the terminal shock is currently of particular interest because of the prospect of one or more spacecraft crossing it in the near future. In attempting to determine the most likely location for the terminal shock, not only must we consider the ram pressure of the supersonic solar wind and the pressure of the interstellar medium into which the wind is expanding, but we must also examine the structure of the postshock subsonic solar wind flow.

Basic to our study of this dynamical interaction is the understanding of mass and momentum balance in a magnetized fluid, which are described by

$$\frac{\partial \rho}{\partial t} + \nabla \cdot (\rho \mathbf{u}) = 0, \qquad 1.$$

$$\frac{\partial}{\partial t}(\rho \mathbf{u}) + \nabla \cdot (\rho \mathbf{u}\mathbf{u}) + \nabla p + \nabla(B^2/8\pi) - (\mathbf{B} \cdot \nabla)(\mathbf{B}/4\pi) = \mathbf{F}, \qquad 2.$$

where ρ, $\mathbf{u}$, and p are the mass density, flow velocity, and thermal pressure of the fluid, respectively; $\mathbf{B}$ is the magnetic field; and $\mathbf{F}$ includes the effects of body forces, such as gravitational, radiative, and frictional forces. We consider below mass and momentum balance both along a streamline in a subsonic flow and across interfaces (either a shock front separating the supersonic and subsonic flow regimes of a fluid or a contact surface separating two nonpenetrating fluids). Across an interface that is parallel to the local magnetic field and perpendicular to the local flow velocity (i.e. for which $\mathbf{B} \cdot \hat{\mathbf{n}} = 0$ and $\mathbf{u} \times \hat{\mathbf{n}} = \mathbf{0}$, where $\hat{\mathbf{n}}$ is the unit vector normal to the surface), we see from Equation 2 that in a steady state the total pressure is conserved, i.e.

$$\rho u^2 + p + B^2/8\pi + p_F = \text{constant}, \qquad 3a.$$

where the total pressure is the sum of the four terms on the left side of Equation 3a, which represent (from left to right) the ram pressure (normal to the surface), the thermal pressure, the magnetic pressure, and the pressures (p_F) associated with the ambient media (e.g. neutral gas and cosmic rays) producing a frictionlike interaction. In a flow that is highly supersonic and super-Alfvénic, like the supersonic solar wind, the ram pressure is much larger than the thermal and magnetic pressures, whereas in a very subsonic or sub-Alfvénic flow, the ram pressure is negligible. Evidently, Equation 3a applies to an interface (like a shock front) across which there is a flow of mass. For a contact surface across which mass does not flow (i.e. $\mathbf{u} \cdot \hat{\mathbf{n}} = 0$), the ram pressure term in Equation 3a disappears and pressure balance across the surface is described by

$$p + B^2/8\pi + p_F = \text{constant}. \qquad 3b.$$

The above descriptions of mass and momentum balance are, of course, incomplete without either a description of energy balance or the imposition of some other condition (e.g. the assumption of incompressibility), and we consider such additional requirements below as the need arises.

Our approach to the following theoretical discussion is guided by Parker's (1961, 1963) description of three fundamental types of interaction between stellar and interstellar plasmas. First, we discuss Parker's results in Sections 3.1–3.3, and then we go on to consider modifications of these results brought about through the effects of the solar magnetic field, the interstellar neutral gas, cosmic rays, and interstellar dust (Sections 3.4–3.7). Finally, in Section 3.8, we conclude by describing the expected structure of the heliospheric cavity formed through the interaction between the solar wind and the interstellar medium.

3.1 Static, Unmagnetized Interstellar Plasma

Let us first consider a steady, radial, spherically symmetric solar wind interacting with a static, unmagnetized interstellar plasma (Parker 1963). In the asymptotic flow regime of the supersonic solar wind, the flow speed u is nearly constant, so according to Equation 1 the density ρ and the ram pressure ρu^2 decline as $1/r^2$, where r is the heliocentric radial distance. Beyond some distance R_t the ram pressure of the supersonic wind falls below the interstellar pressure p_I, and the solar wind can no longer stand off the interstellar medium (cf. Equation 3): Evidently, R_t is given by

$$R_t = (\rho_E u_E^2/p_I)^{1/2}, \qquad 4.$$

where R_t is measured in astronomical units, and the subscript E refers to wind parameters at $r = 1$ AU. In essence, near the distance R_t the supersonic solar wind encounters an interstellar barrier and must undergo a shock transition to a subsonic flow (Clauser 1960, Weymann 1960), which can achieve pressure equilibrium with the interstellar medium. If we assume a strong, adiabatic shock with an adiabatic index (ratio of specific heats) of 5/3, the Rankine-Hugoniot relations [i.e. the equations of mass, momentum, and energy conservation across the shock (cf. Equation 3)] yield the following relations between parameters just upstream (subscript s1) and just downstream (subscript s2) of the shock:

$$u_{s1}/u_{s2} = \rho_{s2}/\rho_{s1} = 4, \qquad 5.$$

$$p_{s2} = 3\rho_{s1}u_{s1}^2/4, \qquad 6.$$

where the magnetic field is neglected (a good approximation in a highly super-Alfvénic flow). In the postshock subsonic region the flow speed declines quite rapidly, and in the absence of a magnetic field (see Section 3.4 for a discussion of magnetic effects in the subsonic region) the flow is very nearly incompressible, so that $\mathbf{u} \cdot \nabla \rho = \nabla \cdot \mathbf{u} = 0$. It then follows from

Equation 2 that a constant of the subsonic flow is $\rho u^2/2 + p \,(= \rho_{s2} u_{s2}^2/2 + p_{s2})$. The rapid decline of the flow speed leads to a dominance of the thermal pressure in the subsonic flow (i.e. $p \gg \rho u^2/2$), so near the heliopause (the boundary of the heliosphere) we have $p \approx \rho_{s2} u_{s2}^2 + p_{s2}$. It follows from Equation 3b that the interstellar pressure p_I must equal $\rho_{s2} u_{s2}^2/2 + p_{s2}$. Hence, using Equations 5 and 6 we can relate the shock location R_s (measured in astronomical units) to the solar wind ram pressure and interstellar pressure by

$$R_s = (7\rho_E u_E^2/8p_I)^{1/2}, \qquad\qquad 7.$$

where we have assumed that $\rho u^2 \propto r^{-2}$ in the supersonic solar wind. As we should expect, comparison of Equations 4 and 7 indicates that R_t and R_s are essentially the same.

Our heliosphere, in this simplest of possible models, is thus characterized by a spherical shock transition at R_s and a spherical heliopause boundary (separating the solar and interstellar plasmas, which are assumed not to interpenetrate) that is very slowly moving outward at a speed $u = u_{s2}(R_s/R)^2$. Evidently, the heliopause radius is a function of time, the solar wind flow speed, and the shock radius, and it can be written

$$r_H = 0.4 \left[\left(\frac{u_E}{4 \times 10^7} \right) \left(\frac{R_s}{100} \right)^2 \left(\frac{t}{10^9} \right) \right]^{1/3} \text{pc}, \qquad 8.$$

where the units of u_E, R_s, and t are centimeters per second, astronomical units, and years, respectively. Over the lifetime of the Sun, assuming a steady solar wind, the heliopause boundary reaches a distance on the order of 1 pc from the Sun. Of course, any flow of the interstellar medium relative to the Sun must drastically modify the shape of this vast region of subsonic solar wind flow, and this is the subject we next address.

3.2 Flowing, Unmagnetized Interstellar Plasma

Let us again consider unmagnetized solar and interstellar plasmas that do not interpenetrate and whose regions of subsonic flow can be approximately treated as incompressible. A flow of the interstellar plasma relative to the Sun produces an asymmetry in the total interstellar pressure, with the maximum pressure exerted at the heliopause along the stagnation line, which runs radially outward from the Sun in the direction upstream in the interstellar wind (cf. Figure 2). This asymmetry of the total interstellar pressure leads to an asymmetry of the terminal shock, with the minimum shock distance occurring along the stagnation line. The resultant deviation of the shock normal from the radial direction produces (at the shock) a turning of the flow away from the stagnation line, which leads to the

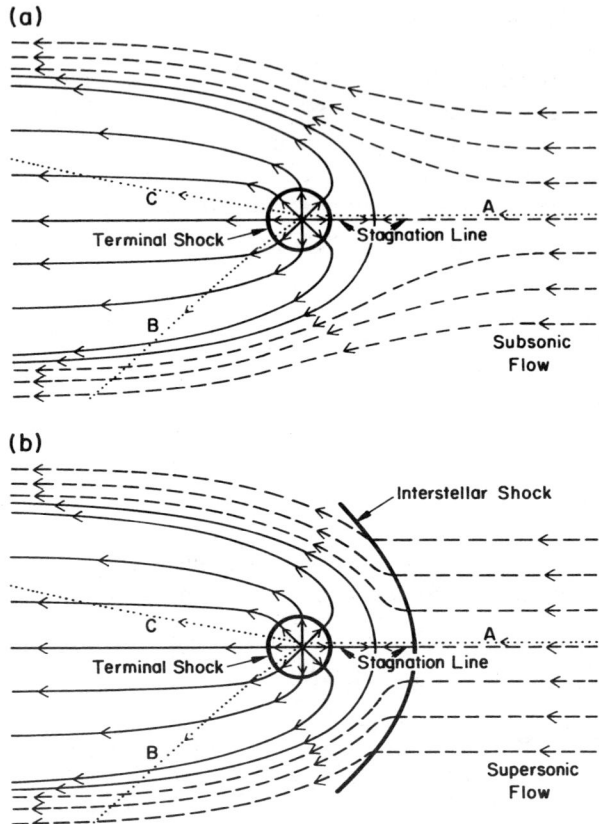

Figure 2 Schematic description of the noninterpenetrating interaction of a flowing, unmagnetized interstellar plasma with an unmagnetized solar wind for the cases in which the interstellar plasma flow relative to the Sun is (*a*) subsonic (adapted from Parker 1963) and (*b*) supersonic [adapted from Baranov et al. (1970) and Parker (1963)]. The curves with arrows are flow lines of the solar (solid) and interstellar (dashed) plasmas. The solid curves without arrows represent the shock front terminating supersonic solar wind flow (panels *a* and *b*) and the bow shock (panel *b*) standing in front of the heliosphere through which the supersonic interstellar flow passes. The dotted curves are trajectories of an interstellar hydrogen atom that is subjected to either a net attractive force (curve AB) or a net repulsive force (curve AC), where the net force is the sum of the solar gravitational force and Ly-α radiation force.

eventual alignment of the postshock subsonic solar wind flow with the interstellar gas flow (as is illustrated in Figure 2).

The flow of the interstellar gas relative to the Sun may be either subsonic (Parker 1963) or supersonic (Baranov et al. 1970, 1976), and the form of the interstellar stagnation pressure π_I is different in these two cases. In the subsonic case ($u_I^2 < 5p_I/3\rho_I$), it is just

$$\pi_1 = p_1 + \tfrac{1}{2}\rho_1 u_1^2, \qquad 9.$$

which is what we would expect from the discussion in Section 3.1. In the supersonic case ($u_1^2 > 5p_1/3\rho_1$), the interstellar gas must pass through a standing bow shock in front of the heliosphere (cf. Figure 2b), and the stagnation pressure becomes

$$\pi_1 = \tfrac{3}{8}p_1 + \tfrac{7}{8}\rho_1 u_1^2, \qquad 10.$$

where ρ_1, u_1 and p_1 are preshock parameters. In either case, the solar wind stagnation pressure remains $p_2 + \rho_2 u_2^2/2$, so the terminal shock position along the stagnation line is given by Equation 7, with p_1 replaced by π_1.

3.3 Magnetically Dominated, Static Interstellar Plasma

In examining the effect of an interstellar magnetic field, let us return to the consideration of a static interstellar plasma (cf. Section 3.1), which this time includes an interstellar magnetic field whose energy density is much larger than that of the other component(s) of the interstellar medium. Such a field can be described in terms of a scalar potential and is characterized by the balance between the magnetic tension and magnetic pressure gradient forces. If a spherical body that excludes magnetic field is inserted into an otherwise uniform interstellar field, the field outside the spherical body remains potential and can be described by (Parker 1963)

$$\psi_1 = -B_0(r + a^\beta/(\beta-1)r^{\beta-1})\cos\theta, \qquad 11.$$

$$\mathbf{B}_1 = -\nabla\psi_1 = B_0[\hat{\mathbf{e}}_r(1 - a^\beta/r^\beta)\cos\theta - \hat{\mathbf{e}}_\theta(1 + a^\beta/(\beta-1)r^\beta)\sin\theta], \qquad 12.$$

where ψ_1 is the magnetic scalar potential, $\hat{\mathbf{e}}_r$ and $\hat{\mathbf{e}}_\theta$ are unit vectors, r is measured from the center of the excluding sphere, $\theta = 0$ is the direction of the undisturbed interstellar field, and $\beta = 3$. If a cylinder, rather than a sphere, excludes the interstellar field, then Equations 11 and 12 still apply, but now $\beta = 2$.

The distorted interstellar field, whose lines of force are defined by $(a/r)[(r/a)^\beta - 1]\sin^{\beta-1}\theta = $ constant, is illustrated in Figure 3, where seven field lines (one the $\theta = 0$ field line) that are equally spaced as $r \to \infty$ (namely, $r\sin\theta \to 0$, $\pm a/2$, $\pm a$, $\pm 3a/2$) are drawn in the r–θ plane for both the sphere (panel a) and the cylinder (panel b). Evidently, the distortion leads to an enhanced interstellar magnetic pressure ($B_1^2/8\pi$), which maximizes for $\theta = \pi/2$. The largest enhancement occurs at the surface of the excluding sphere or cylinder ($r = a$ and $\theta = \pi/2$), where the interstellar magnetic pressure is a factor of 2.25 (4.0) larger than its value far from the sphere (cylinder). Of course, at the two points $r = a$, $\theta = 0, \pi$ the magnetic pressure vanishes; the implications of this effect are discussed in

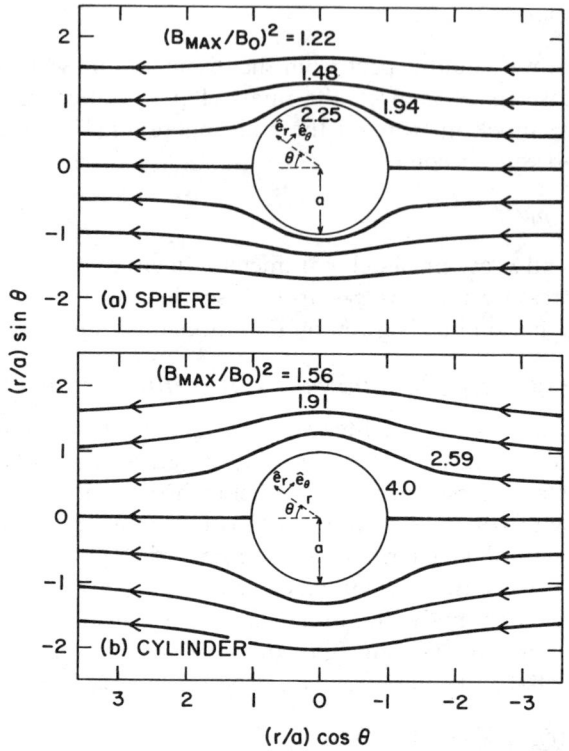

Figure 3 Potential magnetic field lines of an interstellar field distorted by either a spherical (panel *a*) or a cylindrical (panel *b*) heliospheric cavity of radius a that excludes magnetic field. The maximum enhancement of the magnetic pressure along each field line (which occurs at $\theta = 90°$) is given. For a spherical cavity the field is symmetric about the $\theta = 0°$ line, and for a cylindrical cavity the field is symmetric about the $\theta = 0°$ plane. Here r and θ are two of the three coordinates of either a spherical or a cylindrical coordinate system (adapted from Parker 1963).

Sections 3.4 and 3.8, where the consequences of a heliospheric magnetic field are considered.

Clearly, when the magnetic field provides a significant contribution to the total interstellar pressure, we must consider the consequences of the heliospheric distortion of the interstellar magnetic field, and this is done in Section 3.8. Before proceeding, however, we note that any of the magnetic surfaces (cf. Figure 3) that are symmetric about the line $\theta = 0$ (spherical case) or about the plane perpendicular to the paper and including the line $\theta = 0$ (cylindrical case) can be considered the boundary of a region from which the interstellar magnetic field is excluded (i.e. the heliospheric boundary). Parker (1963) made use of this fact in his discussion of the outflow of the subsonic solar wind along the direction of the interstellar magnetic field, a subject to which we return in Sections 3.4, 3.5, and 3.8.

3.4 Effects of a Heliospheric Magnetic Field

The solar magnetic field is drawn out from the Sun by the radially expanding supersonic solar wind, and owing to solar rotation the magnetic field takes the form (on average) of an Archimedes spiral (Parker 1958, 1963), i.e.

$$B_\varphi/B_r = (4.3 \times 10^7/u) R \sin \theta, \qquad 13.$$

where φ and θ are solar azimuth and colatitude, and the units of u and R are centimeters per second and astronomical units, respectively. Evidently, by the time the solar wind reaches the terminal shock ($R \gtrsim 50$ AU) the magnetic field is very nearly azimuthal, except at the highest solar latitudes (cf. Equation 13). For the purpose of the following discussion, we assume that the field is purely azimuthal at and beyond the terminal shock.

As noted above, in the supersonic solar wind the magnetic energy density is negligible ($\lesssim 1\%$) in comparison with the total wind energy density. Across the terminal shock, the magnetic pressure increases by no more than a factor of 16 (for a strong shock) and thus remains small ($<15\%$) in comparison with the gas pressure just downstream of the shock. Hence, as assumed above, the postshock flow is initially very nearly incompressible, which requires that u decrease nearly as r^{-2}. Yet $B \propto (ur)^{-1} \propto r$, so if nearly spherically symmetric subsonic flow extends to a great enough radial distance (which depends on the interstellar neutral gas interactions discussed in Section 3.5), the magnetic energy density becomes dominant, and the magnetic field controls the structure of the flow (e.g. Holzer 1972). In the region of magnetic control of spherically symmetric flow, the magnetic tension and pressure gradient forces nearly balance, so that $B \propto r^{-1}$, which leads to $u =$ constant and $n \propto r^{-2}$. Clearly, the magnetic control of the subsonic flow leads to a decrease with radial distance of the total pressure (e.g. Cranfill 1971, Holzer 1972), and Lee (1988) has argued that this reduction may be significant in determining the terminal shock location [leading, of course, to a smaller shock distance than would be predicted on the basis of the assumption of incompressible flow (cf. Sections 3.1 and 3.2)].

Another important consequence of the magnetic energy density becoming significant in the outer heliosphere involves the structure of the heliospheric boundary (the heliopause) when the interstellar magnetic field dominates the interstellar pressure. Let us consider a situation in which the subsonic solar wind flow is turned (cf. Sections 3.2, 3.5, and 3.8) in a direction perpendicular to the interstellar magnetic field, so that the bulk of the heliosphere takes the shape of the cylinder illustrated in Figure 3b. As noted in Section 3.3, the pressure exerted by the interstellar magnetic field at this boundary decreases substantially from the direction perpendicular to the direction parallel to the interstellar field. If the subsonic solar wind flow were incompressible, then pressure balance at the heliopause would require a substantial distortion of this cylindrical shape, with the heliosphere expanding parallel to the interstellar field and contracting perpendicular to the field until a heliopause shape nearer that described by the outermost interstellar field lines shown in Figure 3b were achieved. Such a shape would clearly be associated with very little enhancement of the interstellar magnetic pressure (cf. Figure 3b) through field distortion resulting from the presence of the heliospheric cavity.

The situation is different, however, if the heliospheric magnetic field is dominant near the heliopause, and the heliospheric field has a significant component in the plane of the paper in Figure 3. This situation is illustrated schematically in Figure 4 (which is a modification of Figure 3), where the heliospheric field is assumed to be circular near the terminal shock and to lie in the plane perpendicular to the axis of the cylinder. Although the heliosphere is distorted from its cylindrical shape, the distortion is relatively minor (owing to the tension of the heliospheric field), and the enhancement of the interstellar magnetic pressure remains relatively large. Note that in the region where the heliopause bulges, the heliospheric

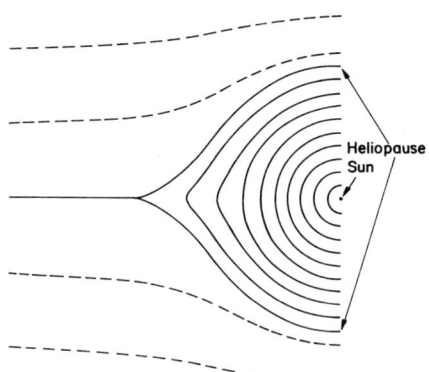

Figure 4 Schematic illustration of the adjustments of the heliospheric and interstellar magnetic fields when the heliospheric cavity is no longer a cylinder (cf. Figure 3), but instead the heliospheric field is taken to be nearly circular near the Sun and potential beyond. Heliospheric field lines are solid, and interstellar field lines are dashed. The spacing of the heliospheric field lines is not intended to indicate field intensity, which declines as r^{-1} near the Sun and much more rapidly near the cusp region.

pressure has decreased below its value at the heliopause in the region near the interstellar magnetic pressure maximum, so that the effect of the interstellar pressure enhancement on the terminal shock distance along flow lines intersecting each region should be essentially the same. We return to this discussion of heliospheric structure, as well as the possible effects of magnetic field line reconnection, in Section 3.8.

3.5 Effects of a Flowing Interstellar Neutral Gas

As indicated in Figure 2, the interstellar neutral gas penetrates relatively freely into the heliosphere. The neutral atoms, particularly H and He, resonantly scatter solar radiation, so that the distribution of interstellar H and He in the heliosphere can be studied by observing sky background radiation in H I $\lambda 1216$ and He I $\lambda 584$. Such studies, coupled with models for the penetration of interstellar neutrals into the heliosphere (Section 3.5.1), allow inferences to be drawn concerning the thermodynamic parameters of the neutral component of the VLISM (cf. Section 2.2). Interstellar neutrals penetrating the heliosphere lead to significant modifications of both the thermal plasma and the energetic particle population in the heliosphere. They slow and heat the supersonic solar wind (Section 3.5.2) and in so doing provide a population of seed particles that can be accelerated at the terminal shock to produce low-energy cosmic rays (Section 3.6.2). Interstellar neutrals also both cool and turn the subsonic solar wind flow and thus play an important role in determining the structure and location of both the terminal shock and the heliopause (Sections 3.5.3 and 3.8).

3.5.1 PENETRATION OF INTERSTELLAR NEUTRALS INTO THE HELIOSPHERE Two possible orbits of a hydrogen atom moving toward the heliosphere with a velocity equal to the interstellar flow velocity and an impact parameter (relative to the Sun) of about 2 AU are shown by the dotted trajectories in Figure 2. Both AB and AC are hyperbolic trajectories, the first corresponding to solar minimum conditions (for which the repulsive solar Ly-α radiation force is smaller than the attractive force of gravity) and the second to solar maximum conditions (for which the Ly-α radiation force exceeds gravity, leading to a net radially outward force on the incoming atom). Since the hydrogen atom travels about 4 AU in a year, the time to traverse the portions of trajectories AB and AC that are shown in Figure 2 is about 200 yr (assuming both that the atom is not ionized in its close encounter with the Sun and that the minimum Sun-heliopause distance is about 100 AU). The real trajectory of an interstellar atom is not smooth like trajectories AB and AC because of the significant variation of the Ly-α radiation force on time scales comparable to and smaller than the time over which the net solar force has a significant effect on the trajectory (e.g. Vidal-Madjar 1977). In addition, the nonnegligible thermal speeds characterizing the interstellar neutral velocity distributions (about 10 km s^{-1} for H I and 5 km s^{-1} for He I, corresponding to $T_1 = 10^4$ K) imply that the incoming neutral atom velocity vector will not generally be closely aligned with the interstellar flow velocity, as it is assumed to be for the illustrative trajectories shown in Figure 2.

Before a hydrogen atom can complete a heliosphere-penetrating trajectory like those shown in Figure 2, it must overcome three obstacles. The first obstacle is the diverging flow of the interstellar ionized gas as it passes the heliosphere. Resonance charge transfer between interstellar hydrogen atoms and protons near the heliosphere can lead to a diversion of some fraction of the neutral gas away from the heliosphere, because through the charge transfer the neutral atom and proton effectively exchange trajectories (Wallis 1975, 1978, 1981, 1984). The region in which this diversion takes place depends on whether the interstellar gas is flowing subsonically (Figure 2a) or supersonically (Figure 2b), and it is characterized by the divergence of the interstellar plasma flow lines from the stagnation line. For subsonic flow (actually subcritical flow, where the critical speed is the hydromagnetic fast mode speed), which is likely the relevant case (cf. Section 3.8), this diversion region extends one to two times the minimum Sun-heliopause distance ahead of the heliosphere. For a charge transfer cross section of 5×10^{-15} cm^2 [appropriate to a 1-eV interaction energy; the cross section decreases to 2×10^{-15} cm^2 for a 1-keV interaction energy (Tawara et al. 1983)], only a fraction of a percent of hydrogen atoms charge exchange if the proton density is 0.01 cm^{-3} (cf. Section 2.2.2), and the minimum Sun-heliopause distance (R_H) is on the order of 100 AU. If the product of the proton density and the Sun-heliopause distance is increased by a factor of 10, then some 30% of the hydrogen atoms charge exchange in the diversion region. Using rather large numbers for the interstellar proton density and the size of the heliosphere, and assuming that a charge transfer implies exclusion of a hydrogen atom from the heliosphere, Ripken & Fahr (1983) have suggested that a significant fraction of interstellar hydrogen atoms are diverted away from the heliosphere. Wallis (1978, 1981, 1984), however, has pointed out that in this particular case the charge transfer process corresponds more closely to a scattering process than to an extinction process, which Ripken & Fahr (1983) have implicitly assumed. The scattering process can be visualized by realizing that the charge transfer effectively produces a new population of interstellar neutrals with the same velocity distribution as the interstellar protons. The exclusion of neutral particles from the heliosphere can then be estimated by comparing the fraction of the newly produced neutral distribution that enters the heliosphere ($\langle f_1 \rangle$) with the fraction of the original neutral distribution ($\langle f_0 \rangle$) that enters. It follows that the interstellar neutral hydrogen density is effectively reduced (through exclusion from the heliosphere by charge transfer) by a factor

$$F = 1 - [1 - (\langle f_1 \rangle / \langle f_0 \rangle)](1 - e^{-\tau_H}), \qquad 14.$$

$$\tau_H = \beta n_e L, \qquad 15.$$

where n_e is in cm^{-3}, L (in astronomical units) is twice the minimum Sun-heliopause distance (R_H), $(1 - e^{-\tau_H})$ is the fraction of atoms undergoing charge transfer, and, in the case at hand, $\beta = 5 \times 10^{-2}$. Since the interstellar flow speed is only a factor of 2 greater than the thermal speed, and since the diversion of the interstellar plasma is relatively gradual (cf. Figure 2), it follows that $[1 - (\langle f_1 \rangle / \langle f_0 \rangle)] \lesssim 0.15$ and, even for the parameters of Ripken & Fahr (1983), that $F \gtrsim 0.95$ (for the parameters given in Sections 2.2.2 and 3.8, it follows that $F > 0.99$).

The second obstacle an interstellar hydrogen atom penetrating the heliosphere faces is charge transfer in the postshock subsonic solar wind. Such charge transfer produces a population of hot hydrogen atoms (formerly solar wind protons) characterized by the subsonic solar wind temperature, which is on the order of 1.5×10^6 K near the terminal shock and decreases gradually with distance from the shock (cf. Section 3.5.3). The scattering process is the same as described in the preceding paragraph, with newly produced atoms taking on the velocity distribution of the protons, which in this case is taken to be a Maxwellian distribution essentially at rest rather than a rapidly drifting Maxwellian. We thus have a larger exclusion factor, $[1 - (\langle f_1 \rangle / \langle f_0 \rangle)] \lesssim 0.4$, and if we choose appropriate representations of β, n_e, and L, Equation 15 becomes

$$\tau_H \approx (9/R_s)[(R_H - R_s)/R_s], \qquad 16.$$

where R_s is the minimum shock distance. Taking $R_H/R_s = 2$, we find from Equations 14 and 16 that $F \gtrsim 0.90$ for $R_s = 50$ AU, and $F > 0.95$ for $R_s = 100$ AU. We must remember, though, that the newly produced atoms in this case have relatively high speeds (> 100 km s^{-1}) and will generally have radial velocity components large enough to shift them into the wings of the solar Ly-α line, thus making them less visible in Ly-α backscatter observations than are interstellar atoms (which have much lower speeds). For the purpose of interpreting Ly-α observations, therefore, it is more appropriate to take $F \gtrsim 0.85$ for $R_s = 50$ AU and $F \gtrsim 0.93$ for $R_s = 100$ AU.

The final obstacle faced by an interstellar hydrogen atom penetrating the heliosphere is the supersonic solar wind, in which the charge transfer process produces a hydrogen atom traveling nearly radially outward from the Sun at the solar wind speed. Such an atom is Doppler shifted so far from the solar Ly-α line center as to become essentially invisible. For our purposes here, therefore, this hydrogen atom can be considered destroyed, so $f_1 = 0$ and Equations 14 and 15 become

$$F = e^{-\tau_H}, \qquad 17.$$

$$\tau_H = 5 \times 10^{-7} \int ds(2.5 \times 10^{-15} nu + 9 \times 10^{-8} r_E^2/r^2), \qquad 18.$$

where the integration is carried out along the trajectory of an interstellar hydrogen atom inside the terminal shock (the units of s are centimeters), nu is the solar wind proton flux density (in square centimeters per second), and the second term in the integrand arises from photoionization. For a hydrogen atom traveling radially inward (toward the Sun) the penetration distance (where $\tau_H = 1$) is

$$R(\tau_H = 1) = \left\{ \left[4.5\left(\frac{n_E u_E}{3 \times 10^8}\right) + 0.7 \right]^{-1} + R_s^{-1} \right\}^{-1} \text{AU}, \qquad 19.$$

which is about 5 AU. In contrast, interstellar helium atoms (for which the primary destruction process is photoionization rather than charge transfer) penetrate to about 0.6 AU (Holzer & Axford 1971).

In the preceding discussion, we have not considered the effects of scattering (through charge transfer in H–H$^+$ collisions and through polarization interaction in He–H$^+$ collisions) on the flow speed and temperature of the interstellar neutral gas that are inferred from UV backscatter observations (cf. Section 2.2.1). Yet when such effects are ignored, there exists a significant discrepancy between the inferred hydrogen and helium temperatures [$0.4 \lesssim T_H/T_{He} \lesssim 0.7$ (Bertaux et al. 1977, 1985, Ajello 1978, Weller & Meier 1981, Dalaudier et al. 1984)]. A number of different treatments of the scattering have been used to produce a wide variety of conflicting results (e.g. Wallis 1978, 1984, 1988, Wallis & Hassan 1978, Wallis & Wallis 1979, Fahr et al. 1985, Chassefiere et al. 1986, 1988a,b, Chassefiere & Bertaux 1987a,b). In selecting the interstellar parameters presented in Section 2.2.1, we have relied primarily on the analysis of Chassefiere et al. (1988a,b).

3.5.2 EFFECTS OF NEUTRALS ON THE SUPERSONIC SOLAR WIND FLOW Charge transfer collisions between interstellar hydrogen atoms and solar wind protons in the supersonic flow regime occur principally outside $R = 5$ AU (Equation 19), where the magnetic field is nearly perpendicular to the solar wind flow (except at high solar latitudes). Thus, a newly produced proton resulting from charge transfer experiences an electric force (associated with solar wind flow perpendicular to the local magnetic field) that accelerates it to an energy approximately twice that of an ambient solar wind proton. Immediately after attaining this energy, the motion of the newly produced proton (which we refer to as a pickup proton or, generically, as a pickup ion) can be described as the superposition of an outward radial motion at the solar wind speed and a circular motion (also characterized approximately by the solar wind speed) in the solar wind rest frame in a plane perpendicular to the local magnetic field. The motion of a newly produced neutral atom is, of course, just an outward radial motion at the solar wind speed. The net effect on the solar wind of a charge transfer collision is, therefore, a reduction of the momentum of the ionized solar wind (equal to the momentum of the newly produced neutral atom) and an increase in the solar wind thermal energy (equal to the energy of circular motion of the newly produced proton). We have avoided in this discussion explicit consideration of the effects of finite solar wind temperature and finite flow speed (relative to the Sun) of interstellar neutrals, but both these effects are quite small.

The two principal effects of interstellar neutral hydrogen on the supersonic solar wind are thus a slowing and heating of the flow (Semar 1970, Wallis 1971a,b, 1973, 1974, Holzer 1972). The aspect of the slowing of the flow in which we are particularly interested is the reduction of the solar wind ram pressure through interaction with the interstellar neutrals, and this reduction can be represented by the factor γ, i.e.

$$\gamma = e^{-\tau_p}, \qquad 20.$$

$$\tau_p = 3 \times 10^{-3} F_t \left(\frac{n_H}{0.1}\right)(R_s - 5), \qquad 21.$$

where the solar wind ram pressure at the terminal shock (upstream in the interstellar wind) is $\gamma\rho u^2$, and F_t is the factor by which the interstellar neutral hydrogen density is decreased from its interstellar value at the terminal shock (cf. Equations 14–16). Away from the direction upstream in the interstellar wind, the hydrogen density inside the terminal shock is a bit lower, so the slowing effect is reduced and γ is correspondingly larger.

Beyond about 5 AU from the Sun, the heating of the supersonic solar wind arising from the interaction with interstellar neutrals is larger than the cooling associated with the spherical expansion of the wind. Thus, one might expect such heating to be observable by spacecraft in the outer solar system. Yet this heating is comparable to that produced by solar wind stream interactions (e.g. Hundhausen 1973, Pizzo 1986, Burlaga 1988) near the ecliptic plane (which is where the spacecraft are located), so distinguishing between the two heating processes would be difficult under the best of circumstances. A further difficulty, however, is presented by the fact that while the pickup ions rapidly pitch-angle scatter to form a spherical shell in velocity space (at a speed several times the proton thermal speed), they diffuse very slowly in energy and thus do not become assimilated into the ambient near-equilibrium solar wind proton velocity distribution (Isenberg 1987). Thus, instruments designed to observe a highly directed proton velocity distribution (characteristic of a highly supersonic flow) are at a distinct disadvantage in seeking that part of the distribution (the pickup ions) for which the random speed is comparable to the flow speed. Fortunately, there have been observations of singly ionized helium pickup ions (Mobius et al. 1985, Mobius 1986), and these observations seem to confirm the description of pickup ions just given. Although the lack of assimilation of pickup ions into the ambient solar wind velocity distribution is largely irrelevant to the description of solar wind dynamics (e.g. the effect of the solar wind pressure gradient force), it is quite important for particle acceleration at the terminal shock, as is discussed in Section 3.6.2.

3.5.3 EFFECTS OF NEUTRALS ON THE POSTSHOCK, SUBSONIC SOLAR WIND FLOW In the region of postshock, subsonic solar wind, where the proton temperature remains high ($T \gtrsim 10^6$ K), charge transfer between a solar wind proton and an interstellar hydrogen atom produces a fast neutral atom (cf. Section 3.5.1) that is very unlikely to undergo another charge transfer collision until it is well outside the heliosphere. As this atom leaves the heliosphere, it carries with it some of the thermal energy and (on average) the momentum of the subsonic wind, so it follows that the charge transfer collision producing the fast neutral atom serves to cool the subsonic solar wind and to turn its flow into the direction of the interstellar neutral gas flow. In the hemisphere toward the incoming interstellar wind, this turning of the flow is aided by the magnetic force directed from solar equator to pole that arises from the more rapid decline of the magnetic pressure at high latitudes and from the component of the magnetic tension force directed from equator to pole (Parker 1958, 1963).

In order to estimate the distance traveled by the subsonic solar wind before it is turned to flow in the direction of the interstellar wind, we need to consider the time required for the subsonic flow to reach a given radial distance. As mentioned earlier (Section 3.4), in the region where the subsonic flow is nearly incompressible the flow speed decreases as r^{-2}, so the time (in seconds) taken to travel from the shock radius R_s to a radius R is

$$t = 5 \times 10^5 \left(\frac{10^7}{u_s}\right) R_s \left[\left(\frac{R}{R_s}\right)^3 - 1\right], \qquad 22.$$

where u_s is the flow speed (centimeters per second) just downstream of the terminal shock. The fraction of solar wind protons that undergo charge transfer with interstellar neutrals during this time t is just $(1 - e^{-\tau_p})$, where

$$\tau_p = 2 \times 10^{-3} \left(\frac{n_H}{0.1}\right) F(R) R_s \left[\left(\frac{R}{R_s}\right)^3 - 1\right], \qquad 23.$$

and $F(R)$ is the factor by which the interstellar hydrogen density is reduced at the radius R. The turning of the flow should be accomplished when τ_p reaches a value of about 2, and for a minimum shock distance of $R_H = 50$ AU, this corresponds to a minimum heliopause distance R_H that is 2 or 3 times R_s.

Once the subsonic solar wind flow is turned into the heliospheric tail (cf. Section 3.8 and Figure 5) it rapidly (within several R_s) reaches velocity and temperature equilibrium with the neutral interstellar gas. Such an adjustment of the heliosphere to the interstellar medium could not be accomplished without the relatively free penetration of interstellar neutral hydrogen into the heliosphere.

3.6 Cosmic-Ray Effects

As noted earlier, Galactic cosmic rays penetrate relatively freely into the heliosphere and thus, like the interstellar neutral gas, can interact directly with both the supersonic and the postshock subsonic solar wind (Section 3.6.1). Yet it appears that the cosmic rays that are most important to our study of the interaction between the solar wind and the interstellar medium are accelerated in the heliosphere itself and comprise ions that were once interstellar neutral atoms that penetrated into the region of supersonic solar wind before being ionized (Section 3.6.2). We now briefly discuss both these interactions.

3.6.1 GALACTIC COSMIC RAYS   As Galactic cosmic rays flow through the heliosphere, they scatter off magnetic fluctuations transported outward by the solar wind (Parker 1956). Through this scattering the solar wind exerts a radially outward force on the cosmic rays, which is balanced by an inward cosmic-ray pressure gradient force. The cosmic-ray pressure gradient force, of course, affects momentum balance in both the subsonic and the supersonic solar wind and thus affects the location of the terminal shock (e.g. Axford & Newman 1965, Jokipii & Parker 1967, Sousk & Lenchek 1969, Wallis 1971a). Recent calculations (Axford & Ip 1986, Ko & Webb 1987, 1988, Ko et al. 1988) indicate that the ram pressure in the supersonic solar

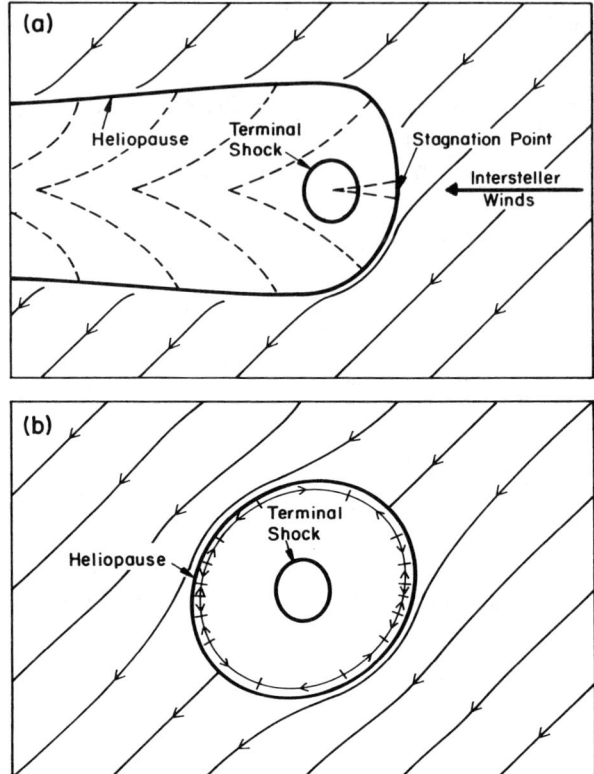

Figure 5 Schematic illustration of the structure of the heliosphere in a magnetically dominated interstellar medium: (*a*) plane containing the interstellar velocity vector and the solar rotation axis (assumed normal to u_I); (*b*) plane perpendicular to the interstellar velocity vector and containing the solar rotation axis. Light arrows outside the heliosphere indicate the direction of the component of the interstellar magnetic field in the plane shown. Dashed curves in panel *a* are loci of solar wind fluid elements emitted from the Sun near the solar equator at the maximum of every third 11-yr solar cycle. The arrows paralleling the heliopause in panel *b* indicate the predominant direction of the component (in the plane of the paper) of the heliospheric magnetic field within a few astronomical units of the heliopause. The 16 places where the field changes direction correspond to boundaries between plasma emitted from the Sun during successive 11-yr solar cycles.

wind is reduced by about 5–10% by the cosmic-ray pressure gradient, while the pressure of cosmic rays excluded from the supersonic region is much less than the total cosmic-ray pressure.

3.6.2 COSMIC-RAY ANOMALOUS COMPONENT An anomalous enhancement is observed in the cosmic-ray spectrum at low energies (5–50 MeV nucleon^{-1}) for elements such as He, N, O, and Ne (e.g. Garcia-Munoz et al. 1973, Hovestadt et al. 1973, McDonald et al. 1974, von Rosenwinge & McDonald 1975, Cummings & Stone 1988). It has been suggested (Fisk et al. 1974) that this cosmic-ray anomalous component arises from interstellar neutral atoms that have penetrated into the region of supersonic solar wind, have been ionized, and have been accelerated to MeV energies. A possible mechanism for the last stage of this process is acceleration of the interstellar pickup ions at the solar wind terminal shock (Fisk 1986, Jokipii 1986). The observed increase in the anomalous component with radial distance, together with the assumption that the anomalous component contains hydrogen in appropriate proportion to its other constituents [anomalous hydrogen is presumably masked by Galactic cosmic-ray protons in the inner heliosphere (Beatty et al. 1985)], implies that if the anomalous component is accelerated at the terminal shock, then the shock must be located inside 50 AU in order that the anomalous hydrogen pressure not exceed the solar wind ram pressure upstream of the shock (Fisk 1986, Jokipii 1986, Lee 1988). Of course, if the anomalous hydrogen pressure is relatively large upstream of the terminal shock, then it should significantly modify the structure of the shock (Drury 1988, Lee & Axford 1988), but it will not affect the shock location (e.g. Lee 1988).

3.7 Effect of Interstellar Dust

The last interstellar component we consider is dust, and we assume that in the VLISM the gas-to-dust mass ratio is 100 and that the typical grain radius is $5 \times 10^{-6} \lesssim a \lesssim 2 \times 10^{-5}$ cm (e.g. Greenberg 1978). Assuming that in the heliosphere a dust grain is charged to 5 V (e.g. Parker 1964, Lamy et al. 1985), we can readily calculate (Parker 1964) the Lorentz force on the grain (G_L), which is the same force exerted on interstellar pickup ions by the magnetized solar wind, and compare it with the solar gravitational and radiative forces exerted on the grain (G_G and G_R):

$$G_L \approx 1.5 \times 10^{-10} \left(\frac{u_E}{10^7}\right) \frac{a}{R} \text{dyn}, \qquad 24.$$

$$G_G \approx 7 \frac{a^3}{R^2} \text{dyn}, \qquad 25.$$

$$G_R \approx 1.4 \times 10^{-4} \frac{a^2}{R^2} \text{dyn}, \qquad 26.$$

where a is the grain radius, R is the heliocentric radial distance, and u_E is the solar wind flow speed at 1 AU. For $a = 10^{-5}$ cm, $R = 100$ AU, and $u_E = 5 \times 10^7$ cm s^{-1}, the Lorentz force is 100 times the gravitational force and 50 times the radiative force, so it is clear that such dust grains will be excluded from the heliosphere (e.g. Levy & Jokipii 1976). Interstellar dust, therefore, will exert a surface force at the heliopause, but this force is negligible in comparison with the other interstellar forces we are considering.

3.8 Structure of the Heliosphere

We have now provided an adequate observational and theoretical basis for discussing the expected large-scale structure of the heliosphere. A schematic view of this structure for a magnetically dominated VLISM[1] is given in Figure 5, where two cross sections of the heliosphere are shown: (a) the plane containing both the interstellar velocity vector (in the solar rest frame) and the solar rotation axis; and (b) the plane perpendicular to the interstellar velocity vector and containing the solar rotation axis. For convenience, the solar rotation axis is taken to be perpendicular to the interstellar wind vector, although the angle between the two vectors actually may be closer to 97° (cf. Section 2.2.1). If the VLISM is dominated by thermal gas pressure rather than magnetic pressure (which we consider unlikely), some aspects of the heliosphere and VLISM illustrated in Figure 2 must be taken into account. Thus, with attention directed to Figures 5 and 2, we proceed with a consideration of heliospheric structure, concentrating on the locations and shapes of the terminal shock (which bounds the region of supersonic solar wind flow) and of the heliopause (which bounds the region of influence of the solar magnetized plasma).

3.8.1 THE TERMINAL SHOCK As in Section 3.1, we first determine the shock distance along the stagnation line (cf. Figure 2), R_s, by equating the total pressures just inside and just outside the heliopause at the stagnation point (cf. Figure 5):

$$\gamma_1\gamma_2\gamma_3\rho_E u_E^2/R_s^2 = \Gamma_1 B_I^2/8\pi + (2n_e + \Gamma_2 n_H)(\Gamma_3 k T_I + \Gamma_4 m_H u_I^2)$$
$$+ \Gamma_5 p_{cr} + \Gamma_6 \rho_d u_I^2. \quad 27.$$

The total pressure just inside the heliopause, which is given by the left side of Equation 27, is written in terms of the solar wind ram pressure at 1 AU

---

[1] Note that when we speak of a magnetically dominated VLISM in the context of heliospheric structure, we are not addressing the issue of the relative energy densities of magnetic field and fluid (thermal gas and cosmic rays) in the interstellar medium. Consider, for example, the case of a 5-μG magnetic field and a negligible thermal gas pressure. The magnetic pressure is 10^{-12} dyn cm^{-2}, which is about the same as the Galactic cosmic-ray pressure. Yet the magnetic field is enhanced near the heliosphere by up to a factor of 4, while only a small fraction of the cosmic-ray pressure plays a role in determining the structure of the heliosphere. Thus, from the standpoint of heliospheric structure, the VLISM is magnetically dominated despite the near equipartition of energy between field and fluid.

($\rho_E u_E^2$), the factors by which the ram pressure is reduced between 1 AU and the terminal shock through spherical expansion (R_s^{-2}, where R_s is in astronomical units) and through interaction with the interstellar neutral gas (γ_1) and Galactic cosmic rays (γ_2), and the factor by which the total pressure at the heliopause is reduced from the ram pressure just inside the terminal shock (γ_3). The total pressure just outside the heliopause, which is given by the right side of Equation 27, is separated into four terms, associated with the interstellar magnetic field, the interstellar thermal gas (neutral and ionized components), Galactic cosmic rays, and interstellar dust. The factor Γ_1 reflects the amplification of the background interstellar magnetic field through distortion by the heliosphere (cf. Section 3.3, and Figures 3 and 4). The factors Γ_2 and Γ_5 are, respectively, the fractions of interstellar neutrals and of Galactic cosmic rays (i.e. cosmic-ray pressure) excluded from the supersonic solar wind. (Note that for simplicity we are treating this exclusion in Equation 27 like an exclusion from the heliosphere.) Finally, the factors Γ_3, Γ_4, and Γ_6 are all of order 1 and reflect the nature of the flow around the heliosphere; illustrative values of Γ_3 and Γ_4 for the two types of flow shown in Figure 2 are given by Equations 9 and 10.

In the discussion of Sections 2 and 3 we have given ranges of possible values for the various parameters that determine the terminal shock distance in Equation 27. We first calculate the shock distances for both low-speed and high-speed solar wind (cf. Table 1) appropriate to the midpoints of these parameter ranges, and then we perform the calculations for extreme values of the parameters in order to produce minimum and maximum values for the shock distance. For the first calculation (using midpoints of the parameter ranges), we take $\gamma_1 = 0.89$, $\gamma_2 = 0.93$, and $\gamma_3 = 0.5$. The value of γ_3 (cf. Section 3.4) is determined using the results of Holzer (1972) and by assuming a heliopause distance that is between 2 and 3 times the shock distance along the stagnation line. A value of $\Gamma_1 = 2.5$ is determined by assuming that the interstellar magnetic field is oriented at 45° to the interstellar wind vector and that a maximum amplification factor[2] of $\Gamma_1 = 4$ is appropriate for an orientation of 90°. [Compare this with the value of $\Gamma_1 = 2.25$ (cf. Section 3.3) normally assumed (e.g. Axford 1972, Axford & Ip 1986, Lee 1988).] For the remaining factors, we choose $\Gamma_5 = 0.23$ (cf. Section 2.2.4), $\Gamma_2 = 1 - F_t = 0.1$ (cf. Section 3.5.1), and $\Gamma_3 = \Gamma_4 = \Gamma_6 = 1$. It follows that the shock distances (for low-speed and high-speed solar wind) appropriate to the midpoints of the parameter ranges are

---

[2] We use the amplification factor appropriate to the cylinder in Figure 3b, because the heliospheric tail is presumably nearly cylindrical, and all flow lines leaving the terminal shock must have an asymptotic pressure corresponding to that of the heliospheric tail.

$R_s(\text{low speed}) = 50 \text{ AU},$ \hfill 28.

$R_s(\text{high speed}) = 60 \text{ AU}.$ \hfill 29.

Taking extreme values of the parameters, we obtain the following maximum and minimum values for the shock distance, again for both low-speed and high-speed wind:

$R_{s\,\text{min}}(\text{low speed}) = 25 \text{ AU},$ \hfill 30.

$R_{s\,\text{max}}(\text{low speed}) = 140 \text{ AU},$ \hfill 31.

$R_{s\,\text{min}}(\text{high speed}) = 27 \text{ AU},$ \hfill 32.

$R_{s\,\text{max}}(\text{high speed}) = 170 \text{ AU}.$ \hfill 33.

An examination of the relative magnitudes of the terms on the right side of Equation 27 reveals that the principal source of uncertainty in the terminal shock location (as reflected in Equations 30–33) is the uncertainty in the interstellar magnetic field. It follows that a direct detection of the shock would indirectly place a significant constraint on the magnetic field of the VLISM and would thus make an important contribution to our understanding of the ISM. (We note, however, that for the very low value of $B_I \approx 2 \times 10^{-6}$ G used in calculating $R_{s\,\text{max}}$, the reduction of γ_1 through slowing of the supersonic wind by interstellar neutral hydrogen quite significantly reduces $R_{s\,\text{max}}$.)

The different shock distances for low-speed and high-speed wind indicate that the terminal shock is not likely to be spherical; rather, it should bulge outward at high solar latitudes, where high-speed wind flows over most of the solar cycle (cf. Section 2.1). The shortest distance to the terminal shock should be along the stagnation line, where the lowest speed wind usually flows and where the modest effect of the interstellar ram pressure is felt most strongly; the antipodal shock distance, however, should only be slightly greater than this shortest distance. Although the distortion in shape of the terminal shock is modest, the density difference at high and low latitudes is large, with the high-latitude (high-speed wind) shock density being about a factor of 4 less than the low-latitude (low-speed wind) shock density. Of course, during the declining phase of the solar cycle, when a mixture of high-speed and low-speed wind flows within some 30° of the solar equator (e.g. Hundhausen 1977), the density ratio will be somewhat less (something like 2.5 to 3). This substantial density difference at low and high latitudes could lead to two spectral peaks in radio emission from the terminal shock and thus might be consistent with the observations of Kurth et al. (1984, 1987; cf. Section 2.3).

The preceding discussion has not touched upon shock motion in response to variations of the solar wind on time scales ranging from 25 days [shorter time-scale variations are filtered out in the inner heliosphere (e.g. Pizzo 1986)] to 11 yr (period of the solar activity cycle). The detailed adjustment of the terminal shock to such variations will generally be relatively complex, often involving the formation and eventual dissipation of multiple shocks (including both forward and reverse shocks). On average, though, the terminal shock should move about 2–3 AU a month in response to solar wind ram pressure changes, so only the relatively long-period solar wind variations should produce a significant change in the shock location. Of course, even quite small changes in the shock location will lead to multiple shock crossings by a spacecraft, and such multiple crossings will have to be carefully distinguished from the crossing of multiple shocks mentioned above.

3.8.2 THE HELIOPAUSE The basic shape of the heliopause is determined by the turning of the subsonic solar wind flow and by the containment of the heliospheric magnetized plasma with the pressure of the interstellar magnetized plasma. The turning of the flow is accomplished in part through the frictionlike interaction between interstellar neutral hydrogen atoms and solar wind protons, in part through the asymmetry of the terminal shock (which turns the flow poleward), and in part through the poleward Lorentz force, which becomes important as the heliospheric magnetic field becomes dominant [well beyond the terminal shock (cf. Section 3.4)]. The interstellar neutrals not only play a major role in turning the postshock solar wind flow, but they also bring the heliospheric plasma toward both flow and temperature equilibrium with the interstellar gas just a few R_s into the heliospheric tail. The evolution toward temperature equilibrium, of course, brings about a modest compression of the heliospheric plasma in the tail, which accounts for the slight narrowing of the heliosphere downstream in the interstellar wind shown in Figure 5.

The distortion of the heliopause by the inherent anisotropy of the interstellar magnetic stress on the heliosphere is shown in Figure 5 (especially panel *b*) as being much more modest than one might expect from Figure 4. This reduction in distortion results from the substantial component of the interstellar field parallel to the axis of the heliospheric tail [$(\mathbf{B}_I \cdot \mathbf{u}_I)^2/(B_I^2 u_I^2) = 0.5$ in Figure 5], which contrasts with the absence of a magnetic field component parallel to the axis of the cylinder in Figure 4. Such a parallel field component in Figure 3*b* would lead to a change in magnetic field pressure along $r = a$ (from $\theta = 90°$ to $\theta = 0°$) from 2.5 to 0.5 times the background pressure, rather than from 4 to 0 times the background pressure. Evidently, the component of the interstellar field parallel to the axis of the heliospheric tail also leads to a reduction of the amplification of the interstellar field caused by the presence of the heliosphere (cf. Section 3.3), which we accounted for when assigning a value to Γ_1 in Section 3.8.1.

One might expect (e.g. Fahr et al. 1986) substantial diffusion across the heliopause, rapidly obscuring completely the boundary between the heliosphere and the VLISM. However, in the magnetically dominated model shown schematically in Figure 5, such diffusion should be quite small. Generation of the Kelvin-Helmholtz instability is suppressed because of the rapid approach to flow equilibrium across the heliopause induced by the interstellar neutrals. Furthermore, magnetic field line reconnection should not be particularly significant for the following reasons.

First, let us consider the predominant direction of the heliospheric magnetic field within several astronomical units of the heliopause. We begin by noting that loci of solar wind fluid elements emitted from the Sun near the solar equator at the maximum of every third 11-yr solar cycle are shown in projection by the dashed curves in Figure 5a. The direction of the solar magnetic field reverses (at solar maximum) from one solar cycle to the next, in the sense that in one solar cycle the field is directed predominantly outward from the Sun in the Northern Hemisphere and predominantly inward toward the Sun in the Southern Hemisphere, while in the preceding and following cycles these directions are reversed. Because the solar rotation axis is nearly perpendicular to the interstellar wind vector and because the heliospheric field is wrapped into a tight spiral (cf. Section 3.4), the component of the predominant heliospheric field in a plane perpendicular to the axis of the heliospheric tail (like that shown in Figure 5b) will be counterclockwise (in both hemispheres) in one solar cycle and clockwise in the preceding and following cycles. Thus, since postshock solar wind originating from the Sun during part or all of five solar cycles appears in the plane of Figure 5b (cf. dashed loci in Figure 5a), there are 16 reversals (of the component in the plane of Figure 5b) of the predominant field direction in the vicinity of the heliopause. [Actually, there will be many more field reversals because of the tilt and distortion of the neutral sheet separating the oppositely directed solar fields in the Northern and Southern Hemispheres (e.g. Hundhausen 1977), but the 16 reversals shown in Figure 5b adequately illustrate our point.] Reconnection can take place at a reasonably rapid rate only when the components of the interstellar and heliospheric fields shown in Figure 5a are oppositely directed, and any reconnection that does take place will eventually lead to the replacement of the reconnected heliospheric field with oppositely directed field from an adjacent heliospheric region. Thus, a skin of heliospheric magnetic field with a direction inappropriate to reconnection with the interstellar field will form at the heliopause, and the reconnection process will be suppressed.

In the absence of significant reconnection or disruption by the Kelvin-Helmholtz instability, it seems likely that the heliopause shown in Figure 5 will maintain its integrity far into the heliospheric tail. Of course, there will be reconnection at current sheets within the heliospheric tail, but this is not likely to affect the heliopause significantly and, indeed, should not be able to maintain the mean heliospheric tail temperature significantly above the interstellar gas temperature.

4. CONCLUDING REMARKS

Given the observational and theoretical information currently available, it appears that the interaction between the solar wind and the VLISM is characterized primarily by the interstellar magnetic containment of the solar magnetized plasma and by the slowing, turning, and cooling of the postshock subsonic solar wind flow through charge transfer with interstellar H atoms. The shock terminating supersonic solar wind flow (and thus accommodating the flow to the interstellar pressure) should be asymmetric, with the greatest shock distance at high latitudes, where the predominantly high-speed solar wind is characterized by a larger ram pressure than the lower speed wind flowing near the solar equator.

The distance to the solar wind terminal shock is very sensitive to the magnitude (and, to a lesser extent, to the direction) of the VLISM magnetic field. Recent indirect inferences of a shock distance of 50 AU (or a bit less) are not inconsistent with current observational estimates of this field, but unfortunately the same could be said of inferences of a shock distance of some 150 AU. If the terminal shock does lie near 50 AU, then the *Voyager 1* spacecraft should cross it within the next few years, and such a crossing would obviously place an important constraint on the interstellar magnetic field.

Clearly, the most important information to acquire in furthering our understanding of the interaction between the solar wind and the VLISM is a considerably improved determination of the VLISM magnetic field, but this is not likely to be obtainable in the near future. It is, therefore, imperative to maintain operation as long as possible of all our deep space missions, in the not-unreasonable hope that we can either directly or indirectly determine the location of the solar wind terminal shock. The observations most likely to be helpful in such a determination are of the plasma, magnetic field, plasma waves, and energetic particles, but it is also important to continue studying the UV radiation scattered by interstellar neutrals that have penetrated into the heliosphere. These UV observations, obtained both from deep space probes and from spacecraft in the inner heliosphere, can provide us not only with information concerning the VLISM ionization state (although only if the observational uncertainties are reduced), but also with valuable information concerning the latitudinal variation of the solar wind mass flux, which is important to our understanding of solar wind acceleration near the Sun (e.g. Lallement et al. 1986).

An obvious lesson to be learned from consideration of the interaction between the solar wind and the interstellar medium is that the community of scientists studying the local interstellar medium and the community of scientists studying the heliosphere should maintain the close contact that has recently been established.

ACKNOWLEDGMENTS

I am particularly grateful to B. C. Low for his comments and criticisms on the manuscript, which have led to a significantly improved review (especially Section 3.3). I also wish to thank Randy Jokipii and Marty Lee for useful discussions, and Liz Boyd for preparing the manuscript. This work was supported in part by NASA Contract W-17,016.

Literature Cited

Ajello, J. M. 1978. *Ap. J.* 222: 1068
Ajello, J. M., Stewart, A. I., Thomas, G. E., Graps, A. 1987. *Ap. J.* 317: 964
Axford, W. I. 1972. In *Solar Wind*, ed. C. P. Sonett, P. J. Coleman Jr., J. M. Wilcox, p. 609. *NASA SP-308*
Axford, W. I. 1973. *Space Sci. Rev.* 14: 582
Axford, W. I. 1976. In *Physics of Solar Planetary Environments*, 1: 270. Washington, DC: Am. Geophys. Union
Axford, W. I., Dessler, A. J., Gottlieb, B. 1963. *Ap. J.* 137: 1268
Axford, W. I., Ip, W.-H. 1986. *Adv. Space Res.* 6(2): 27
Axford, W. I., Newman, R. C. 1965. *Proc. Int. Conf. Cosmic Rays, 9th, London*, p. 173. London: Phys. Soc.
Baranov, V. B., Krasnobaev, K. V., Kulikovsky, A. G. 1970. *Dokl. Acad. Sci. USSR* 194: 41
Baranov, V. B., Krasnobaev, K. V., Ruderman, M. S. 1976. *Astrophys. Space Sci.* 41: 481
Beatty, J. J., Garcia-Munoz, M., Simpson, J. A. 1985. *Ap. J.* 294: 455
Bertaux, J. L., Blamont, J. E. 1971. *Astron. Astrophys.* 11: 200
Bertaux, J. L., Blamont, J. E., Mironova, E. N., Kurt, V. G., Bourgin, M. C. 1977. *Nature* 270: 156
Bertaux, J. L., Lallement, R., Kurt, V. G.,
Mironova, E. N. 1985. *Astron. Astrophys.* 150: 1
Blum, P. W., Fahr, H. J. 1970. *Astron. Astrophys.* 4: 280
Blum, P. W., Grzedzielski, S., Witt, N. 1980. *Astrophys. Space Sci.* 70: 513
Brown, R. L., Chang, C.-A. 1983. *Ap. J.* 264: 134
Burlaga, L. F. 1988. *Proc. Int. Sol. Wind Conf., 6th*, ed. V. J. Pizzo, T. E. Holzer, D. G. Sime, p. 547. *NCAR/TN-306 + Proc*
Chassefiere, E., Bertaux, J. L. 1987a. *Astron. Astrophys.* 174: 239
Chassefiere, E., Bertaux, J. L. 1987b. *Astron. Astrophys.* 176: 121
Chassefiere, E., Bertaux, J. L., Lallement, R., Kurt, V. G. 1986. *Astron. Astrophys.* 160: 229
Chassefiere, E., Bertaux, J. L., Lallement, R., Sandel, B. R., Broadfoot, L. 1988a. *Astron. Astrophys.* 199: 304
Chassefiere, E., Dalaudier, F., Bertaux, J. L. 1988b. *Astron. Astrophys.* 201: 113
Clauser, F. 1960. *Symp. Cosmical Gas Dyn., 4th, Varenna, Italy*, p. 306. Bologna: Nicola Zanichelli
Cox, D. P., Reynolds, R. J. 1987. *Annu. Rev. Astron. Astrophys.* 25: 303
Cranfill, C. 1971. PhD dissertation. Univ. Calif., San Diego
Cummings, A. C., Stone, E. C. 1988. *Proc.*

Int. Sol. Wind Conf., 6th, ed. V. J. Pizzo, T. E. Holzer, D. G. Sime, p. 599. NCAR/TN-306+Proc
Dalaudier, F., Bertaux, J. L., Kurt, V. G., Mironova, E. N. 1984. Astron. Astrophys. 134: 171
Davis, L. Jr. 1955. Phys. Rev. 100: 1440
Davis, L. Jr. 1962. J. Phys. Soc. Jpn. 17A-II: 543
Drury, L. O. 1988. Proc. Int. Sol. Wind Conf., 6th, ed. V. J. Pizzo, T. E. Holzer, D. G. Sime, p. 521. NCAR/TN-306+Proc
Fahr, H. J. 1974. Space Sci. Rev. 15: 483
Fahr, H. J., Nass, H. U., Rucinski, D. 1985. Astron. Astrophys. 142: 476
Fahr, H. J., Neutsch, W., Grzedzielski, S., Macek, W., Ratkiewicz-Landowska, R. 1986. Space Sci. Rev. 43: 329
Feldman, W. C., Asbridge, J. R., Bame, S. J., Gosling, J. T. 1977. In The Solar Output and Its Variations, ed. O. R. White, p. 351. Boulder: Colo. Assoc. Univ. Press
Fisk, L. A. 1986. In The Sun and the Heliosphere in Three Dimensions, ed. R. G. Marsden, p. 401. Dordrecht: Reidel
Fisk, L. A., Kozlovsky, B., Ramaty, R. 1974. Ap. J. Lett. 190: L35
Frisch, P. C. 1986. Adv. Space Res. 6(1): 345
Frisch, P. C., York, D. C., Fowler, J. R. 1987. Ap. J. 320: 842
Garcia-Munoz, M., Mason, G. M., Simpson, J. A. 1973. Ap. J. Lett. 182: L81
Greenberg, J. M. 1978. In Cosmic Dust, ed. J. A. M. McDonnell, p. 187. New York: Wiley
Heiles, C. 1976. Annu. Rev. Astron. Astrophys. 14: 1
Holzer, T. E. 1972. J. Geophys. Res. 77: 5407
Holzer, T. E. 1977. Rev. Geophys. Space Phys. 15: 467
Holzer, T. E., Axford, W. I. 1971. J. Geophys. Res. 76: 6965
Hovestadt, D., Vollmer, O., Gloeckler, G., Fan, C. Y. 1973. Phys. Rev. Lett. 31: 650
Hundhausen, A. J. 1968. Planet. Space Sci. 16: 783
Hundhausen, A. J. 1973. J. Geophys. Res. 78: 1528
Hundhausen, A. J. 1977. In Coronal Holes and High Speed Wind Streams, ed. J. B. Zirker, p. 225. Boulder: Colo. Assoc. Univ. Press
Hundhausen, A. J., Hansen, R. T., Hansen, S. F. 1981. J. Geophys. Res. 86: 2079
Ip, W.-H., Axford, W. I. 1985. Astron. Astrophys. 149: 7
Isenberg, P. A. 1987. J. Geophys. Res. 92: 1067
Jokipii, J. R. 1986. J. Geophys. Res. 91: 2929
Jokipii, J. R., Parker, E. N. 1967. Planet. Space Sci. 15: 1375
Ko, C. M., Jokipii, J. R., Webb, G. M. 1988. Ap. J. 326: 761
Ko, C. M., Webb, G. M. 1987. Ap. J. 323: 657
Ko, C. M., Webb, G. M. 1988. Ap. J. 325: 296
Kojima, M., Kakinuma, T. 1987. J. Geophys. Res. 92: 7269
Kurth, W. S. 1988. Proc. Int. Sol. Wind Conf., 6th, ed. V. J. Pizzo, T. E. Holzer, D. G. Sime, p. 667. NCAR/TN-306+Proc
Kurth, W. S., Gurnett, D. A., Scarf, F. L., Poynter, R. L. 1984. Nature 312: 27
Kurth, W. S., Gurnett, D. A., Scarf, F. L., Poynter, R. L. 1987. Geophys. Res. Lett. 14: 49
Lallement, R., Bertaux, J. L., Kurt, V. G. 1985. J. Geophys. Res. 90: 1413
Lallement, R., Holzer, T. E., Munro, R. H. 1986. J. Geophys. Res. 91: 6751
Lamy, P. L., Lefevre, J., Millet, J., Lafon, J. P. 1985. In Properties and Interactions of Interplanetary Dust, ed. R. H. Giese, P. Lamy, p. 335. Dordrecht: Reidel
Lazarus, A., Belcher, J. 1988. Proc. Int. Sol. Wind Conf., 6th, ed. V. J. Pizzo, T. E. Holzer, D. G. Sime, p. 533. NCAR/TN-306+Proc
Lee, M. A. 1988. Proc. Int. Sol. Wind Conf., 6th, ed. V. J. Pizzo, T. E. Holzer, D. G. Sime, p. 635. NCAR/TN-306+Proc
Lee, M. A., Axford, W. I. 1988. Astron. Astrophys. 194: 297
Levy, E. H. Jokipii, J. R. 1976. Nature 264: 424
McDonald, F. B., Teegarden, B. J., Trainor, J. H., Webber, W. R. 1974. Ap. J. Lett. 187: L105
McKee, C. F., Ostriker, J. P. 1977. Ap. J. 218: 148
McNutt, R. L. Jr. 1988. Geophys. Res. Lett. 15: 1307
Meier, R. R. 1980. Astron. Astrophys. 91: 62
Mobius, E. 1986. Adv. Space Res. 6(1): 199
Mobius, E., Hovestadt, D., Klecker, B., Scholer, M., Gloeckler, G., Ipavich, F. M. 1985. Nature 318: 426
Paresce, F., Bowyer, S. 1973. Astron. Astrophys. 27: 399
Parker, E. N. 1956. Phys. Rev. 103: 1518
Parker, E. N. 1958. Ap. J. 128: 664
Parker, E. N. 1961. Ap. J. 134: 20
Parker, E. N. 1963. Interplanetary Dynamical Processes. New York: Interscience
Parker, E. N. 1964. Ap. J. 139: 951
Patterson, T. N. L., Johnson, F. S., Hanson, W. B. 1963. Planet. Space Sci. 11: 767
Pesses, M. E., Jokipii, J. R., Eichler, D. 1981. Ap. J. Lett. 246: L85
Pizzo, V. J. 1986. Adv. Space Res. 6(1): 353
Ripken, H. W., Fahr, H. J. 1983. Astron. Astrophys. 122: 181
Schwenn, R. 1983. In Solar Wind Five, ed. M. Neugebauer, p. 489. Washington, DC: NASA

Semar, C. L. 1970. *J. Geophys. Res.* 75: 6892
Sime, D. G. 1983. In *Solar Wind Five*, ed. M. Neugebauer, p. 453. Washington, DC: NASA
Sousk, S. F., Lenchek, A. M. 1969. *Ap. J.* 158: 781
Steinitz, R., Eyni, M. 1980. *Ap. J.* 241: 417
Suess, S. T., Dessler, A. J. 1985. *Nature* 317: 702
Tawara, H., Kato, T., Nakai, Y. 1983. *Rep. IPPJ-AM-30*, Inst. of Plasma Phys., Nagoya Univ., Jpn.
Thomas, G. E. 1978. *Annu. Rev. Earth Planet. Sci.* 6: 173
Thomas, G. E., Krassa, R. F. 1971. *Astron. Astrophys.* 11: 218
Thompson, R. C., Nelson, A. H. 1980. *MNRAS* 191: 863
Tinsley, B. A. 1971. *Rev. Geophys. Space Phys.* 9: 89
Torland, T. H., Heiles, C. 1986. *Ap. J.* 301: 339
Vidal-Madjar, A. 1977. In *The Solar Output and Its Variations*, ed. O. R. White, p. 213. Boulder: Colo. Assoc. Univ. Press
von Rosenwinge, T. T., McDonald, F. B. 1975. *Proc. Int. Cosmic Ray Conf., 14th, Munich*, 2: 792
Wallis, M. 1971a. *Rep. TRITA-EPP-71-01*, R. Inst. Technol., Stockholm, Swed.
Wallis, M. 1971b. *Nature* 233: 23
Wallis, M. K. 1973. *Astrophys. Space Sci.* 20: 3
Wallis, M. 1974. *MNRAS* 167: 103
Wallis, M. K. 1975. *Nature* 254: 202
Wallis, M. K. 1978. *Space Res.* 18: 401
Wallis, M. K. 1981. In *Solar Wind Four*, ed. H. Rosenbauer, p. 516. Katlenburg-Lindau, Germ: MPAE
Wallis, M. K. 1984. *Astron. Astrophys.* 130: 200
Wallis, M. K. 1988. *Proc. Int. Sol. Wind Conf., 6th*, ed. V. J. Pizzo, T. E. Holzer, D. G. Sime, p. 687. *NCAR/TN-306+Proc*
Wallis, M. K., Hassan, M. H. A. 1978. *Planet. Space Sci.* 26: 111
Wallis, M. K., Wallis, J. 1979. *Astron. Astrophys.* 78: 41
Webber, W. R. 1987. *Astron. Astrophys.* 179: 277
Weller, C. S., Meier, R. R. 1981. *Ap. J.* 246: 386
Weymann, R. 1960. *Ap. J.* 132: 390
Wu, F. M., Gangopadhyay, P., Ogawa, H. S., Judge, D. L. 1988. *Ap. J.* 331: 1004

SURFACE PHOTOMETRY AND THE STRUCTURE OF ELLIPTICAL GALAXIES

John Kormendy[1]

Dominion Astrophysical Observatory, Herzberg Institute of Astrophysics, Victoria, British Columbia V8X 4M6, Canada

S. Djorgovski[2]

Division of Physics, Mathematics, and Astronomy, California Institute of Technology, Pasadena, California 91125

1. INTRODUCTION

This paper reviews surface photometry of bulges and elliptical galaxies. Work prior to 1982 is discussed by Kormendy (1982a; hereafter K82). Since then, the subject has gone through a revolution. CCD detectors have come into common use, providing photometry accurate enough to measure new classes of subtle properties of ellipticals. Together with improvements in seeing, CCDs have allowed the resolution and study of galaxy cores and nuclei (Section 2). Newly discovered structural details, such as dust, shells, and dynamical subsystems, show the importance of accretion events in galaxy evolution (Sections 3–6). Better measurements of parameter scaling laws have led to improved constraints on galaxy formation (Section 8). Finally, CCDs provide accurate measurements of departures from elliptical isophotes (Section 9) and color gradients (Section 10). These

---

[1] Visiting Astronomer, Canada–France–Hawaii Telescope, operated by the National Research Council of Canada, the Centre National de la Recherche Scientifique of France, and the University of Hawaii.
[2] Alfred P. Sloan Fellow.

observations are currently producing a quantum jump in our understanding of elliptical galaxies.

Some of the present subjects are discussed in more detail in recent reviews by Kormendy (1987a; hereafter K87), Okamura (1988), and Nieto (1988). Techniques are discussed by de Vaucouleurs (1979, 1984), Nieto (1982), Capaccioli & de Vaucouleurs (1983), Capaccioli (1985, 1987, 1988a,b), Okamura (1988), and Djorgovski & Dickinson (1989). Compilations of photometry references for individual galaxies are found in Davoust & Pence (1982) and Pence & Davoust (1985).

Unless otherwise noted, we assume that the Hubble constant is $H_0 = 50$ km s^{-1} Mpc^{-1}.

2. CORES AND NUCLEI

Two kinds of structure are commonly seen at the centers of early-type galaxies. When observed with sufficient resolution, the steep brightness profile of an elliptical usually flattens into a nearly constant surface brightness core. In addition, a nucleus is sometimes seen inside the core. By this, we mean a dynamically distinct cluster of stars that is much smaller and denser than the core. Cores are reviewed by Kormendy (K82, 1984, K87, 1987c) and Lauer (1988b); here we summarize their properties in Section 2.1. Nuclei are less well studied and understood; we review them in Section 2.2.

2.1 *Summary of Core Properties*

Reliable core photometry became available only when problems of seeing and photographic photometry were resolved. All photographic core photometry proves to be unreliable. CCDs do better: Profiles derived by different authors routinely agree to $\lesssim 0.1$ mag arcsec^{-2} (Lauer 1985a, K87). Seeing affected early work on cores so strongly that it was not clear whether most galaxies have cores at all (Schweizer 1979, 1981b). Seeing corrections could be derived if one *assumed* that galaxies have cores, but the derived core parameters were model dependent (K82, Kormendy 1984). Then CCD observations by Lauer (1985a,b) found nearly isothermal cores in about a dozen ellipticals, increasing the sample of resolved cores by a factor of four. More recently, the Canada–France–Hawaii Telescope (CFHT) has provided seeing a factor of about two better than available previously; nearly isothermal cores are now resolved in almost all bright, nearby ellipticals (Kormendy 1985a, 1987c, K87).

CCD data are accurate enough for seeing corrections based on deconvolutions. Simple techniques are used by Djorgovski (1983) and Lauer (1985a,b), and more powerful techniques are discussed by Bendinelli et al.

(1982, 1984a, 1985, 1986, 1988, and references therein). The latter techniques have a mixed record, successfully revealing a stellar nucleus in M81 but finding a similar nucleus in M32 that is not confirmed by higher resolution observations (Section 2.2). Their weakness is that they magnify nonrandom errors, such as kinks in the observed profile. Lauer's simpler technique mines less resolution from the data but finds fewer spurious features. Better still is resolution good enough to require little correction. By the time this article appears, the *Space Telescope* may provide a long-awaited factor-of-five additional improvement in resolution.

Core profile shapes have been examined systematically by Lauer (1985b). His CCD data show that virtually all cores have slightly nonisothermal brightness profiles. Kormendy's (1985a, K87) high-resolution CCD photometry shows further that core profile shape correlates weakly with galaxy luminosity. A few galaxies, including some first-ranked galaxies in clusters, have isothermal profiles. Fainter galaxies have profiles that do not flatten completely into a core. The faintest galaxy with a well-resolved core is M31; it is even less isothermal than the ellipticals (Kent 1983). Such profiles have been interpreted as evidence for central black holes (Young et al. 1978, 1979) or anisotropy (Kormendy 1985a).

Some cores also show kinematic evidence for anisotropy (K87). For example, the core of NGC 1600 falls far below the "oblate line" describing isotropic oblate rotators in the $V_{\max}/\sigma$–ε diagram (Illingworth 1977). Like the overall shapes of ellipticals, the E3 shape of the core of NGC 1600 must be maintained by anisotropy. Dynamical modeling is required to explore the orbital distribution functions implied by these observations.

Considerable effort has also gone into the measurement of characteristic parameters of cores, i.e. central surface brightnesses μ_0, core radii r_c at which the surface brightness has fallen by a factor of two, and central velocity dispersions σ. Structural scaling laws revealed by these data are discussed in Section 8.

2.2 Nuclei

Dense nuclear star clusters superposed on much larger cores have been recognized in at least six bright galaxies. The best example is in M31 (Light et al. 1974); this has $r_c \simeq 0\rlap{.}''4$ and $\mu_{0V} \simeq 12.4$ mag arcsec^{-2}, compared with $r_c = 17''$ and $\mu_{0V} = 15.7$ mag arcsec^{-2} for the bulge. Tremaine & Ostriker (1982) have shown that the nucleus and bulge of M31 are dynamically independent. We refer to these central star clusters as nuclei and distinguish them from bulges with cuspy brightness profiles (e.g. a pure power law like that in M32; Tonry 1984b). We also distinguish nuclei from nonthermal point sources such as those in Seyfert galaxies, quasars, and M87.

Besides M31, nuclei are found in M81 (Kormendy 1985a, Bendinelli et al. 1986) and in other nearby bulges (Kormendy 1985a). Their detection is limited by poor resolution (Lauer 1988b). Nuclei are also seen in many dwarf spheroidal galaxies (Reaves 1977, 1983, Romanishin et al. 1977, Caldwell 1983, 1987, Binggeli et al. 1984, 1985, Ichikawa et al. 1986, van den Bergh 1986, Caldwell & Bothun 1987) and in many disk galaxies that are late enough in type so that they do not contain bulges (e.g. M33; Gallagher et al. 1982, and references therein).

Nuclei are rarely seen in ellipticals. For example, in M87, the nuclear spectrum shows no stellar absorption lines when the spectrum of the underlying core is subtracted (Dressler 1988, Kormendy 1989). A small central excess of brightness above an isothermal core in NGC 3379 (de Vaucouleurs & Capaccioli 1979, K82, Nieto & Vidal 1984) is not due to a stellar nucleus either, but only to a nonisothermal core exactly like those in other ellipticals (Bendinelli et al. 1984b, Kormendy 1985a). Nuclei suspected to exist in M32 and NGC 4649 (Bendinelli et al. 1982, Lauer 1988b) are not confirmed at better resolution (J. Kormendy, in preparation). Nuclei should be relatively easy to see in bright ellipticals because they have large, well-resolved cores. Their apparent scarcity could be due to the existence of a maximum luminosity for nuclei. Also, few ellipticals are as close to us as the bulges that are known to contain nuclei.

Since nuclei are poorly resolved, little further is known about them. Stellar kinematic data are available in M31 (K87, Dressler & Richstone 1988, Kormendy 1988b), NGC 3115 (J. Kormendy & D. O. Richstone, in preparation), and NGC 4594 (Kormendy 1988c). Rapid rotation and velocity dispersions of ~ 100 km s^{-1} (after bulge subtraction) indicate that all three nuclei are disks. (Dressler & Richstone did not come to this conclusion, but they did not subtract the bulge spectrum; then detection of the cold component is difficult.) In NGC 3115 and NGC 4594, the disk structure is also seen in the isophotes.

The available data suggest that disklike nuclei are built out of gas that has fallen into the center (van den Bergh 1976, K82, Kormendy 1982b, 1988b,c, Gallagher et al. 1982, Kormendy & Illingworth 1983). This idea is a natural consequence of the hypothesis that black holes are fueled by infalling gas. If gas can reach the black hole, it may form stars along the way when the density gets high enough in the gravitational funnel. This may even be a necessary step in the formation of nuclear black holes, since collapse times of cores in giant ellipticals are long, whereas nuclei can evolve more rapidly (Kormendy 1988a,b,c). Further discussion is given in Shlosman & Begelman (1987) and in Duschl (1988a,b).

Nuclei may originate in other ways, too. Globular clusters sink toward the center by dynamical friction and may form nuclei (Tremaine et al.

1975). A large galaxy can accrete a small one with a compact core (Section 3). And black holes may produce central density cusps. Accreted nuclei should be distinguishable from black hole cusps: In general they should have smaller σ and a different rotation axis than the rest of the galaxy. However, accreted nuclei and ones grown by gas infall and star formation may be difficult to distinguish.

There are indications that nuclei in dwarf spheroidal and disk galaxies are similar to those seen in bulges. The nucleus of M33 is interpreted by Gallagher et al. (1982) as a composite-age stellar population, consistent with late infall of gas and subsequent star formation. Spectra of nuclei in dwarf spheroidal galaxies suggest that they are $\gtrsim 5$ Gyr old but sometimes contain a contribution from younger (A–F) stars; this is also consistent with secondary formation (Bothun et al. 1985, Caldwell & Bothun 1987, Bothun & Mould 1988). It is also possible that some "nuclei" in dwarf galaxies are really very low-luminosity bulges, since small bulges have small r_c and high μ_0 (see K87 for a review).

3. DYNAMICAL SUBSYSTEMS IN GALAXY CORES: EVIDENCE FOR MERGERS

A major development in recent years has been the realization that galaxies accrete significant amounts of material in the form of gas and small companions. The next three sections discuss some of the evidence. We begin with observations of distinct dynamical subsystems in galaxy cores.

The first clear example was NGC 5813. Efstathiou et al. (1982) found a core-within-a-core brightness profile in this otherwise normal elliptical (i.e. its core contains a second, smaller core of higher surface brightness). The inner core rotates more rapidly than the outer, and, except for the central measurement, has a smaller velocity dispersion. Kormendy (1984) suggested that these observations are the signature of a merger between a low- and a high-luminosity elliptical. Low-luminosity ellipticals have smaller core radii and higher central surface brightnesses than giant ellipticals. Kormendy showed that the robust core of a small elliptical can survive a merger with a giant elliptical and form a distinct subsystem at the center. He predicted that the rotation axis of the subsystem should be oriented randomly with respect to the main galaxy. In practice, observed orientations may be somewhat nonrandom because merger cross sections depend on encounter geometry. Nevertheless, this provides a test of the merger hypothesis. Also, the nucleus should in general rotate more rapidly than the rest of the galaxy because low-luminosity ellipticals are rapid rotators (Davies et al. 1983). Finally, the Faber-Jackson (1976) relation predicts that the velocity dispersion should in some cases decrease toward the center. These effects are also seen in N-body simulations (e.g. Balcells & Quinn 1988).

A number of galaxies that dramatically show this behavior have now been found. Franx & Illingworth (1988), Jedrzejewski & Schechter (1988a), and Bender (1988b) have found seven elliptical galaxies whose cores are kinematically distinct from the rest of the galaxy. In four cases, the inner and outer parts rotate in opposite directions. This is strong evidence for accretion. Further support is provided by the observation of isophote twists between the two subsystems (Efstathiou et al. 1982, Bender 1988b).

In the new cases, no core-within-a-core structure is seen. This is not surprising. Efstathiou et al. (1982) warned us that their NGC 5813 photometry is not of high quality. It should be checked, especially since dust can counterfeit a core-within-a-core structure. Also, distinct cores are only expected in extreme cases, e.g. when a galaxy like M87 eats one like M32 (Figure 3 in Kormendy 1984). Such events should be rare, because faint ellipticals are rare (Binggeli et al. 1985, Sandage et al. 1985b). Mergers between nearly equal galaxies are not likely to leave a signature in the brightness profile.

The merger interpretation is attractive, but alternatives are possible. For example, if the figure rotation velocity in an elliptical is backward with respect to the streaming velocity of the stars, the sum (which is what we observe) can change sign. However, counterstreaming is difficult to achieve (Vietri 1986, 1988) and does not explain subsystems that rotate at right angles to their galaxies (NGC 4406; Bender 1988b, Franx 1988). This interpretation seems improbable. Another possibility is that the inner and outer parts of a galaxy acquire different angular momenta through tidal torques (Binney 1987, Barnes & Efstathiou 1987). This cannot be excluded, although it is least likely near the center. Like the above authors, we conclude that the observed dynamical subsystems result from mergers. These could be mergers of bulges or ellipticals, or ones involving gas infall and star formation (Section 4). IC 1459 and NGC 5322 may be examples of the latter: Their nuclear subsystems appear to be counterrotating stellar disks (Franx & Illingworth 1988, Bender 1988b).

Only a fraction of all merger remnants can be recognized from observations like the above. The fact that about one third of the ellipticals examined so far show nuclear subsystems (Bender 1988b, Jedrzejewski & Schechter 1988a,b) suggests that mergers affect a significant fraction of galaxies.

4. DUST IN ELLIPTICAL GALAXIES

According to classical definition (Hubble 1926, de Vaucouleurs 1959, Sandage 1961), elliptical galaxies contain no dust. Galaxies with dust have usually been given S0 or later-type classifications. Now sensitive searches are finding that even the remaining, classical ellipticals often contain dust. This section summarizes its properties. Other recent reviews have been given by Schweizer (1987), Bertola (1987), and Nieto (1988; hereafter N88). A catalog of dusty ellipticals has been published by Ebneter & Balick (1985).

Progress in this subject has depended critically on the ability to detect subtle absorption features superposed on steep brightness gradients. CCD surveys are especially powerful: Their dynamic range is large, and the data can easily be subjected to digital "unsharp masking" (Sandage & Miller 1964, Malin 1977). This is done by dividing the image by a model of the overall brightness distribution without the fine structure. The model can be a smoothed version of the original image, or a synthetic galaxy image with the best-fitting elliptical isophotes, or an image taken in a redder bandpass. (In the last case, the ratio is a color image.) These techniques show that $\gtrsim 50\%$ of bulges and elliptical galaxies contain dust.

4.1 *Frequency of Occurrence of Dust*

A few dusty ellipticals have been known for years. They received little systematic attention until Bertola & Galletta (1978) pointed out that several ellipticals have dust lanes along their minor axes and therefore may be prolate. This had immediate impact because of the recent discovery (Bertola & Capaccioli 1975, Illingworth 1977) that most bright ellipticals are dynamically supported not by rotation but by velocity dispersion anisotropy, which suggests that they are triaxial (Binney 1976, 1978a,b, 1982a,b).

Systematic surveys for dust followed, and detection rates increased as search techniques improved. Hawarden et al. (1981) examined carefully chosen diskless galaxies on the ESO/SRC IIIa-J and Palomar sky surveys and found a substantial number (40) with dust. Sadler & Gerhard (1985a,b) found dust in $23 \pm 7\%$ of ellipticals with mean diameters of at least 2' on the ESO B survey. Like all such estimates, this is a lower limit. The dust is usually in well-defined, nearly edge-on disks; this implies that many face-on dust distributions are going undetected. Sadler & Gerhard estimated that the true fraction of ellipticals with dust is at least 40%. A CCD survey by Sparks et al. (1985) led to similar conclusions. More recently, Djorgovski & Ebneter (1986) and Ebneter et al. (1988) have detected dust in 36% of the 116 ellipticals they studied. Finally, CCD photometry with the CFHT (Kormendy & Stauffer 1987; J. Kormendy, to be published) shows a still higher detection frequency, because of the excellent seeing on Mauna Kea. Dust distributions are often so small that they are barely detected even with the CFHT. Many more may await discovery with the *Space Telescope*.

This dust was also found by the *Infrared Astronomical Satellite* (*IRAS*). Detection frequencies in co-added *IRAS* survey data on bright, nearby ellipticals are comparable to or larger than those seen optically (Jura et al. 1987). Optical and *IRAS* photometry both imply that typical dust masses are $\sim 10^5$–10^6 $M_\odot$; for canonical gas-to-dust ratios, this corresponds to $\sim 10^7$–10^8 $M_\odot$ of cold gas (e.g. Sadler & Gerhard 1985b, Sparks et al. 1985, Jura 1986, Jura et al. 1987, Véron-Cetty & Véron 1988).

It is now clear that dust in elliptical galaxies is not rare. This is one more piece of evidence that ellipticals contain substantial amounts of interstellar matter [see Schweizer (1987) for a review]. Some gas is acquired by accretion (see the next section), and some is expected from mass loss during stellar evolution (Sandage 1957, Faber & Gallagher 1976). With the discovery that ellipticals generally contain 10^9–10^{10} $M_\odot$ of X-ray–emitting gas (e.g. Forman et al. 1985), the idea that they are surprisingly free of interstellar matter has disappeared.

Although the precise frequency is uncertain because of classification bias, the above surveys show that bulges contain dust still more often than ellipticals. Even prototypical bulges can be riddled with dust (e.g. M31; Johnson & Hanna 1972, Kent 1983, McElroy 1983), as well as ionized (Ciardullo et al. 1988) and other gas.

4.2 Origin of Dust: Further Evidence for Galaxy Mergers

There is strong evidence that many large-scale dust and H I gas distributions are accreted. The most convincing evidence is kinematic: The gas and dust are usually in disks rotating at random orientations with respect to the optical major axis [H I (e.g. Gallagher et al. 1977); H II (Schweizer 1980, 1981a, 1982, Davies & Illingworth 1986, Caldwell et al. 1986); H II associated with dust (Burbidge & Burbidge 1959, Graham 1979, Marcelin et al. 1982, Möllenhoff 1982, Sharples et al. 1983, Caldwell 1984, Bertola et al. 1985, Möllenhoff & Marenbach 1986, Wilkinson et al. 1986, Bland et al. 1987, Varnas et al. 1987, Galletta 1987, Bertola & Bettoni 1988, Bertola et al. 1988a,b, Möllenhoff & Bender 1988)]. Minor-axis dust lanes rotate at right angles to the stars. Sometimes dust lanes and stars even rotate in opposite directions. This gas cannot come from internal mass loss. Accretion is also suggested by the morphology (although dust is not correlated with the presence of ripples and shells; Schweizer & Ford 1985). At large radii, dust lanes often show S-shaped warps or transitions from regular disks to irregular distributions. Such behavior is expected for material just settling into equilibrium, since orbital clocks run slower at larger radii. Note that accretion does not require cannibalism; gas can be donated by a galaxy that gets away (Schweizer 1987).

Small dust lanes are more common than large-scale dust distributions. Whether or not they have the same origin is not clear. They are usually well-defined rings or disks near the center, often oriented parallel to the major axis. Many resemble the inner dust lanes commonly seen in S0 and spiral galaxies (Sandage 1961). Some or even most may have an internal origin. However, a folklore is developing, perhaps prematurely, that dust in ellipticals is always accreted. Kinematic constraints are badly needed on the fraction of inner dust disks that are accreted. Is the fraction of counterrotating cases near 50% (as it is for large-scale dust lanes; Bertola et al. 1988a,b), or is it much smaller? At stake is a better understanding of how much secular evolution results from mergers and how much from internal processes.

4.3 Three-Dimensional Shapes of Ellipticals Containing Dust

Bertola & Galletta's (1978) pioneering paper raised the hope that dust-lane geometry could be used to measure galaxy shapes. However, the large number of free parameters make this complicated. The results provide further evidence that ellipticals are triaxial, and they sometimes favor an oblate or a prolate configuration, but they have not securely told us the shape of any individual galaxy.

This subject is reviewed in detail by Merritt & de Zeeuw (1983) and will also be reviewed in the next volume of this series by de Zeeuw (1990). Thus our summary of the predictions is brief. Gas in a spheroidal or triaxial potential settles into certain preferred planes through differential precession and dissipation (Kahn & Woltjer 1959, Gunn 1979, Lake & Norman 1983). Consider first the simplest case, in which the shape of the potential does not rotate. Then the gas settles into one of two planes, i.e. perpendicular to the shortest or to the longest axis (e.g. Heiligman & Schwarzschild 1979, Tohline et al. 1982, Steiman-Cameron & Durisen 1982). In a spheroidal galaxy, only the equatorial orbits are stable; polar orbits gradually tip over into the equatorial plane. If we knew that ellipticals are spheroids, then those with minor-axis dust lanes would be prolate and those with major-axis dust lanes would be oblate. But ellipticals can be triaxial. Then, for some infall angles and galaxy shapes, gas is captured into polar orbits. Therefore, a dust lane along a particular axis is consistent with either oblate or prolate structure. Already there is no unique relationship between galaxy shape and dust-lane geometry.

The next complication is that the figure can tumble (angular velocity $\Omega_p \neq 0$). However, the angular velocity we measure is that of the stars, and they stream through the figure (as they do through bars and spiral arms). Therefore, we do not know Ω_p; in fact, we are as interested in estimating Ω_p as in finding the shape of the galaxy. Tumbling elliptical galaxies allow additional equilibrium orbits, as summarized in Figure 1.

Half of the configurations shown in Figure 1 may be uncommon. If a galaxy tumbles about its long axis, stellar rotation velocities will be large along the minor axis and zero along the major axis. This has been observed in only a few galaxies (e.g. NGC 4261; Davies & Birkinshaw 1986, Wagner et al. 1988). We assume that ellipticals usually tumble about their short axes. Then stable major-axis dust lanes should be prograde. Minor-axis dust lanes should be perpendicular at small radii and should twist at large radii and show retrograde rotation.

What do we observe? Major-axis dust lanes counterrotate in two of the four ellipticals studied (Bertola et al. 1988a); retrograde gas velocities are also seen in the SB0 galaxy NGC 4546 (Galletta 1987). Of seven minor-axis dust lanes measured so far, three show retrograde-rotating twists [NGC 1316 (Schweizer 1980), NGC 5363 (Sharples et al. 1983, Bertola et al. 1985), and A0609−33 (Möllenhoff & Marenbach 1986)] and four show prograde twists [NGC 4589 (Möllenhoff & Bender 1988), NGC 5128 (e.g.

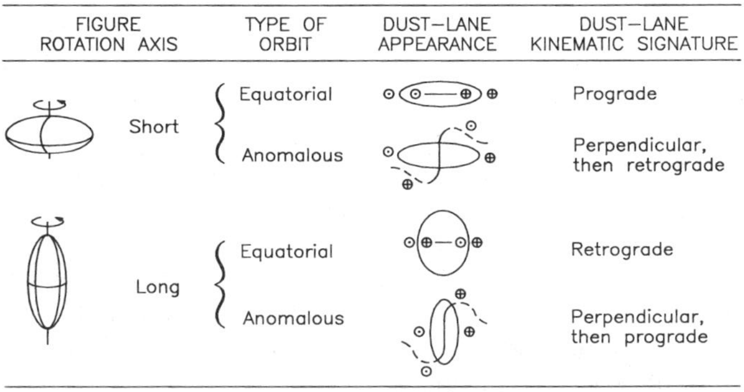

FIGURE ROTATION AXIS	TYPE OF ORBIT	DUST–LANE APPEARANCE	DUST–LANE KINEMATIC SIGNATURE
Short	Equatorial		Prograde
	Anomalous		Perpendicular, then retrograde
Long	Equatorial		Retrograde
	Anomalous		Perpendicular, then prograde

Figure 1 Stable orbits of gas in a rotating triaxial galaxy (adapted from Merritt & de Zeeuw 1983). As illustrated, the figure tumbles in the direction of stellar rotation ($\Omega_p > 0$); if $\Omega_p < 0$, the sense of gas rotation is reversed. Assume that the figure rotates about its shortest or longest axis (*left*). The second column gives the kind of orbit, and the third sketches resulting dust lanes seen edge-on. Anomalous orbits have different orientations at different radii (van Albada et al. 1982). They are the analogues of polar orbits in a stationary potential; at small radii, where Ω_p is unimportant, they are polar. At large radii, the figure rotates several times during an orbit and so is effectively oblate-spheroidal; then the orbit is equatorial (Simonson 1982). In between, the orbits have skew orientations determined by the Coriolis force. The schematic illustrations of dust lanes show the directions of stellar and gas motion; ⊙ indicates approach, and ⊕ indicates recession. The right column states the kinematic signature, i.e. the sense of rotation of the dust lane with respect to the stars.

Davies et al. 1984, Bertola et al. 1985, Wilkinson et al. 1986, Bland et al. 1987), NGC 5266 (Caldwell 1984, Möllenhoff & Marenbach 1986, Varnas et al. 1987), and A0151−49 (Sharples et al. 1983, Bertola et al. 1985)].

The hypothesis that these dust lanes are in equilibrium can be saved if $\Omega_p < 0$ (e.g. Varnas et al. 1987). Although it is difficult (Vietri 1986), Vietri (1988) has succeeded in constructing at least one realistic dynamical model in which retrograde figure rotation is slow enough, and prograde stellar streaming large enough, so that the sum (i.e. the observed galaxy rotation velocity) is opposite to the tumbling direction at some radii (see also Freeman 1966). On the other hand, N-body models that collapse and become bar-unstable have always resulted in $\Omega_p > 0$ [see van Albada (1987) for a review]. It is not clear whether retrograde tumbling is a viable interpretation.

Therefore the observations suggest that many dust-lane warps are transient—that gas has settled to a preferred plane at small radii but still remembers the merger geometry in the warp (Tubbs 1980, Simonson 1982, Bertola et al. 1985, Wilkinson et al. 1986, Schweizer 1987, Schwarzschild 1987, Möllenhoff & Bender 1988). This possibility has existed from the beginning; it was resisted mainly because warps then tell us less about galaxy shapes. But the fact that dust lanes are often regular at small radii and irregular farther out (e.g. NGC 1316; Schweizer 1980) should already have convinced us that settling into principal planes is not always complete.

We dwell on this subject because it has seemed to be the most rigorous new method to measure the shapes of individual ellipticals. It remains promising. But even with photometry and kinematic data, it is difficult to unravel the many unknowns: the amount of triaxiality, the orientation of the galaxy, the pattern rotation speed, and the question of whether dust has settled into equilibrium. There are other uncertainties that we have not discussed. For example, a slowly rotating elliptical may not be exclusively oblate or prolate; it may at some radius change from one to the other. And some conclusions summarized in Figure 1 may be violated in special potentials. Simple deductions seem reasonably secure: (*a*) Ellipticals are generally triaxial; (*b*) some are prolate and others are oblate; and (*c*) some warps imply that $\Omega_p \neq 0$ and others imply nonstationary structure. But more detailed progress has been elusive.

We know of no simple remedies. As Merritt & de Zeeuw (1983) point out, better statistics would help. We may be basing far-reaching conclusions on configurations that turn out to be rare. New discoveries of systematic behavior may reduce the available parameter space. But it appears that the implications of dust-lane geometry and kinematics become statistical. Unless further work sheds new light, we still do not know how to measure the shapes of individual ellipticals.

4.4 Dust and the Distinction Between E and S0 Galaxies

The presence of dust contributes to a blurring of the distinction between E and S0 galaxies. This is partly just a practical problem of classification. When ellipticals were dust free by definition, dusty galaxies were easy to classify. If we now adopt as the main classification criterion the presence (S0) or absence (E) of a disk, then it is difficult to distinguish ellipticals from S0s with faint disks [see K82 and Capaccioli (1987) for reviews]. A significant number of galaxies must be misclassified in the literature. If the E–S0 sequence is continuous, this makes little difference for an individual object (Schweizer 1987). But it can systematically affect galaxy samples selected for physical studies.

There is also a more difficult problem of principle. Since dust lanes and gas can form stars, a galaxy can change our perception of its morphological type. For example, a slowly rotating, bright elliptical may, through judicious cannibalism, grow a disk and come to look like an S0. This would contribute noise to correlations between physical properties and type. For example, even if real bulges rotate rapidly, there would be apparent exceptions because some S0s started life as ellipticals. This is an example of how secular evolution can obscure a physical correlation that was set up during an earlier phase of galaxy formation. Since far-reaching conclusions are often based on a few galaxies with surprising behavior, we need to be careful to understand and allow for secular evolution.

5. SHELLS AND RIPPLES

Shells or ripples are faint, arc-shaped structures in galaxy halos (Arp 1966, Malin 1979, Schweizer 1980, 1983, Malin & Carter 1980, 1983, Malin et al. 1983, Schweizer & Ford 1985, Schweizer & Seitzer 1988, Prieur 1988). They are reviewed by Schweizer (1983), Quinn (1984), Athanassoula & Bosma (1985), Dupraz & Combes (1986), and Quinn & Hernquist (1987); we discuss only their most important implications.

Schweizer (1980, 1982, 1983) was the first to suggest that shells result from the accretion of small galaxies. Numerical simulations by Quinn (1982, 1984), Toomre (1983), Dupraz & Combes (1985, 1986), Quinn & Hernquist (1987), Hernquist & Quinn (1987, 1988), and others make a convincing case that they form through phase and spatial wrapping of cold material sloshing back and forth in the gravitational potential of an elliptical. Supporting this interpretation are the observations that (*a*) shells are made of stars similar in color to or slightly bluer than the underlying elliptical (Schweizer 1980, Carter et al. 1982, 1988, Bosma et al. 1985, Fort et al. 1986, Pence 1986, Clark et al. 1987, Schombert & Wallin 1987); (*b*) shells at successive increasing radii alternate on opposite sides of the center [e.g. NGC 3923; Quinn 1982, 1984, Malin & Carter 1983, Fort et al. 1986, Pence 1986 (but see Prieur 1988)]; and (c) their outer edges are sharp and often edge brightened, like folded sheets (e.g. Malin & Carter 1980, 1983). Such structures form when a small accreted galaxy falls almost radially into a smooth and stationary potential.

Shell detection frequencies show that accretion of small companions is a normal event in the life of a galaxy. Surveys by Malin & Carter (1983) and by Schweizer & Ford (1985) find shells in 17% and 44% of field ellipticals, respectively. Not all encounter geometries and viewing angles produce visible shells, so the real percentage is larger. This suggests that a typical elliptical has experienced one or more accretion events. Disk galaxies also contain shells, albeit less often than ellipticals (Schweizer & Ford 1985, Schweizer & Seitzer 1988). There is no reason to believe that disk galaxies do not accrete companions. However, if the victim is too massive, the disk is destroyed. If the disk is very robust, the resulting flattened potential is not gentle enough to form ordered shells (Quinn 1984, Dupraz & Combes 1986).

At first it was hoped that shells could be used to measure galaxy mass distributions, because the number of shells and their radial distribution depend on the gravitational potential (Quinn 1984, Dupraz & Combes 1986, Quinn & Hernquist 1987, Hernquist & Quinn 1987). The steep potential gradient of an $r^{1/4}$-law mass distribution predicts a large number (100–200) of shells. The reason is that stars in the innermost shells have much shorter orbital periods than those at large r; one new shell forms every time inner stars complete an extra half-oscillation with respect to outer ones. But galaxies typically contain $\lesssim 20$ shells. This can be understood if dark matter is added at large radii to reduce orbital periods there. However, this argument is oversimplified. The predictions were based on the assumption that the accreted galaxy is disrupted instantaneously. If not, it sinks toward the center through dynamical friction. Then inner shell stars have spent less time at small r than we thought, so fewer shells are predicted. Dark matter is no longer required. This also solves the problem that some inner shells are surprisingly close to the center. All of this is discussed by Dupraz & Combes (1987), Hernquist & Quinn (1988), and Prieur (1988), who now conclude that shells cannot be used to measure mass distributions. However, Piran & Villumsen (1987) find that shells are not formed at all if stars are stripped too slowly from the victim. It remains to be demonstrated that we know how to construct regular shell systems over the largest observed radius range.

Shells may tell us something about galaxy shapes. Dupraz & Combes (1985, 1986) suggest that (a) when shells are short arcs bisected by the major axis, the elliptical is likely to be prolate and edge-on; (b) when shells align with the minor axis, the elliptical is oblate and edge-on; and (c) when the shells are randomly distributed in azimuth and the elliptical is nearly round, it is likely to be oblate and face-on (e.g. 0422−476; Wilkinson et al. 1987). As in Section 4.3, these results are statistical, except perhaps in the most regular cases. However, they suggest independently that ellipticals span a wide range of shapes from oblate triaxial to prolate triaxial.

6. SUMMARY: SECULAR EVOLUTION BY ACCRETION

The observations reviewed in Sections 3–5 suggest that galaxy structure may be altered significantly by the accretion of gas and small companions. In the past, galaxies showing obvious effects of mergers were regarded as peculiar. Now we know that accretion happens often enough in a typical galaxy that even modest events can add up to a significant effect. However, this is secular evolution and not like the more violent mergers that may completely destroy a disk and convert two spirals into one elliptical. The present results are independent of the debate about what fraction of ellipticals formed in this way. Accreted material continues to trickle in long after dissipational collapse or merger formation is complete.

The amount of material added and its effects on the structure of a typical galaxy are unknown. However, typical masses of dust and gas (10^7–10^8 $M_\odot$; Section 4.1) are not much less than the mass of stars in a core. Substantial changes in core properties could result; this may increase the scatter in core parameter relations (cf. Lauer 1988b). Also, enough material can be accreted to make disks that would change a galaxy's apparent morphological type (Section 4.4). And many shell galaxies have blue colors and early-type spectra implying recent bursts of star formation (Carter et al. 1988).

These results form part of a gradually changing picture of the formation and evolution of elliptical galaxies. Traditionally, galaxies were thought to form on a gravitational collapse time scale, with little subsequent evolution. Recent results suggest a picture in which formation is more gradual. As in disk galaxies, where gas infall and disk building are still going on today (Gunn 1982), accretion and the rearranging of mass in ellipticals may be a significant evolutionary process that is far from over (Schweizer 1983, 1986, Schweizer & Seitzer 1988).

7. BRIGHTNESS PROFILES AND TIDAL EFFECTS

The shapes of galaxy brightness profiles and their dependence on luminosity contain information about galaxy formation. Systematic trends with environment tell us about tidal effects. And the functional form of the profile determines the best way to derive size and density scale parameters (Section 8).

Studies of profile shapes are affected by a variety of problems, some of which are worse for CCDs than for photographic observations. (a) Some data are not accurate at large radii. CCDs have a reputation for omnipotence that does not apply to measurements of galaxy halos. CCDs are small, so sky estimates are often uncertain (Capaccioli 1987). Observers know this, but poor sky subtraction nevertheless plagues even the best CCD photometry of halos. Capaccioli et al. (1988) and Peletier et al. (1988a) cite examples; errors of 0.2–0.5 mag arcsec^{-2} are common. (b) Most CCD data reach out to only a few de Vaucouleurs (1948) effective radii r_e; the systematic departures from $r^{1/4}$ laws that are discussed below begin at about these radii. (c) Seeing is a problem, especially for low-luminosity galaxies. These have such tiny cores that seeing can completely change their brightness profiles. For example, if M32 were in the Virgo cluster, we would know nothing about its inner power-law profile (Tonry 1984b, 1987). No one has studied enough nearby cases; for example, Schombert's (1986, 1987) low-luminosity galaxies are in the Coma cluster. (d) Tidal effects modify outer profiles upward or downward (Section 7.2). (e) Some "ellipticals" are misclassified S0s. (f) Finally, many ellipticals contain unrecognized dust (Section 4.1).

7.1 Do Elliptical Galaxies Have $r^{1/4}$-Law Brightness Profiles?

De Vaucouleurs' (1948, 1953) $r^{1/4}$ law fits bright elliptical galaxies reasonably well except where tidal effects are important. We do not attach physical significance to this choice of function, although Binney (1982c) and Bertin & Stiavelli (1984, 1989) find reasonable distribution functions whose density profiles are similar to it. The $r^{1/4}$ law is a convenient parameterization that extracts all of the scaling information that we are entitled to derive, given the similarity of profiles to power laws (Kormendy 1980, K82). But how well does it work?

No definitive study has been published. Based on large photometric surveys, Michard (1985), Djorgovski et al. (1985), Djorgovski (1985), Schombert (1986, 1987), Kodaira et al. (1986), Jedrzejewski (1987b), Capaccioli et al. (1988), and de Carvalho & da Costa (1988) conclude that ellipticals have a wide variety of profile shapes. A corollary is that fitting functions with two scale parameters but no shape parameter are not particularly useful. However, these conclusions are undermined by the problems discussed above.

Much of what we know about galaxy halos still comes from photographic data. These show that profiles of isolated ellipticals vary with luminosity (e.g. Kormendy 1980, Michard 1985, Schombert 1986, 1987), although with significant scatter. The $r^{1/4}$ law fits best near $M_B = -21$. Even at this luminosity, profiles are slightly concave upward when plotted against $r^{1/4}$; typical deviations are ± 0.1–0.2 mag arcsec^{-2} over $\gtrsim 6$ mag arcsec^{-2} (Kormendy 1977b, Capaccioli 1985). Galaxies much brighter than $M_B = -21$ have more light at large radii than the extrapolation of $r^{1/4}$ laws fitted further in, and fainter galaxies have less.

We believe that a good approach for future investigation of profile shapes is one suggested by Schombert (1986, 1987). For each luminosity bin, Schombert constructs template profiles by averaging many observed profiles. Two further improvements are needed. First, total luminosities should be used. Schombert's 16-kpc metric absolute magnitudes measure different fractions of the total light in giant and dwarf galaxies: They are total magnitudes for dwarfs but contain only $\sim 50\%$ of the luminosity of first-ranked galaxies (see Figure 8 in Schombert 1986). Second, we need to use isolated galaxies to minimize tidal effects. The resulting templates can then be compared with profiles of galaxies that have companions to study tidal effects.

It remains true that characteristic sizes and densities are well measured using two-parameter fitting functions. These are basically equivalent. None has a special physical interpretation, but among formulas explored so far, the $r^{1/4}$ law is most convenient and fits best. Profile fits can be improved by adding a third parameter, but then the parameters are too coupled to be useful. All this has been reviewed by Kormendy (1980, K82) and Capaccioli (1988b). Parameters can also be derived without using fitting functions. For example, the actual half-light radius and surface brightness can be used. Or scale radii can be derived using dimensionless monotonic functions like Petrosian's (1976) η function, i.e. the ratio of the surface brightness at a given radius to the mean surface brightness within that radius.

7.2 Tidal Effects

Tidal effects are reviewed in K82. In the cores of rich clusters, galaxies are observed to have abnormally small sizes (Strom & Strom 1978, and subsequent papers; see K82). This is particularly true of faint ellipticals, which moreover have outer cutoffs in their profiles (see also Schombert 1986, 1987). These observations are convincingly interpreted as truncation by the mean gravitational field of the cluster, especially during virialization (Merritt 1984, and references therein).

Whether small galaxies are truncated by large ones is less clear. Suggestions that M32 (King 1962) and NGC 4486B (Rood 1965, Kormendy 1977a) are truncated conflict with recent photometry [see Nieto & Prugniel

(1987a,b) and N88 for reviews]. On the other hand, King & Kiser (1973) find that NGC 5846A has an outer cutoff. Examples of both "truncated" and untruncated small companions are given by Prugniel et al. (1987, 1988). Some photometry is uncertain because the galaxies are embedded in the halos of companions (N88). Also, some close pairs must be optical doubles. Thus the implications of these observations are unclear.

Despite possible truncation, it is clear that ellipticals like M32 are not dwarfs only because of tidal effects. They are genuinely the low-L end of the luminosity function of elliptical galaxies (Section 8.1; see also Nieto & Prugniel 1987a,b, N88).

Encounters between ellipticals of nearly the same mass cannot by symmetry produce truncation if the total mass lost to the system is small (Aguilar & White 1985, and references therein). Kormendy (1977b, 1982a) concluded that ellipticals with companions of comparable sizes have distended outer profiles, and he interpreted this as tidal stretching or heating. The effect needs checking: Schombert (1988) and de Carvalho & da Costa (1988) did not see it in their samples. However, distension is seen in N-body simulations. Aguilar & White (1986) find that encounters produce transient tidal waves in the density distribution. An encounter heats each galaxy. Strong encounters steepen the profiles (i.e. make the galaxies smaller) because mass is lost; weak encounters make the profiles shallower. In either case, the final profile is set up first at small radii. As time passes, the transition between the old and new profile moves outward until only the final profile is left. An $r^{1/4}$-law profile shape is approximately preserved; there is no truncation.

Azimuthal distortions produced by tides are also observed (K82, Djorgovski 1985, Borne & Hoessel 1988, Borne 1988, Borne et al. 1988, Porter 1988, Davoust & Prugniel 1988, Prugniel et al. 1988, Lauer 1988a). As Borne notes, these are clear evidence for tidal friction in action.

8. PARAMETER CORRELATIONS AND SCALING LAWS

One of the main astrophysical uses of surface photometry is for the study of parameter correlations and scaling laws. These contain valuable information about galaxy formation and evolution. Also, correlations between distance-dependent and distance-independent quantities are vital for the mapping of large-scale structure and velocity fields.

8.1 *Families of Ellipsoidal Stellar Systems*

A fundamental application of parameter correlations has been the demonstration that diffuse dwarf spheroidal (dSph) galaxies are a family of objects unrelated to ellipticals. Baade (1944) long ago noted that NGC 147, NGC 185, NGC 205, and the Galactic dwarf spheroidals form a low-surface-brightness sequence quite unlike ordinary ellipticals. Until the mid-1980s, most people believed that the transition between these sequences is continuous [e.g. Binggeli et al. (1984), but contrast Michard (1979) and Farouki et al. (1983)]. Then, in important and somewhat neglected papers, Saito (1979a,b) pointed out that dSphs have anomalously low binding energies compared to giant ellipticals. He suggested the now-favored explanation that dwarfs have low densities because supernova-driven winds have removed large amounts of gas. Later, Wirth & Gallagher (1984) were the first to emphasize that there are two *unrelated* sequences of early-type galaxies: the diffuse dwarfs, and an E-galaxy sequence whose low-luminosity end consists of galaxies like M32. They pointed out that the extreme properties of M32-like dwarfs are not due to tidal truncation but are intrinsic to low-luminosity ellipticals (see also Section 7.2). Also, they found additional examples in the Fornax cluster, showing that M32 is not a fluke. The case was further strengthened by Kormendy (1985b, 1987c), who demonstrated a clear separation into two families overlapping in luminosity. The key to this was CFHT seeing good enough to define the low-L end of core parameter scaling laws for ordinary ellipticals. The differences between the families are also seen in global properties[3] (Saito 1979a, Okamura 1985, Dekel & Silk 1986, Ichikawa et al. 1986, 1988, Kormendy 1987c). These results are not due to selection effects, as suggested by Phillipps et al. (1988). The distribution of parameters for diffuse dwarfs is undoubtedly biased by selection; remarkably low-surface-brightness galaxies are still being discovered (Sandage & Binggeli 1984, Impey et al. 1988). But luminosity "icebergs" hidden under the sky brightness (Disney & Phillipps 1987, 1988) only contribute to the distinction between E and dSph galaxies. Further evidence for this distinction includes a large difference in luminosity functions: Ellipticals have a nearly Gaussian luminosity function that peaks at $M_B \simeq -18$, while dSph galaxies

---

[3] Uncertainty about whether there is a discontinuity (Binggeli et al. 1984, Sandage et al. 1985a, Binggeli 1985, Caldwell & Bothun 1987) is based mainly on three problems. (*a*) Some global parameters used (e.g. isophotal mean surface brightnesses) are insensitive structure indicators. (*b*) Bright dSph galaxies, which are close to the E sequence, often contain both an exponential component and (apparently) a bulge (e.g. NGC 5206; Caldwell & Bothun 1987). Inclusion of bulges guarantees convergence with the E sequence; disks and bulges should be plotted separately in these diagrams. (*c*) Seeing effects are so large that it is very difficult to define the faint end of the E sequence using ground-based observations of galaxies as far away as 20 Mpc (Kormendy 1987c). Effects (*a*)–(*c*) appear sufficient to explain the apparent merging of the E and dSph sequences seen by the above authors. Nevertheless, it is important that the galaxies they cite as transition objects be measured with *Space Telescope* for inclusion in the parameter diagrams.

begin to appear at $M_B \simeq -18$ and in Virgo then become more numerous at least as rapidly as $L^{-1.35}$ [Wirth & Gallagher 1984, Sandage et al. 1985a,b, Impey et al. 1988; see Binggeli (1987) for a review]. The distinction between the elliptical and diffuse dwarf galaxy families points to a fundamental difference in formation history.

A clue to the origin of dSph galaxies is provided by the observation that they are structurally similar to dwarf spiral and irregular (dS+I) galaxies (Faber & Lin 1983, Lin & Faber 1983, Caldwell 1983, Wirth & Gallagher 1984, Sandage & Binggeli 1984, Binggeli et al. 1985, Kormendy 1985b, 1987c, Okamura 1985, Binggeli 1985, Ichikawa et al. 1986, 1988, Karachentseva et al. 1987, Impey et al. 1988). They are not merely dS+I galaxies seen between bursts of star formation, because they contain virtually no gas (Bothun et al. 1985, Impey et al. 1988). Two formation mechanisms are discussed at length in the literature. First, the basic low-density structure of dwarf galaxies is probably due at least in part to supernova-driven galactic winds; these can turn some dS+I galaxies into dwarf spheroidals (e.g. Larson 1974, Saito 1979b, Silk 1983, Dekel & Silk 1986, Vader 1986a, 1987, Silk et al. 1987, Yoshii & Arimoto 1987). In addition, there is strong evidence, at least in clusters, that some dwarf spheroidals formed from dS+I galaxies by ram-pressure stripping of their gas [see Lin & Faber (1983), Binggeli (1985), and Kormendy (1987c) for reviews]. There are other possibilities too. Dwarf spheroidals could be dS+I galaxies that turned all of their gas into stars (Kormendy 1985b, Davies & Phillipps 1988, Binggeli et al. 1989). In certain circumstances, dSph galaxies could even turn back into dS+I galaxies by accreting gas (Silk et al. 1987). In the search for simple, unique explanations, we should not forget that all of these things may happen.

8.2 Correlations With Galaxy Luminosity

The results of the previous section were based on correlations of various physical scale parameters with total luminosity L. The best known of these is the Faber-Jackson (1976) relation $L \propto \sigma^n$. The slope is $n \simeq 4$, but with a real variation, depending on the sample definition (Faber & Jackson 1976, Tonry & Davis 1981, Tonry 1981, Terlevich et al. 1981, de Vaucouleurs & Olson 1982, Kormendy & Illingworth 1983, Dressler 1984a). Many of these authors combined their data with estimates of effective radii and found a weak correlation between mass-to-light ratio and luminosity, $M/L \propto L^{0.35 \pm 0.15}$. This is also seen in core mass-to-light ratios (K87, Kormendy 1987c).

A correlation between the de Vaucouleurs (1948) effective radius r_e and surface brightness I_e was found by Kormendy (1977b, 1980, K82); modern data give $r_e \propto I_e^{-0.83 \pm 0.08}$ (e.g. Hoessel & Schneider 1985, Hamabe &

Kormendy 1987, Djorgovski & Davis 1987). This implies that more luminous galaxies have larger r_e and fainter I_e, although with large scatter. Hamabe & Kormendy show that these relations are not significantly affected by the coupling of measurement errors in the parameters. They are also largely independent of how the parameters are defined; e.g. similar relations hold for core parameters (K82, Kormendy 1984, 1985b, 1987c, K87, Lauer 1985b, 1988b).

An important correlation between luminosity and the dynamical importance of rotation was discovered by Davies et al. (1983). They found that low-luminosity ellipticals and bulges rotate rapidly, have nearly isotropic velocity dispersions, and are flattened by rotation. In contrast, bright ellipticals rotate slowly, are pressure supported, and owe their shapes to velocity anisotropy. Let V/σ be the ratio of the maximum rotation velocity to a suitable mean velocity dispersion (see Davies et al. 1983). Then the level of rotational support can be parametrized by the ratio $(V/\sigma)^*$ of V/σ to the value expected for an isotropic oblate spheroid. The result that $(V/\sigma)^* \simeq 1$ for faint galaxies and $\ll 1$ for bright ones could arise if protoellipticals acquired angular momenta through tidal torques, and if mergers then produced brighter ellipticals in which rotation got scrambled.

8.3 Multiparameter Correlations: the "Fundamental Plane" of Elliptical Galaxies

Multiparameter correlations were discovered through studies of correlated residuals from relations like those of Section 8.2. A breakthrough in our understanding of scaling laws required the appearance of large, homogeneous data sets based mainly on CCD photometry and long-slit spectroscopy. Also, this work has benefited from the application of statistical tools like principal component analysis (PCA) (e.g. Brosche 1973, Bujarrabal et al. 1981, Brosche & Lentes 1983, Lentes 1983, Efstathiou & Fall 1984, Whitmore 1984, Murtagh & Heck 1987). However, in PCA, the astrophysics can get lost in too many eigenvectors. Therefore, simple techniques like bilinear least-squares fits remain useful to provide physical insight.

The presence of intrinsic scatter in the Faber-Jackson relation was correctly interpreted as an indication of a "second parameter." In an important paper, Terlevich et al. (1981) proposed that this second parameter is metallicity, measured by the Mg_2 index, and possibly axial ratio. Their results were challenged by Tonry & Davis (1981) and then readdressed by Efstathiou & Fall (1984). However, relatively poor data sets available at the time did not permit a resolution of the problem. Authors agreed that elliptical galaxies are at least a two-parameter family, but the second parameter could not clearly be identified.

More accurate data confirm that the variance of global properties is exhausted almost entirely by two variables (Tonry & Davis 1981, Lauer

1985b, 1987, Burstein et al. 1986, Djorgovski & Davis 1986, 1987, Dressler et al. 1987, Faber et al. 1987, Djorgovski 1987a, Dressler 1987, de Carvalho & da Costa 1989). These data show that bulges and ellipticals lie in an inclined "fundamental plane" in the space of observed parameters (Figure 2),

$$R \propto \sigma^{1.4 \pm 0.15} I^{-0.9 \pm 0.1}. \qquad 1.$$

Here R can be any consistently defined radius derived from surface brightness profiles, such as the core or effective radius, but not an isophotal radius. An equivalent relation is obtained for luminosity. The old Faber-Jackson and radius-surface brightness relations are projections of the fundamental plane. The luminosity-color and mass-metallicity relations are also contained in the fundamental plane. Its tilt with respect to the

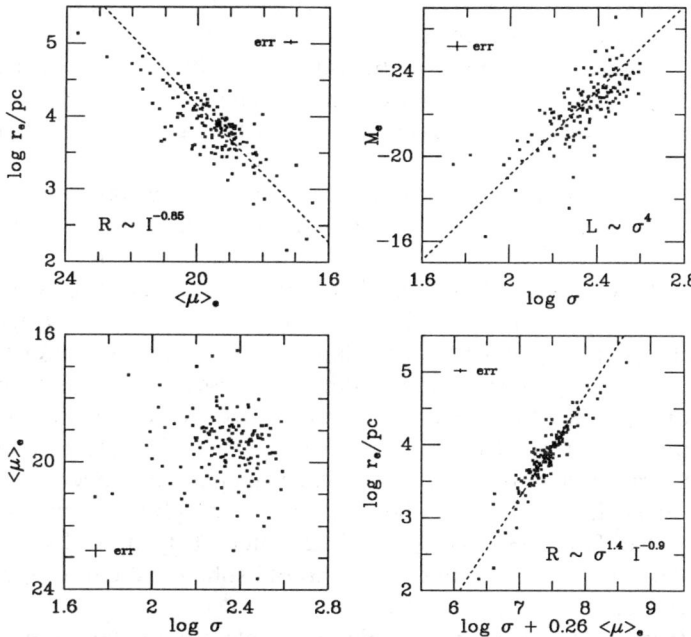

Figure 2 Projections of the fundamental parameter plane of elliptical galaxies. *Top panels*: the one-parameter scaling relations discussed in Section 8.2, i.e. (*left*) the relation between radius and mean surface brightness, and (*right*) that between luminosity and velocity dispersion (the Faber-Jackson relation). *Bottom left*: the surface brightness–velocity dispersion correlation is the fundamental plane seen almost face-on. This is an observer's version of the cooling diagram from theories of galaxy formation. *Bottom right*: this relation between the radius and a combination of surface brightness and velocity dispersion is the fundamental plane seen edge-on. The data are from Djorgovski & Davis (1987). All photometric quantities are in the Lick r_G band and are measured at or within the r_e elliptical isophote. The crosses are median error bars for all points in each panel.

planes of observed parameters produces the correlated intrinsic scatter seen in the projected relations.

An alternate form of Equation 1 is the relation between the modified isophotal diameter and velocity dispersion, $D_n \propto \sigma^{4/3}$ (Burstein et al. 1986, Dressler et al. 1987). Here D_n is defined as the circular diameter within which the mean surface brightness reaches a certain fiducial value, e.g. $\mu_B = 20.75$ mag arcsec^{-2} in the case of Dressler et al. (1987). The above authors show that the D_n–σ relation is equivalent to Equation 1, provided that all elliptical galaxies have brightness profiles of the same shape.

Another alternative representation of the fundamental plane—a relation between radius, surface brightness, and a metallicity indicator (color or Mg_2 index)—has been obtained by de Carvalho & Djorgovski (1989). The two-dimensional nature of the manifold of elliptical galaxies implies that there must be a second parameter in the relation between mass and metallicity; this is identified as the luminosity density (S. Djorgovski & R. R. de Carvalho, in preparation).

The residual scatter about the fundamental plane is $\sim 20\%$ per galaxy (given as the relative error of distance or radius). It is mostly or entirely due to measurement errors. Any cosmic scatter cannot be larger than a few percent.

A group at the Tokyo Astronomical Observatory has obtained and analyzed photographic surface photometry of galaxies of all Hubble types (Kodaira et al. 1983, Okamura et al. 1984, Watanabe et al. 1985). Based on PCA, they also conclude that there are two dominant dimensions in the parameter space of luminosity, isophotal diameter, surface brightness, and central light concentration. Kodaira (1988) proposes that one of the principal components is phase space density. These results are in agreement with work described above when ellipticals are treated separately. They also agree with results obtained separately for spirals (e.g. Whitmore 1984). Spirals and ellipticals have similar principal-component solutions. Nevertheless, it is not clear whether it is meaningful to lump together galaxies of different Hubble types, since different dynamical subsystems and stellar populations (young disks and old spheroids) contribute to the measured quantities.

In retrospect, the fundamental plane was already implicit in papers by Michard (1979), de Vaucouleurs & Olson (1982), Brosche & Lentes (1983), and Lentes (1983), although they did not recognize the full implications of their results.

There is some controversy about whether luminosity is the "first" parameter (i.e. whether it accounts for the greater part of the variance in other parameters). The present authors disagree on this point. SD believes that it is misleading to consider luminosity as the first parameter. The axis perpendicular to the luminosity in the fundamental plane does not correspond to any direct observable. Even if subsystems of galaxies have a first parameter, this does not prove that the same parameter has the same controlling effect on galaxies as a whole. SD therefore believes that it is most profitable to think of the velocity dispersion and surface brightness as the principal variables from which one can derive luminosity, radius, and other quantities of interest. JK is unconvinced. Even though the $\langle \mu \rangle_e$–σ diagram in Figure 2 shows no correlation, he worries that subtle problems may have enlarged the scatter. Large samples were required to explore these issues; then many of the galaxies are far away and suffer from problems like seeing and sample selection. Also, cores of nearby galaxies do show a μ_0–σ correlation (J. Kormendy, in preparation). JK believes that more work is needed on the question of whether one first parameter is more fundamental than the others.

8.4 Uses and Interpretation of the Fundamental Plane

The fundamental plane is a powerful new distance indicator for early-type galaxies. Using it, Lynden-Bell et al. (1988) have discovered large-scale galaxy streaming motions toward the Hydra-Centaurus Supercluster (the "Great Attractor" model).

The plane also contains valuable clues about galaxy formation. Its solutions are very robust, and the residual scatter is very low. The solution has the same form for ellipticals and bulges, it spans about three orders of magnitude in luminosity, it varies little (if at all) in different environments, and it does not depend on how the parameters are measured. It must reflect an important regularity in the process of elliptical galaxy formation or transformation. One useful representation of the fundamental plane is its projection on the log σ–log I plane of observables. This is the "cooling diagram" in theories of galaxy formation [i.e. virial temperature vs. density (e.g. Rees & Ostriker 1977, Faber 1982, Silk 1983, 1985, 1987, Blumenthal et al. 1984)]. The position of a galaxy in this diagram is related to the amount of dissipation during its formation.

The fundamental plane can be understood using the following simple argument (Faber et al. 1987, Djorgovski et al. 1989). The virial theorem implies that galaxies must satisfy a relation that is very similar to Equation 1:

$$R \sim k_S k_E \sigma^2 I^{-1} (M/L)^{-1}. \qquad 2.$$

The parameter k_S reflects the density, luminosity, and kinematic structure of a galaxy; it would be a constant if all galaxies considered had the same dynamical structure. The parameter k_E is the ratio of absolute potential energy to kinetic energy for a galaxy: $k_E > 1$ for a bound system, and

$k_E = 2$ for a virialized one. The deviations in Equation 1 of the coefficients of σ and I from 2 and -1, respectively, reflect the dependence of $k_S k_E (M/L)^{-1}$ on galaxy mass or other fundamental plane variables. If all of the variation is in mass-to-light ratio, this implies the scaling relation $M/L \propto M^{0.2}$ (Faber et al. 1987, Djorgovski 1987b; cf. Section 8.2). A more complete discussion is given by Djorgovski et al. (1989).

The parameters k_S, k_E, and M/L depend on the formation and evolutionary histories of galaxies (Djorgovski et al. 1989). Our present understanding of galaxy formation is that it consists of a series of dissipative merging and infall processes, most of which are affected by environment (e.g. Silk & Norman 1981, Silk 1987). In fact, Vader (1986b) found a marginal but systematic difference between the L–σ–Mg_2 relations in the Virgo and Coma clusters. Also, Djorgovski et al. (1989) find that the D_n–σ relation in different clusters varies with cluster richness class. Within clusters, it varies with distance from the cluster center. Further investigation of the fundamental plane and its dependence on environment is desirable but will require large bodies of high-quality data.

9. ISOPHOTE SHAPES

Fundamental new constraints on the structure of elliptical galaxies are emerging from studies of isophote shapes. This work may resolve a well-known shortcoming of the Hubble classification scheme: While the sequence S0–Im is one of changing physical properties, that from E0 to E6 is not a sequence of anything fundamental (Tremaine 1987). New observations suggest that ellipticals form a physical sequence that is continuous with S0s at one end. Along this sequence, rotation decreases in importance compared with anisotropic velocity dispersions. This subject is developing rapidly; we summarize it as of December 1988.

CCD photometry shows that the isophotes of elliptical galaxies are usually not perfect ellipses. Some are box shaped and others have disk-shaped distortions along their major axes (Carter 1979, 1987, Lauer 1985c, Jedrzejewski 1987a,b, Jedrzejewski et al. 1987, Michard & Simien 1987, 1988, Bender & Möllenhoff 1987, Bender et al. 1987, 1988a, Ebneter et al. 1988, Franx et al. 1988, Peletier et al. 1988a). It is convenient to parametrize these departures by the amplitude $a(4)$ of the $\cos(4\theta)$ term in a Fourier expansion of the isophote radius in polar coordinates [see Carter (1978) and the above papers]. Along the major axis, the fractional radial departures from ellipses are typically $a(4)/a \simeq 1\%$. Positive values of $a(4)/a$ describe disky isophotes; negative values describe boxy isophotes.

Our discussion of $a(4)/a$ measurements follows an excellent paper by

Bender et al. (1988b; hereafter B+88). Figure 3 shows correlations of various parameters with $a(4)/a$. The upper-left panel shows that rotation is dynamically less important in boxy ellipticals than in disky ellipticals (Lauer 1985c, Carter 1987, Bender 1987, 1988a, Nieto et al. 1988, Wagner et al. 1988, B+88, Nieto & Bender 1988). All disky ellipticals show significant rotation, and many are consistent with isotropic models. Boxy ellipticals have a variety of $(V/\sigma)^*$ values but include all of the galaxies with negligible rotation. They are also notable for showing minor-axis rotation (Davies & Birkinshaw 1986, Wagner et al. 1988). Bender, Nieto,

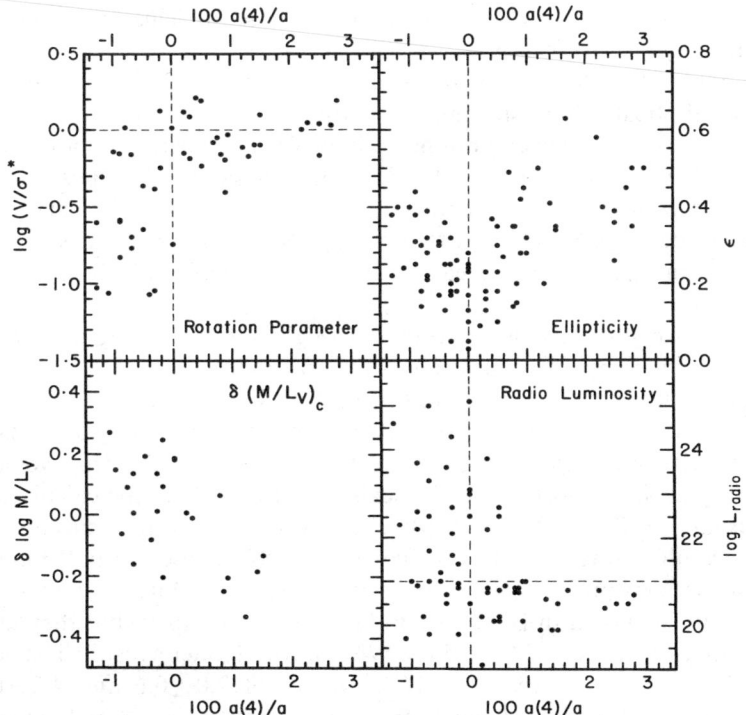

Figure 3 Correlations of selected parameters with isophote shape. Here $100a(4)/a$ is the percent inward or outward perturbation of isophote radii along the major axis (B+88); negative values indicate boxy isophotes, and positive values indicate disky isophotes. Most of the galaxies are ellipticals, but a few galaxies with $100a(4)/a \gtrsim 2$ are S0s. The upper-left panel shows the rotation parameter $(V/\sigma)^*$ (Section 8.2). This panel is adapted from Bender (1988a), but with $a(4)/a$ values from B+88 and with $(V/\sigma)^*$ values added from Davies et al. (1983). The lower-left panel shows deviations of core mass-to-light ratios from the mean relation $M/L_V \propto L^{0.20\pm0.04}$ found in K87; positive values imply that M/L_V is larger than average for the galaxy's luminosity. The right panels (from B+88) show ellipticity ε and radio luminosity L_{radio} at 1.4 GHz (W Hz^{-1}).

and collaborators suggest that ellipticals can be divided into two groups: boxy, slowly rotating, anisotropic ellipticals; and rapidly rotating, disky galaxies that connect with the S0 sequence.

Such a division also segregates ellipticals by other physical properties. For example, the upper-right panel shows that $a(4)/a$ correlates with ellipticity. The distribution of points is V-shaped. Galaxies that appear almost round are almost elliptical. More flattened galaxies tend to be either box- or disk-shaped (see also Jedrzejewski 1987a). Since ellipticals have a preferred shape of E3.8 (Sandage et al. 1970, Binney & de Vaucouleurs 1981), these flattened galaxies are close to edge-on, and most round galaxies are close to face-on. B+88 therefore suggest that essentially all ellipticals are either boxy or disky when seen edge-on. This again suggests a dichotomy (but see below).

Bender et al. (1987) and B+88 also find that X-ray and radio luminosities of ellipticals show striking correlations with $a(4)/a$ (e.g. Figure 3, *bottom right*). With few exceptions, only box-shaped ellipticals are strong radio or X-ray sources. Disky ellipticals have X-ray luminosities that are consistent with emission by compact sources only. Also, they are weak radio sources, like S0 galaxies (Hummel & Kotanyi 1982). These results are remarkably clear-cut, at least in the B+88 sample. We do not know what they mean.

B+88 also note that average mass-to-light ratios are higher in boxy than in disky ellipticals. This is consistent with Heckman's (1983) finding that powerful radio galaxies have higher M/L values than do other ellipticals. B+88 calculate global M/L values using central velocity dispersions and effective radii. However, slow rotation demonstrates that velocity anisotropies are important in ellipticals, and these affect the calculations (e.g. Binney & Tremaine 1987, Merritt 1988). We therefore checked the B+88 results using core M/L values from K87. These are not immune from anisotropy problems but should be more secure. Only 24 objects with $a(4)/a$ values quoted in B+88 are in both galaxy samples, but they are a fair sample of Bender's M/L values. We confirm Bender's result: The boxy and disky galaxies have mean M/L_V values of 7.0 ± 0.6 and 4.2 ± 0.6, respectively. Since M/L_V depends on L, this is not the best way to express the result (although these boxy and disky ellipticals happen to have the same mean luminosity). Rather, the bottom-left panel of Figure 3 suggests that $a(4)/a$ may be a second parameter in the M/L_V–L correlation. Galaxies with low mass-to-light ratios for their luminosities tend to have larger $a(4)/a$ values.

We believe that this is new evidence for velocity anisotropy. Figure 3 (*upper left*) shows that boxy ellipticals are especially anisotropic. If the radial component of σ is larger than the tangential component, then by ignoring this we overestimate M/L_V. Similarly, we underestimate M/L_V in disky ellipticals because we neglect rotational support. This suggests that there should be a correlation between $\delta \log (M/L_V)$ and $(V/\sigma)^*$, and one is observed, but it is not better than the one between $\delta \log (M/L_V)$ and $a(4)/a$. Therefore, anisotropy is not the whole story. A larger galaxy sample is needed to pursue these questions.

The distribution of points in Figure 3 and other similar correlations in B+88 suggest two alternative interpretations. First, it is possible that $a(4)/a$ measures the distribution of ellipticals along a continuous (but not necessarily uniformly populated) sequence that connects smoothly with S0s at one end. As rotation decreases, galaxies become intrinsically more spherical and anisotropic. However, the most anisotropic galaxies must be flattened and turn out to be boxy. Alternatively, perhaps only the disky ellipticals are the continuation of the Im–S0 sequence, and boxy ellipticals are a separate group with a different origin.

It is clear that at least two kinds of boxy structure are seen (e.g. Bender 1988a, Nieto & Bender 1988), because boxy ellipticals include the slowest rotators, whereas box-shaped bulges of disk galaxies rotate rapidly (K82). Interestingly, the few box-shaped ellipticals that rotate rapidly are small companions of much larger galaxies (Nieto & Bender 1988; see also Jedrzejewski 1987a, Peletier et al. 1988a). This suggests to these authors that the boxy structure is related to interactions (May et al. 1985). Accretion events also seem capable of leaving behind an excess of box or tube orbits (ex-polar rings?) that could create slowly or rapidly rotating boxy structure, respectively (Binney & Petrou 1985, Whitmore & Bell 1988, Hernquist & Quinn 1988, Statler 1988). The extreme case IC 3370 may be an example of the latter (Jarvis 1987). Of course, it is also possible that one or both kinds of boxy structure are primordial.

These developments have great potential for clarifying our picture of galaxy formation and structure. However, it is still early. Also, we have ignored complications like dust, variations of $a(4)/a$ with radius, and other Fourier components in the isophotes. The present discussion will undoubtedly prove inadequate; our main aim is to stimulate further work on these important issues.

10. COLORS AND COLOR GRADIENTS

In old stellar systems, colors are a complicated measure of metallicity (Burstein et al. 1984, Aragón et al. 1987) and age (O'Connell 1986). Colors, color gradients, and their correlations with other galaxy properties can therefore be used to test theories of galaxy formation. A few of the many reviews of this subject include Faber (1977), Pagel & Edmunds (1981), Burstein (1985), Norman et al. (1986), and Thomsen & Baum (1988).

The correlation between color and luminosity for early-type galaxies is well known (Sandage 1972, Visvanathan & Sandage 1977, Sandage & Visvanathan 1978a,b, Strom et al. 1976, 1978, Frogel et al. 1978): More luminous galaxies are redder and thus more metal rich. For example, Sandage & Visvanathan find that $\log L = 4.1\,(u-V) + $ constant. Also, the centers of most early-type galaxies are redder than their envelopes; typical gradients are $\Delta(b-V) \simeq -0.03$ and $\Delta(u-V) \simeq -0.10$ magnitudes per decade in radius. Color gradients in bulges of spirals are generally stronger than those in ellipticals (Wirth 1981, Wirth & Shaw 1983). The same effects are seen in spectroscopic metallicity indicators (Faber 1973, Terlevich et al. 1981, Tonry & Davis 1981). From population synthesis models, Tinsley (1978) derives the mass-metallicity relation, $Z \propto M^{0.25}$, where Z is the logarithm of the metallicity. Metallicity variations can partly, but not entirely, explain the observed dependence of M/L on luminosity (Smith & Tinsley 1976).

These results can be understood within the framework of dissipative galaxy formation (Larson 1974, 1975, Silk & Norman 1981, Carlberg 1984a,b, Arimoto & Yoshii 1987, Yoshii & Arimoto 1987, Matteucci & Tornambè 1987). Carlberg's models in particular avoid some technical limitations of Larson's pioneering work and make more detailed predictions. In these models, the removal of enriched gas by supernova-driven galactic winds is more efficient for less massive galaxies (see also Section 8.1). In this spirit, the color-luminosity relation is recast as a metallicity–escape velocity relation by Vigroux et al. (1981): $Z \propto V_{\rm e}^{0.9}$. Similarly, colors and metallicities correlate better with central velocity dispersions than with luminosities. Carlberg (1984b) also predicts the existence of a second parameter in the Faber-Jackson and mass-metallicity relations. Larson and Carlberg both predict that color and metallicity gradients should be stronger in more massive galaxies. Finally, they make testable predictions about the relative shapes of isophotes and isochromes (i.e. isometallicity contours). In Larson's models, isochromes are considerably flatter than isophotes. However, Larson's ellipticals are supported by rotation, which we now know is incorrect (e.g. Illingworth 1981, Davies et al. 1983). Carlberg's models are generally supported by velocity anisotropy; then isochromes are only slightly flatter than isophotes.

CCD photometry has provided high-quality measurements of color gradients for large numbers of galaxies. Boroson et al. (1983), Davis et al. (1985), Cohen (1986), Boroson & Thompson (1987), and Bender & Möllenhoff (1987) present data on relatively small samples of ellipticals, mostly in the Virgo cluster. They conclude that color gradients are common in ellipticals. In the absence of nonthermal emission or recent star formation, colors always get redder toward the center. Interestingly, isophotes and isochromes generally have the same shape. In fact, isochromes are occasionally rounder than isophotes (Boroson et al. 1983). This is consistent with Carlberg's but not Larson's models.

The interpretation of color gradients in terms of stellar population gradients has been discussed recently by Efstathiou & Gorgas (1985), Gorgas & Efstathiou (1987), Davies & Sadler (1987), Couture & Hardy (1988), and references therein. They present extensive evidence for gradients in Mg_2 indices. Assuming the somewhat uncertain conversions between Mg_2 index and metallicity (Terlevich et al. 1981) and between color and metallicity (Strom et al. 1976, 1978, Tinsley 1978), they find that the Mg_2 and color gradients are mutually consistent and imply typical changes of $\Delta[Fe/H] \sim -0.2$ per decade in radius. In excellent papers, Baum et al. (1986) and Thomsen & Baum (1987, 1988) derive metallicity gradients from narrowband surface photometry. They also find that isochromes are not flatter than isophotes, in agreement with spectroscopic results. Similar photometric measurements of the Mg_2 index are reported by Vigroux et al. (1988). Also, Vader et al. (1988) find that Mg_2 gradients correlate well with broadband color gradients. Further constraints are obtained by Peletier and coworkers (Peletier et al. 1987, 1988a,b,c, Peletier & Valentijn 1988, Peletier 1988). They show that observed optical and near-infrared (JHK) color gradients are mutually consistent (i.e. one can be derived from the other using the separate optical and infrared color-luminosity relations). All this suggests that the same change in stellar population produces both the color-luminosity relation and the color gradients. Using the new Yale isochrones (Green et al. 1987), Peletier and coworkers conclude that most color variations and gradients are due to changes in metallicity. In typical ellipticals, these do not exceed a factor of 10 inside r_e. However, age gradients may be present as well; the fraction of young stars may increase at larger radii.

Large data sets are needed to investigate correlations of color gradients with other galaxy properties. The measurements are difficult because color gradients are weak and because differential magnitude measurements are sensitive to systematic errors. Nevertheless, important data for early-type galaxies have been obtained by Jedrzejewski (1987b), Vigroux et al. (1988), Franx (1988), Franx et al. (1988), and Peletier and collaborators (see above).

Vader et al. (1988) analyze data from Vigroux et al. (1988) and obtain several interesting results. Whereas inward reddening is the rule in elliptical galaxies and bulges, they find that dSph galaxies tend to become bluer toward the center. Particularly interesting is the observation that color gradients are correlated with the rotation parameter $(V/\sigma)^*$: Anisotropic, pressure-supported ellipticals have smaller color gradients. We find the same effect, although with more scatter, in the Franx and Peletier et al. data (Figure 4).

The bright ellipticals in the Franx (1988) sample show weak correlations of color gradients with luminosity, velocity dispersion, integrated color, and Mg_2 index: Weaker gradients are seen in brighter, hotter, redder, and more metal-rich galaxies. Gorgas & Efstathiou (1987) also find a marginally significant anticorrelation between Mg_2 gradients and velocity dispersion. However, Peletier et al. (1988a) find no significant correlations with the above quantities. When we combine the Vader et al., Franx, and Peletier et al. samples (Figure 4), color gradients in E and S0 galaxies are weak or absent at low luminosities ($M_B > -20$) and largest near the peak of the luminosity function ($M_B \sim -20$). The scatter exceeds the measurement errors at all luminosities.

We also find a marginal correlation of color gradients with isophote

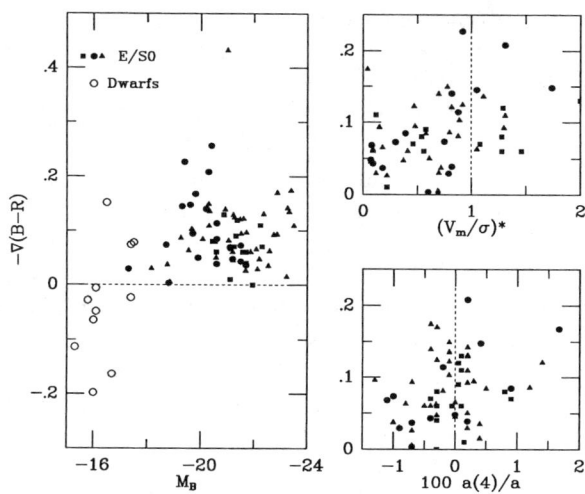

Figure 4 Correlations of color gradients with other galaxy properties. The gradients are defined as $\Delta(\text{Color})/\Delta(\log r)$, in magnitudes per decade in radius; positive values indicate reddening toward the center. The data are from Vader et al. (1988; circles), Franx (1988; squares), and Peletier et al. (1988a; triangles). The left panel shows the dependence of color gradients on luminosity for dSph galaxies (open circles) and for ellipticals and S0s (solid symbols). Ellipticals become redder toward the center, but most dSph galaxies have inverse gradients. The top-right panel shows the relation between color gradients and the level of rotational support; $(V/\sigma)^* = 1$ for an isotropic oblate rotator. Anisotropic galaxies tend to have smaller gradients. The bottom-right panel shows the correlation with isophote shape (measured by B+88). More boxy galaxies [$a(4)/a < 0$] tend to have smaller color gradients. Similar trends are obtained using $(U-R)$ color measurements.

shapes (Figure 4). Color gradients in boxy ellipticals get smaller as boxyness increases [i.e. as $a(4)/a$ decreases further below 0]. There are not enough data to look for variations of color gradients in disky ellipticals.

These results are very preliminary. However, the trends are probably real: All of the parameters are measured independently, and measurement errors only diminish the correlations. If confirmed, these correlations will provide important new information about galaxy formation.

The absence of a strong correlation between the strengths of color gradients and L or σ is contrary to the predictions of the Larson and Carlberg models. However, these do not include the effects of postcollapse mergers, which are important at least for bright ellipticals. The observations suggest that the properties of early-type galaxies are determined by dissipative collapse and then modified by mergers (Vader et al. 1988). Dissipative formation could produce a mass-metallicity relation and metallicity gradients, which are then gradually erased by mergers. The trend with luminosity at $M_B \gtrsim -21$ would then be a fossil of the initial correlation predicted by Larson and Carlberg. Mergers may also produce correlated changes in $(V/\sigma)^*$ and $a(4)/a$ (cf. Figure 3). Normal color gradients are not produced in diffuse dwarfs because of galactic winds (Vader 1986a, Dekel & Silk 1986); their inverse color gradients could be due to recent star formation.

Finally, we discuss color gradients in cooling-flow galaxies. The existence of cooling flows in ellipticals and cluster cores is reasonably well established (e.g. Fabian 1988). Mass flow rates are uncertain, but $\dot{M} \sim 1$–$1000 \ M_\odot \ yr^{-1}$ are believed to be deposited inside a few r_e, with a strong dependence on radius. The most plausible fate of the gas is star formation. Unless the initial mass function (IMF) is strongly biased toward low-mass stars, observable color gradients should result (Sarazin & O'Connell 1983, Silk et al. 1986, O'Connell 1988). Blueing toward the centers of some high-$\dot{M}$ cooling-flow galaxies is seen and interpreted as evidence for age gradients (e.g. Wirth et al. 1983, Romanishin 1986a, 1987, Maccagni et al. 1988). However, the extreme gradients reported by Valentijn (1983) and Valentijn & Moorwood (1985) are not confirmed by subsequent work. A detailed comparison of color gradients in normal and cooling-flow galaxies could provide strong constraints on the fate of gas in cooling flows.

11. cD AND BRIGHTEST CLUSTER GALAXIES

The most luminous galaxies in rich clusters are giant ellipticals. Some of these are classified as D (possessing a large, diffuse envelope) or cD (extra-large D) (Matthews et al. 1964, Morgan & Lesh 1965, Morgan et al. 1975, Albert et al. 1977). The distinction between cD and E galaxies is useful, but we argue below that the term "D galaxy" is not. When we wish to discuss brightest cluster members without regard to their morphology, we will refer to them as BCMs.

The importance of BCMs is twofold. First, because of their large luminosities, they are used as standard candles for cosmological studies. Second, because of their large masses and special locations, they are believed to be the sites of interesting evolutionary phenomena (e.g. dynamical friction, galactic cannibalism, interactions with the intracluster medium, and cooling flows). These subjects are reviewed in detail by Dressler (1984b) and Tonry (1987); here we discuss them briefly in the context of recent photometry. Extensive photometric studies are reported by Oemler (1976), Thuan & Romanishin (1981), Schneider et al. (1983), Lugger (1984), Malumuth & Kirshner (1985), Hoessel & Schneider (1985), Schombert (1986, 1987, 1988), Lauer (1988a), and Porter (1988). Except for Valentijn (1983) and Morbey & Morris (1983), most authors are in fairly good agreement.

The classification of galaxies as D and cD is often done loosely and may be misleading. More luminous ellipticals have shallower brightness profiles (Section 7.1). Also, galaxies can have S0 disks and tidally stretched halos. All of these satisfy the definition of a D galaxy but are not new phenomena, nor even a single class of phenomena. For this reason, we recommend that the term D galaxy not be used (Kormendy 1987b). On the other hand, the cD classification is useful: cDs are physically different from ellipticals.

Finding an objective way to recognize cDs is nontrivial. The most objective way is to look for an inflection in the outer brightness profile that is independent of the way the profile is plotted. This is interpreted as the signature of a halo that is a distinct dynamical subsystem. In practice, not all halos are prominent enough to produce inflection points in the profiles. Then, less objective identification criteria are necessary, such as extra light compared with mean profiles of comparably bright ellipticals. We use the term cD only for bright ellipticals with such extra envelopes. It is not clear whether a cD envelope belongs to the galaxy or whether it was formed by the parent cluster independently of whether there happened to be a bright galaxy at the bottom of the cluster potential well.

BCM and cD galaxies are generally believed to have formed or been modified by mergers (e.g. V Zw 311; Schneider & Gunn 1982, and references therein). The luminosities of BCMs are weakly correlated with some properties of their parent clusters, like Abell richness class or Bautz-Morgan type (Sandage 1972, Sandage & Hardy 1973, Schneider et al. 1983, Morbey 1984, Hoessel & Schneider 1985, Schombert 1987). Schombert (1987) also finds weak correlations with cluster velocity dispersion and X-ray luminosity. Beers & Geller (1983) find that cD galaxies are always found in local density maxima, even if they are not the brightest or central cluster members. Such correlations suggest that environment-dependent processes are responsible for at least some of the luminosity of BCMs. Mergers are a natural candidate, but other options are possible. Examples include a gradual accumulation of cluster tidal debris (Malumuth & Richstone 1984, and references therein) or star formation in (now possibly extinguished) cooling flows (Sarazin 1986, and references therein). Or the large luminosities of BCMs may be a consequence of environment-dependent initial conditions. In the language of biased galaxy formation, BCMs may originate from unusually large primordial fluctuations, which are most likely to occur in dense environments.

An argument often cited in favor of mergers is the high frequency of secondary nuclei in BCMs. These may be semidigested cluster members. Schneider et al. (1983) and Hoessel & Schneider (1985) find that about half of the BCMs in their sample of Abell clusters are multiple-nucleus systems, considerably more than would be expected from chance projections if the clusters have cores. This argument is weakened by the conclusion of Beers & Tonry (1986) that rich clusters do not have cores, but instead have steep number density profiles. Then many nuclei are predicted to be near the central galaxy. Tonry (1984a, 1985; see also Hoessel et al. 1985) shows that many of the secondary nuclei move too quickly to be gravitationally bound to the BCM core. This effect was explained independently by Merritt (1984) and Tonry (1984a) as a natural outcome of the evolution of galaxy orbits in a rich cluster. A possible way to distinguish between bound and unbound secondary nuclei is to decompose BCM images into elliptically symmetric components and look for tidal distortions. Lauer (1986, 1988a) made such a study of 17 multiple-nucleus systems and found that $\sim 50\%$ of the secondary nuclei show isophote distortions. However, these distortions do not correlate as expected with the kinematics. Even nuclei with large relative velocities show distortions. Therefore, the problem of which nuclei are currently being accreted is not settled. Based only on the observed distortions, Lauer estimates that material is being cannibalized at an average rate of $\sim 4 L_*$ (primary galaxy)$^{-1}$ (10 Gyr)$^{-1}$. [Here L_* is the characteristic luminosity of the Schechter (1976) luminosity function.] This is in rough agreement with models by Merritt (1985), which imply accretion rates of $\sim 1 L_*$ (primary galaxy)$^{-1}$ (10 Gyr)$^{-1}$. Since the total luminosity of a cD galaxy is typically $\sim 12 L_*$, this argues that not all of a cD originates through cannibalism.

The structural properties of BCMs are often discussed in the framework of the homologous merger picture (e.g. Ostriker & Tremaine 1975, Ostriker & Hausman 1977, Hausman & Ostriker 1978, Malumuth & Richstone 1984, Merritt 1985; see also White 1982, and references therein). In this picture, the kinetic energy per unit mass is preserved. Then, if the orbital structure of the cannibal galaxy stays the same, its projected central velocity dispersion does not change, even though the mass and luminosity increase. That is, merger products should deviate from the Faber-Jackson relation by being too luminous for their velocity dispersions. Such an effect was found by Malumuth & Kirshner (1985). Because of the conversion of galaxy orbital energy into internal random motions in the merger remnant, the envelope of the cannibal should get shallower after every merger [as measured, say, by the Gunn-Oke (1975) structure parameter α]. Therefore, more luminous merger remnants should have shallower profiles. This is the sense of the luminosity dependence of profile shapes for all ellipticals (Section 7.1; Schneider et al. 1983, Hoessel & Schneider 1985). Finally, compared with the color-luminosity relation for normal galaxies, merger remnants should be too blue for their luminosities. This prediction is not confirmed: Gallagher et al. (1980), Lugger (1984), Lachièze-Rey et al. (1985), and Schombert (1988) find that integrated colors and color gradients in the envelopes of cD galaxies are consistent with those in normal ellipticals of comparable luminosities.

Of more interest is the radius–surface brightness relation (cf. Section 8.2). Kormendy (1980), Thomsen & Frandsen (1983), Lugger (1984), and Romanishin (1986b) find strong correlations consistent with those for normal ellipticals. With larger data sets, Schneider et al. (1983), Schombert (1987), and Hoessel et al. (1987) conclude that the relations between r_e and μ_e or L are steeper for BCM than for non-BCM ellipticals. Hoessel et al. (1987) find that $r_e \propto I_e^{-0.80}$ for BCMs and $r_e \propto I_e^{-0.55}$ for non-BCMs in the Gunn r band. BCMs are larger at a given surface brightness than normal ellipticals. However, since the relation for non-BCM ellipticals that are well resolved is $r_e \propto I_e^{-0.83}$ (Section 8.2), the above difference may be due partly to seeing effects. (The non-BCM ellipticals in the Hoessel et al. sample typically have $r_e \simeq 2-5''$.)

An even more informative comparison of BCMs and other ellipticals can be made using fundamental plane solutions. Hoessel et al. (1987) find that the R–σ–μ solutions for the galaxies in their sample are consistent with solutions for normal ellipticals (Djorgovski & Davis 1987), but with a hint of a different slope. S. Djorgovski & R. R. de Carvalho (in preparation), using data from Malumuth & Kirshner (1985), obtain different solutions for BCMs and normal ellipticals. At a given effective radius or luminosity, the range of velocity dispersions is much smaller than for normal ellipticals; as a result, the scatter in the r_e–μ_e relation is smaller for BCMs than for non-BCM ellipticals. This can be understood in the homologous merger picture, because velocity dispersions are not changed much by mergers, while luminosities and radii increase. Different fundamental planes for BCMs and other ellipticals imply different formation histories.

The r_e–μ_e relation and especially the fundamental plane solutions for BCMs are promising distance indicators (Thomsen & Frandsen 1983, Hoessel et al. 1987; S. Djorgovski & R. R. de Carvalho, in preparation). They may also lead to an improved angular size–redshift cosmological test.

The origin of cD halos remains murky. They are purely a rich cluster phenomenon: Thuan & Romanishin (1981) and Schombert (1986) find that cD halos do not occur in poor clusters. Struble (1988) has discovered what appears to be a cD envelope without a central galaxy in the rich cluster Abell 545. It would be interesting to know whether more such cases exist. They are difficult to find because of their low surface brightnesses ($\mu_V \sim 24$ mag arcsec^{-2}) and unimposing luminosities ($L \sim L_*$ for Struble's "star pile").

The most systematic photometric study of cD envelopes to date is by Schombert (1988), building on work by Oemler (1976). He subtracted template brightness profiles of ellipticals from those of cD galaxies and measured the properties of the envelopes. He found that envelopes have brightness profiles $I(r) \propto r^{-1.6}$ similar to those of their galaxies and any X-ray halos. Envelope luminosities are comparable to those of the central galaxies ($\sim 10^{12} L_\odot$). Therefore, if the theoretical models are correct, mergers are an insufficient source of material to build either cD galaxies or their envelopes. Envelope luminosities correlate with parent galaxy luminosities, which suggests that similar processes may be responsible for both. They also correlate with cluster richness ($L_{env} \propto N^{1.6}$, where N is the Abell galaxy count) and weakly with Bautz-Morgan type. Finally, there is a good correlation with the cluster X-ray luminosity ($L_{env} \propto L_X^{1.06 \pm 0.18}$).

Other connections between BCMs and their clusters include alignment effects. In clusters with well-defined orientations, BCM isophotes tend to align with cluster major axes (Carter & Metcalfe 1980, Binggeli 1982, Porter 1988, Rhee & Roos 1989; see also the review by Djorgovski 1987c). Cluster position angles are uncertain, but Porter (1988) also finds a tendency for alignment with cluster X-ray gas isodensity contours. Ellipticity tends to increase strongly with radius in BCMs. Porter finds that BCMs tend to have larger ellipticities and ellipticity gradients and smaller isophote twists than normal ellipticals.

All of these correlations suggest that cD envelopes are products of their clusters. They may be the accumulated debris of all tidal interactions. Further support for this interpretation comes from the kinematics: The projected velocity dispersions of cDs increase with radius and approach cluster velocity dispersions (Dressler 1979, Carter et al. 1981, 1985).

12. CONCLUSION

This paper is being written at a time of unusually rapid progress in galaxy photometry. Some tentative conclusions, especially in the latter parts of this paper, may even evolve by the time you read this. CCDs have made it possible to measure second-order structure parameters like isophote shape distortions, color gradients, and dust. More difficult physical questions have become accessible. Soon, near-infrared array detectors will provide important new kinds of data. Therefore, the next few years promise substantial progress in our understanding of the structure of elliptical galaxies.

ACKNOWLEDGMENTS

We are most grateful to R. Bender, R. Davies, R. de Carvalho, G. Illingworth, J.-L. Nieto, R. Peletier, F. Schweizer, A. Toomre, and especially T. de Zeeuw for very helpful discussions, comments on the manuscript, or data in advance of publication. Also, it is a pleasure to thank the many people who sent us preprints, including those on subjects that were ultimately not covered because of space limitations. During this work, SD was supported by Caltech and the Alfred P. Sloan Foundation.

Literature Cited

Aguilar, L. A., White, S. D. M. 1985. *Ap. J.* 295: 374
Aguilar, L. A., White, S. D. M. 1986. *Ap. J.* 307: 97
Albert, C. E., White, R. A., Morgan, W. W. 1977. *Ap. J.* 211: 309
Aragón, A., Gorgas, J., Rego, M. 1987. *Astron. Astrophys.* 185: 97
Arimoto, N., Yoshii, Y. 1987. *Astron. Astrophys.* 173: 23
Arp, H. 1966. *Atlas of Peculiar Galaxies.* Pasadena: Calif. Inst. Technol.
Athanassoula, E., Bosma, A. 1985. *Annu. Rev. Astron. Astrophys.* 23: 147
Baade, W. 1944. *Ap. J.* 100: 147
Balcells, M., Quinn, P. J. 1988. *Astrophys. Space Sci.* In press
Barnes, J., Efstathiou, G. 1987. *Ap. J.* 319: 575
Baum, W. A., Thomsen, B., Morgan, B. L. 1986. *Ap. J.* 301: 83
Beers, T. C., Geller, M. J. 1983. *Ap. J.* 274: 491
Beers, T. C., Tonry, J. L. 1986. *Ap. J.* 300: 557
Bender, R. 1987. *Mitt. Astron. Ges. No. 70,* p. 226
Bender, R. 1988a. *Astron. Astrophys.* 193: L7
Bender, R. 1988b. *Astron. Astrophys.* 202: L5
Bender, R., Döbereiner, S., Möllenhoff, C. 1987. *Astron. Astrophys.* 177: L53
Bender, R., Döbereiner, S., Möllenhoff, C. 1988a. *Astron. Astrophys. Suppl.* 74: 385
Bender, R., Möllenhoff, C. 1987. *Astron. Astrophys.* 177: 71
Bender, R., Surma, P., Döbereiner, S., Möllenhoff, C., Madejsky, R. 1988b. *Astron. Astrophys.* In press (B+88)
Bendinelli, O., Di Iorio, A., Parmeggiani, G., Zavatti, F. 1985. *Astron. Astrophys.* 153: 265
Bendinelli, O., Lorenzutta, S., Parmeggiani, G., Zavatti, F. 1984a. *Astron. Astrophys.* 138: 337
Bendinelli, O., Parmeggiani, G., Zavatti, F. 1982. *Astrophys. Space Sci.* 83: 239
Bendinelli, O., Parmeggiani, G., Zavatti, F. 1984b. *Astron. Astrophys.* 140: 174
Bendinelli, O., Parmeggiani, G., Zavatti, F. 1986. *Ap. J.* 308: 611
Bendinelli, O., Parmeggiani, G., Zavatti, F. 1988. *J. Astrophys. Astron.* 9: 17

Bertin, G., Stiavelli, M. 1984. *Astron. Astrophys.* 137: 26
Bertin, G., Stiavelli, M. 1989. *Ap. J.* 338: 723
Bertola, F. 1987. In *Structure and Dynamics of Elliptical Galaxies, IAU Symp. No. 127*, ed. T. de Zeeuw, p. 135. Dordrecht: Reidel
Bertola, F., Bettoni, D. 1988. *Ap. J.* 329: 102
Bertola, F., Buson, L. M., Zeilinger, W. W. 1988a. *Nature* 335: 705
Bertola, F., Capaccioli, M. 1975. *Ap. J.* 200: 439
Bertola, F., Galletta, G. 1978. *Ap. J. Lett.* 226: L115
Bertola, F., Galletta, G., Kotanyi, C., Zeilinger, W. W. 1988b. *MNRAS* 234: 733
Bertola, F., Galletta, G., Zeilinger, W. W. 1985. *Ap. J. Lett.* 292: L51
Binggeli, B. 1982. *Astron. Astrophys.* 107: 338
Binggeli, B. 1985. In *Star-Forming Dwarf Galaxies and Related Objects*, ed. D. Kunth, T. X. Thuan, J. T. T. Van, p. 53. Gif sur Yvette, Fr: Ed. Front.
Binggeli, B. 1987. In *Nearly Normal Galaxies*, ed. S. M. Faber, p. 195. New York: Springer-Verlag
Binggeli, B., Sandage, A., Tammann, G. A. 1985. *Astron. J.* 90: 1681
Binggeli, B., Sandage, A., Tarenghi, M. 1984. *Astron. J.* 89: 64
Binggeli, B., Tarenghi, M., Sandage, A. 1989. Preprint
Binney, J. 1976. *MNRAS* 177: 19
Binney, J. 1978a. *MNRAS* 183: 501
Binney, J. 1978b. *Comments Astrophys.* 8: 27
Binney, J. 1982a. In *Morphology and Dynamics of Galaxies*, ed. L. Martinet, M. Mayor, p. 1. Sauverny: Geneva Obs.
Binney, J. 1982b. *Annu. Rev. Astron. Astrophys.* 20: 399
Binney, J. 1982c. *MNRAS* 200: 951
Binney, J. 1987. In *Dark Matter in the Universe, IAU Symp. No. 117*, ed. J. Kormendy, G. R. Knapp, p. 303. Dordrecht: Reidel
Binney, J., de Vaucouleurs, G. 1981. *MNRAS* 194: 679
Binney, J., Petrou, M. 1985. *MNRAS* 214: 449
Binney, J., Tremaine, S. 1987. *Galactic Dynamics*. Princeton, NJ: Univ. Press
Bland, J., Taylor, K., Atherton, P. D. 1987. *MNRAS* 228: 595
Blumenthal, G. R., Faber, S. M., Primack, J. R., Rees, M. J. 1984. *Nature* 311: 517
Borne, K. D. 1988. *Ap. J.* 330: 61
Borne, K. D., Balcells, M., Hoessel, J. G. 1988. *Ap. J.* 333: 567
Borne, K. D., Hoessel, J. G. 1988. *Ap. J.* 330: 51
Boroson, T. A., Thompson, I. B. 1987. *Astron. J.* 93: 33
Boroson, T. A., Thompson, I. B., Shectman, S. A. 1983. *Astron. J.* 88: 1707
Bosma, A., Smith, R. M., Wellington, K. J. 1985. *MNRAS* 212: 301
Bothun, G. D., Mould, J. R. 1988. *Ap. J.* 324: 123
Bothun, G. D., Mould, J. R., Wirth, A., Caldwell, N. 1985. *Astron. J.* 90: 697
Brosche, P. 1973. *Astron. Astrophys.* 23: 259
Brosche, P., Lentes, F.-T. 1983. In *Internal Kinematics and Dynamics of Galaxies, IAU Symp. No. 100*, ed. E. Athanassoula, p. 377. Dordrecht: Reidel
Bujarrabal, V., Guibert, J., Balkowski, C. 1981. *Astron. Astrophys.* 104: 1
Burbidge, E. M., Burbidge, G. R. 1959. *Ap. J.* 129: 271
Burstein, D. 1985. *Publ. Astron. Soc. Pac.* 97: 89
Burstein, D., Davies, R. L., Dressler, A., Faber, S. M., Lynden-Bell, D., et al. 1986. In *Galaxy Distances and Deviations From Universal Expansion*, ed. B. F. Madore, R. B. Tully, p. 123. Dordrecht: Reidel
Burstein, D., Faber, S. M., Gaskell, C. M., Krumm, N. 1984. *Ap. J.* 287: 586
Caldwell, N. 1983. *Astron. J.* 88: 804
Caldwell, N. 1984. *Ap. J.* 278: 96
Caldwell, N. 1987. *Astron. J.* 94: 1116
Caldwell, N., Bothun, G. D. 1987. *Astron. J.* 94: 1126
Caldwell, N., Kirshner, R. P., Richstone, D. O. 1986. *Ap. J.* 305: 136
Capaccioli, M. 1985. In *New Aspects of Galaxy Photometry*, ed. J.-L. Nieto, p. 53. New York: Springer-Verlag
Capaccioli, M. 1987. In *Structure and Dynamics of Elliptical Galaxies, IAU Symp. No. 127*, ed. T. de Zeeuw, p. 47. Dordrecht: Reidel
Capaccioli, M. 1988a. *Proc. Summer Sch. Extragalact. Astron., 2nd, Cordoba, Argent.* In press
Capaccioli, M. 1988b. In *Le Monde des Galaxies*, ed. H. G. Corwin, L. Bottinelli. New York: Springer-Verlag. In press
Capaccioli, M., de Vaucouleurs, G. 1983. *Ap. J. Suppl.* 52: 465
Capaccioli, M., Piotto, G., Rampazzo, R. 1988. *Astron. J.* 96: 487
Carlberg, R. G. 1984a. *Ap. J.* 286: 403
Carlberg, R. G. 1984b. *Ap. J.* 286: 416
Carter, D. 1978. *MNRAS* 182: 797
Carter, D. 1979. In *Image Processing in Astronomy*, ed. G. Sedmak, M. Capaccioli, R. J. Allen, p. 386. Trieste: Astron. Obs.
Carter, D. 1987. *Ap. J.* 312: 514
Carter, D., Allen, D. A., Malin, D. F. 1982. *Nature* 295: 126
Carter, D., Efstathiou, G., Ellis, R. S., Inglis, I., Godwin, J. 1981. *MNRAS* 195: 15P
Carter, D., Inglis, I., Ellis, R. S., Efstathiou,

G., Godwin, J. G. 1985. *MNRAS* 212: 471
Carter, D., Metcalfe, N. 1980. *MNRAS* 191: 325
Carter, D., Prieur, J.-L., Wilkinson, A., Sparks, W. B., Malin, D. F. 1988. *MNRAS* 235: 813
Ciardullo, R., Rubin, V. C., Jacoby, G. H., Ford, H. C., Ford, W. K. 1988. *Astron. J.* 95: 438
Clark, G., Plucinsky, P., Ricker, G. 1987. In *Structure and Dynamics of Elliptical Galaxies, IAU Symp. No. 127*, ed. T. de Zeeuw, p. 453. Dordrecht: Reidel
Cohen, J. G. 1986. *Astron. J.* 92: 1039
Couture, J., Hardy, E. 1988. *Astron. J.* 96: 867
Davies, J. I., Phillipps, S. 1988. *MNRAS* 233: 553
Davies, R. L., Birkinshaw, M. 1986. *Ap. J. Lett.* 303: L45
Davies, R. L., Danziger, I. J., Fabian, A., Hanes, R., Jones, B. J. T., et al. 1984. *Bull. Am. Astron. Soc.* 16: 410
Davies, R. L., Efstathiou, G., Fall, S. M., Illingworth, G., Schechter, P. L. 1983. *Ap. J.* 266: 41
Davies, R. L., Illingworth, G. D. 1986. *Ap. J.* 302: 234
Davies, R. L., Sadler, E. M. 1987. In *Structure and Dynamics of Elliptical Galaxies, IAU Symp. No. 127*, ed. T. de Zeeuw, p. 441. Dordrecht: Reidel
Davis, L. E., Cawson, M., Davies, R. L., Illingworth, G. 1985. *Astron. J.* 90: 169
Davoust, E., Pence, W. D. 1982. *Astron. Astrophys. Suppl.* 49: 631
Davoust, E., Prugniel, P. 1988. *Astron. Astrophys.* 201: L30
de Carvalho, R. R., da Costa, L. N. 1988. *Ap. J. Suppl.* 68: 173
de Carvalho, R. R., da Costa, L. N. 1989. Submitted for publication
de Carvalho, R. R., Djorgovski, S. 1989. *Ap. J. Lett.* In press
Dekel, A., Silk, J. 1986. *Ap. J.* 303: 39
de Vaucouleurs, G. 1948. *Ann. Astrophys.* 11: 247
de Vaucouleurs, G. 1953. *MNRAS* 113: 134
de Vaucouleurs, G. 1959. *Handbuch der Physik* 53: 275
de Vaucouleurs, G. 1979. In *Photometry, Kinematics and Dynamics of Galaxies*, ed. D. S. Evans, p. 1. Austin: Dept. Astron., Univ. Tex.
de Vaucouleurs, G. 1984. In *Astronomy With Schmidt-Type Telescopes*, ed. M. Capaccioli, p. 367. Dordrecht: Reidel
de Vaucouleurs, G., Capaccioli, M. 1979. *Ap. J. Suppl.* 40: 699
de Vaucouleurs, G., Olson, D. W. 1982. *Ap. J.* 256: 346
de Zeeuw, T. 1990. *Annu. Rev. Astron. Astrophys.* In preparation
Disney, M., Phillipps, S. 1987. *Nature* 329: 203
Disney, M., Phillipps, S. 1988. *New Sci.* 117(1604): 60
Djorgovski, S. 1983. *J. Astrophys. Astron.* 4: 271
Djorgovski, S. G. 1985. PhD thesis. Univ. Calif., Berkeley
Djorgovski, S. 1987a. In *Nearly Normal Galaxies*, ed. S. M. Faber, p. 227. New York: Springer-Verlag
Djorgovski, S. 1987b. In *Structure and Dynamics of Elliptical Galaxies, IAU Symp. No. 127*, ed. T. de Zeeuw, p. 79. Dordrecht: Reidel
Djorgovski, S. 1987c. In *Starbursts and Galaxy Evolution*, ed. T. X. Thuan, T. Montmerle, J. T. T. Van, p. 549. Gif sur Yvette, Fr: Ed. Front.
Djorgovski, S., Davis, M. 1986. In *Galaxy Distances and Deviations From Universal Expansion*, ed. B. F. Madore, R. B. Tully, p. 135. Dordrecht: Reidel
Djorgovski, S., Davis, M. 1987. *Ap. J.* 313: 59
Djorgovski, S., Davis, M., Kent, S. 1985. In *New Aspects of Galaxy Photometry*, ed. J.-L., Nieto, p. 257. New York: Springer-Verlag
Djorgovski, S., de Carvalho, R., Han, M.-S. 1989. In *The Extragalactic Distance Scale*, ed. S. van den Bergh, C. J. Pritchet, p. 329. San Francisco: Astron. Soc. Pac.
Djorgovski, S., Dickinson, M. 1989. In *Highlights of Astronomy*, ed. D. McNally, Vol. 8. In press
Djorgovski, S., Ebneter, K. 1986. In *Instrumentation and Research Programmes for Small Telescopes, IAU Symp No. 118*, ed. J. B. Hearnshaw, P. L. Cottrell, p. 277. Dordrecht: Reidel
Dressler, A. 1979. *Ap. J.* 231: 659
Dressler, A. 1984a. *Ap. J.* 281: 512
Dressler, A. 1984b. *Annu. Rev. Astron. Astrophys.* 22: 185
Dressler, A. 1987. *Ap. J.* 317: 1
Dressler, A. 1988. In *Active Galactic Nuclei, IAU Symp. No. 134*, ed. D. E. Osterbrock, J. S. Miller. Dordrecht: Kluwer. In press
Dressler, A., Lynden-Bell, D., Burstein, D., Davies, R. L., Faber, S. M., et al. 1987. *Ap. J.* 313: 42
Dressler, A., Richstone, D. O. 1988. *Ap. J.* 324: 701
Dupraz, C., Combes, F. 1985. In *New Aspects of Galaxy Photometry*, ed. J.-L. Nieto, p. 151. New York: Springer-Verlag
Dupraz, C., Combes, F. 1986. *Astron. Astrophys.* 166: 53
Dupraz, C., Combes, F. 1987. *Astron. Astrophys.* 185: L1
Duschl, W. J. 1988a. *Astron. Astrophys.* 194: 33

Duschl, W. J. 1988b. *Astron. Astrophys.* 194: 43
Ebneter, K., Balick, B. 1985. *Astron. J.* 90: 183
Ebneter, K., Djorgovski, S., Davis, M. 1988. *Astron. J.* 95: 422
Efstathiou, G., Ellis, R. S., Carter, D. 1982. *MNRAS* 201: 975
Efstathiou, G., Fall, S. M. 1984. *MNRAS* 206: 453
Efstathiou, G., Gorgas, J. 1985. *MNRAS* 215: 37P
Faber, S. M. 1973. *Ap. J.* 179: 731
Faber, S. M. 1977. In *The Evolution of Galaxies and Stellar Populations*, ed. B. M. Tinsley, R. B. Larson, p. 157. New Haven, Conn: Yale Univ. Obs.
Faber, S. M. 1982. In *Astrophysical Cosmology*, ed. H. A. Brück, G. V. Coyne, M. S. Longair, p. 191. Vatican City: Pontif. Acad. Sci.
Faber, S. M., Dressler, A., Davies, R. L., Burstein, D., Lynden-Bell, D., et al. 1987. In *Nearly Normal Galaxies*, ed. S. M. Faber, p. 175. New York: Springer-Verlag
Faber, S. M., Gallagher, J. S. 1976. *Ap. J.* 204: 365
Faber, S. M., Jackson, R. E. 1976. *Ap. J.* 204: 668
Faber, S. M., Lin, D. N. C. 1983. *Ap. J. Lett.* 266: L17
Fabian, A., ed. 1988. *Cooling Flows in Clusters and Galaxies*. Dordrecht: Kluwer. In press
Farouki, R. T., Shapiro, S. L., Duncan, M. J. 1983. *Ap. J.* 265: 597
Forman, W., Jones, C., Tucker, W. 1985. *Ap. J.* 293: 102
Fort, B. P., Prieur, J.-L., Carter, D., Meatheringham, S. J., Vigroux, L. 1986. *Ap. J.* 306: 110
Franx, M. 1988. PhD thesis. Univ. Leiden, Neth.
Franx, M., Illingworth, G. D. 1988. *Ap. J. Lett.* 327: L55
Franx, M., Illingworth, G., Heckman, T. 1988. *Astron. J.* In press
Freeman, K. C. 1966. *MNRAS* 134: 1
Frogel, J. A., Persson, S. E., Aaronson, M., Matthews, K. 1978. *Ap. J.* 220: 75
Gallagher, J. S., Faber, S. M., Burstein, D. 1980. *Ap. J.* 235: 743
Gallagher, J. S., Goad, J. W., Mould, J. 1982. *Ap. J.* 263: 101
Gallagher, J. S., Knapp, G. R., Faber, S. M., Balick, B. 1977. *Ap. J.* 215: 463
Galletta, G. 1987. *Ap. J.* 318: 531
Gorgas, J., Efstathiou, G. 1987. In *Structure and Dynamics of Elliptical Galaxies, IAU Symp. No. 127*, ed. T. de Zeeuw, p. 189. Dordrecht: Reidel
Graham, J. A. 1979. *Ap. J.* 232: 60
Green, E. M., Demarque, P., King, C. 1987. *The Revised Yale Isochrones and Luminosity Functions*. New Haven, Conn: Yale Univ. Obs.
Gunn, J. E. 1979. In *Active Galactic Nuclei*, ed. C. Hazard, S. Mitton, p. 213. Cambridge: Univ. Press
Gunn, J. E. 1982. In *Astrophysical Cosmology*, ed. H. A. Brück, G. V. Coyne, M. S. Longair, p. 233. Vatican City: Pontif. Acad. Sci.
Gunn, J. E., Oke, J. B. 1975. *Ap. J.* 195: 255
Hamabe, M., Kormendy, J. 1987. In *Structure and Dynamics of Elliptical Galaxies, IAU Symp. No. 127*, ed. T. de Zeeuw, p. 379. Dordrecht: Reidel
Hausman, M. A., Ostriker, J. P. 1978. *Ap. J.* 224: 320
Hawarden, T. G., Elson, R. A. W., Longmore, A. J., Tritton, S. B., Corwin, H. G. 1981. *MNRAS* 196: 747
Heckman, T. M. 1983. *Ap. J.* 273: 505
Heiligman, G., Schwarzschild, M. 1979. *Ap. J.* 233: 872
Hernquist, L., Quinn, P. J. 1987. *Ap. J.* 312: 1
Hernquist, L., Quinn, P. J. 1988. *Ap. J.* 331: 682
Hoessel, J. G., Borne, K. D., Schneider, D. P. 1985. *Ap. J.* 293: 94
Hoessel, J. G., Oegerle, W. R., Schneider, D. P. 1987. *Astron. J.* 94: 1111
Hoessel, J. G., Schneider, D. P. 1985. *Astron. J.* 90: 1648
Hubble, E. 1926. *Ap. J.* 64: 321
Hummel, E., Kotanyi, C. G. 1982. *Astron. Astrophys.* 106: 183
Ichikawa, S.-I., Okamura, S., Kodaira, K., Wakamatsu, K.-I. 1988. *Astron. J.* 96: 62
Ichikawa, S.-I., Wakamatsu, K.-I., Okamura, S. 1986. *Ap. J. Suppl.* 60: 475
Illingworth, G. 1977. *Ap. J. Lett.* 218: L43
Illingworth, G. 1981. In *The Structure and Evolution of Normal Galaxies*, ed. S. M. Fall, D. Lynden-Bell, p. 27. Cambridge: Univ. Press
Impey, C., Bothun, G., Malin, D. 1988. *Ap. J.* 330: 634
Jarvis, B. 1987. *Astron. J.* 94: 30
Jedrzejewski, R. I. 1987a. In *Structure and Dynamics of Elliptical Galaxies, IAU Symp. No. 127*, ed. T. de Zeeuw, p. 37. Dordrecht: Reidel
Jedrzejewski, R. I. 1987b. *MNRAS* 226: 747
Jedrzejewski, R. I., Davies, R. L., Illingworth, G. D. 1987. *Astron. J.* 94: 1508
Jedrzejewski, R., Schechter, P. L. 1988a. *Ap. J. Lett.* 330: L87
Jedrzejewski, R. I., Schechter, P. L. 1988b. In *Dynamics of Dense Stellar Systems*, ed. D. Merritt. Cambridge: Univ. Press. In press
Johnson, H. M., Hanna, M. M. 1972. *Ap. J. Lett.* 174: L71

Jura, M. 1986. *Ap. J.* 306: 483
Jura, M., Kim, D. W., Knapp, G. R., Guhathakurta, P. 1987. *Ap. J. Lett.* 312: L11
Kahn, F. D., Woltjer, L. 1959. *Ap. J.* 130: 705
Karachentseva, V. E., Karachentsev, I. D., Richter, G. M., von Berlepsch, R., Fritze, K. 1987. *Astron. Nachr.* 308: 247
Kent, S. M. 1983. *Ap. J.* 266: 562
King, I. 1962. *Astron. J.* 67: 471
King, I. R., Kiser, J. 1973. *Ap. J.* 181: 27
Kodaira, K. 1988. Submitted for publication
Kodaira, K., Okamura, S., Watanabe, M. 1983. *Ap. J. Lett.* 274: L49
Kodaira, K., Watanabe, M., Okamura, S. 1986. *Ap. J. Suppl.* 62: 703
Kormendy, J. 1977a. *Ap. J.* 214: 359
Kormendy, J. 1977b. *Ap. J.* 218: 333
Kormendy, J. 1980. In *ESO Workshop on Two-Dimensional Photometry*, ed. P. Crane, K. Kjär, p. 191. Geneva: ESO
Kormendy, J. 1982a. In *Morphology and Dynamics of Galaxies*, ed. L. Martinet, M. Mayor, p. 113. Sauverny: Geneva Obs. (K82)
Kormendy, J. 1982b. *Ap. J.* 257: 75
Kormendy, J. 1984. *Ap. J.* 287: 577
Kormendy, J. 1985a. *Ap. J. Lett.* 292: L9
Kormendy, J. 1985b. *Ap. J.* 295: 73
Kormendy, J. 1987a. In *Structure and Dynamics of Elliptical Galaxies, IAU Symp. No. 127*, ed. T. de Zeeuw, p. 17. Dordrecht: Reidel (K87)
Kormendy, J. 1987b. In *Structure and Dynamics of Elliptical Galaxies, IAU Symp. No. 127*, ed. T. de Zeeuw, p. 97. Dordrecht: Reidel
Kormendy, J. 1987c. In *Nearly Normal Galaxies*, ed. S. M. Faber, p. 163. New York: Springer-Verlag
Kormendy, J. 1988a. In *Supermassive Black Holes*, ed. M. Kafatos, p. 219. Cambridge: Univ. Press
Kormendy, J. 1988b. *Ap. J.* 325: 128
Kormendy, J. 1988c. *Ap. J.* 335: 40
Kormendy, J. 1989. Submitted for publication
Kormendy, J., Illingworth, G. 1983. *Ap. J.* 265: 632
Kormendy, J., Stauffer, J. 1987. In *Structure and Dynamics of Elliptical Galaxies, IAU Symp. No. 127*, ed. T. de Zeeuw, p. 405. Dordrecht: Reidel
Lachièze-Rey, M., Vigroux, L., Souviron, J. 1985. *Astron. Astrophys.* 150: 62
Lake, G., Norman, C. 1983. *Ap. J.* 270: 51
Larson, R. 1974. *MNRAS* 166: 585
Larson, R. 1975. *MNRAS* 173: 671
Lauer, T. R. 1985a. *Ap. J. Suppl.* 57: 473
Lauer, T. R. 1985b. *Ap. J.* 292: 104
Lauer, T. R. 1985c. *MNRAS* 216: 429
Lauer, T. R. 1986. *Ap. J.* 311: 34
Lauer, T. R. 1987. In *Nearly Normal Galaxies*, ed. S. M. Faber, p. 207. New York: Springer-Verlag
Lauer, T. R. 1988a. *Ap. J.* 325: 49
Lauer, T. R. 1988b. In *Dynamics of Dense Stellar Systems*, ed. D. Merritt. Cambridge: Univ. Press. In press
Lentes, F. T. 1983. *Proc. Stat. Methods in Astron., Strasbourg (ESA SP-201)*, ed. E. J. Wolfe, p. 73. Noordwijk, Neth: ESTEC
Light, E. S., Danielson, R. E., Schwarzschild, M. 1974. *Ap. J.* 194: 257
Lin, D. N. C., Faber, S. M. 1983. *Ap. J. Lett.* 266: L21
Lugger, P. M. 1984. *Ap. J.* 286: 106
Lynden-Bell, D., Faber, S. M., Burstein, D., Davies, R. L., Dressler, A., et al. 1988. *Ap. J.* 326: 19
Maccagni, D., Garilli, B., Gioia, I. M., Maccacaro, T., Vettolani, G., Wolter, A. 1988. *Ap. J. Lett.* 334: L1
Malin, D. F. 1977. *Am. Astron. Soc. Photo Bull. No. 16*, p. 10
Malin, D. F. 1979. *Nature* 277: 279
Malin, D. F., Carter, D. 1980. *Nature* 285: 643
Malin, D. F., Carter, D. 1983. *Ap. J.* 274: 534
Malin, D. F., Quinn, P. J., Graham, J. A. 1983. *Ap. J. Lett.* 272: L5
Malumuth, E. M., Kirshner, R. P. 1985. *Ap. J.* 291: 8
Malumuth, E. M., Richstone, D. O. 1984. *Ap. J.* 276: 413
Marcelin, M., Boulesteix, J., Courtes, G., Millard, B. 1982. *Nature* 297: 38
Matteucci, F., Tornambè, A. 1987. *Astron. Astrophys.* 185: 51
Matthews, T. A., Morgan, W. W., Schmidt, M. 1964. *Ap. J.* 140: 35
May, A., van Albada, T. S., Norman, C. A. 1985. *MNRAS* 214: 131
McElroy, D. B. 1983. *Ap. J.* 270: 485
Merritt, D. 1984. *Ap. J.* 276: 26
Merritt, D. 1985. *Ap. J.* 289: 18
Merritt, D. 1988. *Astron. J.* 95: 496
Merritt, D., de Zeeuw, T. 1983. *Ap. J. Lett.* 267: L19
Michard, R. 1979. *Astron. Astrophys.* 74: 206
Michard, R. 1985. *Astron. Astrophys. Suppl.* 59: 205
Michard, R., Simien, F. 1987. In *Structure and Dynamics of Elliptical Galaxies, IAU Symp. No. 127*, ed. T. de Zeeuw, p. 393. Dordrecht: Reidel
Michard, R., Simien, F. 1988. *Astron. Astrophys. Suppl.* 74: 25
Möllenhoff, C. 1982. *Astron. Astrophys.* 108: 130
Möllenhoff, C., Bender, R. 1988. *Astron. Astrophys.* In press
Möllenhoff, C., Marenbach, G. 1986. *Astron. Astrophys.* 154: 219

Morbey, C. L. 1984. *Publ. Astron. Soc. Pac.* 96: 874
Morbey, C., Morris, S. 1983. *Ap. J.* 274: 502
Morgan, W. W., Kayser, S., White, R. A. 1975. *Ap. J.* 199: 545
Morgan, W. W., Lesh, J. R. 1965. *Ap. J.* 142: 1364
Murtagh, F., Heck, A. 1987. *Multivariate Data Analysis.* Dordrecht: Reidel
Nieto, J.-L. 1982. *Astron. Astrophys. Suppl.* 47: 535
Nieto, J.-L. 1988. *Proc. Summer Sch. Extragalact. Astron.*, 2nd, Cordoba, Argent. In press (N88)
Nieto, J.-L., Bender, R. 1988. *Astron. Astrophys.* In press
Nieto, J.-L., Capaccioli, M., Held, E. V. 1988. *Astron. Astrophys.* 195: L1
Nieto, J.-L., Prugniel, P. 1987a. In *Structure and Dynamics of Elliptical Galaxies, IAU Symp. No. 127*, ed. T. de Zeeuw, p. 99. Dordrecht: Reidel
Nieto, J.-L., Prugniel, P. 1987b. *Astron. Astrophys.* 186: 30
Nieto, J.-L., Vidal, J.-L. 1984. *MNRAS* 209: 21P
Norman, C. A., Renzini, A., Tosi, M., eds. 1986. *Stellar Populations.* Cambridge: Univ. Press
O'Connell, R. W. 1986. See Norman et al. 1986, p. 167
O'Connell, R. W. 1988. In *Cooling Flows in Clusters and Galaxies*, ed. A. Fabian. Dordrecht: Kluwer. In press
Oemler, A. Jr. 1976. *Ap. J.* 209: 693
Okamura, S. 1985. In *ESO Workshop on the Virgo Cluster*, ed. O.-G. Richter, B. Binggeli, p. 201. Garching: ESO
Okamura, S. 1988. *Publ. Astron. Soc. Pac.* 100: 524
Okamura, S., Kodaira, K., Watanabe, M. 1984. *Ap. J.* 280: 7
Ostriker, J. P., Hausman, M. A. 1977. *Ap. J. Lett.* 217: L125
Ostriker, J. P., Tremaine, S. D. 1975. *Ap. J. Lett.* 202: L113
Pagel, B. E. J., Edmunds, M. G. 1981. *Annu. Rev. Astron. Astrophys.* 19: 77
Peletier, R. F. 1988. PhD thesis. Univ. Groningen, Neth.
Peletier, R. F., Davies, R. L., Illingworth, G. D., Davis, L. E., Cawson, M. C. M. 1988a. *Astron. J.* In press
Peletier, R. F., Lauberts, A., Valentijn, E. A. 1988b. *Astron. Astrophys.* In press
Peletier, R. F., Valentijn, E. A. 1988. Preprint
Peletier, R. F., Valentijn, E. A., Davies, R. L., Jameson, R. F. 1988c. *Astron. Astrophys.* In press
Peletier, R. F., Valentijn, E. A., Jameson, R. F. 1987. In *Structure and Dynamics of Elliptical Galaxies, IAU Symp. No. 127*, ed. T. de Zeeuw, p. 443. Dordrecht: Reidel
Pence, W. D. 1986. *Ap. J.* 310: 597
Pence, W. D., Davoust, E. 1985. *Astron. Astrophys. Suppl.* 60: 517
Petrosian, V. 1976. *Ap. J. Lett.* 209: L1
Phillipps, S., Davies, J. I., Disney, M. J. 1988. *MNRAS* 233: 485
Piran, T., Villumsen, J. V. 1987. In *Structure and Dynamics of Elliptical Galaxies, IAU Symp. No. 127*, ed. T. de Zeeuw, p. 473. Dordrecht: Reidel
Porter, A. C. 1988. PhD thesis. Calif. Inst. Technol., Pasadena
Prieur, J.-L. 1988. *Ap. J.* 326: 596
Prugniel, P., Davoust, E., Nieto, J.-L. 1988. *Astron. Astrophys.* In press
Prugniel, P., Nieto, J.-L., Simien, F. 1987. *Astron. Astrophys.* 173: 49
Quinn, P. J. 1982. PhD thesis. Aust. Natl. Univ., Canberra
Quinn, P. J. 1984. *Ap. J.* 279: 596
Quinn, P. J., Hernquist, L. 1987. In *Structure and Dynamics of Elliptical Galaxies, IAU Symp. No. 127*, ed. T. de Zeeuw, p. 249. Dordrecht: Reidel
Reaves, G. 1977. In *The Evolution of Galaxies and Stellar Populations*, ed. B. M. Tinsley, R. B. Larson, p. 39. New Haven, Conn: Yale Univ. Obs.
Reaves, G. 1983. *Ap. J. Suppl.* 53: 375
Rees, M. J., Ostriker, J. P. 1977. *MNRAS* 179: 541
Rhee, G., Roos, N. 1989. *Astrophys. Space Sci.* In press
Romanishin, W. 1986a. *Ap. J.* 301: 675
Romanishin, W. 1986b. *Astron. J.* 91: 76
Romanishin, W. 1987. *Ap. J. Lett.* 323: L113
Romanishin, W., Strom, K. M., Strom, S. E. 1977. *Bull. Am. Astron. Soc.* 9: 347
Rood, H. J. 1965. *Astron. J.* 70: 689
Sadler, E. M., Gerhard, O. E. 1985a. In *New Aspects of Galaxy Photometry*, ed. J.-L. Nieto, p. 269. New York: Springer-Verlag
Sadler, E. M., Gerhard, O. E. 1985b. *MNRAS* 214: 177
Saito, M. 1979a. *Publ. Astron. Soc. Jpn.* 31: 181
Saito, M. 1979b. *Publ. Astron. Soc. Jpn.* 31: 193
Sandage, A. 1957. *Ap. J.* 125: 422
Sandage, A. 1961. *The Hubble Atlas of Galaxies.* Washington, DC: Carnegie Inst. Washington
Sandage, A. 1972. *Ap. J.* 176: 21
Sandage, A., Binggeli, B. 1984. *Astron. J.* 89: 919
Sandage, A., Binggeli, B., Tammann, G. A. 1985a. In *ESO Workshop on the Virgo Cluster*, ed. O.-G. Richter, B. Binggeli, p. 239. Garching: ESO
Sandage, A., Binggeli, B., Tammann, G. A. 1985b. *Astron. J.* 90: 1759

Sandage, A., Freeman, K. C., Stokes, N. R. 1970. *Ap. J.* 160: 831
Sandage, A., Hardy, E. 1973. *Ap. J.* 183: 743
Sandage, A. R., Miller, W. C. 1964. *Science* 144: 382
Sandage, A., Visvanathan, N. 1978a. *Ap. J.* 223: 707
Sandage, A., Visvanathan, N. 1978b. *Ap. J.* 225: 742
Sarazin, C. L. 1986. *Rev. Mod. Phys.* 58: 1
Sarazin, C. L., O'Connell, R. W. 1983. *Ap. J.* 268: 552
Schechter, P. 1976. *Ap. J.* 203: 297
Schneider, D. P., Gunn, J. E. 1982. *Ap. J.* 263: 14
Schneider, D. P., Gunn, J. E., Hoessel, J. G. 1983. *Ap. J.* 268: 476
Schombert, J. M. 1986. *Ap. J. Suppl.* 60: 603
Schombert, J. M. 1987. *Ap. J. Suppl.* 64: 643
Schombert, J. M. 1988. *Ap. J.* 328: 475
Schombert, J. M., Wallin, J. F. 1987. *Astron. J.* 94: 300
Schwarzschild, M. 1987. In *Structure and Dynamics of Elliptical Galaxies, IAU Symp. No. 127*, ed. T. de Zeeuw, p. 123. Dordrecht: Reidel
Schweizer, F. 1979. *Ap. J.* 233: 23 (Erratum. 1980. *Ap. J.* 236: 1056)
Schweizer, F. 1980. *Ap. J.* 237: 303
Schweizer, F. 1981a. *Ap. J.* 246: 722
Schweizer, F. 1981b. *Astron. J.* 86: 662
Schweizer, F. 1982. *Ap. J.* 252: 455
Schweizer, F. 1983. In *Internal Kinematics and Dynamics of Galaxies, IAU Symp. No. 100*, ed. E. Athanassoula, p. 319. Dordrecht: Reidel
Schweizer, F. 1986. *Science* 231: 227
Schweizer, F. 1987. In *Structure and Dynamics of Elliptical Galaxies, IAU Symp. No. 127*, ed. T. de Zeeuw, p. 109. Dordrecht: Reidel
Schweizer, F., Ford, W. K. 1985. In *New Aspects of Galaxy Photometry*, ed. J.-L. Nieto, p. 145. New York: Springer-Verlag
Schweizer, F., Seitzer, P. 1988. *Ap. J.* 328: 88
Sharples, R. M., Carter, D., Hawarden, T. G., Longmore, A. J. 1983. *MNRAS* 202: 37
Shlosman, I., Begelman, M. C. 1987. *Nature* 329: 810
Silk, J. 1983. *Nature* 301: 574
Silk, J. 1985. *Ap. J.* 297: 9
Silk, J. 1987. In *Dark Matter in the Universe, IAU Symp. No. 117*, ed. J. Kormendy, G. R. Knapp, p. 335. Dordrecht: Reidel
Silk, J., Djorgovski, S., Wyse, R. F. G., Bruzual A., G. 1986. *Ap. J.* 307: 415
Silk, J., Norman, C. 1981. *Ap. J.* 247: 59
Silk, J., Wyse, R. F. G., Shields, G. A. 1987. *Ap. J. Lett.* 322: L59
Simonson, G. F. 1982. PhD thesis. Yale Univ., New Haven, Conn.
Smith, H. A., Tinsley, B. M. 1976. *Publ. Astron. Soc. Pac.* 88: 370
Sparks, W. B., Wall, J. V., Thorne, D. J., Jorden, P. R., van Breda, I. G., et al. 1985. *MNRAS* 217: 87
Statler, T. S. 1988. *Ap. J.* 331: 71
Steiman-Cameron, T. Y., Durisen, R. H. 1982. *Ap. J. Lett.* 263: L51
Strom, K. M., Strom, S. E. 1978. *Astron. J.* 83: 73
Strom, K. M., Strom, S. E., Wells, D. C., Romanishin, W. 1978. *Ap. J.* 220: 62
Strom, S. E., Strom, K. M., Goad, J. W., Vrba, F. J., Rice, W. 1976. *Ap. J.* 204: 684
Struble, M. F. 1988. *Ap. J. Lett.* 330: L25
Terlevich, R., Davies, R. L., Faber, S. M., Burstein, D. 1981. *MNRAS* 196: 381
Thomsen, B., Baum, W. A. 1987. *Ap. J.* 315: 460
Thomsen, B., Baum, W. A. 1988. *Ap. J.* In press
Thomsen, B., Frandsen, S. 1983. *Astron. J.* 88: 789
Thuan, T. X., Romanishin, W. 1981. *Ap. J.* 248: 439
Tinsley, B. M. 1978. *Ap. J.* 222: 14
Tohline, J. E., Simonson, G. F., Caldwell, N. 1982. *Ap. J.* 252: 92
Tonry, J. L. 1981. *Ap. J. Lett.* 251: L1
Tonry, J. L. 1984a. *Ap. J.* 279: 13
Tonry, J. L. 1984b. *Ap. J. Lett.* 283: L27
Tonry, J. L. 1985. *Astron. J.* 90: 2431
Tonry, J. L. 1987. In *Structure and Dynamics of Elliptical Galaxies, IAU Symp. No. 127*, ed. T. de Zeeuw, p. 89. Dordrecht: Reidel
Tonry, J. L., Davis, M. 1981. *Ap. J.* 246: 680
Toomre, A. 1983. Quoted in Schweizer 1983, p. 324
Tremaine, S. 1987. In *Structure and Dynamics of Elliptical Galaxies, IAU Symp. No. 127*, ed. T. de Zeeuw, p. 367. Dordrecht: Reidel
Tremaine, S., Ostriker, J. P. 1982. *Ap. J.* 256: 435
Tremaine, S. D., Ostriker, J. P., Spitzer, L. 1975. *Ap. J.* 196: 407
Tubbs, A. D. 1980. *Ap. J.* 241: 969
Vader, J. P. 1986a. *Ap. J.* 305: 669
Vader, J. P. 1986b. *Ap. J.* 306: 390
Vader, J. P. 1987. *Ap. J.* 317: 128
Vader, J. P., Vigroux, L., Lachièze-Rey, M., Souviron, J. 1988. *Astron. Astrophys.* 203: 217
Valentijn, E. A. 1983. *Astron. Astrophys.* 118: 123
Valentijn, E. A., Moorwood, A. F. M. 1985. *Astron. Astrophys.* 143: 46
van Albada, T. S. 1987. In *Structure and Dynamics of Elliptical Galaxies, IAU Symp. No. 127*, ed. T. de Zeeuw, p. 291. Dordrecht: Reidel
van Albada, T. S., Kotanyi, C. G., Schwarzschild, M. 1982. *MNRAS* 198: 303

van den Bergh, S. 1976. *Ap. J.* 203: 764
van den Bergh, S. 1986. *Astron. J.* 91: 271
Varnas, S. R., Bertola, F., Galletta, G., Freeman, K. C., Carter, D. 1987. *Ap. J.* 313: 69
Véron-Cetty, M.-P., Véron, P. 1988. *Astron. Astrophys.* 204: 28
Vietri, M. 1986. *Ap. J.* 306: 48
Vietri, M. 1988. Preprint
Vigroux, L., Chièze, J. P., Lazareff, B. 1981. *Astron. Astrophys.* 98: 119
Vigroux, L., Souviron, J., Lachièze-Rey, M., Vader, J. P. 1988. *Astron. Astrophys. Suppl.* 73: 1
Visvanathan, N., Sandage, A. 1977. *Ap. J.* 216: 214
Wagner, S. J., Bender, R., Möllenhoff, C. 1988. *Astron. Astrophys.* 195: L5
Watanabe, M., Kodaira, K., Okamura, S. 1985. *Ap. J.* 292: 72
White, S. D. M. 1982. In *Morphology and Dynamics of Galaxies*, ed. L. Martinet, M. Mayor, p. 289. Sauverny: Geneva Obs.
Whitmore, B. C. 1984. *Ap. J.* 278: 61
Whitmore, B. C., Bell, M. 1988. *Ap. J.* 324: 741
Wilkinson, A., Sharples, R. M., Fosbury, R. A. E., Wallace, P. T. 1986. *MNRAS* 218: 297
Wilkinson, A., Sparks, W. B., Carter, D., Malin, D. A. 1987. In *Structure and Dynamics of Elliptical Galaxies, IAU Symp. No. 127*, ed. T. de Zeeuw, p. 465. Dordrecht: Reidel
Wirth, A. 1981. *Astron. J.* 86: 981
Wirth, A., Gallagher, J. S. 1984. *Ap. J.* 282: 85
Wirth, A., Kenyon, S. J., Hunter, D. A. 1983. *Ap. J.* 269: 102
Wirth, A., Shaw, R. 1983. *Astron. J.* 88: 171
Yoshii, Y., Arimoto, N. 1987. *Astron. Astrophys.* 188: 13
Young, P. J., Sargent, W. L. W., Kristian, J., Westphal, J. A. 1979. *Ap. J.* 234: 76
Young, P. J., Westphal, J. A., Kristian, J., Wilson, C. P., Landauer, F. P. 1978. *Ap. J.* 221: 721

ABUNDANCE RATIOS AS A FUNCTION OF METALLICITY

J. Craig Wheeler and Christopher Sneden

Department of Astronomy and McDonald Observatory,
University of Texas, Austin, Texas 78712

James W. Truran, Jr.

Department of Astronomy, University of Illinois, Urbana,
Illinois 61801

1. INTRODUCTION

The history of the Galaxy is written in the evolution of its composition. One of the major goals of modern astrophysics is to read that history and thereby reveal the record of the Big Bang; the formation of the Galaxy as a primordial cloud, perhaps ridden with exotic weakly interacting particles; the formation of the first stars, which illuminate the cloud and provide the first injection of heavy elements; the formation of a halo, a bulge, a thick disk, a thin disk, and spiral arms; interactions with companion galaxies; possible phases of nuclear activity; bursts of star formation; and further injection of energy and nucleosynthetic products, infall, and fountains. This is a complex history, coupling a large range in temporal and spatial scales.

The primary tools to read this history are (*a*) the spectroscopy of stars, which reveals their composition; (*b*) the theory of stellar evolution and nucleosynthesis, which guides our understanding of how the elements are synthesized; and (*c*) observations and the theory of Galactic dynamics that helps us to understand the environment in which the stars were born. These tools provide us with a great deal of analytic power. We have a reasonable overall picture of the chemical evolution of the solar neighborhood, in which massive stars have provided the bulk of the heavy-element nucleosynthesis and the accumulation of carbon, oxygen, iron, and a wide variety of other elements and isotopes in the gas and stars as the Galaxy has aged. In the simplest picture (comprised of an isolated volume of gas), stars form from the gas, evolve, and die, injecting some of their synthesized heavy elements back into the gas so that new stars form in the enriched gas and then add their own contribution. In this scenario one need not refer explicitly to the parameter of time, since the accumulation of heavy elements serves as a measure of the evolution of the system. It is popular and pragmatic to adopt this approach to the study of chemical evolution by presenting the evolution of abundances in terms of a single element. Iron is usually chosen to play this role as the fundamental measure of metallicity because it is relatively easy to determine the stellar abundance of Fe.

The chronology can be calibrated by studying the ages of stars. The principal means of doing this is through the study of stellar evolution. Stellar isochrones are fit to clusters or individual stars to compare with theoretical estimates of how stars should age. Stars associated with the halo are found to be among the oldest from theoretical studies of stellar evolution and galaxy formation, and they have the lowest metallicity. The stellar bulge and disk form later in dynamical theories and have higher metallicity.

There are two fundamental time scales involved in this global picture. One is the stellar evolution time scale dictated by the gravity of individual stars and by their nuclear physics. The other is the gravitational/gas-dynamical time scale associated with the formation and evolution of Galactic structure. There is an implicit assumption that these time scales are identical and can be calibrated, but this certainly is not trivially true. The growth of the metallicity is further assumed to be a monotonic measure of the passage of time that links the stellar evolution time scale with global processes through the star formation rate. In practice, this procedure involves the construction of a stellar evolution–metallicity relation, or a relation between metallicity and global kinematics (halo, bulge, thick disk, thin disk), but an actual temporal chronology is not assigned so easily. An important question then is whether implicit assumptions concerning the assignment of time scales is interfering with our reading of the record of chemical evolution. More specifically, does the rise in the heavy-element abundance represent a good clock, how is the clock to be calibrated, and which element is best to use?

There are a large number of additional questions about the chemical evolution of the Galaxy. Among them are the distribution of metal-poor dwarfs and the nature and role of the so-called Population III stars. What was the mass function and yield of the first stars? Has the mass function evolved with time? Does the mass function have multiple components implying different processes of star formation and hence, perhaps, nucleosynthetic history? How was the abundance gradient laid down in the Galaxy, and what was the role of infall, outflow, and fountains in distributing mass, angular momentum, and composition? How has the record of chemical evolution been affected by mixing of stars of different population and origin by stellar and gasdynamical processes? Does chemical enrichment proceed uniformly owing to stellar and gasdynamical mixing, or is it inhomogeneous in globular clusters, the halo, or the disk? How accurate are our theoretical estimates of stellar yields, and what is the role of Type I versus Type II supernovae? Of stars that leave neutron star remnants versus those that produce black holes? What is the astrophysical origin of the r-process elements? What is the effect of temporal irregularities (starbursts) on the chemical evolution?

We cannot hope to answer these questions in this review; however, we do hope to remind the reader that the questions are complex and the interpretations subject to oversimplification. There are a number of patterns that can be discerned—the dynamical classification of a given star, the metallicity [Fe/H], the abundance of a variety of heavy elements. The basic questions concern how these quantities are correlated and how they relate to the chronologies imposed by stellar evolution and dynamics. The correlations can be empirically established, and we attempt to present an up-to-date record of them. The calibration of the chronology and the elucidation of which variables are fundamental and which are secondary are more difficult.

The major part of this review (Sections 2–4) consists of a summary of the observational record of abundances as a function of metallicity, using evidence from field and globular cluster stars of the Galaxy and the Magellanic Clouds. We do not have space to do justice to all the excellent work being done on modeling the chemical evolution of galaxies. Rather, we discuss some of the basic problems of the physical interpretation of the chemical record in terms of nucleosynthesis theory (Section 5), mass functions (Section 6), and age-metallicity relations (Section 7). A summary, with recommendations for both further theoretical and observational work, is given in Section 8.

2. THE ABUNDANCES IN FIELD STARS

In this section we review the current status of abundances in field stars of different metallicity. Our discussion is weighted toward a comparison of solar metallicity stars with those of extreme metal poverty. We generally rely on the standard notation of stellar abundance research—namely

$[X] = \log_{10}(X)_{\text{star}} - \log_{10}(X)_{\text{Sun}}$, where X is any quantity in the stellar atmosphere (usually an abundance ratio such as Fe/H, C/Fe, etc.).

There have been several reviews of stellar abundance trends in recent years, most notably by Spite & Spite (1985) in these volumes, Gustafsson (1987), Lambert (1988), and Gehren (1988). The review by Lambert (1988) provides quantitative intercomparisons of the abundance results from different investigations of the same element abundances, and the review by Gehren (1988) devotes considerable effort to a discussion of the basic techniques of abundance determinations, with emphasis on questions about the reliability of each part of the process (basic line measurement uncertainties, laboratory oscillator strength determinations, non-LTE concerns, etc.).

Below we consider each element group in turn in the usual order of increasing Z, but with the recognition that the order of synthesis of these elements during a star's lifetime is not always a monotonic function of Z, nor is Z necessarily a reasonable indicator of the mass of the star responsible for particular element syntheses. For example, both one of the lightest elements considered (oxygen) and one of the heaviest (europium) apparently are synthesized most efficiently by massive stars, whereas much of the iron probably is manufactured by less massive stars. The nucleosynthesis sites for each element are considered mainly in a later section. Some of the discussion in this section is devoted to the uncertainties in the abundance trends, and one may be left with the impression that the basic observational data and analysis techniques are not yet really trustworthy. It is therefore worth reminding the reader at the start that considerable progress has been made within the last decade on abundance trends in stars of all metallicities, and that for the Sun the abundances determined from photospheric spectra with standard techniques are in excellent agreement with meteoritic values (Anders & Ebihara 1982, Cameron 1982b, Grevesse 1984, Anders & Grevesse 1989) for an impressive number of elements.

2.1 *The CNO Elements*

Our discussion of abundance trends begins with the CNO group. Considerable recent observational effort has extended greatly our knowledge of lithium (and to a lesser extent of beryllium and boron) abundances in stars of all metallicities. Space here does not permit a thorough discussion of these elements, and so we suggest that the interested reader first turn to the excellent review by Boesgaard & Steigman (1985) for an outline of all the lightest element abundances in the context of the overall evolution of the Universe. Then, much of the more recent work on Li, Be, and B may be discovered by reading several articles on these elements that appear in the proceedings of *IAU Symposium No. 132* (e.g. Soderblom 1988, Budge et al. 1988). Finally, a good summary of the Li data as a function of metallicity is given by Rebolo et al. (1988).

The abundance trends of the CNO elements are crucial to Galactic evolution studies because they form the bulk of the heavy elements. As such, they play important roles in stellar interior opacities and energy generation, and thus they affect the lifetimes, Hertzsprung-Russell (HR) diagram positions, and heavy-element yields from stars. The abundances of the CNO elements may be altered at various evolutionary stages by stars of all masses and compositions. Sneden (1985) recently has reviewed the status of CNO abundance determinations, but new surveys of these elements in, especially, halo stars provide some important new information to clarify the abundance trends.

2.1.1   CARBON   It has long been thought that [C/Fe] ≈ 0 in unevolved stars of both the disk and halo (e.g. Peterson & Sneden 1978, Clegg et al. 1981), and the constancy of this ratio has been used as an important constraint on models of Galactic chemical evolution. Is [C/Fe] truly constant? Three recent studies of C in stars spanning a large range of metallicity simultaneously confirm and challenge the standard view of the abundance of this element in different Galactic epochs. All of these studies have employed the CH G-band features in their analyses, for the C I lines become unmeasurably weak in metal-poor ([Fe/H] < -1) stars. This is unfortunate, because there appears to be a systematic offset of about 0.3 dex between C abundances deduced for a star from its CH features and abundances deduced from C I features. (Higher abundances are obtained from the C I features.) It is not known whether the offset is constant or has some trend with metallicity. Also, all of these studies employed LTE spectrum synthesis analyses, since no individual CH lines could be detected in the spectra. Finally, to ascertain the overall trends of [C/Fe] in stars of different epochs, these studies necessarily have concentrated on dwarf stars, since the dredge-ups of CN-cycle hydrogen fusion products alter the surface contents of giant stars. The effects of the CN cycle on the C and N abundances vary from star to star, but they appear to be far more severe in the low-metal, low-mass stars of the halo than in younger, higher mass disk stars (see the brief discussion below). The C-abundance determinations in dwarfs are considered in some detail here to attempt to show the reader to what extent abundance trend claims may be real or may be artifacts of the reduction and analysis procedures. Many of the comments given on C here may apply to the interpretation of other abundance results as well.

First, Laird (1985) published an extensive study of C abundances in 116 dwarfs with metallicities over the range $+0.5 \geq$ [Fe/H] ≥ -2.2, with very few stars below [Fe/H] ≈ -2. The basic data for his work were 1-Å resolution image tube spectra. His analyses, after he applied various corrections for systematic errors in the abundances, revealed no variations in C with metallicity: [C/Fe] ≈ 0 over the whole [Fe/H] range of his sample. Second, Tomkin et al. (1986) analyzed the spectra of 32 halo ($-0.7 \geq$ [Fe/H] ≥ -2.5) dwarfs using coudé digital data of high resolution (≈ 0.25 Å) and low noise ($S/N \geq 100$). They derived [C/Fe] ≈ -0.2 independent of metallicity for stars with [Fe/H] > -2, in agreement with Laird and earlier workers; however, their data showed a rise in C in more metal-poor stars: [C/Fe] $\approx +0.2$ at [Fe/H] ≈ -2.5. The suggested *difference* of $+0.4$ dex between stars with [Fe/H] < -2 and those with [Fe/H] > -2 is more important than the absolute value of [C/Fe], since the overall scale of the ratio is of course sensitive to various analysis assumptions (molecular-line parameters, choice of model atmospheres, how the Fe abundances are derived, etc.). Third, Carbon et al. (1987) derived C abundances for 83 dwarfs from low-resolution (≈ 10 Å), high S/N spectra using the same techniques that they had previously employed in the analyses of halo field and globular cluster giants. They also asserted that a trend of increasing [C/Fe] ratios with decreasing [Fe/H] metallicity existed.

A simple plot of these three data sets together would not give the viewer any confidence about the claims of changing C/Fe ratios in the Galactic halo! It is somewhat misleading, however, to plot these data sets together without further study, for they have been determined with a variety of analysis assumptions. Some of these assumptions would have the main effects of wholesale scale shifts that would be relatively independent of metallicity. These include choices of (*a*) dissociation energies and oscillator strengths for CH, (*b*) the solar C abundance, and (*c*) the solar model atmosphere.

The choice of model solar atmosphere deserves some brief comment here, for it affects nearly all abundance ratios to be discussed in this review. At present, laboratory oscillator strengths have been determined only for a minority of astrophysically interesting transitions, and those that have been published still are subject to internal and scale uncertainties; see, for example, Sneden & Parthasarathy's (1983) attempt to derive absolute abundances for the halo giant HD 122563. Therefore, at least for Fe, many stellar abundance studies begin by deriving empirical oscillator strengths from equivalent widths of the solar features. Very high-quality solar atlases have appeared in recent years [e.g. the center-of-disk intensity spectrum of the Liège group (Delbouille et al. 1973) and the integrated-flux spectrum of the Kitt Peak group (Kurucz et al. 1984)], and thus the chief uncertainty lies simply in the assumption of the model to represent the solar photosphere. Usually, either the empirical solar model derived by Holweger & Müller (1974) or the radiative equilibrium model of the Bell-Gustafsson group (Gustafsson et al. 1975, Bell et al. 1976) or of Kurucz (1979) is employed. Many investigators have adopted the Holweger-Müller model because of its ability (*a*) to predict solar abundances that are in good agreement with meteoritic values, and (*b*) to yield good consistency in abundances of an element from features of different species. On the other hand, the advantage of choosing equilibrium solar models is that they are calculated with the same assumptions as those employed for the program stars in a given study, either through the choice of appropriate grid models or through recomputation using the source codes of Gustafsson et al. (1975) or Kurucz (1970). Therefore, the different abundance ratios should at least partially cancel some of the admitted inadequacies of the models that in any case must be employed for the program stars. The uncertainties induced by the different approaches to this question made by different investigators show most clearly in ratios such as [C/Fe], where the Fe abundance is normally a differential value while the C abundance is derived using laboratory values of the line parameters. The best papers in this field usually will perform numerical experiments to show the reader the results of the different possible choices for solar and stellar model atmospheres.

Typical uncertainties of scale induced in [C/Fe] by the different model assumptions are about 0.15 dex, and line parameter and solar abundance assumptions contribute perhaps an additional 0.1 dex uncertainty. This review is not the place to attempt to sort out all the different effects in the papers that we discuss, so we have considered only the following simple statement: On average, the [C/Fe] abundances determined for a sufficiently large sample of stars in a small metallicity range should be the same in different investigations. With this guidance, we selected stars in the metallicity range $-1.2 \geq$ [Fe/H] ≥ -1.9 for normalization of the three data sets. For stars with [Fe/H] ≤ -1.2, CO formation has a smaller impact on the CH-line strengths than in more metal-rich stars, if we assume that the usually unknown O abundance of a metal-poor dwarf is not too much higher than the Fe abundance (see below). Also, CH-line saturation effects should be less in this metallicity range than in more metal-rich stars. The lower limit of [Fe/H] ≥ -1.9 is set by the desire to avoid the metallicity domain where changes in [C/Fe] are alleged to occur. The average [C/Fe] values of the three data sets considered here are ≈ 0.00 (Laird 1985; 10 stars), ≈ -0.25 (Tomkin et al. 1986; 16 stars), and ≈ -0.10 (Carbon et al. 1987; 36 stars). A plot of [C/Fe] versus [Fe/H] with these scale adjustments makes the proposed upswing in C below [Fe/H] ≈ -2 much more obvious.

Unfortunately, the uncertainties in the [C/Fe] results are not limited simply to wholesale scale shifts. Two problems stand out here: the lack of oxygen abundances for nearly all of the stars in the three studies, and the dependence of C abundances on effective temperatures of the stars. To account for the possible CO formation in these stars, each investigation adopted different assumptions about the O abundances. Laird (1985) chose a mean enhancement of O from previous studies of a few dwarfs and giants: [O/Fe] $\approx$ +0.4. Tomkin et al. (1986) assumed (but did not state in their paper) that [O/Fe] = 0, since their program stars mainly had $T_{\rm eff}$ > 5500 K, and CO formation was expected to be negligible. Carbon et al. (1987) examined the CO formation problem more thoroughly, showing the trends in [C/Fe] for two assumptions about O: [O/Fe] = +0.6 and 0.0. Indeed, examination of the figures presented by Carbon et al. (1987) should give the reader an appropriate sense of caution about claims of variations in [C/Fe] with metallicity. Since the most metal-poor stars in their study also were the coolest stars, these stars were most affected by CO formation, and the trends in [C/Fe] were very dependent on O-abundance assumptions. The necessary labor in determining the [O/Fe] ratios in metal-poor dwarfs has yet to be done, so that this review must leave the CO problem to future investigations.

The effective temperature trends may be dealt with here in approximate ways. First, note that in the metallicity range $-1.2 \geq$ [Fe/H] ≥ -1.9, the trends with temperature Δ[C/Fe]/$\Delta T_{\rm eff}$ are ≈ -0.1 (Laird 1985), $\approx +0.4$ (Tomkin et al. 1986; this trend is most uncertain owing to the small number of stars involved), and ≈ -0.2 (Carbon et al. 1987). The fact that the trends are not all of the same sign suggests that the effects are more analytical than astrophysical, and so here again we adopt the simple view that corrections should be made to the data sets to bring them into general agreement with no pronounced temperature effects. Applying these temperature slopes to the data sets and renormalizing, we obtain the final corrected [C/Fe] abundances and plot them in Figure 1. We notice immediately that the overall trend toward larger [C/Fe] ratios in the most metal-poor stars does not disappear. Before asking theoreticians to accept the reality of C overabundances in the extreme halo, at least two further tasks remain for the observer: to derive O abundances for a substantial fraction of these same stars, and to address the issue of the scatter in [C/Fe] at a given [Fe/H]. Is the scatter real or just another artifact of the observational/analytical technique? If it is not real, then we cannot fairly say at the present time what the [C/Fe] ratio is at, say, [Fe/H] ≈ -2.5. A final plea for the improved determination of C in extreme halo dwarfs is to increase the available sample. The list of Beers et al. (1985) is a promising step in that direction, and detailed C-abundance analyses of their stars should be undertaken.

2.1.2 NITROGEN   A very large data base of nitrogen abundances in un-

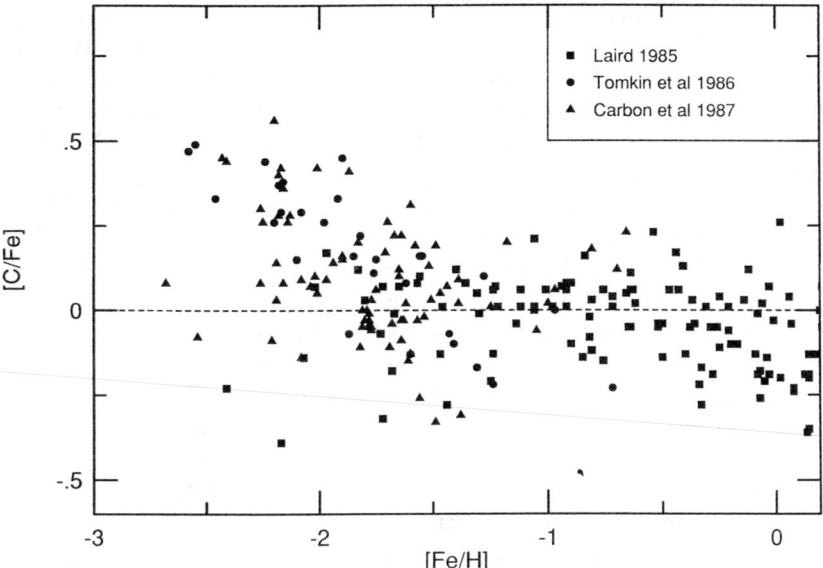

Figure 1 Carbon abundances in halo stars, after application of the T_{eff} and abundance scale corrections described in the text.

evolved stars of various metallicities now exists, but definitive statements about the [N/Fe] trend still are not possible. Difficulties remain in attempts to derive reliable abundances from the NH bands at 3360 Å and 3370 Å, due (*a*) to uncertainties in the continuous opacities at these wavelengths, (*b*) to the severe atomic-line contamination among the NH features, and (*c*) to remaining questions about the value of the dissociation energy of this molecule (see the discussion by Lambert 1988). Attempts to use the CN bands at 4200 Å or 3800 Å to derive N abundances are nearly hopeless, because the dissociation energy of CN is not known to better than a factor of 2 (e.g. see Lambert 1978), the determination of N abundances from CN are dependent on the prior determination of reliable C abundances from other features, and the CN bands become extremely weak in very metal-poor stars as a result of their double dependence on the overall stellar metallicity. Finally, the N I lines may be used in abundance determinations of this element, but only for reasonably metal-rich stars.

It is possible, however, to suggest some general trends based on the current literature for N. The results of several large-scale surveys of dwarf stars (Clegg et al. 1981, Tomkin & Lambert 1984, Laird 1985, Carbon et al. 1987) all produce the same general result: [N/Fe] $\sim$ 0 irrespective of metallicity in the range $+0.3 >$ [Fe/H] > -2. The exact [N/Fe] value is very difficult to determine. For instance, Laird's (1985) raw abundance results from low-resolution image tube spectroscopy of the NH bands indicated that [N/Fe] ~ -0.6, but this offset persisted even in his analyses of stars with solar metallicity ([Fe/H] = 0). Knowing that previous N-abundance studies [e.g. Clegg et al. (1981) from analyses of N I lines; Sneden (1974) from NH and CN bands] had indicated that [N/Fe] ~ 0 in some of these same stars, Laird suggested that this offset was not real and corrected his N abundances upward by 0.65 dex. In contrast, Tomkin & Lambert (1984) derived constant relative deficiencies of N in their sample of 14 stars ([N/Fe] ~ -0.25) from high-resolution, low-noise spectra of the NH bands. Moreover, Carbon et al. (1987), working from low-resolution spectrophotometric scans, found N deficiencies averaging [N/Fe] ~ -0.3. Intriguingly, their results also indicated a slight trend of smaller [N/Fe] at decreasing [Fe/H] values; however, most of that trend could be explained by systematic changes in [N/Fe] with temperature, which they showed also exists both in the Laird and in the Tomkin & Lambert data sets.

If one assumes that cooler metal-poor dwarfs do not in reality have higher surface N content at a given metallicity and corrects the data sets accordingly (as we have attempted for C abundances), then the [N/Fe] trend with [Fe/H] disappears. This is the safest assumption to adopt here, but the question of real correlations of N abundances with T_{eff} values in halo dwarfs should be pursued in the future. More generally, we still are left with considerable uncertainty in the preferred [N/Fe] value, even if it is constant with metallicity. Although Laird (1985) mistrusted his apparent large N deficiencies of -0.6 dex, the other studies quoted here have reached nearly the same conclusion; perhaps his initial N abundances were more correct than he feared, or possibly all extant N results for metal-poor dwarfs are only good to some undetermined additive offset constant. We urge renewed attention to the entire problem of [N/Fe] in stars over the whole metallicity range.

Two footnotes finish our discussion of N in unevolved stars. First, a combination of C and N abundances into N/C ratios is of some interest because abundances of these two elements are correlated through actions of the CN cycle in stellar interiors. Also, if these abundances are determined from features of the largely similar CH and NH molecules, as is often the case, then presenting the abundance results as N/C values should suppress some of the analysis uncertainties. Gehren (1988, his Figure 7) shows the trend of [N/C] with [Fe/H]. The alleged rise of C and fall of N in halo dwarfs combine to create very low values of this ratio ([N/C] ~ -0.6 at [Fe/H] < -2), and it is tempting to extend this trend (as Gehren does) to include the value [N/C] ~ -1.5 of the reigning champion of metal poverty, G77–61, which has an apparent metallicity of [Fe/H] ~ -5.6 (Gass et al. 1988). However, since both [C/Fe] and [N/Fe] are extremely high in G77−61, it is difficult to be sure of the place of this star in Galactic chemical evolution studies.

A second small point is to remind the reader that a handful of halo dwarfs with N overabundances have been found. The data sets of Laird (1985) and Carbon et al. (1987) suggest that the frequency of such nitrogen-rich stars is somewhat less than 5%. Spite & Spite (1986, 1987) have used Li abundance results for the four known N-rich metal-poor dwarfs to argue that the high N abundances in these stars cannot be due to internal CN cycle processing. Specifically, they found that the Li values of the N-rich stars are very consistent with those of all other unevolved halo stars: $\log \varepsilon(\text{Li}) \approx +2$, on the usual number density scale in which $\log \varepsilon(\text{H}) = 12$ (see Rebolo et al. 1988, and references therein). Any internal mixing would be expected to dilute the surface Li content of these stars, and thus it is unlikely that such mixing has occurred. The high N abundances of these stars are likely to be primordial in origin.

2.1.3 OXYGEN Over two decades ago, Conti et al. (1967) presented the first data suggesting the existence of relative oxygen overabundances in metal-poor stars. Since their paper, supersolar [O/Fe] ratios have been confirmed over the whole metallicity range between $-0.5 \geq$ [Fe/H] ≥ -3.0, and current work on this question is concentrated on the determination of the overall level of [O/Fe] and a comparison of this ratio between field and globular cluster metal-poor stars. Lambert et al. (1974) provided the first quantitative O estimate in a very metal-poor star, finding [O/Fe] $\approx +0.6$ for the halo giant HD 122563 ([Fe/H] ≈ -2.7) from the 6300-Å feature. Sneden et al. (1979) followed with a study of the 7774-Å O I triplet in metal-poor dwarfs. They obtained an average value [O/Fe] $\sim +0.5$, but with a fairly large scatter. This is not entirely unexpected, since the 7774-Å lines arise from a very high excitation state, and the line formation is known to be sensitive to luminosity and non-LTE effects (e.g. Sedlmayr 1974, Eriksson & Toft 1979). Clegg et al. (1981) demonstrated the gradual onset of O overabundances in disk stars of decreasing metallicity.

Three recent papers provide the current definition of [O/Fe] in stars of different metallicity. Andersen et al. (1988; see also Nissen et al. 1985) employed careful analyses of the 6150-Å and 7774-Å O I features from very high-resolution spectra of disk population dwarfs to demonstrate that [O/Fe] ≈ -0.5[Fe/H] in the range $0.0 \geq$ [Fe/H] ≥ -1.0. For more metal-poor stars, analyses of similar spectra of the 6300-Å [O I] line by Gratton & Ortolani (1986) and Barbuy (1988) yield a constant [O/Fe] $\approx +0.35$ down to metallicities of [Fe/H] ≈ -3, with very little scatter (± 0.15 dex) in this ratio from star to star. The [O/Fe] ratios for three stars in common between these two studies are in very good agreement. In Figure 2 we show the trends of O as a function of metallicity. As mentioned previously, the traditional metallicity parameter is Fe, but since the site for its synthesis is not yet well established, and since stars of different mass ranges and hence different evolutionary histories may produce Fe, but apparently only high-mass stars are efficient sites of O production, we have chosen to present Figure 2 with O being the metallicity variable. If the O abundance is a better chronometer than the Fe abundance, then the most significant point of this style of plot (as opposed to the more usual [O/Fe] vs. [Fe/H] given in, say, the review by Gehren 1988) is the emphasis of the recent buildup of Fe content in a very short O time scale. We return to this point in Section 8.

These O results seem very clear and consistent at first glance, but it is necessary to note some remaining questions that observers should consider. The above discussion has pointed out that studies using the high-excitation O I lines in metal-poor dwarfs often derive larger O overabundances (≈ 0.2 dex on average) than do studies of [O I] in metal-poor giants. An extreme example of this effect is seen in a comparison of recent results for two extremely metal-poor stars: Magain (1985) found

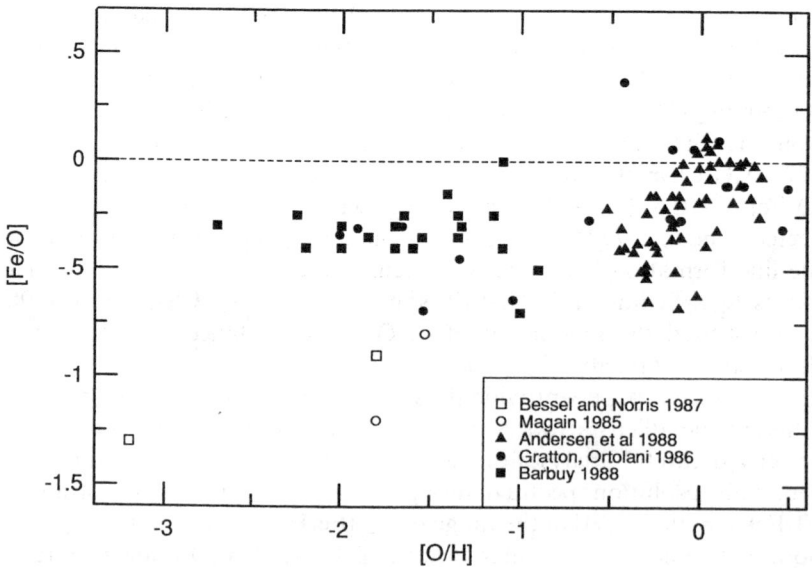

Figure 2 A comparison of O and Fe abundances in field stars, with O taken as the independent metallicity indicator.

[O/Fe] ≈ +1.2 for the dwarf HD 140283, whereas Barbuy (1988) derived [O/Fe] ≈ +0.3 for the giant BD −18°5550. Part of the extremely high [O/Fe] value derived by Magain may be due to his adoption of a very low [Fe/H] value for HD 140283; however, analyses of OH lines near 3150 Å in two stars by Bessell & Norris (1987) support much higher values ([O/Fe] ≈ +1) for both dwarfs and giants. It will be necessary to demonstrate concordance in abundances from OH, [O I], and O I features in individual stellar spectra before absolute confidence can be attached to the level of O overabundances in the halo field. Additional complications are provided by the puzzling O abundances in some globular cluster giants; we consider this question at length in Section 3.

2.1.4 EFFECTS OF THE FIRST DREDGE-UP No discussion of light-element abundances in stars of different metallicity would be complete without some mention of recent work on CNO in evolved stars. Theoretically, it is well known that during the main sequence lifetimes of stars there is a gradual conversion of ^{12}C to ^{13}C and ^{14}C through the actions of the CN-cycle hydrogen fusion reactions. Then, as stars become red giants, their deepening convective envelopes serve to mix these altered interior light-element isotopes with those of the envelope layers, allowing the first indications of interior nucleosynthesis to become visible at the stellar surfaces. Quantitative estimates for the changes in surface C and N contents in Population I solar metallicity stars during first ascent of the giant branch suggested (a) that the C/N values would drop from ≈ 5 in zero-age main sequence stars to ≈ 1 in giants, (b) that the $^{12}C/^{13}C$ ratios would decrease from ≈ 90 to values of 20–30, and (c) that the ^{16}O abundances would not be altered because the ON portion of the CNO cycle would not be effective in the relatively low temperatures of the stellar core regions reachable by the convective envelopes [see Iben & Renzini (1984) for references on first dredge-up predictions]. These theories, of course, have been amply confirmed both qualitatively and quantitatively, chiefly by Lambert and collaborators (e.g. Lambert & Ries 1981, and references therein).

It was realized quite early in these studies, however, that the description of surface abundance changes during first dredge-up are more complex than that given by initial theoretical predictions. Specifically, a class of giants with anomalously low carbon isotope ratios could be identified. The prime example of this class was the old disk, mildly metal-poor star α Bootis, with $^{12}C/^{13}C = 7\pm2$ (Lambert & Dearborn 1972, Day et al. 1973). In fact, very low carbon isotope ratios and high C/N ratios are characteristic of the atmospheres of old disk giants (Cottrell & Sneden 1985). Lambert & Ries (1981) speculated that these ratios would occur in low-mass stars as a result of slow circulation currents (not predicted by standard stellar evolution theory) that would develop during the long main sequence lifetimes of these stars. The initial envelope ^{12}C content would be brought thereby into contact with interior regions, which would allow conversion of ^{12}C into ^{13}C but not ^{14}N.

Masses of field giants are nearly impossible to pin down accurately— witness the debate over the mass of α Boo in the papers by, for example, Mäckle et al. (1975), Blackwell & Willis (1977), and Trimble & Bell (1981). Therefore, for old disk giants it is very difficult to determine whether a low metallicity ($\langle[Fe/H]\rangle \sim -0.5$ in the Cottrell & Sneden study) or a low mass ($\langle M \rangle \sim 1.1\ M_\odot$) of an old disk field giant provides the dominant effect in producing the anomalously high C/N and low $^{12}C/^{13}C$ values. Recently, Gilroy (1988) has determined carbon isotope ratios in giants of 20 open clusters with turnoff masses ranging from 1.2 to 8 $M_\odot$. Her cluster stars, all of which have approximately solar metallicity, divide into two isotope ratio groups: Those with $M > 2.5\ M_\odot$ have $^{12}C/^{13}C = 25 \pm 5$, in good accord with theory; those with lower masses have $^{12}C/^{13}C = 15 \pm 5$, with a trend toward smaller isotope ratios for smaller mass stars.

The disagreements with first dredge-up theory become more dramatic when considering halo giants. For over a decade, Kraft and his colleagues at Lick Observatory have used scanner spectrophotometry of globular cluster and field giants to define alterations of C and N contents along the Population II giant branch. Much of their work is summarized in reviews by Langer & Kraft (1984) and Kraft (1986). Here we confine our remarks on their work to emphasizing that the C/N ratios of very metal-poor giants often are much lower than metal-rich giants (e.g. see the results for field halo giants by Kraft et al. 1982). It is not immediately obvious that the low C/N ratios are a result of internal mixing effects, for in some stars the ratio $(C+N)/Fe$ is constant, whereas in others it appears that N is more abundant than would be expected from conversion of the star's initial C content [see Smith's (1987) summary of the observational evidence for and against mixing and primordial abundance variations in these stars]. However, Sneden et al. (1986) discovered that many field halo giants also possess extremely low carbon isotope ratios: $^{12}C/^{13}C = 4-10$ in the more luminous giants. They also noted that since lower luminosity halo giants had carbon isotope ratios between 30 and 40, the anomalously low values of the brighter giants must have been due to internal mixing effects. Important confirmation of the existence of anomalously low carbon isotope ratios in metal-poor giants is provided by the recent study of the CO spectra of the stars of two globular clusters by Smith & Suntzeff (1989). They showed that the ^{13}CO bands are quite strong and derived low $^{12}C/^{13}C$ values very consistent with those seen in the field halo giants.

These large variations in C, N, and $^{12}C/^{13}C$ are not predicted by standard dredge-up theories—see, for example, the quantitative predictions in Sneden et al. (1986). Ordinary interior theories predict that the convection zone depths become shallower with decreasing mass and metallicity of a star, and thus low-mass Population II stars should show very weak evidence for the first dredge-up. Sneden et al., following earlier suggestions by Sweigart & Mengel (1979), speculated that meridional circulation currents during giant branch evolution allow the convective envelope of a halo star to go very deep to regions of highly depleted ^{12}C and of enhanced ^{13}C and ^{14}N. Indeed, they noted that it might be possible for the convective envelopes to extend down to the O-depleted ON-cycle layers of these stars and thus to create O-poor halo giants. This proposed effect has not been observed in field metal-poor giants, but in Section 3 we discuss two globular cluster giants that *may* show evidence for ON-cycle products at their surfaces. The discussion in this review cannot do justice to this topic, but it is clear that the detailed studies of the CNO elements in giants of different populations do indicate very fundamental differences in the internal evolution of stars of different metallicities.

2.2 The Light Metals

In this element group ($11 \leq Z \leq 22$) we count the observable elements Na, Mg, Al, Si, S, K, Ca, and perhaps Ti. The even-Z elements of this group all are synthesized by α-capture processes. They may be manufactured by a star of any initial metallicity, whereas the production of odd-Z elements is dependent on the neutron excess (and *perhaps*, therefore, on the initial metallicity) of the synthesis region. Thus, following traditional practice, we consider in turn all even-Z and odd-Z elements grouped together.

2.2.1 THE EVEN-Z ELEMENTS Many recent surveys of the even-Z α-element (Mg, Si, S, Ca, and perhaps Ti) abundances in stars over a wide metallicity range have produced results encouragingly in good agreement with each other. These include studies of halo dwarfs (Tomkin et al. 1985, François 1986, 1987, 1988, Magain 1987, 1989, Andersen et al. 1988, Hartmann & Gehren 1988) and halo giants (Gratton & Sneden 1987, 1988). Detailed intercomparison of the abundances of these elements are already extant (Lambert 1987, 1988, Gehren 1988), and so a few remarks should suffice here.

Beginning at the solar metallicity, for each element a well-defined increase in the ratio [α/Fe] occurs somewhere in the range $0 \geq $ [Fe/H] ≥ -0.5. We show this trend in Figure 3, averaging the available α-element results from the quoted studies. The rise in [α/Fe] ratios is defined most clearly by the study of disk stars by Andersen et al. (1988), who show (*a*) that these ratios are very similar in all stars of a given [Fe/H], and (*b*)

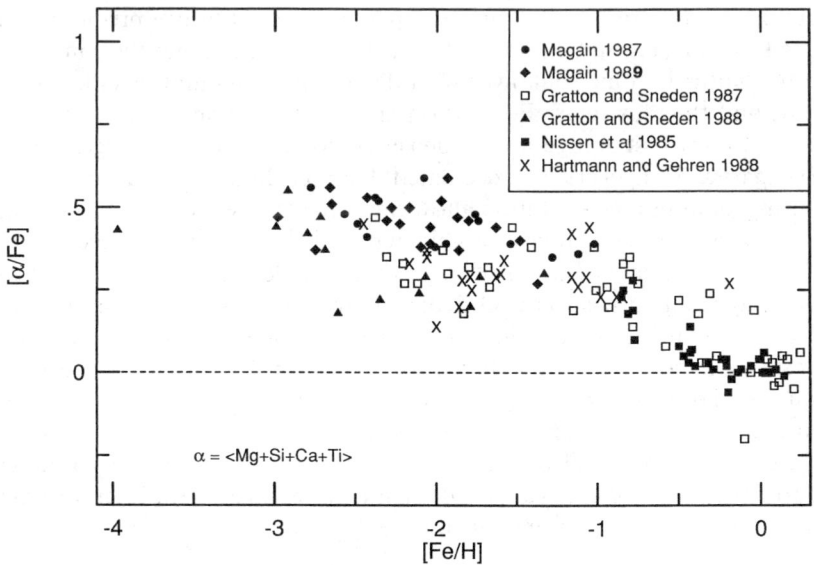

Figure 3 The average α-element abundances as functions of the traditional metallicity indicator Fe. Note that the averaging of the four α-element abundances may blur the possibly real variations with Fe among this group.

that the onset of the [α/Fe] increases may occur at different [Fe/H] values for each element. Specifically, the ratios [O/Fe] and [S/Fe] apparently begin to increase at metallicities very close to [Fe/H] ≈ 0, whereas those of [Mg/Fe] and [Si/Fe] delay their increases until [Fe/H] ≈ −0.5. Does the difference in behavior of these two sets of elements reflect real variations in the production efficiency of these α-elements in the disk, or does it simply occur because of systematic abundance analysis errors? It is impossible to provide here a definitive answer to this question, but a study of the relevant papers fails to suggest any obvious observational and/or analytical uncertainties that would explain away the differences among the α-elements. Obviously, a careful analysis of all assumptions in the various analyses needs to be done (including, especially, full non-LTE line formation predictions for the line strengths) before insisting that the differences are real, but theoreticians should be alert to potential future demands for explanations of the apparent clash in these abundance ratios in disk stars.

In the metallicity range spanning at least $-1 \geq$ [Fe/H] ≥ -2, α-element ratios remain nearly constant and fairly similar from element to element: [α/Fe] = +0.4±0.15, with quite small star-to-star scatter. One notes that these ratios also are very close to those of [O/Fe], discussed previously.

The only potential difference is in S, for which François (1987, 1988) claims that [S/Fe] ≈ +0.6; however, Lambert (1988) suggests that about 0.2 dex of this value may be due to a choice of oscillator strengths for the S I lines by François that are too low, and that the [S/Fe] ratio may well turn out to be ≈ +0.4, in good accord with the other elements of this group. The derived ratios in giants are slightly smaller ([α/Fe] ≈ +0.3; Gratton & Sneden 1987, 1988) than for dwarfs ([α/Fe] ≈ +0.45; Magain 1987, 1989), but we attribute these differences to differences in observational material and analysis assumptions in these studies. The star-to-star scatter appears smallest in Magain's abundances from dwarf stars, so we suggest primary adoption of his [α/Fe] results.

Below [Fe/H] ≈ −2, the trends of [α/Fe] are not quite so obvious. Clearly the values remain supersolar, but there are some small indications that these ratios may actually rise even further in the extreme halo. Both the dwarfs and giants show this effect to some extent. Correcting for the offsets between the two samples, one can see a change Δ[α/Fe] ≈ +0.13±0.04 from [Fe/H] ≈ −2 to [Fe/H] ≈ −3. As for the disk star abundances discussed previously, the reader is left wondering whether to ascribe this apparent change to nucleosynthesis or analysis effects! Here we again urge caution in interpreting these results, and call for more studies of the α-elements in stars with [Fe/H] < −2.5 to understand this problem further.

The element Ti deserves some comment, for it may be viewed as either the heaviest α-element or one of the lightest Fe-peak elements. Magain's (1989) abundances of Ti in very metal-poor dwarfs show a clear overabundance of this element ([Ti/Fe] ≈ +0.4±0.15), while those of Gratton & Sneden (1987, 1988) show a lower overabundance ([Ti/Fe] ≈ +0.25± 0.3, with a larger scatter). The rise in Ti/Fe is apparent in moderately metal-poor disk stars also (Wallerstein 1962, Gratton & Sneden 1987), and thus we tentatively include Ti with the α-elements. More work on this element would be welcome, and we alert the reader to further comments on Ti in the discussion of the very heavy elements in Section 2.4.

2.2.2 THE ODD-Z ELEMENTS The observable odd-Z elements in this domain include Na, Al, and K. For many years observers struggled with the definition of trends in the abundances of these elements. In our discussion, we follow standard convention in which Mg, or some average of Mg, Si, Ca, and Ti, is used (rather than Fe) to compare with the odd-Z elements; this choice tends to emphasize the "odd-even" effect that has been postulated to exist among the light metals. A lively debate over these abundance ratios persisted for many years. Analyses of photographic spectra of strong lines of Na and especially Al in metal-poor dwarfs (e.g.

Peterson 1978) yielded very low [Al/Mg] ratios, whereas those involving giants (e.g. Spite & Spite 1978) seemed to indicate no anomalies in the odd-Z elements. A first hint of a resolution to this problem came from Arpigny & Magain (1983), who showed that previous analyses of the Al I resonance lines in the spectra of giants were doubly suspect: Not only do these lines occur between the Ca II H and K lines, which makes continuum placement difficult, but also one of these Al I lines is badly blended with a CH line, which affects giant star spectra far more than those of dwarfs and leads to spuriously large Al abundances.

Recent attacks on the odd-Z elements have benefited from the ability of silicon diode detectors to obtain high-resolution spectra of the weaker Al I and Na I lines in the red and near-infrared. These data, supplemented by better data and analyses of the resonance lines of these species, seem to settle the issue. The [Na/Mg] ratio declines as a near mirror image of the increase of [Mg/Fe] with decreasing [Fe/H] in the disk and halo; obviously, this means that [Na/Fe] ≈ 0 in stars of all metallicities. Unfortunately, the star-to-star scatter in this ratio remains frustratingly large (e.g. see Figure 8 of Lambert 1988), and at present it is impossible to tell whether the decline in [Na/Mg] is attributable to anything more than the change in Mg. We note that Gratton & Sneden (1988) attempted to use the Na D resonance lines to help define the trend in this element in very metal-poor giants. They derived [Na/Mg] ratios about 0.3 dex higher on average than in their previous analyses, which employed weaker Na lines. These abundances are suspect, probably because the Na D lines still are saturated enough even in these weak-lined stars to be sensitive to microturbulence and damping parameter choices as much as to abundances.

Some recent work has clarified substantially the trends for Al. The ratio [Al/Mg] begins to decline with decreasing metallicity; the trend is noticeable at very mild metal deficiencies and reaches [Al/Mg] ≈ -1.2 at [Fe/H] ≈ -2. As yet we do not have convincing evidence as to whether the ratio continues to decline at even smaller metallicities. Also, the march of Al to the large deficiencies at [Fe/H] ≈ -2 is not precisely defined, as results from Magain (1989) show a precipitous drop in Al abundances beginning at [Fe/H] ≈ -1, whereas those of Gratton & Sneden (1987, 1988) show a more gradual decline; however, all studies now seem to agree that [Al/Mg] really is quite low in metal-poor stars, and the deficiency is far in excess of that amount due to the rise in [Mg/Fe].

Further evidence of the reality of the odd-even effect is provided by several recent studies (Tomkin & Lambert 1980, Barbuy 1987, Barbuy et al. 1987, McWilliam & Lambert 1988) of ^{25}Mg and ^{26}Mg in moderately metal-poor stars. As expected from nucleosynthesis theory (e.g. Arnett 1971), these Mg isotopes are virtually undetectable and thus very deficient with respect to ^{24}Mg below [Fe/H] = −1. This agrees with the overdeficiency of Al in these same stars and makes one suspect that further analyses of Na may show further declines in that element in halo stars.

Finally, we mention the work on K by Gratton & Sneden (1987). This first attempt to derive K abundances in stars over a large metallicity range used the only available transition, the resonance line of K I at 7699 Å. The results suggest that [K/Fe] > 0 in metal-poor stars, as is the case for the α-elements, but with an unacceptably large star-to-star scatter ($\approx \pm 0.4$ dex) at a given [Fe/H]. The reader is referred back to the discussion on the use of the Na D lines for abundance determinations, and thus the K results should be viewed with caution.

2.3 The Iron Group Elements

The iron group elements comprise at least the elements V, Cr, Mn, Fe, Co, and Ni, and possibly the lighter elements Sc and Ti and the heavier elements Cu and Zn. As a group they are important because the ubiquity of their transitions in stellar spectra leads to their adoption as the defining elements for stellar metallicity, even though their bulk contributions to the heavy-element abundances are much less than the CNO group. Some of the early contributions to stellar abundance studies suggested that the odd-Z elements of the Fe group (at least V and Mn) are overdeficient in metal-poor stars. Arnett (1971) collected the data then available for these elements. His summary indicated the existence (barely) of an odd-Z/even-Z effect, although the deficiencies of V and Mn were much less pronounced than for, say, Na and Al; however, in subsequent years doubt was cast on the basic data employed in the early studies (the transitions employed for Mn, in particular, were strong lines susceptible to saturation effects) and on the analysis techniques (the hyperfine structure corrections necessary for the odd-Z elements often either were not done or were done only with approximate methods). Therefore, at the time of Spite & Spite's (1985) review, the prevailing notion of most spectroscopists was that all element ratios among the Fe group elements were solar at all metallicities. Some recent papers have challenged this notion, and in this section we highlight the recent progress on these elements. Following the usual practice, we first consider whether the even-Z elements Cr, Fe, Ni, and Zn have an abundance pattern consistent with the solar system distribution; we then turn our attention to possible deficiencies of the odd-Z elements Sc, V, Mn, Co, and Cu.

2.3.1 IRON We begin by briefly considering the question of Fe itself. What are the basic accuracies in [Fe/H] attainable with current analyses? A full discussion of this issue involves a detailed discussion of the assumptions about atmosphere parameters and analysis techniques and is beyond the scope of the present article. A reasonable starting point, however, would be a simple tabulation of the various [Fe/H] estimates by different observers for some well-studied stars. Much of the labor for this task has been completed by Cayrel de Strobel et al. (1984). Inspection of the entries for most stars in their noncritical catalog of [Fe/H] values suggests that observers have been unable to determine metallicities even for well-studied stars to better than a factor of 2 or so. More discouragingly, the recent literature on [Fe/H] values for many stars shows no tendency toward convergence. We must conclude that uncertainties in [Fe/H] for most stars are not better than ± 0.15 at present, based solely upon the scatter in the determinations and not on considerations of the uncertainties that may affect all analyses together.

As a general rule, the "absolute" analyses based on laboratory and/or theoretical oscillator strengths should be preferable to those that rely on "differential" analyses; however, the entries in the Cayrel de Strobel et al. (1984) catalog do not suggest that the two types of analyses yield substantially different metallicity estimates. Moreover, the modern analyses, with the advantages of much better observational data and model atmosphere techniques, seem often to confirm the results of the pioneering efforts of the first abundance analyses, which of necessity used noisy photographic data and coarse analysis techniques. When all possible sources of error are accounted for, we suggest that the "metallicities" of stars from Fe abundance determinations are now known to no better than factors of 2 or 3.

2.3.2 OTHER EVEN-Z ELEMENTS All studies of chromium in stars of any metallicity have yielded the same result: [Cr/Fe] ~ 0. The recent analyses (e.g. Gratton & Sneden 1988, Gilroy et al. 1988, Hartmann & Gehren 1988, Magain 1989) show (*a*) good agreement between the results for dwarfs and giants, and (*b*) a reasonably small star-to-star scatter at a given metallicity. Note that the scatter in [Cr/Fe] in the Hartmann & Gehren sample is slightly higher than in the other studies cited here, probably as a result of their use of photographic spectra. Also, the star-to-star scatter in the results from all the studies seems to increase at the lowest metallicities. It is suspected that this effect is one of observation and analysis, but future workers should be alert to the possibility that the scatter might reflect a real [Cr/Fe] variation in the extreme halo. Finally, Bessell & Norris (1984) and Gratton & Sneden (1988) derived [Cr/Fe] ~ -0.5 for the ultrametal-deficient giant CD $-38°245$, but their results are based on very few lines and should be viewed with caution.

Nickel abundances have received renewed attention in recent years thanks to the intriguing claim by Luck & Bond (1983, 1985a) that [Ni/Fe] > 0 in the most extremely metal-poor ([Fe/H] < −2.5) stars. Their assertion arose from analyses of their Cassegrain echelle image tube photographic spectra of halo giants. To support their observations of anomalous abundances, Luck & Bond (1983) pointed to theoretical predictions of relative Ni overproductions under certain nucleosynthesis conditions. Specifically, in a study of silicon-burning reactions in stars, Woosley et al. (1973) showed that if the densities in the burning region are too low to support full nuclear statistical equilibrium (the classical e-process; see Burbidge et al. 1957), then an excess of alpha particles per available heavy seed nucleus will occur. Under these conditions, usually labeled the "α-rich freeze-out," the element mix resulting from the capture of these alpha particles will favor the heavy-element end of the Fe group elements, or larger Ni/Fe ratios than from the typical e-process.

The reality of the claim of Ni overabundances is still being debated. In Figure 4 we show [Ni/Fe] ratios from different investigations. It is clear from this figure that other observers have not been able as yet to confirm the trend seen by Luck & Bond. Frustratingly, only a few stars of sufficiently low metallicity have been observed by other workers to conclusively answer this question, but those that have been analyzed do not yield overabundances of Ni. Note in particular that the analysis of CD

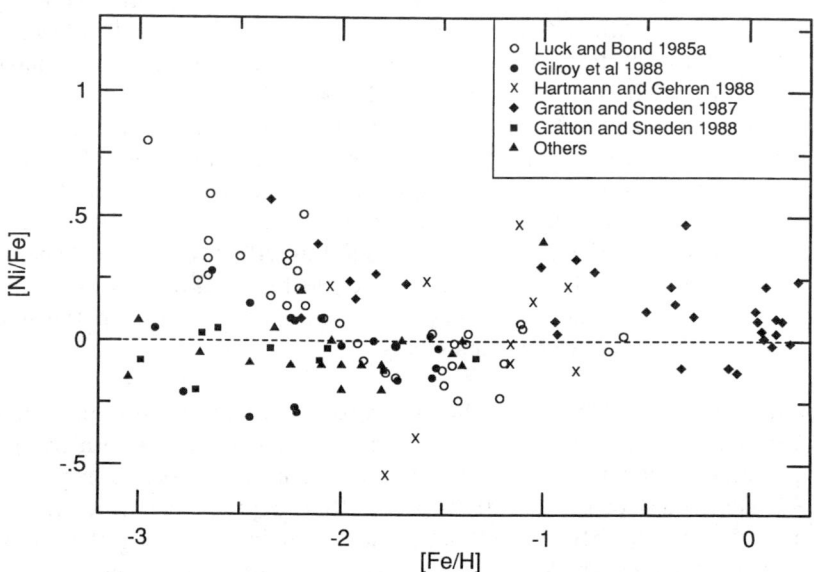

Figure 4 Nickel abundances in stars of different metallicities.

−38°245 by Bessell & Norris yielded an apparent *under*abundance of Ni (unfortunately, this abundance is based on only one Ni I line); however, as Bond & Luck (1988) point out, no definitive arguments have been presented to demonstrate where the data and/or analysis in the Luck & Bond (1983, 1985a) papers could be wrong. A cautious interpretation is that probably [Ni/Fe] ∼ 0 at all metallicities, but that further work needs to be done on this question. This work should involve not only the determination of [Ni/Fe] ratios in more stars with [Fe/H] < −2.5 but also a careful reanalysis of Luck & Bond's data, at least from their published (Luck & Bond 1985b) equivalent widths.

It is often asserted that elements with $Z > 28$ cannot be Fe group elements, since they cannot be synthesized in major nuclear-burning stages in stars. This is not exactly true, for at least copper may be manufactured in explosive neon burning, and so can zinc in explosive silicon burning. Indeed, the realization that Zn is produced in relatively large quantities by a low-density "α-rich freeze-out" condition in silicon burning provides a different way to test the claims of Ni overabundances in extreme halo stars. If both Ni and Zn are products of the α-rich freeze-out in early Galactic nucleosynthesis, then any excess Ni synthesis surely will be matched by the synthesis of copious amounts of Zn (see, for example, the quantitative predictions of Woosley 1986).

New observations of Zn in halo stars show no Zn/Fe anomaly. Sneden & Crocker (1988) analyzed the only two Zn I transitions spectroscopically accessible in five stars with metallicities −1.3 > [Fe/H] > −2.7 and found [Zn/Fe] ∼ 0 for all stars. Very few other investigations of this element have been published. It is worth noting that Luck & Bond (1985a) also found [Zn/Fe] ∼ 0 in their data. Star-to-star scatter was too large to draw firm conclusions on the [Zn/Fe] trend; however, if one averages their Zn abundances for a few stars with similar metallicities, a hint can be seen of a rise of [Zn/Fe] in the most metal-poor stars. As with the case of Ni, further observations of Zn in extreme halo stars are urged. If these two elements are indeed overabundant with respect to Fe below [Fe/H] ∼ −2.5, an important step will have been taken in defining the nature of the first generation of supernovae in the Galaxy.

2.3.3 THE ODD-Z ELEMENTS The observational situation for scandium is not satisfactory at present. Most studies of this element in both dwarfs and giants (Wallerstein 1962, Gratton & Sneden 1987, 1988, Gilroy et al. 1988, Hartmann & Gehren 1988) agree that [Sc/Fe] ∼ 0, although the results by Gratton & Sneden (1988) show a quite large range at the lowest metallicities; however, Magain's (1989) recent analysis of halo dwarfs yields [Sc/Fe] ∼ +0.4. Lambert (1988) has considered the assumptions of the various studies in detail. The interested reader is referred to his discussion, which apportions blame for the spread in Sc abundances nearly equally among the different investigations. Lambert emphasizes that none of these abundance studies have properly accounted for the hyperfine structure of the Sc transitions. Perhaps Sc/Fe has the solar ratio at all metallicities, but definitive statements on this element ratio must await the appearance of more careful analyses.

Little controversy exists at present for vanadium; all studies of this element agree that [V/Fe] ~ 0 in stars over the whole metallicity range. We accept this result at present, but as with Sc the lack of accurate accounting for possible hyperfine structure effects keep the question open to future investigation. We note that since most of the current analyses of V are differential ones (e.g. see Luck & Bond 1985a, or Gilroy et al. 1988), the change in [V/Fe] when hyperfine calculations are made cannot be predicted, because the splitting affects lines in both stellar and solar spectra to varying degrees.

Gratton (1989) has provided the first really trustworthy analysis of manganese in stars over the complete range $-0.2 >$ [Fe/H] > -2.4. His work is more credible than most earlier efforts due to his avoidance of the very strong Mn I lines near 4000 Å, his use of reliable laboratory oscillator strengths, and his proper treatment of hyperfine structure effects. In Figure 5 we show his results, along with those of other selected recent investigations. Most of the other Mn abundance results considered here probably are less accurate than Gratton's. A notable exception seems to be the careful work of Beynon (1978a,b) on selected Mn I transitions in (mostly) disk stars. A clear pattern of Mn underabundances emerges from inspection of this figure. Logical extensions of the current work would be a reanalysis of the other data sets to be consistent with Gratton's work, and especially an extension to stars of even lower [Fe/H] values to ascertain whether [Mn/Fe] remains constant at extreme halo metallicities.

The available data for cobalt are not very consistent. Gilroy et al. (1988) suggest that [Co/Fe] ~ -0.3, while Luck & Bond's (1985a) analyses yield overabundances ([Co/Fe] $\sim +0.2$, with a tendency for larger [Co/Fe] values at smaller metallicities). There is every hope that more reliable Co abundances may be obtained in the near future with observations of carefully chosen Co I lines with high-resolution, low-noise spectra and through use of published accurate laboratory oscillator strengths (Cardon et al. 1982).

The assignment of copper to the Fe group may not be done as easily as in the case of Zn. It is possible to synthesize Cu in explosive neon burning (Woosley & Weaver 1980, Woosley 1986), but traditionally it has been thought to be produced with s-process neutron capture reactions (Burbidge

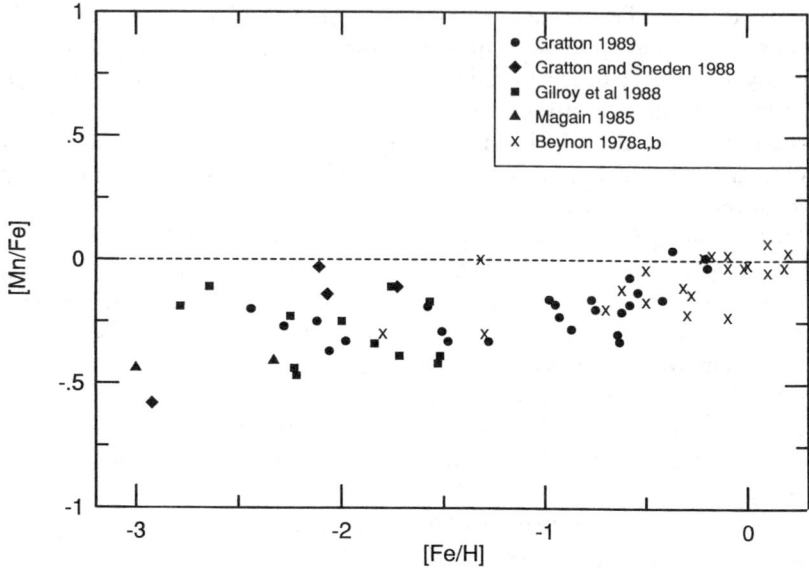

Figure 5 Manganese abundances at different metallicities. Only recent determinations have been included in this figure.

et al. 1957, Clayton et al. 1961). We attach this element to the others of the Fe group simply by atomic number similarity.

Cu is represented in stellar spectra by only a handful of Cu I lines, and in very metal-poor stars all but one of these transitions are vanishingly weak. For this reason, very little work has been done on Cu in halo stars until quite recently. Sneden & Crocker (1988) have provided Cu abundances for five stars based on high-S/N spectra, laboratory oscillator strengths, and correct accounting of hyperfine structure in the two transitions used. Cu is underabundant in metal-poor stars, and Sneden & Crocker (1988) suggest that it declines continuously with decreasing [Fe/H]. The Cu abundances in a few halo giants measured by Gratton & Sneden (1988; no hyperfine structure calculations were done in their analysis) are consistent with Sneden & Crocker's results, although a trend with metallicity is not apparent in the Gratton & Sneden data.

If the continued decline is real, then Cu could be one of the few truly "secondary" elements—that is, an element whose synthesis depends on prior generation of seed nuclei and whose abundance therefore will lag the buildup of other heavy elements. Interestingly, the bulk yield of Cu through either explosive neon burning or s-process neutron capture is sensitive to the overall heavy-element abundance level in the synthesis region. Therefore, the behavior of Cu with Fe does not provide a discriminant between the two possible synthesis mechanisms (see the discussion in Sneden & Crocker's paper). More Cu abundances are needed to establish firmly the suspected secondary nature of this element.

2.4 The Very Heavy Neutron Capture Elements

All elements heavier than the Fe group are synthesized predominantly by neutron capture reactions. For convenience, these are often referred to as s-process or r-process elements, depending on whether slow or rapid neutron capture synthesis chains are responsible for the production of the major fraction of the dominant isotope(s) of these elements in solar system material; however, this breakdown is fairly artificial and can be very misleading, for nearly all very heavy ($Z > 30$) elements can be synthesized in either the r- or s-process. For that reason we simply refer to these elements collectively as very heavy or neutron capture elements, rather than dividing them into possibly indefensible r- or s-process categories. Before the abundances of the very heavy elements are reviewed, it is appropriate to remark briefly on the observational constraints that limit our knowledge of the distribution of these elements.

First, for stars substantially more metal poor than the Sun, only the element groups Sr–Zr ($38 < Z < 40$) and Ba–Yb ($56 < Z < 70$) are spectroscopically accessible. For these elements, only transitions of the singly ionized species are strong enough to be detected with the usual spectrographic equipment. (The exception to this rule is Sr, for which the Sr I 4607-Å line often is measurable.)

Second, for two of these elements, Sr and Ba, the line selection is limited to two or three lines only, and these lines are strong (saturated) even in very metal-poor stars. For the rest of the elements there are more available transitions, but many of them are undetectably weak. For example, in Magain's (1989) study of halo dwarfs, only the lines of Sr, Ba, and Eu proved detectable in stars with [Fe/H] < -2. This is quite unfortunate, for, as we discuss below, the ratio of Ba to its immediate neighboring elements is often crucial to the assignment of the relative contributions of the s- and r-process to very heavy-element abundances.

Finally, the oscillator strengths of these elements are known to various accuracies, from quite precise (e.g. Y, Zr) to barely adequate (e.g. La, Pr). This remains one of the chief sources of uncertainty in determining the abundance distribution of these elements.

It was realized in some of the earliest abundance studies that the distribution of the very heavy elements in Population II stars is not identical to the solar mix. For example, Pagel's (1965) careful reanalysis of the Wallerstein et al. (1963) data for the classical metal-deficient giant HD

122563 ([Fe/H] $\sim -2.60\pm 0.15$) yielded abundance ratios that are characteristic of many halo stars: [Sr-Y-Zr/Fe] ~ -0.3, [Ba/Fe] ~ -1.0, [Ba/La] ~ -0.5, and [Eu/Fe] ~ -0.4. Clearly, in this star the overall level of the very heavy elements is moderately subsolar when compared with Fe. More importantly, however, the Ba abundance is extremely low. Many subsequent analyses of HD 122563 have yielded the same result (e.g. see Sneden & Parthasarathy 1983, and references therein; Luck & Bond 1985a, Gilroy et al. 1988).

Spite & Spite (1978) provided the first systematic study of a few of the very heavy elements in halo stars, finding that the pattern of severe Ba underabundances and nearly normal Eu abundances was characteristic of all five stars of their sample. Spite & Spite, following the usual interpretation of most prior and future authors, interpreted their abundance results as evidence for normal r-process abundances and very deficient s-process abundances in the halo stars. This standard interpretation is based on the fact that the solar system abundance of Ba is dominated by s-process synthesis (88% by mass; Cameron 1982a), whereas the abundance of Eu is predominantly due to the r-process (the s-process contributes only 9% of the mass fraction of this element); however, Truran (1981), realizing that at least some of each heavy-element abundance even in the solar system must be due to the r-process, went a step further in postulating that the low Ba/Eu ratio in metal-poor stars was a natural consequence of the existence of *only* r-process syntheses in the manufacture of the very heavy elements in the early Galaxy.

The assertion that these elements in extreme halo stars are exclusively r-process products has received support through the determination of detailed abundance patterns of these elements in many metal-poor stars (Sneden & Parthasarathy 1983, Sneden & Pilachowski 1985, Gilroy et al. 1988). In Figure 6 we show an abundance distribution characteristic of stars in the metallicity range $-2.0 >$ [Fe/H] > -2.5. This metallicity was chosen to strike a balance between detectability of many species (at the low limit of the range) and severe saturation problems in Ba and Sr (at the high limit) and to avoid contamination of the heavy-element abundances by s-process contributions (at the upper limit also; see the discussion below). The data are averages for five stars from the Gilroy et al. (1988) data; we have included only those stars for which almost all accessible very heavy elements have been analyzed. We compare the observed abundances with (*a*) the solar system r-process fractional abundances deduced by Cameron (1982a), (*b*) a theoretical s-process abundance distribution from Malaney (1987; for the elements in question these are very similar to those of Cowley & Downs 1980), and (*c*) a theoretical r-process abundance distribution computed by J. J. Cowan (private communication; these are

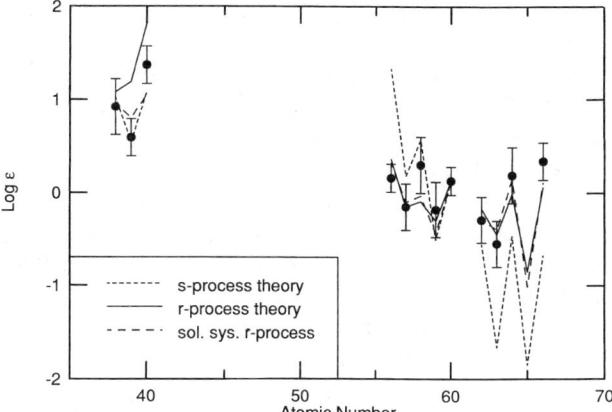

Figure 6 Average abundances of the observable neutron capture very heavy elements in halo stars, compared with several empirical and theoretical abundance curves. See the text for details.

nearly the same as those published by, e.g., Cowan et al. 1982), summed by atomic number instead of mass number. All three of these abundance distributions have been normalized arbitrarily to the observed Nd abundance point.

The *s*-process curve, although chosen to provide a reasonable ratio of the Sr-Y-Zr group to the Ba-····-Yb group from among a family of curves published by Malaney (1987), clearly does not fit the extreme halo star data. In particular, it fails to account for the observed Ba/La, Ba/Eu, and Nd/Eu ratios (or any combination of these abundances). As was emphasized by Sneden & Pilachowski (1985), not only does the *s*-process always lead to a very much larger Ba/Eu ratio than that seen in the halo stars, but it also always produces a Ba/La ratio of about 10, in contrast to the observed ratio of about 2. The Ba/La ratio is as good an *r*/*s*-process discriminant as the more obvious Ba/Eu ratio.

The *r*-process curves fit the whole data set much better. It is apparent that the solar *r*-process fractions account very well for the total very heavy-element abundance distribution in very metal-poor stars. Either the *r*-process yields are insensitive to the different burning conditions that may exist in the explosive events that lead to release of the copious quantities of neutrons for the *r*-process, or else the site of the *r*-process has been basically the same throughout the Galaxy's history.

The overall level of very heavy-element abundances appears to be only loosely correlated with Fe in extreme halo stars. The scatter in [very heavy/Fe] at a given [Fe/H] can be seen in data sets of both dwarfs (Magain 1989) and giants (Gilroy et al. 1988), and it appears to be substantially larger than maximum possible observational errors. This effect can be seen in Figures 1 and 2 of Sneden et al. (1988) and in Figure 17 of Lambert (1988). The scatter obviously suggests that local supernova events were important, at least for the heavy elements, in seeding the local interstellar medium of a poorly mixed early Galactic halo.

If the heavy elements in extreme halo stars are manufactured only by the r-process, at what metallicity does the s-process begin to contribute substantially to the abundance patterns? The available data (Sneden et al. 1988, Magain 1989) suggest that in the metallicity range $-2.5 \leq$ [Fe/H] ≤ -2.0, the ratio [Ba/Eu] is ≈ -0.6, and in the range $-2.0 \leq$ [Fe/H] ≤ -1.0, this ratio climbs as metallicity increases to its solar ($=0$) value [see Figure 13 of Gilroy et al. (1988) and Figure 15 of Lambert (1988)]. This increase of the [Ba/Eu] ratio undoubtedly is due to the onset of bulk s-process contributions to the Ba abundance. Gilroy et al. (1988) suggest that since s-process nucleosynthesis is most likely to occur in stars with $M < 10$ $M_\odot$, the typical evolution time scale of 10-$M_\odot$ stars—a few times 10^7 yr—gives one indication of the time scale for the initial rise of metallicity of the halo to [Fe/H] ≈ -2.

In the most extremely metal-poor stars, there is some weak indication that all of the heavy elements may become extremely depleted with respect to Fe (see Luck & Bond 1985a, Gilroy et al. 1988). In particular, data gathered from several studies by Sneden et al. (1988) seem to show a decline in average [Eu/Fe] ratios below [Fe/H] ≤ -2.5. Perhaps the heavy-element nucleosynthesis in the progenitors of these very metal-poor stars simply could not progress past the Fe peak much at all. [The effect may be most pronounced in the heaviest neutron capture elements (Gilroy et al. 1988).] We note, however, that these tentative conclusions are based on the spectra of giant stars only, since all but Ba and Sr features become almost unmeasurably weak in extremely metal-poor dwarfs. In this same cautionary spirit, note that the precipitous decline of Ba/Fe ratios is seen in both dwarfs and giants, but that it becomes noticeable at somewhat different metallicities: at [Fe/H] ≈ -2.0 in giants and -2.5 in dwarfs (see Figure 2 in François & Brocato 1987). The reason is not well understood, and François & Brocato suggest that the Ba line formation mechanism may not be fully understood in one or both types of stars; however, if one plots [Ba/Ti] instead of [Ba/Fe] as a function of metallicity (using the data of Magain and Gilroy et al.), there is much better concordance between dwarfs and giants. This is of course due to the tendency noted in Section 2.2.1 for the Ti abundances derived in giants to be somewhat lower than those derived in dwarfs. Since Ti is much closer in ionization potential to Ba than is Fe, perhaps the [Ba/Ti] ratio serves better to weaken atmosphere uncertainties affecting the observed Ba depletion.

Finally, in a potentially exciting use of very heavy-element abundances for Galactic evolution, Butcher (1987, 1988) has observed Th and Nd features in stars of the young and old disk components. His results show that the ratio Th/Nd is substantially constant for all stars irrespective of age, and he argues that therefore the Galaxy must be fairly young—about 11 Gyr. Clayton (1987, 1988) has warned that conclusions about the age of the Galaxy from this particular chronometer are quite sensitive to details of the growth in time of the r- and s-process bulk yields of the Galaxy. This is because Nd is produced by both processes, and indeed the solar system abundance of this element is divided almost equally between the r- and s-process (Cameron 1982a). This question, of course, is exactly the type of problem we have been concerned with in the preceding few paragraphs, and clearly the understanding of this problem is just getting underway. We join Clayton in urging caution on the Th interpretation, and wish especially to have Th observations in more metal-poor stars as a first step toward greater understanding of this point. Even detection of Th in more metal-poor stars may not be enough, for actually we need to ascertain the initial abundance of Th of the star in question. This value can only be guessed at, but abundance determinations of other (stable) elements near Th (e.g. Pt) would be of great interest. We shall return to the question of the Th/Nd chronometer in Section 7.4.

3. CHEMICAL HISTORIES OF THE GLOBULAR CLUSTERS

Globular clusters represent perhaps the oldest known objects in our Galaxy. For this reason, detailed information concerning their composition trends and abundance patterns can be expected to provide important clues to and constraints upon the formation and early chemical and dynamical evolution of the Galaxy. In particular, we are concerned with global trends as a function of such parameters as [Fe/H], Galactocentric distance, or age (if real age differences do exist), rather than with the chemical inhomogeneities observed in individual clusters, which presumably (with the exception of ω Cen) reflect the complexities of the evolution of low-mass stars.

The available abundance determinations for Galactic globular clusters, while certainly not complete, may nevertheless be used to guide our thinking with respect to models for their formation and chemical evolutionary history. Here we note some of the interesting trends in overall metallicity levels and in detailed abundance patterns of the clusters. We remind the reader that the faintness of globular clusters makes only the brightest, evolved stars accessible. These are affected by mixing, so primordial C, N, O abundances are not accessible.

3.1 Globular Cluster Metallicities

There exist some strikingly narrow abundance spreads within clusters, except for ω Cen and possibly M22, as reflected in the narrow widths of their giant branches in the color-magnitude diagram (see, for example, the reviews by Kraft 1979, Harris & Racine 1979, Freeman & Norris 1981, and Smith 1987), which suggests that the observed stellar generation formed from thoroughly homogenized gas. This is often viewed as evidence in support of the view that the clusters were born with their observed metallicity, and that the ejecta of its own stellar population has systematically been lost to the system. In this context, the unusual globular cluster ω Cen, which has a giant branch in the color-magnitude diagram that is four times wider than those of typical clusters, may be interpreted as having been sufficiently massive to retain enough of its processed matter to form a subsequent generation of stars (Dopita & Smith 1986, Smith 1987). The alternative view, that the globular clusters were formed with no heavy elements but generated their own metallicities in massive stars and supernovae associated with an early generation (see e.g. Fall & Rees 1985, 1987), remains to be fully examined with detailed numerical models. There are, however, as we shall see, abundance data that support this latter viewpoint (Truran 1983, 1987, 1988, Cayrel 1986).

The frequency distribution of globular cluster abundances, as reflected in [Fe/H], appears distinctly bimodal (Zinn 1985). Most globular clusters are found to be associated with a lower metallicity peak at [Fe/H] $= -1.6$, while there also exists a less populated peak at [Fe/H] $= -0.5$ that is composed of systems with more of a thick disk than a spheroidal distribution. This suggests that there may indeed be two distinct populations— disk and spheroidal globular clusters.

While the disk clusters are clearly more metal rich, there is no strong evidence for an abundance gradient with Galactocentric distance in the spheroidal cluster system (Searle & Zinn 1978, Freeman & Norris 1981, Pilachowski 1984, Zinn 1985). While Pilachowski (1984) most recently argued for a weak tendency for the mean abundances to fall off with Galactocentric distance, the more complete analysis by Zinn (1985) does not support this conclusion. There appears, rather, to be a spread of metallicities for clusters at a given distance. Cayrel (1986; see also Zinn 1985) argued on this basis that these metallicities, resulting from self-pollution, are stochastic variables. This spread certainly does not seem to support the view that the clusters formed from progressively enriched gas during the phase of halo collapse. In fact, the evidence for an abundance gradient in the field halo stars is itself not very strong. Saha (1985) found no clear gradient in metallicity out to 25 kpc from an analysis of local and distant RR Lyraes, and Carney et al. (1988) drew the same conclusion from their analysis of a large sample of halo population stars. It would thus appear that there may be a consistency in the absence of evidence for gradients with Galactocentric distance in the globular cluster and field halo star populations.

The globular cluster stars do not reveal the same metallicity distribution function as the field halo stars (Laird et al. 1988). Although the two populations may share the same mean metallicity, the field stars have a much broader range in metallicity than do the clusters. In particular, a substantial fraction of the field stars are much more metal poor than any of the globular clusters. Zinn (1985) lists NGC 5053 at metallicity [Fe/H] $= -2.58$, but other estimates of [Fe/H] for this cluster are higher [e.g. Suntzeff et al. (1988) obtain [Fe/H] $\simeq -2.2$]. No other cluster in Zinn's sample has [Fe/H] < -2.25, whereas there are a substantial number of field halo stars at metallicities $-4.0 \leq$ [Fe/H] ≤ -2.25. The limit on the metallicities of the clusters seems suggestive of models of self-contamination, in that a natural lower limit on the abundance levels in clusters arises from the fact that a single supernova of a typical massive star yielding $\sim 10\ M_\odot$ of heavy elements is sufficient to contaminate a cluster to a metallicity $\sim 10^{-2}$ of solar (Truran & Brown 1988).

Finally, studies of extragalactic globular cluster systems reveal mean color differences between the clusters and the underlying field populations (Forte et al. 1981, Harris 1983, 1986, Mould 1986), in the sense that the globular clusters are consistently bluer (of lower metallicity) at a specified Galactocentric distance than the mean for the field stars. This would appear to imply that the globular clusters formed first, while the field halo stars formed later in the halo collapse phase with higher metallicities.

3.2 Detailed Abundance Patterns in Globular Clusters

Observational studies of individual stars in globular clusters now permit the determination of individual elemental abundances in these stars. The results of the two most extensive studies to date (Pilachowski et al. 1983, Gratton 1987b, and references therein) reveal specifically that globular cluster stars exhibit abundance patterns that are similar in some ways to those of the extreme metal-deficient field halo stars discussed previously and quite distinct from solar system abundances (Anders & Grevesse 1989). In Table 1 we present some average abundance ratios in globular cluster stars ($-0.7 \geq$ [Fe/H] ≥ -2.2). Most of the data for this table are taken from R. G. Gratton's (private communication) summary of the studies published by his group (e.g. Gratton 1987b) of giants in moderately

Table 1 Abundances in globular cluster stars

	Metal rich[a]	Intermediate[b]	Metal poor[c]
[Fe/H]	−0.80	−1.35	−2.25
[O/Fe]	0.50	0.35	≤0.50
[Na/Fe]	0.25	0.05	0.25
[Mg/Fe]	0.20	0.30	−0.10
[Si/Fe]	0.30	0.25	—
[Ca/Fe]	0.10	0.20	0.45
[Sc/Fe]	0.10	0.05	0.05
[Ti/Fe]	0.30	0.35	0.30
[V/Fe]	0.25	0.25	0.35
[Cr/Fe]	0.10	0.05	−0.10
[Mn/Fe]	−0.10	−0.30	−0.15
[Co/Fe]	−0.00	0.00	—
[Ni/Fe]	−0.20	−0.15	0.05
[Cu/Fe]	−0.20	−0.30	−0.40
[Y/Fe]	0.10	−0.05	−0.20
[Zr/Fe]	−0.15	0.05	—
[Ba/Fe]	−0.35	−0.20	−0.15
[La/Fe]	0.00	0.05	—

[a] The abundances are straight means of the results for NGC 104 (47 Tuc), NGC 6352, and NCG 6838 (M71) (R. G. Gratton, private communication).
[b] The abundances are means of the results for 17 clusters with $-0.8 \geq$ [Fe/H] ≥ -1.9 (R. G. Gratton, private communication).
[c] The abundances are averages of the abundances of M15 and M92 (Cohen 1979).

metal-poor clusters. We have also added to this table a set of average abundances of the very metal-poor globular clusters M92 and M15 (Cohen 1979). The detailed abundance patterns published by Pilachowski et al. (1983, and references therein) are in good accord with those in Table 1. Our general comments on observed trends seen in the entries of this table include the following points.

1. Generally high O/Fe ratios are encountered in giants of some globular clusters, although the observational situation is at present a bit uncertain (see below).
2. The intermediate-mass α-elements Mg, Si, Ca, and Ti are enhanced at nearly the same levels ($[\alpha/\text{Fe}] \approx +0.3$) as are observed in the field stars.
3. In general, the situation with respect to the light odd-Z elements Na and Al can be quite complicated. It has been known for some time that features of Al I and Na I have some significant star-to-star variations in some globular clusters (e.g. see Peterson 1980, Norris et al. 1981, Norris & Pilachowski 1985, and references therein). Also, the Na and Al line strengths often seem to be correlated with the strengths of the

CN molecular bands. At present it is not clear whether these variations are due to primordial abundance variations, to internal mixing events, or simply to atmospheric line formation problems in the different stars of these clusters. For a more extensive description of this issue, the reader should consult the reviews of Freeman & Norris (1981) or Smith (1987) and the recent detailed analysis of several ω Cen giants by François et al. (1988).

4. Some evidence exists for the occurrence of more pronounced odd-even effects in Z for some heavier elements, as predicted for metal-poor stellar populations by calculations of explosive nucleosynthesis (Truran & Arnett 1971). In this regard, we note specifically the case of [Sc/Ti], for which Sc is found consistently to be depleted and a mean value of -0.31 is obtained; however, the limited data for [Na/Mg] and [V/Cr] show no such consistent trends.

5. A slight depletion of Cu is noted, again very consistent with the observed field star trend.

6. The heavy-element ($A > 60$) abundance patterns in globular cluster stars reveal some mild anomalies in the ratios Zr/Fe, Ba/Fe, and La/Fe similar to but less extreme than the trends observed in field halo stars with [Fe/H] < -2. In particular, the systematically high values obtained for [La/Ba] for cluster stars, with a mean value $+0.35$, again seem suggestive of an r-process origin (Truran 1988). The modest deficiencies of Ba itself are not too worrisome here, because the pronounced drop-off in this element in field stars does not become apparent until [Fe/H] < -2. Further indicators that would reflect a pure r-process origin include the following expected excesses (Truran & Brown 1988): [La/Ba] $= +0.48$, [Nd/Ba] $= +0.44$, and [Eu/Ba] $= +0.90$; however, these expectations do not appear to be fulfilled in giants of the globular cluster M55, which seem to exhibit solar mixes of the very heavy elements (Pilachowski et al. 1984). Since M55 has an overall metal deficiency of only [Fe/H] ≈ -1.3, a solar composition, with contributions from both r- and s-processes, is not very surprising from our knowledge of the field star abundance distributions.

7. Overall, none of the detailed element ratios show any discernible trend with overall metallicity of the clusters.

At present it would be imprudent to accept these conclusions at face value, for the state of the art in globular cluster abundance determinations is not yet to the level achieved for the brighter field stars. We stress that the abundance patterns for globular clusters are derived from analysis of giants, which are cooler and more luminous than the typical field stars observed. The spectra therefore are intrinsically more difficult to analyze successfully, and thus the abundance uncertainties may be larger than for field stars. As one specific example of this problem, we consider here the status of oxygen abundance determinations in globular cluster stars.

The early work on O in globular clusters, which utilized relatively high-noise photographic image tube echelle spectra, was summarized by Pilachowski et al. (1983), who suggested that while most cluster giants share the [O/Fe] ≈ +0.3 value of field metal-poor giants (again from analyses of the [O I] lines), a significant minority of the cluster giants showed large relative *under*abundances of O. It was noted that all O-deficient stars are members of clusters with blue horizontal branches, but not all giants from these clusters are O deficient. More recent work with lower noise digital spectra has substantiated the early claims in two cases: star VII-18 in M92 has [O/Fe] ≤ −0.4 (Pilachowski 1988), and star II-67 in M13 shows [O/Fe] ≤ −0.7 (Leep et al. 1986, Hatzes 1987). Pilachowski (1988) notes that the M92 star with a low O abundance has one of the highest N abundances (Carbon et al. 1982), and that [(C+N+O)/Fe] is approximately constant for all M92 giants. Therefore, the O/N anticorrelation in one giant is suggestive of the appearance of ON-cycle products at the surface of that star according to the theory mentioned at the end of Section 2.1.4; however, an extensive study of M13 giants now in progress (C. A. Pilachowski, private communication) shows a more uncertain picture. Not only is $\Sigma(C+N+O)$ variable from star to star in the cluster, but the O abundances seem to show some correlation with the strengths of Na and Al features in each giant. Since there are no obvious nucleosynthetic scenarios to produce these correlations, it is not clear whether the occasional very low O abundances are due to internal processing or to some anomalous atmospheric effects.

This issue cannot be resolved here, but it is worth noting that caution should be used when comparing O abundances of globular cluster and field halo giants. The region of the HR diagram in which the cluster stars with low [O/Fe] values arise is largely unexplored in field giants. This is demonstrated in Figure 7, in which [O/Fe] results from various studies are correlated with effective temperatures. A plot with luminosities would produce essentially the same result, but with increased scatter, since luminosities are much more difficult to determine for the field giants (e.g. see Bond 1980). The low O abundances in cluster giants apparently occur only at $T_{\rm eff} \leq 4200$ K, and only one field star with a published O abundance has an effective temperature that low. The suspicion that the O deficiencies are limited to the coolest globular cluster giants is strengthened by the recent survey of O in six globular clusters by Gratton (1987a). Nearly all of the stars with $T_{\rm eff} \geq 4200$ K that Gratton and other investigators have

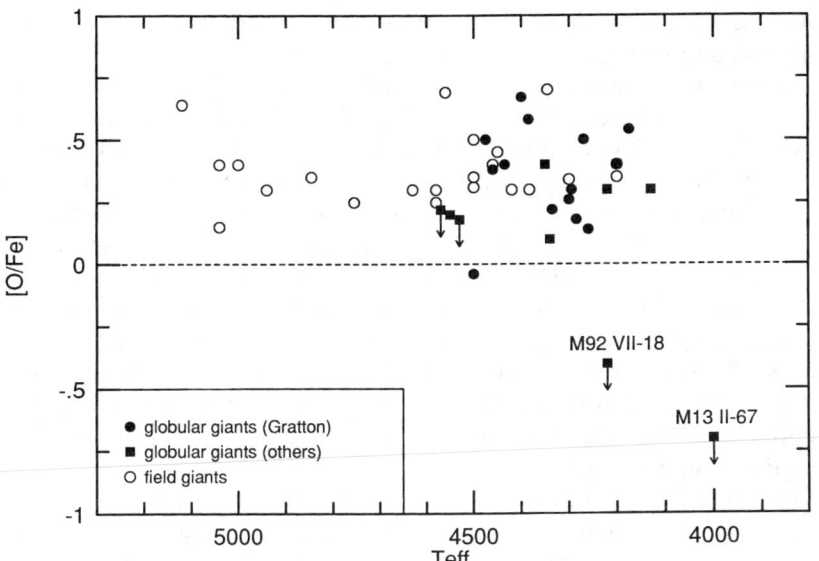

Figure 7 Oxygen abundances in globular cluster and field giants as functions of their effective temperatures. For the globular cluster points labeled "other," we have taken data from Pilachowski (1988) and Leep et al. (1986).

analyzed show substantial O excesses: [O/Fe] ≈ +0.4, in reasonable accord with the field stars. This plot does not suggest a resolution to the puzzle of O underabundances in cluster giants, but it does show that the few anomalous O-deficient stars are not the same sorts of objects considered in studies of field halo giants. It would be of interest to derive O abundances in cooler field giants to see whether any of them can be shown to be O deficient.

Similar cautionary tales could be told of other abundance ratios in globular cluster stars. We look forward to the time when abundances for many species may be determined from high-resolution, high-signal-to-noise spectroscopy of stars over a large range of globular cluster giant branch positions.

4. MAGELLANIC CLOUD ABUNDANCES

The recent occurrence of supernova SN1987A has stimulated interest in abundance determinations for the Magellanic Clouds, particularly for the Large Magellanic Cloud (LMC). The Clouds provide an excellent opportunity for probing the detailed chemical evolution of other galaxies, since they are close enough for high-dispersion spectra to be obtained for giant and supergiant stars. Although this review is concerned almost exclusively with the interpretation of stellar abundance results, our knowledge of the abundances in the LMC and the Small Magellanic Cloud (SMC) has been based largely upon studies of abundances in H II regions. Dufour (1984) has provided a review of the results of a large number of H II region analyses, which provide measures of the relative abundances of He, C, N, O, Ne, S, Cl, and Ar. Dufour (1984) quotes mean abundance levels for oxygen of [O/H] = -0.44 for the LMC, and of [O/H] = -0.85 for the SMC. Comparable depletions are found for the intermediate-mass nuclei Ne, S, Cl, and Ar. This is consistent with the fact that the nuclei O, Ne, S, Cl, and Ar have a common nucleosynthesis origin in massive stars (Truran 1984, Woosley & Weaver 1986a). Similar depletions of these intermediate-mass elements have also been determined for both the SMC and the LMC by Smith (1980).

The H II region data also provide evidence for interesting relative abundances of carbon, nitrogen, and oxygen. In particular, Dufour (1984) quotes depletions of carbon of [C/H] = -0.75 for the LMC, and of [C/H] = -1.49 for the SMC. The corresponding depletions for nitrogen are [N/H] = -0.99 and [N/H] = -1.50, respectively. It follows that both carbon and nitrogen are depleted in the H II regions of the Clouds, relative to oxygen, by factors of ~ 3. This feature cannot be readily explained by the relative youth of the Clouds with respect to the Galaxy, as has been suggested by Russell et al. (1988a). Stars in the Galaxy with a metallicity (as reflected in the oxygen abundance) one third of solar (comparable to the metallicity of the LMC) do not show such relative depletions of carbon and nitrogen.

Improved data concerning the abundances of elements in individual stars in the LMC and SMC are clearly required. While determinations of lower terminal velocities for O-star winds (Garmany & Conti 1985) imply a lower metallicity in the LMC, direct evidence from analyses of photospheric stellar spectra have only recently become available. Kudritzki et al. (1987) have performed quantitative spectroscopy of two early B supergiants in the LMC: SK 21−65 and SK 41−68. The heavy-element abundance patterns in both of these stars confirm a low metallicity for the LMC relative to the Galaxy. Magnesium is depleted by a factor of 10 in both stars, and aluminum is depleted by at least a factor of 4, although the silicon abundance appears to differ in the two objects and in one case is essentially solar. The abundances of several lighter nuclei are also quite different from solar, but the situation is confused by the possibility of the presence of significant mass loss. In particular, both stars show substantial enrichments of helium and of nitrogen and depletions of both carbon and oxygen. Similar anomalies have been identified for Galactic OBN stars by Schönberner et al. (1988) and may be interpreted as resulting from a combination of CN cycling and mass loss effects. It would then appear that the depletions observed for Mg, Al, and Si are the best indicators of the LMC abundances.

Abundance determinations from high-resolution spectroscopy of F supergiants in both the SMC and the LMC have recently become available (Russell et al. 1988b, Russell & Bessell 1988). These authors quote mean iron abundances of [Fe/H] = −0.30 for the LMC, and of [Fe/H] = −0.65 for the SMC. They find the elemental abundances relative to iron in both the LMC and SMC to be similar to those of solar system matter. This is true, in particular, for the abundance of carbon relative to iron, even though the H II region analyses quoted previously indicate a significant relative carbon depletion. This may suggest that grain formation depletes the gaseous carbon in the nebulae. Unfortunately, these studies of stellar spectra do not provide information concerning nitrogen or oxygen that might serve further to clarify the issue. Finally, these authors note the presence of overabundances of the heavy neutron capture products Nd and Sm, which supports earlier findings by Spite et al. (1986) for a globular cluster star in the SMC. On the other hand, their quoted abundances of such other neutron capture elements as Sr, Y, Zr, Ba, La, and Ce, relative to iron, are quite compatible with solar abundances. In our view, therefore, no clear pattern emerges with respect to these heavy-element abundance leads; further studies are required to clarify the situation.

The Magellanic Clouds provide a very accessible testing ground for models of galactic chemical evolution. The data available to date, however, do not provide an unambiguous picture of the chemical enrichment histories of the SMC and LMC. Their respective mean metallicities are [O/H] = −0.85 and [O/H] = −0.44 from H II region analyses, and [Fe/H] = −0.65 and [Fe/H] = −0.30 from analyses of stellar spectra. The relative abundances of the elements in the mass range from oxygen through the iron-peak region seem reasonably compatible with solar abundances in both Clouds, with the possibly very important exceptions of Mg and Al. The pronounced relative depletions of nitrogen and carbon, inferred from H II region analyses, are not easily understood in populations with metallicities as high as those in the Clouds. The stellar spectra reveal, rather, that carbon is not so depleted with respect to iron (and oxygen). This is an important issue that must be resolved before we can proceed to build realistic models for the nucleosynthesis histories of the Magellanic Clouds.

5. NUCLEOSYNTHESIS PROCESSES, SITES, AND TIME SCALES

Nucleosynthesis theory has taught us that the abundances of the elements heavier than helium in the Galaxy are generally attributable to nuclear processes occurring in stellar and supernova environments. In order to provide a framework within which observations of abundance patterns as a function of Galactic age may be interpreted, we first present, in this section, a review of the basic sites of nucleosynthesis that calls attention both to critical dependences on the characteristics of the underlying environment and to the characteristic production time scales. More extensive reviews of various aspects of theories of nucleosynthesis and galactic chemical evolution include, specifically, discussions of nucleosynthesis mechanisms (Trimble 1975, Truran 1984), stellar evolution (Iben & Renzini 1983, Chiosi & Maeder 1986), supernovae (Wheeler 1981, Trimble 1982, 1983, Woosley 1986, Woosley & Weaver 1986b), nucleocosmochronology (Schramm 1974), and the chemical evolution of the Galaxy (Tinsley 1980, Audouze 1986, Chiosi 1986).

Detailed predictions of nucleosynthesis yields are now available for both the quiescent phases of stellar evolution and the matter ejected in supernova events. For the purposes of our discussions of abundance patterns and of chemical evolutionary trends, it is convenient to identify several interesting ranges of stellar mass with which specific nucleosynthesis processes and products are associated: (*a*) very massive stars ($M \gtrsim 100\ M_\odot$), which may undergo the pair formation instability; (*b*) "ordinary" massive stars ($20 \lesssim M/M_\odot \lesssim 100$), which are expected to undergo core collapse; (*c*) the mass range $1 \lesssim M/M_\odot \lesssim 10$ of single, intermediate-mass stars, for which significant nucleosynthesis is expected to occur during the asymptotic giant branch phase of evolution; and (*d*) the subset of intermediate-mass stars in appropriate binary systems, which is now believed to give rise to the outbursts of Type Ia supernovae.

There exists a crucial distinction in the nucleosynthesis production time scales for these sources of heavy elements as dictated by their respective stellar evolutionary lifetimes. The most massive stars ($\sim 100\ M_\odot$) evolve on time scales $\tau \lesssim 4 \times 10^6$ yr (Maeder & Meynet 1987), massive stars ($M \gtrsim 10\ M_\odot$) evolve on time scales $\tau \lesssim 2 \times 10^7$ yr (Cox & Guili 1968), and intermediate-mass stars ($1 \lesssim M/M_\odot \lesssim 10$) evolve on time scales $\tau > 10^8$–10^9 yr. We can therefore anticipate that extreme halo population stars (and globular cluster stars) might be distinguished by "anomalous" abundance patterns relative to solar abundances, which reflect the specific nucleosynthesis processes that occur in massive stars.

We now briefly identify and discuss the particular nucleosynthesis processes that are associated with these various critical ranges of stellar mass. For our present purposes, we confine our attention to the following interesting elements or classes of nucleosynthesis products: carbon, nitrogen, oxygen, the elements in the range from neon to calcium, the iron-peak nuclei, the s-process heavy elements, and the r-process heavy elements. A more complete discussion of these distinct nucleosynthesis products and of the expected astrophysical sites for their production is provided in the review by Truran (1984) and references therein.

5.1 Nucleosynthesis in Very Massive Stars

There are speculations that the first primordial Population III stars were very massive objects. Even if stars of lower mass were formed with a mass function similar to that observed today, there may have been a component of very massive stars that would be the first to explode in any given generation. Such very massive stars are expected to explode by the pair formation process in cores rich in oxygen. The quasi-static production of C and the explosive synthesis of Fe should be relatively small. Some of these stars might leave behind collapsed black hole cores, and others could be disrupted entirely, according to the theory (see Woosley & Weaver 1986b, and references therein). It is not clear that these stars will be an abundant source of r-process matter in either case. The lower boundary to stars that undergo the pair formation process is particularly ill defined. Several studies suggest that pair formation will ensue in stars with oxygen cores in excess of about 30 $M_\odot$ (Barkat et al. 1967, Arnett 1978, Woosley & Weaver 1986a). The open question is the main sequence mass that produces an oxygen core of this mass. If rotationally induced turbulent diffusion or a similar process leads to mixing and homogenization of massive stars on the main sequence (Endal & Sofia 1978, Maeder 1987), then such stars will form much larger helium, and subsequently oxygen, cores for a given main sequence mass than traditional theory would predict. The lower main sequence mass limit for producing pair formation is traditionally thought to be in excess of $\sim 70\ M_\odot$, but this could be as low as 30 $M_\odot$ if strong main sequence mixing is the rule in radiation-pressure-dominated massive stars.

5.2 Nucleosynthesis in Massive Stars

The cores of massive stars during the late stages of presupernova evolution and in the ensuing Type II and perhaps Type Ib supernova events are thought to represent the sites of formation of the bulk of the heavy nuclei present in nature, in the range from carbon to nickel (Arnett 1978, Weaver et al. 1978, Thielemann & Arnett 1985, Woosley & Weaver 1986a,b). Woosley & Weaver suggest that carbon, neon, and magnesium are underproduced in such events relative to oxygen. The outer layers of the heavy-element mantle are rich in O but less so in C, Ne, and Mg, since the $^{12}C(\alpha,\gamma)^{16}O$ reaction favors O as the burning product of helium burning at the higher temperatures in the more massive stars. On the other hand, recent calculations for a 25-$M_\odot$ model by Hashimoto et al. (1989) show Ne/O and Mg/O to be nearly solar. These ratios are clearly sensitive to the $^{12}C(\alpha,\gamma)^{16}O$ rate and other features of the models. The Si to Ca elements are in approximate solar ratios to O in 25-$M_\odot$ models, but enhanced in 15-$M_\odot$ models, for current estimates of the $^{12}C(\alpha,\gamma)^{16}O$ rate (Woosley & Weaver 1986b). Hashimoto et al. find that S, Ar, and Ca are underproduced relative to O for similar assumptions regarding $^{12}C(\alpha,\gamma)^{16}O$, which again reflects the sensitivity of the results to the models. Although the precise mass ejected in the form of iron-peak nuclei is quite uncertain owing to uncertainties associated with the position of the mass cut and the expected remnant mass, there is a tendency for iron to be underproduced. The occurrence of supernova SN1987A in the Large Magellanic Cloud and the confirmation that the energetics of the decay of ^{56}Co to ^{56}Fe dominate the exponential bolometric light curve lead to the conclusion that $\sim 0.075\ M_\odot$ of nuclei of mass $A = 56$ are ejected in this event (Arnett et al. 1989). For a progenitor star of 20 $M_\odot$, this is compatible with an O/Fe ratio about three times solar for the ejected heavy-element-rich matter.

Massive stars ($M \gtrsim 10\ M_\odot$) are thus probably the primary source of oxygen. They may also contribute substantially to the intermediate-mass elements from neon to calcium (but see the discussion of Type Ia events below regarding silicon through calcium). The synthesis of these elements in the upper end of this mass range is rather uncertain, since a significant proportion of these stars may collapse to make black holes and thus swallow all the heavy elements. This point is discussed in more detail below. It may be that no star between about 30 and 100 $M_\odot$ contributes significantly to bulk nucleosynthesis. Such stars may expel an interesting amount of He along with CNO-processed matter before they collapse (Maeder & Meynet 1987, and references therein). The major work of bulk nucleosynthesis in massive stars may thus be done by stars with mass in the range $10 \lesssim M/M_\odot \lesssim 30$, with the lower limit being set by the small amount of heavy-element matter overlying the degenerate O/Ne/Mg core, and the upper end by the total collapse of the star. In this mass range the yields change rapidly with mass and are sensitive to treatment of convection and overshoot as well as nuclear reaction rates.

Another feature of nucleosynthesis in these explosive environments is the occurrence of reduced relative concentrations of odd-Z nuclei for stars of low metallicity (Arnett 1973, Truran 1973). Since the most abundant nuclear constituents of the stellar core immediately prior to the supernova event are self-conjugate nuclei (e.g. ^{12}C, ^{16}O, ^{28}Si), and since explosive burning time scales do not allow significant conversion of protons to neutrons by weak interaction processes, it follows that the final products of these explosive burning episodes must lie along or very near to the $Z = N$ line. The abundances of the odd-Z nuclei Na, Al, P, Cl, K, Sc, V, Mn, Co, and Cu formed in these environments are then sensitive to trace abundances of nuclei with $N > Z$. A significant neutron excess can be achieved during hydrogen and helium burning when first all critical CNO nuclei are converted to ^{14}N, and subsequently helium-burning reactions proceed as

$$^{14}N(\alpha,\gamma)^{18}F(e^+\nu)^{18}O(\alpha,\gamma)^{22}Ne.$$

Since the neutron excess defined in this manner by the presence of ^{18}O and ^{22}Ne is a function of the initial metal content of the star, we might anticipate that nonstellar odd-even effects in Z can appear in extremely metal-deficient stars. Indeed, calculations of explosive nucleosynthesis in metal-poor stars (Truran & Arnett 1971) indicate that the resulting concentrations both of odd-Z nuclei and of the neutron-rich isotopes of even-Z nuclei may be significantly reduced. As we have seen, observations both of field halo stars and of globular cluster stars reveal the possible presence of such trends.

It also seems most likely that these massive stars represent the site of production of the r-process heavy nuclei. Proposed models for r-process nucleosynthesis in these supernova environments range from those involving the expansion and cooling of neutron-rich matter from the vicinity of the neutronized core (see the review by Hillebrandt 1978) to those concerned with capture of neutrons released in the passage of supernova shock waves through the helium and carbon shells (Hillebrandt & Thielemann 1977, Truran et al. 1978, Lee et al. 1979, Cowan et al. 1982). While the detailed features of the r-process environment remain quite uncertain, the general identification of an r-process with massive stars seems reasonably secure. Again, we have argued in Section 2 that such identification is strongly suggested by recent spectroscopic studies, which reveal that the heavy-element ($A > 60$) abundance patterns in extreme metal-deficient stars have a distinctly r-process character.

This identification of the r-process site with massive stars holds implications for nucleocosmochronology. Since the interesting nuclear chronometers ^{232}Th, ^{235}U, ^{238}U, and ^{244}Pu are generally believed to be formed in the r-process, the production history of these nuclei is thereby tied to the history of massive star formation. It follows that the age obtained with the use of these chronometers should be closely related to the onset of the earliest epoch of significant star formation and nucleosynthesis activity in the Galaxy.

5.3 Nucleosynthesis in Intermediate-Mass Stars

The majority of the lowest mass stars surely die by ejecting their envelopes and leaving the degenerate C/O cores to cool as white dwarfs. Stars in the mass range $1 \lesssim M/M_\odot \lesssim 10$ are able to provide significant contributions to the abundances of the heavy elements in the Galaxy as a consequence of the occurrence of thermal pulses in their helium-burning shells on the asymptotic giant branch. The elements thus produced are subsequently ejected along with the envelope (see, for example, the review by Iben & Renzini 1983, and references therein). Estimates of nucleosynthesis yields from these asymptotic giant branch stars (Iben & Truran 1978, Renzini & Voli 1981) indicate that significant production of carbon, nitrogen, and the s-process heavy elements, in solar proportions, can be achieved in this environment. At the upper end of this mass range there is still considerable uncertainty. If stars between about 6 and 8 $M_\odot$ retain their envelopes and explode, they could be a major contributor to the synthesis of Fe and Ca. The longer time scales of evolution for intermediate-mass stars suggest that the nucleosynthesis products of these stars will begin to influence the interstellar gas abundances only relatively late in the process.

Several features of the nucleosynthesis processes that operate in red giant environments deserve further discussion. Schwarzschild & Härm (1967) first identified the environment provided by low- and intermediate-mass stars possessing both hydrogen- and helium-burning shells as a promising site for neutron capture synthesis. They determined that thermal pulses associated with the helium shells in such stars on the asymptotic branch will trigger the growth of a convective shell encompassing most of the matter within the ^{4}He and ^{12}C region between the carbon-oxygen core and the hydrogen envelope. The temperatures achieved in the helium thermal pulses are sufficient to generate neutrons by means of one or the other of the ^{13}C$(\alpha, n)^{16}$O or the ^{22}Ne$(\alpha, n)^{25}$Mg reactions. (It has yet to be established which one of these reactions is the primary source of neutrons.) The capture of these neutrons on preexisting "seed" iron-peak and heavier nuclei then produces s-process heavy nuclei in the mass range $70 \lesssim A \lesssim 209$. Note that the s-process elements here represent secondary nucleosynthesis products, in the sense that their formation demands the presence of seed iron-peak nuclei from the ashes of prior stellar generations.

The existence of a diverse population of peculiar red giants (S stars, carbon stars, etc.), whose surface abundance patterns differ significantly from those of solar system matter, provides evidence for the fact that the products of helium shell nucleosynthesis are ultimately transported to the upper reaches of the hydrogen envelope by convection. The early detection of the presence of the element technetium (which has no stable isotopes) in the atmospheres of red giant stars (Merrill 1952) unambiguously established the occurrence of this behavior. In fact, the helium shell provides a source both of freshly formed s-process elements and of freshly synthesized ^{12}C. Transportation of such enriched matter to the surface regions ultimately gives rise to detectable enrichments (relative to solar abundances) of s-process elements. Furthermore, the realization of carbon enrichments such that C/O exceeds unity leads to the occurrence of carbon stars. Processing of this freshly synthesized ^{12}C into ^{14}N as a consequence of CN-cycle hydrogen burning in the hydrogen-burning shell can result in the formation of primary nitrogen (Truran & Cameron 1971, Pagel & Edmunds 1981).

While both carbon and nitrogen can therefore be synthesized and returned to the interstellar medium in intermediate-mass stars of a range of initial metallicity, the production of significant amounts of s-process nuclei requires the presence of primordial seed iron-peak nuclei on which neutron captures can proceed. The s-process elements therefore represent secondary nucleosynthesis products. Their return to the interstellar gas is therefore expected to be delayed relative to the abundances of the iron-peak nuclei on which they are built (Truran & Cameron 1971, Tinsley 1980, Twarog & Wheeler 1982). It follows that the buildup of significant abundances of s-process heavy elements is delayed in the Galaxy both by the need for the prior production of seed iron-peak nuclei and by the longer lifetimes of the stars in which s-process nucleosynthesis occurs. We note, however, that the synthesis of heavy s-process nuclei through mass $A \sim 90$ only can occur in the helium-burning cores of massive stars (Lamb et al. 1977, Prantzos et al. 1987). We might therefore anticipate that the abundances of light s-process nuclei, through the peak at strontium at mass number $A \sim 88$, can increase faster than those at barium or lead. A possible ambiguity here in the interpretation of observed abundance trends is that the strontium-to-barium ratio in r-processed matter is also high with respect to the solar value (Truran 1981, Cameron 1982b). We return to this question later when we seek interpretations of the abundance trends observed in metal-deficient stars.

5.4 *Nucleosynthesis in Type Ia Supernovae*

Currently favored models for the outbursts of Type Ia supernovae are those involving the growth of a white dwarf to the Chandrasekhar limiting mass as a consequence of accretion in a close binary system, following which carbon or helium ignition occurs under highly degenerate conditions and a thermonuclear runaway ensues. The realization of a critical mass may be accomplished by the merger of two white dwarfs. This is the double-degenerate model described by Iben & Tutukov (1984) and Webbink (1984). Since a significant fraction of the core is ultimately burned to a nuclear statistical equilibrium peak centered on ^{56}Ni, the total thermonuclear energy liberated is sufficient to disperse the entire star at high velocities, leaving no condensed remnant. In this model, the decay of ^{56}Ni ($\tau_{1/2} = 6.1$ days) through ^{56}Co ($\tau_{1/2} = 78.5$ days) to ^{56}Fe serves to power the light curve. A more extensive review of many aspects of this problem, and appropriate references, is provided by Woosley & Weaver (1986b; see also Nomoto 1986).

Theoretical studies currently suggest that nucleosynthesis in Type Ia supernovae complements that of massive stars by forming primarily iron-peak nuclei. Type Ia supernovae are believed to be the major contributors to the abundances of the iron-group nuclei in Galactic matter. Calculations of explosive nucleosynthesis associated with carbon deflagration models of Type I supernovae (Thielemann et al. 1986) predict that sufficient iron-peak nuclei are formed to explain the light curves of Type I supernovae by the decay of ^{56}Ni and ^{56}Co. The question of whether such events are prone to overproducing the observed mass fraction of iron in Galactic matter ($\sim 10^{-3}$) is still open to debate (Twarog & Wheeler 1982, Nomoto et al. 1984). The association of these thermonuclear explosions with kinematically old Type Ia events means that they are expected to occur at large scale heights and to be particularly prone to uncertainties in the amount of the ejecta that is blown from the Galaxy in supernova-driven winds.

The predicted yields of the intermediate-mass nuclei (neon to calcium) are roughly consistent with observations of Type Ia spectra. There is insufficient C, O, Ne, and Mg relative to Fe to contribute significantly to the abundances of these nuclei in the Galaxy. These model explosions predict significant abundances of Si, S, Ar, and Ca compared with Fe (Thielemann et al. 1986), and if the massive stars have a paucity of these elements, the Type Ia events may fill in the gap. Nomoto et al. (1984) point out that if Fe is overproduced in the current models, then these elements (especially Ca) may be also overproduced. A continuing problem with the deflagration model of Type Ia events is the tendency to overproduce neutron-rich isotopes, especially ^{54}Fe and ^{58}Ni, in the relatively slow expansion that allows significant weak interactions. Correcting this flaw may, of course, have significant implications for other nucleosynthesis aspects of the model as well.

There remain critical unanswered questions concerning the nature and evolutionary histories of the putative binary progenitors of Type Ia supernovae and their frequencies of occurrence in different stellar populations.

For the purposes of our present discussion, we assume that such binary models for Type I supernovae produce iron and iron-peak nuclei on time scales compatible with the lifetimes of the intermediate-mass stars that are assumed to characterize these systems, while at the same time recognizing that some proportion of these systems could have moderately short lifetimes.

5.5 Nucleosynthesis Summary

Very massive stars ($M \gtrsim 30$–70 $M_\odot$) will evolve in less than 4×10^6 yr and may produce pair formation explosions rich in oxygen with virtually no C, Fe, or r-process elements. Massive stars of $M \gtrsim 10$ $M_\odot$, which evolve on time scales $\tau \lesssim 2 \times 10^7$ yr, are the major Galactic sources of oxygen and contribute substantially to the intermediate-mass elements from neon to calcium and to the r-process elements. Intermediate-mass stars ($1 \lesssim M/M_\odot \lesssim 10$) evolving on time scales $\tau \gtrsim 10^8$–10^9 yr are the main Galactic sources of carbon, nitrogen, and the s-process elements. Type Ia supernovae produce the iron-peak nuclei and make substantial contributions to the elements from Si to Ca on comparable time scales ($\tau \gtrsim 10^8$ yr).

6. INITIAL MASS FUNCTION, DEATH FUNCTION, AND YIELDS

In models of the chemical history of the Galaxy, a key ingredient is the yield. For massive stars, calculation of the yield involves the integration over the mass function that contributes to the production and ejection of given elements. The most common procedure is to integrate over the initial mass function, but it is not at all obvious that this is appropriate. Rather, one needs to integrate over the *death function*—that is, the mass function that actually contributes to stellar explosions or other mass ejection processes, not necessarily the one that represents the distribution at birth. The death function can differ from the initial mass function by the influence of binary companions on stellar evolution (a star can die with more mass than it had at birth) and by total collapse, which prevents any contribution to the yield. The influence of binary evolution on the death function is beyond the scope of this review, but some discussion of the death function of massive stars is appropriate.

A number of authors (Twarog & Wheeler 1982, 1987, Larson 1986, Matteucci 1986, Olive et al. 1987) argue that with the best current estimates of yields, mass functions, star formation rates, and Galactic gas content, oxygen is overproduced if all massive stars up to ~ 100 $M_\odot$ are included in the mass function. Estimates of the present total oxygen abundance are particularly sensitive to the assumed mass function of the massive stars. For a thorough discussion of the mass function, see Scalo (1986). The key parameter is the slope of the mass function in the relevant range. One can write the initial mass function as

$$\frac{dN(m)}{dm} = km^{-(1+x)},$$

or equivalently $dN(\log m)/d\log m = km^{-x}$, where $N(m)$ is the number of stars born with mass between m and $m+dm$, and $x(m)$ is the effective slope. The slope x is rather small (even negative?) for very low-mass stars, of order unity for solar masses ($x = 1.35$ is the famous Salpeter value), and ~ 2 for massive stars, with large uncertainty.

Twarog & Wheeler (1987) argue that for current estimates of the $^{12}C(\alpha,\gamma)^{16}O$ rate, if all stars up to 50 $M_\odot$ contribute to the oxygen yield, then O is overproduced by about a factor of 5 for $x = 1$ and by a factor of ~ 1.5 for $x = 2$, but that there would be no overproduction if $x = 3$. Matteucci (1986) finds O overproduced by a factor of 2.6 for $x = 1.35$ and by 2.1 for $x = 2.2$ if all stars up to 100 $M_\odot$ are included. Truran et al. (1986) reach similar conclusions. The oxygen overproduction problem does vanish if the slope of the mass function for high-mass stars $\gtrsim 15\ M_\odot$ is $x \gtrsim 3$. Note, however, that one cannot simply invoke the contributions of all stars up to $\sim 100\ M_\odot$ in the oxygen yield and then argue that this cannot be ruled out within the uncertainties in the mass function. Rather, if the yield from massive stars is to be finite, then one must be prepared to argue that the slope of the mass function *is* as steep as $x \sim 3$ and eventually to back that up with observations. These models that predict oxygen overabundance are all based on variants of "infall" models, where infall means a (sometimes time-varying) input of low- or zero-metallicity gas designed to solve the "G-dwarf" problem. This infall could be literal (although there is little direct evidence for it), recycled gas from previous generations of low-mass star formation, or it could be other effects with the same numerical result. The G-dwarf problem is discussed in Section 7.4.

An obvious and interesting way to overcome the problem of oxygen overproduction is to alter the death function. To invoke all stars up to a high mass in the stellar yield is to assume that the initial mass function and the final mass function are the same. There is ample circumstantial evidence from study of binary X-ray sources that some massive stars make black holes. This raises the question of whether some massive stars contribute to the initial mass function but not at all to the final mass function in the context of oxygen nucleosynthesis because they collapse and swallow all their synthesized oxygen. With this idea in mind, Twarog & Wheeler (1982, 1987) argue that the oxygen overproduction problem could be solved if no stars above some mass contribute to oxygen synthesis. They find the upper mass limit to be 24–28 $M_\odot$, depending on the $^{12}C(\alpha, \gamma)^{16}O$ rate. This is an interestingly low mass, since it is often argued that the "typical" star synthesizing heavy elements has a mass in the same range. To the best of our knowledge, there is no direct evidence, nucleosynthetic or otherwise, that demands that stars in excess of 30 $M_\odot$ explode.

Another way of considering the mass function is in terms of bimodal star formation. In this picture there are two physical mechanisms for producing stars, one for low-mass stars and one for high-mass stars. Evidence in favor of such a picture is the apparent bimodal form of the initial mass function. The advantage of the bimodal star formation picture is that the mix of high- and low-mass stars can be altered in Galactic history to better account for observations. In particular, one can speculate that in the very early halo phase (or Pop III?) of the Galaxy, the high-mass mode was the primary mode of star formation, and that this accounts for the large [O/Fe] in the halo. In the current context, this exacerbates the oxygen overproduction problem, because with less mass locked into low-mass stars the yield of oxygen must increase. In the model of Larson (1986), the oxygen yield increases by about a factor of 2, and contemporary oxygen overproduction can be avoided only by truncating the final mass function at $\lesssim 20$ $M_\odot$. The explosion of SN1987A would constrain the final mass function of exploding stars to extend to at least 15–20 $M_\odot$ if it were at all typical.

7. AGE-METALLICITY RELATIONS

It is traditional to discuss the evidence for the chemical evolution of the Galaxy in terms of the changes of ratios of species as a function of metallicity, [Fe/H]. This process implicitly assumes that [Fe/H] represents a clock in some sense. To fully understand the history written in the abundance ratios, we must understand the clock. We must calibrate the flow of time that links the variety of physical processes involved. This in turn requires that we establish age-metallicity relations. As we argue here, this is not a trivial process, and current work may be biased by severe conceptual errors. A related cautionary discussion is given by Sandage (1988).

7.1 *The Nature of Time Scales*

In trying to map out the chemical history of the Galaxy, it is important to distinguish relative and absolute chronometers associated with the astrophysical processes that dictate the evolution of abundances. The increase in [O/H] or [Fe/H] serves as a chronometer of sorts, in the sense that the integral accumulation of either abundance increases monotonically (presuming that no catastrophic Galactic outflow or grain formation process selectively depletes a particular element). The evolution of the ratio of an abundance to that in the Sun is, however, only a relative measure of the passage of time and requires calibration. The color of stars in a certain evolutionary phase is also only a relative measure of time, and Sandage (1988) warns that color may actually measure metallicity, not age, in important cases.

There are absolute time scales determined in various ways. One fundamental time scale is the age of the Sun at 4.6 ± 0.1 Gyr (Wasserburg et al. 1977). A nearly fundamental time scale is that based on stellar evolution. This is not quite as direct as measuring the decay of an unstable species, since it requires theoretical calculations to affect the comparison of observations with the effects of nuclear and atomic data, but the success of stellar evolution theory puts this absolute time scale on a reasonably firm basis. Although there are uncertainties, one is fairly certain of calibrating the age of a given open or globular cluster to within perhaps 50%, and that is certainly adequate to distinguish the age of the Hyades from that of 47 Tuc. A common presumption is that the age scale established from stellar evolution increases uniformly. One may, however, envisage different physical effects that enter into globular cluster evolution, so that the "clock" jumps between the oldest open clusters and the globular clusters. A possible example of this would be the main sequence winds [postulated by Willson et al. (1987) and elaborated on by Guzik et al. (1987)], which are pictured to affect only low-mass stars just above the onset of outer convective layers. There are also nuclear chronometers based on laboratory measurements of decay rates of radioactive species such as U to Th (e.g. Fowler 1987), but these are rather model dependent and the question of whether this technique is as reliable as stellar evolution for dating the oldest epochs is debatable.

7.2 *The Age of the Halo/Disk Transition*

The overall form of the age-metallicity relation of the Galactic disk presented by Twarog (1980) can be reproduced by a simple Galactic infall model that assumes that the disk began to evolve 13 Gyr ago from a metallicity of [Fe/H] = -1. How well is this time scale established?

The ages of the globular clusters continue to be a controversial topic, illustrating the remaining uncertainty in stellar evolution theory. Questions involve the effect of the heavy-element distribution (not just the average metallicity, but the distribution of C, N, O, and Fe), convective overshooting, and more exotic and controversial effects such as main sequence winds (Willson et al. 1987). The Yale group (e.g. see Demarque et al. 1988, King et al. 1988) employ the revised Yale isochrones (Green et al. 1987) to suggest that a possibly significant age spread exists among globular clusters, and that the ages seem to be loosely correlated with cluster metallicity. For example, they find an age of 16.0 ± 1.5 Gyr for M68 with [Fe/H] = -2.15, and 12.5 ± 1.5 Gyr for 47 Tuc with [Fe/H] = -0.76. In contrast, VandenBerg and his colleagues (e.g. see the review by VandenBerg 1988) find that all the globular clusters have about the same age of ~ 14 Gyr, if isochrones computed by VandenBerg (1983, 1985) or VandenBerg & Bell (1985) are employed. Their work on 47 Tuc (Hesser et al. 1987) yields an age of 13.5 Gyr, in good agreement with the Yale group, but their derived age for M68 is only 14 Gyr (McClure et al. 1987). The disagreements do not appear to be heading toward convergence at this time.

As for the disk, Winget et al. (1987) and Liebert et al. (1988) have argued from the failure to detect low-luminosity white dwarfs that the age of the disk is of order 8 Gyr but this seems to contradict the ages of the oldest disk stars established from isochrones. Norris & Green (1989) claim that stars composing the thick disk (itself a topic of controversy) are as much as 6 Gyr younger than 47 Tuc, whereas Janes et al. (1988) argue that the old open cluster NGC 6971 is even older than NGC 188 and may be only 1 Gyr younger than 47 Tuc (Janes 1988). Scalo (1987a,b) states that a number of lines of evidence suggest that the major phase of star formation in the disk began with a burst that occurred only about 5 Gyr ago. Although there seems to have been some star formation in the disk prior to this phase, the amount of nuclear processing may have been relatively small, and such a burst might represent the onset of significant chemical evolution in the disk. Reproducing such an age for the disk evolution is not a matter of a simple scaling of the chemical evolution model. Some stars that can evolve and die in 12 Gyr simply will not have done so in only 5–8 Gyr.

7.3 Is [Fe/H] the Right Stuff?

If there is uncertainty in dating the beginning of the evolution of the disk, what of the other crucial ingredient in the age-metallicity relation, i.e. the metallicity itself? We have noted that the metallicity has traditionally been measured by the iron abundance [Fe/H] because it was relatively easy to measure. The efficacy and appropriateness of this choice has been called into question by the discovery that while C scales with Fe over a large range in [Fe/H], oxygen distinctly does not do so (refer back to Section 2.1.3). Furthermore, there are sound theoretical reasons for believing that

Fe is not an appropriate measure of the "clock" of Galactic chemical evolution. One is that there are probably two distinct sources of Fe production in the Galaxy—stellar core collapse and thermonuclear explosion (see Section 5). The former is generally associated with high mass and the latter with lower mass stars, perhaps in binary systems, so that the time scales for the contributions of Fe from these two sources are distinctly different. In addition there are major uncertainties in both. The amount of Fe produced in a core collapse is very uncertain because the amount of iron produced in the shock and then ejected vs. that which falls back onto a compact remnant is very difficult to compute, given the failure of any dynamical model to produce unambiguous explosions. The association of the thermonuclear explosions with kinematically old Type Ia events means that they are expected to occur at large scale heights and to be particularly prone to uncertainties in the amount of the ejecta that is blown from the Galaxy in supernova-driven winds. Further complications may also arise, since some Type I events may be core collapses (Type Ib) and some Type II events may be thermonuclear explosions (Type II–linear light curve).

As a step toward thinking critically about the fundamental assumptions concerning Galactic chemical evolution, we propose that [O/H] may be a superior variable in which to couch the evolution of relative abundances. The preponderance of O probably comes from one source, namely massive stars. In addition, the bulk synthesis of the oxygen yield is more straightforward to estimate theoretically, since the oxygen is primarily produced in the quasi-static evolution of the star. This is not to say that O is a perfect clock by any means. Its abundance is affected by alterations in the mass function, and its rate of accumulation, while probably monotonic, will reveal the effects of temporal irregularities like massive starbursts only with careful scrutiny. Nevertheless, we feel that important questions such as the nature of the "break" in the abundance curves at [Fe/H] = -1, which is attributed to the onset of "disk" as opposed to "halo" evolution, can be more clearly posed and contemplated if, for instance, [O/H] is regarded as a more fundamental variable than [Fe/H]. The information in a plot of [O/H] vs. [Fe/H] is the same as in a plot of [Fe/H] vs. [O/H], but the perspective is not, and the proper perspective may aid in posing and answering the appropriate questions.

7.4 *The Metal-Poor Dwarf Problem: Closing Some Boxes*

An important aspect of the overall problem of Galactic chemical evolution that has been actively pondered for two decades is the "paucity of metal-poor dwarfs." Many studies have suggested that there are insufficient numbers of metal-poor dwarfs, which should be the remnants of the first epochs of star formation and which should live long enough to reveal that record now. Understanding the distribution of the metal-poor stars is also needed as an ingredient in establishing the age-metallicity relation. This problem has been recently discussed by Pagel (1988), who considers the problem in the disk and halo separately. Pagel shows (using [O/H] as the metallicity indicator!) that despite reanalysis and updating, the problem remains severe in disk stars of even moderate metallicity. A simple "closed-box" model gives a significant overestimate of the number of dwarfs already for $[O/H] \lesssim -0.3$.

In order to account for the perceived paucity of metal-poor dwarfs, a number of proposals have been put forward, including prompt initial enrichment, metal-enhanced star formation, "infall" of primordial material that allows the metallicity to climb more rapidly in early phases but not to exceed the observed metallicity at the current epoch, and bimodal star formation (Larson 1986). Pagel argues that a range of analytic infall models (Clayton 1987, 1988) that adequately fit the disk dwarf metallicity distribution fail to simultaneously match both solar system and stellar nucleocosmochronologies. These infall models demand a large age (~ 15 Gyr) based on solar system data and a small age (~ 10 Gyr) based on stellar data on Th/Nd (Butcher 1987, 1988; recast as Th/Eu by Pagel to avoid problems with the s-process origin of some Nd). Pagel concludes that prompt initial enrichment models may give the best self-consistent fit to the metal-poor dwarf problem and to both the solar system and the stellar radioactivity data with an age of the disk of about 10 Gyr. This age is roughly commensurate with the age of the first burst of r-processing estimated by Fowler (1987), of the oldest white dwarfs by Winget et al. (1987), and of the first burst of star formation in the disk as argued by Scalo (1987a,b). Pagel concludes that if the initial spike in enrichment were the result of infall from the halo to the disk, the disk must have built up to its present mass in a short time. Alternatively, if disk star formation were suppressed in the disk during infall (Larson 1976) and began with a burst, these various pictures might be reconciled, with the time scale so determined being the age of the first major onset of star formation in the disk but not necessarily the age of the halo, nor of the Universe. We return to this possibility in Section 8.

In the simplest "closed-box" models, the distribution of metallicity, dN/dZ, should be roughly constant in the early phases, with small metallicity before gas depletion becomes important. This is the phase that should apply to the halo. The potential problem in the halo was underlined by the study of Bond (1981), who emphasized that his discovery of only two objects with $[Fe/H] < -3.0$ in a survey of halo subdwarfs and globular clusters was significantly less than expected in a closed-box picture. Pagel (1988) reconsidered the distribution of metallicity in halo and disk globular clusters and concludes that the globular clusters are basically consistent with a "closed-box" model. Beers et al. (1985) and Beers (1987) compiled a complete sample of metal-poor Southern Galactic Hemisphere stars with an objective-prism survey. The metal-poor stars, determined by weak or absent Ca II lines, were of spectral type F0 to G5 and included a mix of stars near the halo main sequence turnoff and on the red horizontal branch, giant branch, and asymptotic giant branch. Beers shows that the distribution of metallicity (measured, strictly speaking, by [Ca/H] rather than by [Fe/H]) is flat at low metallicities. Out of a sample of 120 stars with [Fe/H] < -2.0, there are 50 stars with [Fe/H] < -2.5 and 10 stars with [Fe/H] < -3.0, compared with 38 and 12, respectively, expected from a simple closed-box model. Note that at low metallicity, both closed-box and infall models (Audouze & Tinsley 1976) predict cumulative metallicity distributions proportional to metallicity. Thus the results of Beers, while consistent with a closed box, do not rule out more complex models.

These results suggest that it is entirely possible that the halo evolved as a simple "closed box." Note that this must apply piecewise to individual portions of the halo, with different rates of accumulation of Fe if the local scatter in [Fe/H] with age in the solar neighborhood is properly interpreted as a sampling of different parts of the halo. Closed-box halo evolution in turn suggests that the halo stars formed from pristine primordial material, and that there was no significant sudden early enrichment from Population III stars. If one adopts this point of view, the next question is how to merge this picture into the overall view of the age-metallicity relation and of the evolution of the metallicity in the disk. Did the halo form from gas, or was it accreted, perhaps in chunks, as stars that formed in some more remote, but closed, box (Searle & Zinn 1978, Rodgers & Paltoglou 1984, Norris & Ryan 1989)? If the distribution in metallicity up to [Fe/H] = -2.0 is consistent with a simple closed box, what span of absolute time does that represent, and is it different in different parts of the halo? What is the evolution of other basic elements, such as C and O, over this interval? How long does the epoch of globular cluster formation last? Are all halo stars formed in globular clusters, some of which subsequently dissolve, or are some of the present field halo stars born that way? Does the character of the formation of globular cluster or field halo stars change as the metallicity evolves?

7.5 *Current Age-Metallicity Relations*

With these various issues in mind, let us consider the status of current attempts to establish the age-metallicity relation. The age-metallicity relation of Twarog (1980) is based on narrowband photometry to establish a color-magnitude diagram for the sample stars. This technique has the advantage of yielding a large sample, but it also has the disadvantage of not having kinematic information. One thus *assumes* that all stars with [Fe/H] > −1 are disk stars. The ages of individual stars are then determined by fitting isochrones of appropriate metallicity to the positions of individual stars. The raw data show considerable scatter, so that averages are taken in age bins. Within the error bars of the averaged data, a relatively smooth relation is derived in which [Fe/H] = −1 at the beginning of the disk 13 Gyr ago and rises smoothly to the metallicity of the Sun at 4.6 Gyr ago and then climbs another 0.1 dex by the current epoch.

Boesgaard (1989) and Nissen (1988) have recently redetermined the age-metallicity relation from spectroscopy and Strömgren photometry, respectively, of Galactic clusters. The age assignments based on distributions of cluster stars are perhaps more reliable than isochrone fits to individual stars, but clusters may sample the Galactic abundance gradient in an uncertain way, as pointed out by Twarog (1980). The results of Boesgaard and Nissen are in interesting concordance, in that they show a peak in [Fe/H] at about 3×10^8 yr ago, another at $7–9 \times 10^8$ yr ago, and perhaps a third peak at 2×10^9 yr ago (Lambert 1988). The amplitude of the oscillations in this age-metallicity relation is of order ± 0.1 in [Fe/H], which is larger than the formally quoted internal errors of ± 0.04 dex, but systematic errors could be larger. These variations could be evidence that the disk produces chemical inhomogeneities related to radial, longitudinal, or vertical variations in the rate of star formation and nucleosynthesis. Sandage (1988) has argued in favor of an age-metallicity relation that varies with Galactic radius and scale height, in the sense that the increase in metallicity is slower at larger radii and larger scale heights. If we ignore this oscillation in [Fe/H], the resulting age-metallicity relation deduced from open clusters by Boesgaard and by Nissen is flat to within ± 0.1 dex for the last 2×10^9 yr, in agreement with the cluster age-metallicity relation discussed by Twarog (1980).

More interesting for the nature of the age-metallicity relation over the long term is the reanalysis of Twarog's data by Carlberg et al. (1985). In this study, Carlberg et al. use VandenBerg's (1983, 1985) isochrones (which give larger ages for older stars than do the Yale isochrones), a photometric metallicity calibration that tends to give larger metallicity to the cooler, generally older stars, and several other refinements. They find that the age-metallicity declines from $+0.1$ to only -0.2 over the last 15×10^9 yr, in substantial disagreement with the results of Twarog (1980) for the stars in excess of 5 Gyr. In Figure 8 we show the age-metallicity relations proposed by Carlberg et al. and by Twarog. There are several factors involved in this difference. Some of the ages assigned by Twarog according to the hooks in the isochrones may have been ambiguous, since the isochrones

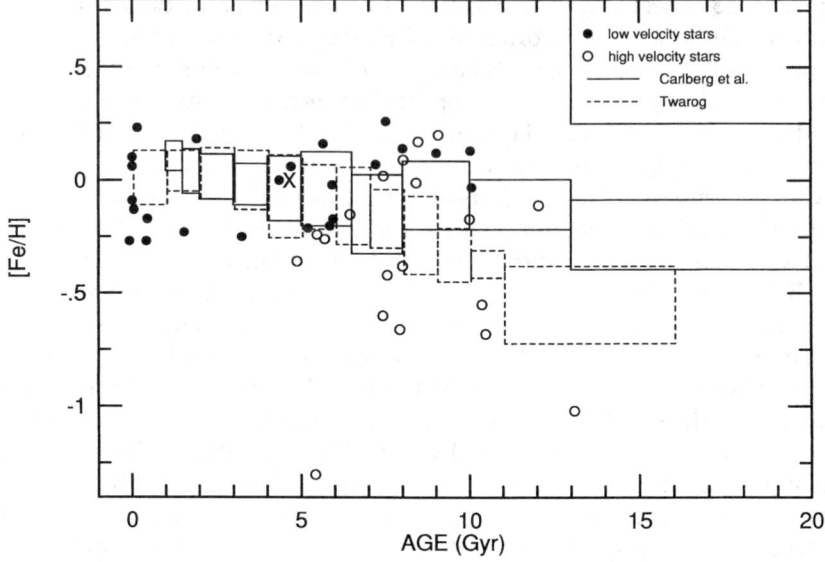

Figure 8 The age-metallicity-kinematics relation of Barry (1988; circles) compared with the age-metallicity relations of Twarog (1980) and Carlberg et al. (1985). The × denotes the location of the Sun in this diagram.

are nearly vertical, are closely spaced, and are sensitive to Z. Carlberg et al. thus assign older ages to some stars of given metallicity than did Twarog. In addition, while Twarog gave relatively large weights to stars of low metallicity, Carlberg et al. omit about 50 stars from the analysis because of perceived bias in the data.

Given this level of disagreement, it is important that other independent age-metallicity relations also be examined. An interesting one is that discussed by Barry et al. (1987) and Barry (1988), in which an empirical relation is established between the decrease in chromospheric activity (as measured by the ratio of chromospheric Ca emission to bolometric flux) with cluster age for solar-type stars ($B-V = 0.60 \pm 0.10$). Barry uses this chromospheric chronometer to examine the age-metallicity relation for a sample of stars for which there are also velocity data. We have added Barry's data to Figure 8. The correlation that he presents shows that the low-velocity stars yield an essentially flat age-metallicity relation within the scatter of ± 0.3 dex for the last 10 Gyr. The high-velocity stars do not show a decline in [Fe/H] with age, but rather a broad real scatter of amplitude perhaps ± 0.5 dex. The lowest metallicity stars ([Fe/H] $\lesssim -0.5$) show a range of ages from about 5.5 to 13 Gyr.

Barry (1988) concludes that the oldest stars tend to be high-velocity stars, that the high-velocity stars have a larger dispersion in metallicity than do the low-velocity stars, and that the high-velocity stars scatter, but do not trend, toward lower metallicity. This scatter may arise from the fact that the older high-velocity stars sample a range in halo birth sites and hence a vertical or radial halo metallicity gradient. It is interesting that some of the low-metallicity, high-velocity stars are young. J. M. Scalo (private communication) suggests that some of the high velocity could be due to gasdynamics rather than stellar dynamics associated with disk vs. halo motion. For instance, stars born in the wake of a violent starburst could be associated with the disk, but with high proper motions imparted to the gas by the attendant ionizing flux and supernova explosions.

Barry notes that 8 of the 19 high-velocity stars in his sample fall on a boundary suggesting that the most recently born, high-velocity stars formed only 5×10^9 yr ago with [Fe/H] $= -0.4$ and that the earliest born, high-velocity, high-metallicity stars formed at about twice this age with four times the metallicity. This is a very difficult trend to explain in any traditional model. The most obvious explanation is that the chromospheric activity is not simply a function of age, but also of metallicity in the sense that higher metallicity is associated with decreasing chromospheric activity and hence a larger apparent "age." It is not clear why such a correlation should exist, and one might even expect an opposite correlation.

Recent studies thus tend to suggest that there is no simple age-[Fe/H] relation. Low-velocity disk stars of age $\lesssim 10-15$ Gyr tend to have the same value ([Fe/H] $= 0.0 \pm 0.3$) and high-velocity stars show a scatter in ages and metallicity. It is thus possible that the age-metallicity relations of Twarog (1980) and Carlberg et al. (1985) are misleading in the absence of correlated dynamical information. The apparent downturn in [Fe/H] with age, mild though it is in the Carlberg et al. study, may represent the scatter among the high-velocity stars that has been inappropriately averaged. The identification of [Fe/H] $= -1$ as the lowest metallicity disk stars may be completely misleading if all stars with [Fe/H] < -0.3 are high-velocity stars. Hartkopf & Yoss (1982) and Yoss et al. (1987) give the distribution of giants with metallicity at different scale heights. For $z < 500$ kpc, the distribution drops precipitately at [Fe/H] ~ -0.4, although there is a small tail extending to -0.8.

A constant metallicity in the disk implies that all the iron in disk stars in the solar neighborhood was produced before the disk began to evolve. This is consistent with, but does not necessarily demand, the notion that there was an appreciable hiatus between the formation of the halo and the onset of active star formation in the disk, during which appreciable numbers of longer lived stars had a chance to finish their synthesis of Fe. In the high-velocity stars, the scatter in age and [Fe/H] suggests that the production of Fe continued in the halo until ~ 5 Gyr ago in both high- (Galactic center) and low- (Galactic outskirts) metallicity regions, and that the solar neighborhood samples all these environments.

If [Fe/H] has been sensibly constant over the age of the disk, then it may represent a useful point of normalization for the trends in other elements, but it should certainly not be thought of as a chronometer. (Galactic clocks that stand still are not even right twice a day!) Clearly a major effort must be made to establish a calibration of the oxygen abundance in a class of stars for which age estimates are independently available in order to determine the degree to which [O/H] might serve an an abundance chronometer to supplement or replace iron.

8. CONCLUSIONS

We have assembled recent data on the evolution of abundances as a function of metallicity and raised a number of conceptual issues relating to the chemical evolution of the Galaxy. Future progress will depend on the acquisition of data, guided by the proper physical framework, and this may require rethinking of some of the basic assumptions regarding chemical evolution.

One lesson that emerges clearly in this review is that great care must be exercised in the *independent* determinations of ages, abundances, and kinematics so as not to ascribe the properties of one population to another. The assumption that all stars with [Fe/H] > -1 are disk stars may give rise to serious misconceptions if the majority of moderately low-metallicity stars are actually high-velocity stars. Current data suggest that there is little or no evidence for a temporal gradient of [Fe/H] in the disk, and hence that there is no evidence that [Fe/H] represents a good clock for disk evolution. Its efficacy for timing the halo is thus also cast in doubt. The notion that [Fe/H] is roughly constant in low-velocity stars but scatters at high velocity suggests a physical picture very different from that of a steady trend to lower [Fe/H] in stars of a single kinematic population.

As a chronometer, Fe has the advantage that it may arise predominantly from lower mass stars with longer response times, and hence its production may integrate and smooth bursts of star formation. The drawback is that Fe is produced in different types of explosions with different intrinsic time scales for production and different theoretical uncertainties. We suggest that as more data accumulate, oxygen may prove more useful. As a chronometer, O has the advantages that it comes from basically a single source (massive stars), and that its production in the quasi-static evolution of the stars is relatively well understood. Potential drawbacks are that O has few features and its abundance is hard to determine, and that it will be produced sporadically if the birthrate of massive stars is sporadic. In any case, careful thought must be given to the time calibration of a given value of a given metal abundance.

It is also clear that one must incorporate dynamical aspects into chemical evolution models. It is simply not correct to make a chemical evolution model of the solar neighborhood with a gas-permeable bag (infall) and ignore the influx of stars from other regions of the Galaxy with different chemical histories. The scatter in the metallicity of the field halo stars suggests that the sample has been drawn from locales with different chemical histories. The field stars exist in the solar neighborhood because they have orbited here from some distant birth site. The Population II field stars may show a different abundance distribution than that of the globular clusters, which suggests that they are not simply dissolved globular clusters, and that the globular clusters have yet another chemical history. Chemical evolution models that purport to follow the buildup of the local abundances from very small metallicities with simple infall of gas into the solar neighborhood are clearly not to be taken literally. Rather, we must develop a class of models that realistically incorporates the dynamical infusion of stars from other chemical environments into the solar neighborhood in order to properly interpret the abundance, age, and kinematic information.

8.1 *A Brief History of Chemical Evolution*

The basic ingredients of the problem we are trying to solve are the structural and star formation history of the Galaxy, coupled with the nucleosynthetic yields of various kinds of stars. We can outline how these ingredients should contribute to chemical evolution and thereby attempt to gain some insight into the meaning of the observed abundance variations.

Starting from the formation of the proto-Galaxy, there may have been some role for the so-called Population III stars. These may have formed throughout the proto-Galactic cloud or just in the nuclear bulge. We do not know the mass function of such stars, nor whether such stars or globular clusters were the basic unit of star formation. If Pop III stars were dominated by very massive stars, then they might have been selective producers of oxygen in the earliest phases of the formation of the Galaxy. Even if the Pop III stars had a mass function typical of the contemporary solar neighborhood, the most massive stars would have evolved first and could have been heavy producers of O to the exclusion of Fe and, to a great extent, C. These stars may also have been deficient in the production of *r*-process elements. If the decline in the *r*-process tracer [Eu/Fe] at very low metallicity is real, it may suggest a role for very massive stars exploding by the pair formation process. If this is the case, then the epoch of the downturn of [Eu/Fe] at [Fe/H] = -2.5 might be calibrated by the lifetimes of such stars to be $\sim 4 \times 10^6$ yr. The hints of an upturn in [C/Fe] at [Fe/H] ~ -2.5 may have a similar explanation, since the very massive stars should produce some C but little Fe and less Eu. When massive stars far enough down the mass function to produce core collapse evolve, they will commence to produce ~ 0.1 $M_\odot$ of Fe and some r-process elements but not much more C. Hence, we should see a rise in [Eu/Fe] and a decline in [C/Fe]. Further evolution down the mass function should bring in relatively more carbon. Using Fe to normalize studies of the most massive stars, which may make little Fe, may not be the most efficacious choice. A careful determination of [C/O] might show the gradual enrichment of C/O with increasing [O/H] that is expected as lower mass stars evolve, but before the C- and Fe-rich ejecta of intermediate- and low-mass stars come into play.

Globular clusters and the halo stars are a primordial component of the Galaxy. The time scales over which they formed is uncertain. The globular clusters show an appreciable range in metallicity. Does this mean that they formed from an increasingly iron-rich gas over an appreciable range in time, or are they self-enriched and formed at more nearly the same epoch? The field stars do occur with very low metallicity, and current data are consistent with them having been formed in a "closed box." But what closed box? Although the notion that low-metallicity stars in the solar neighborhood conform to a simple closed-box model is appealing, this statement actually raises a troubling question, since the box is clearly not closed to the dynamical influx of stars and there are ample reasons to think that distant locales participated in different chemical histories. If the net effect at the solar neighborhood is that of a closed box, then apparently every environment throughout the halo evolved a metallicity distribution as a locally closed box. Even then, if regions that increased their metallicity at different rates contribute with different weights to the solar neighborhood, it is not clear why the net result should resemble a single "closed-box" distribution.

What sort of stars contributed to the evolution of the halo? In Figure 9 we show the distribution of elements as a function of atomic weight for elements from carbon to dysprosium for stars with a metallicity of [Fe/H] ≈ -2.3, or [O/H] ≈ -1.9. The abundance points in this figure represent our best estimates of the abundances of each element from the field star data of all the investigations reviewed in Section 2. The error bars also are our (somewhat subjective) evaluations of the uncertainties associated with the relative abundances. Note that in keeping with our notion that O may represent a better measure of chemical evolution than Fe, we have normalized this figure to O. This normalization emphasizes

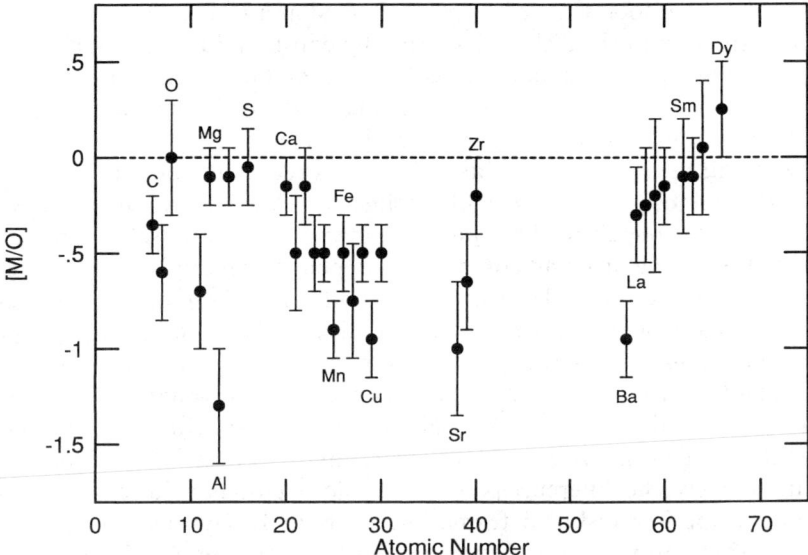

Figure 9 Average abundances, normalized to oxygen, of very metal-poor stars ([Fe/H] ≈ −2.3) as a function of atomic number. The abundances are based on the discussions in Section 2.

that the α-chain nuclei from Ne through V are roughly solar. Carbon and iron are deficient by a factor of about 3, but the very heavy *r*- and *s*-process elements are solar with respect to O. This distribution is roughly consistent with the yields expected solely from massive stars.

The next step in the evolution of the Galaxy seems particularly uncertain. Larson (1976) found that infall from the halo to the disk continued over several billion years, and that he needed to suppress star formation in the disk to form a respectable disk at all, rather than a galaxy that is mostly bulge. Recently, Vasquez & Scalo (1989) have suggested that infall will naturally suppress star formation. They argue that star formation is not just a simple function of the gas density. On the contrary, infall provides a source of kinetic energy to maintain gas motion that causes disruptive collisions of clouds rather than sticky collisions that build more massive clouds. This work leads to the general idea that there could have been a considerable hiatus between star formation in the halo and that in the disk.

The work of Vasquez & Scalo (1989) also suggests that when the infall slackens, the disk will undergo a burst of star formation. Scalo (1987a,b) has argued that the irregularities in the stellar luminosity function in the solar neighborhood can be better interpreted as a record of past bursts of star formation in the disk with essentially constant mass function, rather than a bimodal, or trimodal, mass function and constant star formation rate. He cites evidence for two or more large bursts in the main sequence luminosity function, in the age distribution determined from Ca emission-line strengths, in the lithium abundances in red giant stars, in the white dwarf luminosity function, and in time-correlated bursts in the Large Magellanic Cloud. The oldest burst could represent the onset of rapid star formation in the disk and could have happened as recently as 6 Gyr ago. The age calibrations of Twarog (1980) and Barry (1988) suggest that there was considerable star formation in the disk in the epoch 8–12 Gyr ago, but there might still have been a hiatus in star formation in the disk of about 4 Gyr from the oldest globular clusters at 16 Gyr ago and the onset of star formation in the disk at about 12 Gyr ago. This might account for the sudden increase in white dwarfs and r-processing (Section 7.4). Alternatively, the difference between the oldest globular clusters at 16 Gyr ago and the first disk star formation at about 12 Gyr ago could be the time for the propagation of globular cluster formation from the inner to the outer halo. If so, this would imply that there was minimal delay between the formation of the globular clusters at large Galactic radius and the onset of star formation in the disk in the solar neighborhood at about 12 Gyr, if the Yale chronology for the globular clusters is correct. Yet another possibility is that we are yet not dating the globular clusters correctly, and that both they and the disk are younger than traditional estimates.

The subsequent evolution of the disk could then have been driven by intermittent star formation, interrupted by occasional large global bursts. A picture in which the halo formed over 4 Gyr and appreciable star formation in the disk was significantly delayed and then proceeded in stochastic fashion, punctuated by large bursts, is radically different than one in which the halo formed in about 1 Gyr and the disk formed at about the same time and proceeded with nearly constant star formation for 12 Gyr. The record of metallicity with age and kinematics will likewise be different in such pictures.

8.2 Alternate Histories

To illustrate, consider two extreme examples of possible Galactic evolution. In the first, more traditional picture, the halo forms in about 1 Gyr, followed rather quickly by the disk, which then produces stars at essentially a constant rate for about 12 Gyr. In the opposite extreme, the halo forms in about the same time, but then star formation is suppressed in the disk by infall for several billion years and then proceeds by occasional global bursts.

In Figure 10 we represent schematically some of the differences that could arise in these alternate histories. In Figure 10a we represent the standard model, in which the star formation rate (top panel) is essentially constant in the disk. The halo should be especially enriched in O coming

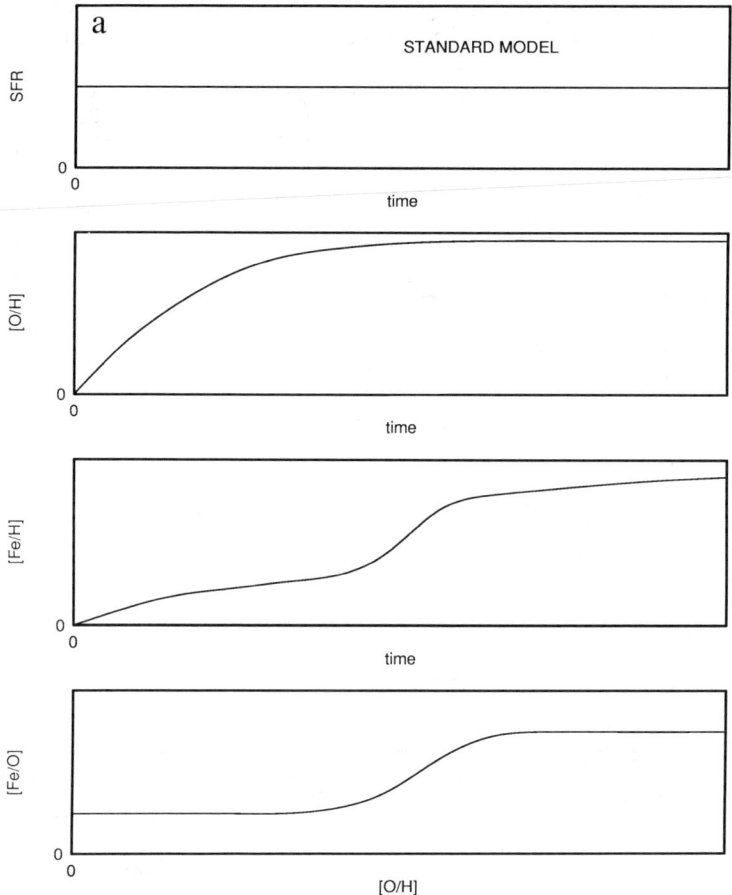

Figure 10 Various star formation histories and their effects on metallicity are shown schematically. For these displays, we have assumed that oxygen comes from massive stars that follow the star formation rate promptly, and that while some iron comes promptly from massive stars, the bulk arises from intermediate- and low-mass stars after some delay. In the panels of part (*a*) a constant star formation rate is illustrated, in (*b*) (p. 340) a hiatus in the star formation rate is shown, and in (*c*) (p. 342) a burst in the star formation rate is assumed.

rapidly from the massive stars. The O abundance should then accumulate steadily over the age of the disk (second panel). At first, the Fe abundance grows slowly as the massive stars produce little Fe, but after some time the production of Fe picks up as the lower mass, iron-producing stars in both the halo and the disk begin to evolve (third panel). The growth as a function of metallicity is thus as shown in the bottom panel, which illustrates [Fe/O] as a function of [O/H]. Although [O/H] is not a linear chronometer (ordinate of panel two), it is better behaved than [Fe/H] (ordinate of panel three).

Figure 10b shows the schematic behavior if there is a hiatus in star

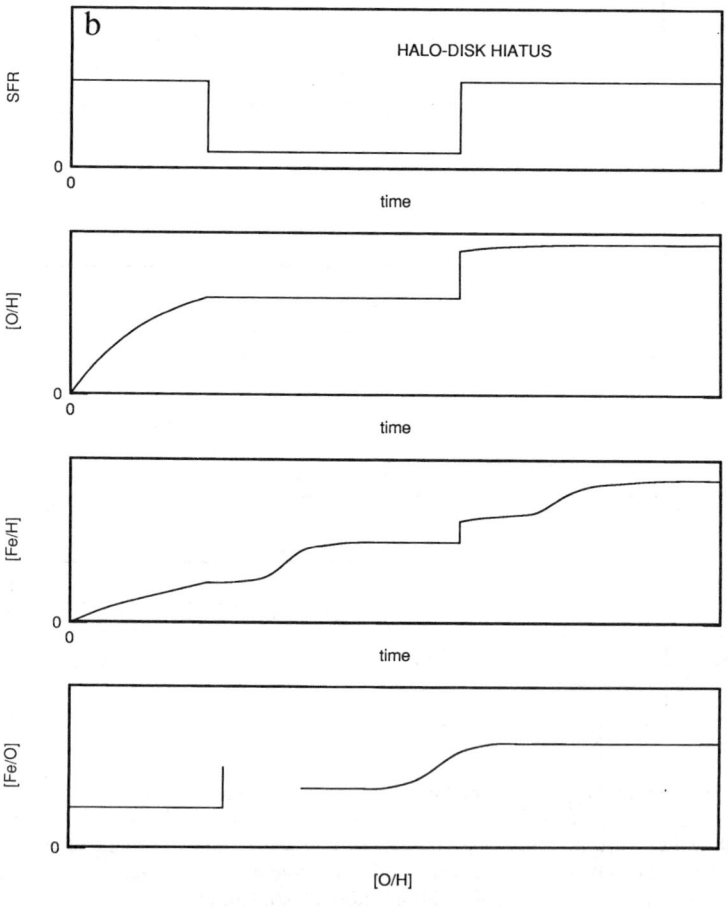

Figure 10b

formation (top panel). In this case, nearly all the halo stars that can contribute to nucleosynthesis will have done so during the long hiatus between the formation of the halo stars and the onset of star formation in the disk. During this hiatus before disk star formation, the evolution of [O/Fe] will depend on the degree to which star formation continues in the halo and the degree to which new generations of halo stars are enriched by the products of previously rapidly released O and slowly released Fe. Alternatively, all the Fe could settle into the disk awaiting the first burst of star formation there. Figure 10b assumes that the intermediate- and low-mass Fe-producing stars have not had a chance to evolve before the onset of the hiatus. Thus the hiatus is reflected directly as a plateau in [O/H] (second panel), but the Fe abundance, which at first shows a plateau due to the cessation of Fe production from massive stars, rises during the hiatus (panel three) as the lower mass stars born before the hiatus finally evolve. After the onset of disk star formation, there should be a rapid increase in the O abundance (panel two) and perhaps a smaller associated step in Fe from the massive stars followed by a slower rise in Fe from the lower mass stars born after the resumption of more rapid star formation (panel three). The resulting profile of abundance as a function of metallicity (panel four) is not pretty. The whole rise in [Fe/H] from the low-mass early-born stars occurs during the hiatus in star formation and hence in massive star O production, and thus it appears as a step in [O/H]. There should be essentially no stars within the range of [O/H] corresponding to the vertical rise in [O/H] in panel two when the star formation rate jumps and the O production rate responds. Thus there should be a corresponding gap in the distribution. On the high side of the gap in [O/H], the massive stars will have contributed their Fe as well as O, but the result should be O-rich compared with the mix just before the gap, which also had the contribution from the low-mass Fe-producing stars that evolved during the hiatus. Thus the value of [Fe/O] should be somewhat lower after the gap than before it, but the ratio should rise as the lower mass stars born after the hiatus finally make their contribution. Clearly, one would read [O/H] as a chronometer in such a case with trepidation.

Figure 10c shows the result of a burst of star formation on top of an otherwise constant star formation rate (top panel). In this case there is a step in [O/H] with time (panel two) and then a slower rise as the lower star formation rate ensues. A similar behavior is reflected in the massive star contribution to Fe (panel three), but then the lower mass, Fe-producing stars born in the burst come in to increase the Fe production rate. The profile of [Fe/O] vs. [O/H] again shows a gap corresponding to the sudden production of O in the burst (panel four). Figure 10c assumes, for purposes of illustration, that prior to the burst the constant star formation had

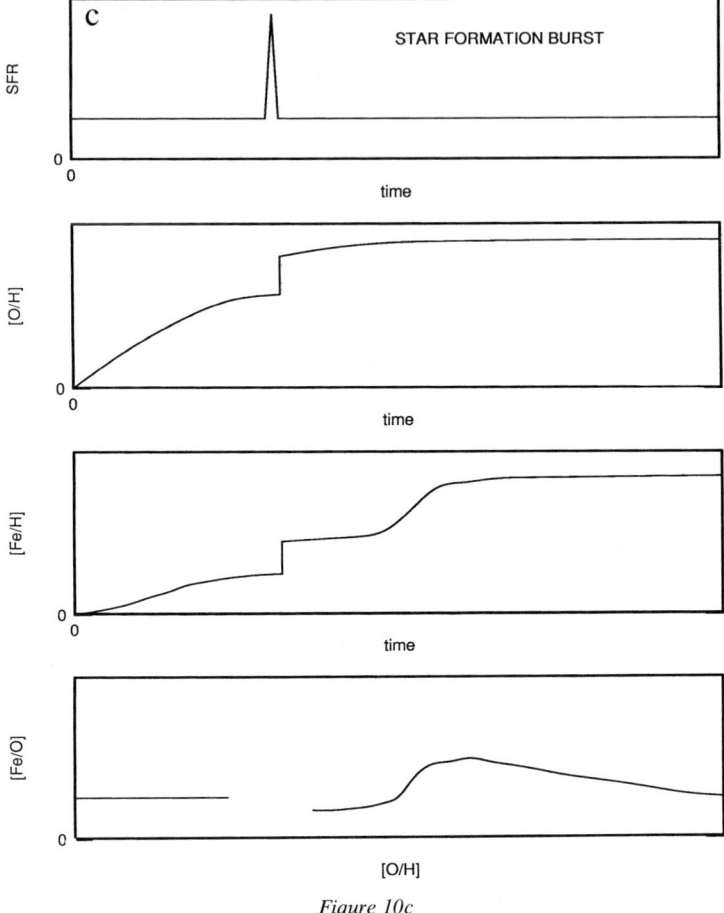

Figure 10c

proceeded long enough for production rates of both high- and low-mass stars to be in steady state. After the burst, and the attendant gap in [O/H], the value of [Fe/O] is thus a bit lower, since it only contains the contribution from massive stars. The lower mass stars produced in the burst then contribute, which leads to the rise in [Fe/O]. If the postburst star formation continues steadily for long enough, the Fe production from lower mass stars could tend back toward steady state and thus lead to the same asymptotic value of [Fe/O] as before the burst, as illustrated in panel four. Once again, the calibration of the clock represented by the evolution of metallicity in such a case would have to be done with great care.

In the real Galaxy with a halo, a star formation hiatus interrupted by a major burst and a succession of subsequent bursts could exhibit a complex combination of the phenomena illustrated in Figure 10. The [O/H] ratio should increase in a "stair-step" mode after each major global burst, with the [Fe/H] ratio increasing more smoothly with a memory of previous bursts due to the longer production time for Fe from moderate-mass stars. There could be other complications introduced by global starbursts, such as winds driven by the large, concentrated ionizing flux, which could selectively affect different elements differently, depending on which stars participate, their scale heights in the Galactic potential, and the yields. If the massive stars formed in a burst produce oxygen, which might be driven out in a wind, what of the iron produced from older stars born in a previous burst?

8.3 *The Next Generation of Observation and Theory*

With these ideas in mind, it is useful to reconsider the class of questions one should be asking about abundance patterns for various species. It has traditionally been argued that there is a break in the [O/Fe] curve that sets in at [Fe/H] ~ -1. The same is also approximately true for Mg and S. This break has been discussed in terms of the onset of Fe production in the disk by longer lived stars. This value of the metallicity ([Fe/H] ~ -1) is also supposed to be the minimum metallicity of the disk. Is this to be regarded as a coincidence, since the time scale for the evolution of the halo and the formation of the disk need have no relation to the time scale for the evolution of stars with masses of $\sim 8\, M_\odot$, or is there a need to examine this case more carefully?

A key question is the absolute Galactic time at which this break occurs. This date is abundantly unclear if the low-velocity stars show constant [Fe/H] back to 10^{10} yr and the high-velocity stars show only scatter in [Fe/H] but no trend with age. In that case, [Fe/H] $= -1$ does not refer to any chronological aspect of the disk. If Sandage (1988) is correct, and thus the age-metallicity relation varies with Galactic radius, then [Fe/H] $= -1$ refers to very different chronological times at different locations in the halo. The fact that [O/Fe] correlates with [Fe/H] despite the scatter of [Fe/H] with age in the high-velocity stars suggests that there are systematic trends, but perhaps these refer to localized portions of the halo, far from the solar neighborhood that we are now restricted to in our samples. In this case, the basic aspects of the picture in which Fe is produced from longer lived stars than those that produce O may still be pertinent. It still seems remarkable that [O/Fe] correlates with [Fe/H] at low [Fe/H] if O and Fe are produced on different time scales and at different rates in different parts of the halo and then are sampled randomly in the solar neighborhood. How sure are we that a break occurs at [Fe/H] ~ −1, and that there is not actually a constant decline in [X/Fe] that happens to appear as a break because of a scatter in the sparse data for [Fe/H] ~ −1 to −1.5, or some other behavior that the scatter in the data has masked?

Consider again Figure 2, which gives [Fe/O] vs. [O/H]. This figure was constructed to see if it gave a different perspective than the standard way of plotting [O/Fe] vs. [Fe/H]. In the standard picture, one would argue that conditions were fixed in some sense, producing a constant ratio of [Fe/O] until the epoch corresponding to [O/H] ~ −0.3, when the production of Fe increased, presumably as a result of the evolution of longer lived stars. If, on the other hand, there were steady star formation interrupted by a burst, as shown schematically in Figure 10c, then several things would be different. One would expect a jump in the oxygen abundance as the massive, short-lived stars rapidly ejected their oxygen. This would produce a gap in the distribution of stars between those with [O/H] equal to the value before the burst and those with the value of [O/H] born after the burst. After some interval of time, and accumulation of [O/H], the slower evolving stars would then come into play and release ejecta that might be particularly rich in Fe. If these stars were concentrated in a narrow mass range (say 6–8 $M_\odot$), then the increase in [Fe/O] at that epoch might be precipitate. If the burst altered the mass function to bring in a greater preponderance of low-mass iron-producing stars, the [Fe/O] ratio might be permanently changed. If not, the [Fe/O] ratio might tend to relax back to the preburst value as all components of the death function came back into equilibrium with the steady postburst star formation rate.

There is certainly an intriguing gap in Figure 2 between [O/H] ~ −1.0 and −0.5. Is this a selection effect, or is it rather some evidence that there was a burst in massive oxygen-producing stars? Ignoring for a moment the eight stars with [Fe/O] < −0.5 at [O/H] ~ −0.2, there does seem to be a precipitate jump in [Fe/O] at an [O/H] of ~ −0.2. Is there a hint that [Fe/O] then declines back toward the value it had at [O/H] < −1.0? We would argue that Figure 2 does not rule out such a different interpretation and thus raises challenges for both observers and theorists.

Observers should try to fill in the gap in the [O/H] distribution. The stars in Figure 2 should all be studied kinematically, bearing in mind that high velocities could be the result of a violent starburst as well as Galactic dynamics. What about the stars with [Fe/O] < −0.5 at [O/H] ~ −0.2, which seem to stand out in this plot? In a traditional plot of [O/Fe] vs. [Fe/H], this group of stars reinforces the notion that there is a break in the slope at [Fe/H] ~ −1.0, despite the fact that they have lower [Fe/O] values than those of lower metallicity stars, and that without them the standard diagram could also be interpreted as showing a constant value of [O/Fe] ~ +0.3 and then a precipitate shift to a mean value of [O/Fe] ~ 0 at [Fe/H] ~ −0.3. Is there something special about these stars in terms of their kinematics or other properties? Another worry is that the O abundance is determined by different techniques in the stars with [Fe/H] > −1 than it is in stars with [Fe/H] < −1. Barry (1988) assigns ages in the full range 0–10 Gyr to stars with [Fe/H] ~ 0±0.3. This corresponds to the clump of stars with [Fe/O] ~ 0 and [O/H] ~ 0±0.3 in Figure 2. Does this bunch of stars represent a large range of ages? Of kinematic properties? What about the stars on the constant band at [Fe/H] ~ 0.3±0.1 that one could imagine extending from [O/H] ~ −3 all the way up to [O/H] ~ 0? What are their ages and kinematic properties?

Theorists should turn their attention to constructing bursting models to see if the properties of Figure 2 can be explained. Can a strong burst account for the increase in [O/H] by a factor of 3 at [O/H] ~ −1.0? Should there be a postburst decline in [Fe/O] if the high-mass stars produce some Fe but the low-mass stars are temporarily left out of the picture? Can the subsequent rise in [Fe/O] be accounted for quantitatively at the appropriate value of [O/H] ~ −0.2? Will there be a tendency to restore the preburst value of [Fe/O] before the next burst? These are questions suggested by Figure 2 that have not arisen in traditional analyses. If star formation has proceeded by bursts in the disk, then many of the traditional assumptions of nuclear cosmochronology must also be reexamined.

The problem of the chemical evolution of the Galaxy remains a vibrant one, with the promise of major new insights into this complex machine.

Literature Cited

Anders, E., Ebihara, M. 1982. *Geochim. Cosmochim. Acta* 46: 2363–80

Anders, E., Grevesse, N. 1989. *Geochim. Cosmochim. Acta* 53: 197–214

Andersen, J., Edvardsson, B., Gustafsson, B., Nissen, P. E. 1988. In *The Impact of Very High S/N Spectroscopy on Stellar Physics, IAU Symp. No. 132*, ed. G. Cayrel de Strobel, M. Spite, pp. 441–43. Dordrecht: Kluwer

Arnett, W. D. 1971. *Ap. J.* 166: 153–73

Arnett, W. D. 1973. *Annu. Rev. Astron. Astrophys.* 11: 73–94

Arnett, W. D. 1978. *Ap. J.* 219: 1008–16

Arnett, W. D., Bahcall, J. N., Kirshner, R. P., Woosley, S. E. 1989. *Annu. Rev. Astron. Astrophys.* 27: 629–700

Arpigny, C., Magain, P. 1983. *Astron. Astrophys.* 127: L7–9

Audouze, J. 1986. In *Nucleosynthesis and Chemical Evolution*, ed. B. Hauck, A. Maeder, pp. 429–530. Sauverny: Geneva Obs.

Audouze, J., Tinsley, B. M. 1976. *Annu. Rev. Astron. Astrophys.* 14: 43–79

Barbuy, B. 1987. *Astron. Astrophys.* 172: 251–56

Barbuy, B. 1988. *Astron. Astrophys.* 191: 121–27

Barbuy, B., Spite, F., Spite, M. 1987. *Astron. Astrophys.* 178: 199–202

Barkat, Z., Rakavy, G., Sack, N. 1967. *Phys. Rev. Lett.* 18: 379–80

Barry, D. C. 1988. *Ap. J.* 334: 436–48

Barry, D. C., Cromwell, R. H., Hege, E. K. 1987. *Ap. J.* 315: 264–72

Beers, T. C. 1987. In *Nearly Normal Galaxies*, ed. S. M. Faber, pp. 41–44. New York: Springer-Verlag

Beers, T. C., Preston, G. W., Schectman, S. A. 1985. *Astron. J.* 90: 2089–2102

Bell, R. A., Eriksson, K., Gustafsson, B.,

Nordlund, Å. 1976. *Astron. Astrophys. Suppl.* 23: 37–95
Bessell, M. S., Norris, J. 1984. *Ap. J.* 285: 622–36
Bessell, M. S., Norris, J. 1987. *J. Astron. Astrophys.* 8: 99–102
Beynon, T. G. R. 1978a. *Astron. Astrophys.* 64: 145–52
Beynon, T. G. R. 1978b. *Astron. Astrophys.* 64: 299–301
Blackwell, D. E., Willis, R. B. 1977. *MNRAS* 180: 169–76
Boesgaard, A. M. 1989. *Ap. J.* 336: 798–807
Boesgaard, A. M., Steigman, G. 1985. *Annu. Rev. Astron. Astrophys.* 23: 319–78
Bond, H. E. 1980. *Ap. J. Suppl.* 44: 517–33
Bond, H. E. 1981. *Ap. J.* 248: 606–11
Bond, H. E., Luck, R. E. 1988. In *The Impact of Very High S/N Spectroscopy on Stellar Physics, IAU Symp. No. 132*, ed. G. Cayrel de Strobel, M. Spite, pp. 477–84. Dordrecht: Kluwer
Budge, K. G., Boesgaard, A. M., Varsik, J. 1988. In *The Impact of Very High S/N Spectroscopy on Stellar Physics, IAU Symp. No. 132*, ed. G. Cayrel de Strobel, M. Spite, pp. 585–88. Dordrecht: Kluwer
Burbidge, E. M., Burbidge, G. R., Fowler, W. A., Hoyle, F. 1957. *Rev. Mod. Phys.* 29: 547–650
Butcher, H. R. 1987. *Nature* 328: 127–31
Butcher, H. R. 1988. *ESO Messenger* 51: 12–15
Cameron, A. G. W. 1982a. *Astrophys. Space Sci.* 82: 123–31
Cameron, A. G. W. 1982b. In *Essays in Nuclear Astrophysics*, ed. C. A. Barnes, D. D. Clayton, D. N. Schramm, pp. 23–43. Cambridge: Univ. Press
Carbon, D. F., Barbuy, B., Kraft, R. P., Friel, E. D., Suntzeff, N. B. 1987. *Publ. Astron. Soc. Pac.* 99: 335–68
Carbon, D. F., Langer, G. E., Butler, D., Kraft, R. P., Trefzger, C. F., et al. 1982. *Ap. J. Suppl.* 49: 207–58
Cardon, B. L., Smith, P. L., Scalo, J. M., Testermann, L., Whaling, W. 1982. *Ap. J.* 260: 395–412
Carlberg, R. G., Dawson, P. C., Hsu, T., VandenBerg, D. A. 1985. *Ap. J.* 294: 674–81
Carney, B. W., Aquilar, L., Latham, D. W., Laird, J. B. 1988. Preprint
Cayrel, R. 1986. *Astron. Astrophys.* 168: 81–88
Cayrel de Strobel, G., Bentolila, C., Hauck, B., Dequennoy, A. 1984. *Astron. Astrophys. Suppl.* 59: 145–86
Chiosi, C. 1986. In *Nucleosynthesis and Chemical Evolution*, ed. B. Hauck, A. Maeder, pp. 197–428. Sauverny: Geneva Obs.
Chiosi, C., Maeder, A. 1986. *Annu. Rev. Astron. Astrophys.* 24: 329–75
Clayton, D. D. 1987. *Nature* 329: 397–98
Clayton, D. D. 1988. *MNRAS* 234: 1–36
Clayton, D. D., Fowler, W. A., Hull, T. E., Zimmerman, B. A. 1961. *Ann. Phys.* 12: 331–408
Clegg, R. E. S., Lambert, D. L., Tomkin, J. 1981. *Ap. J.* 250: 262–75
Cohen, J. G. 1979. *Ap. J.* 231: 751–61
Conti, P. S., Greenstein, J. L., Spinrad, H., Wallerstein, G., Vardya, M. S. 1967. *Ap. J.* 148: 105–27
Cottrell, P. S., Sneden, C. 1985. *Astron. Astrophys.* 161: 314–26
Cowan, J. J., Cameron, A. G. W., Truran, J. W. 1982. *Ap. J.* 252: 348–55
Cowley, C. R., Downs, P. L. 1980. *Ap. J.* 236: 648–57
Cox, J. P., Guili, R. T. 1968. *Principles of Stellar Structure*. New York: Gordon & Breach
Day, R. W., Lambert, D. L., Sneden, C. 1973. *Ap. J.* 185: 213–28
Delbouille, L., Neven, L., Roland, G. 1973. *Photometric Atlas of the Solar Spectrum from 3800 to 10,000 Å*. Liège: Univ. Liège
Demarque, P., Guenther, D. B., King, C. R., Green, E. M. 1988. Preprint
Dopita, M. A., Smith, G. H. 1986. *Ap. J.* 304: 283–94
Dufour, R. J. 1984. In *Structure and Evolution of the Magellanic Clouds, IAU Symp. No. 108*, ed. S. van den Bergh, K. S. de Boer, pp. 353–61. Dordrecht: Reidel
Endal, A. S., Sofia, S. 1978. *Ap. J.* 220: 279–90
Eriksson, K., Toft, S. C. 1979. *Astron. Astrophys.* 71: 178–97
Fall, S. M., Rees, M. J. 1985. *Ap. J.* 298: 18–26
Fall, S. M., Rees, M. J. 1987. In *The Harlow Shapley Symposium on Globular Cluster Systems in Galaxies, IAU Symp. No. 126*, ed. J. E. Grindlay, A. G. D. Philip, pp. 323–32. Dordrecht: Kluwer
Forte, J. C., Strom, S. E., Strom, K. M. 1981. *Ap. J. Lett.* 245: L9–13
Fowler, W. A. 1987. *Q. J. R. Astron. Soc.* 28: 87–108
François, P. 1986. *Astron. Astrophys.* 160: 264–76
François, P. 1987. *Astron. Astrophys.* 176: 294–98
François, P. 1988. *Astron. Astrophys.* 195: 226–29
François, P., Brocato, E. 1987. *ESO Messenger* 50: 47–48
François, P., Spite, M., Spite, F. 1988. *Astron. Astrophys.* 191: 267–77
Freeman, K. C., Norris, J. 1981. *Annu. Rev. Astron. Astrophys.* 19: 319–56
Garmany, C. D., Conti, P. S. 1985. *Ap. J.* 293: 407–13

Gass, H., Liebert, J., Wehrse, R. 1988. *Astron. Astrophys.* 189: 194–98
Gehren, T. 1988. *Rev. Mod. Astron.* In press
Gilroy, K. K. 1988. PhD thesis. Univ. Tex., Austin
Gilroy, K. K., Sneden, C., Pilachowski, C. A., Cowan, J. J. 1988. *Ap. J.* 327: 298–320
Gratton, R. G. 1987a. *Astron. Astrophys.* 177: 177–82
Gratton, R. G. 1987b. *Astron. Astrophys.* 179: 181–92
Gratton, R. G. 1989. *Astron. Astrophys.* 208: 171–78
Gratton, R. G., Ortolani, S. 1986. *Astron. Astrophys.* 169: 201–7
Gratton, R. G., Sneden, C. 1987. *Astron. Astrophys.* 178: 179–93
Gratton, R. G., Sneden, C. 1988. *Astron. Astrophys.* 204: 193–218
Green, E. M., Demarque, P., King, C. R. 1987. *The Revised Yale Isochrones and Luminosity Functions.* New Haven, Conn: Yale Univ. Obs.
Grevesse, N. 1984. *Phys. Scr.* T8: 49–58
Gustafsson, B. 1987. In *Stellar Evolution and Dynamics of the Outer Halo of the Galaxy*, ed. M. Azzopardi, F. Matteucci, pp. 33–45. Garching: Eur. South. Obs.
Gustafsson, B., Bell, R. A., Eriksson, K., Nordlund, Å. 1975. *Astron. Astrophys.* 42: 407–32
Guzik, J. A., Willson, L. A., Brunish, W. H. 1987. *Ap. J.* 319: 957–65
Harris, W. E. 1983. *Publ. Astron. Soc. Pac.* 95: 21–22
Harris, W. E. 1986. *Astron. J.* 91: 822–41
Harris, W. E., Racine, R. 1979. *Annu. Rev. Astron. Astrophys.* 17: 241–74
Hartkopf, W. I., Yoss, K. M. 1982. *Astron. J.* 87: 1679–1709
Hartmann, K., Gehren, T. 1988. *Astron. Astrophys.* 199: 269–90
Hashimoto, M., Nomoto, K., Shigeyama, T. 1989. Submitted for publication
Hatzes, A. 1987. *Publ. Astron. Soc. Pac.* 99: 369–73
Hesser, J. E., Harris, W. E., VandenBerg, D. A., Allwright, J. W. B., Shott, P., Stetson, P. B. 1987. *Publ. Astron. Soc. Pac.* 99: 739–808
Hillebrandt, W. 1978. *Space Sci. Rev.* 21: 639–702
Hillebrandt, W., Thielemann, F.-K. 1977. *Mitt. Astron. Ges.* 43: 234–36
Holweger, H., Müller, E. A. 1974. *Sol. Phys.* 39: 19–30
Iben, I. Jr., Renzini, A. 1983. *Annu. Rev. Astron. Astrophys.* 21: 271–342
Iben, I. Jr., Renzini, A. 1984. *Phys. Rep.* 105: 329–406
Iben, I. Jr., Truran, J. W. 1978. *Ap. J.* 220: 980–95
Iben, I. Jr., Tutukov, A. V. 1984. *Ap. J. Suppl.* 54: 335–72
Janes, K. A. 1988. In *Calibration of Stellar Ages. Van Vleck Obs. Workshop.* In press
Janes, K. A., Tilley, C., Lyngå, G. 1988. *Astron. J.* 95: 771–84
King, C. R., Demarque, P., Green, E. M. 1988. Preprint
Kraft, R. P. 1979. *Annu. Rev. Astron. Astrophys.* 17: 309–43
Kraft, R. P. 1985. In *Production and Distribution of C, N, O Elements*, ed. I. J. Danziger, F. Matteucci, K. Kjär, pp. 21–47. Garching: Eur. South. Obs.
Kraft, R. P., Suntzeff, N. B., Langer, G. E., Carbon, D. F., Trefzger, C. F., et al. 1982. *Publ. Astron. Soc. Pac.* 94: 55–66
Kudritzki, R. P., Groth, H. G., Butler, K., Husfeld, D., Becker, S., et al. 1987. In *SN 1987A*, ed. I. J. Danziger, pp. 39–51. Garching: Eur. South. Obs.
Kurucz, R. L. 1970. *Smithson. Astrophys. Obs. Rep. No. 309*
Kurucz, R. L. 1979. *Ap. J. Suppl.* 40: 1–340
Kurucz, R. L., Furenlid, I., Brault, J., Testermann, L. 1984. *Solar Flux Atlas From 296 to 1300 nm.* Cambridge, Mass: Harvard Univ.
Laird, J. B. 1985. *Ap. J.* 289: 556–69
Laird, J. B., Rupen, M. P., Carney, B. W., Latham, D. W. 1988. *Astron. J.* 96: 1908–17
Lamb, S. A., Howard, W. M., Truran, J. W., Iben, I. Jr. 1977. *Ap. J.* 217: 213–21
Lambert, D. L. 1978. *MNRAS* 182: 249–72
Lambert, D. L. 1987. *J. Astrophys. Astron.* 8: 103–22
Lambert, D. L. 1988. Preprint
Lambert, D. L., Dearborn, D. S. 1972. *Mem. Soc. R. Sci. Liège, Collect. 8°* (6th ser.) 3: 147–65
Lambert, D. L., Ries, L. M. 1981. *Ap. J.* 248: 228–48
Lambert, D. L., Sneden, C., Ries, L. M. 1974. *Ap. J.* 188: 97–100
Langer, G. E., Kraft, R. P. 1984. *Publ. Astron. Soc. Pac.* 96: 339–48
Larson, R. B. 1976. *MNRAS* 176: 31–52
Larson, R. B. 1986. *MNRAS* 218: 409–28
Lee, T., Schramm, D. N., Wefel, J. P., Blake, J. B. 1979. *Ap. J.* 232: 854–62
Leep, E. M., Wallerstein, G., Oke, J. B. 1986. *Astron. J.* 91: 1117–20
Liebert, J., Dahn, C. C., Monet, D. G. 1988. *Ap. J.* 332: 891–909
Luck, R. E., Bond, H. E. 1983. *Ap. J. Lett.* 271: L75–78
Luck, R. E., Bond, H. E. 1985a. *Ap. J.* 292: 559–77
Luck, R. E., Bond, H. E. 1985b. *Ap. J. Suppl.* 59: 249–76
Maeder, A. 1987. *Astron. Astrophys.* 178: 159–69

Maeder, A., Meynet, G. 1987. *Astron. Astrophys.* 182: 243–63
Mäckle, R., Holweger, H., Griffin, R., Griffin, R. 1975. *Astron. Astrophys.* 38: 239–57
Magain, P. 1985. *Astron. Astrophys.* 146: 95–112
Magain, P. 1987. *Astron. Astrophys.* 179: 176–80
Magain, P. 1989. *Astron. Astrophys.* 209: 211–25
Malaney, R. A. 1987. *Ap. J.* 321: 832–45
Matteucci, F. 1986. *Ap. J. Lett.* 305: L81–84
McClure, R. D., VandenBerg, D. A., Bell, R. A., Hesser, J. E., Stetson, P. B. 1987. *Astron. J.* 93: 1144–65
McWilliam, A., Lambert, D. L. 1988. *MNRAS* 230: 573–85
Merrill, P. 1952. *Science* 115: 484
Mould, J. 1986. In *Stellar Populations*, ed. C. A. Norman, A. Renzini, M. Tosi, pp. 9–27. Cambridge: Univ. Press
Nissen, P. E. 1988. *Astron. Astrophys.* 199: 146–60
Nissen, P. E., Edvardsson, B., Gustafsson, B. 1985. In *Production and Distribution of C, N, O Elements*, ed. I. J. Danziger, F. Matteucci, K. Kjär, pp. 131–49. Garching: Eur. South. Obs.
Nomoto, K. 1986. *Ann. NY Acad. Sci.* 470: 294–319
Nomoto, K., Thielemann, F.-K., Wheeler, J. C. 1984. *Ap. J. Lett.* 279: L23–26
Norris, J., Cottrell, P. L., Freeman, K. C., DaCosta, G. S. 1981. *Ap. J.* 244: 205–20
Norris, J., Green, E. M. 1989. *Ap. J.* 337: 272–92
Norris, J., Ryan, S. G. 1989. *Ap. J. Lett.* 336: L17–19
Norris, J., Pilachowski, C. A. 1985. *Ap. J.* 299: 295–302
Olive, K. A., Thielemann, F.-K., Truran, J. W. 1987. *Ap. J.* 313: 813–19
Pagel, B. E. J. 1965. *R. Obs. Bull. No. 104*
Pagel, B. E. J. 1988. In *Evolutionary Phenomena in Galaxies*, ed. J. Beckman, B. E. J. Pagel. Cambridge: Univ. Press
Pagel, B. E. J., Edmunds, M. G. 1981. *Annu. Rev. Astron. Astrophys.* 19: 77–113
Peterson, R. C. 1978. *Ap. J.* 222: 595–99
Peterson, R. C. 1980. *Ap. J. Lett.* 237: L87–91
Peterson, R. C., Sneden, C. 1978. *Ap. J.* 225: 913–18
Pilachowski, C. A. 1984. *Ap. J.* 281: 614–23
Pilachowski, C. A. 1988. *Ap. J. Lett.* 326: L57–60
Pilachowski, C. A., Sneden, C., Green, E. M. 1984. *Publ. Astron. Soc. Pac.* 96: 932–43
Pilachowski, C. A., Sneden, C., Wallerstein, G. 1983. *Ap. J. Suppl.* 52: 241–87
Prantzos, N., Arnould, M., Arcoragi, J.-P. 1987. *Ap. J.* 315: 209–28
Rebolo, R., Molaro, P., Beckman, J. E. 1988. *Astron. Astrophys.* 192: 192–205
Renzini, A., Voli, M. 1981. *Astron. Astrophys.* 94: 175–93
Rodgers, A. W., Paltoglou, G. 1984. *Ap. J. Lett.* 283: L5–7
Russell, S. C., Bessell, M. S. 1988. Preprint
Russell, S. C., Bessell, M. S., Dopita, M. A. 1988a. In *The Impact of Very High S/N Spectroscopy on Stellar Physics, IAU Symp. No. 132*, ed. G. Cayrel de Strobel, M. Spite, pp. 545–50. Dordrecht: Kluwer
Russell, S. C., Bessell, M. S., Dopita, M. A. 1988b. Preprint
Saha, A. 1985. *Ap. J.* 289: 310–19
Sandage, A. R. 1988. In *Calibration of Stellar Ages. Van Vleck Obs. Workshop*. In press
Scalo, J. M. 1986. *Fundam. Cosmic Phys.* 11: 1–278
Scalo, J. M. 1987a. In *Starbursts and Galaxy Evolution*, ed. T. Montmerle, pp. 445–65. Paris: Ed. Front.
Scalo, J. M. 1987b. In *Galaxy Evolution, Proc. Eur. Astron. Meet. IAU, 10th*, ed. J. Palouš, 4: 101–9. Ondřejov: Czech. Acad. Sci.
Schönberner, D., Herrero, A., Becker, S., Eber, F., Butler, K., et al. 1988. *Astron. Astrophys.* 197: 209–22
Schramm, D. N. 1974. *Annu. Rev. Astron. Astrophys.* 12: 383–406
Schwarzschild, M., Härm, R. 1967. *Ap. J.* 150: 961–70
Searle, L., Zinn, R. 1978. *Ap. J.* 225: 357–79
Sedlmayr, E. 1974. *Astron. Astrophys.* 31: 23–35
Smith, G. H. 1987. *Publ. Astron. Soc. Pac.* 99: 67–90
Smith, H. A. 1980. *Astron. J.* 85: 848–52
Smith, V. V., Suntzeff, N. 1989. *Astron. J.* In press
Sneden, C. 1974. *Ap. J.* 189: 493–507
Sneden, C. 1985. In *Production and Distribution of C, N, O Elements*, ed. I. J. Danziger, F. Matteucci, K. Kjär, pp. 1–19. Garching: Eur. South. Obs.
Sneden, C., Crocker, D. A. 1988. *Ap. J.* 335: 406–14
Sneden, C., Lambert, D. L., Whitaker, R. W. 1979. *Ap. J.* 234: 964–72
Sneden, C., Parthasarathy, M. 1983. *Ap. J.* 267: 757–78
Sneden, C., Pilachowski, C. A. 1985. *Ap. J. Lett.* 288: L55–58
Sneden, C., Pilachowski, C. A., Gilroy, K. K., Cowan, J. J. 1988. In *The Impact of Very High S/N Spectroscopy on Stellar Physics, IAU Symp. No. 132*, ed. G. Cayrel de Strobel, M. Spite, pp. 501–6. Dordrecht: Kluwer
Sneden, C., Pilachowski, C. A., VandenBerg, D. A. 1986. *Ap. J.* 311: 826–42

Soderblom, D. R. 1988. In *The Impact of Very High S/N Spectroscopy on Stellar Physics, IAU Symp. No. 132*, ed. G. Cayrel de Strobel, M. Spite, pp. 381–86. Dordrecht: Kluwer
Spite, F., Spite, M. 1986. *Astron. Astrophys.* 163: 140–44
Spite, F., Spite, M. 1987. *J. Astrophys. Astron.* 31: 23–35
Spite, M., Cayrel, R., François, P., Richter, T., Spite, F. 1986. *Astron. Astrophys.* 168: 197–203
Spite, M., Spite, F. 1978. *Astron. Astrophys.* 67: 23–31
Spite, M., Spite, F. 1985. *Annu. Rev. Astron. Astrophys.* 23: 225–38
Suntzeff, N. B., Kraft, R. P., Kinman, T. D. 1988. *Astron. J.* 95: 91–105
Sweigart, A. V., Mengel, J. G. 1979. *Ap. J.* 229: 624–41
Thielemann, F.-K., Arnett, W. D. 1985. *Ap. J.* 295: 604–19
Thielemann, F.-K., Nomoto, K., Yokoi, K. 1986. *Astron. Astrophys.* 158: 17–33
Tinsley, B. 1980. *Fundam. Cosmic Phys.* 5: 287–388
Tomkin, J., Lambert, D. L. 1980. *Ap. J.* 235: 925–38
Tomkin, J., Lambert, D. L. 1984. *Ap. J.* 279: 220–24
Tomkin, J., Lambert, D. L., Balachandran, S. 1985. *Ap. J.* 290: 289–95
Tomkin, J., Sneden, C., Lambert, D. L. 1986. *Ap. J.* 302: 415–20
Trimble, V. 1975. *Rev. Mod. Phys.* 47: 877–976
Trimble, V. 1982. *Rev. Mod. Phys.* 54: 1183–1224
Trimble, V. 1983. *Rev. Mod. Phys.* 55: 511–63
Trimble, V., Bell, R. A. 1981. *Q. J. R. Astron. Soc.* 22: 361–79
Truran, J. W. 1973. *Space Sci. Rev.* 15: 23–49
Truran, J. W. 1981. *Astron. Astrophys.* 97: 391–93
Truran, J. W. 1983. *Mem. Soc. Astron. Ital.* 54: 113–22
Truran, J. W. 1984. *Annu. Rev. Nucl. Part. Sci.* 34: 53–97
Truran, J. W. 1987. In *Relativistic Astrophysics*, ed. M. P. Ulmer, pp. 430–36. Singapore: World Sci.
Truran, J. W. 1988. In *The Impact of Very High S/N Spectroscopy on Stellar Physics, IAU Symp. No. 132*, ed. G. Cayrel de Strobel, M. Spite, pp. 577–83. Dordrecht: Kluwer
Truran, J. W., Arnett, W. D. 1971. *Astrophys. Space Sci.* 11: 430–42
Truran, J. W., Brown, J. H. 1988. Preprint
Truran, J. W., Cameron, A. G. W. 1971. *Astrophys. Space Sci.* 14: 179–222
Truran, J. W., Cowan, J. J., Cameron, A. G. W. 1978. *Ap. J. Lett.* 222: L63–67
Twarog, B. A. 1980. *Ap. J.* 242: 242–59
Twarog, B. A., Wheeler, J. C. 1982. *Ap. J.* 261: 636–48
Twarog, B. A., Wheeler, J. C. 1987. *Ap. J.* 316: 153–61
VandenBerg, D. A. 1983. *Ap. J. Suppl.* 51: 29–66
VandenBerg, D. A. 1985. *Ap. J. Suppl.* 58: 711–69
VandenBerg, D. A. 1988. Preprint
VandenBerg, D. A., Bell, R. A. 1985. *Ap. J. Suppl.* 58: 561–621
Vasquez, E. C., Scalo, J. M. 1989. *Ap. J.* In press
Wallerstein, G. 1962. *Ap. J. Suppl.* 6: 407–43
Wallerstein, G., Greenstein, J. L., Parker, R., Helfer, H. L., Aller, L. H. 1963. *Ap. J.* 137: 280–300
Wasserburg, G. J., Papanastassiou, D. A., Tera, F., Huneke, J. C. 1977. *Philos. Trans. R. Soc. London Ser. A* 285: 7–22
Weaver, T. A., Zimmerman, G. B., Woosley, S. E. 1978. *Ap. J.* 225: 1021–29
Webbink, R. F. 1984. *Ap. J.* 277: 355–60
Wheeler, J. C. 1981. *Rep. Prog. Phys.* 44: 85–138
Willson, L. A., Bowen, G., Struck-Marcell, C. 1987. *Comments Astrophys.* 12: 17–34
Winget, D. E., Hansen, C. J., Liebert, J., Van Horn, H. M., Fontaine, G., et al. 1987. *Ap. J. Lett.* 315: L77–81
Woosley, S. E. 1986. In *Nucleosynthesis and Chemical Evolution*, ed. B. Hauck, A. Maeder, pp. 1–195. Sauverny: Geneva Obs.
Woosley, S. E., Arnett, W. D., Clayton, D. D. 1973. *Ap. J. Suppl.* 26: 231–312
Woosley, S. E., Weaver, T. A. 1980. *Ap. J.* 238: 1017–25
Woosley, S. E., Weaver, T. A. 1986a. In *Radiation Hydrodynamics in Stars and Compact Objects, IAU Colloq. No. 89*, ed. D. Mihalas, K.-H. A. Winkler, pp. 91–120. Berlin: Springer-Verlag
Woosley, S. E., Weaver, T. A. 1986b. *Annu. Rev. Astron. Astrophys.* 24: 205–53
Yoss, K. M., Neese, C. L., Hartkopf, W. I. 1987. *Astron. J.* 94: 1600–15
Zinn, R. 1985. *Ap. J.* 293: 424–44

T TAURI STARS: WILD AS DUST

Claude Bertout

Institut d'Astrophysique de Paris and Laboratoire d'Astrophysique Théorique du Collège de France, 98bis, Boulevard Arago, 75014 Paris, France

> *I know the stars*
> *are wild as dust*
> *and wait for no man's discipline*
> *but as they wheel*
> *from sky to sky they rake*
> *our lives with pins of light*
>
> Leonard Cohen: "Another Night With Telescope"

1. BACKGROUND

T Tauri stars are newly formed low-mass stars that have recently become visible in the optical range, but Alfred H. Joy could hardly have suspected their evolutionary status when he first discovered them in the Taurus-Auriga dark cloud and named the class after its brightest member, T Tauri. He found there a number of faint stars with late spectral types and "emission lines resembling the solar chromosphere." They displayed irregular and large light variations, and were apparently associated with dark or bright nebulae (cf. Joy 1942, 1945, 1949). That T Tauri stars are seen projected onto nebulosities hinted at their young age, but early interpretations instead assumed that they were field stars passing through the nebula and interacting with it. This first false track was by no means the last, so now that we once again have a promising framework for understanding several aspects of these exotic objects, we should keep in mind that the elusiveness of T Tauri stars, while it attracts and challenges, also teaches humility.

Ambartsumian (1947) was first to make some sense out of T Tauri stars.

After finding that they occur in groups that he named T-associations, and that these groups are often found in connection with groups of OB stars (O-associations), he postulated that T Tauri stars were the low-mass counterpart of the short-lived (i.e. recently formed) OB stars. Other early arguments for their youth are summarized in Herbig's (1962) classic review. While it took several years before the pre-main-sequence evolutionary status of T Tauri stars gained wide acceptance in the Western Hemisphere, all subsequent studies have confirmed its validity. Herbig (1977a), for example, found that the radial velocities of 50 T Tauri stars in the Taurus-Auriga stellar formation region are consistent with those of apparently associated molecular clouds, and he thereby demonstrated a kinetic association between the young stars and the dark clouds from which they were presumably born. This result was confirmed by the proper-motion study of Jones & Herbig (1979).

Early studies of the Orion population, which are all stellar objects associated with nebulosity, used both photometric and spectroscopic techniques. Based on their photometric behaviors, several classes of nebular variables were defined within the Orion population (T Tauri variables, RW Aurigae variables, and T Ori variables; cf. Glasby 1974). Early objective-prism surveys of dark clouds and young associations uncovered a large number of stars with prominent Hα emission [i.e. $W_\lambda(H\alpha) \gtrsim 5$ Å], and follow-up observations with slit spectrographs were then conducted to classify them. The spectroscopically defined class of T Tauri stars (see below) includes nebular variables belonging to each of the above categories. The new comprehensive catalog of emission-line stars of the Orion population by Herbig & Bell (1988) lists over 700 pre-main-sequence objects in various star-forming regions, and many of them are T Tauri stars.

While the optical spectroscopic criteria defining T Tauri stars correspond to specific physical conditions common to these objects, the visible range is only one of many spectral domains available to modern astronomy. Not only have other common properties of T Tauri stars become evident in all windows of the electromagnetic spectrum that have opened during the last 20 years, but, more interestingly, the T Tauri phenomenon now appears as one aspect of stellar formation that can no longer be treated independently of others aspects. Consequently, the term T Tauri star has become equivalent over the years to pre-main-sequence low-mass *optical* object; and the concept of *young stellar object*, coined by S. Strom (1972), is now widely used to designate all pre-main-sequence stellar objects, embedded or visible.

The stellar population in the core of the ρ Ophiuchi dark cloud provides an illustration. Only a dozen or so optically visible T Tauri stars are seen at the periphery of the cloud, where extinction is lowest. But observations in the near and mid-infrared reveal more than 70 embedded sources (Vrba et al. 1975, Wilking & Lada 1983, Wilking et al. 1989), many of them with bolometric luminosities in the 0.1–25 $L_\odot$ range. While most mid-infrared sources are concentrated toward the center of the molecular cloud, observations using the *Einstein* X-ray Observatory uncovered a population of pre-main-sequence stars with little activity (besides their X-ray emission) located predominantly at the outskirts of the cloud (Montmerle et al. 1983); and a recent radio continuum survey using the Very Large Array (VLA) detected yet another, albeit smaller, class of objects within the cloud boundaries (André et al. 1987, Stine et al. 1988). Figure 1 displays the spatial relationships between all these objects, which are predominantly low-luminosity pre-main-sequence objects. Although a few sources appear at more than one wavelength, it is remarkable that most stars stand out in only one spectral domain.

Because of the difference in their concentrations toward the cloud's center, the various stellar populations seen in these surveys are believed to represent several evolutionary stages of low-mass stars. Lada & Wilking (1984) and Lada (1986) divide the ρ Ophiuchi infrared sources into three distinct morphological classes based on the shape of their spectral energy distributions in the log λF_λ vs. log λ diagram, and Adams et al. (1987) suggest that this classification corresponds to an evolutionary sequence, from protostars to pre-main-sequence stars (cf. Section 3). In this scheme, *Class I* sources, with flat or rising spectrum toward long wavelengths, are deeply embedded sources (and hence presumably very young) seen only in the infrared range. *Class II* objects have energy distributions typical of T Tauri stars. Since a number of embedded sources show these energy distributions, the small number of optical T Tauri stars detected in spectroscopic surveys probably reflects the molecular cloud's steep extinction gradient. *Class III* sources display reddened blackbodylike energy distributions; they are often found at the cloud periphery and correspond presumably to more evolved (or less active) stars in the cloud. Objects detected primarily in the X-ray range often belong to Class III, and their X-ray emission is presumably related to solar-type magnetic activity (see Section 2). The discovery of radio-emitting objects that apparently are deeply embedded in the cloud but that lack the strong infrared excess of protostellar sources is surprising. While the nature of these sources is unclear, they may provide an initial hint that age is not the only parameter governing the spectral shapes of young objects (André et al. 1987).

T Tauri stars represent a pivotal class between deeply embedded low-luminosity sources, which can be studied only at infrared and radio wavelengths, and solar-type main sequence stars. As such, they display a be-

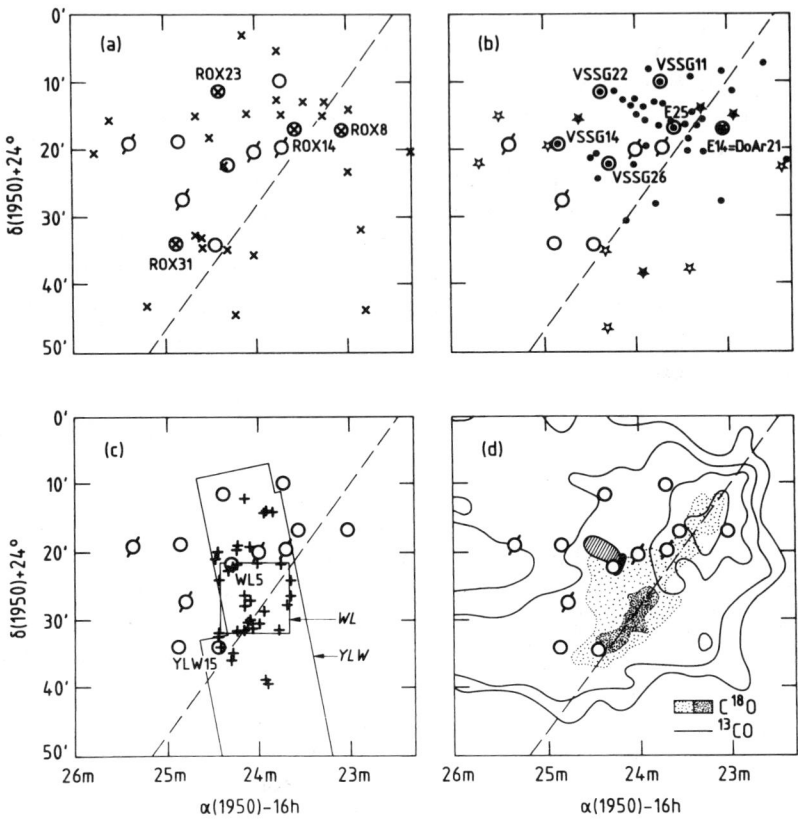

Figure 1 The various populations of young stars seen at different wavelengths in the central part of the ρ Ophiuchi star formation region (from André et al. 1987). Open circles denote the radio sources [their counterparts are named according to the relevant survey (see text for references)], and the dashed line shows the orientation of the high-density gas. (*a*) Crosses are definite X-ray sources. (*b*) Dark circles are bright infrared sources, and star symbols are optical young stellar objects of various masses. (*c*) Plus symbols are faint embedded infrared sources. Two different survey regions (denoted WL and YLW) are delineated. (*d*) Contours show ^{13}CO emission, and increasingly shaded areas represent high density $C^{18}O$. The hatched and black areas are dense regions seen, respectively, in DCO^+ and H_2CO.

wildering variety of phenomena characteristic of earlier as well as of later evolutionary phases. For example, some of them are associated with jets and Herbig-Haro objects, both of which are typical protostellar outflow manifestations (reviewed by Lada 1985, Mundt 1988), whereas others display the large spots commonly found on late-type active dwarfs. Others still (like the "prototype" T Tauri, which now appears to be a complex multiple system) possess most attributes of both embedded young stellar objects and late-type active stars.

This paper provides an overview of low-mass optical young stellar objects, whether they are *classical* T Tauri stars (CTTSs) that match Herbig's (1962) definition and that were discovered from Hα surveys, or whether they are optical low-luminosity pre-main-sequence stars first detected at some other wavelength, e.g. in X-rays (Walter & Kuhi 1981, Feigelson & Kriss 1981, Walter 1986, Feigelson et al. 1987) or in Ca II (Herbig et al. 1986). The second category includes mostly weak emission-line pre-main-sequence stars $[W_\lambda(H\alpha) \lesssim 5 \text{ Å}]$,[1] which, following Herbig & Bell (1988), I call *weak-line* T Tauri stars (WTTSs). I shall refer globally to all these objects as T Tauri stars, but I shall find it useful at times to distinguish between the two subgroups, since each of them displays specific average properties.

Walter et al. (1988) recently presented the first extensive study of WTTSs[2] in the Taurus-Auriga region. They define a WTTS as an X-ray source with an optical counterpart showing pre-main-sequence characteristics. Specifically, Li I $\lambda 6707$ absorption is present with equivalent width larger than 100 mÅ, and stellar radial velocity is consistent with membership in the associated molecular cloud.

Optical spectroscopic criteria that define a CTTS, according to Herbig (1962), are the following: (*a*) Hydrogen Balmer lines and Ca II H and K lines are in emission. (*b*) Anomalous emission of Fe I $\lambda\lambda 4063, 4132$ is often observed. Fluorescent emission in these lines is probably excited either by Ca II H or Hε (Willson 1974, 1975). (*c*) Forbidden emission of O I and S II is observed in many CTTSs. (*d*) Li I $\lambda 6707$ absorption is conspicuously strong.

After the hydrogen and Ca II lines, the strongest emission lines are usually caused by Fe II, Ti II, and He I. The emission-line spectrum is superimposed on a continuous spectrum that may range from a pure continuum (in *extreme* T Tauri stars) through a late-type absorption spectrum with anomalous line strengths (in *veiled* T Tauri stars) to an almost normal absorption spectrum of type F through M in *moderate* T Tauri stars. Figure 2 illustrates the various subgroups of T Tauri stars. There, medium-resolution spectrograms covering the spectral range 3200–

---

[1] This somewhat arbitrary value was used in early objective prism surveys to distinguish T Tauri star candidates [with $W_\lambda(H\alpha) \gtrsim 5$ Å] from dMe field stars. A selection criterion based on the Hα flux would be more meaningful physically.

[2] Because the above definitions of T Tauri stars are phenomenological, the term WTTS is preferred here to the term *"naked" T Tauri star* used by Walter et al. (1988) to designate the same objects.

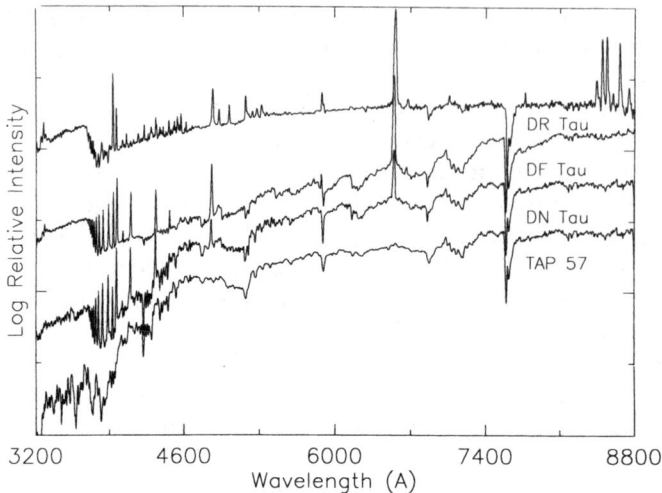

Figure 2 Medium-resolution spectrograms covering the spectral range 3200–8800 Å of four late-K or early-M T Tauri stars, shown in order of increasing emission levels. The relative intensity is displayed in wavelength units.

8800 Å (kindly communicated by G. Basri in advance of publication) are presented for four late-K or early-M stars. TAP 57 (045251+3016) is a WTTS similar in many respects to a standard K7 dwarf (Feigelson et al. 1987, Walter et al. 1988), DN Tau a moderate M0 CTTS, DF Tau a veiled M0 CTTS, and DR Tau an extreme CTTS with probable K5 spectral type. Note the strong Balmer emission present in the spectra of CTTSs. Cohen & Kuhi (1979) and Appenzeller et al. (1983) display low-resolution red spectrograms of many T Tauri stars in their largely spectroscopic catalogs.

Bastian et al. (1983) found Herbig's definition inadequate to define the CTTS in the SIMBAD data base of the Centre de Données de Strasbourg because not all the above emission lines are found in all stars—in particular, [S II] and Fe I lines are visible only in stars with strong emission spectra—and because intrinsic variability implies that not all of the criteria defined above need be fulfilled all the time. They propose instead to use as primary criteria (*a*) association with a region of obscuration, (*b*) presence of Hα and Ca II H and K emission with $W_\lambda(H\alpha) > 5$ Å, and (*c*) spectral type later than F, which altogether are the de facto criteria used in searches for faint T Tauri stars (Herbig 1962).

High-resolution profiles of emission and absorption lines provide evidence for complex gaseous flows in T Tauri star envelopes. Several studies of Hα show a variety of line profiles (Kuhi 1964, 1978, Ulrich & Knapp 1979, Schneeberger et al. 1979, Hartmann 1982, Mundt & Giampapa 1982,

Mundt 1984). According to Kuhi (1978), a majority of T Tauri stars display Type III P Cygni profiles [in Beals' (1950) classification] at Hα; in these profiles, the emission is broad (FWHM ≈ 200 km s^{-1}) and symmetric, and a blueshifted absorption feature at typically 80 km s^{-1} (Herbig 1977a) is superimposed on the emission. While symmetric emission profiles are also common, only a few CTTSs display Type I P Cygni profiles, with blueshifted absorption components reaching below the photospheric continuum. Type I profiles unambiguously indicate that mass loss is taking place, unlike Type III profiles (Ulrich 1976, Wagenblast et al. 1982). Inverse P Cygni profiles indicative of mass accretion are not observed at Hα (Herbig 1977a), although a subclass of T Tauri stars named YY Orionis stars after their prototype by Walker (1972) displays such profiles at the higher members of the Balmer series. The spectroscopic properties of a few bright YY Orionis stars were studied in some detail by Appenzeller & Wolf (1976), Wolf et al. (1977), Edwards (1979), Bertout et al. (1982), and Appenzeller et al. (1986). In some extreme CTTSs (most notably DR Tau), Balmer-line profiles can change from P Cygni to inverse P Cygni in a matter of days, and both P Cygni and inverse P Cygni profiles can occasionally be seen in the same spectrogram (Krautter & Bastian 1980). Ca II and Na D lines of active CTTSs and YY Orionis stars can also exhibit both blue- and redshifted absorptions, and sharp blueshifted absorption components with velocities in the range 50–100 km s^{-1} are observed in a number of stars (Mundt 1984). All these spectroscopic data indicate that complex mass motions are taking place in T Tauri stars' atmospheres, and prominent mass ejection appears as a major characteristic of the CTTS class. The low luminosities and moderate rotational velocities of T Tauri stars put severe efficiency constraints on possible mass-loss mechanisms (Section 3).

FU Orionis objects make up another related class of exotic pre-main-sequence low-mass stars characterized by a strong brightening ($\gtrsim 5$ mag) on a time scale of several months, followed by a slow decay on time scales of years to decades. One such well-studied object is V1057 Cyg, for which a low-resolution preoutburst spectrogram displays T Tauri characteristics (Herbig 1977b). The light-curve decay is accompanied by a spectral type change from F–G after the outburst to late G. Although statistics based on only a few such outbursts are not too reliable, they seem to imply that FU Orionis eruptions may be frequent events during the T Tauri phase of evolution (Herbig 1977b). Whether there exists a continuum of FU Orionis–like events of various strengths in T Tauri stars is also an open question; the recent brightenings of DR Tau (cf. Chavarria 1979) and RY Tau (Herbst 1986) are cases in point (see also Parsamyan & Gasparyan 1987).

When compared with standard stars, the spectral energy distributions of T Tauri stars display a wide range of both ultraviolet excesses (e.g. Kuhi 1974, Herbig & Goodrich 1986, Bertout et al. 1988) and infrared excesses (Mendoza 1966, 1968). Recently published data bases of optical and infrared photometry of T Tauri stars include those of Rydgren et al. (1984b), Bouvier et al. (1988), Strom et al. (1989), and Cohen et al. (1989). Figure 3 displays the observed spectral energy distributions from 3600 Å to 100 μm of the stars whose spectra are shown in Figure 2. The spectral energy distribution of the WTTS TAP 57, which is basically that of a normal

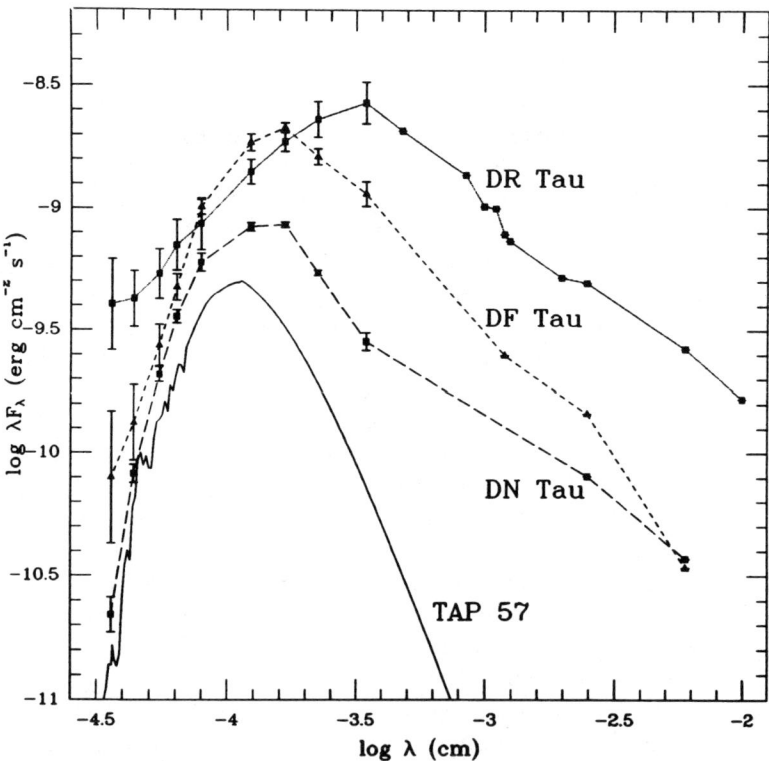

Figure 3 Observed spectral energy distributions from 3600 Å to 100 μm of the stars whose spectra are shown in Figure 2. The energy distribution of the K7V WTTS TAP 57, shown as a solid line, has been displaced downward by 0.3 dex. The filled symbols are simultaneous (for DN Tau and DF Tau) or averaged (for DR Tau) photometric data (cf. Bertout et al. 1988) supplemented by *IRAS* data (Rucinski 1985). When available, observed variability is indicated by error bars. When compared with WTTSs such as TAP 57, CTTSs display prominent ultraviolet and infrared excesses. Excess continuum flux and optical emission-line activity are often correlated.

K7V star, was displaced downward by 0.3 dex in order to facilitate the comparison with the CTTSs, shown at their actual flux level. Visual extinction values toward these objects are approximately 0 for TAP 57, 0.4 mag for DN Tau, and 1.3 mag for DF Tau. Extinction toward DR Tau, an extreme CTTS, is not reliably known but is probably between 1 to 2 mag (Basri & Bertout 1989). Note that the excess luminosity of these objects increases with activity, as measured by optical-line emission, even when the energy distributions are not corrected for extinction.

This brief overview of basic observational properties provides a sufficient background to begin a discussion of inferred T Tauri properties, but more observational data are later introduced when needed. Observations of CTTSs were reviewed by both Bertout (1984) and Cohen (1984), and the reader is referred to these two (often complementary) papers for extensive references prior to 1983.

This review ignores the implications of observations of T Tauri stars for such questions as their importance for the initial mass function or their influence on the surrounding molecular clouds, both of which were addressed recently by Shu et al. (1987). It focuses instead on the stars themselves, their surroundings, and their common evolution toward the main sequence. Aspects of current interest that are discussed in some detail are the photospheric properties of T Tauri stars and their solar-type outer atmospheres (Section 2), recent evidence for circumstellar (possibly protoplanetary) disks around CTTSs, and the mass outflows of CTTSs (Section 3). While this outline may seem to imply both that there *are* such (almost independent) stellar and circumstellar layers and that we are able to distinguish between them, neither is really the case. The adopted presentation is merely convenient when comparing T Tauri stars with other late-type stars and with embedded objects. And even though a framework has emerged over the last few years that now offers more hope of eventually understanding a much wi(l)der range of T Tauri properties than earlier models have been capable of, allowance is of course made for deviance in both behavior and interpretation throughout this review.

2. PHOTOSPHERES AND SOLAR-TYPE OUTER ATMOSPHERES

It is now well established that a solar-type outer atmosphere (i.e. chromosphere, transition region, and corona; cf. Linsky 1981) is formed around main-sequence late-type stars by dissipation of acoustic and magnetic energy above the photosphere, even if details of heating mechanisms are still debated. While there is a wide consensus about the presence of a chromosphere and related regions around T Tauri stars as well, there still is considerable disagreement about the degree of solar-type activity[3] in these stars.

The *International Ultraviolet Explorer (IUE)* and *Einstein* X-ray Observatory have provided us with a wealth of information about the transition regions and hot coronae of T Tauri stars over the last decade. Properties of pre-main-sequence stars in the ultraviolet range were reviewed by Imhoff & Appenzeller (1987). Besides the prominent Mg II h and k lines, emission spectra of CTTSs at *IUE* wavelengths typically show lines of Fe II, He II, C I–IV, Si II–IV, and N II–V (e.g. Appenzeller et al. 1980, Imhoff & Giampapa 1980). The ultraviolet continuum of CTTSs appears as the extension toward short wavelengths of the Balmer continuum emission observed in the optical range (Basri & Bertout 1989). The ultraviolet line fluxes are 10^4–10^6 times stronger than in the Sun, which indicates a correspondingly larger emission measure, apparently because the volume of the emitting regions is larger (Cram et al. 1980). The variation of emission measure with electron temperature was derived by Jordan et al. (1982) for T Tauri. It shows a fast decline from 10^4 K to about 3×10^4 K, followed by a slower decline up to 3×10^6 K (X-ray domain). While WTTSs are quite inconspicuous at *IUE* wavelengths, typical X-ray luminosities of both CTTSs and WTTSs span the range 10^{-4}–10^{-3} L_{bol}, and strong variability is also observed in that range. (T Tauri star properties in X-rays are discussed in more detail below.)

Spectral Peculiarities

T Tauri stars have late-type photospheres that range from F to M. Cohen & Kuhi (1979) demonstrated that a typical CTTS is a K7 star with moderate emission characteristics, which contrasts with extreme CTTSs as defined above. Among several peculiarities in a typical CTTS absorption spectrum, spectral classification is somewhat wavelength dependent, and absorption lines often appear shallower than in comparable standards. The origin of this spectral "veiling" is the subject of much current work.

Although systematic work has yet to be reported on this topic, it is known among workers in the field that the spectral type of a given star depends somewhat upon the spectral range used for its classification (e.g. Herbig & Bell 1988). More specifically, the spectral type can differ by as much as about three subclasses, in the sense that the type determined from the blue region is earlier than that found from the red region. This spectral ambiguity, which seems to occur even in stars with low levels of activity

---

[3] *Solar-type activity* refers here to manifestations of stellar activity commonly observed in late-type dwarfs (e.g. chromospheric and coronal emission, stellar spots) and driven by nonradiative mechanisms presumably related to magnetic dynamo processes.

(Bouvier & Appenzeller 1989), is one manifestation of the so-called veiling of spectral lines first noted by Joy. The term veiling is somewhat confusing because it refers in fact to two distinct phenomena that cannot easily be disentangled: (a) selective filling-in of spectral *lines*, and (b) overlying *continuous* emission. The net effect of either process is to decrease the absorption-line equivalent widths with respect to a standard star. Only a few analyses of observed veiling have been reported so far (Strom 1983, Finkenzeller & Basri 1986), mainly because high-resolution data necessary to study it in detail became available only very recently. While interpretations of veiling are discussed below, a first conclusion to be drawn from its mere existence is the presence of a nonphotospheric emission region in T Tauri atmospheres. Veiling today appears to be the key to understanding the contribution to the overall spectrum of the several atmospheric regions in a T Tauri star.

The spectroscopic definitions of T Tauri stars make it clear that conspicuous lithium absorption is also a typical spectral feature of young stars (cf. Herbig 1965). Lithium, the abundance of which was shown to decrease with age (Skumanich 1972), is thought to be destroyed when convective mixing brings it in contact with high-temperature zones. From analysis performed on a few CTTSs, Zappala (1972) confirmed that Li abundance is close to its interstellar value in these objects. Recent advances in the understanding of lithium abundance evolution (Duncan 1981, Boesgaard & Tripicco 1987) indicate that processes leading to lithium destruction are much more complex than anticipated. The interplay between rotation and lithium abundance is also unclear (Butler et al. 1987). Because of these questions, a reexamination of lithium abundances on the basis of the high-resolution data now available for a cross section of the T Tauri population appears overdue.

Rotation and Duplicity

Veiling makes it difficult to use standard spectroscopic techniques to determine the projected rotational velocity of T Tauri stars with copious emission activity. Available rotation data are thus biased toward WTTSs and low-activity CTTSs. There is, however, work in progress at various places to derive $v \sin i$ in strong emission-line CTTSs from the weak, presumably photospheric absorption lines apparent in the high-dispersion spectra of these objects (e.g. Appenzeller et al. 1988); preliminary evidence so far indicates that projected rotational velocities are comparable in moderate and extreme CTTSs (G. Basri, private communication).

Vogel & Kuhi (1981) made the initial attempt to determine $v \sin i$ for a large sample of pre-main-sequence stars, and it yielded the rather surprising result that most low-mass stars are rotating much slower than their strong activity would seem to predict. Implicit in the expectation of fast rotation had been the ideas that (a) newly formed stars would possess high specific angular momentum and would therefore rotate at close to breakup velocity (about 200 km s^{-1}), and that (b) the T Tauri phenomenon is exaggerated solar-type activity powered by fast rotation (e.g. Mestel 1968). The rotation data, therefore, tell us that newly formed stars dispose of their angular momentum prior to the T Tauri phase. Since the upper limits on $v \sin i$ reached by Vogel & Kuhi were only about 25 km s^{-1}, several groups improved upon these results and measured projected rotational velocities for a large number of T Tauri stars (Bouvier et al. 1986a, Hartmann et al. 1986, 1987, Franchini et al. 1988, Walter et al. 1988).

Current results are illustrated in Figure 4, where all CTTSs and WTTSs with known $v \sin i$ are plotted in the Hertzsprung-Russell diagram (HRD). WTTSs are represented by open circles, and CTTSs by dark circles. In both cases, circle area[4] is proportional to the stellar $v \sin i$, which ranges from 5 to 100 km s^{-1}. Approximate pre-main sequence quasi-static evolutionary tracks for various masses, taken from Cohen & Kuhi (1979), are also plotted in Figure 4, and they show that the more massive stars ($M \gtrsim 1.2\ M_\odot$) rotate significantly faster than less massive ones. Walter et al. (1988) find that distributions of projected rotational velocities are comparable in low-mass CTTSs and WTTSs. The lowest $v \sin i$ value reached by current surveys using either Fast Fourier Transform or cross-correlation techniques of a high-resolution spectrogram is about 10 km s^{-1}. Smaller rotational velocities are, however, attainable with CORAVEL (Benz & Mayor 1981); using this instrument, Bouvier et al. (1986a) positively detected all T Tauri stars that they measured. Their $v \sin i$ range from 6 to about 70 km s^{-1}, but if the sample of T Tauri stars is restricted to masses lower than 1.2 $M_\odot$, the corresponding $v \sin i$'s then encompass a much smaller range (from 6 to about 30 km s^{-1}, with a mean at about 15 km s^{-1}).

Rotational data set constraints on the initial values of rotation rates and on the spread in angular momentum at the beginning of quasi-static evolution toward the main sequence. Restricting the discussion to low-mass T Tauri stars, one finds that they will rotate at velocities ranging from 30 to more than 150 km s^{-1} upon their arrival on the main sequence if (a) loss of angular momentum due to the wind is negligible during this phase, and (b) the stellar moment of inertia evolves along the radiative track, as in models computed by Gilliland (1986). While this range of

---

[4] For practical purposes, all stars for which $v \sin i$ upper limits of 10 km s^{-1} are available have been arbitrarily assigned a 5 km s^{-1} $v \sin i$ in Figure 4 and are thus denoted by the smallest circles.

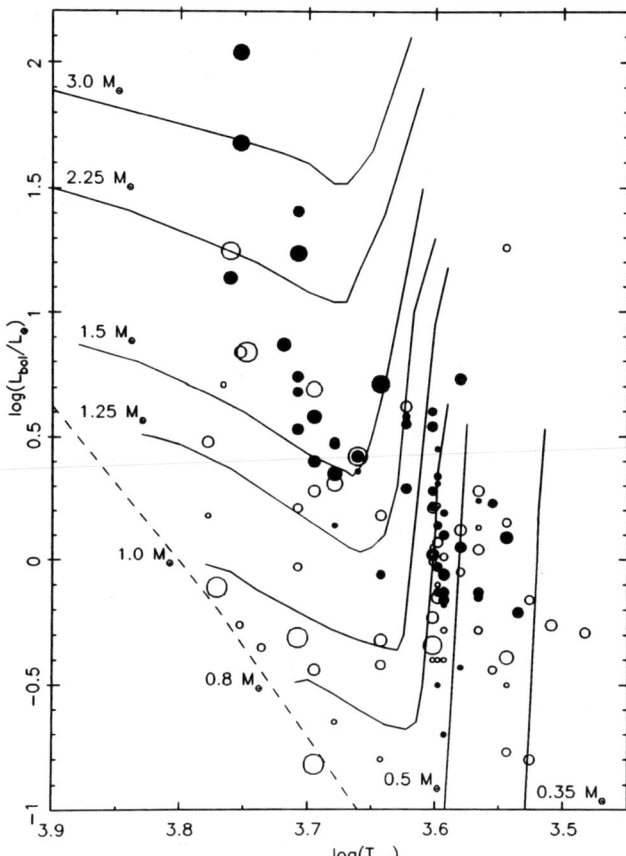

Figure 4 Position in the Hertzsprung-Russell diagram of all CTTSs and WTTSs with known $v \sin i$. WTTSs are represented by open circles, and CTTSs by dark circles. In both cases, the circle area is proportional to the stellar $v \sin i$. Approximate pre-main-sequence quasi-static evolutionary tracks for various masses are also plotted together with the zero-age main sequence (dashed line).

rotational velocities is in agreement with rotation rates of fast rotators in young clusters, members with low rotational velocities (e.g. about one half of the Pleiades K dwarfs) are not accounted for by T Tauri stars. Several possibilities, reviewed by Stauffer (1987) and Hartmann & Noyes (1987), have been proposed to explain this discrepancy, but none of them appears to be entirely satisfying.

The apparent deficiency of spectroscopic binaries among CTTSs (Herbig 1962) has sometimes been attributed to the difficulty of measuring accurate radial shifts caused by veiling and rotation. New high-precision surveys are, however, becoming available that seem to confirm the small number of spectroscopic binaries in CTTSs; in contrast, the statistics of spectroscopic binaries among WTTSs appears to be consistent with that of field stars (Mathieu 1988). There are a large number of close optical pairs among both CTTSs and WTTSs, most of which are probably wide binaries (Cohen & Kuhi 1979, Walter et al. 1988). Spectroscopic and visual binaries sample the two extremes of the distribution of binary star separations. At the distance of the closest star formation regions (150 pc), spectroscopic binaries have separations of typically less than a few astronomical units, and visual pairs are separated by 150 AU or more. Lunar occultations and speckle interferometry can both detect binaries with intermediate separations (cf. Reipurth 1988). Current results based on a small number of detected multiple systems seem to indicate no significant deviations from the multiple-system statistics of field stars in systems with intermediate separations (Simon et al. 1987).

The position of T Tauri stars in the HRD and comparison with theoretical quasi-static evolutionary tracks provide further indirect evidence for their youth. Note that while some WTTSs appear closer to the main sequence than the bulk of CTTSs, and thus may be called post–T Tauri stars (cf. Herbig 1978), many WTTSs are located in the same region of the HRD as CTTSs and therefore do not appear older. Use of quasi-static evolutionary tracks for deriving precise ages and masses of these stars has been severely questioned in the past (e.g. Appenzeller 1983), and more doubts about their usefulness are introduced by the probable presence of disks around CTTSs, which implies that their observed bolometric luminosity includes nonstellar contributions. I stressed earlier, however, that the T Tauri stars whose rotation rates are known today are predominantly stars with low-level activity[5] and thus are more likely to evolve along classical convective-radiative tracks than are very active CTTSs (Section 3). It is therefore probably reasonable to use these tracks in order to get a rough idea of their mass and perhaps of their age.

Photospheric Spots

While CTTSs are noted for their irregular variability on many time scales, there were some early detections of quasi-periodicities on time scales of a few days (Herbig 1962, Hoffmeister 1965). Since the average T Tauri star $v\sin i$ of 15 km s^{-1} translates to a rotational period of 5.3 days when assuming a typical 2-$R_\odot$ radius, one therefore suspects that regions with different temperatures present on their photospheres cause these light-

---

[5] Notable exceptions include T Tau and DF Tau.

curve modulations. This makes it possible to derive the photospheric rotational period, which is a critical parameter in studies of magnetically driven stellar activity.

Several groups have made a huge effort over the last five years to observe the photometric variability of T Tauri stars on time scales comparable to the rotational period, and strict periodicities have been detected in a number of WTTSs and CTTSs (Rydgren & Vrba 1983, Schaefer 1983, Rydgren et al. 1984a, Bouvier et al. 1986b, Vrba et al. 1986, Holtzmann et al. 1986, Herbst et al. 1986, Herbst & Koret 1988, Bouvier & Bertout 1989). Derived rotation periods span the range from about 2 to 9 days. A lower limit of 1 to 2 days for the rotational period P_{rot} is expected on the basis of the $v \sin i$ distribution mentioned above, but the upper limit of 9 days reflects only current observational limitations. (Many periods were found from data gathered during observing runs lasting less than 2 weeks.) These studies also reveal alternate phases of periodic and aperiodic light curves with similar color behaviors in some stars, which suggests a common origin for both regular and irregular light variations taking place on time scales of days to weeks. A likely explanation of the observed rotational modulation, discussed below, is the presence of dark spots at the stellar surface.

Main-sequence late-type stars display various forms of magnetic activity caused by dynamo processes, which in turn probably result from the interaction of differential rotation with convective motions (Durney & Latour 1978). One manifestation of magnetic activity is the formation of dark spots on the stellar surface. They occur when the dynamo amplifies the magnetic field under a portion of the stellar surface. The enhanced magnetic flux partially blocks convective energy transport in the subphotospheric layers, and a flux deficiency (i.e. a dark spot) results at the stellar surface (cf. Hartmann & Rosner 1979). Sufficiently large spots lead to rotational modulation of the stellar light curve, as observed, for example, in many Pleiades fast rotators (Alphenaar & van Leeuwen 1981).

Detailed models by Bouvier & Bertout (1989) reveal that spots on T Tauri stars are often large (typically 10% of the stellar surface) and much colder than the surrounding photosphere. (The temperature difference is typically in the 700–1000 K range.) A spot lifetime of at least a few years is implied by the work of Vrba et al. (1988) on the WTTS V410 Tauri. The properties of its main spotted region have stayed remarkably similar over more than 1000 rotation periods. So far, there is no evidence of periodic variations of either Ca II flux or TiO band strengths within the rotation period in the few spotted stars surveyed spectroscopically (at moderate resolution) by Bouvier & Bertout (1989). Average properties of spots on T Tauri stars derived from the observed light curves compare well with what is expected from a 1300-G stellar magnetic field emerging at the surface of a pre-main-sequence star model, as computed by Appenzeller & Dearborn (1984) for a magnetic flux of 5×10^{25} Mx. And the presence of a magnetic field with strength $1-3 \times 10^3$ G over a substantial part of the stellar surface does not appear unreasonable for late-type active stars (Saar & Linsky 1985, Basri & Marcy 1988).

A comparison of the derived properties of cold spots found on both CTTSs and WTTSs with those of spots observed on other late-type active stars indeed demonstrates the similarity of spotted regions in T Tauri stars, RS CVn stars, and BY Dra (Bouvier & Bertout 1989). Although these stars are at different evolutionary stages, they share one common property: their rotation rates. Rotation has indeed emerged over the past decade as a major parameter governing magnetic activity (see the review by Hartmann & Noyes 1987), and studies of T Tauri star spots support this idea further.

A few CTTSs show clear evidence for spots hotter than the stellar photosphere, which are unlikely to result from solar-type activity. Main examples are BP Tau (Vrba et al. 1986) and DF Tau (Bouvier & Bertout 1989). Both of these objects are most likely surrounded by accretion disks, and hot spots may well be related to accretion phenomena (cf. Section 3).

The Solar Analogy

Joy (1945) was first to note the similarity between the emission spectrum of strong-line CTTSs and the chromospheric (flash) spectrum of the Sun. Finkenzeller & Basri (1987) recently demonstrated that this is also true of T Tauri stars with low-emission characteristics by constructing *ratio* plots of high-resolution T Tauri star spectrograms divided by standard star spectrograms (with the same spectral type and rotation rate) that they qualitatively compared with a solar chromospheric spectrum. Although only the Ca II H and K lines are obviously in emission in the original spectrograms, many strong photospheric absorption lines appear in *pseudoemission* in the ratio plots. This demonstrates that nonradiative heating contributes more heavily to the excitation of these features in T Tauri stars than in inactive standard stars (at a given continuum optical depth). Conversely, it confirms Herbig's (1970) proposition that the photospheric temperature minimum may occur at greater continuum optical depth in T Tauri stars than in the Sun. Weaker photospheric lines that are formed deeper in the atmosphere than the strong lines usually divide out in the ratios; this supports the implicit assumption—made in all empirical attempts to compare atmospheric properties of T Tauri stars and other late-type stars—that the deep photosphere of T Tauri stars is the same as that of the standard star to which it is compared.

Several groups followed Herbig's (1970) suggestion and performed several detailed investigations of continuum and line formation in a chromosphere located at larger optical depths than in the Sun (Dumont et al. 1974, Lago 1979, Cram 1979, Calvet et al. 1984). The last two groups assumed a plane-parallel, homogeneous atmosphere in hydrostatic equilibrium and appended a chromosphere to a late-type photosphere. Both the chromospheric temperature stratification and the optical depth of the chromosphere's anchorage in the atmosphere are free parameters that are adjusted until an acceptable fit to a set of spectral diagnostics is obtained. Their results demonstrate that the metallic emission spectrum and the two manifestations of veiling discussed above can be reproduced by this model. They also show that an extended, optically thin region is needed to reproduce both the strong Hα flux and Balmer decrement of CTTSs.

While the Balmer continuum (in emission in many CTTSs; see Section 1 and Figure 2) could also be formed, at least in principle, in a chromospheric region, this would require a large optical depth at the bottom of the transition region that also drives the Paschen continuum in emission, thus producing a strong veiling in the optical region. In contrast, many CTTSs (e.g. DN Tau in Figure 2) display strong Balmer continuous emission but little veiling. Others, like DF Tau (Figure 2), display both Balmer emission and veiling that could conceivably be of chromospheric origin. In such strong chromospheres, the chromospheric luminosity then becomes comparable to the photospheric luminosity.

A major drawback of these semiempirical models is that evidently they do not directly address the question of the origin of the nonradiative energy flux responsible for the heating of the deep-lying chromosphere. While solar-type magnetic activity is often mentioned as the most likely possibility, the stringent energy requirements for covering observed radiative losses in CTTSs such as DF Tau are a formidable challenge. Calvet & Albarrán (1984) estimated the total magnetohydrodynamic-wave energy flux emerging from stars entirely covered by active regions and showed that the resulting flux, plotted as a function of the stellar luminosity, forms a lower envelope to the observed radiative losses in CTTSs. They conclude that solar-type dynamo-driven activity might drive the activity of those T Tauri stars that show the lowest activity levels, but that an additional energy source is needed for more active stars. They then go on to suggest that flares may be responsible for the excess activity.

That the T Tauri phenomenon could essentially be continuous flaring activity is hardly a new idea. Haro (1968, 1976) emphasized the similarities between T Tauri stars and flare stars. Besides having both been classified by Joy—whose classic 1960 paper defines the UV Ceti class of dMe and dKe flare stars (Joy 1960)—the two groups of objects have in common a propensity for irregular, short-term light variations reminiscent of solar flares, although on different scales. The outburst of energy in the optical continuum is typically 10 times greater in a UV Ceti flare, and 10^{5-6} times greater in a strong T Tauri star flare, than in the strongest solar flare (Haro 1968). In contrast to T Tauri stars, flare stars show little emission during quiescent phases, but the Balmer series and Ca II go into emission during flares and the optical spectrum is veiled by a strong continuum. The width of emission lines is, however, much smaller in UV Ceti stars than in T Tauri stars, which hints at a different widening mechanism. In any case, the similarities of these two classes of late-type stars were more apparent than their differences in early low-resolution studies, and an evolutionary sequence was proposed in which T Tauri stars were progenitors of UV Ceti stars. It is now clear that T Tauri stars are up to one order of magnitude more massive than UV Ceti stars, and that the latter class is present in the solar vicinity as well as in young stellar clusters, while T Tauri stars are confined to the vicinity of star-forming regions. So while the least massive young stellar objects will indeed evolve to become dMe stars, the bulk of T Tauri stars will become solar-type stars instead. Nevertheless, flarelike events at different wavelengths are a common occurrence in T Tauri stars, as a series of recent observations has shown.

While optical and U-band variability on a time scale of 15 min or more is commonly observed in T Tauri stars, there is no convincing evidence so far for ultraviolet flares over shorter time scales. Both Kuan (1976) and Worden et al. (1981) reported such events and claimed that T Tauri stars were flaring continuously, but their data analysis method was recently criticized by Warner (1988). More high-speed photometry of bright T Tauri stars is therefore urgently needed to examine further the continuous-flaring hypothesis. No short time-scale variability has been reported in the infrared range so far. Sizable radio variability (on an unknown time scale) has been observed for the fast-rotating WTTS V410 Tau (Cohen & Bieging 1986). Repeated observations of the ρ Ophiuchi cloud with the VLA (Stine et al. 1988) led to detections of strong radio variability in pre-main-sequence objects—e.g. DoAr 21, which underwent a radio flare (cf. Feigelson & Montmerle 1985)—but none of the CTTSs were observed (cf. Section 1 and Figure 1). Montmerle et al. (1983) repeatedly observed the same cloud with the *Einstein* X-Ray Observatory and discovered about 50 sources characterized by widespread variability. There are 10 CTTSs among these sources, and many of the remaining objects turned out to be WTTSs.

The main argument supporting the idea that this variability represents solarlike flaring is the cumulative flux distribution of X-ray flares derived by Montmerle et al. (1983); it is approximately a power-law with index −1.4, the same as in the Sun. Other X-ray flares have been observed in a number of T Tauri stars (Feigelson & DeCampli 1981, Walter & Kuhi 1984), and the most powerful observed X-ray flare, which released 10^{32} erg s^{-1} at maximum light (i.e. larger by a factor of 10^5 than the strongest X-ray solar flare), originated from a region with temperature of about 1 keV, density of 10^{10} cm^{-3}, and size of about 10^{11} cm (Montmerle et al. 1983). This event is quite similar to a solar flare in its temperature and density, but the difference in luminosity is caused by its comparatively enormous size.

The widely differing energy scales involved in T Tauri stars and solar flares make it difficult to extend—as is done for UV Ceti stars—theoretical models of solar events to T Tauri stars (see, for example, the discussion by Gershberg 1983). The observations reported above, as well as comparisons with flares observed on other active late-type stars (Montmerle et al. 1983), do nevertheless indicate that enhanced solar-type flaring is one component of T Tauri activity. It is, however, obvious that the large-scale variations observed over time scales of months to years, such as the brightening of DR Tau (Chavarria 1979), must have quite a different origin. In fact, several mechanisms using magnetic fields have been proposed for explaining the large-scale variations. For instance, Appenzeller & Dearborn (1984) assume that variable magnetic fields are redistributed over the stellar surface and show that brightness variations of up to about 3 magnitudes (typical of moderately active CTTSs) can be produced in this way. Sizable effective temperature variations would then occur concurrently and lead to significant changes in the stellar spectral type during the eruption. Although it is known that FU Orionis–type eruptions involve large effective temperature changes, there is little information so far about changes in spectral type during observed CTTS light variations, probably because coordinated observations require heavy logistics. Both Walker (1987) and Gahm et al. (1989) do report such observations for a few CTTSs and find no changes of spectral type with photometric variations. Gahm et al. invoke variable occultation by dust to explain the light curve of a particularly wild object, RY Lupi, while Walker finds that changes in a source of continuous emission located somewhere in the system can explain many of his data. Such a mechanism, which involves interaction of the star with its circumstellar disk, is discussed in Section 3.

The difficulty of quantitatively assessing the role of magnetic dynamo activity in the T Tauri phenomenon arises mainly because of the present lack of magnetic field detections in these faint objects (e.g. Johnstone & Penston 1988). An absence of hard facts leaves open the possibility to speculate that large convective regions (Simon et al. 1985), strong differential rotation (Calvet & Albarrán 1984), and fossil magnetic field (Tayler 1987) may all enhance surface magnetic field strengths to values large enough to drive the entire range of activity witnessed in these objects. But a major parameter governing solar-type magnetic activity, the rotation period, is now known for a reasonably large sample of T Tauri stars; thus, it is now possible to compare known activity diagnostics of T Tauri stars with those of other active late-type stars.

Rotation-Activity Connections

Walter & Bowyer (1981) were first to demonstrate a correlation between rotation rate and coronal luminosity in late-type stars. Since then, this connection has been largely confirmed and was shown to extend to chromospheric and transition region diagnostics as well (Basri 1987, Hartmann & Noyes 1987).

Figure 5 (from Bouvier 1989) displays relationships between the rotation periods of a sample of both T Tauri stars and other late-type stars, and the observed flux in two major diagnostics: the X-ray flux and Hα. The observed correlation between the X-ray flux of T Tauri stars and their rotation period closely matches the relationship derived from other late-type stars; this is the strongest argument so far in favor of solar-type coronal heating in T Tauri stars (Bouvier et al. 1985, Bouvier 1987, 1989). The lack of T Tauri stars with periods larger than about 10 days is probably due to observational selection effects, as indicated earlier.

The sample of T Tauri stars on which the rotation-activity relation of Figure 5a is based contains only one extreme CTTS.[6] This is the result of a double observational bias: (a) several of them were not detected in the X-ray range in pointed surveys of CTTSs (e.g. Gahm 1981, Walter & Kuhi 1981), and (b) there is little information so far about their rotation periods. Failure to detect these objects in the X-ray range may be due to the lack of repeated observations, since flaring is an important component of their X-ray activity. On the other hand, solar-type magnetic activity is perhaps quenched by the physical processes responsible for the strong activity of extreme CTTSs. Inclusion of more extreme stars in the rotation-activity plot would be particularly useful for distinguishing between these possibilities. But because extreme CTTSs represent at most 10% of the CTTS class—and thus maybe as little as a few percent of the total T Tauri population—the earlier conclusion that the X-ray emission of T Tauri

---

[6] This is RW Aur, for which there is both an X-ray flux upper limit (Gahm 1981) and a suspected rotation period of 5.39 days (Grinin et al. 1983). The upper limit is consistent with the observed correlation, as it is in other extreme stars for which available $v \sin i$ values permit determinations of most probable periods (J. Bouvier, private communication).

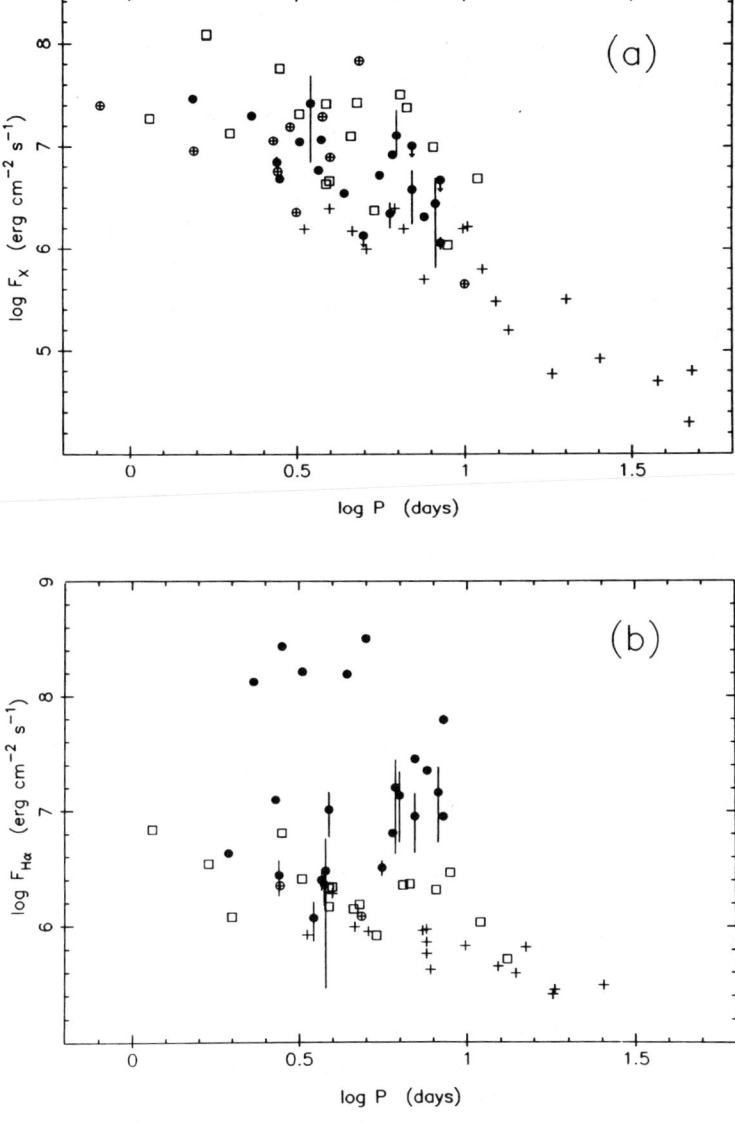

Figure 5 (*a*) Relationship between rotation period and observed X-ray flux for a sample of T Tauri stars and other late-type stars (from Bouvier 1989). T Tauri stars are shown as black dots, and their observed variability (when available) is indicated by a vertical bar. RS CVn stars are displayed as open squares, dKe and dMe stars as open crossed circles, and late-type dwarfs as crosses. The observed correlation between X-ray flux of T Tauri stars and their rotation period closely matches the relationship derived from other late-type stars. (*b*) Relationship between rotation period and observed Hα flux for the same sample of T Tauri stars and other late-type stars as in (*a*) (from Bouvier 1989). While WTTSs follow the same relation as other late-type active stars, CTTSs have much larger Hα fluxes, which are not correlated with the rotation period.

stars is generally consistent with solar-type coronal emission driven by magnetic dynamo processes remains valid.

The relationship between Hα flux and rotation period is, however, another story. A weak correlation is observed for late-type dwarfs and RS CVn stars, but the behavior of T Tauri stars in Figure 5b is quite peculiar. While stars with low Hα flux (the WTTSs) display emission levels comparable to those of dwarfs with similar rotation rates, the bulk of the CTTSs are way above the correlation line. Thus, chromospheric nonradiative losses as measured by Hα appear quantitatively similar in WTTSs and other late-type active stars, but they can be up to 100 times larger in CTTSs; and there is no connection in CTTSs between this activity indicator and rotation rates. It is therefore probable that Hα is not formed in a solar-type chromosphere in CTTSs.[7] In contrast, the Hα line of WTTSs could be entirely of chromospheric origin. One could perhaps argue that the Hα flux is not a good chromospheric indicator, and that the Ca II K line is preferable. While Hα flux was chosen only because it is a widely available quantity in T Tauri stars, a strong correlation exists between Hα, Ca II H and K, and Mg II h and k line fluxes for T Tauri stars of all activity levels (Calvet et al. 1985, Bouvier 1987, 1989), so that the lack of correlation between line flux and rotation observed in the Hα line flux of CTTSs extends to these lines as well. The conclusion that none of these strong lines is formed primarily in a solar-type chromosphere is therefore difficult to escape. These empirical comparisons support Calvet & Albarrán's (1984) conclusion (based on theoretical estimates; see above) that only low-level WTTS activity could possibly be driven by magnetohydrodynamic wave dissipation. This leaves open the question of the origin of these strong emission lines in CTTSs, a problem that is dealt with in Section 3.

Although optical and ultraviolet emission lines appear to be much stronger in moderately active CTTSs than in late-type stars with comparable rotation periods, their X-ray fluxes are quite similar. This is interpreted as *X-ray flux deficit* by those who implicitly assume that the T Tauri phenomenon is *only* scaled-up solar-type activity. In this framework, the strong lines are chromospheric, and correspondingly strong coronal losses should be observed if the solar analogy is valid (e.g. Giampapa et al. 1981). Two approaches were devised to weaken the X-ray flux with respect to the chromospheric and transition region flux. The first one assumes that X rays are partially absorbed in the dense circumstellar medium of T Tauri stars (Gahm 1981, Walter & Kuhi 1981), while the second one proposes

---

[7] Herbig (1985) reached a similar conclusion from his study of the evolution in time of the Hα flux.

that strong CTTS Alfvén-wave driven winds do not reach coronal temperatures (Hartmann et al. 1982). The opposing view postulates that there is an *excess* of optical-line emission compared with X-ray emission, which is supported by recent results reviewed above on the activity-rotation connection in T Tauri stars. Since the activity level seen in WTTSs now appears to be consistent with solar-type magnetic activity as observed in late-type stars with comparable rotation rates, one is inclined to conclude that the WTTS activity level indicates the share of solar-type magnetic activity in the overall CTTS activity.

3. CIRCUMSTELLAR DISKS AND OUTFLOW PHENOMENA

At the distance of the closest star formation regions, one arcsecond corresponds to 150 AU. Subarcsecond resolution is attainable today at radio wavelengths (e.g. with the VLA), and in the near-infrared and optical through direct imaging, lunar occultations, and speckle interferometry. Resolution of T Tauri into three components (Dyck et al. 1982, Nisenson et al. 1985) is indeed a major achievement of speckle work, and more data on T Tauri star binarity and extended structures around these objects are rapidly becoming available (e.g. Simon et al. 1987, Zinnecker et al. 1988, Zinnecker 1988). Direct imaging at optical wavelengths has recently been used very successfully to improve our knowledge of the circumstellar environment of young stellar objects.

The Circumstellar Environment of HL Tau

A CCD image kindly communicated by R. Mundt (Figure 6) illustrates the complexity of the circumstellar environment of some T Tauri stars. It displays the region of HL/XZ Tau as observed in [S II] by Mundt et al. (1988). Both HL Tau and XZ Tau are CTTSs with strong infrared excesses, and a weak radio continuum source (VLA 1 in Figure 6) was detected by Brown et al. (1985) at about 12″ northeast of HL Tau. Several extremely well-collimated jets are seen in [S II] light. One of them originates from HL Tau [position angle (p.a.) 42°]. A second one originates from VLA 1, and the opposite signs of the heliocentric velocity on each side of this jet indicate that the collimated flow is bipolar. A third outflow originates from a continuum source denoted HH 30-star; several aligned knots of this jet, extending up to 2 arcmin (i.e. 0.1 pc) from the star in the northeast direction, can be seen in the large-scale figure. While jets are quite impressive in their extreme collimation, inferred mass-loss rates are usually small to moderate ($0.05-2 \times 10^{-8}\ M_\odot\ \mathrm{yr}^{-1}$; Mundt et al. 1987).

HL Tau was long suspected to be surrounded by a prominent disklike

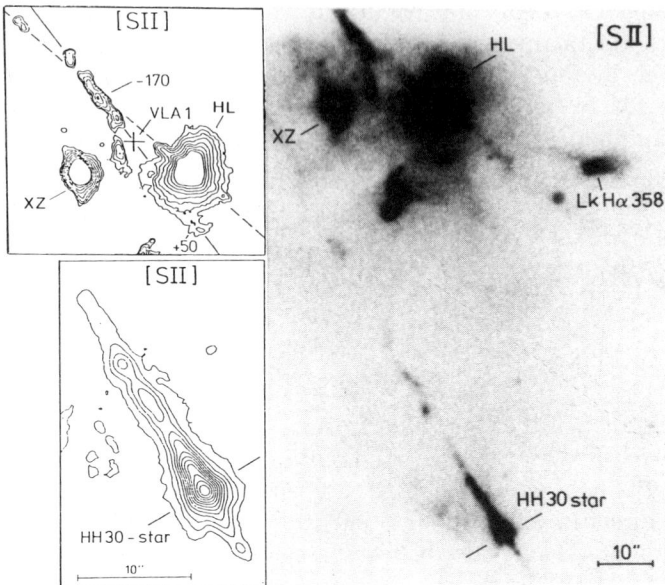

Figure 6 [S II] image of the HL/XZ Tauri region (from Mundt et al. 1988). Two inserts containing enlarged intensity contours of the most interesting details are shown, together with a larger scale image of the entire region. The solid and dashed lines drawn on the upper contour plot show the directions of two independent jets emanating, respectively, from the radio continuum source VLA 1 and from HL Tauri. The lower contour plot displays a blowup of a jet originating from the star associated with HH 30 and extending to at least 0.1 pc.

molecular condensation seen almost edge-on (Cohen 1983). Direct near-infrared imaging (Grasdalen et al. 1984), speckle interferometry (Beckwith et al. 1984), and recent interferometric observations at millimeter wavelengths all confirm this expectation. Beckwith et al. (1986) detected around HL Tau a ^{12}CO unresolved structure (at a resolution of about 6 arcsec) that is elongated in p.a. 150°, i.e. almost parallel to the linear polarization vector (p.a. 135°) of HL Tau and almost perpendicular to its jet (Figure 6). Sargent & Beckwith (1987) gained some information on this structure's size and velocity field from ^{13}CO observations and find it consistent with Keplerian rotation. The overall size of this suspected molecular disk is about 4000 AU. This molecular condensation might conceivably be feeding the inner disk with size of perhaps 100 AU, whose presence is inferred from the infrared energy distribution of HL Tau (see below).

Understanding the exact relationship between circumstellar disks and the various outflow phenomena associated with young stellar objects offers a major challenge to today's stellar astronomy. The most spectacular outflow manifestations (i.e. ionized jets such as those associated with HL/XZ Tau, Herbig-Haro objects, and bipolar molecular outflows) are more typically found in the vicinity of embedded young stellar objects than around optically visible T Tauri stars (Mundt et al. 1987, Levreault 1988). There are, however, various indications that a significant fraction of CTTSs are still losing mass at an appreciable rate, and evidence is growing that the same CTTSs are still surrounded by dusty disks. In contrast, WTTSs do not seem to have strong winds, nor do they display evidence for extended disks. A picture is thus emerging that relates disks to outflows. The disk provides the energy reservoir needed to account for energetic mass and radiative losses observed in young stellar objects; if the disk then dissipates on a time scale smaller than or comparable to the duration of the T Tauri phase, a gradual transition from T Tauri to main-sequence stars can be expected. While several aspects of this tentative scenario are explored in recent papers discussed below, it is far from being firmly established. In particular, the connections between disks and winds are not yet understood, nor is there a widely accepted theoretical mechanism for driving protostellar winds. In contrast, the physics of (Keplerian) disks appears—at first sight—somewhat better understood.

Models for T Tauri Disks

Adams & Shu (1986) recognized that circumstellar dusty disks with masses larger than 10^{-3} $M_\odot$ are optically thick to their own thermal radiation. They computed emitted spectra from passive disks that merely reprocess stellar light (i.e. stellar optical and ultraviolet photons are absorbed in the dusty disk and reemitted thermally in the infrared). Although the star is the only photon source in this picture, the observed bolometric flux of the star/disk system can be up to 50% larger than the stellar photospheric flux because stellar photons that would not reach the observer if the disk were absent are "scattered" by the disk in the observer's direction. Adams & Shu showed that the resulting disk temperature distribution is proportional to $r^{-3/4}$ at large distances r from the star, and that resulting infrared spectra depend only on the disk viewing angle. The main signature of a circumstellar disk around young stellar objects, therefore, is primarily its infrared spectrum. Of course, since any dusty shell surrounding a star will emit primarily in the infrared, the disk hypothesis may not be all that unique. A first line of evidence for the location of dust in a flat disk rather than in a more or less spherical shell is provided by observations of sources discovered with the *Infrared Astronomical Satellite* (*IRAS*) near dense cores by Myers et al. (1987), who show that the amount of dust necessary to explain the infrared spectrum is irreconcilable with the observed low optical extinction values. Polarization data provide additional evidence for disk geometry. Bastien & Ménard (1987) recently demonstrated that observed linear polarization maps around young stellar objects rule out grain alignment (by a magnetic field or otherwise) as a polarization mechanism. They also showed that both the linear polarization maps and the circular polarization recently detected in a few CTTSs (Nadeau & Bastien 1986, Bastien et al. 1989) can be reproduced by a model in which a stellar photon is scattered first in an optically thin dusty envelope and then in an optically thick (but geometrically thin) disk.

Comparisons of computed spectra with observed infrared spectral energy distributions of young stellar objects led Adams et al. (1987) to propose the evolutionary sequence already mentioned in Section 1. Spectra of young stellar objects that rise steeply in the infrared (Class I) can be reproduced when assuming that the young star is surrounded by both a disk and a dusty, infalling envelope. They therefore seem to point to young stellar objects still in their protostellar accretion phase. Class II objects with spectra falling off in the infrared are typical of CTTSs surrounded by disks. Finally, Class III objects with blackbodylike spectral energy distribution are either WTTSs or post–T Tauri stars. While quite successful in this enlightening interpretation, models of thin reprocessing disks fail to account for several details of T Tauri star energy distributions. Specifically, they do not account for (a) the strong ultraviolet excess of CTTSs, (b) CTTSs with infrared luminosities obviously larger than the limit mentioned above, and (c) stars with "flat" infrared spectra.

We know from Figure 3 that ultraviolet and infrared excesses are often correlated, and Section 2 touched on difficulties involved in trying to attribute large ultraviolet (and infrared) excesses to solar-type chromospheric emission. When dealing with purely reprocessing disks, we are confronted with the very same problem. Whereas the main features of infrared spectral energy distributions can be attributed to disk emission, we must postulate still another physical process to produce an ultraviolet excess luminosity comparable, in some cases, to the photospheric luminosity. This problem and the one presented by stars with infrared excess luminosity in excess of half the photospheric luminosity can be overcome if the disk is not only reprocessing photons from its central star, but is also self-luminous. Since accretion of disk matter onto the star is a natural consequence of energy dissipation within the disk, it will occur in the active disks envisioned here. Lynden-Bell & Pringle (1974) were first to suggest that a viscous accretion disk might account for the continuous excesses of CTTSs, but this proposal was heeded only recently (Rucinski 1985, Beall 1986, Bertout 1987, Kenyon & Hartmann 1987, Bertout et al. 1988, Basri

& Bertout 1989). The delay was probably caused by a lack of relevant data; ultraviolet and far-infrared data only became available with the launchings of *IUE* in 1978 and of *IRAS* in 1983. The accretion disk hypothesis has been worked on extensively over the last three years, so that an assessment of this model's successes and shortcomings is now possible.

Current work generally assumes that the viscosity is parametrized according to the so-called α prescription (Shakura & Sunyaev 1973), which relates the local viscosity v in the disk to the sound speed c_s and to the disk height scale H (which determines the maximum eddy size) by $v = \alpha c_s H$. The theory of optically thick steady-state quasi-Keplerian accretion disks predicts an asymptotic temperature distribution proportional to $r^{-3/4}$, i.e. identical to that of reprocessing disks. The emitted far-infrared spectra therefore have the same shape in these two classes of models ($\lambda F_\lambda \propto \lambda^{-4/3}$). But one half of the accretion luminosity ($L_{acc} = GM_*\dot{M}/R_*$, where R_* is the stellar radius, M_* the stellar mass, and $\dot{M}$ the constant mass accretion rate within the disk) is now emitted from the disk together with the reprocessed luminosity. As long as the accretion luminosity is smaller than or comparable to the reprocessed luminosity (i.e. typically for $\dot{M} \lesssim 10^{-8}$ $M_\odot$ yr^{-1}), the infrared spectrum alone does not allow one to distinguish between these two models.

Models of accretion disks surrounding young stellar objects were first worked out by Hartmann & Kenyon (1985, 1987a,b) in order to explain the spectral peculiarities of FU Orionis objects (cf. Section 1). Because of the resemblance to dwarf novae outbursts, Hartmann & Kenyon hypothesized that FU Orionis outbursts were caused by a rapid mass accretion increase (up to 10^{-4} $M_\odot$ yr^{-1} at maximum light) with a correspondingly large disk luminosity increase. Disk instabilities that may lead to strong variations of the mass accretion rates (Lin & Papaloizou 1985) are poorly understood owing to the present lack of knowledge about viscosity. The variation of spectral type with wavelength typical of FU Orionis systems is well reproduced by accretion disk models in which the optical spectrum arises in the inner parts of the disk and the infrared spectrum is formed in the outer-disk regions. Furthermore, the different rotational velocities of optical and infrared absorption lines predicted by the models were observed in FU Ori and V1057 Cyg (Hartmann & Kenyon 1987a,b). Thus, mass accretion through a disk appears quite successful in explaining several aspects of the FU Ori phenomenon [see Goodrich (1986) for a discussion of other models].

Optically visible T Tauri stars have projected rotation rates in the range 10–30 km s^{-1}, i.e. much smaller than breakup velocities (typically 200 km s^{-1}). A quasi-Keplerian accretion disk extending down to the stellar photosphere must therefore join the star in a hydrodynamically complex boundary layer, as envisioned by Lynden-Bell & Pringle (1974). There, disk matter is slowed down to photospheric velocity, and about half of the accretion luminosity is dissipated (cf. Regev 1983, Pringle 1989). Depending on physical conditions prevailing in the shocks that bring the accreted matter's rotational velocity down to subsonic values, gravitational energy will be transformed into internal energy of the gas and/or into heat. Both Pringle & Savonije (1979) and Tylenda (1981) find that the radiative cooling time of the boundary layer is shorter than its adiabatic expansion time as long as its size is small compared with the stellar radius. In that case, boundary layer radiation is expected to emerge from a hot zone with extent comparable to the local disk scale height at the stellar equator (Pringle 1977). For typical parameters of T Tauri disks, the boundary layer will then radiate mainly in the ultraviolet range (Bertout 1987). Additionally, nonradiative processes due to magnetohydrodynamic interactions between the turbulent boundary layer and the photospheric layers may be important; Kenyon & Hartmann (1987) hypothesize that waves generated in this interaction could be responsible for heating a boundary layer chromosphere.

Since very little is known at present about the details of processes taking place in the boundary layer, one must turn to observations for guidance. Bertout et al. (1988) stressed the need for simultaneous ultraviolet and optical data for deriving mass accretion rates. They compared quasi-simultaneous sets of data in the ultraviolet/optical and optical/near-infrared ranges (similar to those displayed in Figure 3) to synthetic spectra emitted by models of a T Tauri *system* made up of a late-type active star, an accretion disk, and its boundary layer. They found that typical T Tauri disks are optically thick over most of their surface, and that the spectral energy distribution of typical CTTSs can be reproduced from about 0.2 to 10 μm if emission from the boundary layer is confined to an equatorial region with width comparable to the local disk scale height (typically 2% of the stellar radius). The isothermal boundary layer temperature, determined from the condition that half the accretion luminosity be emitted from its surface, is then in the 7000–12,000 K range. The disk temperature varies from about 3000 K near the star to typical interstellar temperatures in its outer parts. Positive aspects of this simple model are its self-consistency and its small number of free parameters (the disk mass accretion rate and view angle). A major drawback, however, is the assumption of an optically thick boundary layer; observed Balmer jumps (Figure 2) indicate that the Paschen continuum is at least partially optically thin.

This assumption was relaxed by Basri & Bertout (1989), who used a classical atmosphere code to compute monochromatic gas opacities in the boundary layer, which was again assumed to be isothermal. They chose the boundary layer width (rather than α) to be the free parameter needed to control the optical depth, and computed emergent spectral energy distributions to be compared with observations of the Balmer and Paschen continua region. Figure 7 illustrates current results of their computations for BP Tauri, which is a moderately active star quite representative of

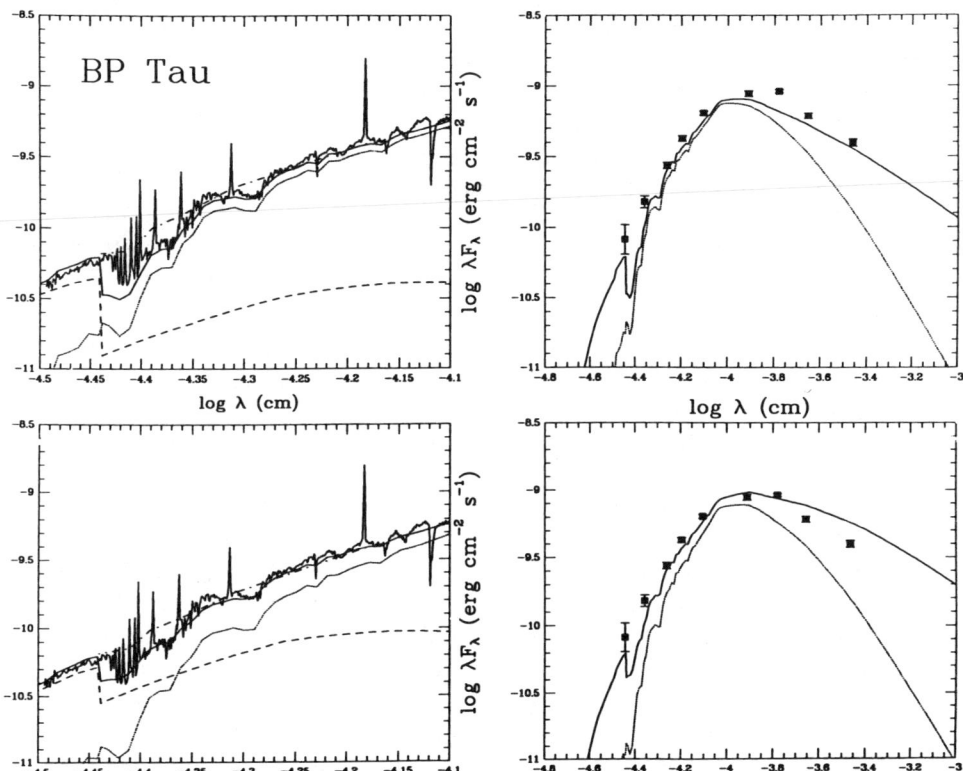

Figure 7 Comparison of observations and models for the spectral energy distribution of the CTTS BP Tauri (from Basri & Bertout 1989). The optical spectrum is shown in the left panels, and the squares in the right panels are simultaneous photometric measurements, with error bars denoting observed variability. The solid smooth line in all panels is the final composite model, the dashed lines in the left panels indicate the boundary layer and disk contributions, the dash-dotted line shows the expected locus of optically thick Balmer line peaks, and the dotted line is the photospheric flux. The right panels show the stellar and overall spectral energy distributions over the full spectral range. The two models shown in the upper and lower panels bracket the range of acceptable model parameters, which differ mainly by their mass accretion rates (2×10^{-8} $M_\odot$ yr^{-1} in the upper panels, and 6×10^{-8} $M_\odot$ yr^{-1} in the lower panels).

CTTSs. The optical spectrum is shown in the left panels together with computed Balmer and Paschen continua for two sets of parameters that differ mainly in their assumed mass accretion rates. The dash-dotted line indicates optically thick Balmer line emission expected from the boundary layer. While the head of the Balmer continuum is optically thick in these models computed with $\alpha = 1$, the Paschen continuum is partially optically thin and the Balmer jump consequently appears in emission. Line emission from Balmer lines with high quantum number appears consistent with optically thick emission from the boundary layer, the width of which is again comparable to the local disk scale height; but there is obviously a more extended and more optically thin region of emission that contributes to the flux in the lowest members of the Balmer series.

The right panels of Figure 7 display the observed spectral energy distribution of BP Tauri in the range 0.36–3.5 μm, together with overall spectral energy distributions computed with the same parameters as before. Because both the projected rotational velocity and the rotation period of BP Tauri are known, the inclination angle of the system can be derived independently with fair accuracy. Consequently, the disk parameters are more reliably known than in most other cases. Basri & Bertout (1989) conclude from a discussion of uniqueness of disk models that a unique set of computational parameters can be found for a given object provided (*a*) that the degree of veiling is determined from high-resolution data, and (*b*) that the inclination angle can be determined as above. While high-resolution spectrograms now available for a large sample of stars make veiling determination possible, the number of CTTSs with known rotation periods is still small (cf. Bouvier & Bertout 1989). Extended photometric monitoring of T Tauri stars is therefore valuable for further modeling.

Disk model predictions can be compared with properties of YY Orionis stars, i.e. those CTTSs that display spectroscopic evidence for accretion (cf. Section 1). Early models for these objects, based on spherical infall, met with problems when it was realized that these stars also exhibit evidence for outflow in high-resolution spectrograms (e.g. Mundt 1984). Both types of flows appear to occur simultaneously, which suggests that they take place in different spatial regions. The presence of inverse P Cygni absorptions indicates that we are viewing the optical continuum source through the infalling region, and most YY Ori stars display a strong ultraviolet excess. The accretion disk hypothesis thus seems to provide an adequate framework and suggests that YY Ori stars are accreting T Tauri stars that are viewed nearly equator-on, as originally proposed by Walker (1972). Inverse P Cygni absorptions are in fact observed at the higher members of the Balmer series, and not at Hα, in agreement with model predictions. But the limitations of the simple disk model envisioned earlier become obvious when we look at the observational data more closely. For example, the strong observed variability of the inverse P Cygni profiles (e.g. Bertout et al. 1982) suggests that accretion is nonsteady. Also, velocities of the infalling gas in typical YY Ori stars are 300–400 km s^{-1}, i.e. much larger than expected for the disk accretion speed of a few kilometers per second if $\alpha \leq 1$. At least two possible explanations of these large accretion velocities can be found. One assumes that $\alpha \gg 1$, which also means that turbulence is supersonic in the inner disk regions. Since supersonic turbulence is usually thought to become rapidly subsonic through the formation of internal shocks in the turbulent medium, one wonders how such a large α could be maintained over long periods of time.[8] The second, more likely possibility assumes that the inner disk is disrupted by the stellar magnetic field, and that accretion occurs along field lines at close to free-fall velocities (Bertout et al. 1988); it, in turn, hints at a complex interaction between disk and star.

Current Limitations of the Disk Hypothesis

There are other worries besides the obvious shortcomings mentioned above. For instance, one can argue that the present lack of understanding of chromospheric activity and of extinction toward T Tauri stars undermines the usefulness of simple disk and boundary layer models (Hartmann & Kenyon 1988). This view is probably overly pessimistic, since solar-type activity of T Tauri stars is at least partially understood (as Section 2 has shown) and since problems with the extinction curve are acute only below 2000 Å (Herbig & Goodrich 1986). Lack of knowledge about magnetic and turbulent processes in protostellar disks is a more serious issue. Although the complexity of the T Tauri phenomenon makes it clear that the simple disk models envisioned so far cannot be the whole story, one feels encouraged by its obvious success at explaining in a self-consistent way several heretofore mysterious properties of CTTSs.

One subclass of CTTSs that cannot be explained by classical accretion disk theory is that of T Tauri stars with flat infrared spectra, which show constant or rising far-infrared flux in the log λF_λ vs. log λ diagram. In fact, the far-infrared flux of many CTTSs is often somewhat larger than predicted by the thin accretion disk model; Rydgren & Zak (1987) show that the infrared flux λF_λ of typical CTTSs is proportional to $\lambda^{-3/4}$ rather than to $\lambda^{-4/3}$, as expected in classical thin-disk models. Bertout et al. (1988) noticed, however, that the observed infrared flux excess (over the computed disk infrared flux) is usually smaller than the optical and ultraviolet flux

---

[8] If, however, viscosity were caused by magnetic rather than dynamic stress, the condition $\alpha \leq 1$ need not apply (cf. Pringle 1989).

absorbed by the grains responsible for extinction toward the star. Since interstellar extinction toward the closest star-forming regions is small, extinction takes place in the vicinity of the star and the infrared flux excess thus probably corresponds to thermally reemitted optical and ultraviolet photons. In contrast, the far-infrared flux excess observed in stars with flat infrared spectra is greater than the ultraviolet and optical flux deficit due to intracloud extinction. One should note, however, that visual extinction toward these often peculiar sources is not always reliably known.

HL Tauri (Figure 6) belongs to this category. While this object is also an extreme T Tauri star (with a very diluted photospheric spectrum), not all flat-spectrum sources are extreme (e.g. T Tauri and DK Tauri). Conversely, although all extreme stars have prominent infrared excesses, not all of them display flat spectra, as defined above (e.g. DR Tauri, RW Aurigae). Among the 61 Taurus-Auriga stars of the Rydgren et al. (1984b) catalog, there are 6 extreme CTTSs and 9 flat infrared spectrum stars, 3 of which are also extreme. Flat infrared spectrum sources, in spite of their small number, have attracted much attention in recent years, and several modified disk models have been proposed to account for their properties, with only limited success.

Kenyon & Hartmann (1987) envision a reprocessing disk whose thickness is given by $H \propto r^\delta$, $\delta \geq 9/8$. This particular value (9/8) is expected if hydrostatic equilibrium determines the disk's vertical structure. Because of the "flaring" of outer parts of the disk, more stellar photons are absorbed by these cool regions than in a flat disk, and enhanced emission results, particularly in the 60–100 μm region. For this model to work, one must assume, however, that absorbing dust is suspended high above the disk's midplane during its lifetime (several 10^6 yr), whereas estimates of the dust sedimentation time are usually 100 times shorter (Adams et al. 1988). Convection in the disk could lead to the desired effect, but Hartmann & Kenyon (1988) point out that the purely reprocessing disk that they favor is stable against convection. Adams et al. (1988) noted that a flat spectrum will emerge from an accretion disk with temperature distribution proportional to $r^{-1/2}$, and they computed spectra using this assumption. But this temperature law also implies that either the disk mass or the mass accretion rate must be proportional to the disk radius (Hartmann & Kenyon 1988). The latter authors reject both possibilities as unrealistic because they imply either a disk much more massive than the central object or a mass accretion rate 1000 times larger in the outer disk than in the inner disk. Hartmann & Kenyon suggest instead that the far-infrared flat spectrum is caused by radiative transfer effects due to scattering of the stellar light by flattened dust distributions. Linear polarization data indeed indicate that dust around CTTSs cannot be distributed spherically (Bastien

& Landstreet 1979). From an extensive survey of the polarimetric properties of the T Tauri class, Bastien (1987, 1988, and references therein) concludes that the average degree of linear polarization increases with the amount of infrared excess. In fact, the flat infrared spectrum sources have the strongest polarization (up to 12% in HL Tau; Vrba et al. 1976). While these data support Hartmann & Kenyon's (1988) suggestion, a model based on their hypothesis must still be constructed.

Disks and Stellar Evolution

Crucial to our understanding of pre-main-sequence evolution and planet formation is the time scale of evolution for disks surrounding young stellar objects. Comparative studies of CTTSs and WTTSs are interesting in this respect, since most WTTSs, whether or not they are located in the same part of the HR diagram as CTTSs, lack the strong infrared excess characteristic of CTTSs (Walter et al. 1988). The infrared excess distribution in the pre-main-sequence stars of the Taurus-Auriga star formation region was recently examined by Strom et al. (1989) using averaged broadband optical and infrared photometry as well as *IRAS* data. In order to derive the excess infrared flux, they fit a blackbody curve of appropriate temperature at the effective wavelength of the *R*-band (0.64 μm). While this procedure overestimates the photospheric flux (and underestimates the infrared excess) of veiled stars, it is not expected to significantly alter the conclusions of Strom et al., as these are based on a large number of T Tauri stars. Strom et al. find that a significant fraction of WTTSs ($>1/3$) have an excess in the *K*-band larger than 0.1 dex and also often display an excess at wavelengths larger than 10 μm, which suggests that dusty disks, while less prominent than in CTTSs, are still present around these objects. There is generally no spectroscopic evidence for mass accretion in WTTSs (i.e. no Balmer continuous emission; cf. Basri & Bertout 1989). The small infrared excess observed in some WTTSs is thus likely produced by the reprocessing of stellar photons. Based on their statistics, Strom et al. find that the time scale for disk dissipation is typically in the range 3×10^6–10^7 yr, and Basri & Bertout (1989) suggest that some extreme CTTSs with prominent disks may be even older. Of course, this assumes that isochrones derived from the quasi-static evolutionary tracks are approximately correct.

Estimates of mass accretion rates are useful in determining the impact of disks on the stellar evolution of CTTSs. Mass accretion rates smaller than a few times $10^{-8} M_\odot$ yr^{-1} cannot be detected from infrared continuum data alone, but Balmer continuous emission now offers an accretion diagnostic that allows detection of mass accretion rates as small as a few times $10^{-9} M_\odot$ yr^{-1}. At the same time, disks must account for infrared excess without dominating the optical spectrum, since the stellar photosphere is visible in all but the most extreme CTTSs. For reasonable values of the stellar radius, this implies a maximum mass accretion rate of a few times 10^{-7} $M_\odot$ yr^{-1}. Thus, observed infrared and ultraviolet diagnostics constrain the mass accretion rates of CTTSs to a relatively small range. Even if disk accretion at a rate of 10^{-9} $M_\odot$ yr^{-1} during 10^7 yr is not expected to significantly alter the star's evolution, accretion at a rate of 10^{-7} $M_\odot$ yr^{-1} on a low-mass star during the same period of time will. For purposes of illustration, I now assume that the pre-main-sequence evolution of a solar-mass star will be significantly modified by the presence of the disk if half of the stellar mass is accreted from the disk, i.e. if $\dot{M}_{acc} \geq 5 \cdot 10^{-8}$ $M_\odot$ yr^{-1} over a 10^7-yr disk lifetime.

Estimating the fraction of T Tauri stars whose evolution could be affected by disk accretion is, however, difficult. First, accretion probably does not proceed in a steady state. Observed long-term light variations of at least some CTTSs may be due to changes in the mass accretion rate, and the mass accretion rate of FU Orionis objects at maximum light may be up to 1000 times larger than typical CTTS mass accretion rates. Disk instabilities, therefore, may well play a role in the long-term irregular variability of CTTSs and FU Orionis objects (Hartmann & Kenyon 1985, Bertout et al. 1988). Even if we assume a constant $\dot{M}_{acc}$, determining the percentage of T Tauri stars with significant mass accretion is difficult. From estimates based on the infrared excess, Kenyon & Hartmann (1987) conclude that at most 20% of CTTSs show evidence for accretion at rates larger than about 10^{-7} $M_\odot$ yr^{-1}. Bertout (1988) compares the $U-B$ and $B-V$ colors of T Tauri stars with theoretical colors of optically thick disk models and finds that up to 50% of the Taurus-Auriga T Tauri stars of the Rydgren et al. (1984b) data base may be accreting at rates larger than a few times 10^{-8} $M_\odot$ yr^{-1}. A similar result is found by Cabrit et al. (1989) from a detailed study of infrared excesses in a large sample of CTTSs.

In any case, none of the WTTSs appears to accrete at a sizable rate, and Walter et al. (1988) show that WTTSs may largely outnumber CTTSs. Hence, the overall fraction of T Tauri stars whose evolution may be altered by accretion from a disk (i.e. those stars for which classical convective-radiative quasi-static tracks are inappropriate) is probably small anyway. But it appears nevertheless likely that a significant fraction (between 20% and 50%) of CTTSs may not be following usual pre-main-sequence evolutionary tracks. In these systems, the star may grow partly from viscous transport of disk matter onto the protostellar core (Cameron 1978), but the evolution of stars formed in this way is not well understood. Tscharnuter (1985) computed the approximate evolution of a protostellar disk under somewhat comparable conditions but did not follow in detail the evolution of the central protostar.

The finding that accretion of matter with high specific angular momentum can occur during a substantial fraction of the pre-main-sequence phase of evolution in CTTSs might explain why they keep sizable rotation rates in spite of their relatively strong winds (Kenyon & Hartmann 1987). While there is no evidence that WTTSs located in the same region of the HRD as CTTSs display smaller rotation rates (Walter et al. 1988), WTTSs have much weaker winds, so that magnetic braking should not be nearly as efficient in these stars as in CTTSs. Further quantitative and observational work on the topic of rotational evolution during the T Tauri phase should test for the possibility that rotation rates evolve differently in CTTSs and in WTTSs. The apparent deficiency of spectroscopic binaries among CTTSs (cf. Section 2) is a related issue; Walter et al. (1988) suggested that close CTTS systems may evolve into WTTSs because of tidal disruption of their disks. The existence of at least two CTTS spectroscopic binaries does not support this hypothesis (Mathieu 1988), but tidal effects leading to disk instabilities, increased mass accretion, and therefore large-scale variability may be important in spectroscopic binaries as well as in multiple systems with wider separations. Other processes, such as the accretion of the entire disk onto the star or the formation of planets within the disk, could also make a diskless WTTS out of a CTTS. Whether most WTTSs go through an earlier CTTS phase is an issue that deserves further investigation.

Wind-Disk Connections: The Twilight Zone

We now focus on the connections between the winds and disks of T Tauri stars. This is an area of extremely active research today, and a major overhaul of past ideas about protostellar and T Tauri winds will probably result from current work on this topic. But the conventional wisdom on T Tauri mass loss is still relevant and needs summarizing before dealing with current ideas [see also the extensive reviews by Hartmann (1986) and Panagia (1988)].

Mass-loss diagnostics include the permitted emission lines (most notably $H\alpha$), the forbidden lines, and in some cases radio continuum emission, associated Herbig-Haro objects, and optical jets. Because of its prominence in T Tauri stars, the $H\alpha$ line has traditionally been used for mass loss rate determination. The $H\alpha$ line formation problem is, however, a complex one, and the P Cygni Type III profiles characteristic of CTTSs are particularly ambiguous (Section 1). In any case, early estimates yielded mass loss rate values up to several times 10^{-7} $M_\odot$ yr^{-1} (Herbig 1962, Kuhi 1964, Kuan 1975). As first discussed in detail by DeCampli (1981), it is quite difficult to understand how such strong mass losses can originate from T Tauri stars: The ratio of wind kinetic energy to stellar luminosity is typically 0.01

for a T Tauri star losing mass at a rate as small as 10^{-8} $M_\odot$ yr^{-1}. This implies a high efficiency for the wind production mechanism if the wind kinetic energy flux is related to the stellar object's bolometric flux. Applying the theory of Alfvén wave–driven wind to T Tauri stars, Lago (1979, 1984), DeCampli (1981), and Hartmann et al. (1982) take advantage of the efficiency of conversion of linear Alfvén wave flux into mass flux (10–20%) to construct T Tauri winds models with mass loss rates of up to 3×10^{-8} $M_\odot$ yr^{-1}. DeCampli also proposes that Balmer-line widening might not entirely be caused by organized outflow but rather by turbulent velocity fields, so that actual mass loss rates need not be as high as previously thought. Hartmann et al. (1982) find from their detailed models of turbulent Alfvén wave–driven winds that a wide range of observed Hα equivalent widths can indeed be reproduced with mass loss rates $\leq 10^{-8}$ $M_\odot$ yr^{-1}. They also find that the variation of emission measure with temperature observed in CTTSs (cf. Section 2) can be understood in the same framework. Holzer et al. (1983) caution, however, that the scale length over which wave damping takes place is a major parameter of Alfvén wave–driven wind models—it determines both the mass loss rate and the asymptotic flow speed, as well as the wind temperature structure—but find no physical basis for the constant damping length assumed in the models of Hartmann et al. (1982).

Possibilities offered by the presence of an equatorial disk were not considered in any of the above wind models, which were proposed before evidence for disks around CTTSs became compelling. These models are nevertheless instructive because they demonstrate so clearly how difficult it is to find a theoretical mechanism for driving gaseous outflows with mass loss rates larger than 10^{-8} $M_\odot$ yr^{-1} from a CTTS. Since spherical symmetry was assumed, this value is an upper limit to the T Tauri mass loss that models produce today. In contrast, models of centrifugally driven winds emanating from stellar objects rotating at close to breakup velocities or from their disks can produce strong mass losses, and several variants of this mechanism have been proposed to drive outflows from embedded sources, the rotational velocities of which are unknown (Mestel 1968, 1984, Hartmann & MacGregor 1982, Draine 1983, Pudritz & Norman 1983, Uchida & Shibata 1985, Sakurai 1987, Shu et al. 1988, Pringle 1989).

Radio continuum observations (Cohen et al. 1982, Bieging et al. 1984, Cohen & Bieging 1986, André et al. 1987, Montmerle & André 1988) also indicate that CTTSs must have a mass loss rate typically smaller than about 10^{-8} $M_\odot$ yr^{-1} in the ionized component of their winds (or even less if T Tauri winds are strongly collimated; cf. Reynolds 1986). There is, however, growing concern that the ionized wind might not be the only wind component emerging from protostellar systems. Snell & Bally (1986) conclude from a VLA survey of embedded infrared sources driving molecular outflows that mass loss rates derived from the radio continuum emission from low-luminosity sources are usually much smaller than those inferred from the molecular flow. It has then been suggested that protostellar winds might be largely made up of atomic material (Lizano et al. 1988, Natta et al. 1988), and this property could well extend to CTTS winds.

With recent advances in high-resolution spectroscopy, it has become possible to obtain resolved profiles not only of Hα but also of weaker emission lines, in particular the forbidden [O I] and [S II] lines. As it turns out, their profiles provide indirect evidence for the presence of disks around CTTSs. Forbidden lines, formed in regions of low electronic densities, are thought to primarily probe the outer parts of the stellar wind. Initial observations by Jankovics et al. (1983) reveal that the intensity-weighted systemic velocity of forbidden lines is blueshifted. Appenzeller et al. (1985) show that an opaque screen (i.e. an optically thick circumstellar disk) must "hide" the redshifted emission, a conclusion that does not depend on details of disk or wind models. Further work by Edwards et al. (1987) confirms these results and suggests that the typically double-peaked profiles (with one peak close to rest velocity and the second one blueshifted) can be explained by a latitude-dependent, loosely collimated bipolar velocity field.

Since forbidden lines have low absorption probabilities, they are optically thin, which considerably simplifies the line formation problem. Thus, mass loss rates computed from these lines should be more reliable in principle than those computed from Hα, but Edwards et al. (1987) find that they span the range from 10^{-8} to a few 10^{-7} $M_\odot$ yr^{-1}. These high values are puzzling, as is their similarity to those originally found from Hα (e.g. Kuhi 1964). One should perhaps investigate whether a turbulent velocity component could contribute, together with the organized velocity field, to the broadening of forbidden lines. Alternatively, the forbidden lines might come from several atmospheric regions. Hartmann & Raymond (1989) suggest that [O I] may be formed partly in a wind and partly together with [S II] in oblique shocks at the interface between the wind and a "flaring" disk, while Kwan & Tademaru (1988) propose that the two profile components arise in two different wind components (a jet and a low-velocity wind from the disk). High spectral resolution studies of forbidden lines are currently underway to distinguish between all these possibilities (S. Edwards, private communication).

A recent study of forbidden emission in a large sample of T Tauri stars by Cabrit et al. (1989) shows that if the [O I] λ6300 line flux is larger than

10^{-13} erg s^{-1} cm^{-2}, then [S II] $\lambda 6731$ is detected, [O I] and [S II] line fluxes are correlated, and the intensity-weighted [O I] velocity is largely blueshifted. If, on the other hand, the [O I] flux is smaller than 10^{-13} erg s^{-1} cm^{-2}, then [S II] $\lambda 6731$ is not detected, [S II] and [O I] fluxes are not correlated, and the intensity-weighted [O I] velocity is less blueshifted than in the first case. All this suggests different emission mechanisms for the two subgroups. Cabrit et al. (1989) propose a qualitative model in which a warm stellar wind provides a significant fraction of the [O I] flux, while internal shocks in unresolved jets or Herbig-Haro objects embedded in the wind emit in both [O I] and [S II] and contribute an increasing fraction of the total [O I] flux as the wind velocity (and presumably the mass loss rate) increases.

Since many low-luminosity embedded protostellar sources as well as the most extreme CTTSs display resolved jets in [S II] (Figure 6), this phenomenon may well extend to less active stars on a smaller scale. Mundt (1988) indeed suggests that the sharp blueshifted absorption components found in the Na D lines of CTTSs (cf. Mundt 1984) signal a slow (50–100 km s^{-1}), poorly collimated wind, while the ionized jets seen in [S II] represent a second, fast (300 km s^{-1}) and well-collimated but also less massive wind component. While Mundt points out the difficulty of finding out the origin of the Na D sharp absorption components, it is tantalizing that the average velocity of the Hα reversal in CTTSs is 80 km s^{-1} (Herbig 1977a). Furthermore, the velocity of the Hα reversal is correlated with the [O I] $\lambda 6300$ blueshifted peak (Edwards et al. 1987). It therefore seems probable that Na D and Hα absorption components as well as one component of [O I] emission are all probing different regions of the outflow. While awaiting further investigation of this topic, one should keep in mind that the wind terminal velocity might be 2 to 4 times smaller than the canonical 200 km s^{-1} velocity assumed in most mass loss rate determinations.

At the low end of the mass loss scale, WTTSs show little evidence of strong winds. With few exceptions, no forbidden-line emission is observed in these objects (Walter et al. 1988, Strom et al. 1989). And yet, like moderately active CTTSs, WTTSs also display X-ray luminosities equal to about 10^{-3} times their bolometric luminosities; these radiative losses are comparable to those of a solar-type wind with a mass loss rate of a few times $10^{-9} M_\odot$ yr^{-1}. That massive outflows appear confined to CTTSs raises the question of the role of disks in driving their outflows.

Cohen et al. (1989) and Cabrit et al. (1989) offer first evidence that the disks and winds of CTTSs must indeed be somehow causally connected. Both teams try to disentangle photospheric flux from disk flux by fitting an appropriate stellar energy distribution at a chosen wavelength, either in the optical range (Cohen et al. 1989) or 0.64 μm (Cabrit et al. 1989). Comparison with disk models shows that the fit procedure of Cohen et al. may lead to a much larger overestimation of the photospheric luminosity in the case of boundary layer emission. Both groups find correlations between [O I] λ6300 emission-line flux and the infrared excess in a large sample of stars; the tighter correlation found by Cabrit et al. (1989) probably reflects a closer estimate of the photospheric contribution. Since Hα and [O I] are known to be correlated (Cohen & Kuhi 1979), the Hα flux is also correlated with the infrared excess. None of these quantities, however, is correlated with photospheric luminosity, even for stars along the same convective track. Both groups are therefore led to conclude that the infrared excess is not age dependent. This result, combined with the presence of WTTSs on convective tracks (Figure 4), suggests that the mass and extent of the circumstellar disks formed with the young stellar objects during the protostellar phase are largely determined by the initial conditions of star formation (i.e. by physical conditions in the molecular core from which the star was born). This, in turn, means that the most prominent T Tauri characteristics, due largely to the presence of a circumstellar disk, do not depend primarily on the age of the central star. This finding, together with the discovery of an embedded population of radio-emitting Class III objects in the ρ Ophiuchi cloud (André et al. 1987), suggests that an evolutionary scheme based only on the spectral appearance of young stellar objects (Adams et al. 1987) cannot give a complete view of protostellar evolution and should be refined in order to take into account the wide range of initial properties implied by the data.

If the [O I] flux reflects the wind mass loss rate, the correlation between [O I] flux and infrared excess indicates that T Tauri winds might be powered by mass accretion. Estimated ratios of mass loss to mass accretion rates— computed by assuming that the entire [O I] flux originates from the wind— range from a few percent to almost 100% (with considerable uncertainty).[9] If taken at face value, these numbers would imply that the conversion of mass accretion to mass outflow must be quite efficient. Their exact significance is, however, unclear owing to the uncertainties discussed above about the origin of [O I] emission. While these results provide preliminary suggestive evidence that disk accretion may be driving T Tauri winds, the mere presence of optically thick disks around CTTSs also has several consequences for Hα formation that future wind models must take into account. Typical Hα emission profiles are broad and symmetric, quite

---

[9] Edwards & Strom (1988) find similar conversion ratios for embedded protostellar objects driving molecular outflows, but derivation of mass loss rates from CO observations is also uncertain (cf. Bertout 1988).

unlike those of forbidden lines, which are systematically blueshifted. This means that occultation effects, while important for the forbidden lines, are unimportant for Hα.

Two possibilities then suggest themselves within the framework of the accretion disk hypothesis. On the one hand, the disk may be optically thin from the stellar photosphere up to say five stellar radii or more, in which case there would be little occultation of the Hα-forming region, and P Cygni Type III Hα profiles could result from a bipolar wind as in models computed by Bertout (1985). This scenario requires, however, either large values of α, corresponding to uncomfortably supersonic turbulence, or disruption of the inner part of the disk by a magnetic field leading to accretion along field lines. If, on the other hand, the circumstellar disk is optically thick down to the stellar surface, as in current models, with moderate α values, this in turn implies that the broad Hα emission base must be formed in a region dominated by turbulence; we have noted above that turbulent Hα broadening is also desirable from a theoretical point of view. In this picture, the blueshifted absorption component is the feature of Hα directly related to mass loss. The boundary layer is expected to be turbulent and to generate acoustic and magnetic waves, which raises the possibility that the wind might originate from that region. Both Torbett (1984) and Uchida & Shibata (1985) have discussed somewhat related models in which the wind is driven primarily by thermal pressure, but inferred radiative losses do not appear compatible with observations of T Tauri stars (Hartmann 1986). Other plausible wind-driving mechanisms [e.g. the hydromagnetic wind originating from the boundary layer proposed by Pringle (1989)] depend on the interplay of stellar and disk magnetic fields. Future work in this direction, while complicated, should be quite fruitful.

4. CONCLUSION

This review has emphasized the similarities and differences between T Tauri stars and stars of comparable masses during both earlier and later phases of evolution. T Tauri stars are tentatively depicted as complex systems whose properties depend mostly on (*a*) the initial conditions of star formation, which apparently determine the properties of their circumstellar disks; and (*b*) their rotation rates, which appear to control the magnetic dynamo activity in T Tauri stars as they do in other late-type stars. The most exotic traits of CTTSs are primarily due to the disk and its interaction with the star, and the properties of WTTSs are mainly manifestations of the enhanced solar-type magnetic activity expected from their rotation rates. While some WTTSs may be born without sizable circumstellar disks, CTTSs are expected to become WTTSs when their disks dissipate. One possible cause of disk dissipation might be planet formation. Several aspects of this paradigm await extensive observational and theoretical testing, and topics urgently needing work have been noted throughout the review. Although this global picture appears seductive, several of its fundamental aspects are still not understood: the evolution of stars undergoing mass accretion through a disk, the generation and collimation of winds from CTTSs, the transport of matter in disks, and the stability properties of protostellar disks.

ACKNOWLEDGMENTS

It is a pleasure to thank Joli Adams, Immo Appenzeller, Gibor Basri, Pierre Bastien, Jérôme Bouvier, Sylvie Cabrit, Nuria Calvet, Suzan Edwards, Thierry Montmerle, Reinhard Mundt, and Jean-Claude Pecker. Their comments on early versions of this review led to significant improvements of both its contents and presentation. I am also indebted to Gibor Basri, Jérôme Bouvier, Reinhard Mundt, and Thierry Montmerle for providing illustrations.

Literature Cited

Adams, F. C., Lada, C. J., Shu, F. H. 1987. *Ap. J.* 312: 788
Adams, F. C., Lada, C. J., Shu, F. H. 1988. *Ap. J.* 326: 865
Adams, F. C., Shu, F. H. 1986. *Ap. J.* 308: 836
Alphenaar, P., van Leeuwen, F. 1981. *Inf. Bull. Variable Stars No. 1957*
Ambartsumian, J. A. 1947. *Stellar Evolution and Astrophysics.* Erevan: Acad. Sci. Armen. SSR
André, P., Montmerle, T., Feigelson, E. D. 1987. *Astron. J.* 93: 1182
Appenzeller, I. 1983. *Rev. Mex. Astron. Astrofis.* 7: 151
Appenzeller, I., Chavarria, C., Krautter, J., Mundt, R., Wolf, B. 1980. *Astron. Astrophys.* 90: 184
Appenzeller, I., Dearborn, D. S. R. 1984. *Ap. J.* 278: 689
Appenzeller, I., Jankovics, I., Krautter, J. 1983. *Astron. Astrophys. Suppl.* 53: 291
Appenzeller, I., Jankovics, I., Jetter, R. 1986. *Astron. Astrophys. Suppl.* 64: 65
Appenzeller, I., Jankovics, I., Oestreicher, R. 1985. *Astron. Astrophys.* 141: 108
Appenzeller, I., Reitermann, A., Stahl, O. 1988. *Publ. Astron. Soc. Pac.* 100: 815
Appenzeller, I., Wolf, B. 1977. *Astron. Astrophys.* 54: 713
Basri, G. 1987. *Ap. J.* 316: 377
Basri, G., Bertout, C. 1989. *Ap. J.* In press
Basri, G., Marcy, G. W. 1988. *Ap. J.* 330: 274
Bastian, U., Finkenzeller, U., Jaschek, C., Jaschek, M. 1983. *Astron. Astrophys.* 126: 438
Bastien, P. 1987. *Ap. J.* 317: 231
Bastien, P. 1988. In *Polarized Radiation of Circumstellar Origin*, ed. G. V. Coyne. Vatican City: Vatican Obs. In press
Bastien, P., Landstreet, J. D. 1979. *Ap. J. Lett.* 229: L137
Bastien, P., Ménard, F. 1987. *Ap. J.* 326: 334
Bastien, P., Robert, C., Nadeau, R. 1989. *Ap. J.* In press
Beall, J. H. 1986. *Ap. J.* 316: 227
Beals, C. S. 1950. *Publ. Dominion Astrophys. Obs.* 9: 1
Beckwith, S., Sargent, A. I., Scoville, N. Z., Masson, C. R., Zuckerman, B., Phillips, T. G. 1986. *Ap. J.* 309: 755
Beckwith, S., Zuckerman, B., Skrutskie, M. F., Dyck, H. M. 1984. *Ap. J.* 287: 793
Benz, W., Mayor, M. 1981. *Astron. Astrophys.* 93: 235

Bertout, C. 1984. *Rep. Prog. Phys.* 47: 111
Bertout, C. 1985. In *Nearby Molecular Clouds*, ed. G. Serra, p. 161. Berlin: Springer-Verlag
Bertout, C. 1987. In *Circumstellar Matter*, ed. I. Appenzeller, C. Jordan, p. 23. Dordrecht: Reidel
Bertout, C. 1988. In *Formation and Evolution of Low-Mass Stars*, ed. A. K. Dupree, p. 45. Dordrecht: Reidel
Bertout, C., Basri, G., Bouvier, J. 1988. *Ap. J.* 330: 350
Bertout, C., Carrasco, L., Mundt, R., Wolf, B. 1982. *Astron. Astrophys. Suppl.* 47: 419
Bieging, J. H., Cohen, M., Schwartz, P. R. 1984. *Ap. J.* 282: 699
Boesgaard, A. M., Tripicco, M. J. 1987. *Ap. J.* 307: 389
Bouvier, J. 1987. In *Protostars and Molecular Clouds*, ed. T. Montmerle, C. Bertout, p. 189. Gif-sur-Yvette, Fr: CEN
Bouvier, J. 1989. Submitted for publication
Bouvier, J., Appenzeller, I. 1989. In preparation
Bouvier, J., Bertout, C. 1989. *Astron. Astrophys.* 211: 99
Bouvier, J., Bertout, C., Benz, W., Mayor, M. 1985. In *Nearby Molecular Clouds*, ed. G. Serra, p. 222. Berlin: Springer-Verlag
Bouvier, J., Bertout, C., Benz, W., Mayor, M. 1986a. *Astron. Astrophys.* 165: 110
Bouvier, J., Bertout, C., Bouchet, P. 1986b. *Astron. Astrophys.* 158: 149
Bouvier, J., Bertout, C., Bouchet, P. 1988. *Astron. Astrophys. Suppl.* 75: 1
Brown, A., Drake, S. A., Mundt, R. 1985. In *Radio Stars*, ed. R. M. Hjellming, D. M. Gibson, p. 105. Dordrecht: Reidel
Butler, R. P., Cohen, R. D., Duncan, D. K., Marcy, G. W. 1987. *Ap. J. Lett.* 319: L22
Cabrit, S., Edwards, S., Strom, S. E., Strom, K. M. 1989. Submitted for publication
Calvet, N., Albarrán, J. 1984. *Rev. Mex. Astron. Astrofis.* 9: 35
Calvet, N., Basri, G., Imhoff, C. L., Giampapa, M. S. 1985. *Ap. J.* 293: 575
Calvet, N., Basri, G., Kuhi, L. V. 1984. *Ap. J.* 277: 725
Cameron, A. G. W. 1978. In *Protostars and Planets*, ed. T. Gehrels, p. 453. Tucson: Univ. Ariz. Press
Chavarria, C. 1979. *Astron. Astrophys.* 79: L18
Cohen, M. 1983. *Ap. J. Lett.* 270: L69
Cohen, M. 1984. *Phys. Rep.* 116: 173
Cohen, M., Bieging, J. H. 1986. *Astron. J.* 92: 1396
Cohen, M., Bieging, J. H., Schwartz, P. R. 1982. *Ap. J.* 253: 707
Cohen, M., Emerson, J. P., Beichman, C. A. 1989. *Ap. J.* In press
Cohen, M., Kuhi, L. V. 1979. *Ap. J. Suppl.* 41: 743

Cram, L. E. 1979. *Ap. J.* 234: 949
Cram, L. E., Giampapa, M. S., Imhoff, C. L. 1980. *Ap. J.* 238: 905
DeCampli, W. M. 1981. *Ap. J.* 244: 124
Draine, B. 1983. *Ap. J.* 270: 519
Dumont, S., Heidmann, N., Kuhi, L. V., Thomas, R. N. 1974. *Astron. Astrophys.* 29: 199
Duncan, D. K. 1981. *Ap. J.* 248: 651
Durney, B. R., Latour, J. 1978. *Geophys. Astrophys. Fluid Dyn.* 9: 241
Dyck, H. M., Simon, T., Zuckerman, B. 1984. *Ap. J. Lett.* 243: L89
Edwards, S. 1979. *Publ. Astron. Soc. Pac.* 91: 329
Edwards, S., Cabrit, S., Strom, S. E., Heyer, I., Strom, K. M., Anderson, E. 1987. *Ap. J.* 321: 473
Edwards, S., Strom, S. E. 1988. In *Cambridge Cool Star Workshop, 5th*, ed. J. L. Linsky, R. Stencel, p. 443. Berlin: Springer-Verlag
Feigelson, E. D., DeCampli, W. M. 1981. *Ap. J. Lett.* 243: L89
Feigelson, E. D., Jackson, J. M., Mathieu, R. D., Myers, P. C., Walter, F. D. 1987. *Astron. J.* 94: 1251
Feigelson, E. D., Kriss, G. A. 1981. *Ap. J. Lett.* 248: L35
Feigelson, E. D., Montmerle, T. 1985. *Ap. J. Lett.* 289: L19
Finkenzeller, U., Basri, G. 1987. *Ap. J.* 318: 823
Franchini, M., Magazzù, A., Stalio, R. 1988. *Astron. Astrophys.* 189: 132
Gahm, G. F. 1981. *Ap. J. Lett.* 242: L163
Gahm, G. F., Fischerström, C., Liseau, R., Lindroos, K. P. 1989. *Astron. Astrophys.* 211: 115
Gershberg, R. E. 1983. In *Activity in Red Dwarf Stars*, ed. P. B. Byrne, M. Rodono, p. 487. Dordrecht: Reidel
Giampapa, M. S., Calvet, N., Imhoff, C. L., Kuhi, L. V. 1981. *Ap. J.* 251: 113
Gilliland, R. L. 1986. *Ap. J.* 300: 339
Glasby, J. S. 1974. *The Nebular Variables.* Oxford: Pergamon
Goodrich, R. W. 1986. *Publ. Astron. Soc. Pac.* 99: 116
Grasdalen, G. L., Strom, S. E., Strom, K. M., Capps, R. W., Thompson, D., Castelaz, M. 1984. *Ap. J. Lett.* 283: L57
Grinin, V. P., Petrov, P. P., Shakhovskaya, N. I. 1983. In *Activity in Red Dwarf Stars*, ed. P. B. Byrne, M. Rodono, p. 513. Dordrecht: Reidel
Haro, G. 1968. In *Nebulae and Interstellar Matter (Stars and Stellar Systems*, Vol. 7), ed. B. M. Middlehurst, L. H. Aller, p. 141. Chicago: Univ. Chicago Press
Haro, G. 1976. *Bol. Inst. Tonatzintla* 2: 3
Hartmann, L. 1982. *Ap. J. Suppl.* 48: 109

Hartmann, L. 1986. *Fundam. Cosmic Phys.* 11: 279
Hartmann, L., Edwards, S., Avrett, A. 1982. *Ap. J.* 261: 279
Hartmann, L., Hewett, R., Stahler, S., Mathieu, R. D. 1986. *Ap. J.* 309: 275
Hartmann, L., Kenyon, S. J. 1985. *Ap. J.* 299: 462
Hartmann, L., Kenyon, S. J. 1987a. *Ap. J.* 312: 243
Hartmann, L., Kenyon, S. J. 1987b. *Ap. J.* 322: 393
Hartmann, L., Kenyon, S. J. 1988. In *Formation and Evolution of Low-Mass Stars*, ed. A. K. Dupree, p. 163. Dordrecht: Reidel
Hartmann, L., MacGregor, K. B. 1982. *Ap. J.* 259: 180
Hartmann, L. W., Noyes, R. W. 1987. *Annu. Rev. Astron. Astrophys.* 25: 271
Hartmann, L., Raymond, J. C. 1989. *Ap. J.* In press
Hartmann, L., Rosner, R. 1979. *Ap. J.* 230: 802
Hartmann, L., Soderblom, D. R., Stauffer, J. R. 1987. *Astron. J.* 93: 907
Herbig, G. H. 1962. *Adv. Astron. Astrophys.* 1: 47
Herbig, G. H. 1965. *Ap. J.* 141: 588
Herbig, G. H. 1970. *Mem. Soc. R. Sci. Liège* 19: 13
Herbig, G. H. 1977a. *Ap. J.* 214: 747
Herbig, G. H. 1977b. *Ap. J.* 217: 693
Herbig, G. H. 1978. In *Problems of Physics and Evolution of the Universe*, p. 171. Yerevan: Acad. Sci. Armen. SSR
Herbig, G. H. 1985. *Ap. J.* 289: 269
Herbig, G. H., Bell, K. R. 1988. *Lick Obs. Bull. No. 1111*
Herbig, G. H., Goodrich, R. W. 1986. *Ap. J.* 309: 294
Herbig, G. H., Vrba, F. J., Rydgren, A. E. 1986. *Astron. J.* 91: 575
Herbst, W. 1986. *Publ. Astron. Soc. Pac.* 98: 1088
Herbst, W., Booth, J. F., Chugainov, P. F., Zajtseva, G. V., Barksdale, W., et al. 1986. *Ap. J. Lett.* 310: L71
Herbst, W., Koret, D. L. 1988. *Astron. J.* 96: 1949
Hoffmeister, C. 1965. *Veröff. Sternwarte Sonneberg* 6: 97
Holtzmann, J. A., Herbst, W., Booth, J. 1986. *Astron. J.* 92: 1387
Holzer, T. E., Fla, T., Leer, E. 1983. *Ap. J.* 275: 808
Imhoff, C. L., Appenzeller, I. 1987. In *Scientific Accomplishments of the IUE*, ed. Y. Kondo, p. 295. Dordrecht: Reidel
Imhoff, C. L., Giampapa, M. S. 1980. In *The Universe at Ultraviolet Wavelengths: The First Two Years of IUE*, ed. R. D. Chapman, p. 185. *NASA CP-2171*

Jankovics, I., Appenzeller, I., Krautter, J. 1983. *Publ. Astron. Soc. Pac.* 95: 883
Johnstone, R. M., Penston, M. V. 1988. *MNRAS* 227: 797
Jones, B. F., Herbig, G. H. 1979. *Astron. J.* 84: 1872
Jordan, C., de Ferraz, M. C., Brown, A. 1982. *Proc. Eur. IUE Conf., 3rd. ESA Publ. ESA-SP 176*, p. 83
Joy, A. H. 1942. *Publ. Astron. Soc. Pac.* 54: 15
Joy, A. H. 1945. *Ap. J.* 102: 168
Joy, A. H. 1949. *Ap. J.* 110: 424
Joy, A. H. 1960. In *Stellar Atmospheres (Stars and Stellar Systems*, Vol. 6), ed. J. L. Greenstein, p. 653. Chicago: Univ. Chicago Press
Kenyon, S. J., Hartmann, L. 1987. *Ap. J.* 323: 714
Krautter, J., Bastian, U. 1980. *Astron. Astrophys.* 88: L6
Kuan, P. 1975. *Ap. J.* 202: 425
Kuan, P. 1976. *Ap. J.* 210: 129
Kuhi, L. V. 1964. *Ap. J.* 140: 1409
Kuhi, L. V. 1974. *Astron. Astrophys. Suppl.* 15: 47
Kuhi, L. V. 1978. In *Protostars and Planets*, ed. T. Gehrels, p. 708. Tucson: Univ. Ariz. Press
Kwan, J., Tademaru, E. 1988. *Ap. J. Lett.* 332: L41
Lada, C. J. 1985. *Annu. Rev. Astron. Astrophys.* 23: 267
Lada, C. J. 1986. In *Star Forming Regions*, ed. M. Peimbert, J. Jugaku, p. 1. Dordrecht: Reidel
Lada, C. J., Wilking, B. A. 1984. *Ap. J.* 287: 610
Lago, M. T. V. T. 1979. DPhil. thesis. Univ. Sussex, Engl.
Lago, M. T. V. T. 1984. *MNRAS* 210: 323
Levreault, R. M. 1988. *Ap. J. Suppl.* 67: 283
Lin, D. N. C., Papaloizou, J. 1985. In *Protostars and Planets II*, ed. D. C. Black, M. S. Matthews, p. 981. Tucson: Univ. Ariz. Press
Linsky, J. 1981. In *Solar Phenomena in Stars and Stellar Systems*, ed. R. M. Bonnet, A. K. Dupree, p. 99. Dordrecht: Reidel
Lizano, S., Heiles, C., Rodriguez, L. F., Koo, B.-C., Shu, F. H., et al. 1988. *Ap. J.* 328: 763
Lynden-Bell, D., Pringle, J. E. 1974. *MNRAS* 168: 603
Mathieu, R. D. 1988. In *Highlights of Astronomy*, ed. R. McNally. In press
Mendoza, V. E. E. 1966. *Ap. J.* 143: 1010
Mendoza, V. E. E. 1968. *Ap. J.* 151: 977
Mestel, L. 1968. *MNRAS* 138: 359
Mestel, L. 1984. In *Cool Stars, Stellar Systems, and the Sun*, ed. S. L. Baliunas, L. Hartmann, p. 49. Berlin: Springer-Verlag

Montmerle, T., André, P. 1988. In *Formation and Evolution of Low-Mass Stars*, ed. A. K. Dupree, M. T. V. T. Lago, p. 225. Dordrecht: Kluwer

Montmerle, T., Koch-Miramond, L., Falgarone, E., Grindlay, J. E. 1983. *Ap. J.* 269: 182

Mundt, R. 1984. *Ap. J.* 280: 749

Mundt, R. 1988. In *Formation and Evolution of Low-Mass Stars*, ed. A. K. Dupree, p. 257. Dordrecht: Reidel

Mundt, R., Brugel, E. W., Bührke, T. 1987. *Ap. J.* 319: 275

Mundt, R., Giampapa, M. S. 1982. *Ap. J.* 256: 156

Mundt, R., Ray, T., Bührke, T. 1988. *Ap. J. Lett.* 333: L69

Myers, P. C., Fuller, G. A., Mathieu, R. D., Beichman, C. A., Benson, P. J., et al. 1987. *Ap. J.* 319: 340

Nadeau, R., Bastien, P. 1986. *Ap. J. Lett.* 307: L5

Natta, A., Giovanardi, C., Palla, F., Evans, N. J. 1988. *Ap. J.* 327: 817

Nisenson, P., Stachnik, R. V., Karouska, M., Noyes, R. 1985. *Ap. J. Lett.* 297: L17

Panagia, N. 1988. In *Galactic and Extragalactic Star Formation*, ed. R. E. Pudritz, M. Fich, p. 25. Dordrecht: Kluwer

Parsamyan, E. S., Gasparyan, K. G. 1987. *Astrofizika* 27: 447

Pringle, J. E. 1977. *MNRAS* 178: 95

Pringle, J. E. 1989. *MNRAS* 236: 107

Pringle, J. E., Savonije, G. J. 1979. *MNRAS* 187: 777

Pudritz, R. E., Norman, C. A. 1983. *Ap. J.* 274: 677

Regev, O. 1983. *Astron. Astrophys.* 123: 146

Reipurth, B. 1988. In *Formation and Evolution of Low-Mass Stars*, ed. A. K. Dupree, M. T. V. T. Lago, p. 305. Dordrecht: Kluwer

Reynolds, S. P. 1986. *Ap. J.* 304: 713

Rucinski, S. M. 1985. *Astron. J.* 90: 2321

Rydgren, A. E., Schmelz, J. T., Zak, D. S., Vrba, F. J. 1984b. *Publ. US Nav. Obs.*, Vol. 25, Part 1

Rydgren, A. E., Vrba, F. J. 1983. *Ap. J.* 267: 191

Rydgren, A. E., Zak, D. S. 1987. *Publ. Astron. Soc. Pac.* 99: 141

Rydgren, A. E., Zak, D. S., Vrba, F. J., Chugainov, P. F., Zajtseva, G. V. 1984a. *Astron. J.* 89: 1015

Saar, S. H., Linsky, J. L. 1985. *Ap. J. Lett.* 299: L47

Sakurai, T. 1987. *Publ. Astron. Soc. Jpn.* 39: 821

Sargent, A. I., Beckwith, S. 1987. *Ap. J.* 323: 294

Schaefer, B. 1983. *Ap. J. Lett.* 266: L45

Schneeberger, T. J., Worden, S. P., Wilkerson, M. S. 1979. *Ap. J. Suppl.* 41: 369

Shakura, N. I., Sunyaev, R. A. 1973. *Astron. Astrophys.* 24: 337

Shu, F. H., Adams, F. C., Lizano, S. 1987. *Annu. Rev. Astron. Astrophys.* 25: 23

Shu, F. H., Lizano, S., Ruden, S. P., Najita, J. 1988. *Ap. J. Lett.* 328: L19

Simon, M., Howell, R. R., Longmore, A. J., Wilking, B. A., Peterson, D. M., Cheng, W.-P. 1987. *Ap. J.* 320: 344

Simon, T. S., Herbig, G. H., Boesgaard, A. M. 1985. *Ap. J.* 293: 551

Skumanich, A. 1972. *Ap. J.* 171: 565

Snell, R. L., Bally, J. 1986. *Ap. J.* 303: 683

Stauffer, J. R. 1987. In *Cool Stars, Stellar Systems, and the Sun*, ed. J. L. Linsky, R. E. Stencel, p. 182. Berlin: Springer-Verlag

Stine, P. C., Feigelson, E. D., André, P., Montmerle, T. 1988. *Ap. J.* 335: 940

Strom, K. M., Strom, S. E., Edwards, S., Cabrit, S., Strutskie, M. F. 1989. *Astron. J.* In press

Strom, S. E. 1972. *Publ. Astron. Soc. Pac.* 84: 745

Strom, S. E. 1983. *Rev. Mex. Astron. Astrofis.* 7: 201

Tayler, R. J. 1987. *MNRAS* 227: 553

Torbett, M. 1984. *Ap. J.* 278: 318

Tscharnuter, W. M. 1985. In *Birth and Infancy of Stars*, ed. R. Lucas, A. Omont, R. Stora, p. 601. Amsterdam: Elsevier

Tylenda, R. 1981. *Acta Astron.* 31: 127

Uchida, Y., Shibata, K. 1985. *Publ. Astron. Soc. Jpn.* 37: 515

Ulrich, R. K. 1976. *Ap. J.* 210: 377

Ulrich, R. K., Knapp, G. R. 1979. *Ap. J. Lett.* 230: L99

Vogel, S. S., Kuhi, L. V. 1981. *Ap. J.* 245: 960

Vrba, F. J., Herbst, W., Booth, J. F. 1988. *Astron. J.* 96: 1032

Vrba, F. J., Rydgren, A. E., Chugainov, P. F., Shakovskaya, N. I., Zak, D. S. 1986. *Ap. J.* 306: 199

Vrba, F. J., Strom, K. M., Strom, S. E., Grasdalen, G. L. 1975. *Ap. J.* 197: 77

Vrba, F. J., Strom, S. E., Strom, K. M. 1976. *Astron. J.* 81: 962

Wagenblast, R., Bertout, C., Bastian, U. 1982. *Astron. Astrophys.* 120: 6

Walker, M. F. 1972. *Ap. J.* 175: 89

Walker, M. F. 1987. *Publ. Astron. Soc. Pac.* 99: 392

Walter, F. M. 1986. *Ap. J.* 306: 573

Walter, F. M., Bowyer, C. S. 1981. *Ap. J.* 245: 677

Walter, F. M., Brown, A., Mathieu, R. D., Myers, P. C., Vrba, F. J. 1988. *Astron. J.* 96: 297

Walter, F. M., Kuhi, L. V. 1981. *Ap. J.* 250: 254

Walter, F. M., Kuhi, L. V. 1984. *Ap. J.* 284: 194

Warner, B. 1988. In *High Speed Astro-*

nomical Photometry, p. 106. Cambridge: Univ. Press
Wilking, B. A., Lada, C. J. 1983. *Ap. J.* 274: 698
Wilking, B. A., Lada, C. J., Young, E. T. 1989. Preprint
Willson, L. A. 1974. *Ap. J.* 191: 143
Willson, L. A. 1975. *Ap. J.* 197: 365
Wolf, B., Appenzeller, I., Bertout, C. 1977. *Astron. Astrophys.* 58: 163
Worden, S. P., Schneeberger, T. J., Kuhn, J. R., Africano, J. L. 1981. *Ap. J.* 244: 250
Zappala, R. R. 1972. *Ap. J.* 172: 57
Zinnecker, H. 1988. In *Formation and Evolution of Low-Mass Stars*, ed. A. K. Dupree, M. T. V. T. Lago, p. 111. Dordrecht: Kluwer
Zinnecker, H., Perrier, C., Chelli, A. 1988. *Proc. NOAO/ESO Conf. High Resolution Imaging by Interferometry*. Garching: ESO. In press

ASTROPHYSICAL CONTRIBUTIONS OF THE INTERNATIONAL ULTRAVIOLET EXPLORER[1]

Yoji Kondo

Laboratory for Astronomy and Solar Physics, Code 684, NASA/Goddard Space Flight Center, Greenbelt, Maryland 20771

Albert Boggess

Space Telescope Project Office, Code 440, NASA/Goddard Space Flight Center, Greenbelt, Maryland 20771

Stephen P. Maran

Laboratory for Astronomy and Solar Physics, Code 680, NASA/Goddard Space Flight Center, Greenbelt, Maryland 20771

1. INTRODUCTION

The *International Ultraviolet Explorer* (*IUE*) was launched from the Kennedy Space Center on January 26, 1978 (Boggess et al. 1978a,b). The launch was the culmination of nearly ten years of joint study and development by the National Aeronautics and Space Administration (NASA), the European Space Agency (ESA), and the UK's Science and Engineering Research Council (SERC). The mission was conceived from the beginning as an ultraviolet spectroscopic observatory that could be used by astronomers as they were accustomed to observing from the ground, without special training in the mysteries of orbital operations. This dictated that the satellite be placed in a high orbit, and in fact an elliptical geosynchronous orbit was chosen to allow the observing to proceed at the diurnal rate and to simplify the logistics of ground stations. The disadvantage that only a

---

[1] The US Government has the right to retain a nonexclusive, royalty-free license in and to any copyright covering this paper.

small telescope could be lifted to geosynchronous orbit was offset by the multiplex advantages offered by the two-dimensional SEC (secondary electron conduction) vidicon cameras that were selected as detectors for the spectrographs. The success of the concept has been demonstrated amply by the continuing intense demand for observing time from astronomers all over the world.

The spacecraft is controlled from the NASA/Goddard Space Flight Center (GSFC) in Greenbelt, Maryland. The scientific instruments on board are operated 16 hours daily from the science operations center there and 8 hours daily from the ground station in Villafranca, near Madrid, Spain.

IUE was planned from the beginning to be a long-lived space observatory, but in the early 1970s, *IUE*'s three-year design lifetime requirement seemed like a very long life indeed to the responsible scientists and engineers. The fact that the observatory is still maintaining a full and productive observing schedule with undiminished sensitivity after more than 10 years in orbit is a tribute to the skill and commitment of those who built and have operated the satellite and ground system. The spacecraft, telescope, instrumentation, and US ground system were built by GSFC. The SEC vidicon cameras were produced by the UK's Appleton/Rutherford Laboratories, and ESA's ESTEC provided solar arrays and the European ground system.

The satellite observatory consists of a 45-cm Cassegrain telescope equipped with two spectrographs, for the 1900–3200 Å and 1150–1950 Å spectral ranges. There are two SEC vidicon detectors for each spectrograph, and primary and redundant cameras, designated LWP, LWR for the longer wavelength range and SWP, SWR for the shorter wavelength range. The observational characteristics of *IUE* are summarized in Table 1.

The scientific operations of *IUE* are normally performed in real time, with the observer at the console assisted by a Resident Astronomer and a Telescope Operator. The data are telemetered to the ground after an exposure is completed, and the reduced data are routinely available within 24 hr. At the GSFC *IUE* science operations center, a regional data analysis facility is available to facilitate prompt evaluation of the observations.

During the first 10 years of *IUE* operations, 832 different individuals used the GSFC observing facilities, while approximately 770 astronomers used the Villafranca ground station. Although observers have come from all over the world, these totals in fact represent significant fractions of the astronomers active in research in North America and Europe.

As a measure of the scientific productivity of *IUE*, we note that as of late 1987, a few weeks before the tenth anniversary of launch, 1571 papers in refereed journals were identified as based on *IUE* results. Of course, the

Table 1 Characteristics of *IUE* instrumentation

Telescope	
Clear aperture	0.45 m
Figure	Ritchey-Chretien
Focal ratio	15
Image quality	3 arcsec
Long-wavelength spectrograph	
Spectral range	1900–3200 Å
Echelle resolving power	13,000
Low-dispersion resolution	8 Å
Short-wavelength spectrograph	
Spectral range	1150–1950 Å
Echelle resolving power	12,000
Low-dispersion resolution	6 Å

most important measure of an observatory is the quality of the work; here we describe some highlights of the *IUE* research. The scope of that work is so great that we have had to omit many very significant individual findings and even several major subject areas. For more detailed information and for reviews of *IUE* results over the full range of astrophysical disciplines, the reader is referred to the compendium of review articles edited by Kondo et al. (1987) and to the proceedings of nine *IUE* symposia, especially the most recent one, *A Decade of UV Astronomy With the IUE Satellite* (*ESA SP-281*, two volumes), which contains the proceedings of a symposium held in April 1988. The dates, locations, and sponsors of the *IUE* symposia were as follows:

1979	London	SERC, ESA, and NASA
1980	Greenbelt	NASA
	Tübingen	ESA and SERC
1982	Greenbelt	NASA
	Madrid	ESA and SERC
1984	Greenbelt	NASA
	Rome	ESA and SERC
1986	London	SERC, ESA, and NASA
1988	Greenbelt	NASA, ESA, and SERC

2. STELLAR CHROMOSPHERES AND TRANSITION REGIONS

Prior to the advent of space observations in the UV, the study of chromospheres in stars other than the Sun was limited mostly to the study of the Ca II H and K resonance doublet emission lines (Wilson & Bappu 1957). In the early 1970s, a series of balloon-borne UV observations provided the spectra of the Mg II resonance doublet emission lines at 2795 and 2802 Å (e.g. Kondo et al. 1972, 1976). Observations of the Mg II doublet from *Copernicus* (e.g. Weiler & Oegerle 1979) expanded the data base. It is *IUE*, however, that has made practical an in-depth, comprehensive investigation of stellar chromospheres by extending the spectral coverage to the entire mid- and far-UV region and by enabling observations of much fainter objects. For recent reviews of *IUE* results on this subject, see Jordan & Linsky (1987) and Dupree & Reimers (1987). In addition, there are valuable reviews on chromospheres and related subjects in the proceedings of the above-mentioned *IUE* symposia. Following these review papers, some principal findings from the *IUE* on stellar chromospheres can be summarized as follows:

1. Observations of chromospheric emission have been expanded to include stars as early as spectral type A and as late as M dwarfs (e.g. Brown & Jordan 1981, Ayres et al. 1981); such emission also has been detected in evolved stars in this spectral range (e.g. Linsky & Haisch 1979, Dupree et al. 1979, Brown et al. 1979, Carpenter & Wing 1979). This has permitted the correlation of various stellar parameters that pertain to subphotospheric convection zones with chromospheric emission. Empirical relationships between the stellar age and rotation and the chromosphere and transition regions have also been established. Rotation enhances the level of emission, and as the star evolves, stellar winds tend to brake the rotation, so that the emission fluxes generally decrease (e.g. Hartmann et al. 1984). As a star evolves, the decay of magnetic activity in the outer atmosphere can occur more rapidly at lower levels than at higher ones (Simon et al. 1985).

2. Detailed studies of emission-line fluxes and line widths suggest that for stars later than mid-F, chromospheres are heated primarily by magnetic fields or stellar dynamo processes (e.g. Parker 1970, Ayres & Linsky 1980), but that for stars earlier than mid-F and for weak chromospheres, the heating may be dominated by acoustic waves.

3. The chromospheric properties of single stars can be considered in terms of solar-type, non-solar-type, and hybrid stars. A solar-type star has a hot transition region (temperatures in the range of a few hundred thousand degrees) and a coronal temperature in the range of millions of degrees. A non-solar-type star lacks a hot transition region but has a massive, cool stellar wind. A hybrid star has both a hot transition region and a massive, cool wind. The demarcation line between the solar-type and non-solar-type stars in the Hertzsprung-Russell (HR) diagram is somewhere

near spectral class K1 III (Linsky & Haisch 1979, Mullan & Stencel 1982).

4. Chromospheres in close binary stars are often markedly different from those in single stars. A late-type star in a close binary system rotates more rapidly than a single star owing to its tidally locked rotation. Loss of angular momentum through winds leads to slower rotation in a single star, but in a binary the loss of the angular momentum is compensated for by the large orbital angular momentum. Faster rotations in binaries invigorate the stellar dynamo, giving rise to more active chromospheres.

5. Three years of continuous monitoring of Betelgeuse both in the UV and visible light revealed a periodic variation in the light of the star and a lag between the photospheric disturbances and the chromospheric variations, suggesting the presence of a propagating wave (Dupree et al. 1987). The results indicate that pulsation may be at work in heating and extending the atmosphere and driving the stellar wind. Identification of pulsation in luminous normal stars would be of interest for studies of stellar evolution. Certain regions of instability in the HR diagram, such as those of the Cepheid and long-period variable stars, have long been known. Pulsation may be more ubiquitous than previously thought.

6. The structure of the outer atmospheres of cool stars has been explored in detail through observations of binary systems in which such stars have hot companions (e.g. RS CVn–type systems and ζ Aur–type binaries). The cool-star atmosphere can be observed projected against the hot star at appropriate orbital phases. Some references on this topic are given in the section on close binary stars.

3. EVOLUTIONARY PROCESSES IN INTERACTING BINARIES

It has been estimated that two thirds or more of all stars are members of double or multiple systems. In addition, a large fraction of astrophysically intriguing objects, such as novae and X-ray binaries, involve binary systems. Hence, understanding the evolution in binary stars is important in understanding overall stellar evolution.

Mass flow in close binaries is thought to play an important role in evolution. The temperatures and densities of the flowing gas are such that the spectral signatures are most readily observable in the UV, particularly in the resonance absorption lines of singly or multiply ionized atoms, such as Mg II, Fe II, Al III, C IV, Si IV, and N V. Some metastable lines (e.g. those of Fe II and C III) have also proven valuable in providing information on the gas densities.

Gas flow is present in practically all close binaries observed by *IUE* in the high-spectral-resolution mode. Judging from the radial velocities of the absorption lines observed, a fraction of the gas is being lost from the binary. At the same time, a portion of the gas flowing out of the mass-losing star is apparently accreting onto the companion, as evidenced by high-temperature spectral features (such as Si IV, C IV, and N V) that would be too hot for single stars of the same spectral types.

Listed in the following sections are some of the significant results on close binary stars from the *IUE*.

3.1 *Discovery of High-Temperature Regions and Accretion*

IUE observations show high-temperature regions typically in terms of the absorption lines of highly ionized atoms such as Si IV, C IV, and N V in the case of Algol-type binaries. In the case of β Lyr, high-temperature lines are seen both in absorption and in emission (Hack et al. 1975). In Algol-type systems, the high-temperature regions are observed as absorption lines (Kondo et al. 1979, Peters & Polidan 1984). In W UMa–type binaries, the high-temperature regions are seen as emission lines of multiply ionized species (Rucinski 1985a). Rucinski (1985b) finds that the chromospheres of these binaries are exceedingly active, but that their coronal activities are less than those found in single stars. He also suggests that the gas in the common envelope in earlier type W UMa systems may escape because there are no magnetic loops to confine them.

Enhanced emission lines have been reported during and sometimes outside eclipse in a number of binaries, such as SX Cas and W Ser (Plavec et al. 1982, Plavec 1988), and have been attributed by Plavec (1988) to an accretion disk. Such lines were first observed in β Lyr (Hack et al. 1975). It is still not clear where such lines originate. The possibilities include the energized extended atmosphere of the mass-losing star, a plasma cloud extraneous to both of the stars, and an accretion disk.

The high-temperature absorption and emission lines cannot be accounted for simply in terms of the radiation field of the stars involved; they indicate the existence of a nonthermal energy source. In the case of the hot absorption features superposed on the B star in an Algol-type system, the high temperature is probably a result of accretion onto the B star.

3.2 *Mass Loss From Binary Systems*

The Doppler velocities in the short-wavelength-shifted absorption lines observed in most close binary systems indicate that matter is being lost from the systems. This indicates strongly that mass flow is not caused simply by the gravitational overflow from the critical Roche equipotential surface. Nongravitational forces must be responsible for propelling the gas out of the binaries (e.g. Modisette & Kondo 1980).

3.3 Discovery of Hot Companions

Due primarily to the efforts of Parsons (1981, 1983), hot companions have been found and studied in a number of binary systems. Of particular interest are HD 207739 and 22 Vul. The former contains an optically thick, variable gaseous envelope, and the latter is a short-period (249.083 days) ζ Aur–type binary with an atmospheric eclipse of the G3 supergiant star.

3.4 Highlights of Work on Individual Binary Systems

3.4.1 GRAVITATIONAL REDSHIFT DUE TO A NEUTRON STAR? V Sagittae, which was studied by Koch et al. (1986), exhibits emission lines with a constant redshift of 700 km s^{-1}. Since the redshift persists, it is difficult to explain it in terms of matter falling toward one of the components. One possible explanation is that the emission lines arise in the gaseous envelope surrounding a neutron star, and that the redshift is thus gravitational. If so, this is the first case in which a neutron star has been detected spectroscopically.

3.4.2 LARGE-SCALE STRUCTURES IN AN ACTIVE K DWARF. *IUE* observations of the eclipsing binary V471 Tau (K2 V+DA) revealed the three-dimensional structure of atmospheric loops and their associated starspots (Guinan et al. 1986). Just before and after total eclipse of the white dwarf component, when it is seen projected near the limb of the K dwarf, absorption lines such as C II, C III, C IV, and Si IV appear superposed on the continuum of the hot white dwarf. These lines probably arise in cool coronal loops overlying spots in the atmosphere of the K dwarf. Occasionally, a loop may extend almost one stellar radius from the K star.

3.4.3 DYNAMIC MASS FLOW IN U CEPHEI The Algol-type binary U Cep (B7–8 V+G8 III–IV) undergoes periodic episodes of extremely active mass flow, in which optically thick plasma engulfs the less evolved B component. Other binaries that involve optically thick plasma, which is concomitant to the short-lived dynamic mass flow phase, include β Lyr, R Ara, and HD 207739 (Kondo et al. 1985). Study of U Cep, which may be near the end stage of this phase, provides valuable information on what actually happens in this stage of binary evolution. During U Cep's latest active gas flow episode in 1986, high-resolution UV spectra were obtained for the first time. They showed that an optically thick plasma apparently emanating from the G giant covered much of the B companion, suppressing much of the UV flux from the hot star (McCluskey et al. 1988). Figure 1 shows the Mg II resonance doublet, in which gas streams were observed as multiple absorption features. Figure 2 shows an extremely broad absorption, due probably to overlapping multivelocity streams. Note that the flat bottom does not reach the zero flux level, indicating that the gas streams cover some 70% of the B-star hemisphere.

3.4.4 ADDITIONAL WORK ON BINARIES Further information on *IUE* results on close binaries is given by McCluskey & Sahade (1987). Additional reviews on this subject include those of Shore (1988), Sahade (1986), Rahe (1984), and McCluskey (1982). *IUE* observations have contributed significantly to the improved understanding of another category of binary systems, the cataclysmic variables, as shown by the role of *IUE* results in a recent IAU colloquium on cataclysmic variables (Drechsel et al. 1987). The reader is also referred to reviews on novae by Starrfield & Snijders (1987), on accretion onto compact objects by Cordova & Howarth (1987), and on eclipses of the extended atmospheres of supergiants by Hack & Strickland (1987). *IUE* work on RS CVn stars, another important subgroup of close binary stars, is reviewed by Catalano (1986).

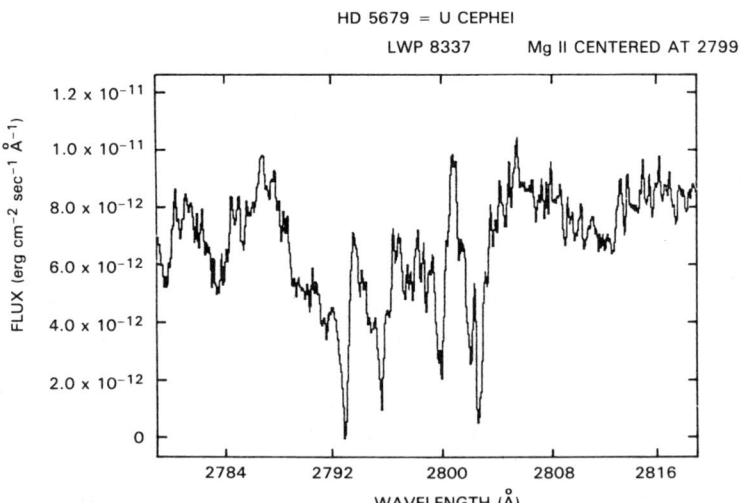

Figure 1 Mg II resonance doublet features of U Cep during a dynamic mass flow event (observation made on 4 June 1986). The rest wavelengths of the doublet are 2795.5 and 2802.7 Å. Multiple gas streams produce the numerous absorption lines at different velocities (McCluskey et al. 1988).

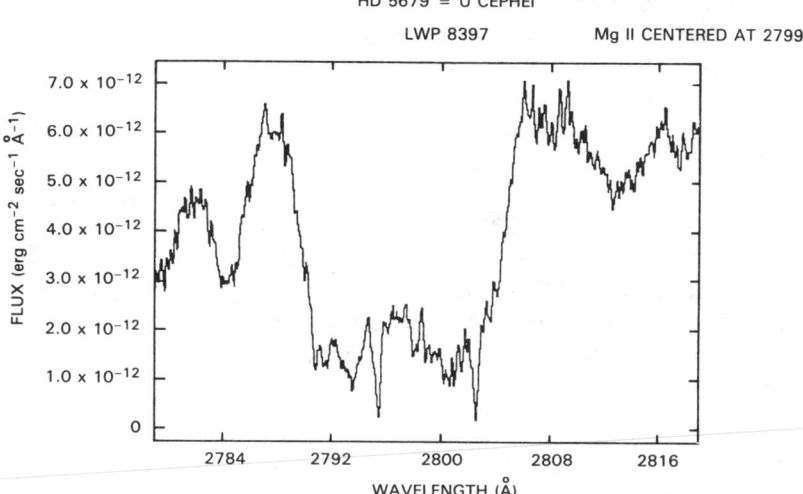

Figure 2 Mg II resonance doublet features in U Cep during the dynamic mass flow event of June 1986. This observation was made 10 days after that shown in Figure 1. The rate of mass flow has increased, and the numerous velocity components overlap to produce the appearance of an extremely broad absorption band (McCluskey et al. 1988).

4. WINDS FROM EARLY-TYPE STARS

IUE has contributed significantly to the investigation of winds in hot stars through its ability to reach much fainter stars than previous UV observatories. This has made a wider variety of stars accessible to observation, including the central stars of planetary nebulae and even the more luminous hot stars in a few galaxies of the Local Group, all with sufficient resolution to identify and measure wind components in spectral line profiles. The longevity of the spacecraft is also a major factor, for within the framework of an observatory administration that must allow for work in many other astronomical disciplines, it has enabled *IUE* users to (*a*) obtain spectra of literally hundreds of hot stars, allowing reliable inferences to be drawn about population properties and trends in wind properties with stellar characteristics; (*b*) make repeated observations of many stars, to establish and investigate variability in stellar winds; and (*c*) have enough time to design, propose, and execute new investigations in reaction to earlier *IUE* findings.

Inspection of reviews written at intervals over the first decade of *IUE* flight operations (e.g. Boggess & Maran 1980, Lamers 1980, Garmany 1984, Henrichs 1984, Kudritzki et al. 1986, Cassinelli & Lamers 1987,

Willis 1988) shows how initial conclusions on the winds from hot stars based on *IUE* spectra were debated, reinforced, or revised as the observations continued. Early indications, for example, that the variation of wind parameters with stellar type is inconsistent with radiative driving gave way to a picture in which most workers are satisfied that radiative acceleration is at work. Initial evidence from *IUE* that the mass loss rates and wind terminal velocities are systematically lower in the OB stars of the Magellanic Clouds (LMC and SMC) than in corresponding stars of the Galaxy, presumably reflecting lower abundances of metals in the stars of the LMC and SMC, have also evolved. It is no longer clear that there are substantial differences in the mass loss rates when stellar types are accurately determined, but the differences in terminal velocities may be real.

4.1 *Winds in Extragalactic Stars*

Astronomers working with *IUE* have obtained UV spectra of hot stars and their winds in the LMC, SMC, M31, and M33. This subject was recently reviewed by Hutchings et al. (1987). The wind velocities are systematically lower in the observed OB stars of the LMC and SMC than in the Galaxy, and the intensities of the strong emission lines that drive the winds in the radiative acceleration model are also weaker. The differences are consistent with the trend of metal abundances, from highest to lowest: Galaxy, LMC, SMC. However, this simple picture clashes with the observation that OB supergiants in M31 and M33 also have weaker lines and slower winds than corresponding stars in the Galaxy. Also, in the Magellanic Clouds, it appears that the mass loss rates are not as small as predicted by the line-driven wind model (Garmany et al. 1988). Whether further progress demands nonradiative momentum transfer or simply more detailed consideration of the respective properties of individual stars remains to be seen.

4.2 *Winds in Planetary Nebulae Nuclei and Hot Subdwarfs*

Early *IUE* observations revealed the presence of strong winds in the central stars of planetary nebulae (Heap 1978). Subsequent work is reviewed by Köppen & Aller (1987) and Vauclair & Liebert (1987). The related discovery and follow-up work on winds in hot subdwarfs is reviewed by Willis (1988). Central stars of low-excitation and many hot-excitation planetary nuclei have fast winds; hot subdwarfs above some limiting luminosity appear to all have such winds. Reported terminal velocities for central stars and subdwarfs range from about 1000 to 5000 km s^{-1}, allowing them to be readily detected in low-dispersion *IUE* spectra.

Indicative of the extent of *IUE* investigations in this field, Cerruti-Sola & Perinotto (1985) report on a sample of 60 planetary nebula central stars. They found that the spectra of 42 stars had continua suitable for the detection of P Cyg profiles; of these, 22 stars actually showed such profiles. Analysis of the sample revealed that central stars with effective temperatures lower than about 63,000 K "almost always have a wind" and those with radii larger than 0.32 $R_\odot$ invariably have winds. Cerruti-Sola & Perinotto concluded that when the stellar surface gravity is below about $\log g = 5.2$, a planetary nebula central star always produces a wind.

The observation by *IUE* that high-speed winds are prevalent in central stars of planetary nebulae is widely interpreted as evidence for the colliding-winds theory for the origin of these nebulae, proposed in original form by Kwok et al. (1978).

4.3 *Stellar Winds and Stellar Evolution*

An especially significant result of the observations of stellar winds with *IUE* is increased knowledge of the dependence of mass loss rate on evolutionary state. Direct observational evidence of this phenomenon is crucial for progress in realistic calculations of stellar evolution. The review by Lamers (1980) reported that the rate of mass loss in young, massive stars seems to increase by two orders of magnitude from the main sequence to the hydrogen-shell-burning phase. Wolf-Rayet stars, which presumably are at a more advanced evolutionary stage, exhibit mass loss rates that may be still another order of magnitude greater. In these very luminous stars, this considerable increase in mass loss rate during evolution is accompanied by relatively little increase in the luminosity. Later, Garmany & Conti (1984) found that the ratio of the terminal wind velocity to the stellar escape velocity in O stars depends systematically on evolutionary condition. Specifically, near the zero age main sequence, this ratio is usually less than 3.0, while stars in which the velocity ratio is larger than 3.0 are usually evolved.

In a more recent review, Cassinelli & Lamers (1987) combine *IUE* data (which are available on many hot stars, but in which the derived mass loss rates are model dependent) with radio observations of a modest number of luminous hot stars, which yield the mass loss rate directly. They demonstrate that as massive stars evolve at nearly constant luminosity, the rate of mass loss increases with temperature and decreases with surface gravity. The data are fit by the relation

$$\log \dot{M} = -5.5 + 1.6 \log(L/10^6) - 1.0 \log(T/T_{\text{ZAMS}}),$$

where T_{ZAMS} is the effective temperature of the star on the zero age main sequence.

5. THE INTERSTELLAR MEDIUM

5.1 Earlier Studies

Ultraviolet studies of the interstellar dust and gas were important objectives of several satellite projects and many rocket investigations prior to the *IUE*, and a great deal was learned about properties of the dust and gas that could be inferred from UV data. In particular, extinction was carefully studied with *OAO-2* and *TD-1* (Code et al. 1976, Nandy et al. 1976), and our understanding of the gaseous component was revolutionized by many fundamental results from *Copernicus*, such as the detection of O VI along many lines of sight (Jenkins & Meloy 1974), which led to a comprehensive new model of the interstellar medium (McKee & Ostriker 1977) consisting of a high-temperature, low-density substrate with scattered clouds that are at least partially ionized on the outside, but with cold interiors containing neutral material. *IUE*'s faint limiting magnitude has made possible refinements of this picture, as well as important improvements in our knowledge of the gas in the solar neighborhood and in the halo of the Galaxy.

5.2 Survey Results

In a high-resolution survey of OB stars, Shull & Van Steenberg (1985) measured H I column densities for 205 stars earlier than B2.5 and found the mean column density/r to be 0.46 cm^{-3}. This value is found to decrease for $r > 1$ kpc, perhaps owing to the effects of interarm gas. Van Steenberg & Shull (1988) have used the same survey to measure abundances of iron and silicon and find that both elements are substantially depleted (by factors of 10 to 1000), and that the depletion factor correlates with hydrogen column density, as one might expect if the heavy elements are being spent in grain formation in dense clouds. Substantial depletions have been measured along lines of sight encountering the ρ Oph cloud (Crutcher & Chu 1984) and the cloud toward HD 147889 (Snow & Joseph 1985).

5.3 Local Interstellar Medium

The local interstellar medium has been probed, using several types of stars as background sources. Spectra of nearby white dwarfs have been obtained for this purpose by Bruhweiler & Kondo (1981, 1982a) and by Dupree & Raymond (1982). White dwarfs are among the few nearby objects suitable for interstellar studies because they have strong continua, particularly in the UV, with very little photospheric structure to complicate the analysis. Consequently, column densities can be measured from features of N I, O I, Si II, C II, Fe II, and Mg II. Other studies have used nearby fast-rotating A and B stars (Frisch 1981, Bruhweiler & Kondo 1982b, Bruhweiler et al.

1984a,b, Freire Ferrero 1984, Skuppin et al. 1987) and late-type stars (Böhm-Vitense 1981, Drake et al. 1984, France et al. 1984, Vladilo et al. 1985). For both types of sources, data are obtained primarily using the Mg II doublet.

Although hydrogen densities cannot be measured directly, good values can be inferred from N I, which is known from *Copernicus* data (Ferlet 1981) to correlate closely with H I. Table 2 shows densities calculated for five white dwarfs, showing a strong anticorrelation between hydrogen density and distance. These data, taken together with *Copernicus* and backscattering results as plotted in Figure 3, indicate a high-density region (or cloud) in the vicinity of the Sun and Sirius B, with substantially reduced densities at greater distances. This picture is substantiated when Mg II data from the more numerous A, B, and late-type stars are considered. Genova et al. (1986) provide results for 91 cool stars, and 40% of them show no evidence of an interstellar feature.

Mg II data from stars within 25 pc indicate that the Sun is located near the outer edge of the Local Cloud, which is between us and the Galactic center. Lines of sight in the hemisphere toward the Galactic center show densities an order of magnitude higher than those towards the anticenter. As pointed out by Bruhweiler (1982), the location of this cloud correlates well with a dust patch discovered by Tinbergen (1982), who measured polarization in nearby stars. Values of N(Mg II)/N(Mg I) derived from *IUE* and *Copernicus* data for nearby stars indicate a temperature near the Sun of 7500–10,000 K (Bruhweiler et al. 1984a,b), consistent with the Sun's position near the edge of the cloud. On the other hand, Bruhweiler & Kondo (1982b) have found C I toward α PsA, in the direction of the polarization patch and Mg II concentration, which indicates a much lower temperature near the cloud center.

It is interesting that there is no evidence for any other cloud or high-

Table 2 Interstellar neutral hydrogen toward nearby white dwarfs

Star	d (pc)	log N(H I) (cm^{-2})	n(H I) (cm^{-3})
Sirius B[a]	2.7	17.93	0.087
W1346	13	18.04	0.024
HD 149499B	34	18.09	0.012
G191-B2B	48	17.92	0.006
Feige 24	90	18.30	0.008

[a] N I data are not available for Sirius B. Value deduced from average of Mg II and Si II.

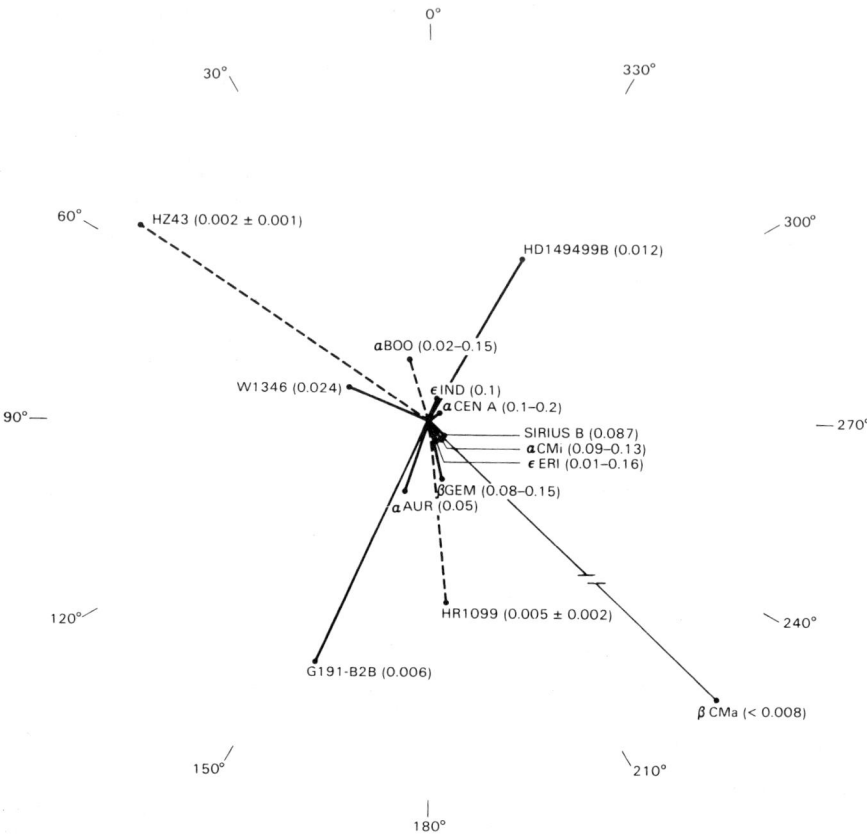

Figure 3 Average neutral hydrogen number densities toward nearby stars (Bruhweiler & Kondo 1982a). Vectors for each object lie in the direction of the object's Galactic longitude, with the length proportional to its actual distance. Dashed lines indicate paths with angles more than 45° out of the Galactic plane with lengths greater than 5 pc. Notice the dropoff in $\bar{n}_{H\,I}$ at larger distances. The values for W1346 ($d = 13$ pc), HD 149499B ($d = 34$ pc), Sirius B ($d = 2.7$ pc), and G191-B2B ($d = 48$ pc) are from Bruhweiler & Kondo (1982a). The $\bar{n}_{H\,I}$ value for HZ 43 is from the *Voyager* EUV experiment, while the other results are from *Copernicus*.

density feature within at least 50 pc of the Sun. This is in contrast with the McKee & Ostriker (1977) model, where the mean free path between cloudlets is about 12 pc. This may indicate either that the model needs modification or that the solar neighborhood is anomalous. Recent supernova activity in the solar vicinity may have evaporated many clouds.

5.4 Interstellar Medium at High Galactic Latitudes

One of the early discoveries of *IUE* was the existence of broad, strong absorption troughs due to C II and other species in the spectra of hot stars in the Magellanic Clouds (de Boer et al. 1978, 1980, Gondhalekar et al. 1980, Savage & de Boer 1981). These observations indicated that ionized gas existed to considerable distances above the Galactic plane and perhaps along the entire line of sight to the Magellanic Clouds. Subsequently, other lines of sight were explored using quasars at appropriate redshifts (York et al. 1984), extragalactic supernovae (Pettini et al. 1982, Jenkins et al. 1984), and high-latitude stars bright in the UV (e.g. de Boer & Savage 1984, Savage & Massa 1987).

Most of the material exists in neutral and low-ionization absorption systems that are found at distances of up to a few hundred parsecs above the Galactic plane. The gas exhibits metal abundances typical of interstellar gas in the solar neighborhood, and so it may be material that originated in the disk and was levitated by a Galactic fountain flow, as described by Shapiro & Field (1976), or by support from cosmic rays (Chevalier & Fransson 1984). It seems likely that it is also similar to the high-velocity clouds seen in 21-cm emission (van Woerden et al. 1985).

The more highly ionized gas shows a very different velocity distribution from the cool material and probably does not coincide with it. The C IV gas has a scale height of a few kiloparsecs (de Boer 1985, Savage & Massa 1987) and may result from the cooling of the hot, 10^6 K coronal material as well as from direct ionization due to cosmic rays and the short-UV radiation field. A few Magellanic Cloud stars have been analyzed to investigate abundances of the interstellar gas in the Clouds (e.g. Fitzpatrick & Savage 1983, de Boer et al. 1985). The depletion pattern is similar to that found in the Milky Way, but the overall abundance levels are lower, reflecting the lower metal content.

6. SUPERNOVA 1987A

The brightest supernova since Kepler's star of 1604 was discovered by Ian Shelton on 24 February 1987 in the Large Magellanic Cloud. Within three hours of notification of the discovery, *IUE* was pointed to the supernova in an existing target-of-opportunity program (Kirshner et al. 1987).

The first exposure showed that SN 1987A was extremely bright in the far UV. However, the UV flux declined rapidly hour by hour. Within three days, the far-UV flux, which was mainly from the exploding gas, declined by three orders of magnitude. Then, *IUE* observations revealed the spectra of two stars adjacent to the supernova. In readily available preexplosion photographs, only two stars were apparent. It was speculated that the progenitor was an unknown object and that since the exploding envelope was relatively thin, SN 1987A might be a Type I supernova involving a white dwarf and a companion.

However, careful analysis of a predetonation plate showed that there had really been three stars in the direction of SN 1987A. The supernova was still brightening in the visible range, and there was no way to take another plate from the ground to identify the missing star. The *IUE* spectra of the two remaining stars were examined by Sonneborn et al. (1987) at Goddard and by Gilmozzi et al. (1987) at Villafranca. They independently concluded that the star that disappeared was a blue supergiant. The high-resolution spectra obtained immediately after the discovery of SN 1987A proved invaluable also in the study of the interstellar media in the Milky Way, the Large Magellanic Cloud, and the intervening halo or haloes. We may never have such a bright UV background source available again for this purpose!

The *IUE* Fine Error Sensor (FES) used for tracking the target provided continuous coverage of the visual light curve of SN 1987A. The maximum brightness near magnitude 2.9 was attained three months after the explosion. An expanding gas cloud heated by radioactivity from isotopes created in the explosion accounted for the brightening. An exponential decline set in several weeks after maximum light. This was strong evidence that the radioactive decay of ^{56}Co to ^{56}Fe was then the main source of energy for heating the ejecta (Sonneborn 1988).

Several months after the explosion, high-temperature emission lines became observable in the far-UV. Detailed analysis of these features revealed that the emitting gas has an abnormal nitrogen-to-carbon ratio, which indicates that the composition of the gas was altered by nuclear processes inside the progenitor star prior to its release in a wind. These lines presumably arose in the gaseous shell formed by the stellar wind present in the red supergiant phase that preceded the immediate pre-supernova blue supergiant state. The lines were very narrow, which indicates that the expansion velocity of the emitting shell is less than 20 km s^{-1}, consistent with a red supergiant wind origin (Fransson et al. 1988). The brightness of the narrow emission lines peaked about 400 days after the supernova explosion. The radius of the circumstellar shell is therefore about 400 light-days or less.

7. ACTIVE GALACTIC NUCLEI

Prior to the launch of *IUE*, 3C 273 was the only active galactic nucleus that had been observed at UV wavelengths—first by the *Astronomical*

Netherlands Satellite (*ANS*; Wu 1977) and then by a rocket spectrograph of Davidsen et al. (1978). Since 1978, however, *IUE* has produced data on all types of active galactic nuclei (AGNs). Since it can only detect relatively bright AGNs (generally brighter than magnitude 17), observations have been restricted to low- to intermediate-redshift objects. Even so, *IUE* has played a valuable role in extending the wavelength range over which studies of the AGN spectrum can be made and in providing UV data for nearby, low-z sources to compare with ground observations of intrinsic UV spectra in high-redshift sources.

AGN spectra are characterized by a nonthermal continuum, usually ascribed to synchrotron emission, and broad emission lines. The emission lines are of two types: permitted resonance lines of abundant ions, many of which occur only in the UV, with Doppler widths of order 10,000 km s^{-1}; and forbidden lines with Doppler widths of order 1000 km s^{-1}. The continuum is thought to originate from a very small nucleus, perhaps powered by accretion. This nucleus is thought to be surrounded by a so-called broad-line region (BLR), of order 0.1 pc in diameter, that contains numerous high-density, high-velocity clouds that produce the broad, permitted emission lines. The BLR is supposedly surrounded by a larger narrow-line region where the forbidden lines originate.

Early efforts to understand the nonthermal continuous spectra of AGNs were complicated by attempts to intercompare data from low- and high-redshift objects based only on those wavelengths that were shifted into the bandpass of ground-based instruments. It was difficult to know if observed discrepancies were due to evolution effects, transmission effects in the intergalactic medium, or were intrinsic differences with wavelength. When it became possible to observe the spectra of individual sources over several decades of wavelength, from the UV with *IUE* into the far-infrared, it became clear that different phenomena dominate in the different wavelength regimes. In particular, many AGNs exhibit a so-called blue bump at wavelengths below 3000 Å (Oke & Zimmerman 1979, Oke & Goodrich 1981), which makes it impossible to describe the nonstellar continuum by a single power law. It now appears that this blue excess comprises most of the energy output for many AGNs.

For many AGNs the blue bump can be modeled satisfactorily by a single-temperature blackbody curve with $T = 26,000$ K (Edelson & Malkan 1986). Figure 4 shows the Malkan et al. (1987) model for the Seyfert 1 galaxy Mrk 335, indicating that the UV energy may be explained by a summation of the underlying power law, plus recombination radiation from BLR clouds and the 26,000-K blue-bump blackbody. In other sources the data can be fit better by models involving a geometrically thin, but optically thick, accretion disk (Malkan 1983).

MKN 335

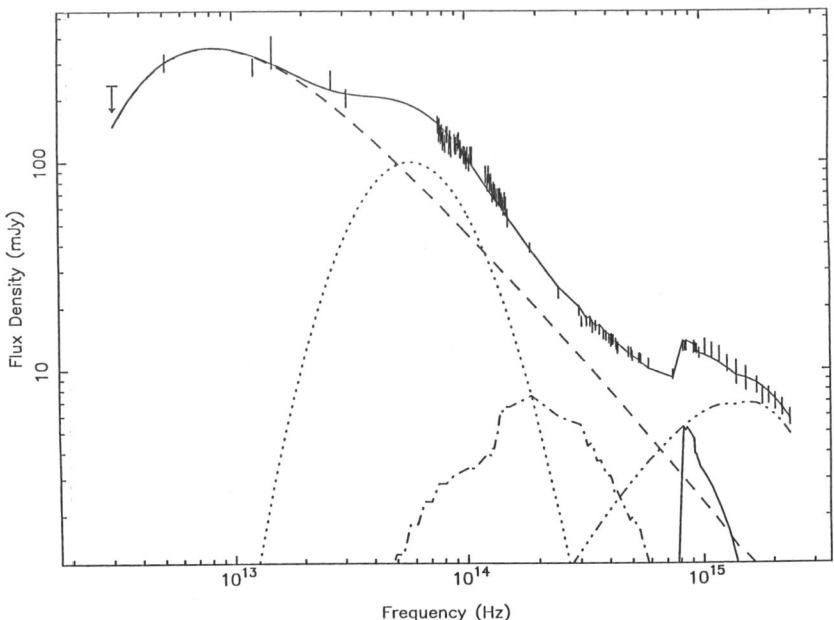

Figure 4 Decomposition of Edelson & Malkan's (1986) fit to the multiwavelength continuum of the Seyfert 1 galaxy Markarian 335. The dashed line is a power law with slope −1.2; the dot-dash line is the flux from starlight; the triple-dot–dash line is a blackbody with temperature 26,000 K; the short solid curve in the ultraviolet is recombination radiation from the BLR; and the dotted curve is a parabola centered at 5.8×10^{12} Hz with a FWHM of 3.3 octaves. Their sum is the long solid line, which fits the spectrophotometric data, shown by vertical error bars. From Malkan et al. (1987).

High signal-to-noise *IUE* spectra in the 2000–3000 Å region have revealed low-contrast structure due to the blending of many Fe II emission lines. These lines are so closely spaced that they can produce a pseudocontinuum containing an integrated flux comparable to that in Ly-α (Wills et al. 1985, Netzer et al. 1985). Models of the BLR clouds will have to be adjusted to allow for these unexpectedly high Fe II line fluxes.

Other UV results pose problems for modeling the BLR region: Although it is usually assumed that the broad emission lines are produced by photoionization because their strengths correlate well with the continuous spectrum, Baldwin (1977a) has pointed out that the C IV 1550-Å luminosity is anticorrelated with the nearby continuum at 1450 Å. *IUE* data were used to examine this relation over the full extent of AGN luminosities

(Wu et al. 1983), and it appears to hold not only statistically but also for individual sources as they vary in luminosity (Barr et al. 1983). Wu et al. (1983) suggest that the covering factor, or solid angle of the BLR clouds as seen from the nucleus, may be larger for less luminous AGNs; Malkan & Sargent (1982) suppose that the 1450-Å continuum radiation may not be proportional to the ionizing flux; and Mushotzky & Ferland (1984) suggest that the ionization parameter is larger in low-luminosity AGNs.

Measures of the Ly-α/H β intensity ratio are substantially less than should be expected from radiative recombination, as first pointed out by Baldwin (1977b). The average value of this ratio is 5 (Wu et al. 1980, Lacy et al. 1982), whereas the predicted value is 40. Examination of other Balmer and Paschen recombination lines suggests that the discrepancy is due to an enhancement of the Balmer series, which can give clues regarding the ionization structure of the BLR clouds (Kwan & Krolik 1981, Canfield & Puetter 1981).

A remarkable body of data has been assembled on NGC 4151, the nearest and brightest Seyfert 1 galaxy. Starting in the first year of *IUE* operations, a group of European astronomers pooled their observing time to obtain an extended series of observations on NGC 4151 in order to examine variability and other phenomena that could not be studied from isolated spectra of randomly selected sources. These data have been discussed by various groups (e.g. Penston et al. 1981, Bromage et al. 1985, Ulrich et al. 1985). In several Seyfert 1 galaxies, including NGC 4151 (Perola et al. 1982), luminosity variations have considerably larger amplitudes in the UV than in the visible or infrared. This may be due to the greater contamination by starlight at the longer wavelengths, which would reduce the apparent percentage of variation of the nuclear luminosity. Typical times for a doubling of the continuum brightness in NGC 4151 range from 5 to 30 days, implying a source size less than 0.01 pc. This is in contrast to X-ray flares, which can develop in less than 12 hr. It may be that the X radiation is more confined than the UV and optical radiation, possibly being produced by a different mechanism.

The temporal variations of the continuum and emission lines in the Seyfert galaxy Fairall 9 were observed with *IUE* and with visible and infrared instruments. The results show clearly that there is a time lag between maximum light in the continuum, which presumably arises in the inner part of the accretion disk surrounding the putative supermassive black hole, and maximum light in the lines, which are supposed to originate from the outer regions of the disk. The UV and visible light emission lines reached maximum 155 days after the UV continuum maximum, while the IR lines peaked at 400 days after UV continuum maximum. The time lags yield information on the width and physical properties of the accretion disk (Clavel et al. 1989).

The bright, high-redshift ($V = 16.1$, $z = 2.72$) quasar HS 1700+6416 was observed with *IUE* by Reimers et al. (1989) at low and high resolution to yield the first information on a quasar at rest wavelengths as short as 330 Å. Although intergalactic Lyman-line and continuum absorption depresses the continuum significantly between rest wavelengths 850–450 Å, the continuum rose sharply toward shorter wavelengths to at least 330 Å. If this result applies to other quasars and AGNs, the continuum maximum probably lies beyond about 300 Å.

8. SOLAR SYSTEM OBJECTS

The *IUE* has been used to study solar system objects, including planets, natural satellites, asteroids, and comets. These observations are challenging from the point of spacecraft operations, as these objects show motions relative to the inertial frame. Among the planets studied are Venus, Mars, Jupiter, Saturn, and Uranus (Moos & Encrenaz 1987). The planetary satellites observed are the Galilean satellites of Jupiter (namely Io, Europa, Ganymede, and Callisto) and four large satellites of Saturn (namely, Tethys, Dione, Rhea, Iapetus; Nelson & Lane 1987). Observations of the remarkable Io torus and its interactions with the Jovian magnetosphere were recently reviewed by Ballester et al. (1988).

Over two dozen comets have been observed with the *IUE* (Festou & Feldman 1987). Observations of Comet IRAS-Araki-Alcock were particularly noteworthy. The comet came within 5×10^6 km of the Earth, which made the tracking quite a challenge but which also made the point-like circumnuclear region a bright object at closest approach. The unexpected discovery of the sulfur dimer emissions in the far-UV spectrum of this comet is one of the most significant cometary results of the *IUE* (A'Hearn & Feldman 1985). A systematic record of the varying UV emissions of Comet Giacobini-Zinner allowed *IUE* to support the first intercept of a comet, the visit of the *International Cometary Explorer* (*ICE*) to that object (McFadden et al. 1987). *IUE* made the most extensive space-based telescopic observations of Comet Halley during its sweep through the inner solar system in 1985–86. The *Challenger* accident prevented the planned March 1986 launch of the UV *Astro* mission payload, and *IUE* accordingly became the workhorse UV support facility as several interplanetary probes (ESA's *Giotto*, Japan's *Sakigake* and *Suisei*, USSR's *Vegas 1* and *2*, and USA's *ICE*) flew through or past the comet. *IUE* observations of the OH emission from Comet Halley showed that it was ejecting several tons of water per second near its perihelion passage (Rahe 1988).

9. OTHER TOPICS

There are 31 chapters on distinct research topics in the compendium *Exploring the Universe With the IUE Satellite* (Kondo et al. 1987), but even those chapters were too few to portray the full scope of the astrophysical research performed with this satellite observatory. Likewise, some of the themes not covered in this review include symbiotic stars (Nussbaumer & Stencel 1987), intrinsic variable stars (Böhm-Vitense & Querci 1987), pre-main-sequence stars (Imhoff & Appenzeller 1987), white dwarfs (Vauclair & Liebert 1987), interstellar dust and extinction (Mathis 1987), supernovae and their remnants (Blair & Panagia 1987), H II regions (Dufour 1987), starburst galaxies (Kunth & Weedman 1987), globular clusters (Castellani & Cassatella 1987), blazars (Bregman et al. 1987), and quasar absorption lines and galaxy haloes (Bergeron et al. 1987).

ACKNOWLEDGMENTS

We acknowledge helpful discussions with Drs. Thomas R. Ayres, Andrea K. Dupree, Matthew A. Malkan, George E. McCluskey, and C. Megan Urry.

Literature Cited

A'Hearn, M. F., Feldman, P. D. 1985. In *Ices in the Solar System*, ed. D. Benest, A. Dollfus, R. Smoluchowski, p. 463. Dordrecht: Reidel
Ayres, T. R., Linsky, J. L. 1980. *Ap. J.* 241: 279
Ayres, T. R., Marstad, N. C., Linsky, J. L. 1981. *Ap. J.* 247: 545
Baldwin, J. A. 1977a. *MNRAS* 178: 67
Baldwin, J. A. 1977b. *Ap. J.* 214: 679
Ballester, G. E., Moos, H. W., Feldman, P. D., Strobel, D. F., Skinner, T. E., et al. 1988. In *A Decade of UV Astronomy With the IUE Satellite*, ed. E. Rolfe, 1: 79. *ESA SP-281*
Barr, P., Willis, A. J., Wilson, R. 1983. *MNRAS* 202: 453
Bergeron, J., Savage, B., Green, R. F. 1987. See Kondo et al. 1987, p. 703
Blair, W. P., Panagia, N. 1987. See Kondo et al. 1987, p. 549
Boggess, A., Carr, F. A., Evans, D. C., Fischel, D., Freeman, H., et al. 1978a. *Nature* 275: 372
Boggess, A., Bohlin, R. C., Evans, D. C., Freeman, H. R., Gull, T. R., et al. 1978b. *Nature* 275: 377
Boggess, A., Maran, S. P. 1980. *Phys. Today* 33(9): 40

Böhm-Vitense, E. 1981. *Ap. J.* 244: 504
Böhm-Vitense, E., Querci, M. 1987. See Kondo et al. 1987, p. 223
Bregman, J., Maraschi, L., Urry, C. M. 1987. See Kondo et al. 1987, p. 685
Bromage, G. E., Boksenberg, A., Clavel, J., Elvius, A., Penston, M. V., et al. 1985. *MNRAS* 215: 1
Brown, A., Jordan, C. 1981. *MNRAS* 196: 757
Brown, A., Jordan, C., Wilson, R. 1979. In *The First Year of IUE*, ed. A. Willis, p. 232. London: Univ. College London
Bruhweiler, F. C. 1982. In *Advances in Ultraviolet Astronomy: Four Years of IUE Research*, ed. Y. Kondo, R. Chapman, J. Mead, p. 125. *NASA CP-2238*
Bruhweiler, F. C., Kondo, Y. 1981. *Ap. J. Lett.* 248: L123
Bruhweiler, F. C., Kondo, Y. 1982a. *Ap. J.* 259: 232
Bruhweiler, F. C., Kondo, Y. 1982b. *Ap. J. Lett.* 260: L91
Bruhweiler, F., Oegerle, W., Weiler, E., Stencel, R., Kondo, Y. 1984a. In *Future of Ultraviolet Astronomy Based on Six Years of IUE Research*, ed. J. M. Mead, R. D. Chapman, Y. Kondo. p. 200. *NASA CP-2349*

Bruhweiler, F., Oegerle, W., Weiler, E., Stencel, R., Kondo, Y. 1984b. In *The Local Interstellar Medium*, ed. F. Bruhweiler, Y. Kondo, B. Savage, p. 64. *NASA CP-2345*
Canfield, R. C., Puetter, R. C. 1981. *Ap. J.* 243: 390
Carpenter, K. G., Wing, R. F. 1979. *Bull. Am. Astron. Soc.* 11: 419
Cassinelli, J. P., Lamers, H. J. G. L. M. 1987. See Kondo et al. 1987, p. 139
Castellani, V., Cassatella, A. 1987. See Kondo et al. 1987, p. 637
Catalano, S. 1986. *Proc. RAL Workshop Astron. and Astrophys.*, ed. P. M. Gondhalekar, p. 105. *RAL-86-085*
Cerruti-Sola, M., Perinotto, M. 1985. *Ap. J.* 291: 237
Chevalier, R. A., Fransson, C. 1984. *Ap. J. Lett.* 279: L31
Clavel, J., Wamsteker, W., Glass, I. 1989. *Ap. J.* 337: 236
Code, A. D., Davis, J., Bless, R. C., Hanbury Brown, R. 1976. *Ap. J.* 203: 417
Cordova, F. A., Howarth, I. D. 1987. See Kondo et al. 1987, p. 395
Crutcher, R. M., Chu, Y.-H. 1984. *Ap. J.* 290: 671
Davidsen, A. F., Hartig, G. F., Fastie, W. G. 1978. *Nature* 269: 203
de Boer, K. S. 1985. *Mitt. Astron. Ges.* 63: 21
de Boer, K. S., Fitzpatrick, E. L., Savage, B. D. 1985. *MNRAS* 207: 115
de Boer, K. S., Koorneef, J., Savage, B. D. 1978. *Bull. Am. Astron. Soc.* 10: 726
de Boer, K. S., Koorneef, J., Savage, B. D. 1980. *Ap. J. Lett.* 136: L7
de Boer, K. S., Savage, B. D. 1984. *Astron. Astrophys.* 136: L7
Drake, S. A., Brown, A., Linsky, J. L. 1984. *Ap. J.* 284: 774
Drechsel, H., Kondo, Y., Rahe, J. eds. 1987. *Cataclysmic Variables*. Dordrecht: Reidel
Dufour, R. J. 1987. See Kondo et al. 1987, p. 577
Dupree, A. K., Baliunas, S. L., Guinan, E. T., Hartmann, L., Nassiopoulos, G. E., Sonneborn, G. 1987. *Ap. J. Lett.* 317: L85
Dupree, A. K., Black, J. H., Davis, R. J., Hartmann, L., Raymond, J. C. 1979. In *The First Year of IUE*, ed. A. Willis, p. 217. London: Univ. College London
Dupree, A. K., Raymond, J. 1982. *Ap. J. Lett.* 263: L63
Dupree, A. K., Reimers, D. 1987. See Kondo et al. 1987, p. 321
Edelson, R. A., Malkan, M. A. 1986. *Ap. J.* 308: 59
Ferlet, R. 1981. *Astron. Astrophys.* 98: L1
Festou, M. C., Feldman, P. D. 1987. See Kondo et al. 1987, p. 101
Fitzpatrick, E. L., Savage, B. D. 1983. *Ap. J.* 267: 93

Franco, M. L., Crivellari, L., Molaro, P., Vladilo, G., Ramella, M., et al. 1984. *Astron. Astrophys. Suppl.* 58: 693
Fransson, C., Cassatella, A., Gilmozzi, R., Kirshner, R. P., Panagia, N., et al. 1988. *Ap. J.* 336: 429
Freire Ferrero, R. 1984. *Proc. Eur. IUE Conf., 4th*, ed. E. Rolfe, B. Battrick, p. 133. *ESA SP-218*
Frisch, P. 1981. *Nature* 293: 377
Garmany, C. D. 1984. In *Future of Ultraviolet Astronomy Based on Six Years of IUE Research*, ed. J. M. Mead, R. D. Chapman, Y. Kondo, p. 17. *NASA CP-2349*
Garmany, C. D., Conti, P. S. 1984. *Ap. J.* 284: 705
Garmany, C. D., Kudritzki, R. P., Husfeld, D. 1988. In *A Decade of UV Astronomy with the IUE Satellite*, ed. E. Rolfe, 2: 137. *ESA SP-281*
Genova, R., Beckman, J. E., Molaro, P., Vladilo, G. 1986. In *Symp. 7, XXVI COSPAR Meet. Toulouse, Fr.* Pap. 7.1.7
Gilmozzi, R., Cassatella, A., Clavel, J., Fransson, C., Gonzales, R., et al. 1987. *Nature* 328: 318
Gondhalekar, P. M., Willis, A. J., Morgan, D. H., Nandy, K. 1980. *MNRAS* 193: 875
Guinan, E. F., Wacher, S. W., Baliunas, S. L., Loeser, J. G., Raymond, J. C. 1986. In *New Insights in Astrophysics: 8 Years of UV Astronomy with IUE*, ed. E. Rolfe, p. 197. *ESA SP-263*
Hack, M., Hutchings, J. B., Kondo, Y., McCluskey, G. E., Plavec, M., Polidan, R. S. 1975. *Ap. J.* 198: 453
Hack, M., Stickland, D. 1987. See Kondo et al. 1987, p. 445
Hartmann, L., Baliunas, S. L., Duncan, D. K., Noyes, R. W. 1984. *Ap. J.* 279: 778
Heap, S. R. 1978. In *Mass Loss and Evolution of O-Type Stars*, ed. P. S. Conti, C. W. H. de Loore, p. 99. Dordrecht: Reidel
Henrichs, H. 1984. *Proc. Eur. IUE Conf., 4th*, ed. E. Rolfe, B. Battrick, p. 43. *ESA SP-218*
Hutchings, J. B., Lequeux, J., Wolfe, B. 1987. See Kondo et al. 1987, p. 605
Imhoff, C. L., Appenzeller, I. 1987. See Kondo et al. 1987, p. 295
Jenkins, E. B., Meloy, D. A. 1974. *Ap. J. Lett.* 193: L121
Jenkins, E. B., Wallerstein, G., Silk, J. 1984. *Ap. J.* 278: 649
Jordan, C., Linsky, J. L. 1987. See Kondo et al. 1987, p. 259
Kirshner, R. P., Sonneborn, G., Crenshaw, D. M., Nassiopoulos, G. E. 1987. *Ap. J.* 320: 602
Koch, R. H., Corcoran, M. F., Holenstein, B. D., McCluskey, G. E. 1986. *Ap. J.* 306: 618

Kondo, Y., Giuli, R. T., Modisette, J. L., Rydgren, A. E. 1972. *Ap. J.* 176: 153
Kondo, Y., McCluskey, G. E., Parsons, S. B. 1985. *Ap. J.* 295: 580
Kondo, Y., McCluskey, G. E., Stencel, R. E. 1979. *Ap. J.* 233: 906
Kondo, Y., Morgan, T. H., Modisette, J. L. 1976. *Ap. J.* 207: 167
Kondo, Y., Wamsteker, W., Boggess, A., Grewing, M., de Jager, C., et al., eds. 1987. *Exploring the Universe With the IUE Satellite*. Dordrecht: Reidel
Köppen, J., Aller, L. H. 1987. See Kondo et al. 1987, p. 589
Kudritzki, R. P., Pauldrach, A., Puls, J. 1986. In *New Insights in Astrophysics: 8 Years of UV Astronomy with IUE*, ed. E. Rolfe, p. 247. *ESA SP-263*
Kunth, D., Weedman, D. 1987. See Kondo et al. 1987, p. 623
Kwan, J., Krolik, J. 1981. *Ap. J.* 250: 478
Kwok, S., Purton, C. R., Fitzgerald, P. M. 1978. *Ap. J. Lett.* 219: L125
Lacy, J. H., Soifer, B. T., Neugebauer, G., Matthews, K., Malkan, M., et al. 1982. *Ap. J.* 256: 75
Lamers, H. J. G. L. M. 1980. In *The Universe at Ultraviolet Wavelengths*, ed. R. D. Chapman, p. 93. *NASA CP-2171*
Linsky, J. L., Haisch, B. 1979. *Ap. J. Lett.* 229: L27
Malkan, M. A. 1983. *Ap. J.* 268: 582
Malkan, M. A., Sargent, W. L. 1982. *Ap. J.* 254: 22
Malkan, M. A., Alloin, D., Shore, S. 1987. See Kondo et al. 1987, p. 655
Mathis, J. S. 1987. See Kondo et al. 1987, p. 517
McCluskey, G. E. 1982. In *Advances in Ultraviolet Astronomy: Four Years of IUE Research*, ed. Y. Kondo, R. Chapman, J. Mead, p. 102. *NASA CP-2238*
McCluskey, G. E., Kondo, Y., Olson, E. C. 1988. *Ap. J.* 332: 1019
McCluskey, G. E., Sahade, J. 1987. See Kondo et al. 1987, p. 427
McFadden, L. A., A'Hearn, M. F., Feldman, P. D., Bohnhardt, H., Rahe, J., et al. 1987. *Icarus* 69: 329
McKee, C. F., Ostriker, J. P. 1977. *Ap. J.* 218: 148
Modisette, J. L., Kondo, Y. 1980. *Ap. J.* 240: 180
Moos, H. W., Encrenaz, T. 1987. See Kondo et al. 1987, p. 45
Mullan, D. J., Stencel, R. E. 1982. In *Advances in Ultraviolet Astronomy: Four Years of IUE Research*, ed. Y. Kondo, R. Chapman, J. Mead. p. 235. *NASA CP-2238*
Mushotzky, R., Ferland, G. J. 1984. *Ap. J.* 278: 558
Nandy, K., Thompson, G. I., Jamar, C., Monfils, A., Wilson, R. 1976. *Astron. Astrophys.* 51: 63
Nelson, R. M., Lane, A. L. 1987. See Kondo et al. 1987, p. 67
Netzer, H., Wamsteker, W., Wills, B. J., Wills, D. 1985. *Ap. J.* 292: 143
Nussbaumer, H., Stencel, R. E. 1987. See Kondo et al. 1987, p. 203
Oke, J. B., Goodrich, R. W. 1981. *Ap. J.* 243: 445
Oke, J. B., Zimmerman, B. 1979. *Ap. J. Lett.* 231: L15
Parker, E. N. 1970. *Annu. Rev. Astron. Astrophys.* 8: 1
Parsons, S. B. 1981. *Ap. J.* 247: 560
Parsons, S. B. 1983. *Ap. J. Suppl.* 53: 553
Penston, M. V., Boksenberg, A., Bromage, G. E., Clavel, J., Elvius, A., et al. 1981. *MNRAS* 196: 857
Perola, G. C., Boksenberg, A., Bromage, G. E., Clavel, J., Elvius, M., et al. 1982. *MNRAS* 200: 293
Peters, G. J., Polidan, R. S. 1984. *Ap. J.* 283: 745
Pettini, M., Benvenuti, P., Blades, J. C., Boggess, A., Boksenberg, A., et al. 1982. *MNRAS* 199: 409
Plavec, M. 1988. In *Algols, IAU Colloq. No. 123*, ed. A. Batten. Dordrecht: Kluwer. In press
Plavec, M., Weiland, J. L., Koch, R. H. 1982. *Ap. J.* 256: 206
Rahe, J. 1984. In *Future of Ultraviolet Astronomy Based on Six Years of IUE Research*, ed. J. M. Mead, R. D. Chapman, Y. Kondo. p. 51. *NASA CP-2349*
Rahe, J. 1988. *Trans. IAU* 20A: 622
Reimers, D., Clavel, J., Groote, D., Engels, D., Hagen, H. J., et al. 1989. *Astron. Astrophys.* In press
Rucinski, S. M. 1985a. *MNRAS* 215: 615
Rucinski, S. M. 1985b. In *Interacting Binary Stars*, ed. J. E. Pringle, R. A. Wade, p. 13. Cambridge: Univ. Press
Sahade, J. 1986. In *New Insights in Astrophysics: 8 Years of UV Astronomy with IUE*, ed. E. Rolfe, p. 267. *ESA SP-263*
Savage, B. D., de Boer, K. S. 1981. *Ap. J.* 243: 460
Savage, B. D., Massa, D. 1987. *Ap. J.* 314: 380
Shapiro, P. R., Field, G. B. 1976. *Ap. J.* 205: 762
Shore, S. N. 1988. In *A Decade of UV Astronomy with the IUE Satellite*, ed. E. Rolfe, 1: 67. *ESA SP-281*
Shull, J. M., Van Steenberg, M. E. 1985. *Ap. J.* 294: 599
Simon, T., Herbig, G., Boesgaard, A. M. 1985. *Ap. J.* 293: 551
Skuppin, R., Grewing, M., Bianchi, L., de Boer, K. S. 1987. *Astron. Astrophys.* 177: 228

Snow, T. P., Joseph, C. L. 1985. *Ap. J.* 288: 277
Sonneborn, G. 1988. In *A Decade of UV Astronomy with the IUE Satellite*, ed. E. Rolfe, 1: 111. *ESA SP-281*
Sonneborn, G., Altner, B., Kirshner, R. P. 1987. *Ap. J. Lett.* 323: L35
Starrfield, S., Snijders, M. A. 1987. See Kondo et al. 1987, p. 377
Tinbergen, J. 1982. *Astron. Astrophys.* 105: 53
Ulrich, M. H., Altamore, A., Boksenberg, A., Bromage, G. E., Clavel, J., et al. 1985. *Nature* 313: 747
Van Steenberg, M. E., Shull, J. M. 1988. *Ap. J.* 330: 942
van Woerden, H., Schwarz, U. J., Hulsbosch, A. N. M. 1985. In *The Milky Way Galaxy, IAU Symp. No. 106*, ed. H. van Woerden, R. J. Allen, W. B. Burton, p. 387. Dordrecht: Reidel
Vauclair, G., Liebert, J. 1987. See Kondo et al. 1987, p. 355
Vladilo, C., Beckman, J. E., Crivelli, L., Franco, M. L., Molaro, P. 1985. *Astron. Astrophys.* 144: 81
Weiler, E. J., Oegerle, W. W. 1979. *Ap. J. Suppl.* 39: 357
Willis, A. J. 1988. In *A Decade of UV Astronomy with the IUE Satellite*, ed. E. Rolfe, 2: 53. *ESA SP-281*
Wills, B. J., Netzer, H., Wills, D. 1985. *Ap. J.* 288: 94
Wilson, O. C., Bappu, M. K. V. 1957. *Ap. J.* 125: 661
Wu, C.-C. 1977. *Ap. J. Lett.* 217: L117
Wu, C.-C., Boggess, A., Gull, T. R. 1980. *Ap. J.* 242: 14
Wu, C.-C., Boggess, A., Gull, T. R. 1983. *Ap. J.* 266: 28
York, D. G., Ratcliff, S., Blades, J. C., Cowie, L. L., Morton, D. C., Wu, C.-C. 1984. *Ap. J.* 276: 92

CLASSIFICATION OF SOLAR FLARES

T. Bai and P. A. Sturrock

Center for Space Science and Astrophysics, Stanford University, Stanford, California 94305

INTRODUCTION

Solar flares exhibit diverse phenomena: They cause electromagnetic radiations ranging from kilometric radio waves to tens of MeV gamma rays, and they accelerate energetic particles interacting in the solar atmosphere as well as those escaping into interplanetary space. They are also associated with shock waves that sometimes propagate to several astronomical units and with ejections of magneto-plasma configurations into interplanetary space. A point of controversy has been whether there is only one class of solar flares or many classes of flares. According to one school of thought, flares are basically the same and must intrinsically involve all flare phenomena, the relative strengths of which vary from flare to flare. During very energetic flares all flare phenomena are strong enough to be detected, but during less energetic flares many flare phenomena are too weak to be detected although they take place [see Kahler (77) concerning the "big-flare syndrome"]. According to the other school of thought, there are different classes of flares, and some flare phenomena take place during one class of flares but not during other classes of flares. In this view, additional physical processes take place during flares exhibiting complex flare phenomena. It is important to determine which view is closer to reality.

The problem with classifications in general is that objects or phenomena do not fall into neatly arranged boxes. There are always exceptions, and some objects fall in the gray areas [see (1) for a summary of a discussion on solar flare classification]. However, a good classification scheme can serve useful purposes by organizing seemingly bewildering phenomena or objects. A good classification scheme should be closely related to the physical processes involved, and it is desirable that the names of different classes should indicate important characteristics of the classes (70). Our aim is to base a classification of flares on energy release processes, including the important process of particle acceleration.

Through observational data collected during the last solar cycle, especially those obtained by the *Solar Maximum Mission* (*SMM*) (131) and the Japanese *Hinotori* satellite (174), we now have a more complete understanding of what is going on in solar flares. Thus this is a good time to review the thoughts of solar physicists on flare classification schemes. Because solar flare physics is such a diverse field, we confine the scope of our paper to observations and theoretical ideas directly related to flare classification. Hence, many important papers that have contributed to our general understanding of solar flares are perforce not discussed here.

Since we are trying to understand the differences between different classes of flares, we do not dwell much on the primary (sometimes called the "first-phase") acceleration process, which is responsible for nonthermal electrons with energies up to a few hundred keV; instead, we concentrate on those additional processes that distinguish various classes of flares, emphasizing high-energy phenomena and interplanetary phenomena involving energetic particles. Energetic particles play important roles in the development of flares; not only do they transport energy from the energy release site to the lower atmosphere and cause almost all the detectable radiations, but also some of them escape into interplanetary space to influence the near-Earth space environment. Gamma-ray emission from solar flares has been reviewed by Chupp (34), Hudson (73), and Ramaty & Murphy (144), and hard X-ray emission has been reviewed by Dennis (46).

Although we frequently refer to radio emission in discussing the observational properties of flares, it is not possible here to consider theoretical aspects of flare radio emission. However, comprehensive reviews of radio emission have already been published in this series (50, 193, 194). For more recent reviews on radio emissions from solar flares, see (102, 120). Melrose & Dulk (120) and Trottet (178), in particular, discuss particle acceleration in terms of radio-burst observations. A recent monograph (119) also deals with solar radio bursts. Coronal mass ejections (CMEs), as is shown, are well associated with a certain class of flares. For reviews of CMEs, see (69, 80, 188). For details on other aspects of flares, see the *SMM* workshop monograph (103).

We organize this paper as follows. In Section 2 we discuss the historical background before the *SMM* launch. In Section 3 we review the recent developments made by observations with *SMM*, *Hinotori*, and other contemporary satellites and ground-based observatories. Based on the observations discussed in Sections 2 and 3, in Section 4 we classify solar flares into the following five classes: *thermal hard X-ray flares, nonthermal hard X-ray flares, impulsive gamma-ray/proton flares, gradual gamma-ray/proton flares,* and *quiescent filament-eruption flares.* Also in Section 4 we examine the roles of filament eruptions in flare development. In Section 5 we discuss theoretical ideas related to processes occurring in different classes of flares. Closing remarks are given in Section 6.

2. HISTORICAL BACKGROUND

2.1 *Proton Flares and Radio Bursts*

Long before soft and hard X rays and gamma rays from solar flares were observed, energetic protons accelerated by flares had been detected. As early as 1946, protons above 1 GeV were detected from their ground-level effects (GLEs; 56a), and subsequently protons with energy of order 10 MeV were detected from the resulting "polar cap absorption" events (189). Solar flares that cause polar cap absorption events are called "proton flares," and flares that cause ground-level effects are often called "GLE events" or "cosmic-ray flares." However, the term "proton flares" in general refers to both groups.

Ellison et al. (54) first noticed that cosmic-ray flares are typically two-ribbon flares, with two large Hα ribbons that slowly drift apart. Later studies have shown that the same is true for the majority of proton flares (7). In the meantime, radio astronomers found that nearly all proton flares are associated with type IV metric bursts, and they claimed that type IV emission is the most reliable indicator of proton acceleration by the parent flare (194). Type IV bursts are almost always preceded by type II bursts [see Pick (138) for emission mechanisms for type IV]. According to Wild et al. (194), type II radio bursts, which are narrowband meter-wave emission, the frequency of which slowly drifts to lower frequencies, are due to energetic electrons accelerated by shock waves propagating in the corona. It has been proposed (194) that the same shocks also accelerate energetic protons arriving at Earth.

As indicated above, some researchers have for some time considered that proton flares may differ from most flares, and that they may involve additional physical processes. On the other hand, McCracken & Rao (118) have expressed the view that it depends only on the detector sensitivity as to whether or not we detect energetic interplanetary (IP) protons after a flare, implying that flares producing energetic IP protons do *not* comprise a unique class of flares. With the advent of energetic particle detectors aboard spacecraft, many flares have been found to accelerate protons detectable in the neighborhood of the Earth, so that the characteristics of proton flares can be studied in more detail. Lin (103a) showed that the majority of proton events were associated with type II and type IV radio bursts. Svestka & Fritzova (170) even suggested that almost all flares producing type II bursts were proton flares. However, later studies (9, 78) have shown that only a small fraction of flares with type II radio bursts are associated with proton events, although the majority of proton events are associated with type II and type IV radio bursts. By studying solar flare soft X rays and energetic particles with detectors aboard the same satellites, Sarris & Shawhan (151) found that soft X-ray decay times of proton flares, ranging from about 80 to 100 min, are longer than those of electron flares (with high-energy electrons but without high-energy protons), and that the ratios between rise and decay times of proton flares are smaller than those of electron flares. This has been confirmed by many other studies (e.g. 76, 128), and Kahler (76) has called flares with slowly decaying soft X-ray emission "long decay events." Sturrock (161) proposed that the slow separation of the Hα ribbons of a two-ribbon flare and the associated long-lasting soft X-ray emission may be attributed to the progressive reconnection of the oppositely directed field lines of an open field configuration. He considered that an initially closed flux tube may be slowly opened by the same stress that drives the solar wind. Kopp & Pneuman (99) later considered a similar model, proposing that an erupting prominence would lead to an opening of the overlying flux system. However, neither of these articles discussed the location or mechanism of proton acceleration.

2.2 Skylab Observations

The next important contribution to flare classification came from *Skylab* observations. Analyzing *Skylab* soft X-ray images (2–60 Å) of flares observed near the solar limb, Pallavicini et al. (132) proposed that there are two classes of flares—compact and extended. Extended flares do exhibit a large array of distinct characteristics: large and diffuse soft X-ray sources (with volumes in the range 10^{28}–10^{29} cm^3), high soft X-ray sources (heights of order 5×10^9 cm), long-enduring soft X-ray emission (with e-folding times of hours), low energy density (10–10^2 erg cm^{-3}, as compared with 10^2–10^3 erg cm^{-3} for compact flares), association with type II and type IV radio bursts, and association with prominence eruption and white-light coronal transients (now often called CMEs). Although Pallavicini et al. did not study energetic IP protons, one can see that many of the above-mentioned properties of extended flares are also the properties of proton flares, with some important new additions.

In view of the foregoing, it is natural that by the end of the *Skylab* Workshop, the majority of solar flare physicists had reached the following view (162, 163): Solar flares consist of two classes—simple, compact flares, for which the flare energy is released mainly during a "first phase"; and complex, extended flares, for which additional energy release and acceleration take place during a "second phase." The second-phase acceleration energizes protons to high energies and electrons to relativistic energies, possibly via shocks propagating in the corona (42, 152, 168, 194). Evidence for continuous energy release by Kopp-Pneuman-Sturrock–type reconnection after the first phase of extended flares is that the top is the brightest part of a loop (110, 132), and that the decay time is much longer than the cooling time (110). Svestka's (169) review paper, although written comparatively recently, maintained basically this pre-SMM view by dividing flares into confined flares and dynamic flares.

2.3 First and Second Phases

Wild et al. (194) used the terms "first phase" and "second phase" of particle acceleration. Analyzing the hard X-ray (> 10 keV) observations by the *Fifth Orbiting Geophysical Observatory*, Kane (90) reported the discovery of impulsive and slow components of hard X rays. The impulsive component is due to bremsstrahlung of nonthermal electrons, while the slow component represents thermal X rays. Since then, the terms "impulsive phase" and "gradual phase" have been widely used. The term "impulsive phase" has often been used as a synonym for "first phase." However, the term "impulsive phase" has become ambiguous because the hard X-ray emission of some flares shows gradual behavior. Some people use the term to refer to the whole period of hard X-ray emission, and others use it to refer only to the period in which the hard X-ray flux changes impulsively. Therefore, in describing observational characteristics, we use the term "first phase" instead of "impulsive phase." (Later, when we come to propose a theoretically based terminology, we introduce other terms.) First-phase phenomena include acceleration of nonthermal electrons in the first phase, and direct consequences of this acceleration (such as hard X-ray and microwave emission and chromospheric responses).

Frost & Dennis (61) and Hudson et al. (75) interpreted gradual hard X rays as resulting from electrons accelerated during the second phase. However, Kahler (79) and Tsuneta et al. (182) have shown that the source height of the gradual hard X-ray emission is well below the location of the type II–producing shock. We also regard the gradual hard X-ray emission as a first-phase phenomenon, for the reasons discussed in the following sections.

Type II and type IV radio bursts start several minutes later than the start of hard X-ray emission; therefore, in short-duration flares, type II and type IV bursts start after the hard X-ray emission has decayed completely. However, in long-duration flares, type II and type IV bursts start while hard X-ray emission is still in progress. Thus, first-phase and second-phase phenomena are not cleanly separated in time, but there are indications that first-phase and second-phase phenomena are separated in space, with the latter occurring in the high corona.

2.4 Nuclear Gamma Rays

Energetic protons and heavy ions interacting in the solar atmosphere produce excited nuclei, which are in turn promptly deexcited to produce gamma-ray lines. The lines at 4.4 MeV and 6.1 MeV, which are due to excited ^{12}C and ^{16}O, are the two most prominent nuclear lines. Nuclear interactions also produce neutrons, which are thermalized in the solar atmosphere and combine with protons to produce deuterons and the 2.2-MeV gamma-ray line. Pions produced by nuclear interactions decay to muons, which in turn decay to electrons or positrons. Positrons are also produced by the β-decay of radioactive nuclei produced by nuclear interactions. Positrons annihilate with electrons to produce 0.511-MeV gamma rays. The 2.2-MeV line is delayed with respect to nuclear gamma rays because neutrons take time to be captured. The 0.511-MeV line is also delayed because positrons from energetic pions take time to slow down and annihilate, and because radioactive nuclei have finite half-life times for β-decay. For the above reasons, we can learn about high-energy protons and ions by studying gamma rays originating from nuclear interactions (143, 144). In particular, the time profiles of a nuclear line or gamma rays in the 4–7 MeV band (where nuclear gamma rays are dominant over the bremsstrahlung continuum) can tell us when protons and heavy ions are accelerated.

On the basis of observations of energetic IP protons, it was expected that solar flares would emit detectable nuclear gamma rays (108), and it was also expected that such gamma rays would be detected after the first phase. Before the launch of *SMM*, solar nuclear gamma rays had been observed by means of the *Seventh Orbiting Solar Observatory* (*OSO 7*) (35), the *First High Energy Astronomical Observatory* (*HEAO 1*) (74), and the *Third High Energy Astronomical Observatory* (*HEAO 3*) (141). Such observations showed that nuclear gamma rays are emitted nearly contemporaneously with hard X rays, with some delays of tens of seconds. Such delays were interpreted as evidence for the "second-phase" acceleration of protons (16, 74).

3. RECENT DEVELOPMENTS

3.1 *SMM* Results

One of the main achievements of *SMM* has been the detection of nuclear gamma rays from many flares—45 flares up to January 1985 (39)—

far more than the modest prelaunch expectation of the Gamma-Ray Spectrometer (GRS) group. Figures 1 and 2 show time profiles of hard X rays and 4.1–6.4 MeV gamma rays for two flares detected with the GRS. Observations such as these run counter to the traditional notion of two phases of acceleration. It is now clear that protons and relativistic electrons are accelerated during the first phase before the start of a type II radio burst. Furthermore, for impulsive flares such as that of 7 June 1980, protons must be accelerated promptly with time scales of a second or so. Although some observations before *SMM* showed that nuclear gamma rays are produced during the first phase, contemporaneously with hard X rays, the GRS detection of nuclear gamma rays from many flares made solar physicists come to terms with this fact. The delay of nuclear gamma rays with respect to hard X rays can be seen in these figures, and it ranges from less than 2 s up to 100 s (34, 91, 199).

The detection of nuclear gamma rays during the first phase raises the following two questions. Since it is known that the first phase can accelerate protons to high energies, do we need a second phase of acceleration (32, 33, 58, 59, 141)? And are protons accelerated during all flares or only during some classes of flares? These two questions are discussed in this section in reverse order.

CHARACTERISTICS OF GAMMA-RAY LINE FLARES   In addition to the impulsive behavior of nuclear gamma-ray time profiles and the very short delays of gamma rays with respect to hard X rays, the GRS group (34, 57, 184) found a good correlation between gamma-ray continuum fluences > 270

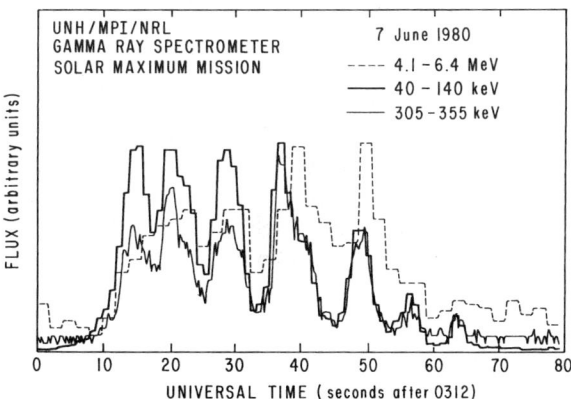

Figure 1   Hard X-ray and gamma-ray time profiles of the 7 June 1980 flare. Gamma rays in the 4.1–6.4 MeV range are emitted simultaneously with hard X rays. Nevertheless the gamma-ray time profile shows a delay of about 2 s with respect to the hard X-ray time profiles. A type II radio burst started at 0313 UT and lasted until 0332 UT. From Chupp (32).

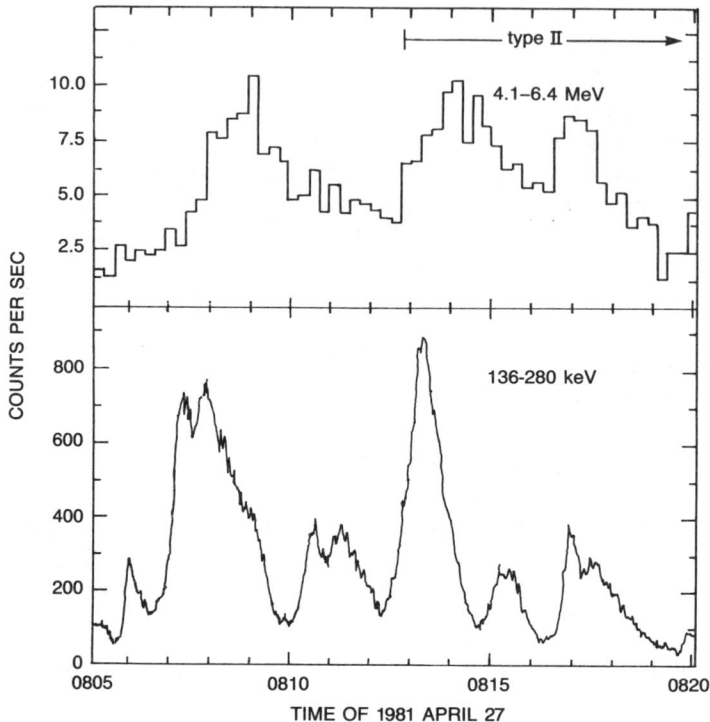

Figure 2 Hard X-ray and gamma-ray time profiles of the 27 April 1981 flare. Gamma rays in the 4.1–6.4 MeV interval are emitted simultaneously with hard X rays, but the gamma-ray time profile is delayed by about 1 min. A type II radio burst started at 0812.8 UT and lasted until 0836.3 UT. The gamma-ray time profile is from Chupp (34), and the hard X-ray time profile is from HXRBS data.

keV and excess 4–8 MeV fluences (excess over the power-law continuum). From these discoveries, the GRS group at the University of New Hampshire (32, 58, 184) proposed that a single, primary acceleration may be responsible not only for low-energy electrons (< 200 keV), but also for relativistic electrons and energetic ions. It is implicit in this view that as far as acceleration is concerned, there is only one class of flares.

In a dissenting opinion, Bai (8) proposed at a meeting at La Jolla in 1981 that in addition to the classical second-phase acceleration (42, 194), an additional process energizes protons to gamma-ray-producing energies and electrons to relativistic energies during the first phase. Bai & Ramaty (17) had initially proposed such an acceleration, which they termed "second-step acceleration," for the further acceleration of electrons during the first phase.

It is implicit in this view that gamma-ray-producing flares differ from other flares in that an additional acceleration process operates during the first phase. In order to test this hypothesis, Bai et al. (13) and Bai & Dennis (12) compared gamma-ray line (GRL) flares (which emit detectable nuclear gamma rays) with flares having peak Hard X-Ray Burst Spectrometer (HXRBS) rates $> 10^4$ counts s^{-1} but without detectable nuclear gamma rays. Because the comparison group has peak HXRBS rates as large as those of GRL flares, we may regard the comparison group as "non-GRL flares" or "gamma-ray-poor flares." Bai et al. and Bai & Dennis found the following results:

1. GRL flares have flatter hard X-ray spectra (the average spectral index being 3.5).
2. A large fraction (75%) of GRL flares produce either type II or type IV radio bursts or both, whereas only a small fraction (30%) of the comparison group do so.
3. The hard X-ray spectra of some GRL flares harden as a spike burst progresses, whereas the majority of flares show a soft-hard-soft behavior.

Bai (9) has confirmed these findings by more detailed statistical analyses. Cliver et al. (40) also have confirmed that GRL flares are well associated with type II and type IV bursts by studying a larger sample of GRL flares that had been observed with the GRS up to January 1985.

The progressive hardening of hard X-ray spectra observed in GRL flares is shown in Figures 3 and 4. Such spectral hardening is equivalent to a

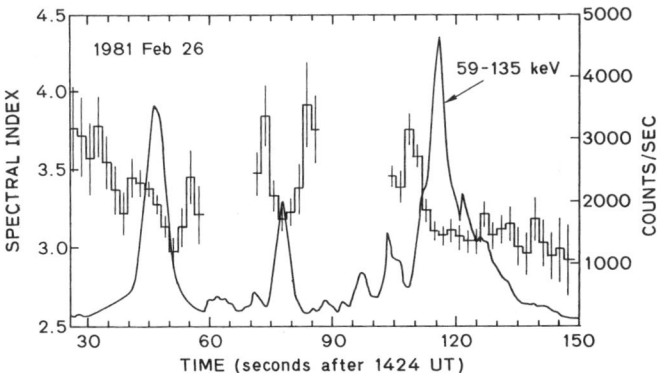

Figure 3 HXRBS hard X-ray time profile and spectral evolution of the 26 February 1981 flare. This is an example of an impulsive gamma-ray/proton (GR/P) flare. The spectrum hardens during the first spike burst (around 1424:45 UT) and the third spike burst (around 1425:55 UT), but it shows "soft-hard-soft" behavior during the second spike burst. From Bai & Dennis (12).

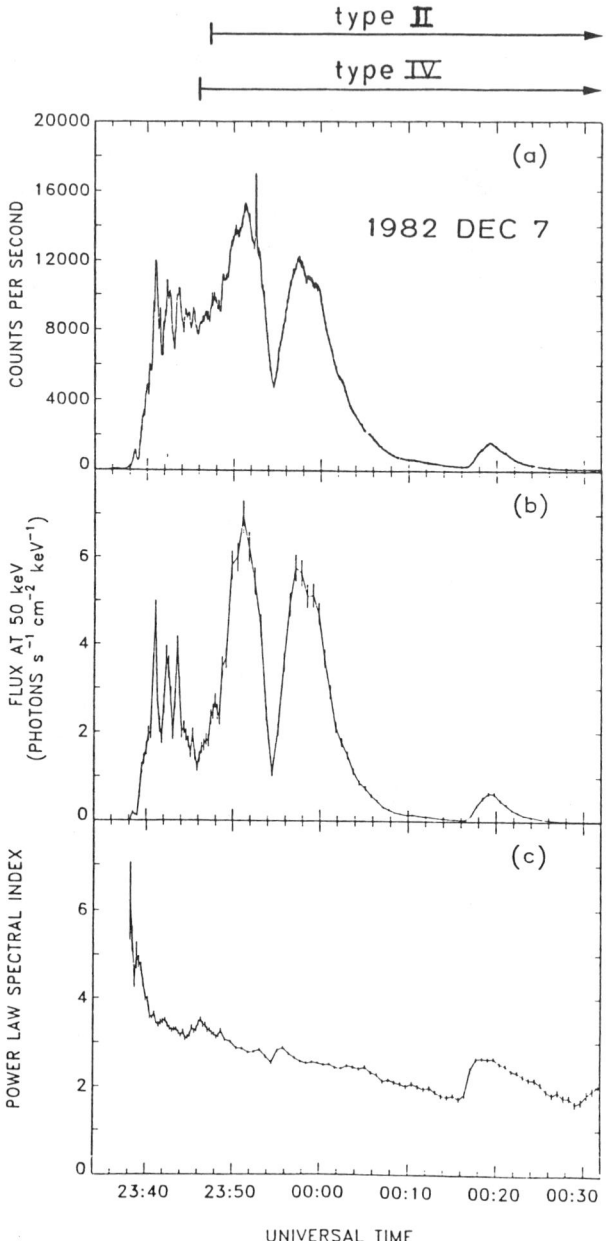

Figure 4 HXRBS hard X-ray time profile and spectral evolution of the 7 December 1982 flare. Panel (*a*) shows the total counts of HXRBS (32–559 keV). This flare, which emitted hard X rays for 50 min, is an example of a gradual GR/P flare. The hard X-ray time profile is impulsive in the beginning but becomes gradual. A short-duration spike at 2352:30 UT is from a flare at another active region. From Bai (9).

delay of high-energy hard X-ray time profiles with respect to lower energy hard X rays (12). Such delays or spectral hardening can be due to the fact that the energy loss time of energetic electrons trapped in a coronal loop increases with increasing energy (17, 179, 186). Inclusion of the precipitation of electrons from a hypothetical trap to the chromosphere does not change this property, although it shortens the delay times (112, 150). Even if the trap interpretation is correct, GRL flares are still different in that such delays are observed in almost all cases in GRL flares. The second-step acceleration (8, 9, 12, 14, 119a, 126) was invoked to explain not only the delays but also the acceleration of relativistic electrons and gamma-ray-producing protons.

SECOND PHASE ACCELERATION Now we address the question of whether there is a second phase of acceleration. One of the puzzling results of *SMM* was that there is a very poor correlation between the number of energetic protons deduced from nuclear gamma-ray observations and the number of energetic protons estimated from IP observations (37, 39, 135, 187). Some flares that produced large fluxes of IP protons produced hardly any nuclear gamma rays; on the other hand, some other flares with large gamma-ray counts produced only small fluxes of IP protons. Two possibilities were proposed: (*a*) The same acceleration mechanism accelerates both gamma-ray-producing protons and energetic IP protons, but the escape rate varies from flare to flare. (*b*) There are two acceleration mechanisms for protons—one for protons that are mostly confined in the solar atmosphere and produce gamma rays, and the other for protons that mostly escape into IP space. In retrospect, considering that impulsive GRL flares did not exhibit the characteristics of proton flares that were well known before *SMM*, it seems easy to opt for the second possibility. But this was not so obvious at the time: One of the reasons may be that the first well-publicized gradual GRL flare was that of 27 April 1981, which did not produce a noticeable change in the high background flux of IP energetic protons owing to a big event 3 days earlier. Bai (8) advanced the second possibility, and Bai et al. (15) subsequently proposed that there are two classes of gamma-ray/proton (GR/P) flares—impulsive and gradual. (Here GR/P flares refer to flares that produce nuclear gamma rays and/or energetic IP protons. GRL flares refer specifically to flares for which nuclear gamma rays are detected, so that GRL flares are a subset of GR/P flares.) According to Bai et al. (15), in gradual GR/P flares, additional (second-phase) acceleration produces energetic protons that mostly escape into IP space, whereas in impulsive GR/P flares only a "second-step" mechanism is operative. This proposal has been further supported by a later, more detailed study by Bai (9). By studying the properties of flares producing energetic IP particles, Cane et al. (23) also have classified IP particle events into impulsive and long-duration events, based on the duration of the associated soft X-ray emission.

Bai (9) studied 17 GRL flares (both impulsive and gradual) observed with the GRS in 1980 and 1981, and, in addition, 23 "HXRBS gradual flares" selected on the basis of HXRBS observations in the interval 1980 through 1982. If the total hard X-ray duration of a flare is longer than 10 min and the duration of the strongest hard X-ray spike burst is longer than 1.5 min, the HXRBS flare was regarded as a gradual flare (Figures 2, 4). Although the selection of these gradual flares was based solely on the characteristics of hard X-ray emission (i.e. on "first-phase" characteristics), they were found to exhibit a large set of characteristics [16 are listed in Table 6 of (9); see Table 1], and many of these characteristics are also the known characteristics of extended (or proton) flares.

Out of these 23 HXRBS gradual flares, the 9 flares with the largest peak HXRBS counting rates were found to emit detectable nuclear gamma rays. Therefore, it seems reasonable to infer that the other gradual flares also emitted nuclear gamma rays, but that their emission levels were below the GRS threshold. The large majority of the HXRBS gradual flares west of E20°, which is the region well connected to Earth by magnetic field lines, turned out to produce detectable energetic IP protons.

In view of the evidence that these HXRBS gradual flares must have produced nuclear gamma rays as well as energetic IP protons, they were called "gradual GR/P flares." Compared with gradual GR/P flares, impulsive GR/P flares do not exhibit characteristics pertaining to phenomena occurring in the high corona and IP medium, with the exception that they may produce type II and type IV radio bursts. In fact, the majority of impulsive GR/P flares produced metric type II bursts, but none of them was found to produce kilometric type II bursts, which are indicative of IP shocks (26). Also, it was found that the estimated ratio between the number of energetic IP protons and the number of gamma-ray-producing protons is small ($\ll 1$) for impulsive GR/P flares, while it is relatively large (> 1) for gradual GR/P flares. For these reasons, it was concluded that energetic IP protons from gradual GR/P flares are mostly produced in open magnetic field configurations in the high corona, possibly by shocks in that region. It is proposed that, in both impulsive GR/P flares and gradual GR/P flares, protons are accelerated during the first phase (or hard X-ray phase) in closed magnetic loops by the second-step mechanism, and that only a small fraction of the energetic protons escape into IP space.

Bai's studies (9, 12) have led to the following conclusions that are relevant to the problem of flare classification: (*a*) Gradual GR/P flares are sufficiently distinct to be selected solely on the basis of the characteristics of their hard X-ray emission (a first-phase phenomenon). (b) Gradual and impulsive GR/P flares share some common characteristics, in addition to nuclear gamma-ray emission. In addition, these studies throw light on the relationship between gamma-ray-producing protons and energetic IP protons.

Cane et al. (23) have studied 67 solar particle events detected with *IMP 8* (the *Eighth Interplanetary Monitoring Platform*) and *ISEE 3* (the *Third International Sun-Earth Explorer*, now *International Cometary Explorer*) in the period from September 1978 to December 1983. (Although this study draws heavily from results obtained with other satellites, it is discussed in this section because it is related to *SMM* results.) These events were divided into two groups—impulsive events and long-duration events. Flare-associated events were considered impulsive if the 1–8 Å *X-ray duration* (not the duration of proton flux enhancement) is less than 1 hr at the 10% level of the peak; otherwise, they were considered to be long-duration events. These two groups were found to be distinct in many respects; in particular, the long-duration events have exactly the same characteristics as extended flares (proton flares, or gradual GR/P flares). This study also found several new and interesting points. (a) The parent flares of impulsive events are mainly in the Western Hemisphere, which is well connected to Earth by magnetic field lines, whereas a considerable fraction of the parent flares of long-duration events are in the Eastern Hemisphere. (b) The decay times of proton and electron fluxes are short (< 10 hr) for impulsive events, whereas the decay times of long-duration events are long. (c) The proton fluxes of long-duration events are, on the average, much larger than those of impulsive events. However, the electron fluxes of both groups are similar.

From these findings, we can draw the following inferences: (a) In long-duration events, protons are accelerated mainly by shock waves propagating in large volumes of the corona [cf. (18, 115)]. (b) In impulsive events, protons and electrons are accelerated in closed loops during the first phase of flares, and only small fractions of these particles escape along open magnetic field lines near the flare site. (c) The shocks accelerating protons in long-duration events are not efficient in accelerating relativistic electrons, so that long-duration events are "electron poor." Evenson et al.'s (56) finding that GRL flares are "electron rich" is applicable only to impulsive GRL flares, since the GRL flares in Evenson et al.'s study all happen to be impulsive (9, 23). In this connection, we recall the result of Sarris & Shawhan (151) that proton events are associated with long-duration soft X-ray emissions, whereas electron events are associated with short-duration soft X-ray emissions.

In a recent paper, Cliver et al. (39) have compared the number of energetic protons deduced from gamma-ray observations with that deduced from IP observations, for flares observed in the period from February 1980 through January 1985, and arrived at essentially the same conclusion as Bai (9) and Cane et al. (23). We have mentioned that energetic protons accelerated during the second phase mostly escape into IP space and do not produce substantial gamma rays. However, Chupp et al. (31) and Ramaty et al. (145) have found that the gamma rays showing gradual variations after 1143.5 UT of the 3 June 1982 flare are due to protons with a flatter spectrum than the protons of the first phase, and Ramaty et al. have proposed that if a small fraction (4%) of protons accelerated by the second-phase acceleration interacted at the Sun, they could account for the observed gradual gamma rays.

We have mentioned that for impulsive GR/P flares, the ratio of IP protons to gamma-ray-producing protons is small. In further detail, Hua & Lingenfelter (71) have found that for impulsive GR/P flares, this ratio increases as the total number of energetic protons increases. However, this result is based on a small number of flares.

In contrast to the interpretations of Bai (9), Cane et al. (23), and Cliver et al. (39), Zaitsev & Stepanov (201) have proposed that one mechanism accelerates both gamma-ray-producing protons and IP protons, and that the ratio between the plasma pressure and the magnetic pressure, $\beta = 8\pi P/B^2$, determines the escape rate. If $\beta \ll 1$, energetic protons are mostly confined in the flare loops and produce gamma rays. If, on the other hand, $\beta > \beta_* = 0.3–1.0$, the hot plasma and high-energy protons escape from the loops by means of the flute instability. Lin & Hudson (105) have proposed an idea similar to that of Zaitsev & Stepanov (201).

The above interpretation is implausible from the following consideration. We would expect an impulsive GR/P flare to have a high plasma pressure because it involves a large energy deposition by energetic electrons in a short time interval in a small volume. In a gradual GR/P flare, on the other hand, energetic electrons deposit their energy more slowly in a larger volume. Furthermore, many gradual GR/P flares are quite weak in hard X-ray emission, which implies that they involve little heating by energetic electrons (9, 38). In support of this view, we may note that Pallavicini et al. (132) found that gradual (or extended) flares have lower energy densities than impulsive flares.

Kocharov (98) has proposed that the density of the ion interaction region is low (10^{10}–10^{11} cm^{-3}) for gradual GRL flares and high (10^{12}–10^{13} cm^{-3}) for impulsive GRL flares. He proposed that the long lifetimes of ions interacting in the low-density medium of gradual flares lead to higher escape probabilities. However, IP observations of energetic protons from gradual events are consistent with protons being accelerated in a large volume for a long time (23, 115).

3.2 Hinotori Results

The *Hinotori* satellite, which was in operation from February 1981 to October 1982, led to important advances in our understanding of flare classification. *SMM* failed to obtain X-ray images of gradual flares [except for imaging a diffuse soft X-ray source long after the main flare of 21 May 1980 (44)], partly because its Hard X-Ray Imaging Spectrometer (HXIS) ceased to operate in November 1980 and was not rendered operational at the time of the Space Shuttle repair mission of April 1984. *Hinotori*, on the other hand, made soft and hard X-ray images of many gradual GR/P flares (129, 173, 181, 182). Although soft X-ray images of gradual GR/P flares had been obtained by *Skylab*, and although there had been some indication [from hard X-ray observations of over-the-limb flares (61, 72, 75, 94, 149)] that hard X-ray sources extend to great heights, *Hinotori* results directly showed for the first time that the hard X-ray sources of gradual GR/P flares are located high above the photosphere ($h > 10^9$ cm). *Hinotori* also made valuable gamma-ray observations (197, 198, 200), which supplemented those obtained by means of the *SMM* gamma-ray detector.

Mainly relying on *Hinotori* observations of soft and hard X-ray images and spectra, Tanaka (175, 176) and Tsuneta (180) classified flares into three types. According to Tanaka (176), these are as follows.

Type A (Hot Thermal Flares): These flares are effective in producing a superhot component of temperature $T = (3-4) \times 10^7$ K that emits hard X rays in the range $E < 40$ keV and strong Fe XXVI lines. The hard X-ray time profile shows a gradual rise and fall similar to the soft X-ray profile, and the hard X-ray source is compact (< 5000 km). The spectrum above 40 keV is soft (effective power law index $\gamma = 7-9$). Radio emission is weak.

Type B (Impulsive Flares): These show typical impulsive hard X-ray bursts consisting of rapidly varying spikes emitted from the low corona, including the loop footpoints. The spectral index is in the range $\gamma = 3-7$. The later phase of some flares evolves to a more gradual time profile with a softer spectrum ($\gamma = 5-8$) and to a more compact source structure located at a higher altitude.

Type C (Gradual Hard Flares): These flares show a long-lived (> 30 min) burst with a broad peak or peaks showing no impulsive variation. The source is located high in the corona ($h > 4 \times 10^9$ cm) and can be identified with large extended loops. The spectrum is very hard, is well characterized by a power law ($\gamma = 2.5-4$), and shows systematic hardening with time. Microwave emission is very strong.

Hinotori made unique contributions to flare classification, including the imaging of hard X-ray sources, the identification of type A flares as another class, and a detailed study of the superhot component ($\sim 3.5 \times 10^7$ K) initially identified by Lin et al. (106). One may argue that during the first phase of all flares, heating and acceleration of nonthermal electrons take place simultaneously, and thus that it is only a matter of degree as to whether heating or acceleration is more efficient. However, it is important to find out what parameters determine the outcome. *Hinotori* made an important contribution in this regard.

3.3 Other Results

Cliver et al. (36) have studied the properties of gradual hard X-ray bursts (GHBs). They selected flares for which the FWHM durations of 2.8-GHz radio bursts (as measured at the Ottawa River Solar Observatory and the Dominion Radio Astrophysical Observatory) exceed 10 min and studied their properties in hard X rays, microwaves, and other aspects. Their events seem in some respects to be different from the gradual GR/P flares of Bai, although they are similar in other respects. [Three of the 10 events examined by Cliver et al. are in Bai's (9) list of gradual GR/P flares.] Except for two cases, all GHBs were preceded by impulsive bursts about 30 min before the GHB. Nevertheless, these GHBs show flattening of their hard X-ray spectra, "microwave richness," and association with CMEs. Kai et al. (89), working with their 17-GHz microwave data, have studied gradual bursts occurring 10–30 min after impulsive bursts, and they also found that the hard X-ray spectra of these delayed gradual bursts flatten with time.

4. A NEW CLASSIFICATION SCHEME

The new observations discussed in the preceding sections enable us to identify the following four classes of flares: *thermal hard X-ray flares*, *nonthermal hard X-ray flares*, *impulsive GR/P flares*, and *gradual GR/P flares*. The *Hinotori* group (175, 176, 180) first proposed thermal hard X-ray flares as a separate class, and Bai and his co-workers (9, 12, 13) first proposed that GRL flares should be considered as a distinct class. In addition to the above four classes, flares resulting from quiescent filament eruptions are regarded as constituting their own class, as shown in Section 4.5. We discuss the properties and relative frequencies of these five classes of flares:

1. thermal hard X-ray flares,
2. nonthermal hard X-ray flares,
3. impulsive GR/P flares,

4. gradual GR/P flares,
5. quiescent filament-eruption flares.

4.1 *Thermal Hard X-Ray Flares*

As shown in Figure 5, for these flares, the hard X-ray emission below 40 keV is dominated by thermal bremsstrahlung due to plasmas with a

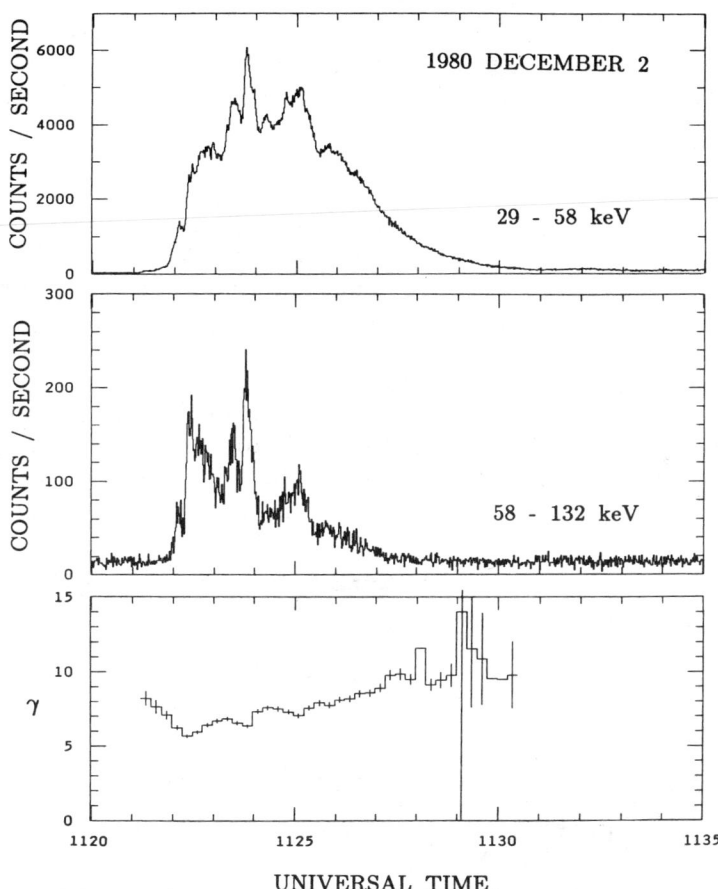

Figure 5 HXRBS time profiles and spectral evolution of the thermal hard X-ray flare on 2 December 1980. The time profile of the 29–58 KeV range is gradual, with some contribution from a spiky nonthermal component, but that of the 58–132 keV interval is more spiky. For power law fits, the spectral indices are large. For thermal fits, the temperature is about 6×10^7 K. Both pure power law fits and pure thermal fits result in large χ^2 values, which indicates that the X rays might be a combination of thermal and nonthermal emissions (figure courtesy of Larry Orwig).

temperature of order 30×10^6 K. At high energies (>40 keV), hard X-ray time profiles reveal impulsive variations. For thermal hard X-ray flares, there is no clear-cut separation between the impulsive component and the gradual thermal component (176). The impulsive component is embedded in the gradual thermal component. The light curve of 30-keV hard X rays is similar to that of 5-keV X rays, and both are considered to be thermal bremsstrahlung, although from plasmas of different temperatures. The hard X-ray spectrum above 40 keV is very steep, with spectral indices $\gamma = 7$–9. Microwave emission is weak because of the paucity of high-energy electrons. To date, none of the flares of this class has been known to emit type II or type IV radio bursts, nor do they show any other second-phase activity.

4.2 Nonthermal Hard X-Ray Flares

Among intense flares (peak HXRBS rates > 1000 counts s^{-1}), the majority belong to this class. Flares of this class show impulsive variations with time scales ranging from 0.1 to 30 s and intermediate hard X-ray spectral indices ($\gamma = 3.5$–6). Figure 6 shows the light curves and the spectral evolution of hard X-ray emission.

Energy is released impulsively during the first phase in both impulsive acceleration of electrons and in situ heating of the plasma in the loop. According to the thick-target model (2, 20, 104, 105), during the first phase, the chromosphere is suddenly heated by the precipitating electrons to coronal temperatures. The following evidence supports the idea that

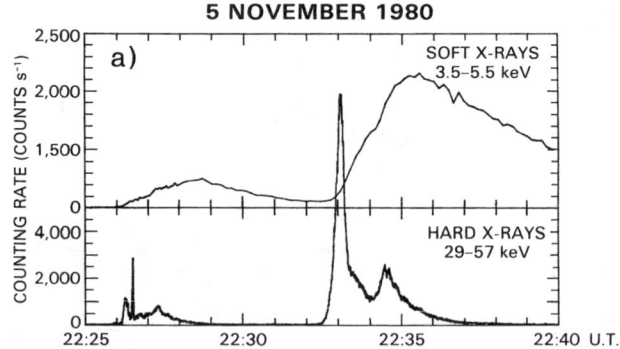

Figure 6(a) Soft and hard X-ray time profiles of the 5 November 1980 flare. The soft X-ray light curve behaves like the time integral of the hard X-ray count rate. (*b*) Time profiles and spectral evolution of a nonthermal hard X-ray flare on 5 November 1980. Notice the "soft-hard-soft" spectral behavior and the similarity of the UVSP (UV Spectrometer on *SMM*) counting rate of the O V line to the hard X-ray time profiles. Both Figures 6*a* and 6*b* are from Dennis (46).

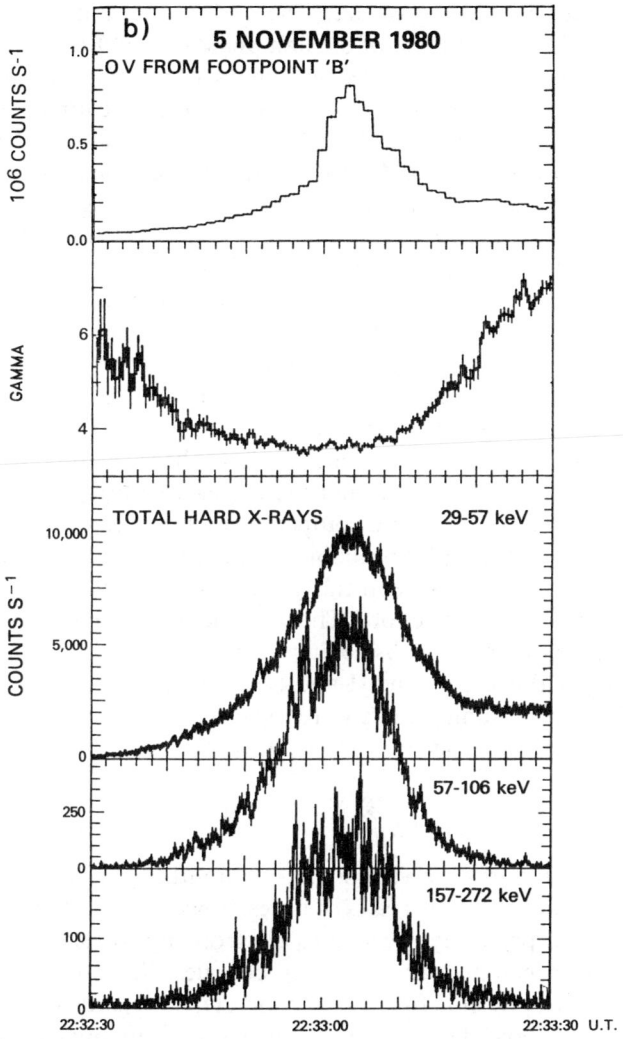

Figure 6—(continued)

this model is basically correct for this class of flares. (*a*) UV and optical radiation show good timing coincidence with hard X rays (55, 92, 195). (*b*) The estimated energy content of nonthermal electrons is quite well correlated with that of thermal plasmas (46, 159), although uncertainties are large. (*c*) The suddenly heated uprising plasma ("chromospheric evaporation") that radiates blueshifted X-ray lines seems to account for the increase of the emission measure. The turbulence and the blueshifted component subside at the end of the first phase (5). (*d*) For some flares the light curve of soft X rays resembles the integral of hard X-ray or microwave emission (43, 46, 127, 174; Figure 6*a*). (*e*) Although the directivity of low-energy (30-keV) electrons is hard to observe, high-energy electrons (> 300 keV) may be directional (11, 48, 137, 185). (*f*) The momentum carried by the uprising plasma is balanced by the downward-moving cool plasma that is radiating redshifted Hα radiation (202). (*g*) There is a good correlation between hard X-ray spikes and type III emission (146).

During the first phase of nonthermal hard X-ray flares, electrons are accelerated up to several hundred keV, but the efficiency of high-energy electron acceleration is low, which results in rather steep hard X-ray spectra. Flares of this class do not accelerate protons to gamma-ray-producing energies in any appreciable quantities. For this class of flares, energy release occurs mainly during the first phase. Although we cannot rule out the possibility of additional energy release during the decay phase of this class of flares, there is no compelling evidence for it (196). After the first phase the thermal plasma cools by radiation and conduction. The increase of emission measure in this gradual decay phase is due to evaporation caused by conduction. The turbulent motions of the plasma observed in the first phase subside during the decay phase (5, 6). For some of the flares of this class, type II and type IV radio bursts are observed. Traditionally, such radio bursts were regarded as second-phase phenomena. However, no other second-phase phenomena are observed during this class of flares.

4.3 *Impulsive GR/P Flares*

Impulsive GR/P flares are very similar to nonthermal hard X-ray flares. During the first phase of this class of flares, however, an additional process takes place and promptly accelerates electrons up to relativistic energies and protons to gamma-ray-producing energies. We do not know what mechanism is responsible for this "second-step acceleration," but there is no shortage of candidates. Various mechanisms involving shocks have also been proposed, such as shock drift acceleration, shock-resonant acceleration, and others (30, 45, 53, 130). Under suitable conditions, these mechanisms seem to be able to accelerate very rapidly. However, they are hard pressed to accelerate electrons to 20 MeV in a few seconds, as some observations require (91, 148).

Hard X-ray spectra are relatively flat because of the second-step acceleration of relativistic electrons. Figure 3 shows the time profile and the spectral evolution of the hard X-ray emission of the impulsive GR/P flare on 26 February 1981. During the majority of impulsive GR/P flares, type II and type IV radio bursts are observed, but seldom are IP shocks produced. Only small fractions (10^{-3}–10^{-2}) of electrons and protons accelerated in the first phase escape into IP space, so that they can be detected near the Earth. The fluxes of IP particles from this class of flares rise and decay rapidly, with a characteristic time of a few hours. Kane et al. (93) have shown that relativistic electrons detected with *ISEE 3* (now *ICE*) on 14 August 1982 were accelerated during the first phase of the flare at 0509 UT, which produced gamma rays above 2 MeV. [Cane et al. (23) proposed the same for energetic IP protons coming from this flare.] Therefore, impulsive flares producing relativistic electrons detected in IP space (23, 56, 151) are thought to emit nuclear gamma rays as well as bremsstrahlung by relativistic electrons.

Since impulsive GR/P flares are well associated with type II and type IV radio bursts, the second-step acceleration and type II and type IV radio bursts must somehow be related. Initially it was proposed (14) that shock waves propagating in the flare loop might act as a second-step acceleration mechanism and produce type II bursts when they propagate to the high corona. However, a close investigation of the timing of type II bursts and hard X-ray emission makes it difficult to maintain this view (40).

4.4 *Gradual GR/P Flares*

Gradual GR/P flares (Figures 2, 4) show a large array of characteristics, and therefore these flares were recognized as a separate class from quite early on (54, 132, 194). These flares have been called *two-ribbon flares, proton flares, extended flares, long-decay events, gradual flares,* etc., depending on the method of observation. Associations between various characteristics of gradual GR/P flares have been studied by many researchers. Instead of going into details of such studies, we present the results in Table 1. In the left column, 19 characteristics of gradual GR/P flares are listed. In the top row numbers 1 to 19 are assigned corresponding to the 19 characteristics in the left column. An example of how to interpret this table is given in the following: We see letters "B" and "M" at the location of the 18th row and the 17th column. This means that according to studies by Cane et al. (23) and Kahler et al. (86) (see the reference code in the footnotes to the table) the majority of flares with characteristic (18) (production of energetic IP protons) exhibit characteristic (17) (high-speed CMEs). This table is not diagonally symmetric: For example, the majority of flares with type II bursts are not proton flares, even though the majority of proton flares produce type II bursts.

All of the 19 characteristics are not present in all gradual GR/P flares. It is difficult to determine which characteristics are most commonly present and which characteristics are most rarely present, because some of the

Table 1 Characteristics of gradual GR/P flares and associations between them[a]

Characteristic[b]	(1)	(2)	(3)	(4)	(5)	(6)	(7)	(8)
(1) Gradual hard X-ray (HXR) emission	—	AYZb	AYZ	A	ASZ	ASZ	ANZ	A
(2) Long-duration HXR emission	AYZb	—	AEOZ	AEOZ	AEZ	AEOZ	AEOZ	AEO
(3) Hardening of HXR spectrum			—					
(4) Flat HXR spectrum				—				
(5) High HXR source					—			
(6) Gradual and long-duration microwaves						—		
(7) Large microwave richness index							—	
(8) Delay of microwaves w.r.t. HXR emission								—
(9) Nuclear gamma-ray emission								
(10) Spreading two Hα ribbons								
(11) Long-duration soft X-ray (SXR) emission								
(12) Large SXR source								
(13) Type II radio burst		Z						
(14) Type IV radio burst								
(15) Active-region filament eruption								
(16) Interplanetary shocks								
(17) High-speed CMEs								
(18) Energetic IP protons								
(19) Slow decay of the IP proton flux								

[a] Reference code: A:9, B:23, C:24, D:25, E:36, F:40, G:54, H:76, I:78, J:82, K:83, L:85, M:86, N:88, O:89, P:97, Q:125, R:128, S:129, T:132, U:153, V:154, W:155, X:156, Y:175, Z:176, a:177a, b:180, c:191, d:194, e:151.

[b] Characteristics (1) through (9) pertain to the first- (or main) phase phenomena. Blanks do not necessarily mean lack of association. In many cases, they indicate a lack of comprehensive study on the association.

characteristics are either difficult to observe or require instruments aboard spacecraft to be detected. Gradual GR/P flares are all two-ribbon flares. However, two-ribbon flares that do not exhibit other characteristics of gradual GR/P flares are quite common, and some two-ribbon flares show very weak hard X-ray emission (52).

Type II and type IV radio bursts are also quite common (87). Many

Table 1 (*continued*)

(9)	(10)	(11)	(12)	(13)	(14)	(15)	(16)	(17)	(18)	(19)	
A	AZ	AZ	YZb	AP	A		A	A	A		(1)
A	AEO	AZ		AEO	AEO		A	A	A		(2)
											(3)
											(4)
											(5)
											(6)
											(7)
											(8)
—									F		(9)
—											(10)
		—				U		H		B	(11)
T						T		T			(12)
				—							(13)
		C	D			—			Ia		(14)
						—					(15)
							—	X	B	B	(16)
		Wc		JL		Qc	VX	—	KM		(17)
	G	BHRe		d	d		B	BM	—		(18)
		B					B	B		—	(19)

flares that are not gradual GR/P flares have been found to emit such bursts. In particular, type II radio emission is a poor indicator of gradual GR/P flares. Only 30% of type II bursts are associated with type IV bursts (25). Therefore, it is evident that some shocks responsible for energetic electrons producing type II bursts do not accelerate protons efficiently. Maxwell & Dryer (116) proposed that shocks causing only type II bursts might be blast-wave shocks, while shocks that accelerate protons and heavy ions are piston-driven shocks. Kahler (78) and Bai (9) concur with this proposal.

All gradual flares do not produce energetic IP protons; therefore, production of IP protons should not be regarded as a necessary condition for a gradual GR/P flare. For example, the GRL flare of 13 May 1981 has all the characteristics of gradual GR/P flares, including IP shocks, although there are no data on CMEs (9). Although this GRL flare emitted intense soft X rays with its X-ray class X1, it failed to produce noticeable changes in the background of the energetic IP proton flux due to a gradual flare on May 10. (If the 1–8 Å peak flux of a flare is in the $1-2 \times 10^{-4}$ W m^{-2} interval, its X-ray class is X1; if the peak flux is in the $2-3 \times 10^{-4}$ W m^{-2} interval, its class is X2. An M1 class flare is 10 times less intense than an X1 flare, and so forth). On the other hand, two preceding gradual flares, on May 8 and May 10, produced large fluxes of energetic IP protons, although their X-ray classes were only M8 and M1, respectively. Filament eruptions are also not observed from all gradual GR/P flares (see, however, Section 4.6).

In the 1960s, type IV emission was regarded as the best indicator of gradual GR/P flares (194). At the present time, long-duration soft X-ray emission seems to be a better indicator of gradual GR/P flares, perhaps due to the availability of continuous soft X-ray observations. However, the distribution of the duration of soft X-ray emission is continuous, and a dichotomy is not obvious. Certain hard X-ray characteristics (long durations of >10 min, gradual variation with spike-burst durations longer than 1.5 min, spectral hardening) seem to be an even better indicator of gradual GR/P flares. Long-lasting, gradual microwave emission is also a good indicator of such flares. In forecasting energetic protons arriving at the Earth from solar flares, characteristics appearing in hard X-ray emission and microwave emission are more useful because one can identify a gradual GR/P flare in the early phase.

Even though gradual GR/P flares show a large set of characteristics, and indeed they can be identified on the basis of hard X-ray characteristics alone, one nevertheless finds that if the time scale of a gradual GR/P flare is reduced by a factor of 5 to 10, the characteristics of the hard X-ray emission phase are not distinguishable from those of an impulsive GR/P flare (9).

Almost all *gradual GR/P flares* show impulsive behavior in the beginning of their hard X-ray emission and gradual behavior later on. None of them is found to show impulsive behavior *after* the gradual hard X-ray emission. At least two gradual GR/P flares (8 May 1981, 13 May 1981) do not show impulsive behavior at all.

After the hard X-ray phase of gradual GR/P flares, the soft X-ray flux begins to decrease gradually, with decay times ranging up to hours (Figure 7). It is thought that reconnection of open magnetic field lines is the source of continued energy input (4, 28, 99, 123, 161). In gradual GR/P flares,

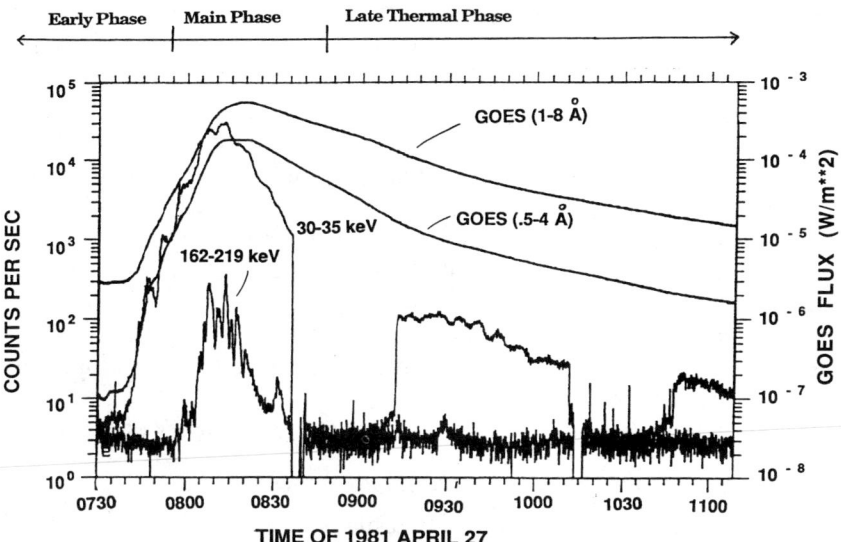

Figure 7 Phases of a gradual GR/P flare. The soft X-ray fluxes decay very slowly. The e-folding decay time of the 1–8 Å soft X rays is about 20 min in the time interval from 0820 to 0857 UT. The soft X-ray light curves decreased to the background levels more than 8 hr after the start of the flare. Two gaps in the hard X-ray data are due to the satellite's nights. Compare with Figure 2, (figure courtesy of Larry Orwig).

full-fledged second-phase phenomena occur in the high corona, including second-phase particle acceleration and generation of shocks.

Hinotori observations (129, 173, 181) show that for gradual GR/P flares, soft (5–10 keV) and hard (16–38 keV) X-ray sources are both high (1–4 × 10^9 cm) and almost cospatial. Observations by Kane et al. (94), on the other hand, show that the bulk of 150-keV emission originates less than 2.5 × 10^8 cm from the photosphere. However, hard X-ray observations from limb-occulted flares (72, 75) show that energetic electrons accelerated during gradual GR/P flares can reach great heights (10^{10} cm), and that some of these electrons can reach even greater heights (4 × 10^{10} cm) and radiate meter-wave continuum (35a, 89, 97). Therefore, we can infer that high-energy electrons of gradual GR/P flares are accelerated in loops with great heights (1–4 × 10^9 cm), and that some of them then precipitate to the chromosphere, while others escape into loops of greater heights and into IP space.

4.5 *Quiescent Filament-Eruption Flares*

Although eruptions of quiescent filaments do not lead to impulsive flare activity, they often lead to the development of pairs of faint Hα ribbons, together with IP shocks and energetic IP protons and heavy ions (22, 81). Dwivedi et al. (52) have found that two-ribbon flares without hard X-ray emission are due to eruptions of filaments at the outskirts of active regions. Because such events cause Hα brightening, a gradual rise and fall of microwaves, and soft X-ray emission, they should be considered to be flares. According to Dodson & Hedeman (49), during the 1956–67 period, 83 "spotless flares" with Hα class > 1 were observed in plages without large sunspots, and the majority of them were related to activation of quiescent filaments. However, only a small fraction of them produced energetic IP particles. Such flares resulting from quiescent filament eruptions constitute an additional class. It is thought that an eruption of a quiescent filament can produce IP shocks and energetic IP protons only when the eruption is fast enough.

Table 2 summarizes the characteristics of different classes of flares.

4.6 Roles of Erupting Filaments in Flares

From the early days, it was known that filament activity is closely related to the occurrence of flares (95, 203). Studying 297 flares with Hα class 1

Table 2 Characteristics of different classes of flares

	Thermal hard X-ray flares	Nonthermal hard X-ray flares	Impulsive GR/P flares	Gradual GR/P flares	Quiescent filament-eruption flares
Hard X-ray light curves and spectrum	Gradual and thermal for <40 keV; spiky and very steep for >40 keV	Spiky, steep; $\langle\delta\rangle = 4.5$	Spiky, flat; $\langle\delta\rangle = 3.5$	Gradual, flat; $\langle\delta\rangle = 3.5$	Not observed
Hard X-ray spectral evolution	Soft-hard-soft	Soft-hard-soft	Soft-hard-harder	Soft-hard-harder	Not known
Hard X-ray source	Small, low	Small, low	Small, low	Large, high ($>10^9$ cm)	Expected to be high
Nuclear gamma rays	No	No	Yes	Yes	No
Type II or IV	No	Rare	Often	Very often	Rare
IP shocks	No	No	Very rare	Often	Rare
IP particles	No	No	Often (low flux)	Very often (high flux)	Rare
Soft X-ray duration	Short	Short	Short	Long	Long
Microwave to hard X-ray flux ratio	Small	Normal	Normal	Large	Not known

or higher, Martin & Ramsey (114) showed that 53% of these flares were associated with some kind of filament activity, such as rapid darkening, expansion or apparent outward motion, breakup into more than one segment, transition to emission, ejection of at least one segment, complete disappearance or point of minimum visibility, and appearance of absorption in a new location. In particular, it has been noted that erupting filaments (or eruptive prominences seen at the limb) are associated with two-ribbon flares or extended flares (7, 132). There are many well-documented cases, particularly the 7 August 1972 (204) and 29 July 1973 flares (122, 123). However, some gradual flares do not show a filament eruption. For example, the filament of the gradual GR/P flare on 10 April 1981 did not erupt but remained intact, although a postflare loop system developed later (F. Tang, private communication, 1984). There is no extensive statistical study of what fraction of extended flares or gradual flares are associated with filament eruption.

In addition to gradual GR/P flares, some impulsive flares are also associated with erupting filaments. Kahler et al. (84) have studied filament eruptions in four impulsive flares. They find that the eruptive motion commences before the onset of the flare, and that its acceleration evolves smoothly through the first phase. However, it is often reported that the eruptive motion shows a rapid acceleration near the flare onset times (114). It is not clear why flares associated with erupting filaments are sometimes gradual and other times impulsive.

A large fraction of CMEs are associated with flares, filament eruptions, or both (125, 191). Among such CMEs with solar association, 66% of *SMM* CMEs and 78% of *Skylab* CMEs are found to be associated with either eruptions of active-region filaments (in such cases flares occur) or eruptions of quiescent filaments (191).

Summarizing the above results, we suggest that all gradual GR/P flares are associated with eruptions of magnetic fields. An Hα filament eruption occurs only when the material in the magnetic field configuration is cool [see (168, p. 216)]; therefore, during some gradual GR/P flares, if the field configurations that erupt contain hot material, no Hα filament eruption is observed. Tang (177) has shown that in some two-ribbon flares, only the top layer of the filament erupts while the bottom layer remains intact. What follows an eruption of a filament may be determined by the properties of the surrounding medium (10, 111). If the overlying magnetic field is strong enough, the filament is prevented from fully erupting, in which case an impulsive flare is produced. If the magnetic field strength of the overlying arcade is weak, the filament erupts, distending the overlying field lines and leading to reconnection and gradual GR/P flares (4, 99, 167). If the field of the overlying magnetic loop (or arcade) is very weak and its plasma density is low, as in the case of a quiescent filament, an eruption of the filament does not cause energetic flare activity; instead, it merely causes two faint Hα ribbons and weak X-ray and/or microwave emission. However, such filament eruptions can lead to CMEs, IP shocks, and/or energetic IP particles.

4.7 Relative Frequencies of Different Classes of Flares

What are the relative frequencies of the different classes of flares? No one has done a systematic study on this, but the question is important enough to convey whatever meager understanding of this is available in the literature.

From a correlation diagram between the spectral index and the peak HXRBS rate (9, 46), we find that during the period from February 1980 to January 1980 only 13 out of 126 HXRBS flares with peak rates greater than 1000 counts s^{-1} had a power law spectral index greater than 6.5. These flares have weak microwave emission compared with their hard X-ray fluxes, and none of them are associated with type II or type IV bursts. Granting that all these flares are thermal hard X-ray flares, we find that thermal hard X-ray flares are rare among intense hard X-ray flares. We do not have any information on less intense hard X-ray flares. In agreement with this, Kosugi et al. (100) identified only 3 flares as thermal hard X-ray flares among the 400 flares commonly detected by HXRBS and the 17-GHz radiometer at Nobeyama, Japan.

It is difficult to estimate what fraction of flares belong to the impulsive GR/P flares because the detection of nuclear gamma rays and energetic IP particles is limited by the detector threshold. Among 42 very intense hard X-ray flares with HXRBS peak rates $> 10^4$ counts s^{-1} observed during 1980 and 1981, 9 flares (21%) turned out to be impulsive GR/P flares (9). Among less intense hard X-ray flares, the fraction seems to be smaller, judging from hard X-ray spectral indices. Short-duration soft X-ray flares producing energetic protons or relativistic electrons detected in IP space (23) are impulsive GR/P flares. Among 31 such events [Table 1*B* of (23)], 13 flares were observed by HXRBS during their peak of hard X-ray emission. Only two of these 13 flares have peak HXRBS rates less than 5000 counts s^{-1}. Again, the detection of relativistic electrons is subjected to the threshold, but we do not expect a large fraction of flares with peak HXRBS rates < 5000 counts s^{-1} to be impulsive GR/P flares.

Among HXRBS flares of 1980 and 1981, only 18 are regarded as gradual GR/P flares (9). Fourteen of them have HXRBS peaks > 1000 counts s^{-1} (9). If we consider that 266 HXRBS flares of 1980 and 1981 have peak rates > 1000 counts s^{-1}, it is clear that gradual GR/P flares are rare. Because the selection of gradual GR/P flares is based on time scales of hard X-ray emission, it does not suffer from any threshold effect among flares with HXRBS peak rates > 300 counts s^{-1}. Figure 8 shows the relative frequencies and dynamic ranges of impulsive and gradual GR/P flares for 1980 and 1981. Impulsive GR/P flares shown in this figure are identified from nuclear gamma-ray emission detected by the GRS. Hence, the identification of impulsive GR/P flares is *limited by the GRS threshold effect*. However, as mentioned earlier, the fraction of impulsive GR/P flares is expected to decrease with decreasing HXRBS peak rates.

The frequency of quiescent filament-eruption flares is not well known;

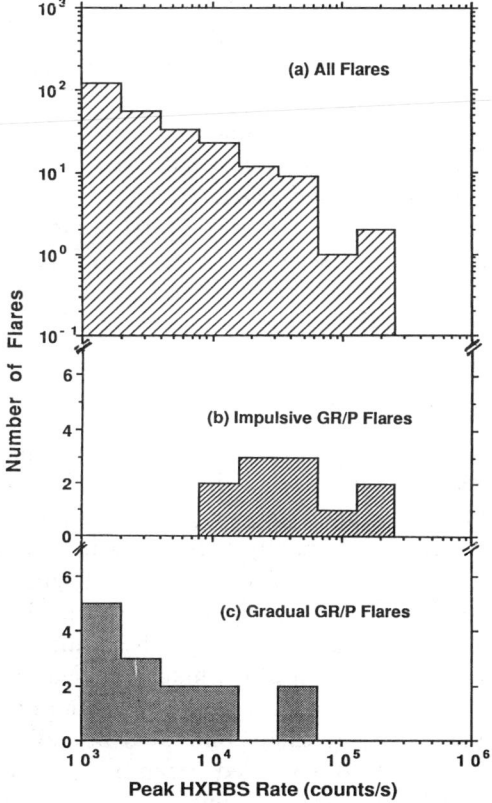

Figure 8 Size distributions of different classes of flares. The vertical axis represents the number of flares in each bin of equal logarithmic interval. Panel (*a*) shows the size distribution of all flares with HXRBS count rates > 1000 counts s^{-1} for 1980 and 1981 [data from (47)]. Panel (*b*) shows the frequency of the *impulsive GR/P flares*, and (*c*) of the *gradual GR/P flares*. The identification of impulsive GR/P flares is limited by the GRS threshold. The majority of flares in panel (*a*) are thought to be nonthermal hard X-ray flare. Data from Bai (9).

the six reported by Cane et al. (22) for the 1978–84 period are the ones associated with IP shocks. Dodson & Hedeman (49) reported 83 "spotless flares" with Hα class > 1 for the 1956–67 period. Since the majority of these spotless flares are associated with quiescent filament eruptions, we can form an idea about the frequency of this class of flares. Among intense hard X-ray flares with peak HXRBS rates > 1000 counts s^{-1}, the majority of these are *nonthermal hard X-ray flares*. It is possible that we will end up subdividing this class of flares as our knowledge of flares increases.

5. THEORETICAL INTERPRETATION

We now face the problem of trying to offer a theoretical interpretation of the fact that flares may be divided into distinct classes, adopting for our purposes the five classes proposed in the previous section. From a theoretical point of view, we need to understand what it is that flares from different classes have in common, and what it is that distinguishes flares in one class from flares in another class.

We address these questions in two ways. First, we try to understand the primary energy release processes that can occur in flares, as well as pay a little attention to some secondary energy conversion processes. Second, we try to decide which of these processes play a role in the various classes of flares.

5.1 *Energy Release Processes*

It is generally agreed that the energy released in a flare is initially embodied in the magnetic field of the active region in which the flare occurs. However, it is not possible to release the total energy of such a field configuration, since the photosphere is sufficiently highly conducting and sufficiently massive that, on the time scale of a solar flare, magnetic field lines are effectively "frozen" into the photosphere. This means that the only magnetic field rearrangements that can occur as the result of a flare are those that leave the normal component of the magnetic field, B_n, unchanged.

The minimum energy state of a magnetic field with prescribed values of the normal magnetic field at each point of a bounding surface is the current-free "potential" field, so called because it can be represented as the gradient of a scalar potential. A simple way to understand this is to consider the "thought experiment" of replacing the highly conducting corona by a medium with finite resistivity. Wherever there are currents, there would then be joule heating, so that the energy of the magnetic field would slowly decline. This decline will continue as long as there is a nonzero current density anywhere in the coronal region. Hence the field will end up in a state with zero current density everywhere in the corona (i.e. it will end up as a potential field), and that state will be the state of minimum energy.

These considerations lead to the useful concept of the "available energy" or "free energy" of the magnetic field configuration of an active region. This is the difference between the energy of the field and the energy of the potential field that has the same photospheric boundary condition, i.e. the same value of B_n, at each point of the photosphere. The basic problem of solar flares is therefore to determine the various types of nonpotential magnetic field configurations that can exist in active regions, and to understand the processes that can lead to the release of some or all of the free energy of these configurations.

It was first proposed by Giovanelli (63, 64) that a flare represents an electromagnetic process of energy conversion. Dungey (51) was the first person to propose that magnetic "neutral points," where the field strength is zero but the current density is nonzero, are representative of configurations that can be unstable and so lead to a catastrophic energy release. Sweet (171) first drew attention to the potential importance of current sheets, across which there is either a reversal of magnetic field direction or at least a sharp change in the direction. Parker (133) analyzed in detail the rate at which magnetic reconnection (the process that eliminates the current in a current sheet) might occur, and he concluded that it is too slow to explain solar flares. However, Petschek (136) subsequently presented a new analysis and argued that reconnection could occur at a rate approaching the rate at which an Alfvén wave could propagate through the region.

At about the same time, Furth et al. (62) published the first comprehensive analysis of resistive instabilities, including the "tearing mode" that can lead to reconnection of a current sheet. According to their linear theory, reconnection would occur in a time that is the geometric mean of the time it takes for magnetic field to diffuse (owing to finite resistivity) across the current layer and the time it takes for an Alfvén wave to propagate across the layer. Whether or not that time is sufficiently short to explain a solar flare depends on the thickness of the current layer, which is an unknown and unobservable quantity. However, what appear to be reasonable values of this quantity do not lead to sufficiently rapid energy release to explain solar flares.

Since that time, the tearing-mode process has been investigated by many individuals and groups, and certain modifications have been found to speed up the reconnection process. Spicer (158) pointed out that nonlinear mode-mode coupling could speed up the reconnection rate, and his argument has been supported by the study of numerical models by Carreras et al. (29). Steinolfson & van Hoven (160) have incorporated radiative effects in a tearing-mode model and find that this too speeds up the reconnection rate. Sakai & Tajima and their collaborators [see, for instance, (172)] have shown that after the tearing mode develops an array of current filaments in a current-sheet configuration, another process, which they call the "coalescence instability," will lead to the rapid merging of those current filaments to enhance and speed up the energy release process. It is interesting to note that this process is very similar to an early proposal by Gold & Hoyle (65), although the latter authors did not present a rigorous analysis of the coalescence process.

It is fair to say that the concept of magnetic reconnection has dominated theoretical investigation of solar flares over the past 25 years. Moreover, there has been the implicit assumption that the tearing-mode concept is equivalent to the reconnection concept. Alfvén & Carlqvist (3) registered an early dissenting opinion when they proposed that energy release might be due to a process that they called "current disruption." If a process were to occur in the corona that interrupted—or attempted to interrupt—the current of a current-carrying flux tube (for instance, by a two-stream type of instability), the stored energy of the current-carrying magnetic field (the energy to be ascribed to the inductance of the system) would lead to the development of a large induced electric field that would tend to maintain the current. The proposal of Alfvén & Carlqvist attracted some [see, for instance, (157)] but not great interest at that time, partly because it was presented in terms of circuit rather than plasma physics concepts, and partly because they did not give a detailed comparison of their theory with observational data.

The situation has changed as the result of *SMM* experiments, which, as we have seen, show that particle acceleration up to many MeV can occur on a time scale of only 1–2 s. Such rapid acceleration is difficult to understand on the basis of the tearing-mode instability, since it is not that rapid, and since most of the energy released during that instability goes into mass motion and into joule-type heating, not into particle acceleration (192). The tearing mode is likely to produce a high level of MHD turbulence in the surrounding region, and this turbulence can lead to stochastic acceleration. Some authors have argued that this acceleration could be sufficiently rapid to explain the *SMM* results [see, for instance, (142)].

Various mechanisms involving shocks have also been proposed, such as shock drift acceleration, shock-resonant acceleration, and others (30, 45, 53, 130). Under suitable conditions these mechanisms seem to be able to accelerate very rapidly. However, such acceleration should affect electrons and protons quite differently, giving preferential acceleration to protons and comparatively little acceleration to electrons. This expectation does not square with the observational fact (91, 148) that in GR/P flares, both electrons and protons are accelerated to high energies at the same time, and that electrons are rapidly accelerated to tens of MeV.

For these reasons, the concept of current interruption has recently been revived. Haerendel (66) has proposed that the ion-acoustic instability (109) may play a key role, although under some conditions the electrostatic ion-cyclotron instability [see, for instance, (41)] is more likely to occur. Whether or not the ion acoustic instability occurs depends on two factors: (a) the ratio of the electron drift velocity (due to the current) to the ion sound speed, and (b) the ratio of the electron temperature to the ion temperature. Both ratios must exceed unity by a small factor for the instability to occur. Sturrock (164) has recently proposed that these conditions are likely to be met if we make the following assumption and argument. The assumption is that there is no steady coronal heating, only a flarelike impulsive heating, so that most of the flux tubes in the corona are filled with cool gas at a temperature that could be as low as the chromospheric temperature or even the photospheric temperature. The argument is that if a flux tube suddenly expands (as the result of a filament eruption, for instance), the ions will tend to cool adiabatically, whereas the electrons will tend to remain at the same temperature as the boundary, since electron thermal conductivity is much higher than ion thermal conductivity, and since heat exchange between the two species is quite slow. Hence a sudden expansion would suddenly decrease the ion sound speed and increase the ratio T_e/T_p. This argument offers an explanation of why a filament eruption should lead to the sudden acceleration that is known to occur in (and be responsible for) impulsive and gradual GR/P flares.

As we pointed out in Section 4, in some flares there is such a close connection between the filament eruption and the flare itself that Kiepenheuer (95) and, more recently, Moore (121) have argued that the two should be regarded as part and parcel of the overall flare phenomenon. S. F. Martin (1988, private communication) has expressed the view that most—but not all—flares show evidence of some kind of rapid mass motion. It seems likely that in flares involving filament eruption or a similar process, the initiation and perhaps the driving process of the flare are processes that lead to the sudden motion of a massive structure. This might be just a sudden change in configuration or position of the structure, but it also might be the complete eruption and ejection of the structure such as we witness when a filament erupts completely and is associated with a coronal mass ejection. Hence the most basic flare problem is perhaps not that of determining how and why energy is released during the impulsive phase, but that of understanding why sudden mass motions occur that are quickly followed by processes that lead to the visible and otherwise detectable manifestations of a flare.

In order to understand what dynamic—including magnetohydrodynamic (MHD)—processes could affect filaments, we need to have an accepted model of the structure of a filament. Unfortunately, no detailed model has been presented and received wide acceptance. It is agreed that the cool hydrogen, which is visible (in Hα) as a dark feature on the disk or as a bright feature on the limb, must be supported in regions of a magnetic field where the field lines are essentially horizontal, and that an upwardly directed curvature of the field lines in those regions is favorable for long-term retention of the gas, as in the early Kippenhahn & Schlüter (96) model. It is also agreed that filaments occur above polarity reversal lines, and that high-resolution Hα photographs (60) and vector-magnetograms by Hagyard et al. (67) indicate that the magnetic field in or near a filament is almost parallel to the direction of the polarity reversal line. This has led to the concept that the magnetic field of a filament comprises a flux tube that runs above and parallel to the polarity reversal line (121). If this is an appropriate model, then the eruption of a filament may be interpreted as a purely MHD instability of such a magnetic flux tube that, in its initial state, is held in place by the stress of an array of overlying magnetic loops (139).

Such a scenario for the structure and instability of a model filament would explain the sudden disruption of a filament, but it offers no explanation for the associated heating [sometimes called "preflare heating" (183)] that is detectable by its UV or soft X-ray radiation or by Hα brightenings. An alternative is that depicted schematically in Figure 9a, in which a filament is taken to be a ropelike structure, made up of many intertwined magnetic "strands." Most of these strands are tied to the photosphere at points along the length of the filament, rather than merely at the ends of the filament. Such a picture seems closer to the Hα appearance of filaments (60). It also helps explain the observational result of Tang (177) that, on some occasions, a part of a filament may erupt, leaving the rest of the filament intact.

The balance of forces of such a model is somewhat similar to that of a large helium-filled balloon that is held close to the ground by many thin ropes. What may happen in such a situation is that two contiguous magnetic strands, which have opposite polarities where they meet the photosphere, may reconnect at a location such as that indicated in Figure 9a. This would produce impulsive heating low in the atmosphere, which could explain the Hα brightenings that occur shortly before a flare, in association with filament activation, and perhaps the small X-ray brightenings that are sometimes detected and referred to as "preheating" of the flare site (183). As the result of such a reconnection, two flux strands that tied the filament to the photosphere have been severed. This is like cutting two of

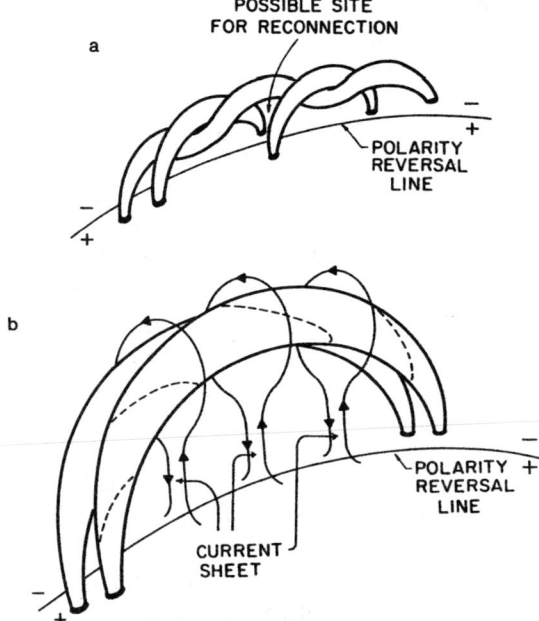

Figure 9 This is a schematic representation of the magnetic field configuration associated with a filament and its development during a filament eruption and flare. Panel (*a*) shows the preflare and preeruption filament configuration as an association of "magnetic strands" comprising a "magnetic rope" with multiple connections to the photosphere. Where strands with magnetic field of opposite polarity come into contact, reconnection can occur. This gives rise to energy release and also to a partial disconnection of the filament magnetic field from the photosphere. Panel (*b*) shows the magnetic field configuration that will develop if a "runaway" series of disconnections occur, leaving the flux system tied to the photosphere only at the two extremities. The new flux tube will tend to erupt (possibly expanding indefinitely into interplanetary space, if it is sufficiently stressed), distorting the overlying magnetic field lines as shown to form an extended vertical current sheet.

the thin ropes holding down the helium balloon. This change puts more stress on the remaining strands. If the configuration is near a critical state, there can be a runaway process in which all the remaining strands are progressively severed (except that the filament must remain magnetically tied to the photosphere at its ends), and the filament lifts off, as indicated in Figure 9*b*, just as the balloon would lift off if the ropes were to break one after another (166).

Whether a filament merely erupts to a stable configuration at a higher level (somewhere in the corona) or is completely ejected from the Sun depends on the strength of the overlying field and the degree to which the erupting magnetic field is stressed. A highly twisted flux rope can have more magnetic energy than the energy of the corresponding open magnetic field with the same boundary conditions at its ends. If this is the case, we must expect the filament to expand indefinitely. It would then appear to be ejected completely from the Sun; this ejection is driven by its own magnetic energy. Moore (121) has recently applied such considerations to an analysis of the flare of 7 August 1972.

If a filament were to erupt as outlined above, one must expect that much of the magnetic energy that is being released would be converted into the kinetic energy of the moving filament material. Analysis of the energy budgets of certain flares [see, for instance, (27, 190)] indicates that the kinetic energy of mass motion can indeed exceed the total energy emitted by the flare in the form of electromagnetic radiation of all types. However, the eruption of a filament has secondary effects that can lead to different types of energy conversion. One of these has already been mentioned: An erupting filament will lead to the sudden expansion of magnetic flux tubes in the neighborhood of the filament, and this can lead to a sudden increase in the electron-to-ion-temperature ratio, so precipitating the ion-acoustic or ion-cyclotron instability that leads to sudden particle acceleration.

Another important effect of filament eruption is that it may generate a shock wave that propagates through the corona. Such a shock wave can influence the chromosphere in such a way as to produce a Moreton wave (124). It can also accelerate electrons in such a way as to produce a type II radio burst and an associated type IV radio burst (101). The shock wave may have the nature either of a blast wave, caused by the sudden motion of the filament, or of a bow shock that runs ahead of the filament, if the filament moves sufficiently rapidly. In the latter case, the shock would persist as the plasmoid moves through interplanetary space, where it would be detectable by the generation of an interplanetary kilometric type II burst.

It is known that as the result of MHD turbulence and particle scattering, shock waves can accelerate particles (ions preferentially) to high energies [see, for instance, (19)]. Hence it is likely that strong interplanetary particle events are due to shock acceleration, as was indeed suggested by Wild et al. (194) many years ago. This is the conventional interpretation of "second-phase" acceleration.

We have referred earlier to another consequence of filament eruption that has important consequences for energy conversion. As has been pointed out by Kopp & Pneuman (99), Anzer & Pneuman (4), Sturrock et al. (167), Cliver et al. (36), and others, the eruption of a filament distorts and distends the overlying magnetic field configuration in such a way that we expect the overlying field to form a current sheet between the filament and the chromosphere (see Figure 9b). This magnetic field topology is of the form proposed by Sturrock (161) as an explanation of two-ribbon flares. The interpretation is that the current sheet slowly reconnects and that the released energy, when thermalized, provides the energy content of the X-ray-emitting flare plasma. The same heating process heats the upper layers of the chromosphere to high temperature (of order 10^7 K) so that it "evaporates" to fill the magnetic flux system with hot, dense, X-ray-emitting plasma. The various physical processes that take place at this stage are shown schematically in Figure 10, taken from Cliver et al. (36).

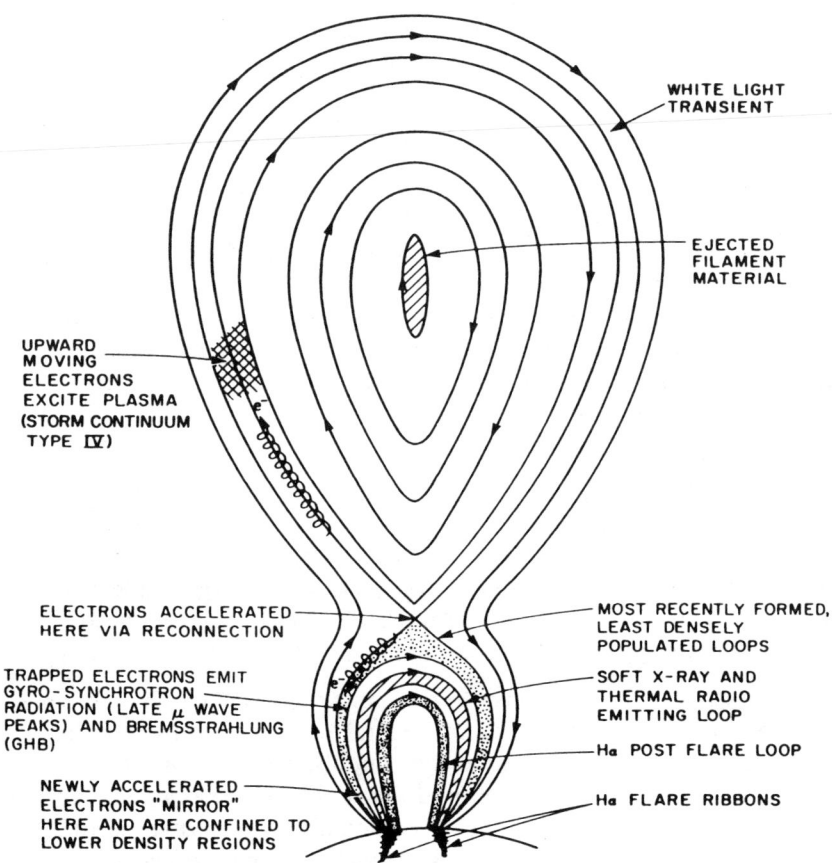

Figure 10 This shows the processes that may occur in the vertical current sheet shown in Figure 9b as the result of progressive reconnection. From Cliver et al. (35a, 36). Similar figures are found in Anzer & Pneuman (4) and Sturrock et al. (167). Although this figure depicts that CMEs are driven by erupting filaments, there are contradictory observations (68, 69, 80).

5.2 Phases of Flares

We can now attempt to use the above ideas concerning energy release for the elucidation of the similarities and differences of the classes of flares proposed in Section 4. We carry out this analysis by supposing that any flare involves one or more phases of energy release or conversion. We then attempt to understand each class of flares as involving certain phases, and we also try to understand each phase in terms of the energy release mechanisms listed above. In the literature, the terms "first phase" and "second phase" have been widely used, and we have followed the convention in Sections 2 through 4. However, here we devise new terms because these two phases are inadequate to describe all the energy release phases of solar flares. We first discuss flare phases.

EARLY PHASE   We use this term to denote the phase of energy release that precedes the main phase. This phase of energy release may be subdivided into two subclasses, as follows.

EARLY THERMAL PHASE   Since we choose to include filament behavior (where appropriate) in the overall flare event, this category of early phase would include filament activation and eruption, as well as processes that have been termed "preheating." We have proposed an interpretation of filament activation and eruption that incorporates the operation of reconnection in the lower atmosphere, and we suggest that heating due to this reconnection is the cause of preheating.

EARLY NONTHERMAL PHASE   It is well known that there are manifestations of nonthermal energy release shortly before the principal initiation of a flare, such as Type III radio emission (18a). The usual interpretation of a Type III radio burst is that it is caused by plasma oscillations in the coronal plasma that are excited by an electron beam with energy in the range 10–100 keV. This beam may be produced as a result of reconnection in a current sheet high in the solar corona.

MAIN PHASE   We use this term to represent the principal phase of energy release in a flare. On the basis of Hα and hard X-ray observations, this phase is sometimes referred to as the "impulsive phase" or "flash phase." On the basis of observations of hard X-ray emission that are indicative of electron accelerations, this phase is sometimes referred to as the "first phase." We feel that the term " first phase" is inappropriate, since (as noted above) there is often an "early phase" that precedes the phase now under discussion. This class of energy release may be subdivided into two subclasses: nonthermal and thermal.

MAIN NONTHERMAL PHASE   This is used to describe the main phase of energy release when this phase involves particle acceleration, as would for instance be evident from the detection of hard X-ray emission. In most flares, we may select one of two subcategories: *main nonthermal impulsive*, or *main nonthermal gradual*, depending upon the profile of the time curve for hard X-ray emission, as discussed in Section 4. If the flare seems to involve a phase with the combined features of both the impulsive and gradual hard X-ray phases, then we simply use the term *main nonthermal phase*.

Our proposed theoretical interpretation is that the main nonthermal phase hinges on the operation of current interruption. The sudden onset of current interruption may be due to a sudden change in the plasma parameters (especially the ratio T_e/T_p) due to a sudden expansion, which may in turn be caused by a filament eruption. We propose that the main nonthermal impulsive phase involves energy release in a compact system of low-lying magnetic loops that are adjacent to, or perhaps part of, the magnetic field structure associated with the filament. A main nonthermal gradual phase is likely to originate in an extended flux system that suddenly expands, such as the magnetic field lines that overlie the filament and are stretched outward as the filament erupts.

MAIN THERMAL PHASE   This term is used to refer to the main phase of energy release operative in those flares for which the energy is converted primarily into a hot plasma that emits soft X rays and some hard X rays. In these flares, the time scale for energy release is longer than that of a typical impulsive phase. The main thermal phase may be caused by reconnection in a current sheet, such as the sheet that we expect to form in an emerging-flux configuration [see, for instance, (140)].

The third main subdivision of flare energy release phases is the *late phase*, which may in turn be subdivided into two categories:

LATE THERMAL PHASE   This term is used to describe, for instance, the late and long-lived phase of two-ribbon flares while the ribbons are slowly drifting apart. As Moore et al. (123) have shown, the X-ray emission during this phase sometimes seems to require continued energy input, rather than being due simply to the decay of the hot, dense flare plasma produced during the main phase. We ascribe this phase to reconnection of the extensive current sheet that forms below an erupting filament. The reason that this phase produces lower energy radiation than the main thermal phase is that the magnetic field is weaker, since it is generated by an extended flux system high in the corona rather than a compact flux system low in the corona.

LATE NONTHERMAL PHASE   This term is used to describe the operation of nonthermal processes late in the development of a flare, such as type II radio emission and "second-phase" particle acceleration. As described earlier, we ascribe this behavior to particle acceleration caused by a shock wave—either a blast wave or a bow shock—produced as the result of sudden mass motion (for instance, filament eruption). Martens (113) has proposed that a DC electric field developed in the reconnection phase of a two-ribbon flare can accelerate protons that produce nuclear gamma rays. However, nuclear gamma rays are emitted during the main phase long before the reconnection phase. Furthermore, impulsive GR/P flares that do not develop the reconnection phase emit nuclear gamma rays. Therefore, we may regard such acceleration as late-phase acceleration rather than main-phase acceleration. Sometimes gradual hard X-ray emission is observed about 30 min after the main phase (36, 89), and this and stationary type IV radio bursts may be due to electrons accelerated during the reconnection.

5.3 Classes of Flares

We now turn to the five classes proposed in Section 4 and offer an interpretation of these classes in terms of the above proposed phases.

THERMAL HARD X-RAY FLARES We attribute this class to the operation of the main thermal phase of energy release. We suppose that there is no filament eruption, and that this explains why the other phases do not occur.

NONTHERMAL HARD X-RAY FLARES In these flares, the dominant phase is the main nonthermal impulsive phase. There is little evidence of any other phase taking place, except in those events that produce type II and type IV bursts, which we attribute to the late nonthermal phase. It may be that all of these flares involve some kind of mass motion, but in most cases the only consequence is the main nonthermal impulsive phase. On the other hand, it may be that most of these flares involve a simple, more or less stable, magnetic geometry, and that the current interruption process occurs more or less spontaneously. For instance, it may be that the cooling phase of a coronal loop can lead to a situation in which the protons have cooled more rapidly than the electrons, so setting the stage for the ion-acoustic instability or the ion-cyclotron instability.

IMPULSIVE GR/P FLARES These flares primarily involve the main nonthermal impulsive phase—that is, current interruption in a compact flux system. It is likely that these flares exhibit sudden mass motion and filament activation. Hence it is possible that the current interruption process may be attributed to a sudden rearrangement of magnetic field caused by this mass motion. The second-step acceleration may be due either to the current interruption process itself or to stochastic acceleration caused by the turbulence resulting from the filament activation. We proposed that for this class of flares the magnetic field configuration is such that the overlying magnetic loops are not significantly distended, so that there is no main nonthermal gradual phase and, of course, no late thermal phase, since no extended current sheet is formed.

GRADUAL GR/P FLARES These flares involve several phases. They typically display an early thermal phase when a filament begins to erupt and when there is preheating, as seen by X-ray and Hα emission. Such a flare will involve a main nonthermal gradual phase of energy release, perhaps in combination with a main nonthermal impulsive phase. As indicated earlier, we attribute the main nonthermal phase to current interruption: A main, nonthermal impulsive phase is produced in a compact flux system that suddenly expands, whereas a main nonthermal gradual phase is created in an extended flux system that suddenly expands, such as the magnetic field lines that overlie the filament and are stretched outward as the filament erupts. As these field lines begin to close under the erupting filament to form a current sheet, reconnection will occur that represents the late phase (thermal and nonthermal) of the flare. If the filament eruption is sufficiently rapid, it will generate a shock wave that is responsible for a late nonthermal phase.

QUIESCENT FILAMENT-ERUPTION FLARES These flares appear to involve some of the phases of a gradual GR/P flare. There is an early thermal phase, when a filament is erupting, that may involve some weak Hα emission. It appears that a flare of this class does not exhibit a main phase (either nonthermal or thermal), but it does exhibit both the late thermal phase and the late nonthermal phase that we attribute to the slow reconnection of a newly formed current sheet and to shock waves generated by the eruption.

6. CLOSING REMARKS

The recent great advances in our observational knowledge of solar flares, due in large measure to the international program of space exploration, has had a dramatic effect on our understanding of solar flares. On the one hand, this great increase in our knowledge has made life more difficult for solar physicists, since a complex problem has become even more complicated, and since some new observations are clearly incompatible with some fondly held ideas. On occasion, we may even pine for the days in the early 1950s when the theorists were challenged to explain only two facts in constructing a flare theory—the total energy of a flare, and the time scale of the impulsive phase!

On the other hand, solving the flare problem requires that we identify and understand the many physical processes that occur in flares. Since flares are intrinsically complicated, this clearly calls for more and better data, such as those that we have obtained in recent years. We hope and expect that even more sophisticated equipment on rockets, balloons, and future space missions will yield even more detailed information about the many physical processes that occur in flares.

However, it would be wrong to give the impression that observational advances all stem from space experiments. Advanced new ground-based equipment is being planned and will no doubt yield exciting new information when it is in operation. This includes both a new optical observatory and a new radio telescope in Japan, and a new vector-magnetograph in the United States.

In the last two decades, our knowledge of solar physics has benefited greatly from an explosive increase in our understanding of the plasma state, due primarily to the rapid development of theoretical plasma physics fostered by the fusion reactor program. More recently, our knowledge of the plasma state has been greatly expanded by the advent of numerical simulation as a branch of science that supports both experimental and analytical studies.

Finally, it is becoming recognized, more clearly and more forcefully, that plasma processes that occur in the Sun are likely to occur in other astronomical bodies as well. Now much of solar physics can be regarded as part of the larger study of plasma astrophysics. It has long been agreed that flares occur on stars other than the Sun, and that many of these flares are far more energetic than solar flares. It is also possible that gamma-ray bursts from sources in our Galaxy may be due to flares in neutron-star magnetospheres. In addition, astrophysicists are now exploring the possibility that activity in galaxies and quasars may be caused by flarelike processes.

In view of all of these interconnections, it is to be hoped that an increased understanding of solar flares will contribute not only to solar physics, but also in some measure to the wider field of astrophysics.

ACKNOWLEDGMENTS

Our research has been supported by NASA grant NGL 05-020-272 and ONR contract N00014-85-K-0111. We thank Ed Cliver, Hugh Hudson, Steve Kahler, Jim Klimchuk, and Ron Moore for reading an early version of this paper and giving extensive comments. We also thank Larry Orwig for providing some of the figures.

Literature Cited

1. Acton, L. W. 1982. *Observatory* 102: 123–24
2. Acton, L. W., Canfield, R. C., Gunkler, T. A., Hudson, H. S., Kiplinger, A. L., Leibacher, J. W. 1982. *Ap. J.* 263: 409–22
3. Alfvén, H., Carlqvist, P. 1967. *Sol. Phys.* 1: 220–28
4. Anzer, U., Pneuman, G. W. 1982. *Sol. Phys.* 79: 129–47
5. Antonucci, E., Gabriel, A. H., Dennis, B. R. 1984. *Ap. J.* 287: 917–25
6. Antonucci, E., Gabriel, A. H., Acton, L. W., Culhane, J. L., Doyle, J. G., et al. 1982. *Sol. Phys.* 78: 107–23
7. Avignon, Y., Martres, M. J., Pick, M. 1964. *Ann. Astrophys.* 27: 23–28
8. Bai, T. 1982. See Ref. 107, pp. 409–17
9. Bai, T. 1986. *Ap. J.* 308: 912–28
10. Bai, T. 1986. *Adv. Space Res.* 6: 203–6
11. Bai, T. 1988. *Ap. J.* 334: 1049–53
12. Bai, T., Dennis, B. R. 1985. *Ap. J.* 292: 699–715
13. Bai, T., Dennis, B. R., Kiplinger, A. L., Orwig, L. E., Frost, K. 1983. *Sol. Phys.* 86: 409–19
14. Bai, T., Hudson, H. S., Pelling, R. M., Lin, R. P., Schwartz, R. A., von Rosenvinge, T. T. 1983 *Ap. J.* 267: 433–41
15. Bai, T., Kiplinger, A. L., Dennis, B. R. 1984. *Bull. Am. Astron. Soc.* 16: 535
16. Bai, T., Ramaty, R. 1976. *Sol. Phys.* 49: 343–58
17. Bai, T., Ramaty, R. 1979. *Ap. J.* 227: 1072–81
18. Beeck, J., Mason, G. M., Hamilton, D. C., Wibberenz, G., Kunow, H., et al. 1987. *Ap. J.* 322: 1052–72.
18a. Benz, A. O., Barrow, C. H., Dennis, B. R., Pick, M., Raoult, A., et al. 1983. *Sol. Phys.* 83: 267–83
19. Blandford, R. D. 1979. In *Particle Acceleration Mechanisms in Astrophysics, AIP Conf. Proc. No. 56*, ed. J. Arons, C. McKee, C. Max, pp. 333–55. New York: AIP
20. Brown, J. C. 1971. *Sol. Phys.* 18: 489–502
21. Deleted in proof
22. Cane, H.V., Kahler, S. W., Sheeley, N. R. Jr. 1986. *J. Geophys. Res.* 91: 13,321–29
23. Cane, H. V., McGuire, R. E., von Rosenvinge, T. T. 1986. *Ap. J.* 301: 448–59
24. Cane, H. V., Reames, D. V. 1988. *Ap. J.* 325: 895–900
25. Cane, H. V., Reames, D. V. 1988. *Ap. J.* 325: 901–4
26. Cane, H. V., Stone, R. G. 1984. *Ap. J.* 282: 339–44
27. Canfield, R. C., Cheng, C. C., Dere, K. P., Dulk, G. A., McLean, D. J., et al. 1980. See Ref. 163, pp. 451–69
28. Cargill, P. J., Priest, E. R. 1983. *Ap. J.* 266: 383–89
29. Carreras, B. A., Hicks, H. R., Holmes, J. A., Waddell, B. V. 1980. *Phys. Fluids* 23: 1811–26
30. Chiueh, T. 1988. *Ap. J.* 333: 366–85
31. Chupp, E. L., Debrunner, H., Fluckiger, E., Forrest, D. J., Golliez, F., et al. 1987. *Ap. J.* 318: 913–25
32. Chupp, E. L. 1982. See Ref. 107, pp. 363–81
33. Chupp, E. L. 1983. *Sol. Phys.* 86: 383–93
34. Chupp, E. L. 1984. *Annu. Rev. Astron. Astrophys.* 22: 359–87
35. Chupp, E. L., Forrest, D. J., Higbie, P. R., Suri, A. N., Tsai, C., et al. 1973. *Nature* 241: 333–35
35a. Cliver, E. W. 1983. *Sol. Phys.* 84: 347–59
36. Cliver, E. W., Dennis, B. R., Kiplinger, A. L., Kane, S. R., Neidig, D. F., et al. 1986. *Ap. J.* 305: 920–35
37. Cliver, E. W., Forrest, D. J., McGuire, R. E., von Rosenvinge, T. T. 1983. *Proc. Int. Cosmic Ray Conf., 18th, Bangalore*, 10: 342–45
38. Cliver, E. W., Kahler, S. W., McIntosh, P. S. 1983. *Ap. J.* 264: 699–707
39. Cliver, E. W., Forrest, D. J., Cane, H. V., McGuire, R. E., Reames, D. V., et al. 1989. *Ap. J.* In press
40. Cliver, E. W., Forrest, D. J., McGuire, R. E., von Rosenvinge, T. T., Reames, D. V., et al. 1987. *Proc. Int. Cosmic Ray Conf., 20th, Moscow*, 3: 61–64
41. Dakin, D. R., Tajima, T., Blenford, G., Rynn, N. 1976. *J. Plasma Phys.* 15: 175–95
42. de Jager, C. 1969. In *Solar Flares and Space Research, Proc. Symp. Plenary Meet. COSPAR, 11th*, pp. 1–15. Amsterdam: North-Holland. 419 pp

43. de Jager, C. 1987. *Proc. Int. Cosmic Ray Conf.*, *20th, Moscow*, 7: 66–76
44. de Jager, C., Svestka, Z. 1985. *Sol. Phys.* 100: 435–63
45. Decker, R. B., Vlahos, L. 1986. *Ap. J.* 306: 710–29
46. Dennis, B. R. 1985. *Sol. Phys.* 100: 465–90
47. Dennis, B. R., Orwig, L. E., Kiplinger, A. L., Gibson, B. R., Kennard, G. S., et al. 1985. *NASA TM 86236*
48. Dermer, C. D., Ramaty, R. 1986. *Ap. J.* 301: 962–74
49. Dodson, H. W., Hedeman, E. R. 1970. *Sol. Phys.* 13: 401–19
50. Dulk, G. A. 1985. *Annu. Rev. Astron. Astrophys.* 23: 169–224
51. Dungey, J. W. 1953. *Philos. Mag., Ser. 7* 44: 725–38
52. Dwivedi, B. N., Hudson, H. S., Kane, S. R., Svestka, Z. 1984. *Sol. Phys.* 90: 331–41
53. Ellison, E. C., Ramaty, R. 1985. *Ap. J.* 298: 400–8
54. Ellison, M. A., McKenna, S. M. P., Reid, J. H. 1961. *Dunsink Obs. Publ.* 1: 53
55. Emslie, A. G., Nagai, F. 1985. *Ap. J.* 288: 779–88
56. Evenson, P., Meyer, P., Yanagita, S., Forrest, D. J. 1984. *Ap. J.* 283: 439–49
56a. Forbush, S. E. 1946. *Phys. Rev.* 70: 771–72
57. Forrest, D. J. 1983. In *Positron-Electron Pairs in Astrophysics, AIP Conf. Proc. No. 101*, ed. M. L. Burns, A. K. Harding, R. Ramaty, pp. 3–14. New York: AIP
58. Forrest, D. J., Chupp, E. L. 1983. *Nature* 305: 291–92
59. Forrest, D. J., Chupp, E. L., Reppin, C., Rieger, E., Ryan, J. M., et al. 1981. *Proc. Int. Cosmic Ray Conf.*, *17th*, *Paris*, 10: 5–8
60. Foukal, P. V. 1971. *Sol. Phys.* 19: 59–71
61. Frost, K. J., Dennis, B. R. 1971. *Ap. J.* 165: 655–59
62. Furth, H. P., Killeen, J., Rosenbluth, M. N. 1963. *Phys. Fluids* 6: 459–84
63. Giovanelli, R. G. 1946. *Nature* 158: 81–82
64. Giovanelli, R. G. 1947. *MNRAS* 107: 338–55
65. Gold, T., Hoyle, F. 1960. *MNRAS* 120: 89–105
66. Haerendel, G. 1987. *Proc. ESLAB Symp., 21st*, pp. 205–14. *ESA SP-275*
67. Hagyard, M. J., Moore, R. L., Emslie, A. G. 1984. *Adv. Space. Res.* 4: 71–80
68. Harrison, R. A., Waggett, P. W., Bentley, R. D., Philips, K. J. H., Bruner, M., et al. 1985. *Sol. Phys.* 97: 387–400
69. Hildner, E., Bassi, J., Bougert, J. L., Duncan, R. A., Gary, D. E., et al. 1986. See Ref. 103, pp. 6-1–71
70. Hodge, P. W. 1966. *Physics and Astronomy of Galaxies and Cosmology*. New York: McGraw-Hill
71. Hua, X.-M., Lingenfelter, R. E. 1987. *Sol. Phys.* 107: 351–83
72. Hudson, H. S. 1978. *Ap. J.* 224: 235–40
73. Hudson, H. S. 1985. *Sol. Phys.* 100: 515–35
74. Hudson, H. S., Bai, T., Gruber, D. E., Matteson, J. L., Nolan, P. L., et al. 1980. *Ap. J. Lett.* 236: L91–95
75. Hudson, H. S., Lin, R. P., Stewart, R. T. 1982. *Sol. Phys.* 75: 245–61
76. Kahler, S. W. 1977. *Ap. J.* 214: 891–97
77. Kahler, S. W. 1982. *J. Geophys. Res.* 87: 3439–48
78. Kahler, S. W. 1982. *Ap. J.* 261: 710–19
79. Kahler, S. W. 1984. *Sol. Phys.* 90: 133–38
80. Kahler, S. W. 1987. *Rev. Geophys.* 25: 663–75
81. Kahler, S. W., Cliver, E. W., Cane, H. V., McGuire, R. E., Stone, R. G., et al. 1986. *Ap. J.* 302: 504–10
82. Kahler, S. W., Cliver, E. W., Sheeley, N. R. Jr., Howard, R. A., Koomen, M. J., et al. 1985. *J. Geophys. Res.* 90: 177–82
83. Kahler, S. W., Hildner, E., van Hollebeke, M. A. I. 1978. *Sol. Phys.* 57: 429–43
84. Kahler, S. W., Moore, S. R., Kane, S. R., Zirin, H. 1988. *Ap. J.* 328: 824–29
85. Kahler, S. W., Sheeley, N. R. Jr., Howard, R. A., Koomen, M. J., Michels, D. J. 1984. *Sol. Phys.* 93: 133–41.
86. Kahler, S. W., Sheeley, N. R. Jr., Howard, R. A., Koomen, M. J., Michels, D. J., et al. 1984. *J. Geophys. Res.* 89: 9683–93
87. Kai, K. 1979. *Sol. Phys.* 61: 187–99
88. Kai, K., Kosugi, T., Nitta, N. 1984. *Publ. Astron. Soc. Jpn.* 37: 155–62
89. Kai, K., Nakajima, H., Kosugi, T., Stewart, R. T., Nelson, G. J., et al. 1986. *Sol. Phys.* 105: 383–98
90. Kane, S. R. 1969. *Ap. J. Lett.* 157: L139–42
91. Kane, S. R., Chupp, E. L., Forrest, D. J., Share, G. H., Rieger, E. 1986. *Ap. J. Lett.* 300: L95–98
92. Kane, S. R., Donnelly, R. F. 1971. *Ap. J.* 164: 151–63
93. Kane, S. R., Evenson, P., Meyer, P. 1985. *Ap. J. Lett.* 299: L107–10

94. Kane, S. R., Fenimore, E. E., Klebesadel, R. W., Laros, J. G. 1982. *Ap. J. Lett.* 254: L53–57
95. Kiepenheuer, K.O. 1964. In *The Physics of Solar Flares*, ed. W. N. Hess, pp. 323–31. *NASA SP-50*
96. Kippenhahn, R., Schlüter, A. 1957. *Z. Astrophys.* 43: 36–62
97. Klein, L., Anderson, K., Pick, M., Trottet, G., Vilmer, N. 1983. *Sol. Phys.* 84: 295–310.
98. Kocharov, G. E. 1983. *Proc. Eur. Cosmic Ray Conf., 8th, Bologna, Invited Talks*, pp. 51–67
99. Kopp, R. A., Pneuman, G. W. 1976. *Sol. Phys.* 50: 85–98
100. Kosugi, T., Dennis, B. R., Kai, K. 1988. *Ap. J.* 324: 1118–31
101. Kundu, M. R. 1965. *Solar Radio Astronomy*. New York: Interscience. 660 pp.
102. Kundu, M. R., Vlahos, L. 1982. *Space Sci. Rev.* 32: 405–62
103. Kundu, M. R., Woodgate, B., eds. 1986. *Energetic Phenomena on the Sun, The Solar Maximum Mission Workshop Proc.* Washington, DC: NASA
103a. Lin, R. P. 1970. *Sol. Phys.* 12: 266–303
104. Lin, R. P., Hudson, H. S. 1971. *Sol. Phys.* 17: 412–35
105. Lin, R. P., Hudson, H. S. 1976. *Sol. Phys.* 50: 153–78
106. Lin, R. P., Schwartz, R. A., Pelling, R. M., Hurley, K. M. 1981. *Ap. J. Lett.* 251: L109–14
107. Lingenfelter, R. E., Hudson, H. S., Worrall, D. M., eds. 1982. *Gamma Ray Transients and Related Astrophysical Phenomena, AIP Conf. Proc. No. 77* New York: AIP. 500 pp.
108. Lingenfelter, R. E., Ramaty, R. 1967. In *High Energy Nuclear Reactions in Astrophysics*, ed. B. S. P. Shen, pp. 99–158. New York: Benjamin
109. Lotko, W. J. 1986. *J. Geophys. Res.* 91: 191–203
110. MacCombie, W. J., Rust, D. M. 1978. *Sol. Phys.* 61: 69–88
111. Machado, M. E., Moore, R. L., Hernandez, A. M., Rovira, M. G., Hagyard, M. J., et al. 1988. *Ap. J.* 326: 425–50
112. MacKinnon, A. L., Brown, J. C., Trottet, G., Vilmer, N. 1983. *Astron. Astrophys.* 119: 297–300
113. Martens, P. C. H. 1988. *Ap. J. Lett.* 330: L131–33
114. Martin, S. F., Ramsey, H. E. 1972. In *Solar Activity Observations and Predictions*, ed. P. S. McIntosh, M. Dryer, pp. 371–87. Cambridge, Mass: MIT Press
115. Mason, G. M., Gloeckler, G., Hovestadt, D. 1984. *Ap. J.* 280: 902–16
116. Maxwell, A., Dryer, M. 1982. *Space Sci. Rev.* 32: 11–25
117. Deleted in proof
118. McCracken, K. G., Rao, U. R. 1970. *Space Sci. Rev.* 11: 155–233
119. McLean, D. J., Labrum, N. R., eds. 1985. *Solar Radiophysics* Cambridge: Univ. Press
119a. Melrose, D. B., Brown, J. C. 1976. *MNRAS* 176: 15–30
120. Melrose, D. B., Dulk, G. A. 1987. *Phys. Scri.* T18: 29–38
121. Moore, R. L. 1988. *Ap. J.* 324: 1132–37
122. Moore, R. L., La Bonte, B. J. 1980. In *Solar and Interplanetary Dynamics, IAU Symp. No. 91*, ed. M. Dryer, E. Tandberg-Hanssen, pp. 207–11. Dordrecht: Reidel. 558 pp.
123. Moore, R. L., McKenzie, D. L., Svestka, Z., Widing, K. G., Antiochos, S. K., et al. 1980. See Ref. 163, pp. 341–409
124. Moreton, G. E., Ramsey, H. E. 1960. *Publ. Astron. Soc. Pac.* 72: 357–58
125. Munro, R. H., Gosling, J. T., Hildner, E., MacQueen, R. M., Poland, A. I., et al. 1979. *Sol. Phys.* 61: 201–15
126. Nakajima, H., Kosugi, T., Kai, K., Enome, S. 1983. *Nature* 305: 292–94
127. Neupert, W. M. 1968. *Ap. J. Lett.* 153: L59–64
128. Nonnast, J. H., Armstrong, T. P., Kohl, J. W. 1982. *J. Geophys. Res.* 87: 4327–37
129. Ohki, K., Takakura, T., Tsuneta, S., Nitta, N. 1983. *Sol. Phys.* 86: 301–11
130. Ohsawa, Y., Sakai, J.-I. 1988. *Ap. J.* 332: 439–46
131. Orwig, L. E., Frost, K. J., Dennis, B. R. 1980. *Sol. Phys.* 65: 25–37
132. Pallavicini, R., Serio, S., Vaiana, G. S. 1977. *Ap. J.* 216: 108–22
133. Parker, E. N. 1963. *Ap. J. Suppl.* 8: 177–211
134. Deleted in proof
135. Pesses, M. E., Klecker, B., Gloeckler, G., Hovestadt, D. 1981. *Proc. Int. Cosmic Ray Conf., 17th, Paris*, 3: 36–39
136. Petschek, H. E. 1964. In *The Physics of Solar Flares*, ed. W. N. Hess, pp. 425–39. *NASA SP-50*
137. Petrosian, V. 1985. *Ap. J.* 299: 987–93
138. Pick, M. 1986. *Sol. Phys.* 104: 19–32
139. Priest, E. R. 1988. *Ap. J.* 328: 848–55
140. Priest, E. R., Heyvaerts, J. 1974. *Sol. Phys.* 36: 433–42
141. Prince, T. A., Ling, J. C., Mahoney, W. A., Riegler, G. R., Jacobson, A. S 1982. *Ap. J. Lett.* 255: L81–84
142. Ramaty, R. 1979. In *Particle Accel-*

eration Mechanisms in Astrophysics, AIP Conf. Proc. No. 56, ed. J. Arons, C. McKee, C. Max, pp. 135–54. New York: AIP
143. Ramaty, R., Kozlovsky, B., Lingenfelter, R. E. 1979. *Ap. J. Suppl.* 40: 487–526
144. Ramaty, R., Murphy, R. J. 1987. *Space Sci. Rev.* 45: 213–68
145. Ramaty, R., Murphy, R. J., Dermer, C. D. 1987. *Ap. J. Lett.* 316: L41–44
146. Raoult, A., Pick, M., Dennis, B. R., Kane, S. R. 1985. *Ap. J.* 299: 1027–35
147. Deleted in proof
148. Rieger, E., Reppin, C., Kanbach, G., Forrest, D. J., Chupp, E. L., et al. 1983. *Proc. Int. Cosmic Ray Conf., 18th, Bangalore*, 4: 79–82
149. Roy, R.-J., Datlowe, D. W. 1975. *Sol. Phys.* 40: 165–82
150. Ryan, J. M. 1986. *Sol. Phys.* 105: 365–82
151. Sarris, E. T., Shawhan, S. D. 1973. *Sol. Phys.* 28: 519–32
152. Sawyer, C. 1984. *Proc. Meudon Sol.-Terrest. Prediction Workshop*, ed. G. Heckman, M. Shea, P. Simon, pp. 1–3. Boulder: Colo. Assoc. Univ. Press
153. Sheeley, N. R., Jr., Bohlin, J. D., Brueckner, G. E., Purcell, J. D., Scherrer, V. E., et al. 1975. *Sol. Phys.* 45: 377–92
154. Sheeley, N. R. Jr., Stewart, R. T., Robinson, R. D., Howard, R. A., Koomen, M. J., et al. 1984. *Ap. J.* 279: 839–47
155. Sheeley, N. R. Jr., Howard, R. A., Koomen, H. M., Michels, D. J. 1983. *Ap. J.* 272: 349–54
156. Sheeley, N. R. Jr., Howard, R.A., Koomen, H. M., Michels, D. J. 1985. *J. Geophys. Res.* 90: 163–75
157. Smith, D. F., Priest, E. R. 1972. *Ap. J.* 176: 487–95
158. Spicer, D. S. 1977. *Sol. Phys.* 53: 249–54
159. Starr, R., Heindl, W. A., Crannell, C. J., Thomas, R. J., Batchelor, D. A., Magun, A. 1988. *Ap. J.* 329: 967–81
160. Steinolfson, R. S., van Hoven, G. 1984. *Ap. J.* 276: 391–98
161. Sturrock, P. A. 1968. In *Structure and Development of Solar Active Regions, IAU Symp. No. 35*, ed. K. O. Kiepenheuer, pp. 471–80. Dordrecht: Reidel
162. Sturrock, P. A. 1980. See Ref. 163, pp. 411–49
163. Sturrock, P. A., ed. 1980. *Solar Flares*. Boulder: Colo. Assoc. Univ. Press. 513 pp.
164. Sturrock, P. A. 1989. *Sol. Phys.* In press
165. Deleted in proof
166. Sturrock, P. A. 1989. *Outstanding Problems in Solar System Plasma Physics*. Am. Geophys. Union. In Press
167. Sturrock, P. A., Kaufman, P., Moore, R. L., Smith, D. F. 1984. *Sol. Phys.* 94: 341–57
168. Svestka, Z. 1976. *Solar Flares*. Dordrecht: Reidel. 399 pp.
169. Svestka, Z. 1986. In *The Lower Atmosphere of Solar Flares*, ed. D. F. Neidig, pp. 332–55. Sunspot, N. Mex: Natl. Sol. Obs.
170. Svestka, Z., Fritzova, L. 1974. *Sol. Phys.* 36: 417–31
171. Sweet, P. A. 1958. In *Electromagnetic Phenomena in Cosmical Physics. IAU Symp. No. 6*, ed. B. Lehnert, pp. 123–34. Cambridge: Univ. Press
172. Tajima, T., Sakai, J.-I., Nakajima, H., Kosugi, T., Brunel, F., et al. 1987. *Ap. J.* 321: 1031–48
173. Takakura, T., Tsuneta, S., Ohki, K., Nitta, N., Makishima, K., et al. 1983. *Ap. J. Lett.* 270: L83–87
174. Tanaka, K. 1983. *Sol. Phys.* 86: 3–6
175. Tanaka, K. 1983. In *Activity in Red-Dwarf Stars, IAU Colloq. No. 71*, ed. P. B. Bryne, M. Rodono, pp. 307–20. Dordrecht: Reidel
176. Tanaka, K. 1987. *Publ. Astron. Soc. Jpn.* 39: 1–45
177. Tang, F. 1986. *Sol. Phys.* 105: 399–412
177a. Thompson, A. R., Maxwell, A. 1960. *Nature* 185: 89–90
178. Trottet, G. 1986. *Sol. Phys.* 104: 145–64
179. Trottet, G., Vilmer, N. 1984. *Adv. Space Res.* 4: 153–56
180. Tsuneta, S. 1984. *Proc. Jpn.-Fr. Semin. Active Phenomena in the Outer Atmosphere of the Sun and Stars*, ed. J.-C. Pecker, Y. Uchida, pp. 243–60. Meudon: Obs. Paris
181. Tsuneta, S., Takakura, T., Nitta, N., Ohki, K., Makishima, K., et al. 1983. *Sol. Phys.* 86: 313–21
182. Tsuneta, S., Takakura, T., Nitta, N., Ohki, K., Tanaka, K., et al. 1984. *Ap. J.* 280: 887–91
183. van Hoven, G., Anzer, U., Barbosa, D. D., Birn, J., Cheng, C. C., et al. 1980. See Ref. 163, pp. 17–116
184. Vestrand, W. T. 1988. *Sol. Phys.* 118: 95–121
185. Vestrand, W. T., Forrest, D. J., Chupp, E. L., Rieger, E., Share, G. H. 1987. *Ap. J.* 322: 1010–22
186. Vilmer, N., Kane, S. R., Trottet, G. 1982. *Astron. Astrophys.* 108: 306–13
187. von Rosenvinge, T. T., Ramaty, R., Reames, D. V. 1981. *Proc. Int. Cosmic Ray Conf., 17th, Paris*, 3: 28–31

188. Wagner, W. J. 1984. *Annu. Rev. Astron. Astrophys.* 22: 267–89
189. Warwick, C. S., Haurwitz, M. W. 1962. *J. Geophys. Res.* 67: 1317–32
190. Webb, D. F., Cheng, C. C., Dulk, G. A., Edberg, S. J., Martin, S. F., et al. 1980. See Ref. 163, pp. 471–99
191. Webb, D. F., Hundhausen, A. J. 1987. *Sol. Phys.* 108: 383–401
192. White, R. B. 1983. In *Basic Plasma Physics I*, ed. A. A. Galeev, R. N. Sudan, pp. 611–76. Amsterdam: North-Holland
193. Wild, J. P., Smerd, S. F. 1972. *Annu. Rev. Astron. Astrophys.* 10: 159–96
194. Wild, J. P., Smerd, S. F., Weiss, A. A. 1963. *Annu. Rev. Astron. Astrophys.* 1: 291–366
195. Woodgate, B. E., Shine, R. A., Poland, A. I., Orwig, L. E. 1983. *Ap. J.* 265: 530–34
196. Wu, T. S., de Jager, C., Dennis, B. R., Hudson, H. S., Simnett, G. M., et al. 1986. See Ref. 103, pp. 5-1–73
197. Yoshimori, M. 1984. *J. Phys. Soc. Jpn.* 53: 4499–4506
198. Yoshimori, M. 1985. *J. Phys. Soc. Jpn.* 54: 1205–13
199. Yoshimori, M. 1989. Submitted for publication
200. Yoshimori, M., Okudaira, K., Hirasima, Y., Kondo, I. 1983. *Sol. Phys.* 86: 375–82
201. Zaitsev, V. V., Stepanov, A. V. 1985. *Sol. Phys.* 99: 313–21
202. Zarro, D. M., Canfield, R. C., Strong, K. T., Metcalf, T. R. 1988. *Ap. J.* 324: 582–89
203. Zirin, H. 1970. In *Upper Atmosphere Geophys. Rep. UAG-8*, ed. J. V. Lincoln, pp. 30–33. Boulder, Colo: World Data Cent.
204. Zirin, H., Tanaka, K. 1973. *Sol. Phys.* 32: 173–207

DIFFUSE GALACTIC GAMMA-RAY EMISSION

Hans Bloemen

Leiden Observatory, P.O. Box 9513, 2300 RA Leiden, The Netherlands

1. INTRODUCTION

This review concentrates on gamma radiation that originates from cosmic-ray (CR) interactions with gas and photons in the interstellar medium (ISM). The γ-ray energy band of interest here ranges from ~ 1 MeV to ~ 100 GeV, but this work is restricted to the energy interval between about 50 MeV and 5 GeV (generally referred to as "high-energy" γ rays), which is mainly dictated by data availability. The experimental development is most advanced at these energies. Also, high-energy γ rays are of particular interest from an astrophysical point of view because they are partly produced by the CR particles that carry the bulk of the CR energy. The feasibility of detecting high-energy γ quanta from the ISM was predicted over 35 years ago (101, 120), but the experimental development of γ-ray astronomy has been slow because space-borne telescopes with high technical requirements are needed in order to detect the rare γ rays in a high background of CR particles. After several pioneering balloon and a few satellite experiments [reviewed by Fazio (80)], a real breakthrough in observational γ-ray astronomy was achieved with the NASA satellite *SAS-2* (83), followed soon after by the ESA satellite *COS-B*. Since mid 1982, when the *COS-B* experiment was switched off, no high-energy γ-ray satellite has been operative, and it may not be obvious at first glance that considerable progress has been made in this field recently. This article emphasizes new results obtained from analyses of the final *COS-B* database (168). Descriptions of the instrument and of the prelaunch and in-flight calibrations can be found in (21, 31, 109, 209, 233).

Both *SAS-2* and *COS-B* viewed essentially the entire Galactic disk up to $|b| \approx 20°$, with several extensions to higher latitudes. The characteristics of the two instruments were not vastly different (geometrical area of the detector ~ 500 cm^2, field of view ~ 0.3 sr, average angular resolution above 100 MeV $\sim 2.5°$ FWHM), but the counting statistics of the *COS-B* observations are higher by a factor of ~ 25 owing to the long duration of the *COS-B* mission—namely, about $6\frac{1}{2}$ years compared with about 7 months for *SAS-2*. In addition, *COS-B* could measure energies beyond the approximate 200-MeV upper limit of the *SAS-2* instrument. These advantages have been fully exploited in recent analyses. On the other hand, the *SAS-2* observations have a lower instrumental background owing to a differing orbit of the satellite, which is important for studies of the low-intensity emission observed at high latitudes. Figure 1 shows a *COS-B* map of the Milky Way for the energy range 100 MeV–6 GeV, which is best suited for pictorial purposes because it gives a good compromise between counting statistics and angular resolution (which degrades with decreasing energy).

During the 10-yr time interval between the first *COS-B* observations and the release of the final data base, our understanding of the instrument and the data has clearly improved in several aspects. Not surprisingly, the results have evolved. I have included here some unpublished updates and extensions of earlier work. All figures were extracted from the final *COS-B* data base.

Specific aspects or more general properties of diffuse gamma radiation have been reviewed in the monographs by Stecker in 1971 (222), Chupp in 1976 (61), Fichtel & Trombka in 1981 (90), Ginzburg in 1984 (93), and Ramana Murthy & Wolfendale in 1986 (196) and in the review papers by Fazio in 1967 (79), Fichtel in 1977 (82), Van der Walt & Wolfendale in 1988 (246) and Dogiel & Ginzburg in 1989 (72). In the 1983 volume of this series, Bignami & Hermsen (24) have reviewed the γ-ray point sources that have been detected.

In Section 2, some basic principles are described and the main observational results are summarized. The interpretation of these findings is discussed further in Sections 3 and 4. Two important types of inference can be distinguished: the relevance to molecular-gas studies (Section 3), and the relevance to CR studies (Section 4). In Section 5 the status regarding γ-ray point sources is briefly discussed, with emphasis on their possibly diffuse nature.

Unless stated otherwise, the radius of the solar circle is assumed to be 10 kpc.

2. GAMMA RAYS AND THE INTERSTELLAR MEDIUM

2.1 *Gamma-Ray Emission Processes*

Radiative transfer in γ-ray astronomy is relatively simple because the Galaxy is practically transparent to γ rays with energies up to $\sim 10^{14}$ eV.

DIFFUSE GALACTIC GAMMA-RAY EMISSION 471

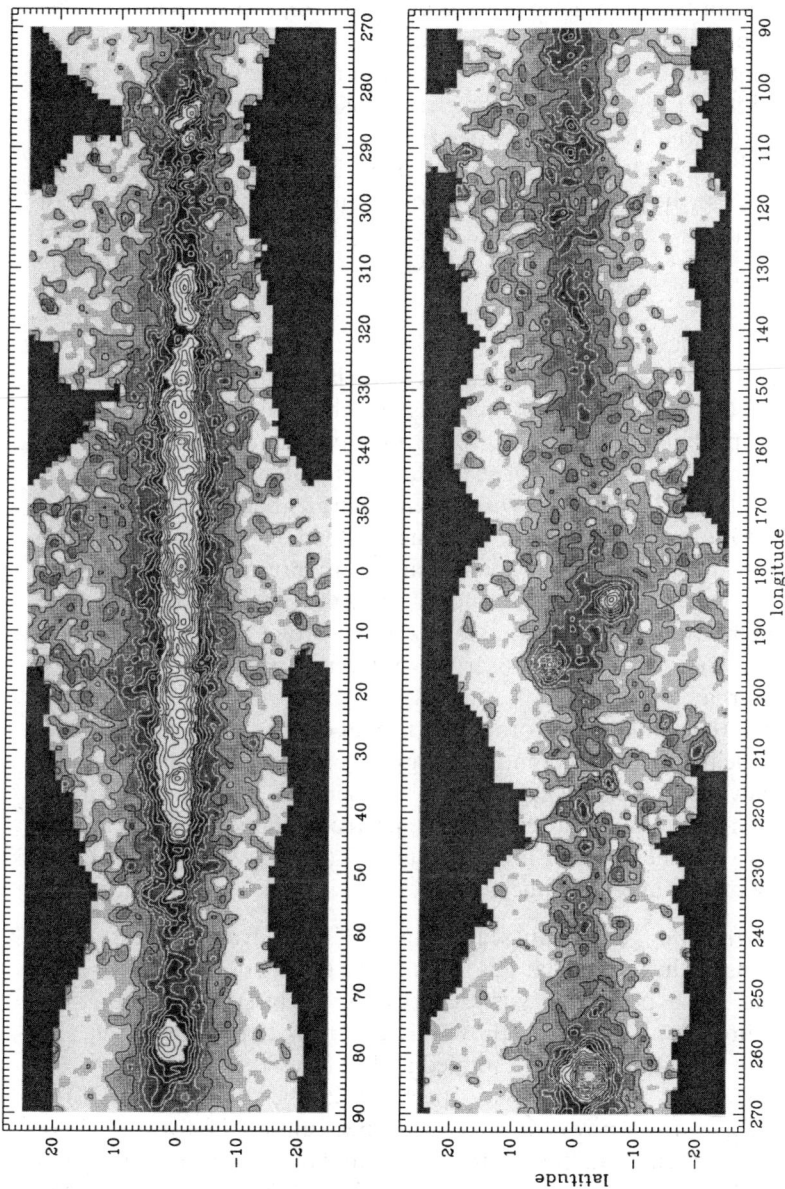

Figure 1 Gamma-ray intensity map of the Galactic disk derived from the final *COS-B* data base. The energy range is 100 MeV–6 GeV. Contour lines are indicated at multiples of 5×10^{-5} photon cm^{-2} s^{-1} sr^{-1}. An isotropic background level of 8×10^{-5} photon cm^{-2} s^{-1} sr^{-1} ["on axis" (233)], which is largely of instrumental origin, has been subtracted. The dark areas at the edges of the map correspond to regions with an exposure $< 4 \times 10^7$ cm^2 s (corresponding to roughly 15 days of observing time).

The γ-ray production mechanisms related to CR interactions with the ISM are well understood and have been described in detail by a number of authors [e.g. in the monographs by Chupp (61) and Stecker (222)]. The important production processes for the 50 MeV–5 GeV range are π°-decay, bremsstrahlung, and inverse-Compton (IC) scattering, of which a brief description is given below. The γ-ray emissivity per unit of volume is denoted by Q_γ, and the emissivity per H atom by $\varepsilon_\gamma \equiv Q_\gamma/4\pi n$ [photon (H atom)$^{-1}$ s^{-1} sr^{-1}], where n is the number density of H atoms (or, more precisely in case the gas is ionized, hydrogen nuclei). The quantity ε_γ is, of course, particularly meaningful when the γ-ray emission is directly related to the gas particles, which is the case for the π°-decay and bremsstrahlung processes.

(a) *Nuclear interactions* (π°-*decay*) Nuclear interactions between CR particles and nuclei of the interstellar gas lead via various decay chains (69, 223, 227) to the production of π°-mesons, which decay rapidly ($\sim 10^{-16}$ s) and with a probability of almost 100% into two γ quanta. The bulk of the π°-mesons is produced by those cosmic rays that carry most of the CR energy density, i.e. CR protons with kinetic energies between about 1 GeV and a few tens of GeV. (The term "protons" instead of "nuclei" is therefore mostly used throughout this paper.) Each γ quantum has an energy of $m_{\pi^\circ}c^2/2 \approx 68$ MeV in the rest frame of the π°-meson (where m_{π° is the mass of the π° meson), which transforms into a broad energy distribution centered on ~ 68 MeV in the observer's reference system (see Figure 2). At γ-ray energies above ~ 1 GeV, the shape of the π°-decay γ-ray spectrum converges to that of the CR proton spectrum at energies above a few GeV. More specifically, if the differential energy spectrum of the CR protons is a power-law spectrum, $I_p(E_p) = K_p E_p^{-\Gamma}$ (where K_p is a normalization factor), then the differential emissivity spectrum of the π°-decay γ rays converges to a similar power-law spectrum, $Q_\gamma(E_\gamma) \propto n K_p E_\gamma^{-\Gamma}$, where n is the density of the target gas nuclei.

(b) *Bremsstrahlung* Coulomb scattering of CR electrons on the nuclei and electrons of the interstellar gas leads to the production of bremsstrahlung γ rays. At a particular γ-ray energy the bremsstrahlung emission originates predominantly from those CR electrons that have energies in the range of about one decade immediately above the γ-ray energy, so that $E_e \approx 3 E_\gamma$. The γ-ray emissivity spectrum can be written as $Q_\gamma(E_\gamma) \propto n I_e(> E_\gamma)/E_\gamma$, so for a power-law electron spectrum of the form $I_e(E_e) = K_e E_e^{-\Gamma}$, the production spectrum is a similar power-law spectrum $Q_\gamma(E_\gamma) \propto n[K_e/(\Gamma-1)]E_\gamma^{-\Gamma}$, where n is the density of the gas nuclei. For the energetic electrons that are of interest here ($E_e \gtrsim 100$ MeV), screening by electrons in the target atoms plays a role in the scattering process (41).

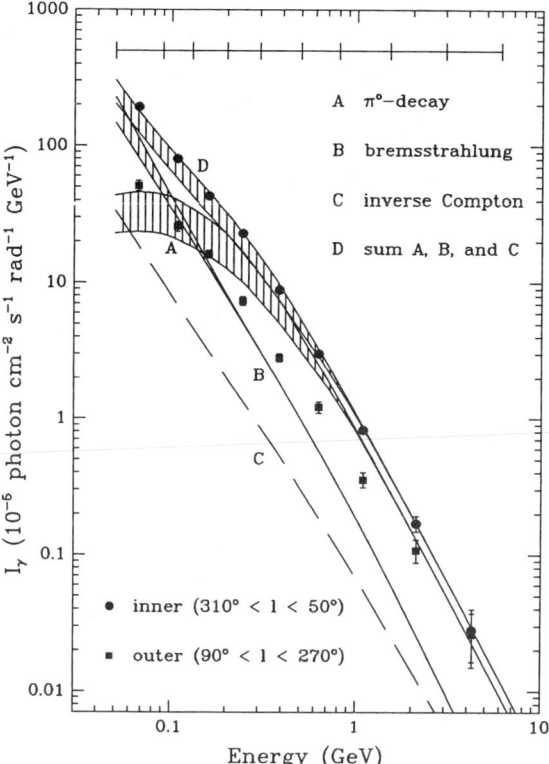

Figure 2 Spectra of the γ-ray emission in the general directions of the inner and the outer Galaxy ($|b| < 30°$) as observed by the COS-B satellite, and spectra derived from the local interstellar CR electron and proton spectra (33). The sum of the spectra for the three individual γ-ray production processes is normalized to the observed inner-Galaxy spectrum at ~1 GeV. The shaded areas indicate uncertainties in the predictions that result largely from uncertainties in the corrections for solar modulation of the observed CR spectra (Section 2.2). The hint for a flatter γ-ray spectrum in the outer-Galaxy direction is discussed at the end of Section 2.3.

Consequently, the bremsstrahlung emissivity for electrons incident on an atomic gas is not identical to that for electrons incident on an ionized gas. In addition, electron-electron scattering needs to be taken into account in the latter case; the bremsstrahlung cross sections for an electron incident on an unshielded free proton and an unshielded free electron are roughly equal (96). The relevant parameter here is the so-called screening factor, given by $\Delta = 68.5 m_e c^2 E_\gamma / [2 E_e (E_e - E_\gamma)]$ for atomic hydrogen; screening is important if $\Delta \ll 1$, which is the case for high-energy electrons. At $E_\gamma = 100$ MeV, for example, $\Delta \approx 0.03$ and the total ε_γ for ionized hydrogen (including electron-electron scattering) is about 30% higher than ε_γ for atomic hydrogen (41). Although the effect is small and in fact negligible for studies of the Galactic disk, where the fraction of the gas that is in ionized form is on average small, it may be an important clue for studies of medium-latitude regions (Section 2.4).

(c) *Inverse Compton* The scattering of high-energy ($E_e \gtrsim 10$ GeV) CR electrons on soft photons of energy $\varepsilon \ll m_e c^2$ (mainly optical and infrared photons and the 2.7-K universal background radiation) leads to the production of IC γ rays. If the incident electrons have a power-law spectrum, $I_e(E_e) = K_e E_e^{-\Gamma}$, then the IC emissivity has approximately a power-law spectrum of the form $Q_\gamma(E_\gamma) \propto n_{\rm ph} \langle \varepsilon \rangle^{(\Gamma-1)/2} K_e E_\gamma^{-(\Gamma+1)/2}$, where $\langle \varepsilon \rangle$ and $n_{\rm ph}$ are the average energy of the target photons and their number density, respectively. For the γ-ray energies of interest here, the mean energy of the produced γ quantum is given by $\langle E_\gamma \rangle \approx \frac{4}{3}\gamma^2 \langle \varepsilon \rangle$, where $\gamma = E_e/m_e c^2$. For example, at $\langle E_\gamma \rangle = 100$ MeV, electrons with very high energies of typically $E_e \approx 200$ GeV are responsible for IC emission from the 2.7-K background ($\langle \varepsilon \rangle \approx 6 \times 10^{-4}$ eV).

A few general remarks can be made based on the above.

First, the volume emissivities Q_γ of the three γ-ray production processes are proportional to the CR density (in a certain energy range) and to either the gas density or the photon density. The γ-ray intensity I_γ can be written as $\int Q_\gamma(E_\gamma, \mathbf{r})/4\pi \, d\mathbf{r}$, where the integration is along the line of sight, or

$$I_\gamma(E_\gamma) = \frac{1}{4\pi} \int_l d\mathbf{r}\, n(\mathbf{r}) \int_{E_\gamma}^\infty dE\, \sigma(E_\gamma, E) I_{\rm CR}(E, \mathbf{r}),$$

where $\sigma(E_\gamma, E)$ is the cross section of the process, and $n(\mathbf{r})$ is the gas or photon density at position $\mathbf{r}$. Following this notation, the γ-ray emissivity per H atom for the π°-decay and bremsstrahlung processes is given by

$$\varepsilon_\gamma(E_\gamma, \mathbf{r}) = \frac{1}{4\pi} \int_{E_\gamma}^\infty dE\, \sigma(E_\gamma, E) I_{\rm CR}(E, \mathbf{r}).$$

Second, the relative importance of the emission processes is a strong function of γ-ray energy. The most distinct change of importance occurs around 100 MeV, which is evident from Figure 2; the relative contributions of the bremsstrahlung and IC processes increase strongly with decreasing energy because they have roughly power-law emissivity spectra, whereas the π°-decay spectrum shows the characteristic bump. In addition, for $E_\gamma \gtrsim 1$ GeV, the importance of IC emission can be expected to increase. The reason for this is that the IC emissivity follows approximately an E_γ^{-2} power law [at least in the solar vicinity, to which our knowledge of the high-energy electron spectrum is limited: $\Gamma \approx 3\text{--}3.4$ for $E_e \gtrsim 100$ GeV (179, 180)], whereas the π°-decay and bremsstrahlung emissivities are both approximately proportional to $E_\gamma^{-2.7}$ [$\Gamma \approx 2.7$ for the local interstellar electron spectrum at GeV energies (e.g. 250) and for the proton spectrum above 10 GeV (e.g. 206)].

Third, the relative contributions of the three processes can be expected to be a function of position in the Galaxy. Most obvious is the impact of different distributions of interstellar matter and soft photons, which lead to spatial variations of the contribution from CR-photon and CR-matter interactions. Another reason for variations of the contributions is the production of so-called secondary electrons by the interaction of the CR proton-nuclear component with interstellar gas, which may lead to enhanced bremsstrahlung emission in high-density regions (such as molecular clouds). Finally, a variety of effects influences the spectral distribution of CR particles during propagation and hence, possibly, the relative contributions of the γ-ray production processes.

2.2 Gamma-Ray Intensity Versus Gas Column Density

Several authors have reached the conclusion that the IC contribution to the measured γ-ray intensities near the Galactic plane does not exceed 5–10% (25, 32, 131, 191, 214, 224). As an example, Figure 2 (from 33) shows estimates of the relative importance of the three main emission processes in the general direction of the inner Galaxy. For further discussions in this paper it is illustrative to know how these estimates were obtained. Only the IC spectrum in the figure is a "real" prediction of absolute intensities (32). It is based on a variety of studies of the interstellar radiation field; the density of high-energy electrons was assumed to increase toward the inner Galaxy (a factor of two between $R = 10$ and 5 kpc—see Section 4.3) and to have a spectral shape as measured near Earth and a scale height of 750 pc. Also, the shapes of the predicted π°-decay and bremsstrahlung γ-ray spectra are based on measurements of the CR electron and proton spectra near Earth. The relative γ-ray intensities of these two production mechanisms are fixed (defined by the local CR electron/proton ratio), but the sum was normalized such that the total predicted γ-ray intensity at 1 GeV is equal to that observed in the inner-Galaxy direction. The shaded areas indicate uncertainties in the predictions that result largely from uncertainties in the corrections for solar modulation of the observed CR spectra. The upper bound of the shaded area for the π°-decay curve represents the spectrum calculated by Stephens & Badhwar (227; their "M_U" curve), and the lower bound corresponds to the spectrum given by Dermer (69). The upper and lower bounds of the shaded area for the bremsstrahlung spectrum correspond to an $E_e^{-2.4}$ and $E_e^{-2.1}$ electron spectrum for $E_e \lesssim 200$ MeV, respectively, which covers the uncertainties at these energies (250).

The IC contribution is expected to increase with increasing Galactic latitude because the scale height of the interstellar gas is smaller than that of optical photons and, of course, the isotropic 2.7-K background. At $E_\gamma \gtrsim 1$ GeV, the IC contribution may even be very substantial. Owing to uncertainties in the electron scale height, however, it is very hard to make a precise estimate. Further discussion on this point is given in Section 2.4.

The γ-ray intensity is a measure of the product of CR density and total gas density, integrated along the line of sight, if the emission can be largely attributed to CR-matter interactions. I_γ, for instance, is a measure of the total gas column density N, practically irrespective of the composition or the physical state of the gas, if the CR density can be assumed to be constant: $I_\gamma = \varepsilon_\gamma N$. Vice versa, the CR distribution can be studied if independent information on the gas distribution is available. Several γ-ray studies have used H I 21-cm line observations and millimeter-wave line observations of the CO molecule to trace the column densities of atomic [N(H I)] and molecular hydrogen [$N(H_2)$], respectively, and maps of selected sky areas were compared with the observed γ-ray intensity distributions. The relationship that is evaluated in these comparisons is of the form $I_\gamma = \varepsilon_\gamma \{N(\text{H I}) + 2XW_{CO}\}$, where W_{CO} is the velocity-integrated CO antenna temperature and $X \equiv N(H_2)/W_{CO}$ is the CO-to-H_2 calibration factor in units of molecule cm^{-2} (K km s^{-1})$^{-1}$. Both ε_γ and X can be estimated from a correlation analysis of the H I, CO, and γ-ray maps. This basic principle has been extended to determine large-scale Galactic variations of X and ε_γ (Section 2.3).

In the remainder of this section, the main observational results are presented. Further discussion on the interpretation of the findings is given in Sections 3 and 4.

2.3 Gamma Rays From the Galactic Disk

Ever since the first firm detection of celestial γ rays by the *OSO-3* experiment (137), but particularly since the *SAS-2* observations have become available (83), it has been clear that the γ-ray sky is dominated by emission from the Galactic disk. The total luminosity of the Galactic disk above 100 MeV is $(1-2) \times 10^{39}$ erg s^{-1}, of which $\sim 50\%$ originates beyond the solar circle. Early studies already showed that most of the radiation could be understood, at least qualitatively, in terms of the diffuse processes described above (e.g. 22, 133, 226). Because of the limited angular resolution of *SAS-2* and *COS-B*, however, γ-ray emission from starlike objects might be hidden and would appear part of the diffuse emission to these experiments. Although there were no firm predictions, this possibility had to be considered very seriously after the first *COS-B* observations showed evidence for several point sources (114) in addition to the few bright ones that could be seen by *SAS-2* [the Crab (134, 243) and Vela (241, 244) pulsars, the puzzling source "Geminga" in the Galactic anticenter (243), and Cyg X-3 (142)—the latter is under debate (89, 111)]. The second *COS-B* catalog (109, 239) contains 19 such sources at low Galactic latitudes, none of which could be unambiguously identified. These point sources were defined as significant peaks in the γ-ray intensity distribution ($E_\gamma >$ 100 MeV) with a shape that is consistent with the *COS-B* point-spread function. Owing to the poor angular resolution, however, it cannot be excluded that these sources are actually extended; most of the unidentified sources might have an angular diameter up to $\sim 2°$.

In the meantime, correlation studies of the observed γ-ray emission and gas tracers (H I and CO observations) have shown with increasing confidence that most of the γ-ray emission is probably of diffuse origin. The observed and expected intensity distributions show generally good agreement (40, 147, 232), and even several unidentified γ-ray sources in the *COS-B* catalog can be attributed to peaks in the gas column density distribution (170, 193, 194, 202). Some other unidentified γ-ray sources can probably be explained by CR-irradiated clouds as well, if one allows for an enhanced CR density inside the cloud (116, 158, 159). On the other hand, evidence for some new sources is found when the expected diffuse γ-ray emission is subtracted from the observed emission. This point is discussed in Section 5.

The main objective of γ-ray studies of the Galactic disk has usually been to determine the CR distribution in the Galaxy and thereby to extract information on the origin of cosmic rays. This is feasible if independent information is available on the distribution of the target gas particles with which the cosmic rays interact. Early studies suffered in this respect from uncertainties in the molecular-hydrogen distribution in the Galaxy. The atomic-hydrogen distribution was well mapped by H I 21-cm line observations, although there were (and still are) some uncertainties due to optical depth effects (e.g. 48). The sky coverage of the observations of the CO molecule, which is probably the best tracer of the large-scale distribution of H_2, was limited to a very narrow band along the Galactic plane. Furthermore, the relation between measured CO intensity and corresponding H_2 column density was uncertain. The small latitude extent of the CO surveys was inadequate for a direct comparison with the γ-ray observations because of the low angular resolution of the latter. The situation has improved significantly with the availability of the large-scale CO surveys from the Columbia 1.2-m telescopes in New York City and on Cerro Tololo in Chile—the combined survey covers the entire Milky

Way up to $|b| \approx 7-10°$, with several large extensions to higher latitudes (68).

The importance of H_2 for the interpretation of the γ-ray data was, however, certainly realized in early work as well. The H I data and the sparse CO observations (or other information on the H_2 distribution) were generally converted into a model of the gas distribution in the Galaxy, although the CO-to-H_2 conversion factor was an uncertain parameter in the modeling. For the second and third Galactic quadrants, the H_2 contribution to the gas column densities has long been known to be small on average, so the lack of a complete CO survey did not seriously hamper γ-ray studies of the outer Galaxy. The *SAS-2* team generally constructed a spiral-arm model of the gas distribution; based on the assumption that the CR density is proportional to the matter density on the scale of spiral arms, agreement between the predicted and observed γ-ray intensities was obtained (22, 23, 86, 100, 131, 132). Paul et al. (187) reached a similar conclusion; they followed a somewhat different approach, however, taking into account the observed nonthermal radio emission of the Galaxy. From follow-up work, Cesarsky et al. (57) concluded that a radial CR density gradient may exist, in addition to CR-matter coupling on the scale of spiral arms (but cosmic rays are assumed to see only half of the molecular gas in this modeling). Fuchs et al. (91) considered CR-matter coupling on larger scales, using a hydrostatic equilibrium model for the ISM. Others did not make a priori assumptions on the proportionality between CR and gas density. The Galacto-centric distribution of γ-rays was determined from a geometrical unfolding technique applied to the inner-Galaxy γ-ray data; a comparison with estimates of the radial gas distribution showed that the γ-ray emissivity ε_γ increases toward the inner parts of the Galaxy. The latter approach was generally followed by Stecker and coworkers (99, 224, 226) and Wolfendale and coworkers (122, 157).

The new Columbia CO surveys enabled use of a third method to derive information on the Galactic CR distribution from the γ-ray data, and this method was applied by the *COS-B* group. It is an extension of the basic principle described at the end of Section 2.2, namely a multivariate correlation analysis of γ-ray sky maps with H I and CO maps for selected Galacto-centric distance intervals (typically a few kiloparsecs wide). The distance information needed to construct the H I and CO maps was derived from the H I and CO line velocities, together with an adopted Galactic rotation curve. The method has the advantage that it takes into account all spatial structures in the data, enabling not only a study of the Galactic CR distribution but also of the $N(H_2)$-to-W_{CO} ratio X. Basically, this method consists of fitting the observed γ-ray intensity distributions by the relation $I_\gamma = \Sigma \, \varepsilon_\gamma(R_i) \{N(\text{H I})_i + 2YW_{\text{CO},i}\}$, where the sum is over the rings

i, and $\varepsilon_\gamma(R_i)$ and Y are the free parameters (which were determined by a maximum-likelihood procedure). The fitted model included other components, such as an estimate of the IC contribution, the intense γ-ray point sources, and an isotropic γ-ray background. The above relation contains a parameter Y (instead of X) to remind us of the fact that the CR density inside molecular clouds may differ from that in the ambient medium: $Y = X \cdot \varepsilon_{\gamma,H_2}/\varepsilon_{\gamma,HI}$, so if CR particles are not excluded from, or concentrated in, molecular clouds, then Y equals X independent of γ-ray energy. Also, the $YW_{CO,i}$ term covers to some extent possible arm/interarm emissivity contrasts because CO is at least in part a tracer of spiral structure. A first (less complete) study by Lebrun et al. (147) showed that the angular resolution and counting statistics of *COS-B* are indeed sufficiently good to put a meaningful γ-ray constraint on X. Bloemen et al. (40) and Strong et al. (232)—the former concentrating mainly on the first and second Galactic quadrants and the latter studying the entire Milky Way—extended this work and derived both X and the Galacto-centric emissivity distribution $\varepsilon_\gamma(R)$ for three γ-ray energy intervals (70–150 MeV, 150–300 MeV, and 300 MeV–5 GeV—these intervals were chosen, among other reasons, because they have approximately equal counting statistics). Longitude distributions of the observed γ-ray emission and the fitted model are presented in Figure 3. Figure 4 shows the model sky map for the 100 MeV–6 GeV range, based on the fit parameter values derived by Strong et al. [although these were obtained for a more limited latitude range ($|b| < 10°$)] and the H I and CO surveys used in that work. The map can be compared with the *COS-B* map shown in Figure 1. (Note, however, that the model map is not convolved with the *COS-B* point-spread function.)

The resulting distributions of $\varepsilon_\gamma(R)$ (Figure 5) show only a weak Galacto-centric gradient. The same result was obtained from a similar analysis (35, 36) restricted to the outer Galaxy (second and third quadrants), for which it is sufficient to use only H I observations as a gas tracer. In fact, for the high-energy range no significant decrease with Galacto-centric radius was found at all beyond the solar circle—although the radial distribution from the recent work described above is the same within uncertainties, it does show a weak gradient, as can be seen in Figure 5. Possible energy dependencies are discussed below in detail. These findings seemed surprising, particularly for the outer Galaxy, because analyses of the *SAS-2* data had shown evidence for a strong gradient; a steep emissivity decrease with increasing distance beyond the solar circle was first found by Dodds et al. (71) and confirmed by, e.g., Cesarsky et al. (57) and Higdon (115). Later studies of the *SAS-2* data (2, 35, 238), however, showed much less evidence for such a strong gradient ($E_\gamma > 100$ MeV). It turned out that this discrepancy can be attributed to an improvement in the calibration of the

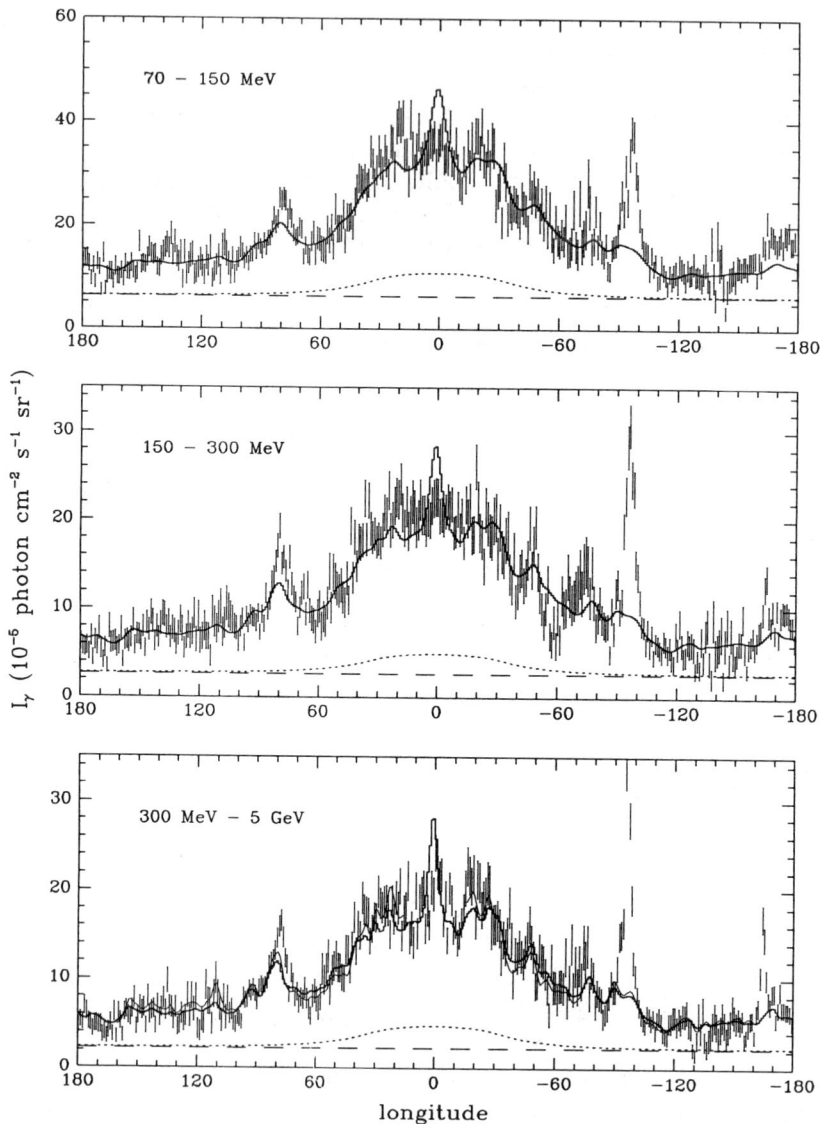

Figure 3 Longitude distributions of the observed ($\pm 1\sigma$ error bars) and modeled (histogram) γ-ray intensity, averaged over $|b| < 5°$ (232). The dotted line indicates the individual contribution in the model from inverse-Compton emission. The isotropic background is indicated by the dashed line. The specific model shown here has an energy-independent shape of radial emissivity distributions (circles in Figure 5) but appears almost identical in this presentation for an energy-dependent shape (black dots in Figure 5). The thin full line in the figure for the 300 MeV–5 GeV range indicates the prediction from a model with CR-matter coupling (172), discussed in Section 4.4.

DIFFUSE GALACTIC GAMMA-RAY EMISSION 481

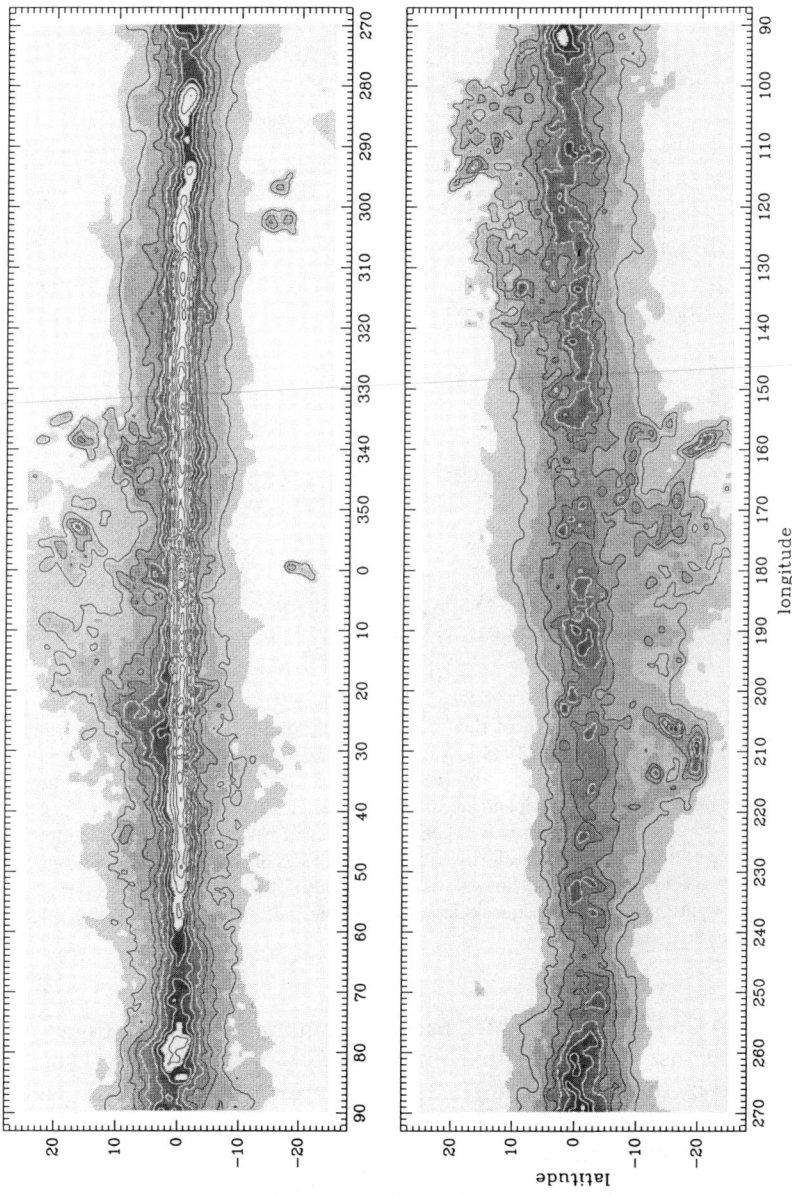

Figure 4 Map of the expected γ-ray intensity distribution (100 MeV–6 GeV) from CR-matter interactions, derived from H I and CO surveys and fit parameter values obtained by Strong et al. (232). The H I and CO data have *not* been convolved with the *COS-B* point-spread function, although a little smoothing was done for pictorial purposes. Contour levels: 4, 8, 12, 16, 24, 32, 40, 52, 68, 90, 120, 150, 180, 240, 300, 400, ..., $\times 10^{-5}$ photon cm^{-2} s^{-1} sr^{-1}.

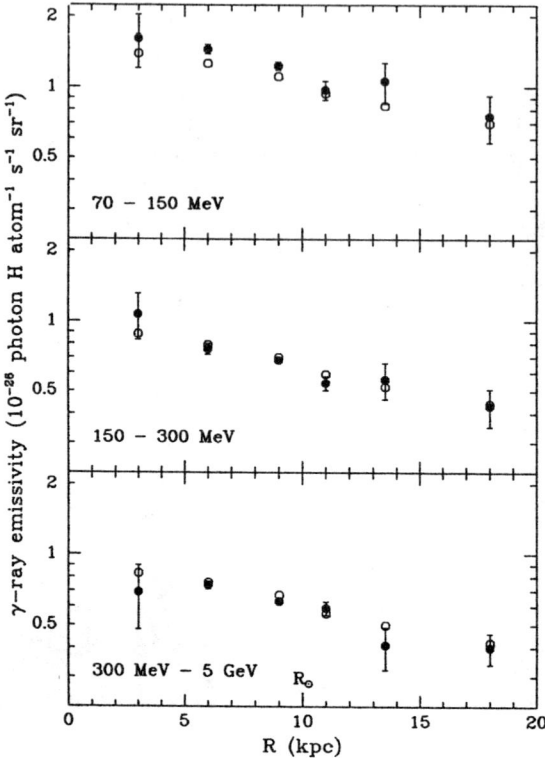

Figure 5 Galacto-centric distribution of the γ-ray emissivity obtained from a correlation analysis of H I, CO, and γ-ray data (232). (*Circles*) Shape of the radial distribution is adopted to be independent of γ-ray energy—the observed large-scale spectral variation along the Milky Way is attributed to an energy dependence of the parameter Y (see Section 2.3.1). (*Black dots*) Vice versa: Y adopted to be energy independent—spectral variation attributed to an energy dependence of the shape of the radial distribution. Both sets of emissivity distributions lead to an equally good fit of the data; early statistical tests (40) of a subset of the data used to construct this figure preferred the energy-dependent shape, but recent ones (232) prefer the energy-independent shape.

SAS-2 data around 1978 [C. Fichtel, personal communication; see Figure 4 in (34)]. As a consequence, the final outer-Galaxy γ-ray intensities (>100 MeV) released by the *SAS-2* group in 1978 (85) are about a factor of two higher than those presented in 1975 (83), which were used in the studies showing the steep gradient.

The Durham group of Wolfendale and collaborators has advocated for many years the presence of a strong gradient. This was to a large extent based on the early *SAS-2* data, but several of their recent papers (15, 16,

166, 240, 246, 255) suggest that there is still no real consensus on the gradient issue, i.e. that in their opinion the emissivity gradient is not as weak as claimed by the *COS-B* group. Their present best estimate of the radial emissivity distributions [first presented by Bhat et al. (15) and shown in most of the other papers] was, however, actually taken from Bloemen et al. (40), with the exception that they increased the emissivity values for the smallest radius by 25%, as described by Bhat et al. The point of debate, therefore, appears to be rather minor.

Although most of the recent γ-ray work on the CR distribution has concentrated on radial gradients, there is still the possibility that this is not an adequate description of the CR distribution in the Galaxy. It was already noted above that the *SAS-2* group generally assumed proportionality between the CR density and gas density on the scale of spiral arms, which led to a good description of the γ-ray data. Melisse & Bloemen (172) have recently reconsidered this possibility, using the *COS-B* data and the same H I and CO surveys used for the gradient studies. Their conclusion is that the option of CR-matter coupling is still viable. Details are given in Section 4.4.

2.3.1 SPECTRAL INFORMATION There is evidence for a large-scale variation of the γ-ray spectrum along the Milky Way. The first clear indication of this effect came from the *COS-B* work of Mayer-Hasselwander (167). His longitude distribution ($|b| < 10°$) of the ratio between the intensity in the energy range 70–150 MeV and that in the range 150 MeV–5 GeV indicates a softer γ-ray intensity spectrum toward the inner Galaxy than for the remainder of the disk. This result is obviously sensitive to background corrections; in earlier *COS-B* work (169), for instance, when the γ-ray background was less well understood, no spectral variation was found. Since Mayer-Hasselwander did not use the final *COS-B* data base, I have repeated his analysis. Another reason for repeating it is the choice of the latitude interval; in order to minimize the influence of γ rays that originate in the local ISM and possibly the Galactic halo (see Section 2.4), it is preferable to restrict the interval to $|b| < 5°$. However, the difference between the angular resolutions of the 70–150 MeV and 150 MeV–5 GeV energy ranges is not negligible in this case. This difference leads to a stronger broadening of the γ-ray-emitting Galactic plane for the 70–150 MeV range, and thus to a larger "spill-over" to $|b| > 5°$ for this energy range. First-order corrections as a function of longitude can be obtained from the model sky maps of Strong et al. (232) described above [i.e. from comparisons of the model sky maps before and after the convolution with the *COS-B* point-spread functions—the method is described in some further detail in (39)]. The correction factors for the spectral ratio turn out to be typically $\sim 10\%$. Figure 6 shows the resulting spectral-ratio distribution and confirms Mayer-Hasselwander's findings.

Indeed, the correlation analyses of Bloemen et al. (40) and Strong et al. (232), discussed above, showed that the interpretation of the *COS-B* data requires an energy-dependent model. In these studies, the energy dependence can be attributed to either a higher Y value for low energies than for high energies (which produces a softer spectrum toward the inner Galaxy because most of the molecular gas is located inside the solar circle), or to a steeper γ-ray emissivity gradient for low energies. These two options can be tested, in principle, because the angular distributions of H I and H_2 are different. In practice, however, such a test is hampered by the degradation of the angular resolution of *COS-B* with decreasing energy. This is evident from the results: In the work of Bloemen et al. (40), the likelihood tests favored energy dependence of $\varepsilon_\gamma(R)$, whereas the tests by Strong et al. (232) favored energy dependence of Y. In both works the two options could not be distinguished at a high level of significance, although the latter is probably to be preferred because Strong et al. used more data. The energy dependence of Y is of course unrelated to X, which is by definition energy independent, but it may be related to CR propagation

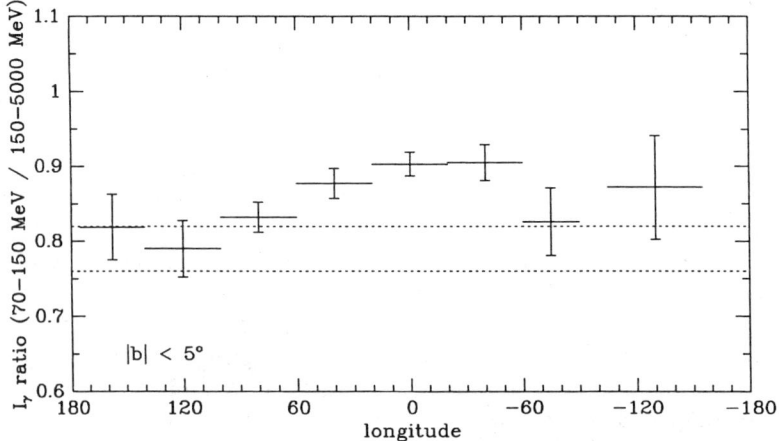

Figure 6 Longitude distribution ($|b| < 5°$) of the spectral ratio between the observed γ-ray intensity in the 70–150 MeV and 150 MeV–5 GeV energy ranges, showing a tendency of softer γ-ray spectra in the inner-Galaxy direction. The correction for differing angular resolutions is described in the text. Regions containing the strong γ-ray point sources Vela, Crab, and Geminga are excluded. The two dotted lines indicate the range of spectral ratios for typical estimates of local γ-ray emissivities (see Section 4.2).

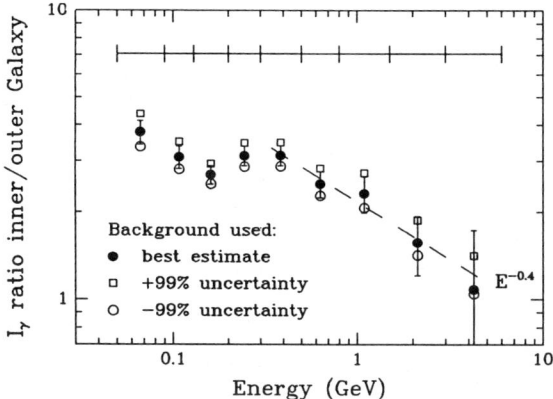

Figure 7 Ratio between the γ-ray spectra toward the inner (310° < l < 50°) and the outer Galaxy (90° < l < 270°), integrated over |b| < 30° (33). The black dots and error bars correspond to the spectra shown in Figure 2. The squares and circles indicate that uncertainties in the γ-ray background have only a minor impact.

and production in or near molecular clouds or due to the presence of sources with steep spectra distributed like CO (Section 4.3). The constraints on X are addressed in Section 3.1; the interpretations with regard to the Galactic CR distribution are discussed in Section 4.3.

In addition to the broad-band spectral difference described above (70–150 MeV versus 150–300 MeV + 300 MeV–5 GeV), a large-scale spectral variation exists within the 300 MeV–5 GeV range. This effect is evident from Figure 7 (33), which shows the ratio between the γ-ray intensity spectrum in the general direction of the inner Galaxy (310° < l < 50°; |b| < 30°) and in the direction of the outer Galaxy (90° < l < 270°; |b| < 30°). (The actual spectra are shown in Figure 2.) The outer-Galaxy spectrum above a few hundred MeV is flatter, with a spectral-index difference of ∼0.4. Later work (39, 204) showed, however, that the integration of the emission over the broad latitude range of |b| < 30° had masked an important effect: The relative flatness of the average outer-Galaxy spectrum results from a gradual spectral hardening with increasing latitude (see Figure 8). Owing to the limited counting statistics, this follow-up work had to be restricted to two energy ranges, namely 300–800 MeV and 800 MeV–6 GeV. Toward the inner Galaxy, this spectral flattening with latitude is not seen, which is a strong indication that the findings cannot be attributed to systematic uncertainties in the intensity level and spectrum of

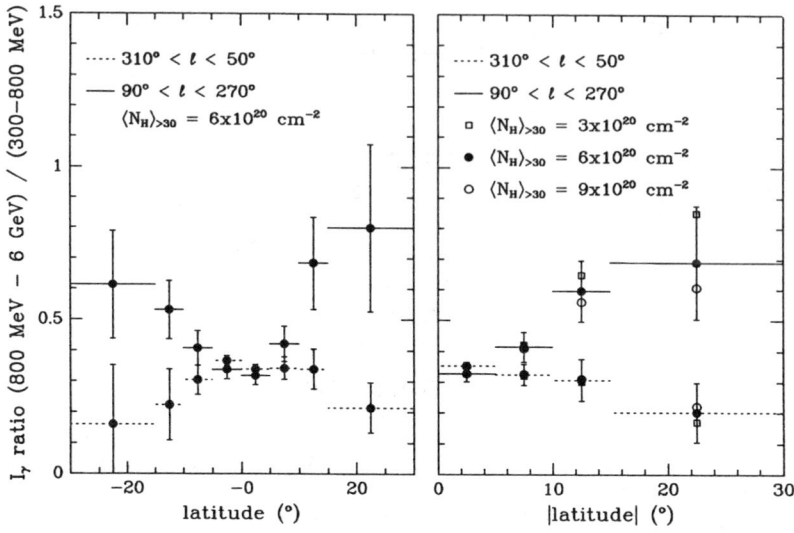

Figure 8 Latitude distribution of the ratio between the intensities for the energy ranges 800 MeV–6 GeV and 300–800 MeV in the general directions of the inner and the outer Galaxy (39). (*Left*) Regions below and above the Galactic plane are shown separately. (*Right*) Regions on both sides of the plane are combined. The results are shown for different choices of the average gas column density at $|b| > 30°$, which is related to the background correction made—the figure shows that uncertainties in this correction cannot explain the spectral difference seen toward the inner and outer Galaxy. Details are given in (39).

the isotropic γ-ray background. The interpretation of this rather surprising result is discussed in Section 4.5.

2.4 Away From the Galactic Plane: Local ISM and Halo

Most of the γ-ray emission observed beyond a few degrees from the Galactic plane is almost certainly of diffuse nature because the γ-ray point sources detected by *COS-B* have a very narrow latitude distribution ($\langle|b|\rangle \approx 1.5°$; 109, 239), despite the greater detectability away from the intense disk emission. Early analyses of the *SAS-2* data by Fichtel et al. (84, 88) showed that the latitude distribution of the γ-ray intensity at $|b| \gtrsim 10°$ can be described by a two-component model: a component that resembles the latitude distribution of $N(\text{H I})$, and an isotropic component, possibly of extragalactic origin. In later work, the Lick galaxy counts catalog (211, 212) was used to derive total gas column densities; studies of both the *SAS-2* data (149, 150, 237, 242) and the *COS-B* data (148, 228, 230, 231, 234) showed that the correlation of γ-ray intensity with total gas column density is an improvement over the correlation with $N(\text{H I})$ alone.[1] Toward intermediate-latitude regions ($10° \lesssim |b| \lesssim 20°$) in the general direction of the inner Galaxy, however, the observed intensities were found to be significantly larger than expected, particularly at positive latitudes. Both the *SAS-2* data (149, 150) and the *COS-B* data (16, 228, 230, 231, 234) show this effect. After a recent improvement in the calibration of the galaxy counts, it turned out that the discrepancy is even larger than found previously and that a similar excess is indeed clearly present below the Galactic plane (150). This improvement in the galaxy-counts calibration— which reduced the gas column-density estimates for some regions of the sky—resulted from the discovery by Lebrun (145, 146) of an observational bias in counting galaxies. The γ-ray excess is discussed below in further detail.

It is generally assumed that the intermediate-latitude γ-ray emission originates mainly from the ISM in the solar vicinity. There is strong evidence that this is to a large extent correct, such as the γ-ray counterparts of Gould's Belt (83, 100, 169) and possibly Dolidze's Belt (18, 19) and the clear detection of γ rays from the Orion (37, 53, 117, 124) and Ophiuchus (113, 123) molecular-cloud complexes in Gould's Belt. There are indications, however, that a significant fraction of the γ-ray emission at medium latitudes may not be of local origin, which brings us to the long-standing question of the existence of a γ-ray halo or "thick disk."

First, for the general direction of the outer Galaxy (second and third Galactic quadrants), it can be estimated from the new Leiden–Green Bank H I survey (49, 51) that $\sim 40\%$ of the H I column density at $|b| \approx 15-20°$ originates beyond 1 kpc from the Sun (39, 52) and thus beyond a few hundred parsecs from the midplane, which is a result of the flaring of the H I disk. This significant contribution of nonlocal gas was less evident from previous medium-latitude H I surveys (105, 106) owing to the limited sensitivity of these observations. *Second*, the medium-latitude γ-ray excess toward the inner Galaxy may not be of local origin, as discussed below. A *third* indication of a nonlocal component in the observed γ-ray emission at medium latitudes is the spectral flattening above 300 MeV with increasing latitude in the outer-Galaxy direction, as described in Section 2.3. Whatever the precise reason for this flattening may be, the observed symmetry of this effect—it occurs systematically on both sides of the

---

[1] Owing to the highly eccentric orbit of the *COS-B* satellite, unlike the near-Earth one chosen for *SAS-2*, the instrumental background of the *COS-B* observations is significantly larger than that of the *SAS-2* observations and dominates the observed emission at high Galactic latitudes. Most *COS-B* studies of regions away from the Galactic disk were therefore restricted to medium-latitude regions, up to $|b| \approx 20°$.

Galactic plane for the entire second and third Galactic quadrants (39)—suggests that it has a large-scale origin. Nevertheless, a local origin cannot be excluded (204). This point is taken up again in Section 4.5.

2.4.1 THE MEDIUM-LATITUDE EXCESS The discrepancy between the observed γ-ray emission at medium latitudes in the general direction of the inner Galaxy and the expectation from galaxy-counts data is confirmed if one uses H I and CO observations as gas tracers instead of galaxy counts. This was recently shown from a study restricted to the 300 MeV–5 GeV range (39), but the same holds for lower energies, which is illustrated in Figures 9 and 10. This alternative approach is attractive because it enables usage of only one model for both low- and medium-latitude regions. The γ-ray expectation shown in Figure 9 is the extension of the empirical model of the disk emission presented by Strong et al. (232), as discussed in Section 2.3; this extension is based on the fit parameter values obtained for $|b| < 10°$ (i.e. the γ-ray emissivity values for different Galacto-centric annuli, the CO-to-H_2 conversion factor, and the background levels). The medium-latitude region is certainly not properly covered by the Columbia CO surveys, but this incompleteness cannot explain the excess (39). The longitude range from $l \approx 330°$ to $l \approx 40°$, for instance, is almost completely mapped at positive latitudes up to $b \approx 20°$.

The spectrum of the excess is remarkably soft compared with the emission from the Galactic disk. This was noted by Lebrun & Paul (150) and can also be seen in the work of Bhat et al. (14), both using *SAS-2* data. Figures 9 and 10 suggest the same for the *COS-B* data, which can be illustrated as follows. Integrating the intensities in Figure 10 over the first and fourth quadrants gives excess fluxes above the Galactic plane ($5° < b < 20°$) of 25.3 ± 1.1, 7.8 ± 0.7, and $7.0 \pm 0.6 \ 10^{-6}$ photon cm^{-2} s^{-1} (70–150 MeV, 150–300 MeV, and 300 MeV–5 GeV, respectively) and 25.1 ± 1.2, 7.2 ± 0.8, and $5.2 \pm 0.6 \ 10^{-6}$ photon cm^{-2} s^{-1} below the plane ($-20° < b < -5°$). Combining the two regions, one obtains flux ratios of $4.1:1.2:1$ (low:medium:high energies). These values can, for instance, be compared with emissivity ratios derived for the solar vicinity, which are typically $1.6:1.1:1$ (40, 228, 232; see Section 4.2). The latter values are in good agreement with the expectation from the sum of $\pi°$-decay and bremsstrahlung emission for the local electron/proton ratio; using the average of the spectra presented in Figure 2, one obtains $1.63:1.03:1$. The spectral softness of the excess is evident. Figure 10 suggests that the excess for the 70–150 MeV interval extends over a wider longitude range than the excess at higher energies. If the zero levels for the three energy ranges are shifted such that optimum fits are obtained for the second and third quadrants (although there is a priori no reason for such a shift), then the

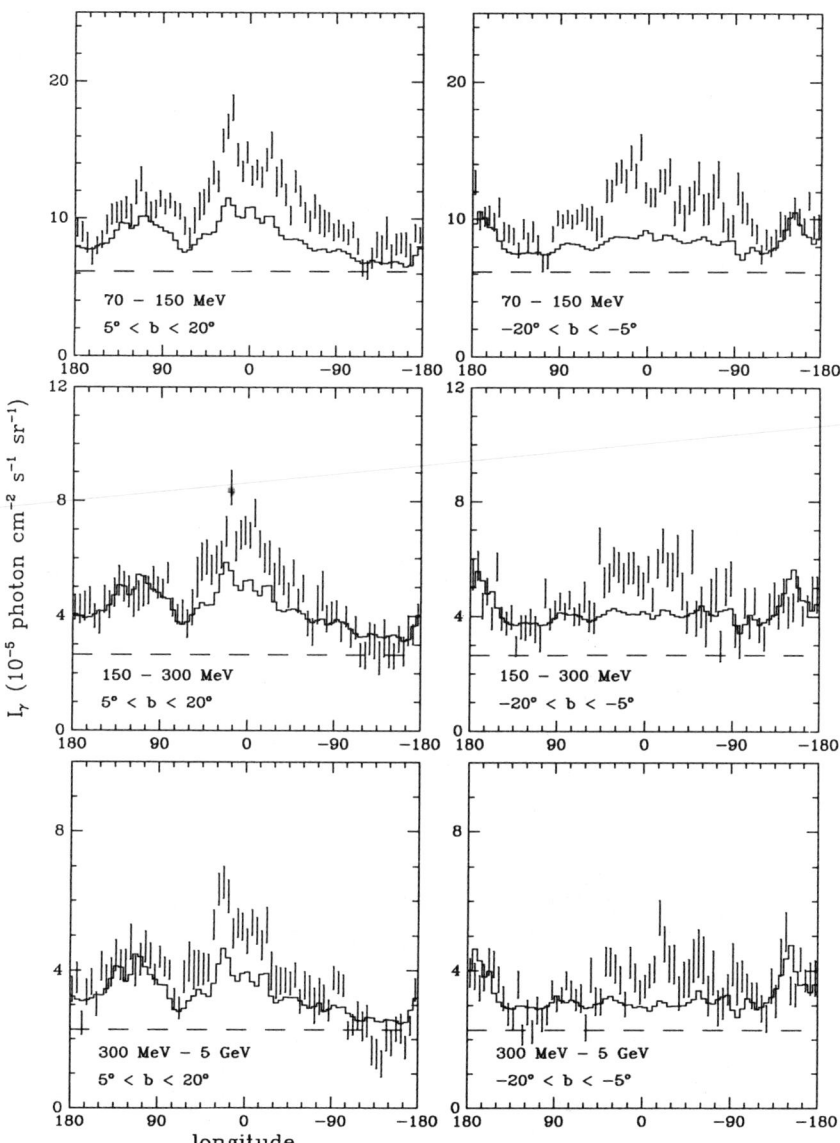

Figure 9 Longitude distributions of the observed ($\pm 1\sigma$ error bars) and modeled (histogram) γ-ray intensities at intermediate latitudes. (*Left*) Above the Galactic plane, $5° < b < 20°$. (*Right*) Below the plane, $-20° < b < -5°$. The model predictions include only CR interactions with atomic and molecular gas and were derived from an extension of the model of the Galactic disk emission presented by Strong et al. (232), which is shown in Figure 3. The dashed line indicates the isotropic background level, also from (232).

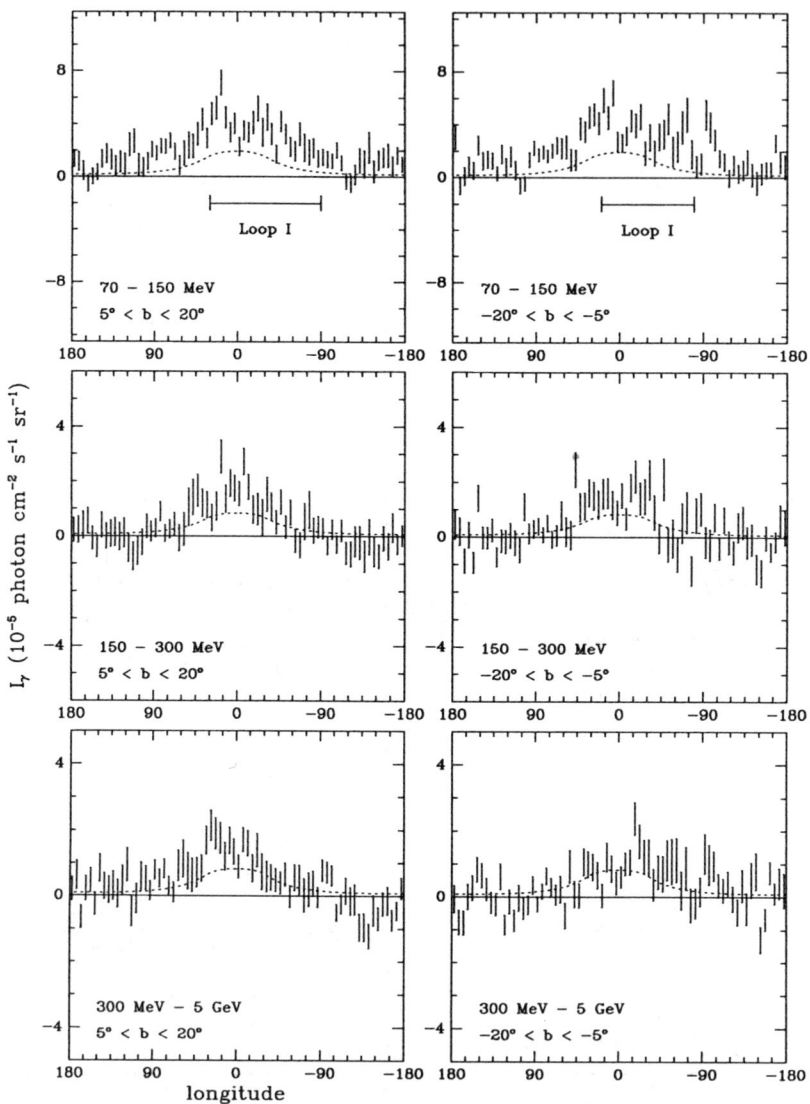

Figure 10 Longitude distributions of the difference between the observed and modeled γ-ray intensities at intermediate latitudes, as shown in Figure 9. (*Left*) Above the Galactic plane, $5° < b < 20°$. (*Right*) Below the plane, $-20 < b < -5°$. The dotted line shows the prediction from the IC model used by Strong et al. (232). In the upper two figures, the horizontal bars centered on $l \approx 330°$ indicate the extent of the Loop I region.

excess fluxes for the first and fourth quadrants are 16.2, 8.4, and 6.6 10^{-6} photon cm^{-2} s^{-1} above the plane and 16.0, 7.8, and 4.9 10^{-6} photon cm^{-2} s^{-1} below the plane. Combining the two regions gives flux ratios of 2.8 : 1.4 : 1; the softness is less pronounced but still very clear.

Several possibilities have been put forward to explain the γ-ray excess:

1. Some authors (228, 234) have attributed a significant fraction of the excess to IC emission. Figure 10 shows that the prediction from the IC model used by Strong et al. (232) explains about half of the excess intensities—most others estimated even lower IC intensities at medium latitudes. On the other hand, there appears to be enough uncertainty in the high-energy electron spectrum and distribution to attribute the entire excess to IC emission (38). In addition, an IC interpretation can account for the soft spectrum. For an E^{-3} electron spectrum (as used in Figure 10), which leads to approximately an E^{-2} IC γ-ray spectrum (Section 2.1), the IC emissivity ratios are 2.4 : 1.1 : 1. For an $E^{-3.4}$ spectrum, the ratios are 3.6 : 1.3 : 1.

2. Others (14, 150) noted that the excess is possible evidence for an enhanced CR density inside Loop I, which has a radius of $\sim 60°$ centered on $l \approx 330°$, $b \approx 18°$ (11). Figure 10 suggests, however, that the angular extent of the excess is larger than that of Loop I, particularly at low energies, but it remains a very attractive explanation. The Durham group [Bhat et al. (14), with a recent update given by Van der Walt & Wolfendale (246)] has considered this possibility in detail. They derive an excess γ-ray intensity for Loop I of 4.9×10^{-6} photon cm^{-2} s^{-1} sr^{-1} above 100 MeV, which agrees with the predicted value of 7×10^{-6} as given by Blandford & Cowie (29). However, the intensity of the excess in Figure 10 is on average almost an order of magnitude larger, namely $\sim 4.5 \times 10^{-5}$ photon cm^{-2} s^{-1} sr^{-1} (>100 MeV); the Durham group attributed about 90% of the total excess emission to a local CR gradient.

Rogers & Wolfendale (205) have made the interesting claim that also Loop III and the Vela supernova remnant are visible in the *COS-B* data, and that the observed emission is close to theoretical predictions. The detections are not firm, particularly for Loop III, which is claimed to be detected at an intensity level of 10% of the instrumental background. Detailed studies of the Vela pulsar by Kanbach et al. (128) and Grenier et al. (97) show no evidence of steady γ-ray emission from the pulsar environment.

3. Another possible reason for the excess is the omission of ionized gas in the γ-ray prediction (39). There is evidence, particularly from pulsar dispersion-measure data, for the existence of a widespread ionized medium with a scale height of the order of 1 kpc and a midplane density of ~ 0.03

cm^{-3} (63, 139, 160). The contribution of this gas to the total gas column density near the midplane is only a few percent, but it may be comparable to N(H I) at medium latitudes and thus sufficient to explain the excess flux. [A uniform layer with density 0.03 cm^{-3} and a thickness of 2 kpc would provide $\sim 0.5N$(H I) at $|b| \approx 15°$.] There is not much information on the Galacto-centric distribution of this medium. Obviously, in order to explain the γ-ray excess this gas should not extend far beyond the solar circle; such a scenario seems plausible because the gas is probably ionized by O stars (164), which are mainly concentrated inside the solar circle. Unless the CR electron-proton ratio is relatively high in this ionized medium (i.e. the bremsstrahlung contribution is relatively high), the spectral softness is hard to understand. The absence of screening in the bremsstrahlung process helps (Secion 2.1) but seems insufficient. It is possible that the ionized gas did not show up in the galaxy counts because of the relatively high destruction rates of interstellar grains in this medium due to the high shock frequencies that can be expected here (74, 75, 171).

It is not clear whether this soft γ-ray excess is physically related to the hard γ-ray spectra (above 300 MeV) found at medium latitudes in the outer-Galaxy direction. The counting statistics at these high energies are not good enough to investigate whether the intense emission above ~ 800 MeV correlates with the gas distribution or whether it is actually an excess.

3. RELEVANCE TO STUDIES OF MOLECULAR GAS

3.1 *The CO-H_2 Calibration*

There has been considerable debate on the use of CO ($J = 1 \rightarrow 0$) observations as a tracer of molecular gas [see, e.g., the discussions by Lequeux (151), Liszt (156), Dickman et al. (70), Verter (247), and Maloney & Black (162)]. It is not evident that this CO transition can be used at all to obtain H_2 column densities because it is optically thick for most molecular clouds. Apart from the γ-ray evidence discussed below, probably the main empirical justification for using velocity-integrated CO emission (W_{CO}) as a tracer of cloud masses is that the line width has been observed to increase with cloud size (67, 143, 174, 207, 220). For the Galactic cloud ensemble, W_{CO} may then trace $N(H_2)$ because shadowing of individual clouds is rather unimportant as a result of differential Galactic rotation. Although the $N(H_2)$–W_{CO} conversion factor X may vary from cloud to cloud owing to temperature and abundance variations [see, for instance, the discussion by Maloney & Black (162)], a measurement of an average value is useful for the determination of H_2 masses on a Galactic scale. Gamma-ray observations have this potential and provide some insight in a possible large-scale (radial) variation of X throughout the Galaxy. Such a gradient may exist because abundance gradients have been found (e.g. 154, 183), although X may not be a strong function of CO abundance (141, 162, 220), and a temperature gradient may be present.

It was already illustrated in Sections 2.2 and 2.3 how γ-ray observations provide a way of determining X. Although this method has the virtue that it does not require any assumptions on excitation, abundance, optical depth, or virial equilibrium, there are two potential problems—namely, (*a*) the possible existence of unresolved γ-ray point sources (although their contribution seems small; see Section 2.3), and (*b*) the possibility that the CR density in molecular clouds may differ from that in the ambient medium. With regard to (*a*), if a population of unresolved (genuine) γ-ray point sources exists, with an angular distribution similar to that of CO, then X is overestimated (i.e. in the terminology of the previous section, $X < Y$). With regard to point (*b*), there are the possibilities of incomplete CR penetration of dense clouds and, alternatively, CR production/concentration in molecular clouds. Theoretical work indicates that only low-energy CR protons [estimates range from $E_p \lesssim 300$ MeV (218) to $\lesssim 50$ MeV (59)], which are irrelevant for the production of high-energy γ rays, fail to penetrate a dense cloud completely. The alternative, an enhanced CR density in clouds (particularly CR electrons), cannot be excluded (see Section 4.3). So in this case, again, $X < Y$.

Two main groups have used the *COS-B* observations to calibrate the CO-H$_2$ relationship, namely the Durham group of Wolfendale and co-workers, and the *COS-B* Collaboration together with the Columbia CO group. The methods of analysis are different and are discussed here separately. In this discussion "Y" is reserved for the parameter actually determined from the γ-ray data; "X" is used when either the Y value determined is claimed to be a good estimate of the real $N(H_2)$-to-W_{CO} ratio or some interpretation is involved (such as corrections for CR enhancements or point sources).

3.1.1 APPROACH OF THE COS-B GROUP   The applied method has been presented in Section 2.3. In summary, Y and γ-ray emissivities were determined simultaneously from a correlation analysis of H I, CO, and γ-ray sky maps. A first paper by Lebrun et al. (147) concentrated on the first Galactic quadrant. It was restricted to the energy range 300 MeV–5 GeV (for this work, a constant γ-ray emissivity was assumed) and gave $Y \approx 3 \times 10^{20}$ molecule cm^{-2} (K km s^{-1})$^{-1}$. The next work [by Bloemen et al. (40)] dealt mainly with the first and second quadrants, giving $Y = (2.75 \pm 0.35) \times 10^{20}$ (70 MeV–5 GeV). A third paper [by Strong et al.

(232)] dealt with the whole Galaxy, giving $Y = (2.3 \pm 0.3) \times 10^{20}$ (150 MeV–5 GeV). The difference between the last two estimates, although statistically insignificant, results mainly from the different γ-ray energy intervals used; the third paper, in particular, showed evidence for a larger Y value in the low-energy interval (70–150 MeV), namely $\sim 3.3 \times 10^{20}$, as discussed in Section 2.3. For several reasons (232) the Y value obtained for these low energies is the least reliable indicator of X, so the authors stated that the 150 MeV–5 GeV range is to be preferred for estimating X. Recently, Melisse & Bloemen (172) have extended the work of Strong et al. by allowing for a possible CR-matter coupling. They found that such a coupling cannot be excluded by the data (see Section 4.4), but interestingly, approximately the same Y value results.

Fitting a more extended model to the data, Bloemen et al. and Strong et al. found that the ratio $Y(2-8 \text{ kpc})/Y(>8 \text{ kpc})$ is not significantly different from unity (40, 232). Hence, there is no indication for a significant difference between the average X value for $R = 2-8$ kpc and that found for $R > 8$ kpc. Variations on smaller scales cannot be excluded, and the Galactic-center region may deviate (see Section 3.3).

For the reasons given above, the best Y estimate of $\sim 2.3 \times 10^{20}$ should strictly be regarded as an upper limit on X. Although the γ-ray data do not provide a stringent lower limit, there are several reasons to believe that X is not much smaller than 2.3×10^{20}—namely, (a) the good agreement between model and data; (b) the agreement between the resulting Y value and the one obtained from a similar analysis of the Orion region (37), where the source contribution is probably negligible; and (c) the similar result obtained by using the model that allows for CR-matter coupling.

The Y estimates correspond to the Columbia CO data, so systematic differences between the Columbia data and other surveys should be taken into account if these estimates are applied to other surveys. Furthermore (and probably needless to say), it is important to keep in mind that the best Y estimate of 2.3×10^{20} molecule cm^{-2} (K km s^{-1})$^{-1}$ is an average value for the Galactic cloud ensemble and may not be directly applicable to individual clouds and other galaxies.

3.1.2 APPROACH OF THE DURHAM GROUP Bhat et al. (17) have given a comprehensive account of the Durham analyses regarding the CO-H$_2$ relationship. An update is given by Wolfendale (254). This section is restricted to their γ-ray studies. With regard to studies of the Galaxy at large, the most important difference between the Durham approach and that of the COS-B group is that the Durham group does not determine simultaneously $\varepsilon_\gamma(R)$ and Y. They prefer to use independent information on $\varepsilon_\gamma(R)$, which is largely dictated by the fact that the Durham group does not use the large-scale Columbia survey but rather CO observations restricted to regions within $\sim 1°$ from the Galactic plane. A complete correlation analysis, as done by the *COS-B* group, is therefore not feasible. The radial distribution of ε_γ used in their work is mainly based on a correlation analysis of γ-ray data and H I data for selected disk regions at $|b| \gtrsim 2.5°$ where the H_2 contribution to the gas column density was expected to be small (122). This method is necessarily restricted to Galactocentric radii between ~ 7 and ~ 13 kpc, but the resulting emissivity distribution was argued to be similar to that of the surface density of supernova remnants, and the latter was therefore taken to be representative of $\varepsilon_\gamma(R)$ throughout the Galaxy. Combining this emissivity distribution with radial unfoldings of the γ-ray, H I, and CO data [the latter taken from Sanders et al. (208)], Bhat et al. (13) found $Y \approx 0.7 \times 10^{20}$ molecule cm^{-2} (K km s^{-1})$^{-1}$ at $R \approx 6$ kpc. Bhat et al. (13, 17) claim independent support for this low X value in the inner Galaxy from X-ray absorption, infrared observations, and the virial theorem (update in 161), which led to their best estimate of $X \approx 1.0 \times 10^{20}$ molecule cm^{-2} (K km s^{-1})$^{-1}$ at $R \approx 5$–6 kpc.

The Durham group has derived an average X value for the solar vicinity from studies of several individual molecular clouds and cloud complexes. The γ-ray emissivity was mostly adopted to be the mean value of a number of determinations at intermediate latitudes, obtained from correlation analyses of γ-ray and galaxy-counts data [as summarized in (237)]. Bhat et al. (13) compared LTE-mass estimates of molecular clouds with the fluxes of γ-ray excesses at or near the positions of these clouds [an update of the work of Issa & Wolfendale (125)] and found $Y \approx 1.2 \times 10^{20}$. For Orion clouds A and B, Houston & Wolfendale (117) found $Y = (1.9 \pm 0.5) \times 10^{20}$ from a correlation study of γ-ray intensity and W_{CO}. Richardson & Wolfendale (199, 200), dealing with the molecular clouds in Taurus, Cepheus, and Orion, have reported that Y shows a falloff with increasing γ-ray energy, ranging from ~ 2–2.5×10^{20} at 100 MeV to $\sim 1.5 \times 10^{20}$ at 1 GeV. The evidence for this effect is at a marginal level of significance, but it is in qualitative agreement with their earlier work on the inner Galaxy (201) and with the relatively high Y value for the 70–150 MeV range found by the *COS-B* group. The best Durham estimate of the local X value obtained from this work is $\sim 1.5 \times 10^{20}$ molecule cm^{-2} (K km s^{-1})$^{-1}$.

3.1.3 COMPARISON Although the results from the two groups are not drastically different, there is some discrepancy, at least in the conclusions. Irrespective of the differences, however, it is clear that γ-ray observations favor rather low X values [other principal determinations are compiled in, e.g., (40)]. The most important discrepancy between the findings of the two groups corresponds to the so-called molecular ring at 4–7 kpc from the Galactic center, where most of the molecular gas in the Galaxy is located. The Durham group obtained a significantly smaller Y value here than the *COS-B* group. In addition, the Durham group claims that Y (as well as X) is smaller in the inner Galaxy than in the solar vicinity, whereas the *COS-B* group finds no significant evidence for such a radial dependence. The reason for the differing results appears to be twofold: (*a*) The radial γ-ray emissivity gradient adopted by the Durham group for this work is steeper than the one derived by the *COS-B* group, and (*b*) the specific CO surface densities used by the Durham group [namely those presented by Sanders et al. (208)] are relatively high in comparison with the Columbia results. With regard to (*a*), the Durham group assumes an emissivity increase by a factor of ~ 2.2 between $R = 10$ and 5 kpc (13), whereas the *COS-B* group finds an increase of only $\sim 25\%$ (Figure 5). Point (*b*) is discussed in detail by Bronfman et al. (44), who showed that the high CO surface densities presented by Sanders et al. can be attributed to the following two facts. Firstly, the W_{CO} values of the survey used by Sanders et al. are simply $\sim 20\%$ higher than those of the Columbia survey. Secondly, there is some debate among CO observers on the radial-unfolding procedure to be applied; using the procedure advocated by Bronfman et al. would lead to an average CO surface density in the inner Galaxy that is $\sim 40\%$ lower than that derived by Sanders et al. The method used by the *COS-B* group does *not* depend on such unfolding procedures. Altogether, these effects account for a correction factor of $(2.2/1.25) \times 1.2 \times 1.4 \approx 3.0$ (although the factor of 1.4 corresponding to the unfolding procedures may, of course, not be applicable). This renders the difference between the derived Y values fully understandable.

3.2 The Molecular Gas Content of the Galaxy

The good correlation between the observed γ-ray intensity distribution of the Milky Way and the expectation from H I and CO observations, with a simple linear $N(H_2)$–W_{CO} relationship, indicates that CO luminosity is a good large-scale mass tracer. The *COS-B* group estimated a total H_2 mass in the inner Galaxy (2–10 kpc) of $\sim 1.0 \times 10^9 \ M_\odot$ (40, 232), which is, strictly speaking, an upper limit; the Durham group found $\sim 0.6 \times 10^9 \ M_\odot$ (13, 17). For comparison, Henderson et al. (108) derived an H I mass in the inner Galaxy of $\sim 0.9 \times 10^9 \ M_\odot$, and thus the γ-ray observations favor rough equality in the H I and H_2 masses in the inner Galaxy (within a few hundred parsecs from the midplane). The *COS-B* mass estimates from the Columbia data have been derived in two different ways, both of which led to approximately the same result—namely, by scaling to the H I mass in the inner Galaxy [for details, see (40)] and by scaling the H_2 mass derived by Bronfman et al. (44) from the Columbia data with the updated X value. In both cases, the obtained H_2 mass is not affected by uncertainties in the absolute CO and γ-ray intensity calibrations.

The γ-ray observations clearly favor the lowest H_2 masses obtained in previous studies [see, e.g., the compilation given in (40)]. Bloemen et al. (40) found that the H_2 mass even in the 2–8 kpc annulus exceeds the H I mass by not more than $\sim 35\%$. The H_2/H I mass ratio may be higher or lower at the peak of the molecular ring depending on the actual variation of X within the 2–8 kpc annulus. Outside the solar circle, for $R < 15$ kpc, the γ-ray results indicate that $M(H_2)$ is $\sim 20\%$ of the total gas mass, namely $\sim 2.5 \times 10^8 \, M_\odot$ (36, 40), compared with $M(H\,I) \approx 1.1 \times 10^9 \, M_\odot$ (108). Altogether, the γ-ray observations show that the total H_2 mass in the Galaxy does not exceed $1.2 \times 10^9 \, M_\odot$, compared with a total H I mass of $\sim 4.8 \times 10^9 \, M_\odot$ (108).

3.3 *The Galactic Center*

The central 400 pc of the Milky Way (here referred to as the Galactic center) contains a very large column density of CO and other molecules that is about 5–10 times greater than that of a typical line of sight through the entire Galactic disk (e.g. 181, and references therein). This concentration of molecules can be expected to show up as a strong peak in the γ-ray observations, if X and ε_γ near the Galactic center are similar to the values measured for the Galactic disk. (This assumption was made in Figure 3.) Blitz et al. (30) analyzed the *COS-B* data and showed that such an excess is lacking, or, more specifically, that the γ-ray flux ($E_\gamma > 300$ MeV) from the Galactic center is nearly an order of magnitude smaller than expected on this basis. Bhat et al. (13) have reached a similar conclusion. The derived (3σ) upper limit of 4×10^{-7} photon cm^{-2} s^{-1} (300 MeV–5 GeV) on the γ-ray flux from the Galactic center is lower than the values obtained from earlier studies of the *SAS-2* and *COS-B* data (4, 122), primarily because of the much larger CO and γ-ray data bases that are available now. The new result implies that in the Galactic center either the γ-ray emissivity is anomalously low or H_2 is nearly an order of magnitude less abundant than estimates made from CO observations with an average Galactic X value. The former suggests that the density of CR protons with GeV energies is anomalously small relative to the local value, or perhaps that these cosmic rays do not efficiently penetrate the molecular clouds. The latter implies that the 3σ upper limit to the H_2 mass of the Galactic center is only $\sim 6 \times 10^7 \, M_\odot$, which is at least an order of magnitude smaller than that previously advocated by millimeter-line observers (e.g. 107, 155, 208).

Although it could not be decided from these γ-ray analyses whether the

Galactic center γ-ray deficit is due to a low γ-ray emissivity or a low X value, there is some circumstantial evidence in favor of a low X value, as discussed by Blitz et al. (30) and Stacy et al. (221). On the other hand, for the highest energies (1–5 GeV) observable with *COS-B*, Silk & Bloemen (215) did not find evidence for a γ-ray deficiency, but this finding needs to be studied in further detail. Since the π°-decay γ-ray emission at $E_\gamma > 1$ GeV originates largely from CR protons with energies $E_p \gtrsim 5$ GeV (see, e.g., 69), these results may indicate that the CR spectrum near the Galactic center is "normal" above ~ 5 GeV but flattens off at lower energies in comparison with the average spectrum in the Galactic disk. A possible explanation is that the Galactic center produces a wind, which tends to exclude low-energy particles and modifies the spectrum of the cosmic rays propagating near the Galactic center, just as the solar modulation affects local cosmic rays. This possibility requires further attention.

4. RELEVANCE TO STUDIES OF COSMIC RAYS

4.1 *Some Background*

The CR particle flux measured near Earth consists principally of protons, to which helium nuclei contribute about 10% and heavier nuclei and electrons (+positrons) each add less than a few percent. A comprehensive review of the CR composition and its energy dependence is given by Simpson (217). The abundance distribution of cosmic rays is similar to that of the solar system and the local region of the Galaxy, but there are also some significant deviations. The most obvious deviations are the overabundances in cosmic rays of the spallation products lithium, beryllium, boron, and the subiron elements ($20 < Z < 26$), which are largely produced by CR interactions with interstellar matter. These secondary particles play an important role for the understanding of CR propagation and confinement [see, e.g., the review by Cesarsky (55)]. Several other deviations remain after the fragmentation processes in the ISM have been taken into account. These differences give important clues on the origin of cosmic rays. An extensive review is given by Meyer (173).

At least in the solar vicinity, the energy density of cosmic rays is comparable to that of the magnetic field, the interstellar gas (both thermal and macroscopic motions), and the interstellar photon field (all ~ 1 eV cm^{-3}). Ever since the classical work of Parker (184–186), it has been well known that the cosmic rays and magnetic field may be of essential importance for understanding the structure and dynamics of the ISM. The CR energy density can largely be ascribed to protons with energies between about 1 GeV and a few tens of GeV, which also generate probably the bulk of the observed high-energy γ rays at $E_\gamma \gtrsim 100$ MeV (Section 2.1). Clearly, γ-ray observations trace an interesting part of the CR spectrum. One important aspect is the unique possibility to study the Galactic distribution of these particles. Furthermore, γ-ray spectra reflect the CR spectral distribution, which contains information on CR propagation characteristics. At γ-ray energies below ~ 200 MeV, the observations trace the electron spectrum below ~ 1 GeV, which adds useful information to CR electron studies based on low-frequency radio observations of the diffuse Galactic synchrotron emission and to direct CR electron measurements near Earth.

Shock acceleration (7, 10, 27, 138), i.e. first-order Fermi acceleration in strong shocks induced mainly by supernova remnants, is at present the favored mechanism for the production of cosmic rays with total energies up to $\sim 10^5$ GeV [see, e.g., the review by Axford (6)]. It produces quite naturally a power-law CR spectrum with a spectral index of 2 or somewhat larger, which is consistent with the CR source spectrum inferred from observations, and it can account for the power required to produce the observed CR energy density. Cosmic rays may get reaccelerated when encountering shocks in the ISM (28) or by second-order Fermi acceleration in interstellar turbulence (81); continuous acceleration, occurring solely during propagation, is, however, very hard to reconcile with the observed abundance ratios of secondary and primary particles and with the observed power-law shape of CR spectra (64, 65, 77, 102). Cesarsky (56) has reviewed the state of the art of this topic. Although the nonlinear aspects of shock acceleration are not fully understood yet (e.g. 249), the mechanism is very attractive and the problem of the origin of cosmic rays ($\lesssim 10^5$ GeV) seems to be shifting towards finding the CR injectors, i.e. the sources that provide the CR material and speed it up to moderate suprathermal energies. On the basis of energetics as well as CR composition, stellar flares and winds are considered to be likely candidates (3, 54, 173), although others have argued that CR injectors are not needed and that shocks can (or have to) accelerate particles directly out of an interstellar plasma (5, 76, 77).

There is no conclusive proof of the confinement region of cosmic rays. At least the CR electrons have a Galactic origin, because they cannot survive the Compton losses in the microwave background. It is in principle possible, however, that (part of) the CR proton-nuclear component fills a much larger volume than that of a galaxy, such as a (super)cluster, as advocated by Burbidge (42, 46, 47). This possibility has frequently been criticized, particularly by Ginzburg and collaborators (e.g. 94, 95). As cosmic rays are closely attached to the interstellar magnetic field lines because of their relatively small gyroradii (for 10-GeV protons in a field of a few μG, for instance, $r_g \approx 10^{-6}$ pc), their propagation is largely determined by the characteristics of the field. Starting with the work of

Pikelner (192), Shklovskii (213), and Ginzburg (92) in the early 1950s, many observational and theoretical studies have indicated that the distributions of the magnetic field and cosmic rays extend far beyond the Galactic disk. Direct evidence for the existence of such halos or thick disks follows from low-frequency radio-continuum observations of our Galaxy (8, 12, 43, 189, 190) and some edge-on galaxies (1, 9, 78, 118, 119, 129, 130), with indirect support from the observed composition and spectra of cosmic rays. Several studies of the polarization of starlight and radio synchrotron emission and of the rotation measures of pulsars and extragalactic radio sources [reviewed by Sofue et al. (219) and Heiles (103)] have shown that the Galactic magnetic field consists of a systematic component, preferentially aligned parallel to the Galactic plane, and an irregular component of comparable strength. Owing to the presence of this tangled component, the CR particles will not only diffuse along the field lines, but also will spread around to other field lines (195). It has been argued that CR propagation can therefore be described in first-order approximation as isotropic diffusion (93). Alternatively, in a so-called dynamical halo model, the magnetic field and CR particles can be convected away from the disk in a Galactic wind (126, 127, 136, 182); cosmic rays might even help to power a wind (121). In addition to the CR density gradients on a Galactic scale, which can be expected from these processes, several scenarios that lead to small-scale gradients have been proposed, such as trapping of cosmic rays near their sources (66), in spiral arms (188), or in tunnels (210). Cesarsky (55) has discussed these transport and confinement models in detail. We see below that γ rays give some insight into the role of the diffusion and convection processes and the confinement of cosmic rays.

4.2 *Cosmic Rays in the Solar Vicinity*

Given the γ-ray emissivity spectrum of the local ISM, some information can be deduced on the local CR electron and proton spectra. This is particularly useful for the CR electrons, because the low energies that can be studied this way (50 MeV $\lesssim E_e \lesssim$ 500 MeV) can barely be addressed by direct CR measurements (as a result of strong solar modulation for $E_e <$ 1 GeV) and radio measurements (as a result of strong free-free absorption for frequencies below $\sim$5 MHz, corresponding roughly to $E_e \lesssim$ 300 MeV). The shape of the electron spectrum between about 1 GeV and 10 GeV is reflected in the spectral shape of radio continuum observations between about 50 MHz and 5 GHz [the directions of the Galactic poles and anticenter are mostly chosen for these analyses (203, 251)] and can be normalized to the directly observed electron spectrum at $\sim$10 GeV.

The importance of γ-ray observations for the determination of the low-energy electron spectrum was first discussed in detail by Fichtel et al. (87). The constraints have successively been studied by Strong et al. (238), Strong & Wolfendale (236), Cesarsky et al. (58), Lebrun & Paul (149), Lebrun et al. (148), Gualandris & Strong (98), and Strong (229). It is not possible to discuss here the different approaches, but the basic idea should be clear from Section 2.1—namely, that given an estimate of the π°-decay γ-ray spectrum (based on CR proton measurements near Earth), the bremsstrahlung γ-ray spectrum and thus the electron spectrum can be obtained for a given measurement of the local γ-ray emissivity spectrum. A common conclusion drawn from these studies is that the γ-ray estimate of the low-energy electron flux at $E_e \simeq 100$ MeV lies above plausible extrapolations of the electron spectrum constructed from direct electron measurements and radio observations. In the course of these studies, however, the discrepancy has become smaller:

- The estimate of the low-energy electron flux from radio and direct electron measurements went up. The reason, as discussed by Webber (250), is that an uncertainty by a factor of about two existed in trying to normalize the demodulated electron spectrum and the radio spectrum at $E_e \simeq 10$ GeV, but the measurements have converged to the previous "high" spectrum.
- The γ-ray estimate of the electron flux went down. The main reason is that the γ-ray emissivity value for low energies has decreased; in the *COS-B* energy range 70–150 MeV, for instance, the values have changed from $\varepsilon_\gamma \simeq 1.4 \times 10^{-26}$ photon (H atom)$^{-1}$ s^{-1} sr^{-1} (35, 231) to (1.0–1.1) $\times 10^{-26}$ (40, 232, 234), largely because of the availability of more data and improvements in the analyses.

Table 1 summarizes the present best estimates of local γ-ray emissivities

Table 1 Local γ-ray emissivities

E (MeV)	Total[a]	Total[b]	π°-decay[c]	π°-decay[d]	π°-decay[e]	Bremsstrahlung[f]
			[10^{-26} photon (H atom)$^{-1}$ s^{-1} sr^{-1}]			
70–150	1.10±0.08	1.02±0.10	0.44	0.52	0.40±0.10	0.48
150–300	0.76±0.09	0.65±0.06	0.37	0.49	0.41±0.08	0.20
300–5000	0.68±0.09	0.62±0.06	0.45	0.49	0.56±0.09	0.10

[a] *Measurement*: From correlation study at medium latitudes ($10^\circ < |b| < 20^\circ$) (234).
[b] *Measurement*: From correlation study of the Galactic disk ($|b| < 10^\circ$) (232).
[c] *Prediction*: Derived by Stephens & Badhwar (227) (their B&S curve).
[d] *Prediction*: Derived by Stephens & Badhwar (227) (their M_U curve).
[e] *Prediction*: Derived by Dermer (69).
[f] *Prediction*: Derived from electron spectrum given by Webber (250, his Figure 1) with $E_e^{-2.4}$ extrapolation for $E_e < 300$ MeV.

from two different studies (near the Galactic plane and at medium latitudes), together with π°-decay and bremsstrahlung predictions. The π°-decay predictions are based on the demodulated proton spectrum; the differences originate from the different demodulation procedures and cross sections used. The bremsstrahlung estimates were derived from the maximum low-energy electron spectrum that Webber (250) considers acceptable. Measurements and predictions are clearly very close now.

4.3 *Radial Gradients*

If cosmic rays are of Galactic origin, then a significant radial gradient of the CR intensity can be expected on a large scale in the Galaxy, simply because more potential CR sources are present in the inner regions. Effective mixing of cosmic rays in the Galaxy (as in a halo diffusion model) may, however, lead to a much weaker falloff for the particle distribution than for the source distribution, so an absence of a strong gradient does not necessarily imply an extragalactic origin. Several studies of the radial CR distribution have been performed in the past, using the γ-ray emissivity as a measure of the CR intensity. The basic findings were discussed in Section 2.3. It is clear from that discussion that there has been some confusion due to uncertainties in the H_2 content of the Galaxy and to differences between recent results and those obtained from early studies of the *SAS-2* survey, before the release of the final data base. At present, the principal conclusion that can be drawn is that radial emissivity gradients are present over the Galaxy as a whole, for the entire γ-ray energy range studied, but the gradients are much weaker than found in early work. The radial emissivity distributions obtained from the latest and most robust analysis (232) were shown in Figure 5. Thus, the CR intensity appears to decrease weakly with Galacto-centric radius, although the electron and proton-nuclear components of relevance here may behave somewhat differently, as discussed below.

We saw in Section 2.3 that the emissivity gradient may be stronger for the low-energy band studied (70–150 MeV) than for the two other bands, but there was a question mark. To recall the situation: There is good evidence that the interpretation of the γ-ray data requires an energy-dependent model such that the γ-ray intensity spectrum toward the inner Galaxy is softer (i.e. a surplus of 70–150 MeV γ rays) than for the remainder of the disk, suggesting that the γ-ray emissivity spectrum is softer in the inner regions. However, it is still debatable whether (*a*) this holds only for the molecular clouds or (*b*) it is a ubiquitous phenomenon. Also, it cannot be excluded that a concentration of unresolved steep-spectrum γ-ray point sources in the inner Galaxy is responsible for the soft spectrum.

(a) ENERGY-INDEPENDENT GRADIENT/SOFT γ-RAY SPECTRUM FOR MOLECULAR CLOUDS If the gradient is assumed to be energy independent, then it follows that the radial (exponential) scale length for both the electrons and protons is as large as ~15 kpc for $R \gtrsim 4$ kpc (40, 232). In contrast, the objects that are generally considered to be potential CR sources, injecting the particles and/or accelerating them, such as supernova remnants, early-type stars, and in fact disk stars in general, all have radial scale lengths of typically ~5 kpc for $R \gtrsim 4$ kpc (135, 160, 165). In order to investigate whether this difference can be understood in a CR diffusion model, Dogiel & Uryson (73) and Bloemen & Dogiel (38) have recently numerically modeled the situation with a source distribution resembling that of supernova remnants and different choices for the dimensions of the CR halo. The diffusion coefficient ($\kappa = \kappa_0 E^\delta$, with $\delta \approx 0.5$) was assumed to be independent of position in the Galaxy. Their findings are summarized in Figure 11. It is clear that the difference between the strong radial gradient of the source distribution and the observed weak CR gradient can only be explained if a very large CR halo is present. Stecker & Jones (225) have

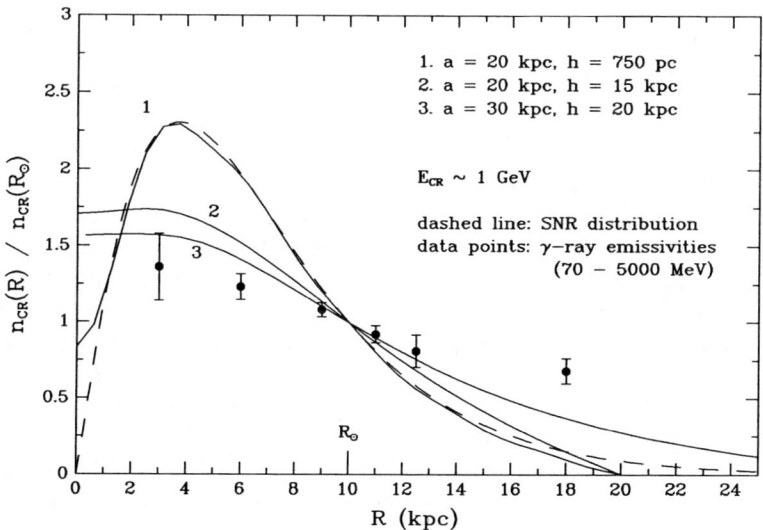

Figure 11 Examples of the Galacto-centric CR density distribution in a diffusion model with a supernova remnant–like CR source distribution and different halo dimensions (*a*: halo radius in the Galactic plane; *h*: halo extent outside the plane). For details, see Bloemen & Dogiel (38). The data points indicate the γ-ray emissivity values derived by Strong et al. (232) (the energy-independent case, shown in Figure 5). All quantities are normalized at the radius of the solar circle.

performed a similar analysis but concluded that thin halo models or source-dominated diffusion models provide a good fit to the data. This conclusion was based, however, on the stronger radial gradient deduced from the early *SAS-2* data base.

Wolfendale (253) has considered the possibility of an increase in the mean CR lifetime with increasing Galacto-centric distance as an explanation for the weak CR gradient, which may be related to the flaring of the gas disk in the outer Galaxy. In this scenario the shape of the CR confinement volume resembles that of the gas distribution, with the nearest escape boundary in the z-direction, so the distance to the CR escape boundary D increases with increasing R; hence, in simple terms, the effective diffusion velocity ($v \propto \kappa/D$) is smaller in the outer Galaxy and the mean CR lifetime ($\tau \propto D/v \propto D^2/\kappa$) is larger.

As noted above, the energy dependence of the γ-ray model is in this case attributed to phenomena related to molecular clouds (or steep-spectrum γ-ray sources, distributed like the molecular gas). A soft γ-ray spectrum for molecular clouds is most simply explained by an enhanced electron-to-proton ratio due to the production of secondary electrons [Brown & Marscher (45, 163)], which may be trapped [Cesarsky & Völk (59)] and even further accelerated [Morfill (175, 176)]. A useful review is given by Völk (248).

(*b*) ENERGY-DEPENDENT GRADIENT   A possible energy dependence of the radial emissivity distribution can be attributed to a stronger gradient for the electrons than for the protons. In this case, the proton distribution is even harder to understand in the framework of a diffusion model. Also, there is no straightforward explanation for the energy dependence. Again, as in the scenario depicted above, differing CR lifetimes in the inner and outer Galaxy may play a role, but in this case owing to the fact that the relevant electrons have lower energies than the protons. This possibility was considered by Bhat et al. (15) and Bloemen (33). Measurements of the abundance ratios between CR secondaries and primaries indicate that locally the CR lifetime is approximately proportional to $E^{-0.5}$ above a few GeV/nucleon and flattens off at lower energies [see, e.g., the review by Wefel (252), and references therein]. Thus the stronger gradient for electrons than for protons (i.e. a stronger gradient for low-energy particles) may indicate that the CR lifetime falls with E increasingly more slowly with increasing Galacto-centric distance; that is, in the inner Galaxy τ falls with E more rapidly than $E^{-0.5}$, and in the outer Galaxy it falls more slowly. In view of the probably widely different structures of the ISM in the inner and outer Galaxy [e.g. Heiles (104)], such a different energy dependence of τ might be explicable. For instance, nonlinear wave damping

(60), which produces a stronger energy dependence for the diffusion coefficient than the linear damping mechanism (140), may be important for $R \lesssim R_\odot$ but not in the outer Galaxy; this can possibly be ascribed to a radial falloff in the abundance of the highly ionized ($\sim 10^4$ K) medium (e.g. 139) in which the nonlinear mechanism may operate.

The γ-ray observations themselves, however, provide some counterevidence. If the energy dependence of the CR lifetime is indeed weaker in the outer Galaxy than in the inner Galaxy, this should show up in the γ-ray emission at high energies ($\gtrsim 300$ MeV), which reflects directly the shape of the CR proton spectrum (Section 2.1). In other words, the γ-ray spectrum at high energies should be steeper in the inner-Galaxy direction than in the outer-Galaxy direction. As discussed at the end of Section 2.3, Bloemen (33) indeed found evidence for this effect (see Figure 7) but had to conclude from follow-up work (39) that the relative flatness of the average outer-Galaxy spectrum above 300 MeV results from a gradual spectral flattening with increasing latitude (see Figure 8); near the Galactic plane, the high-energy γ-ray spectra of the inner and outer Galaxy are not significantly different.

Alternatively, the CR source spectrum may flatten with R, but here also a physical explanation is not evident.

It is clear from the above that option (b) is hard to understand physically. Fortunately, as discussed in Section 2.3, the most recent tests prefer option (a), although only marginally.

The term "CR gradient" should be used with care because the observed γ-ray intensities trace predominantly CR particles in those regions along the line of sight that contribute significantly to the total gas column density. For instance, in the local Galactic environment more than 80% of the gas mass seems to be concentrated in clouds with a filling factor <20%, so practically no CR information can be obtained here from γ-ray observations for the major part of space. The CR density may be lower as well as higher in the intercloud region without being noticed by γ-ray observations. Hence, the radial gradient of the *volume-averaged* CR density may be stronger (or weaker) than that derived from γ-ray observations.

4.4 Cosmic-Ray–Matter Coupling

It has been argued on theoretical grounds that the CR density is correlated with the matter density, but the scale on which this correlation may occur is uncertain. Bignami & Fichtel (22) have summarized some fundamental theoretical considerations and concluded that it may at least occur on the scale of spiral arms and large clouds. As discussed in Section 2.3, follow-up work showed that a model with CR-matter coupling on the scale of arms is indeed in agreement with γ-ray observations. Mainly the *SAS-2* data have been analyzed this way, although Fichtel & Kniffen (86) found reasonable agreement for the *COS-B* data as well. A radial gradient was not needed.

One can ask whether the above implies that the available γ-ray observations cannot distinguish between a "gradient model" and a "coupling model." Until recently, it was not very meaningful to compare the predictions from the two approaches because the coupling option was applied only to a model of the gas distribution, whereas studies of the gradient option made use of detailed H I and CO surveys, taking into account all structures in the data. New insight follows from the preliminary results of a study by Melisse & Bloemen (172), who are reconsidering CR-matter coupling models based on these surveys. They investigate whether the *COS-B* intensity distribution of the Milky Way can be described by a relation of the form

$$I_\gamma = \varepsilon_\gamma(\mathbf{r}_\odot) \int_l n_H(\mathbf{r}) \left(\frac{n_H(\mathbf{r})}{n_H(\mathbf{r}_\odot)}\right)^\alpha d\mathbf{r},$$

where $n_H(\mathbf{r})$ is the gas distribution, derived from CO and H I surveys, and the integration is along the line of sight. (Some modeling is required because of the distance-ambiguity problem.) The full velocity resolution of the surveys is used for the derivation of the gas densities, corresponding to length scales of typically 10–100 pc (i.e. scales of large clouds and cloud complexes). Free parameters are $\varepsilon_\gamma(\mathbf{r}_\odot)$, $n_H(\mathbf{r}_\odot)$, α, and the $N(H_2)$-to-W_{CO} ratio X. One of the main aims of this work is to put constraints on the value of the α exponent in the ρ_{CR}–ρ_{gas} relationship ($\rho_{CR} \propto \rho_{gas}^\alpha$), but this analysis has not been completed yet. As an example, a longitude profile of the modeled intensity distribution for $\alpha = 0.5$ is included in Figure 3; in order to enable a direct comparison with the gradient model of Strong et al. (232), the same IC model has been added. This figure suggests that the coupling model fits the data not significantly worse than the gradient model—statistical tests will show whether this is indeed the case. On the other hand, the preliminary results indicate that fits with $\alpha \gtrsim 0.5$ become increasingly worse. It is too early for further discussion on this point.

4.5 *Large-Scale Spectral Variations*

We saw in Section 4.3 that the large-scale variation of the γ-ray spectrum along the Milky Way does not necessarily imply that the shape of the CR spectrum changes: The most plausible explanation for the relatively high 70–150 MeV intensities in the inner-Galaxy direction is an enhanced electron density in molecular clouds. However, from the observed spectral variations of the γ-ray emission within the 300 MeV–5 GeV band (Section 2.3), it is very hard to avoid the conclusion that spectral variations for cosmic rays (protons) do exist. The hardening of the γ-ray spectrum with increasing latitude in the general direction of the outer Galaxy (39) suggests that the CR proton spectrum at GeV energies is flatter at medium latitudes, with a change in spectral index of ~ 0.4–0.6. Toward the inner Galaxy, this effect is not seen and the spectra are similar to that near the plane in the outer Galaxy (and consistent with direct measurements of the proton spectrum near Earth).

Bloemen & Dogiel (38) have investigated an alternative possibility— namely, whether the presence of an intense IC halo can explain the spectral behavior of the γ-ray data—triggered by the facts that a large CR halo may account for the weak CR gradient and that the IC γ-ray spectrum is relatively flat (Section 2.1). They found that the presence of an IC halo indeed leads to a hardening of the γ-ray spectrum with latitude; however, the observed effect is significantly stronger, and they could not explain the fact that it is only seen toward the outer Galaxy.

It is very interesting that a recent spectral analysis of radio-continuum surveys of the Galaxy (at 408 and 1420 MHz) by Reich & Reich (197, 198) shows a similar spectral flattening with latitude in the outer-Galaxy direction. In this case, GeV electrons are traced; the change in the radio spectral index is about 0.2–0.25, so the change in the electron spectral index is ~ 0.4–0.5 (if most of the radio emission has a nonthermal origin). Reich & Reich found some evidence for a similar effect toward the inner Galaxy, but it was not possible to perform a reliable analysis here because only Northern Hemisphere data are available at 1420 MHz and a large fraction of the available sky coverage is dominated by emission from Loop I. On the basis of radio-continuum data alone, a thermal origin of the flattening could not be excluded; however, this interpretation seems unlikely, since a similar effect is visible in the γ-ray data.

Although the interpretation of these findings is not straightforward, it is clear that a standard CR diffusion model, as described by Ginzburg & Syrovatskii (95), encounters problems: A spectral steepening with distance from the Galactic plane would be expected for the electrons, due to synchrotron and Compton radiation losses, and no spectral changes for the protons. At low radio frequencies a weak spectral steepening with latitude was indeed found in previous work [see, e.g., the review by Lawson et al. (144)]. A dynamical halo model (Section 4.1) may provide a viable explanation. The competition of spatial diffusion, convection, adiabatic deceleration, and (electron) radiation losses in such a model may in principle lead to the observed effects, but the problem is very complex and requires robust modeling. Reich & Reich (198) and Bloemen et al. (39)

noted that the findings are to a large extent in agreement with the asymptotic spectral predictions of the Galactic wind model given by Lerche & Schlickeiser (152, 153). Rogers et al. (204; see also 245, 246) have suggested that the spectral changes may have a local origin, related to our location on the inner edge of the Orion spiral arm and there being acceleration mechanisms at work within the arm such as to give a flatter CR spectrum inside the arm than in the interarm region. At the moment there certainly seems room for further speculation on the nature of this peculiar spectral behavior. Radio studies of other galaxies may shed new light on this topic in the near future.

5. GAMMA-RAY SOURCES

This section briefly reports on the status of searches for γ-ray point sources in the *COS-B* data base, mainly illustrating how considerable progress could (and still can) be made with the improved understanding of the diffuse emission from the Milky Way. A complete account of studies of individual sources is beyond the scope of this paper—a comprehensive review is given by Bignami & Hermsen (24). I skip all discussion of the work on source models—it would be hopelessly incomplete in the space remaining—but remind the reader that γ-ray sources may be a manifestation of the presence of CR sources, such as the interaction of a supernova remnant with an adjacent gas cloud [see, e.g., the review by Morfill & Tenorio-Tagle (177)]. The identification of such sources would enhance significantly our confidence in the interpretation of the large scale γ-ray distribution.

After some early claims based on balloon experiments, firm detection of γ-ray sources was first achieved with *SAS-2*: It confirmed the detection of γ rays from the Crab pulsar (134, 243), discovered the intense γ-ray emission from the Vela pulsar (241, 244) and an hitherto puzzling source in the Galactic anticenter (243; now known as "Geminga"), and reported periodic emission from Cyg X-3 (89, 142). The latter result is under debate—it is not seen by *COS-B* (111). Already the first *COS-B* observations showed evidence for several additional sources (114), which at that stage were simply defined as statistically significant peaks in the γ-ray intensity distribution, with an angular shape consistent with the *COS-B* point-spread function. An extensive account of the search for sources following this definition is given by Hermsen (109, 110); analysis of the data obtained during roughly the first half of the *COS-B* mission led to the so-called 2CG catalog (239), containing 25 γ-ray sources, including the Crab and Vela pulsars and Geminga. Crab and Vela are unambiguously identified by their timing signatures; their temporal and spectral characteristics have been analyzed in great detail [see (62) and (97), respectively, and references therein]. There have also been searches for pulsed emission from other sources, but no firm detections have been obtained. All other identifications have to be based solely on positional coincidence. Given the limited angular resolution, it is difficult to differentiate between intrinsically compact objects and objects that might be up to a few degrees in size. Uniquely identifying counterparts at other wavelengths is therefore practically impossible, but two proposed identifications are very convincing: the quasar 3C 273 (e.g. 20, 112) and the ρ Oph molecular cloud, which was in fact shown to be an extended γ-ray excess in subsequent work (113).

The unidentified sources of the 2CG catalog, except one (so 20 in total), are all located very close to the Galactic plane. This raised the question [one, in fact, already raised by Black & Fazio (26) long before the sources were detected] as to whether a significant fraction of the γ-ray sources are possibly of diffuse origin, i.e. are associated with CR-irradiated clouds, or, more precisely, with peaks in the gas column-density distribution [see, e.g., Li Ti pei & Wolfendale (159)]. With the availability of CO surveys it became feasible to take this diffuse emission into account in the search for sources. Therefore, the data obtained during the second half of the *COS-B* mission have never been used to update the 2CG catalog, applying the search method presented by Hermsen (109, 110). The new approaches are presented by Pollock et al. (193, 194) and Simpson & Mayer-Hasselwander (170, 216); they consist of searching for γ-ray excesses, consistent with the *COS-B* point-spread function, superimposed on a model for the diffuse emission based on H I and CO surveys, as described in Section 2.3. This implies a change in the original definition of a γ-ray source: Those original sources that find an explanation in terms of the same CR-matter interactions as the underlying diffuse emission are excluded. Remaining sources may of course still be of diffuse origin, possibly indicating an enhanced CR density somewhere along the line of sight.

Unfortunately, these new searches for γ-ray sources have not been completed yet. The results published so far suggest that about 30–50% of the 2CG sources can be attributed to structure in the gas distribution without requiring any CR density enhancement. On the other hand, there is some evidence for new sources. A 3CG catalog of γ-ray sources can soon be expected—possibly even more than one version, based on different methods for determining the statistical significance of the sources and the extraction of source parameters.

In order to leave the interested reader not entirely with empty hands at this stage, I have produced a "finding chart" of potential γ-ray sources for the energy range 100 MeV–6 GeV (Figure 12). Basically this map repre-

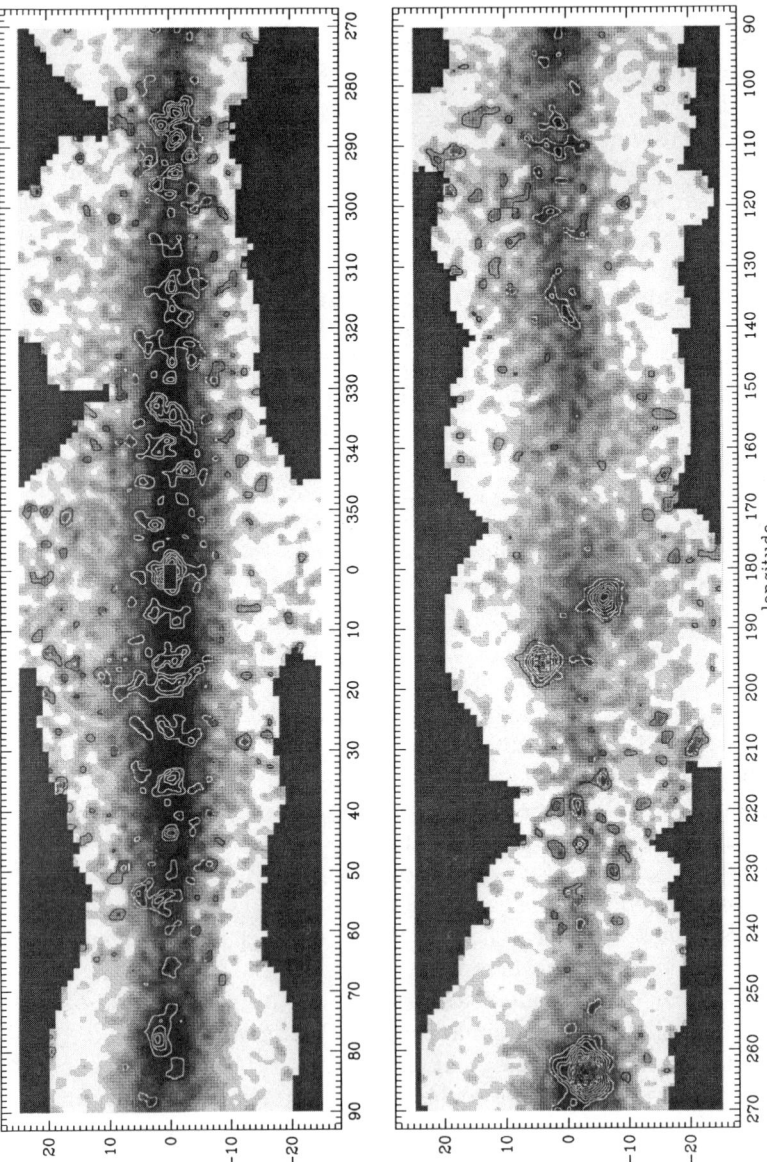

Figure 12 "Finding chart" for potential γ-ray sources (indicated by contours), derived from the *COS-B* γ-ray sky map in the energy range 100 MeV–6 GeV (Figure 1) and the map of predicted diffuse emission, based on H I and CO surveys and the model parameters deduced by Strong et al. (232) (Figure 4). The subtraction method is described in Section 5. The underlying gray-scale map represents the observed γ-ray sky. The small box in the Galactic-center direction indicates a region that was excluded from the analysis; the feature surrounding it is probably an artifact. The strong excesses near $l = 185°$, $195°$, and $263°$ correspond to the Crab, Geminga, and Vela γ-ray sources. Note the extended features visible in the map. Contour values: 8, 13, 18, ..., $\times 10^{-5}$ photon cm^{-2} s^{-1} sr^{-1}.

sents the γ-ray excesses that are left when the expected diffusion emission [based on the modeling of Strong et al. (232) (i.e. the map shown in Figure 4, convolved with the *COS-B* point-spread function)] is subtracted from the observed γ-ray intensity map shown in Figure 1. However, it is not a pure subtraction. In order to correct for intermediate-scale deviations (10–20°) between model and observations, the model intensities have first been readjusted to the observed intensities on this scale: For a given position, the predictions in a strip of 21 pixels in longitude direction and 1 pixel in latitude (pixel size is 0.5° × 0.5°), centered on this position, were scaled upward or downward, such that for only 20% of the pixels are the observed intensities below the predicted intensities. Clearly, this is rather arbitrary, and no formal statistical significance can be assigned to the excesses visible in the finding chart. Interestingly, in addition to pointlike features, extended "sources" can be seen, with the most pronounced ones being the structures near $l \simeq 19°$ (not coinciding with Loop I) and $l \simeq 334°$, both of which extend over at least 10–15°. I leave speculations on their origin to the reader. These features may turn out to be the cream of the *COS-B* data base, just hinting at what the next generation of γ-ray telescopes may show us. By the time this review appears in print, the Soviet-French γ-ray telescope *GAMMA-1* will hopefully be in orbit, and NASA's *Gamma Ray Observatory* should follow within half a year.

ACKNOWLEDGMENTS

I am grateful to V. A. Dogiel, V. L. Ginzburg, W. Hermsen, F. P. Israel, P. Maloney, A. W. Strong, and A. W. Wolfendale for critically reading the manuscript. In addition, I acknowledge the receipt of a Fellowship from the Royal Netherlands Academy of Arts and Sciences (KNAW) and partial support by the Laboratory for Space Research, Leiden.

Literature Cited

1. Allen, R. J., Baldwin, J. E., Sancisi, R. 1978. *Astron. Astrophys.* 62: 397
2. Arnaud, K., Li Ti pei, Riley, P. A., Wolfendale, A. W., Dame, T. M., et al. 1982. *MNRAS* 201: 745
3. Arnaud, K., Cassé, M. 1985. *Astron. Astrophys.* 144: 64
4. Audouze, J., Lequeux, J., Masnou, J. L., Puget, J. L. 1979. *Astron. Astrophys.* 80: 276
5. Axford, W. I. 1981. In *Origin of Cosmic Rays*, ed. G. Setti, G. Spada, A. W. Wolfendale, p. 339. Dordrecht: Reidel
6. Axford, W. I. 1987. *Proc. Int. Cosmic Ray Conf., 20th, Moscow*, 8: 120
7. Axford, W. I., Leer, E., Skandron, K. G. 1977. *Proc. Int. Cosmic Ray Conf., 15th, Plovdiv*, 11: 132
8. Baldwin, J. E. 1976. In *The Structure and Content of the Galaxy and Galactic Gamma Rays*, ed. C. E. Fichtel, F. W. Stecker, p. 206. Greenbelt, Md: Goddard Space Flight Cent.
9. Beck, R., Biermann, P., Emerson, D. T., Wielebinski, R. 1979. *Astron. Astrophys.* 77: 25
10. Bell, A. R. 1978. *MNRAS* 182: 147
11. Berkhuijsen, E. M. 1971. *Astron. Astrophys.* 14: 359
12. Beuermann, K., Kanbach, G., Berkhuijsen, E. M. 1985. *Astron. Astrophys.* 153: 17

13. Bhat, C. L., Issa, M. R., Houston, B. P., Mayer, C. J., Wolfendale, A. W. 1985. *Nature* 314: 511
14. Bhat, C. L., Issa, M. R., Mayer, C. J., Wolfendale, A. W. 1985. *Nature* 314: 515
15. Bhat, C. L., Mayer, C. J., Rogers, M., Wolfendale, A. W., Zan, M. 1986. *J. Phys. G* 12: 1087
16. Bhat, C. L., Mayer, C. J., Wolfendale, A. W. 1984. *Astron. Astrophys.* 140: 284
17. Bhat, C. L., Mayer, C. J., Wolfendale, A. W. 1986. *Philos. Trans. R. Soc. London Ser. A* 319: 249
18. Bignami, G. F. 1981. *Philos. Trans. R. Soc. London Ser. A* 301: 555
19. Bignami, G. F., Barbareschi, L., Bloemen, J. B. G. M., Buccheri, R., Caraveo, P. A., et al. 1981. *Proc. Int. Cosmic Ray Conf., 17th, Paris*, 1: 182
20. Bignami, G. F., Bennett, K., Buccheri, R., Caraveo, P. A., Hermsen, W., et al. 1981. *Astron. Astrophys.* 93: 71
21. Bignami, G. F., Boella, G., Burger, J. J., Keirle, P., Mayer-Hasselwander, H. A., et al. 1975. *Space Sci. Instrum.* 1: 245
22. Bignami, G. F., Fichtel, C. E. 1974. *Ap. J. Lett.* 189: L65
23. Bignami, G. F., Fichtel, C. E., Kniffen, D. A., Thompson, D. J. 1975. *Ap. J.* 199: 54
24. Bignami, G. F., Hermsen, W. 1983. *Annu. Rev. Astron. Astrophys.* 21: 67
25. Bignami, G. F., Piccinotti, G. 1977. *Astron. Astrophys.* 59: 233
26. Black, J. H., Fazio, G. G. 1973. *Ap. J. Lett.* 185: L7
27. Blandford, R. D., Ostriker, J. P. 1978. *Ap. J. Lett.* 221: L29
28. Blandford, R. D., Ostriker, J. P. 1980. *Ap. J.* 237: 793
29. Blandford, R. D., Cowie, L. L. 1982. *Ap. J.* 260: 625
30. Blitz, L., Bloemen, J. B. G. M., Hermsen, W., Bania, T. M. 1985. *Astron. Astrophys.* 143: 267
31. Bloemen, J. B. G. M. 1985. PhD thesis. Leiden Univ., Neth.
32. Bloemen, J. B. G. M. 1985. *Astron. Astrophys.* 145: 391
33. Bloemen, J. B. G. M. 1987. *Ap. J. Lett.* 317: L15
34. Bloemen, J. B. G. M. 1988. In *Genesis and Propagation of Cosmic Rays*, ed. M. M. Shapiro, J. P. Wefel, p. 163. Dordrecht: Reidel
35. Bloemen, J. B. G. M., Bennett, K., Bignami, G. F., Blitz, L., Caraveo, P. A., et al. 1984. *Astron. Astrophys.* 135: 12
36. Bloemen, J. B. G. M., Blitz, L., Hermsen, W. 1984. *Ap. J.* 279: 136
37. Bloemen, J. B. G. M., Caraveo, P. A., Hermsen, W., Lebrun, F., Maddalena, R. J., et al. 1984. *Astron. Astrophys.* 139: 37
38. Bloemen, J. B. G. M., Dogiel, V. A. 1989. In preparation
39. Bloemen, J. B. G. M., Reich, P., Reich, W., Schlickeiser, R. 1988. *Astron. Astrophys.* 204: 88
40. Bloemen, J. B. G. M., Strong, A. W., Blitz, L., Cohen, R. S., Dame, T. M., et al. 1986. *Astron. Astrophys.* 154: 25
41. Blumenthal, G. R., Gould, R. J. 1970. *Rev. Mod. Phys.* 42: 237
42. Brecher, K., Burbidge, G. R. 1972. *Ap. J.* 174: 253
43. Brindle, C., French, D. K., Osborne, J. L. 1978. *MNRAS* 184: 283
44. Bronfman, L., Cohen, R. S., Alvarez, H., May, J., Thaddeus, P. 1988. *Ap. J.* 324: 248
45. Brown, R. L., Marscher, A. P. 1977. *Ap. J.* 212: 659
46. Burbidge, G. R. 1974. *Philos. Trans R. Soc. London Ser. A* 277: 481
47. Burbidge, G. R. 1983. In *Composition and Origin of Cosmic Rays*, ed. M. M. Shapiro, p. 245. Dordrecht: Reidel
48. Burton, W. B. 1976. *Annu. Rev. Astron. Astrophys.* 14: 275
49. Burton, W. B. 1985. *Astron. Astrophys. Suppl.* 62: 365
50. Deleted in proof
51. Burton, W. B., te Lintel Hekkert, P. 1985. *Astron. Astrophys. Suppl.* 62: 645
52. Burton, W. B., te Lintel Hekkert, P. 1986. *Astron. Astrophys. Suppl.* 65: 427
53. Caraveo, P. A., Bennett, K., Bignami, G. F., Hermsen, W., Kanbach, G. 1980. *Astron. Astrophys.* 91: L3
54. Cassé, M., Goret, P. 1978. *Ap. J.* 221: 703
55. Cesarsky, C. J. 1980. *Annu. Rev. Astron. Astrophys.* 18: 289
56. Cesarsky, C. J. 1987. *Proc. Int. Cosmic Ray Conf., 20th, Moscow*, 8: 87
57. Cesarsky, C. J., Cassé, M., Paul, J. A. 1977. *Astron. Astrophys.* 60: 139
58. Cesarsky, C. J., Paul, J. A., Shukla, P. G. 1978. *Astrophys. Space Sci.* 59: 73
59. Cesarsky, C. J., Völk, H. J. 1978. *Astron. Astrophys.* 70: 367
60. Chin, Y., Wentzel, D. G. 1972. *Astrophys. Space Sci.* 16: 465
61. Chupp, E. L. 1976. *Gamma Ray Astronomy*. Dordrecht: Reidel
62. Clear, J., Bennett, K., Buccheri, R., Grenier, I. A., Hermsen, W., et al. 1987. *Astron. Astrophys.* 174: 85
63. Clifton, T. R., Frail, D. A., Kulkarni, S. R., Weisberg, J. M. 1988. *Ap. J.* 333: 332

64. Cowsik, R. 1980. *Ap. J.* 241: 1195
65. Cowsik, R. 1986. *Astron. Astrophys.* 155: 344
66. Cowsik, R., Wilson, L. W. 1975. *Proc. Int. Cosmic Ray Conf.*, *14th, Munich*, 2: 659
67. Dame, T. M., Elmegreen, B. G., Cohen, R. S., Thaddeus, P. 1986. *Ap. J.* 305: 892
68. Dame, T. M., Ungerechts, H., Cohen, R. S., de Geus, E., Grenier, I., et al. 1987. *Ap. J.* 322: 706
69. Dermer, C. D. 1986. *Astron. Astrophys.* 157: 223
70. Dickman, R. L., Snell, R. L., Schloerb, F. P. 1986. *Ap. J.* 309: 326
71. Dodds, D., Strong, A. W., Wolfendale, A. W. 1975. *MNRAS* 171: 569
72. Dogiel, V. A., Ginzburg, V. L. 1989. *Space Sci. Rev.* In press
73. Dogiel, V. A., Uryson, A. V. 1988. *Astron. Astrophys.* 197: 335
74. Draine, B. T., Salpeter, E. E. 1979. *Ap. J.* 231: 77
75. Draine, B. T., Salpeter, E. E. 1979. *Ap. J.* 231: 438
76. Eichler, D. 1979. *Ap. J.* 229: 419
77. Eichler, D. 1980. *Ap. J.* 237: 809
78. Ekers, R. D., Sancisi, R. 1977. *Astron. Astrophys.* 54: L973
79. Fazio, G. G. 1967. *Annu. Rev. Astron. Astrophys.* 5: 481
80. Fazio, G. G. 1973. In *X- and Gamma-Ray Astronomy*, ed. H. Bradt, R. Giacconi, p. 303. Dordrecht: Reidel
81. Fermi, E. 1954. *Ap. J.* 119: 1
82. Fichtel, C. E. 1977. *Space Sci. Rev.* 20: 191
83. Fichtel, C. E., Hartman, R. C., Kniffen, D. A., Thompson, D. J., Bignami, G. F., et al. 1975. *Ap. J.* 198: 163
84. Fichtel, C. E., Hartman, R. C., Kniffen, D. A., Thompson, D. J., Ögelman, H. B., et al. 1977. *Ap. J. Lett.* 217: L9
85. Fichtel, C. E., Hartman, R. C., Kniffen, D. A., Thompson, D. J., Ögelman, H. B., et al. 1978. *NASA Tech. Memo. 79650*
86. Fichtel, C. E., Kniffen, D. A. 1984. *Astron. Astrophys.* 134: 13
87. Fichtel, C. E., Kniffen, D. A., Thompson, D. J., Cheung, C. Y. 1976. *Ap. J.* 208: 211
88. Fichtel, C. E., Simpson, G. A., Thompson, D. J., 1978. *Ap. J.* 222: 833
89. Fichtel, C. E., Thompson, D. J., Lamb, R. C. 1987. *Ap. J.* 319: 362
90. Fichtel, C. E., Thrombka, J. I. 1981. *Gamma-Ray Astrophysics, New Insight into the Universe. NASA SP-453*
91. Fuchs, B., Schlickeiser, R., Thielheim, K. O. 1976. *Ap. J.* 206: 589
92. Ginzburg, V. L. 1953. *Usp. Fiz. Nauk* 51: 343
93. Ginzburg, V. L., ed. 1984. *Cosmic-Ray Astrophysics*. Moscow: Nauka (Engl. translation: 1989, North-Holland)
94. Ginzburg, V. L. 1987. *Proc. Cosmic Ray Conf.*, *20th, Moscow*, 7: 7
95. Ginzburg, V. L., Syrovatskii, S. I. 1964. *The Origin of Cosmic Rays*. Oxford: Pergamon
96. Gould, R. J. 1969. *Phys. Rev.* 185: 72
97. Grenier, I. A., Hermsen, W., Clear, J. 1988. *Astron. Astrophys.* 204: 117
98. Gualandris, F. L., Strong, A. W. 1984. *Astron. Astrophys.* 140: 357
99. Harding, A. K., Stecker, F. W. 1985. *Ap. J.* 291: 471
100. Hartman, R. C., Kniffen, D. A., Thompson, D. J., Fichtel, C. E., Ögelman, H. B., et al. 1979. *Ap. J.* 230: 597
101. Hayakawa, S. 1952. *Prog. Theor. Phys.* 8: 517
102. Hayakawa, S. 1969. *Cosmic Ray Physics*. New York: Wiley-Interscience
103. Heiles, C. 1987. In *Interstellar Processes*, ed. D. J. Hollenbach, H. A. Thronson, p. 171. Dordrecht: Reidel
104. Heiles, C. 1988. *Ap. J.* 324: 321
105. Heiles, C., Cleary, M. N. 1979. *Aust. J. Phys. Suppl.* 47: 1
106. Heiles, C., Habing, H. J. 1974. *Astron. Astrophys. Suppl.* 14: 1
107. Heiligman, G. 1982. PhD thesis. Princeton Univ., Princeton, N.J.
108. Henderson, A. P., Jackson, P. D., Kerr, F. J. 1982. *Ap. J.* 263: 182
109. Hermsen, W. 1980. PhD thesis. Leiden Univ., Neth.
110. Hermsen, W. 1983. *Space Sci. Rev.* 36: 61
111. Hermsen, W., Bennett, K., Bloemen, J. B. G. M., Buccheri, R., Jansen, F. A., et al. 1987. *Astron. Astrophys.* 175: 141
112. Hermsen, W., Bennett, K., Bignami, G. F., Bloemen, J. B. G. M., Buccheri, R., et al. 1981. *Proc. Int. Cosmic Ray Conf.*, *17th, Paris*, 1: 230
113. Hermsen, W., Bloemen, J. B. G. M. 1983. In *Surveys of the Southern Galaxy*, ed. W. B. Burton, F. P. Israel, p. 65. Dordrecht: Reidel
114. Hermsen, W., Swanenburg, B. N., Bignami, G. F., Boella, G., Buccheri, R., et al. 1977. *Nature* 269: 494
115. Higdon, J. C. 1979. *Ap. J.* 232: 113
116. Houston, B. P., Wolfendale, A. W. 1983. *Astron. Astrophys.* 126: 22
117. Houston, B. P., Wolfendale, A. W. 1985. *J. Phys. G* 11: 407
118. Hummel, E., Sancisi, R., Ekers, R. D. 1984. *Astron. Astrophys.* 133: 1
119. Hummel, E., Smith, P., van der Hulst,

J. M. 1984. *Astron. Astrophys.* 137: 138
120. Hutchinson, G. W. 1952. *Philos. Mag.* 43: 847
121. Ipavich, F. 1975. *Ap. J.* 196: 107
122. Issa, M. R., Riley, P. A., Strong, A. W., Wolfendale, A. W. 1981. *J. Phys. G* 7: 973
123. Issa, M. R., Strong, A. W., Wolfendale, A. W. 1981. *J. Phys. G* 7: 565
124. Issa, M. R., Wolfendale, A. W. 1981. *J. Phys. G* 7: L187
125. Issa, M. R., Wolfendale, A. W. 1981. *Nature* 292: 430
126. Jokipii, J. R. 1976. *Ap. J.* 208: 900
127. Jones, F. C. 1979. *Ap. J.* 229: 747
128. Kanbach, G., Bennett, K., Bignami, G. F., Buccheri, R., Caraveo, P. A., et al. 1980. *Astron. Astrophys.* 90: 163
129. Klein, R., Urbanik, M., Beck, R., Wielebinski, R. 1983. *Astron. Astrophys.* 127: 177
130. Klein, R., Wielebinski, R., Beck, R. 1984. *Astron. Astrophys.* 133: 19
131. Kniffen, D. A., Fichtel, C. E. 1981. *Ap. J.* 250: 389
132. Kniffen, D. A., Fichtel, C. E., Thompson, D. J. 1977. *Ap. J.* 215: 765
133. Kniffen, D. A., Hartman, R. C., Thompson, D. J., Bignami, G. F., Fichtel, C. E., et al. 1973. *Ap. J. Lett.* 186: L105
134. Kniffen, D. A., Hartman, R. C., Thompson, D. J., Bignami, G. F., Fichtel, C. E., et al. 1974. *Nature* 251: 397
135. Kodaira, K. 1974. *Publ. Astron. Soc. Jpn.* 26: 255
136. Kóta, J., Owens, A. J. 1980. *Ap. J.* 237: 814
137. Kraushaar, W. L., Clark, G. W., Garmire, G. P., Borken, R., Higbie, P., et al. 1972. *Ap. J.* 177: 341
138. Krymsky, G. F. 1977. *Dokl. Akad. Nauk SSSR* 234: 1306
139. Kulkarni, S. R., Heiles, C. 1987. In *Interstellar Processes*, ed. D. J. Hollenbach, H. A. Thronson, p. 87. Dordrecht: Reidel
140. Kulsrud, R. M., Pearce, W. P. 1969. *Ap. J.* 156: 445
141. Kutner, M. L., Leung, C. M. 1985. *Ap. J.* 291: 188
142. Lamb, R. C., Fichtel, C. E., Hartman, R. C., Kniffen, D. A., Thompson, D. J. 1977. *Ap. J. Lett.* 212: L63
143. Larson, R. B. 1981. *MNRAS* 194: 809
144. Lawson, K. D., Mayer, C. J., Osborne, J. L., Parkinson, M. L. 1987. *MNRAS* 225: 307
145. Lebrun, F. 1984. In *Nearby Molecular Clouds. Lect. Notes Phys.*, ed. G. Serra, 237: 3. Berlin: Springer-Verlag
146. Lebrun, F. 1986. *Ap. J.* 306: 16
147. Lebrun, F., Bennett, K., Bignami, G. F., Bloemen, J. B. G. M., Buccheri, R., et al. 1983. *Ap. J.* 281: 634
148. Lebrun, F., Bignami, G. F., Buccheri, R., Caraveo, P. A., Hermsen, W., et al. 1982. *Astron. Astrophys.* 107: 390
149. Lebrun, F., Paul, J. A. 1983. *Ap. J.* 266: 276
150. Lebrun, F., Paul, J. A. 1985. *Proc. Int. Cosmic Ray Conf., 19th, La Jolla*, 1: 309
151. Lequeux, J. 1981. *Comments Astrophys.* 9: 117
152. Lerche, I., Schlickeiser, R. 1982. *MNRAS* 201: 1041
153. Lerche, I., Schlickeiser, R. 1982. *Astron. Astrophys.* 107: 148
154. Lester, D. F., Dinerstein, H. L., Werner, M. W., Watson, D. M., Genzel, R., Storey, J. W. V. 1987. *Ap. J.* 320: 573
155. Linke, R. A., Stark, A. A., Frerking, M. A. 1981. *Ap. J.* 243: 147
156. Liszt, H. S. 1984. *Comments Astrophys.* 10: 137
157. Li Ti pei, Riley, R. A., Wolfendale, A. W. 1982. *J. Phys. G* 8: 1141
158. Li Ti pei, Wolfendale, A. W. 1981. *Astron. Astrophys.* 100: L26
159. Li Ti pei, Wolfendale, A. W. 1981. *Astron. Astrophys.* 103: 19
160. Lyne, A. G., Manchester, R. N., Taylor, J. H. 1985. *MNRAS* 213: 613
161. MacLaren, I., Richardson, K. M., Wolfendale, A. W. 1988. *Ap. J.* 333: 821
162. Maloney, P., Black, J. 1988. *Ap. J.* 325: 389
163. Marscher, A. P., Brown, R. L. 1978. *Ap. J.* 221: 583
164. Mathis, J. S. 1986. *Ap. J.* 301: 423
165. Mathis, J. S., Mezger, P. G., Panagia, N. 1983. *Astron. Astrophys.* 128: 212
166. Mayer, C. J., Richardson, K. M., Rogers, M. J., Szabelski, J., Wolfendale, A. W. 1987. *Astron. Astrophys.* 180: 73
167. Mayer-Hasselwander, H. A. 1983. In *Kinematics, Dynamics, and Structure of the Milky Way*, ed. W. L. H. Shuter, p. 223. Dordrecht: Reidel
168. Mayer-Hasselwander, H. A. 1985. *Explanatory Supplement to the COS-B Database* (available from K. Bennett, SSD, ESTEC, Noordwijk, Neth.)
169. Mayer-Hasselwander, H. A., Bennett, K., Bignami, G. F., Buccheri, R., Caraveo, P. A., et al. 1982. *Astron. Astrophys.* 105: 164
170. Mayer-Hasselwander, H. A., Simpson, G. 1989. *Adv. Space Res.* In press
171. McKee, C. F., Hollenbach, D. J., Seab,

C. G., Tielens, A. G. G. M. 1987. *Ap. J.* 318: 674
172. Melisse, J., Bloemen, J. B. G. M. 1989. In preparation
173. Meyer, J.-P. 1985. *Ap. J. Suppl.* 57: 173
174. Meyers, P. C. 1983. *Ap. J.* 270: 105
175. Morfill, G. E. 1982. *MNRAS* 198: 583
176. Morfill, G. E. 1982. *Ap. J.* 262: 749
177. Morfill, G. E., Tenorio-Tagle, G. 1983. *Space Sci. Rev.* 36: 93
178. Deleted in proof
179. Müller, D., Tang, J. 1981. *Proc. Int. Cosmic Ray Conf., 17th, Paris*, 9: 142
180. Nishimura, J., Fujii, M., Taira, T., Aiza, E., Hiraiwa, H., et al. 1980. *Ap. J.* 238: 394
181. Oort, J. H. 1977. *Annu. Rev. Astron. Astrophys.* 15: 295
182. Owens, A. J., Jokipii, J. R. 1977. *Ap. J.* 215: 677
183. Pagel, B. E. J., Edmunds, M. G. 1981. *Annu. Rev. Astron. Astrophys.* 19: 77
184. Parker, E. N. 1966. *Ap. J.* 145: 811
185. Parker, E. N. 1969. *Space Sci. Rev.* 9: 651
186. Parker, E. N. 1976. In *The Structure and Content of the Galaxy and Galactic Gamma Rays*, ed. C. E. Fichtel, F. W. Stecker, p. 320. Greenbelt, Md: Goddard Space Flight Cent.
187. Paul, J. A., Cassé, M., Cesarsky, C. J. 1976. *Ap. J.* 207: 62
188. Peters, B., Westergaard, N. J. 1977. *Astrophys. Space Sci.* 48: 21
189. Phillipps, S., Kearsey, S., Osborne, J. L., Haslam, C. G. T., Stoffel, H. 1981. *Astron. Astrophys.* 98: 286
190. Phillipps, S., Kearsey, S., Osborne, J. L., Haslam, C. G. T., Stoffel, H. 1981. *Astron. Astrophys.* 103: 405
191. Piccinotti, G., Bignami, G. F. 1976. *Astron. Astrophys.* 52: 69
192. Pikelner, S. B. 1953. *Dokl. Acad. Sci. USSR* 88: 229
193. Pollock, A. M. T., Bennett, K., Bignami, G. F., Bloemen, J. B. G. M., Buccheri, R., et al. 1985. *Astron. Astrophys.* 146: 352
194. Pollock, A. M. T., Bennett, K., Bignami, G. F., Bloemen, J. B. G. M., Buccheri, R., et al. 1985. *Proc. Int. Cosmic Ray Conf., 19th, La Jolla*, 1: 338
195. Ptuskin, V. S. 1979. *Astrophys. Space Sci.* 61: 259
196. Ramana Murthy, P. V., Wolfendale, A. W. 1986. *Gamma-Ray Astronomy*. Cambridge: Univ. Press
197. Reich, P., Reich, W. 1988. *Astron. Astrophys. Suppl.* 74: 7
198. Reich, P., Reich, W. 1988. *Astron. Astrophys.* 196: 211
199. Richardson, K. M., Wolfendale, A. W. 1988. *Astron. Astrophys.* 201: 100
200. Richardson, K. M., Wolfendale, A. W. 1988. *Astron. Astrophys.* 203: 289
201. Riley, P. A., Wolfendale, A. W. 1984. *J. Phys. G* 10: 1269
202. Riley, P. A., Wolfendale, A. W., Xu, C.-x., Manchester, R. N., Robinson, B. J., Whiteoak, J. B. 1984. *MNRAS* 206: 423
203. Rockstroh, J. M., Webber, W. R. 1978. *Ap. J.* 224: 677
204. Rogers, M. J., Sadzinska, M., Szabelski, J., van der Walt, D. J., Wolfendale, A. W. 1988. *J. Phys. G* 14: 1147
205. Rogers, M. J., Wolfendale, A. W. 1987. *Proc. Int. Cosmic Ray Conf., 20th, Moscow*, 1: 81
206. Ryan, M. J., Balasubrahmanyan, V. K., Ormes, J. F. 1972. *Phys. Rev. Lett.* 28: 985
207. Sanders, D. B., Scoville, N. Z., Solomon, P. M. 1985. *Ap. J.* 289: 373
208. Sanders, D. B., Solomon, P. M., Scoville, N. Z. 1984. *Ap. J.* 276: 182
209. Scarsi, L., Bennett, K., Bignami, G. F., Boella, G., Buccheri, R., et al. 1977. *Proc. 12th ESLAB Symp. ESA SP 124*, p. 3
210. Scott, J. S. 1975. *Nature* 258: 58
211. Seldner, M., Siebers, B., Groth, E. J., Peebles, P. J. E. 1977. *Astron. J.* 82: 249
212. Shane, C. D., Wirtanen, C. A. 1967. *Publ. Lick Obs.* 22: 1
213. Shklovskii, I. S. 1952. *Astron. Zh.* 29: 418
214. Shukla, P. G., Paul, J. A. 1976. *Ap. J.* 208: 893
215. Silk, J., Bloemen, J. B. G. M. 1987. *Ap. J. Lett.* 313: L47
216. Simpson, G., Mayer-Hasselwander, H. A. 1987. *Proc. Int. Cosmic Ray Conf., 20th, Moscow*, 1: 89
217. Simpson, J. A. 1983. *Annu. Rev. Nucl. Part. Sci.* 33: 323
218. Skilling, J., Strong, A. W. 1976. *Astron. Astrophys.* 53: 253
219. Sofue, Y., Fujimoto, M., Wielebinski R. 1986. *Annu. Rev. Astron. Astrophys.* 24: 459
220. Solomon, P. M., Rivolo, A. R., Barrett, J., Yahil, A. 1987. *Ap. J.* 319: 730
221. Stacy, J. G., Dame, T. M., Thaddeus, P. 1987. *Proc. Int. Cosmic Ray Conf., 20th, Moscow*, 1: 117
222. Stecker, F. W. 1971. *Cosmic Gamma Rays*. Baltimore: Mono Book Corp.
223. Stecker, F. W. 1973. *Ap. J.* 185: 499
224. Stecker, F. W. 1977. *Ap. J.* 212: 60
225. Stecker, F. W., Jones, F. C. 1977. *Ap. J.* 217: 843
226. Stecker, F. W., Solomon, P. M., Scoville, N. Z., Ryter, C. E. 1975. *Ap. J.* 201: 90

227. Stephens, S. A., Badhwar, G. D. 1981. *Astrophys. Space Sci.* 76: 213
228. Strong, A. W. 1985. *Astron. Astrophys.* 145: 81
229. Strong, A. W. 1985. *Proc. Int. Cosmic Ray Conf., 19th, La Jolla*, 1: 333
230. Strong, A. W., Bennett, K., Bignami, G. F., Bloemen, J. B. G. M., Buccheri, R., et al. 1983. *Proc. Int. Cosmic Ray Conf., 18th, Bangalore*, 9: 90
231. Strong, A. W., Bignami, G. F., Bloemen, J. B. G. M., Buccheri, R., Caraveo, P. A., et al. 1982. *Astron. Astrophys.* 115: 404
232. Strong, A. W., Bloemen, J. B. G. M., Dame, T. M., Grenier, I., Hermsen, W., et al. 1988. *Astron. Astrophys.* 207: 1
233. Strong, A. W., Bloemen, J. B. G. M., Hermsen, W., Lebrun, F., Mayer-Hasselwander, H. A. 1987. *Astron. Astrophys. Suppl.* 67: 283
234. Strong, A. W., Bloemen, J. B. G. M., Hermsen, W., Mayer-Hasselwander, H. A. 1985. *Proc. Int. Cosmic Ray Conf., 19th, La Jolla*, 1: 317
235. Deleted in proof
236. Strong, A. W., Wolfendale, A. W. 1978. *J. Phys. G* 4: 1
237. Strong, A. W., Wolfendale, A. W. 1981. *Philos. Trans. R. Soc. London Ser. A* 301: 541
238. Strong, A. W., Wolfendale, A. W., Bennett, K., Wills, R. D. 1978. *MNRAS* 182: 751
239. Swanenburg, B. N., Bennett, K., Bignami, G. F., Buccheri, R., Caraveo, P. A., et al. 1981. *Ap. J. Lett.* 243: L69
240. Szabelski, J., van der Walt, D. J., Wdowczyk, J., Wolfendale, A. W. 1989. *Adv. Space Res.* In press
241. Thompson, D. J., Bignami, G. F., Fichtel, C. E., Kniffen, D. A. 1974. *Ap. J. Lett.* 190: L51
242. Thompson, D. J., Fichtel, C. E. 1982. *Astron. Astrophys.* 109: 352
243. Thompson, D. J., Fichtel, C. E., Hartman, R. C., Kniffen, D. A., Lamb, R. C. 1977. *Ap. J.* 213: 252
244. Thompson, D. J., Fichtel, C. E., Kniffen, D. A., Ögelman, H. B. 1975. *Ap. J. Lett.* 200: L79
245. van der Walt, D. J., Wolfendale, A. W. 1988. *J. Phys. G* 14: L159
246. van der Walt, D. J., Wolfendale, A. W. 1988. *Space Sci. Rev.* 47: 1
247. Verter, F. 1987. *Ap. J. Suppl.* 65: 555
248. Völk, H. J. 1983. *Space Sci. Rev.* 36: 3
249. Völk, H. J. 1984. In *High Energy Astrophysics*, ed. J. Audouze, J. Tran Thanh Van, p. 281. Gif-sur-Yvette, Fr: Editions Frontières
250. Webber, W. R. 1983. In *Composition and Origin of Cosmic Rays*, ed. M. M. Shapiro, p. 83. Dordrecht: Reidel
251. Webber, W. R., Simpson, G. A., Cane, H. V. 1980. *Ap. J.* 236: 448
252. Wefel, J. P. 1988. In *Genesis and Propagation of Cosmic Rays*, ed. M. M. Shapiro, J. P. Wefel, p. 1. Dordrecht: Reidel
253. Wolfendale, A. W. 1986. In *Cosmic Radiation in Contemporary Astrophysics*, ed. M. M. Shapiro, p. 135. Dordrecht: Reidel
254. Wolfendale, A. W. 1988. In *Molecular Clouds in the Milky Way and External Galaxies*, ed. R. Dickman, R. Snell, J. Young, p. 76. Heidelberg: Springer-Verlag
255. Wolfendale, A. W. 1989. In *Cosmic Gamma Rays and Cosmic Neutrinos*, ed. M. M. Shapiro, J. P. Wefel. Dordrecht: Reidel. In press

QUASI-PERIODIC OSCILLATIONS AND NOISE IN LOW-MASS X-RAY BINARIES

M. van der Klis

EXOSAT Observatory, Astrophysics Division, Space Science Department of the European Space Agency, ESTEC, Postbus 299, 2200 AG Noordwijk, The Netherlands, and Astronomical Institute "Anton Pannekoek," University of Amsterdam, Roetersstraat 15, 1018 WB Amsterdam, The Netherlands

1. INTRODUCTION

Power spectra of the X-ray intensity variations of low-mass X-ray binaries (LMXB) show dramatic changes in shape as a function of X-ray intensity and spectral hardness. At least five different power spectral components have now[1] been identified whose characteristics change in complex, but consistent, ways in correlation with other source parameters.

These power spectral variations are for the first time providing an observational handle on the bright LMXB, Galactic X-ray binary systems that, in spite of being the brightest and longest-known (25) extrasolar celestial X-ray sources, have long defied interpretation owing to a lack of diagnostic properties. An intricate phenomenology has been revealed, probably involving two different populations among bright LMXB, each having its own pattern of correlated X-ray/power spectral behavior. Comparative power spectral studies of other types of X-ray binaries inspired by and modeled on techniques developed for bright LMXB suggest links with "classical" accreting X-ray pulsars and black hole candidates.

The large amplitudes and short time scales of the observed intensity variations suggest that they are a diagnostic of the accretion process in the innermost regions of LMXB, close to the compact object. It seems likely

---

[1] This review was completed in November 1988, a few weeks after the Fifth Los Alamos Space Physics/Astrophysics Workshop "Quasi-Periodic Oscillations in Luminous Galactic X-Ray Sources" in Santa Fe, New Mexico.

that "power spectroscopy" of these sources, combined with X-ray spectral studies, is providing insight into such topics as neutron star magnetic field decay, disk-magnetosphere interaction, inner radiation-pressure-dominated accretion disks near the Eddington limit, and the evolutionary connection between LMXB and millisecond radio pulsars.

All power spectral components seen in bright LMXB are broad, which indicates aperiodic variability. Some exhibit a very wide power distribution over typically several decades in frequency; others form a more localized peak (see Figure 1). It is generally assumed (and there is evidence; see 136) that the variations underlying all of these power spectral components are stochastic, but only those with very wide power distributions are called "noise" components, whereas those causing peaks are called quasi-periodic oscillations (QPO). Note that this definition differs from the mathematical

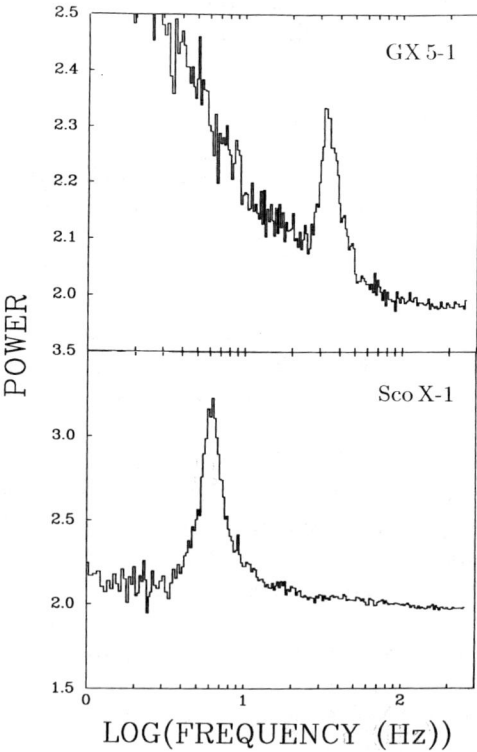

Figure 1 Power spectra of the X-ray intensity variations of two bright LMXB. Note the presence of low-frequency noise in addition to the QPO peak in GX 5−1 and its absence in Sco X-1.

one (e.g. 19). The reason for a separate treatment of power spectral peaks is that with their well-defined preferred frequency (the peak centroid frequency), they are supposedly more amenable to interpretation than the noise components. Some components always form clear peaks; others are more ambiguous and only occasionally appear peaked. Thus, sometimes the choice as to what to call QPO and what to call noise is somewhat arbitrary. In this review I refer to QPO in the case of roughly symmetric peaks with widths not much exceeding their centroid frequencies, and to noise in all other cases. Noise of which the power spectrum monotonically decreases as a function of frequency is, in an obvious analogy, called "red noise."

The occurrence of a broad power spectral component (QPO peak or otherwise) does not tell us what is the underlying signal or physical process. Therefore, QPO and noise could be caused by a variety of phenomena, and indeed observational evidence suggests that more than one process is at work. In spite of this, the X-ray intensity variations are discussed in terms of their power spectra; the reason is that photon counting statistics usually make the data too noisy, at the millisecond time scales most interesting in the context of stellar-mass compact objects, to observe the variations directly, and thus it is necessary to calculate the average power spectrum of large amounts of data in order to detect the signals.

The potential ambiguities inherent in this approach are illustrated by the confusion that arose when, after the discovery of QPO in GX 5−1, Sco X-1, and Cyg X-2 (see Section 1.3), attempts were made to explain all observed power spectral peaks in the same terms. In hindsight, we know that two qualitatively different QPO phenomena had been found in rapid succession, very likely requiring two entirely different explanations. The resolution of the puzzle came from an approach in which QPO properties were considered in relation to (*a*) other power spectral components and (*b*) X-ray spectral properties. This approach—to look for patterns of correlated power spectral and X-ray spectral behavior in order to remove ambiguities remaining when power spectra or X-ray spectra are considered in isolation—has proven to be a very powerful one, and its results have provided a large amount of new insight into LMXB phenomenology.

The QPO peaks that are the main subject of this review (Figure 1; Section 2) typically display the following characteristics:

- centroid frequencies between 5 and 60 Hz,
- peak widths of roughly half the centroid frequency,
- amplitudes between 1 and 10% of the total intensity,
- persistence for typically $> 10^5$ QPO cycles,
- recurrence.

Note that these characteristics, typical of QPO in persistently bright LMXB, may not explore the entire parameter space accessible to the underlying physical phenomena; see, in particular, Section 4 for a discussion of some (possibly closely) related power spectral peaks with properties that are somewhat different from the above.

The study of aperiodic variability in bright LMXB is a field that is very much in flux. In this review, I stress the general patterns of QPO phenomenology and attempts at understanding these patterns rather than detailed differences between individual sources and models for one single aspect of the observed behavior of QPO sources. A comprehensive review of the properties of QPO sources complete up to December 1987 is provided in (58). Discussions of early QPO models can be found in (48, 53), and methods of Fourier spectral analysis relevant to QPO sources are discussed in Chapter 2 of (58) and in (136). For earlier reviews on aspects of QPO sources and models, see also (30, 31, 104–106, 132–135).

1.1 Signal Analysis Aspects of QPO and Noise

Many different types of stochastic signals could in principle give rise to the power spectral components observed in LMXB (see, e.g., 136). The type of signal that on physical grounds has been by far the most popular in modeling X-ray binary power spectra in general is random shot noise (127; see also 110, and references therein), and this preference has propagated into the description of the recent results on QPO and noise (1, 2, 20, 21, 47, 50, 100–102).

Power-law red noise with a power spectrum $P(v)$ described by $P(v) = Av^{-\alpha}$, where v is the frequency, α the power-law slope, and A a normalization constant, is often encountered in practice. Physically, to avoid infinite integrated power, when $\alpha > 1$ such a spectrum must turn over (become flat) below some frequency v_{to}, and when $\alpha < 1$ it must break (become steeper) above some frequency v_{break}. A shot-noise signal consisting of sharp-edged shots has a power-law spectrum with $\alpha = 2$ for large v (128).

QPO spectra can be modeled with signals containing wave trains or "oscillating shots" (50, 143)—short-lived oscillations whose amplitude as well as frequency can vary during the lifetime of the shot (50). The width of the QPO peak is determined by both the range of frequency variation and the shot's lifetime. An example of an oscillating shot is a signal of the form $I(t) = [A + B\cos(2\pi v_{QPO} t)] \exp(-t/\tau_{decay})$ ($t > 0$), where A and B are positive constants. For such a shot, oscillating with a constant frequency v_{QPO} and decaying exponentially on a time scale τ_{decay}, the QPO peak has a Lorentzian shape centered on v_{QPO} with a full width at half maximum of $\Delta v = 1/\pi\tau_{decay}$ (102). The coherence time of the QPO is usually defined as

$\tau_{\text{coh}} \equiv 1/\pi\Delta v$. It reflects frequency changes as well as the shot's lifetime. Unless oscillating shots are fine-tuned to have a zero mean [positive and negative contributions of the oscillation cancel out on time scales longer than $1/v_{\text{QPO}}$ (in our example, $A \equiv 0$)], they cause not only a QPO peak but also an associated red noise component, the shape of which is determined by the oscillating shot's envelope (in our example, an exponential). For random positive-definite shots ($B < A$), the integrated power in red noise and in QPO peak are similar (50).

The integrated power in a power spectral component is a measure of the variance of the underlying signal in the time series. This means that we can estimate the strength of the signal in terms of its root-mean-square (rms) deviation from the mean of the time series. Usually this "rms amplitude" is expressed as a fraction of the mean intensity. When the signal is not always present, the largest deviations from the mean that actually occur can be much larger than the measured rms amplitude.

The rules governing the detection of broad power spectral components are different from those in the case of coherent oscillations, such as from pulsars (see 58, 136). In particular, there is no advantage in using very high power spectral resolutions; in fact, spreading out the power of a broad feature over many power spectral resolution elements makes the feature harder to find (see 132), although formally its significance is not affected. Usually, advantage is taken of this by calculating many low-resolution power spectra of successive short data segments, which saves computing time and in addition allows the monitoring of spectral variations (143). The significance n_σ (in standard deviations) by which a broad feature stands out against the noise is given by $n_\sigma = \frac{1}{2}Ir^2(T/\Delta v)^{1/2}$, where I is the mean intensity, r the fractional rms amplitude, T the observing time, and Δv the feature's width. This means that a small drop in r can be sufficient to make a very significant feature drop below the detection limit, so that observational reports of a power spectral component suddenly appearing or disappearing can well be consistent with only moderate variations in its strength.

1.2 Low-Mass X-Ray Binaries and Millisecond Radio Pulsars

Observationally, LMXB are distinguished from massive X-ray binaries (MXB) in a number of ways (see 55, 95, 146). LMXB are part of an older Galactic population than MXB (see also 16), usually lack the X-ray pulsations and eclipses often seen in MXB, show X-ray bursts (always absent in MXB), have soft X-ray spectra, and emit optical light dominated by an accretion disk rather than by a massive companion star. These differences are attributed to LMXB containing low-mass ($\lesssim 1\ M_\odot$) rather than massive ($\gtrsim 10\ M_\odot$) mass-losing companion stars, which allows them to be on average older (usually $> 10^9$ yr) than MXB ($\lesssim 10^7$ yr) (e.g. 130); this in turn has allowed the magnetic field of the accreting neutron star to decay to values much below the 10^{12}–10^{13} G believed to be characteristic of MXB (see 120). That neutron star surface magnetic field strengths decay on time scales of $\lesssim 10^7$ yr is suggested by observations of radio pulsars (see 43, and references therein). A few LMXB systems may contain a black hole rather than a neutron star (see 67).

The brightest LMXB tend to be located in the Galactic bulge and are less likely to show X-ray bursts than fainter ones (e.g. 156), but there is no evidence for a bimodal luminosity (150) nor, from existing sky surveys (69, 153, 154), for a bimodal *intensity* distribution. Table 1 lists the brightest LMXB in order of decreasing time-averaged X-ray intensity. Except for the black hole candidate GX 339−4, these sources are all "persistently bright" (9). Among the top ten sources in Table 1, eight are located in the Galactic bulge and only one (GX 3+1) has shown regular X-ray bursts, on one single occasion when its luminosity dropped considerably (62). It is among these ten brightest LMXB that most QPO have so far been found. None of them regularly shows X-ray bursts; "reliable" bursters are only found among the bottom eight in the list. The distances to the brightest LMXB are usually not well known, but as a group the X-ray luminosities of the "top ten" are of the order of the Eddington limit (1.5–3 $\times$ 10^{38} erg s^{-1} for hydrogen accretion onto a 1–2 $M_\odot$ neutron star).

The type of companion star may influence the average luminosity of a LMXB (119, 155). Main sequence companions transferring mass through Roche lobe overflow caused by orbital angular momentum loss induced by gravitational radiation and/or magnetic braking would provide low rates of mass transfer and would hence be found in less luminous LMXB. Companions evolving off the main sequence toward the giant branch could drive much larger mass transfer rates by their evolutionary expansion and produce highly luminous LMXB (see 149 for a review). If this is correct, then the most luminous sources should have long ($\gtrsim 10$ hr) orbital periods. This is not easy to check observationally. Fainter LMXB often show periodic X-ray dips (84) that reveal their orbital periods to be mostly below ~ 10 hr. However, the brightest LMXB do not show X-ray dips, and as most of them cannot be observed optically owing to their location in the bulge, the orbital period is not known. Two of the brightest LMXB located outside the bulge (Sco X-1 and Cyg X-2) have optically been found to have long orbital periods; GX 9+9 has a short period (see Table 1).

The low magnetic field strength of the neutron star in a LMXB allows the star to be spun up to millisecond rotation rates by the torques exerted by the accreting matter, and it is believed that radio pulsars with millisecond spin rates are produced through spin-up in binary systems (see 131 for a

Table 1 Bright low-mass X-ray binaries[a]

Source name(s)	l^{II}, b^{II} (°)	I_x^b Mean (μJy)	Min. (μJy)	Max. (μJy)	P_{orb}^c (hr)	Type[d]	Phenomenology[e]
Sco X-1 (1617−155)	359+24	12,400	9300	16,300	19.2	Z	QPO
GX 5−1 (1758−250)	5−1	1200	1070	1410	—	Z	QPO
GX 349+2 (1702−363)[f]	349+3	780	620	980	—	Z	QPO
GX 17+2 (1813−140)	16+1	680	600	780	19.8?[g]	Z	QPO, (bu)
GX 9+1 (1758−205)	9+1	650	550	720	—	A	—
GX 340+0 (1642−455)	340−0	490	400	620	—	Z	QPO
GX 3+1 (1744−265)	2+1	430	230	550	—	A	QPO, (Bu)
Cyg X-2 (2142+380)	87−11	430	290	730	235	Z	QPO, (bu), Mo
GX 13+1 (1811−171)	14+0	340	240	430	—	A	—
GX 9+9 (1728−169)	8+9	290	230	340	4.2	A	Mo
4U 1820−30 (NGC 6624)	3−8	260	94	360	0.2	A	QPO, (Bu), Mo
4U 1705−44	343−2	260	39	440	—	A	Bu
4U 1636−53	333−5	220	100	320	3.8	A	Bu
Ser X-1 (1837+049)	36+5	200	150	290	—	—	Bu
GCX-1 (1742−294)	0−0	170	130	270	—	—	Bu?
4U 1728−33	354−0	170	140	190	—	A	Bu
GX 339−4 (1659−487)	339−4	160	36	250	14.8?	—	QPO, BH?[h]
4U 1735−44	346−7	160	110	210	4.6	A	Bu

[a] All variable objects in 3A Catalogue (69, 153) with an average flux ≥ 100 μJy not identified with an early-type star (excluding Cyg X-3).
[b] Converted from *Ariel V* ASM counts into μJy (2–11 keV) according to 1 ASM c/s = 2.6 μJy (9).
[c] See (84).
[d] Z or A(toll) source; see text. After (36).
[e] QPO: all reported quasi-periodic oscillations are indicated here (see Section 3 for an evaluation of QPO reports in atoll sources); Bu: regular X-ray bursts; (Bu): has shown an episode of regular X-ray bursts; (bu): occasional X-ray bursts reported; BH?: black hole candidate, Mo: shows periodic X-ray modulation (9, 55, 64).
[f] "Sco X-2."
[g] Reference: (37).
[h] References: (77, 157).

review). Likely progenitors of some millisecond radio pulsars are LMXB containing an evolved companion star (see 131; the evolution of these systems may involve the accretion-induced collapse of a white dwarf into a neutron star). Millisecond radio pulsars have magnetic fields in the 10^8–10^{10} G range; however, in apparent conflict with an evolutionary connection, the magnetic field of an LMXB neutron star, decaying on the canonical time scale of $\lesssim 10^7$ yr, would be expected to drop much below these values within the system's lifetime. It has been argued on various grounds [see 131 and (in particular) 43, and references therein] that neutron star magnetic field decay may "bottom out" at field strengths somewhere in the 10^8–10^{10} G range rather than continuing all the way to zero.

The related issues of magnetic field decay in accreting and nonaccreting neutron stars, neutron star formation, neutron star spin-up in LMXB, and the presence and possible observational consequences of evolved companion stars in LMXB are subjects of lively debate (see, e.g., 44). Direct observational information (neutron star surface fields and spin rates, evidence for different populations of LMXB) would help enormously in settling these issues, and consequently the observational diagnostic provided by QPO and noise could be useful here.

In MXB the magnetic field of the neutron star disrupts the accretion disk far ($\sim 10^8$ cm) above the surface; in LMXB, with their low magnetic field, the disk extends much farther down, and an "inner" radiation-pressure-dominated accretion disk can form (see 46, 160, and references therein for views of accretion disk–magnetic field interactions). For luminosities near the Eddington limit these inner disks are likely unstable (e.g. 94), and it is not clear from theory how they behave. Possibly the inner disk is puffed up by radiation pressure and forms a thick accretion torus; relativistic jets could form, maybe producing a triple radio source such as observed in Sco X-1 (38). Bright LMXB are a laboratory of inner-disk physics, and QPO likely are a direct diagnostic of what happens there. In Section 2.4.1 some attempts at interpreting QPO in these terms are described.

Perhaps the most fundamental reason why we would like to obtain information about the conditions in the inner regions of LMXB is that these sources provide us with a different view of neutron star physics. The neutron stars in LMXB probably have low magnetic field strengths, are the only neutron stars spinning at millisecond rates undergoing accretion torques, and [having accreted considerable amounts of matter (up to 0.8 $M_\odot$; e.g. 120)] likely exceed the canonical mass of 1.4 $M_\odot$ (in MXB; 95). In all of these aspects they differ from neutron stars in massive X-ray binaries, and observation of the consequences of these differences would help in our understanding of neutron star physics.

1.3 *History and Prehistory of QPO and Noise in LMXB*

Noise in the X-ray intensity variations of what we now call the persistently bright LMXB has been studied since the early days of X-ray astronomy, initially mainly in the extremely bright source Sco X-1 (54; see also 40, 79, 89) but later also in many other sources (see 23, 85, 90 for surveys; 8 for a review; also 110). However, after the discovery of the rapid variability in Cyg X-1 (82) and periodic X-ray pulsations in Cen X-3 (24), the bulk of the effort with respect to *stochastic* variations went to black hole candidate Cyg X-1 (see 60 for a review) and other "rapidly variable" sources believed to be related (but see also Section 4.3), such as Cir X-1

(Section 4.1.1), GX 339−4 (Section 4.3), and V0332+53 (see also Section 4.3); consequently, the persistently bright LMXB were somewhat neglected. It was not until after the discovery, with *EXOSAT*, of QPO and associated red noise from these sources that systematic power spectral studies of the *shape* of the noise components got underway.

Slow QPO, in the sense of broad power spectral peaks and/or recurrent flares in the 1–20 mHz range, were reported from a number of sources [∼1 mHz in 4U 1626−67 (42, 59); 0.7–20 mHz in Cyg X-3 (138); 20 mHz in 4U 1820−30 (110); 5–10 mHz in GX 349+2 (66; see also 77)]. In some cases these reported oscillations may have been the result of chance occurrences in a (red) noise signal (127). It seems unlikely that they are related to the QPO that are the main subject of this review, although this possibility cannot yet be totally excluded. Slow QPO have recently also been reported from LMC X-1 (see Section 4.3).

Short oscillation trains during or near several Type I (thermonuclear flash; see 55) X-ray bursts have been reported [36 Hz (65); 82 Hz (96); 1.5 Hz (78); 7.6 Hz (R. L. Kelley[2])]. The unique character of these oscillation events makes comparison with QPO seen in the persistently bright LMXB difficult; "Type I burst QPO" usually are more coherent.

Oscillations with a frequency of ∼2 Hz were seen by Tawara et al. (122) during two flat-topped Type II (accretion event; see 55) X-ray bursts in the peculiar "Rapid Burster" (Section 4.1.2). The frequency of the QPO differed between the bursts and drifted within each burst, so it was concluded that they could not be a direct manifestation of neutron star rotation. The relation of Rapid-Burster QPO to QPO in persistently bright LMXB is not clear. In general, Rapid Burster QPO are more coherent and less persistent. It has been argued on the basis of later extensive *EXOSAT* studies (108, 109; see Section 4.1.2) that a relation [with "normal branch" QPO (see Section 2.3)] exists.

In 1985 it was announced that 20–36 Hz QPO and associated red noise, designated "low-frequency noise" (LFN), had unexpectedly been observed from GX 5−1 during a search with *EXOSAT* for periodic millisecond pulsations; the QPO frequency was strongly correlated with X-ray intensity (140, 143). Alpar & Shaham (3, 4) proposed that these QPO were caused by an interaction between a rapidly spinning neutron star's magnetic field and matter circulating in an accretion disk, the characteristic frequency of this interaction being the beat frequency between the Keplerian frequency at the disk's inner edge and the star's spin frequency.[3] They pointed out

---

[2] Paper presented at Santa Fe Workshop (see Note 1).

[3] A similar idea had been considered previously for QPO in cataclysmic variables (86; see also 152). Apart from some very general considerations (143), research on LMXB QPO has not benefited much from the earlier cataclysmic variable work. QPO in these accreting white dwarf systems appear to be even more complex than in LMXB.

that spin rate and magnetic field strength derived in this "beat-frequency" model (Section 2.4.1) fitted with the idea that bright LMXB evolve into millisecond radio pulsars (Section 1.2). Modulation of the accretion rate through magnetic gating was proposed as a feasible mechanism for causing the X-ray intensity variations, and it was shown that the observed red noise was a natural consequence of this accretion-modulation mechanism (50).

However, QPO discovered subsequently from Sco X-1 by Middleditch & Priedhorsky (70, 71) had quite different properties.[4] Their frequency was sometimes low (near 6 Hz) and approximately constant or even slightly anticorrelated with X-ray intensity (71); at other times it was high (between 10 and 20 Hz) and either positively correlated or varying erratically as a function of X-ray intensity, depending on the source state (144). Red noise was weak as compared with the QPO (71, 139). Both results seemed to pose a problem for the beat-frequency model.

In a third source, Cyg X-2, QPO were discovered (33, 34) with very similar properties to GX 5−1, but further QPO observations [Cyg X-2 (80, 81); GX 349+2 (14, 57); GX 17+2 (111); Rapid Burster (113); 4U 1820−30 (115)] presented a confused picture. Stella (104) proposed, based on the analogy with the QPO in Sco X-1, to distinguish between "low-frequency" (<10 Hz), "high-frequency" (>10 Hz), and "erratic" QPO, having, respectively, a negative (or no), a strongly positive, and an erratic frequency-intensity correlation.

The observational situation was rather bewildering at this point, as a large variety of behavior seemed possible for the QPO phenomenon. As the beat-frequency model was unable to explain everything observed, alternative QPO models were being considered (7, 26, 68, 76, 145), and ways were worked out to broaden the range of phenomena that could be explained within the modulated-accretion beat-frequency model (1, 2, 10, 20, 98, 102, 118).

Progress in unraveling the complex QPO phenomenology came from comparing power spectral and X-ray spectral data and from paying attention to the noise components in addition to the QPO peaks. Three of the newly discovered QPO sources were known to exhibit two different X-ray spectral states (11, 103, 158), distinguishable as branches in an X-ray hardness vs. intensity plot, and *EXOSAT* observations confirmed this (Figure 2). For Sco X-1, the suspicion that QPO properties might be

---

[4] As a historical aside, the first observed consequence of the 6-Hz QPO in Sco X-1 may have been the transient narrow power spectral spikes in the 1–10 Hz range reported in 1971 by Angel et al. (5). These were likely statistical fluctuations pushed over the detection limit by the increased average power in the broad QPO peak, which remained unnoticed.

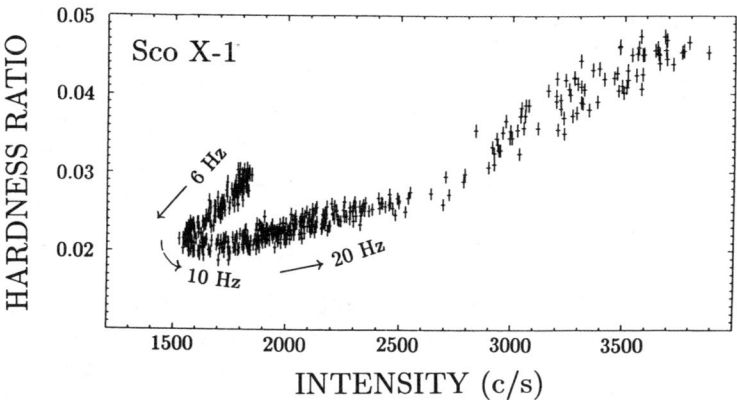

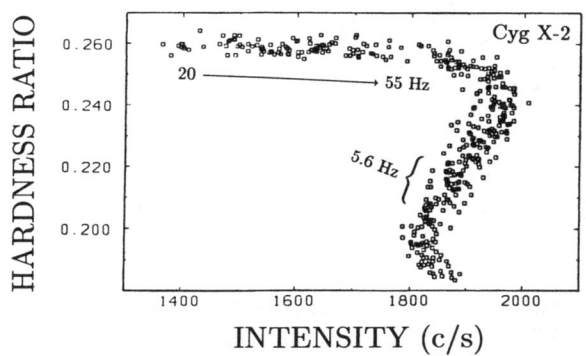

Figure 2 Spectral branches in X-ray spectral hardness vs. intensity diagrams. QPO frequencies in each branch are indicated. After Priedhorsky et al. (93) and Hasinger (28).

related to these X-ray spectral states arose during the initial discoveries and crystallized at the 1985 Workshop on X-ray Time Variability in Taos, New Mexico, when comparing the apparently conflicting data on the source (71, 72, 139, 144). Further work on GX 5−1 and Sco X-1 (93, 141, 142; this work was discussed at the 1985 Cambridge Discussion Meeting on *EXOSAT* Results) and on Cyg X-2 (29, 32) showed conclusively that QPO modes and X-ray spectral states were strongly correlated. The red noise in GX 5−1 was shown to consist of two components: the above-mentioned LFN, which only appeared in the spectral state where the intensity-dependent QPO were seen; and a second, power-law component dominating at frequencies below 0.01–1 Hz that was seen in all states, independent of the QPO, and was designated "very low frequency noise" (VLFN; 142). It was noted that the red noise in Sco X-1 seemed to contain no LFN but apparently only VLFN. [After the first reports on QPO, an extensive study of the red noise in a number of sources was made by Makishima (61); it referred mostly to this VLFN, as it dealt with frequencies below ~ 1 Hz.]

However, the situation was still confusing. True, bimodal patterns of QPO and X-ray spectral behavior were observed in GX 5−1, Cyg X-2, and Sco X-1, with high-frequency, intensity-dependent QPO in one spectral state and ~ 6-Hz (nearly) intensity-independent QPO in the other, but Sco X-1 did not fit in with the other two sources. Its X-ray spectral hardness vs. intensity diagram was different (Figure 2), and LFN was absent even when intensity-dependent QPO appeared. The proposed classification of Stella (104), which combined the high-frequency QPO of GX 5−1 and Cyg X-2 with those of Sco X-1 into one class, was criticized on these grounds (132, 142).

The resolution of the puzzle came from Hasinger (28), who, on the basis of the data summarized in Figure 2 and using the ~ 6-Hz QPO seen in all three sources as the main clue, proposed that the behavior of the QPO sources was in fact *trimodal* and characterized by a Z-shaped pattern in the X-ray hardness vs. intensity diagram, of which we had been seeing the upper half in GX 5−1 and Cyg X-2 and the lower half in Sco X-1. Furthermore, he noted that each of the three branches of the Z had its own characteristic power spectral behavior. These conjectures were beautifully confirmed from further observations, and the class of the "Z sources" is now very well established (see Section 2).

With this clarification of the phenomenology, modeling QPO became a more feasible task. Attempts to explain all observed QPO behavior within the framework of the modulated-accretion magnetospheric beat-frequency model, which had been criticized for being uncompelling (53, 109, 132–134, 142) were abandoned, and the model became the dominant one for the phenomenon it had been conceived for: the intensity-dependent high-frequency QPO and associated LFN originally found in GX 5−1 and later also observed in other sources (Section 2.2).

Millisecond time lags were discovered in these high-frequency QPO signals between X-ray spectral bands (29, 134, 137), suggesting that the X-rays were Compton scattered in a hot plasma before emerging from the vicinity of the neutron star. This was in line with earlier suspicions (159, and references therein) based on the shape of the X-ray spectrum. This result was pleasing, as the temporal smearing caused by such a scattering process had been quoted earlier as one of the possible reasons why periodic pulsations were not observed from these rapidly rotating magnetic neutron stars (50). The problem of accounting for the effect of the scattering on QPO and (hypothetical) beamed pulsar radiation stimulated a great deal of theoretical effort (10, 12, 45, 100, 151). The suggestion from the beat-frequency model that the magnetosphere was very small in these systems (tens of kilometers in radius, rather than thousands as in the accreting pulsars) stimulated a reassessment of magnetospheric theories, and it was pointed out that magnetospheric formation might be qualitatively different in these systems from that in massive X-ray binaries, as for such a small magnetosphere the magnetospheric boundary was located in the radiation-pressure-dominated part of the disk (160).

Further observations produced further puzzles, such as sources that should have shown QPO according to the Z scheme but did not (56, 87), and persistently bright LMXB that exhibited QPO with properties that did not seem to fit well into the scheme (17, 51, 52, 116). Recently, it was proposed (36) that in addition to the Z sources, there exists a second class of slightly fainter bright LMXB, including some of the brightest nonbursting LMXB as well as some bright X-ray burst sources (see Table 1; Section 3), that shows another pattern of correlated X-ray spectral and power spectral behavior. It was argued that these so-called atoll sources do not exhibit QPO similar to those seen in the Z sources. All persistently bright LMXB may be classifiable as either a Z or an atoll source; however, some peculiar LMXB that are not persistently bright (Section 4.1) exhibit different QPO properties.

Recent power spectral studies of other types of X-ray binaries have produced interesting results. In two accreting X-ray pulsars (Section 4.2) (A. Tennant, see Note 2; L. Angelini, see Note 2) QPO peaks were detected at frequencies consistent with the magnetospheric beat-frequency model (but other models are also possible), noise components in accreting pulsars were found to be very strong but otherwise quite similar to those in LMXB (T. Belloni, see Note 2), and QPO were detected with the *Ginga* satellite from two black hole candidates (18, 63; Section 4.3).

2. THE Z SOURCES

The Z sources are named after the pattern (Figure 3, *top*) they trace out, when varying, in an X-ray hardness vs. intensity (H–I) or X-ray color-color diagram. Recently, the use of the color-color diagrams, in which two different measures for X-ray spectral hardness are used (83, 157) has been preferred, as in these diagrams the source tracks are more repeatable

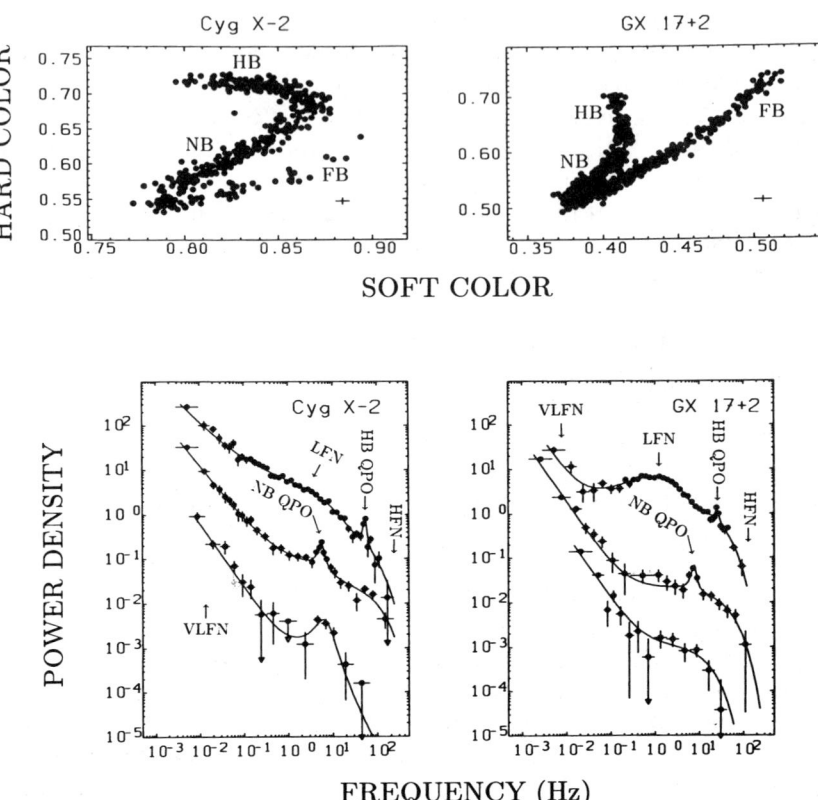

Figure 3 Z source behavior. X-ray color-color diagrams (*top*) are roughly Z shaped; horizontal branch (HB), normal branch (NB), and flaring branch (FB) are indicated. Note the difference in slope of the HB between the two sources. Power spectra (*bottom*) of the X-ray intensity variations refer to HB, NB, and FB (downward), respectively. Various power spectral components are indicated (see Section 2); note the difference in shape of the LFN. After Hasinger & van der Klis (36).

(29, 97). The three branches in the Z correspond to three distinct X-ray spectral/power spectral states. The upper branch is named the "horizontal branch" (HB; 103), the middle one the "normal branch" [NB (61); the positive H–I correlation in this branch was considered "normal" as opposed to the more exceptional HB behavior], and the lower one the "flaring branch" [FB (28), after the flaring shown by Sco X-1 in this branch]. This terminology is not particularly appropriate to the phenomenology as known today; whether the HB is really horizontal depends on source and spectral band, NB behavior is not more common than HB or FB behavior, and in the FB some sources' intensity actually *decreases* rather than flares [Cyg X-2 (30, 97); GX 340+0 (W. Penninx, see Note 2)]. However, the names are generally used, and therefore I shall employ them here as well.

Figure 3 (*top*) shows two examples of the color-color diagrams of Z sources. Note that the HB is well developed and really horizontal in Cyg X-2 (*left*) and short and almost vertical in GX 17+2 (*right*). The sources always follow the Z (11, 97, 103); they do not "jump" between branches. Typically, sources move up and down irregularly in a given branch or around the HB-NB or NB-FB transition point for hours to days before transitioning to another section of the Z (31, 72, 93, 97, 145). There is no evidence that a source spends more time in a particular section of the Z. The parameter varying along the Z is most likely the accretion rate $\dot{M}$, but there is no direct evidence for this, nor for the sense in which $\dot{M}$ varies (see Section 2.4). Optical, UV, and radio emission correlate to X-ray spectral state [see 79 (and references therein), 88; G. Hasinger, see Note 2).

On each branch the power spectra show characteristic structures (Figure 3, *bottom*), discussed in detail below. A total of six power spectral components has been detected. Two ("very low frequency noise" and "high-frequency noise") occur in all three spectral branches. These universal components are discussed in Section 2.1. In Section 2.2, horizontal branch QPO (HB-QPO) and the associated LFN are discussed. These components are strongest in the HB but have also been reported from the NB. Section 2.3 deals with normal branch QPO and flaring branch QPO (NB-QPO and FB-QPO). These two components clearly differ, but there is strong evidence that they are in fact manifestations of the same phenomenon. Universally *absent*, up to now, have been periodic pulsations. Upper limits of fractions of a percent have been obtained (58, and references therein; see also 35).

Table 2 lists the six sources that show Z behavior. References listed there are not repeated below when parameter ranges are quoted. For five of the sources, the Z character seems indisputable in that at least two branches together with the associated power spectral behavior have been observed; the sixth (GX 349+2) has so far only been observed in the FB and in the lowest portion of the NB. Table 3 lists typical power spectral properties of these sources.

2.1 *Universal Noise Components: VLFN and HFN*

VLFN ("very low frequency noise"; 142) and HFN ("high-frequency noise"; 36) are detected in all three branches (35, 36, 145), with no clear variations in strength or shape as a function of position in the Z. Com-

Table 2 Evidence for Z behavior[a]

Source	Any branch		Horizontal branch			Normal branch		Flaring branch		References
	VLFN	HFN	Branch[b]	QPO	LFN	Branch[b]	QPO	Branch[b]	QPO	
GX 17+2	●	●	●	●	●	●	●	●	●	36, 51, 87a, 88, 112
GX 340+0	●	●	●	●	●	●	●	—	—	36, 147[c]
Cyg X-2		●	●	●	●	●	●	●	—	17, 27, 29, 33, 36, 73, 81, 107, 137
Sco X-1	●	●	●	—	●	●	●	●	●	35, 36, 71, 93, 145
GX 5-1	●	●	●	●	●	●	●	—	—	17, 22, 36, 74, 137, 142, 143
GX 349+2		●	—	—	—	●	—[d]	●	—[d]	36, 91

[a] ●: detected; —: not detected.
[b] Refers to detection of spectral branch in X-ray hardness-intensity or color-color diagram.
[c] W. Penninx (see Note 2).
[d] Very broad peaks were reported (36, 91).

QUASI-PERIODIC OSCILLATIONS 533

Table 3 Power spectral properties of Z sources[a]

Component	Shape	Strength[b] (rms)	Other properties		Time lags
All branches					
VLFN	Power law[c]	0.5–3%	α	1.5–2	
HFN	Cutoff power law[d]	1.5–4%	ν_{cut}	40–80 Hz	
HB					
LFN	Cutoff power law[d]	3–4%	α	−1 to +0.3	
			ν_{cut}	5–10 Hz[i]	10-ms soft lags
QPO	Lorentz[e]	2–7%	ν_{QPO}	15–55 Hz[f]	4-ms hard lags
NB					
QPO	Lorentz[e]	1–5%	ν_{QPO}	5–7 Hz[g]	150° hard lags
FB					
QPO	Lorentz[e]	?[h]	ν_{QPO}	10–20 Hz[f]	

[a] Typical values. See Table 2 for references.
[b] $\sim$(1–20 keV).
[c] $\nu^{-\alpha}$.
[d] $\nu^{-\alpha} e^{-\nu/\nu_{cut}}$.
[e] $[(\nu - \nu_{QPO})^2 - (\Delta\nu/2)^2]^{-1}$.
[f] Strongly correlated with position in spectral branch.
[g] Approximately constant.
[h] Only reported in 5–35 keV range in Sco X-1, then 5–10% (rms) (93, 145).
[i] 1–2 Hz for $\alpha < 0$ (peaked LFN).

ponents of similar shape are also seen in atoll sources (Section 3) and even in accreting pulsars (Section 4.2) and black hole candidates (Section 4.3).

VLFN is a power-law component with a total strength of a few percent, usually dominating the power spectra below 0.01–0.1 Hz. It has mostly been studied at frequencies above 1 mHz, and therefore it is not yet clear whether it can be identified with the well-known variability of bright LMXB on time scales of hours to days (23, 85, 90). No evidence for a turnover has been reported. VLFN power-law slopes have been reported between 1.2 and 2.3. It is likely that both the lowest and the highest values in this range are unreliable. Some of the lower values were derived from power spectra not extending to sufficiently low frequencies or from analyses not properly distinguishing between VLFN and other components; the highest values may have suffered from low-frequency leakage (see 16a).

HFN has a shape that can be described as a flat power law with an exponential cutoff: $P(\nu) = \nu^{-\alpha} \exp(-\nu/\nu_{cut})$. A similar shape usually fits the LFN component in the HB (Section 2.2), but in HFN the cutoff frequency is higher: 25 to >100 Hz. The distinction between HFN and LFN becomes apparent in the HB, where both components are sometimes detected simultaneously (35, 36). Strengths between 1.6 and 8% (rms) have been reported; a similar component in atoll sources and accreting pulsars can be considerably stronger (Sections 3 and 4.2).

2.2 The Horizontal Branch: QPO and LFN

Among the five sources exhibiting a HB in their X-ray color-color diagram, four have shown intensity-dependent QPO. [The exception is Sco X-1, whose HB is least developed (35).] The total observed QPO frequency range is 15–55 Hz, with some variation from source to source. QPO frequency v_{QPO} and X-ray intensity I_x are strongly positively correlated; their relation is consistent with being either linear or a power law (Figure 4). Power-law slopes $-(d \log v_{QPO}/d \log I_x)$ of between 0.8 and 2 have been reported. [The slope of 4 quoted in (112) is defined differently.] Obviously, this result depends on the X-ray spectral band (here roughly 1–20 keV) and interstellar absorption. In view of the difficulties in finding the correct model for the X-ray spectra, not much work has been done yet in relating HB-QPO frequency to physically meaningful X-ray spectral components (see, however, 27, 33) or even X-ray luminosity. In all cases, QPO frequency increases along the HB toward the NB.

HB-QPO peaks fit to Lorentzian or Gaussian shapes; the quality of the data usually is not sufficient to strongly reject either of the two [or more complex shapes (20, 101, 102; see, however, 73)]. Determining the precise shape of the peak is complicated by its variations with intensity. The relative peak widths ($\Delta v/v_{QPO}$) are between 0.12 and 0.4, implying coherence times of 2.6 to only 0.8 cycles (see Section 1.1). No clear relation between peak width and HB-QPO frequency applying to all sources has been established.

Fractional rms amplitudes between 2 and 6.5% (1–20 keV) have been reported for HB-QPO, where the lower limit is probably set by observational limitations. In GX 5−1 and Cyg X-2 there is a tendency for fractional amplitude to drop as a function of frequency (and intensity; 33,

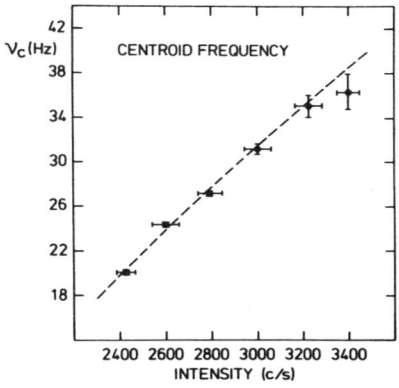

Figure 4 HB-QPO frequency vs. X-ray intensity in GX 5−1. After van der Klis et al. (143).

143), while in absolute terms, QPO amplitude has a maximum halfway across the HB.

Possible second harmonics to the main HB-QPO peak, indicating non-sinusoidal oscillation profiles, have been reported (29, 74, 107, 112; W. Penninx, see Note 2). They are relatively weak and not much is known about them.

LFN is seen in all sources when they are in the HB. Its shape has been described with an exponential (13, 33, 143), which fits well only at frequencies above ~ 1 Hz, and with power laws (107) and Lorentzians centered on zero frequency (22, 74, 112). However, over a wide frequency range, a flat power law with an exponential cutoff (similar to that used for HFN) usually provides better fits (28, 36). Cutoff frequencies (and the roughly equivalent Lorentzian half-widths) vary between 5 and 15 Hz. LFN is not always "red" (monotonic); sometimes it shows a broad peak [e.g. in GX 17+2 (51; Figure 3, *bottom right*) and Sco X-1 (35)]. In a cutoff power-law description, this LFN shape is accommodated by allowing ascending power laws (36). There seems to be a correlation between peak LFN shapes, low HB-QPO amplitudes, and short, vertical HBs (36).

The LFN strength is similar to that of HB-QPO, with their rms amplitude ratio (LFN/QPO) in the range 1-2. In GX 5-1, LFN and HB-QPO strength go up and down hand in hand; their ratio remains close to 1 (143). In Cyg X-2, HB-QPO and LFN strength are anticorrelated (33). On time scales of seconds, HB-QPO and LFN strength in GX 5-1 may vary more or less independently (J. Norris, see Note 2).

Not much detail has emerged as yet about the X-ray *spectrum* of HB-QPO and LFN, other than that it is harder than that of the time-averaged flux (29, 132; K. Mitsuda, see Note 2).

Time lags of up to 4 ms exist in the HB-QPO of Cyg X-2 (29, 137) and GX 5-1 (73, 134, 137) between very roughly the 1-6 and 6-20 keV bands. (Hard X rays lag the soft ones.) Cross-spectral analysis allows the measurement of time lag as a function of frequency and shows that this "hard lag" is a property of the HB-QPO only; in the LFN, *soft* lags are seen (137). It is not clear yet how the time lags depend on frequency inside each of the broad power spectral components. In Cyg X-2 time lag and oscillation frequency are anticorrelated (29, 74, 137). No measurements were reported of the dependence of time lag on photon energy.

HB-NB transitions involve only minor changes in X-ray intensity and spectrum but are quite spectacular in the power spectra: In particular the LFN appears and disappears very suddenly (within 100-200 s; 31) when the source changes state. HB-QPO have been reported to persist together with NB-QPO in Cyg X-2 (73, 81). As state changes can be rapid, confusion between branches could occur. However, recent work (G. Hasinger, see

Note 2 and personal communication) makes it likely that HB-QPO indeed occur in the NB together with NB-QPO.

2.3 Normal and Flaring Branch QPO

QPO in the NB with frequencies between 5 and 7 Hz have been reported from all five Z sources that have shown a full NB in the color-color diagram. Very broad peaks near 6 Hz have also been seen in GX 349+2 (91). (QPO in the 4–8 Hz range reported from other source types may or may not be similar; see Sections 3 and 4.) In Cyg X-2, GX 5−1, and GX 340+0, NB-QPO are only seen in the middle of the NB (29, 73, 147). In Sco X-1, QPO are absent in the upper part of the NB (35), appear in the middle of it, and persist all the way into the FB. Reported NB-QPO relative peak widths are usually between 0.2 and 0.6, but wider peaks are also seen (91, 112). No higher harmonics to the main peak were reported.

QPO frequency ν_{QPO} varies only a small amount in the NB. In Sco X-1 a slight increase (by $<10\%$; 71, 145) is observed along the NB in the direction of the FB; the X-ray intensity I_x *decreases* in the NB in this direction, and therefore ν_{QPO} and I_x are here anticorrelated.

In Sco X-1, ν_{QPO} gradually increases from ~ 6 to ~ 10 Hz when the source moves from the NB through the NB-FB transition region (Figure 5). In the FB, ν_{QPO} varies strongly between 10 and 20 Hz correlated with position of the source in the branch (higher frequency as the source moves "up" the FB away from the NB; see Figure 2). Because intensity increases up the FB, ν_{QPO} correlates positively with I_x here. Up the FB, the QPO peak becomes wider and finally dissolves into the HFN (35; see also 71).

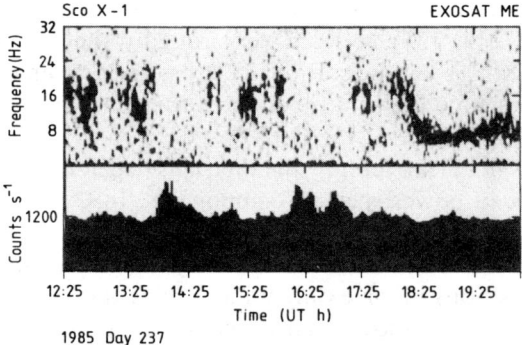

Figure 5 Dynamic power spectrum (*top*; darker pixels indicate higher power) and X-ray light curve (*bottom*) of Sco X-1 exhibiting NB-FB behavior. A gradual transition between FB-QPO (~ 20 Hz) and NB-QPO (~ 6 Hz) is observed near 18:15 UT. After Priedhorsky et al. (145).

In a description of the X-ray spectrum in terms of two spectral components [a "blackbody" and a "Comptonized" one (see 161)], it was found that v_{QPO} in the NB and FB jointly relate approximately linearly with the luminosity of the blackbody component (145). FB-QPO with intensity-correlated frequencies up to 20 Hz have recently also been seen from GX 17+2 (W. Penninx, see Note 2); from the same FB observation, an additional peak near 125 Hz was reported (see also Section 4.1.1).

An X-ray spectrum of NB-QPO in Cyg X-2 was reported by Mitsuda (73, 75; Figure 6, *top*). The fractional QPO amplitude reaches a minimum near 6 keV. This observation reconciles earlier reports about NB-QPO being soft in Cyg X-2 in the 1–6 keV band (29) and hard in Sco X-1 between 1 and 35 keV (93, 145). Large time delays of up to 80 ms occur in Cyg X-2 between QPO in spectral bands above and below 6 keV (hard photons lag the soft ones by $\sim 150°$, or precede them by 210° in phase; 73, 75). The time lag spectrum shows a large jump between 5 and 10 keV (Figure 6, *bottom*).

2.4 Understanding the Z Sources

2.4.1 PHENOMENOLOGY
Sufficient data have been accumulated for some general patterns in the properties of Z sources to become evident.

HB-QPO are qualitatively different from NB-QPO. In the HB, QPO frequency is high and strongly correlated with position in the branch (and hence with intensity); in the NB, it is low and hardly varies. In the HB, there is a strong noise component (LFN) associated with the QPO. NB-QPO have no such associated component, which implies a different shape for the signal underlying the power spectrum. There is a discontinuous change in QPO frequency when moving from the HB into the NB: from high and rising up to ~ 55 Hz in the HB to low and approximately constant

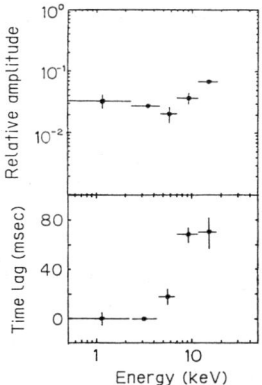

Figure 6 Relative amplitude (*top*) and time lag (*bottom*) spectra of the NB-QPO in Cyg X-2. A large discontinuity in the time lag spectra near 6 keV coincides with a minimum in relative amplitude. After Mitsuda (73).

at ~6 Hz in the NB. HB-QPO may persist into the NB and coexist with NB-QPO—this confirms that they are different phenomena.

The FB-QPO, on the other hand, are probably the same phenomenon as NB-QPO, even though their frequency (like that of HB-QPO) is higher than in the NB and strongly correlated with position in the branch. The main argument for identifying them is that gradual transitions between them are observed in Sco X-1 (Figure 5). A secondary argument is that no LFN is observed in association with either of them.

For these reasons, it is now nearly universally accepted that we are dealing with two physically distinct phenomena: HB-QPO/LFN and NB/FB-QPO (but see 106). Observations provide a number of direct clues to the nature of each of these two types of rapid time variability.

In the HB, the simultaneous presence of QPO and LFN fits in with oscillating-shot models (Section 1.1) in which the shots are positive definite. The "hard" millisecond time lag in HB-QPO could be caused by a simple time shift between hard and soft signals (e.g. caused by Comptonization; see below), whereas X-ray spectral softening during a shot could cause the simultaneous soft lag in the LFN (22, 135, 137).

In the NB the fractional amplitude spectrum of the QPO has a minimum in the 5–10 keV range, and there is a phase lag of close to half a cycle between NB-QPO below 5 keV and above 10 keV (Figure 6). This suggests that NB-QPO can be described in terms of "rocking" of the X-ray spectrum around a pivot point in the 5–10 keV range (49; see, however, 73 for a different interpretation). When the high-energy end of the spectrum goes up, nearly simultaneously the low-energy end goes down, and vice versa, causing larger QPO amplitudes at high and low energies than near the pivot point and a large phase shift between high and low energy QPO (as observed). The absence of red noise implies that this rocking occurs symmetrically around an equilibrium position (see Section 1.1).

Much attention has been given to the relation of QPO frequency v_{QPO} with X-ray intensity I_x. However, if the parameter varying along the Z is accretion rate $\dot{M}$, then clearly I_x is not proportional to $\dot{M}$, and thus the correlation of v_{QPO} with position in the Z seems more relevant. In some sources [Cyg X-2 (30); GX 340+0 (W. Penninx, see Note 2)], X-ray intensity begins to decrease when the source moves up the FB. In such a case, one predicts v_{QPO} to strongly *anti*correlate with X-ray intensity. However, no FB-QPO have been reported from these sources yet.

2.4.2 MODEL SCENARIOS Attempts to model the geometry of the emitting regions in bright LMXB using their X-ray spectrum only have been inconclusive. Two-component spectral models have been discussed (161, and references therein), but no universally accepted physical model has emerged. The recognition of the Z sources as a separate class, as well as the additional constraint provided by their rapid temporal variations, may make modeling these sources a more hopeful enterprise. Recently, attempts were begun to provide scenarios for the entire Z behavior (31, 49). Quantitative modeling within the framework of such scenarios has started (49). In the following, I sketch some considerations relevant to such scenario building.

Considering Z phenomenology, the key questions seem to be the following: What drives source motion through the Z? What happens at the branch transitions? What causes HB-QPO/LFN and NB/FB-QPO? Answering these questions likely entails resolving the older problem of explaining the color-color diagrams in terms of X-ray spectral models.

As a model for HB-QPO/LFN, the modulated-accretion magnetospheric beat-frequency model (4, 50) stands strong. In this model, plasma in near-Keplerian orbits at the inner edge of an accretion disk interacts with a neutron star magnetosphere. Plasma inhomogeneities ("clumps") enter the magnetosphere more readily at some point(s) (e.g. near the magnetic poles) than at others. As the magnetosphere spins (with the star) at a rate v_S, a given clump in the plasma, circulating with a frequency v_K, will periodically pass by a given point of easier entry into the magnetosphere with a frequency $v_B = |v_K - v_S|$ (the beat freqency between Keplerian disk and neutron star spin frequencies). The clump's accretion, modulated at v_B by its interaction with the field, contributes an oscillating shot (Section 1.1) to the X-ray intensity I_x; the time it takes for the clump to accrete corresponds to the duration of the shot, the oscillation frequency v_B is the QPO frequency, and because the shots are positive-definite, QPO and LFN strength are predicted to be similar (Section 1.1; this is also why the model is hard to apply to NB/FB-QPO).

The magnetosphere is defined as the region where the Keplerian disk flow is disrupted and the plasma is forced to corotate. Its radius r_M is determined by a balance between material and magnetic stresses originating in accreting plasma and magnetic field, respectively. In most models (see 46, 160; and references therein) the magnetosphere is quite "stiff," so that the change of r_M in response to variations in $\dot{M}$ is small. The fact that v_{QPO} varies by factors of 2 or 3 whereas $\dot{M}$ likely only varies by factors of order unity, too (the sources remain at luminosities of the order of the Eddington luminosity L_{Edd}), then suggests that v_S and v_K are relatively closely matched, so that a small variation in r_M (and hence, with Kepler's law, in v_K) can cause a relatively large change in their difference v_B (4). This means that v_K must be at least several times as large as v_B, so that for frequencies as observed in HB-QPO (again, with Kepler's law) r_M can be at most several tens of kilometers. The magnetospheric boundary is therefore in the inner, radiation-pressure-dominated region of the disk (160). Magnetospheric formation there is not well understood, but that such radii imply neutron star field strengths of $\lesssim 10^{10}$ G seems very likely. Such field strengths are also expected from evolutionary considerations (Section 1.2).

Within r_M the plasma corotates, so it is likely that $v_S < v_K$; otherwise, centrifugal forces would probably inhibit accretion. The neutron star spin frequency v_S is practically constant, so an observed increase in $v_{QPO} = |v_K - v_S|$ must imply an increase in v_K. With Kepler's law, this implies a decrease in r_M. It is very plausible on general grounds that r_M decreases when $\dot{M}$ increases (but see 160). From the sense of v_{QPO} in the HB, we then derive that $\dot{M}$ increases in the HB toward the NB and is therefore positively correlated with I_x in this branch. This seems consistent, as deviations from a positive $\dot{M}-I_x$ correlation are expected for the highest accretion rates (where more accreting matter might get into the line of sight), while in the picture sketched here, $\dot{M}$ would be lowest in the HB.

In the absence of a definitive theory of accretion onto a magnetized neutron star, the relations between r_M and $\dot{M}$ and between $\dot{M}$ and I_x are uncertain, and it is not possible to estimate the neutron star spin frequency v_S from the slope of the v_{QPO} vs. I_x relation. However, the "matching" argument given above implies that v_S is at least several tens of hertz. Thus the question arises [and has indeed been asked from the early days of the beat-frequency model (e.g. 4, 143)]: If there is a rapidly spinning magnetized neutron star, why are we not seeing a rapid pulsar in these systems?

Some mechanisms might suppress pulsar modulation (50, 143). A particularly effective one may be scattering. A region of scattering plasma can suppress pulsar modulation caused by beamed radiation while *transmitting* much of the isotropic QPO signal, even if their frequencies are similar (10, 45, 151). Such plasma can also cause millisecond time lags in the QPO signal between energy bands: Compton scattering in a hot plasma would result in a situation where photons that have undergone many scattering events emerge with higher energies, but delayed (owing to a longer light travel time) with respect to those scattered only a few times (29; see, e.g., 99). The size of the scattering region required to explain the observed time lags of up to 4 ms is relatively large (10^2-10^3 km; 12, 117, 162). Since most gravitational energy is released in the inner few kilometers from the neutron star surface, it is unlikely that a large fraction of the luminosity could be provided by hot electrons in a cloud of this size (100, 117). However, if one allows the oscillating flux incident on the cloud (before Comptonization) to have a different X-ray spectrum from the non-oscillating incident flux, there may be ways out of this energy problem

(49). Simultaneous measurements of X-ray flux, QPO amplitude, and time lags (all as a function of photon energy) combined with quantitative modeling will provide strong constraints on models of this kind.

The modulated-accretion magnetospheric beat-frequency model apparently allows one to account for most of the observed properties of the HB-QPO/LFN. Its basic strong points are (a) a natural preferred frequency (v_B) whose value (and range of variation) fits with observations without stretching model parameters, and (b) the fact that the model involves modulation of the basic accretion process and has therefore no problem in producing strong oscillations. These strong points have not been matched by competing models for HB-QPO (Section 1.3). However, the model should be confirmed by detection of the predicted low-level periodic pulsations. Provisionally accepting the model, we know the parameter varying along the Z and its sense: It is $\dot{M}$, increasing toward the FB. This points to where we may have to look for answers to the other key questions: What happens at the branch transitions, and what is the nature of the NB/FB-QPO?

In early proposals, the HB-NB transition has been attributed to a switch from magnetically channeled polar cap to more uniform accretion (6), and the NB-FB transition to a drop in the amount of rotational energy fed back from a rapidly spinning neutron star into the disk (92; see, however, 35). In all recent scenarios radiation pressure plays the key role. As a group, Z sources are near the Eddington luminosity L_{Edd}, so when $\dot{M}$ increases in a given source, one expects to see the effects of radiation pressure at some point. The NB-QPO frequency is nearly the same between sources; this suggests that it is determined by some parameter not varying much between sources, e.g. L_{Edd} (28). Two possible consequences of approaching L_{Edd}—a transition to a thick accretion disk and a spherical accretion regime (28, 30, 31, 35, 48, 49, 145), and a decrease in inflow velocity (28, 49)—have been exploited in recent explanations for spectral and temporal behavior in the NB and FB.

The HB-NB transition may be caused by the onset of an inner disk instability that leads to a geometrically thick disk (30, 48, 49). The X-ray spectrum seemingly pivots around a point between 3 and 6 keV when the source moves up and down the NB, which suggests that the X-ray spectral changes in the NB are due to changes in the degree of Comptonization, caused by a changing scattering optical depth of the plasma in the line of sight (29, 49). A transition to a super-Eddington regime may explain the NB-FB transition (30, 31, 35, 49), with outflow and lower scattering optical depth causing spectral changes in the FB similar to those in the NB, but in the opposite direction along the branch (49). Sco X-1 flares and Cyg X-2 dips up the FB; this behavior may be due to differences in inclination (35).

HB-QPO/LFN become much weaker at the HB-NB transition point, which might indicate that the plasma clumps disappear there (31). This disappearance might be caused by a slowdown of the fall velocity of the accreting plasma sufficient for processes occurring at the speed of sound to have the opportunity to smooth out plasma inhomogeneities (49).

Oscillations in a geometrically thick, toruslike inner accretion disk partially obscuring the X rays were proposed to explain NB/FB-QPO [145 (cf. 7), but see also 48]. Hasinger (28) proposed that a standing sound wave in a radiation-pressure-supported plasma layer close to the neutron star surface provides the correct frequency for NB-QPO (see, however, 49, 91). This proposal may have been the first to exploit the fact that at near-Eddington luminosities, radiation pressure slows down the falling plasma sufficiently for sound-speed effects to become important (see also 69a).

Lamb (49) recently proposed that an instability can occur in the sloweddown accretion flow, which originates in the fact that the flow needs a finite time to react (with its reduced fall speed) to changes in luminosity. The resulting optical depth changes in the accreting plasma cause the quasi-periodic rocking of the X-ray spectrum particular to NB-QPO (Section 2.4.1). Preliminary results from quantitative modeling of the flow (G. Miller & F. Lamb, see Note 2) show spectral variations that in an X-ray color-color diagram are reminiscent of HB-NB behavior. The angular momentum of the accreting plasma is neglected in this model; it is likely that more realistic descriptions would involve accretion tori, as earlier invoked in more qualitative scenarios for the spectral changes (47) and QPO (145). In models of the type discussed above, the X-ray luminosity on the NB is always close to L_{Edd}, so that sources in the NB might provide standard candles (28, 49).

Much of the preceding discussion is in a sketchy state. However, a general framework seems to be emerging that has the magnetospheric beat-frequency mechanism and Eddington-limited flows as its two key ingredients. Possibly, for the first time we are dealing with models for the brightest LMXB that can be tested with data obtainable with presently available instrumentation.

3. ATOLL SOURCES

The proposal that there exists a class of bright LMXB that has a pattern of correlated X-ray and power spectral behavior fundamentally different from that of the Z sources is recent (36). Not much is known as yet about the relation of this new class of "atoll" sources to Z sources, and it is still conceivable that the classification is not one of sources, but of source states. However, as a framework within which to discuss the phenomenology of bright LMXB, the division between Z and atoll sources is extremely useful, and this is why I use it here. Sources identified as exhibiting atoll behavior are indicated with an "A" in Table 1. As a group, they are somewhat fainter than the Z sources and, contrary to these, some of them are reliable X-ray bursters. Table 4 summarizes which aspects of atoll behavior were detected in each source.

3.1 Islands and Bananas

Two different structures, provisionally labeled "islands" and "bananas" (36), occur in the color-color diagrams of the atoll sources.

Curved, relatively fuzzy branches such as in Figure 7 (*top left*) are the bananas; the isolated clump in Figure 7 (*top right*) is an example of an island. Both types of structure can occur in a single source, and together they sometimes form a kind of atoll in the color-color diagram. The reason for these fragmented color-color diagrams is observational: In an island state a source apparently does not move much during a typical observational interval of hours to days. Thus, atoll sources in the island state are less active in the color-color diagram than are Z sources, which in the same time trace out large segments of the Z. One transition from an island to the left end of a banana has been observed. It is not clear that islands appear at any fixed position in the diagrams with respect to the bananas, but the bananas recur in a given source in roughly the same place. The X-ray intensity is low in the island state, and this state is characterized by regular X-ray burst behavior; less regular bursting can also occur in the bananas.

Two power spectral components have been identified in atoll sources.

Table 4 Atoll behavior[a]

Source	"Banana" state			"Island" state		
	Branch[b]	VLFN	HFN	Branch[b]	VLFN	HFN
4U 1820−30	●	●	●	●	●	●
4U 1705−44	●	●	●	●	—	●
4U 1636−53	●	●	●	●	—	●
GX 9+1	●	●	●	—	—	—
GX 3+1	●	●	●	—	—	—
4U 1735−44	●	●	●	—	—	—
GX 9+9	●	●	●	—	—	—
GX 13+1	●	●	—	—	—	—
4U 1728−33	—	—	—	●	—	●

[a] ●: detected; —: not detected (36).
[b] Refers to detection of spectral branch in X-ray hardness vs. intensity or color-color diagram.

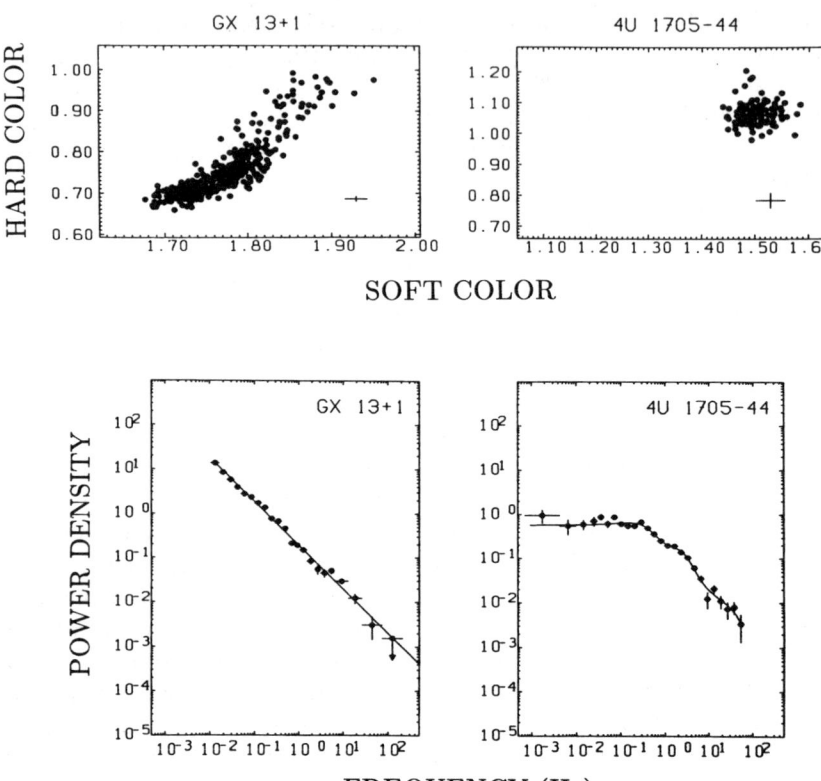

Figure 7 Atoll source behavior. Color-color diagrams (*top*) show elongated "banana" branches (*left*) and isolated "islands" (*right*); power spectra (*bottom*) are nearly pure power laws in the banana state (*left*) and show a clear break in the island state (*right*). Both types of behavior can be observed in the same atoll source at different epochs. After Hasinger & van der Klis (36).

They are similar in shape to the VLFN and HFN discussed in Section 2 and bear the same names; there is no formal evidence that they are related—in fact, in some respects the atoll HFN may be more similar to Z source LFN (see below).

VLFN is strongest in the bananas. Its power-law slope is less than in Z sources; usually it is close to 1 ("1/f noise"). Its strength is typically 2–3% (rms) in the bananas and less than 1% in the islands. HFN dominates the island power spectra; strengths can be large [up to 20% (rms)]. The HFN shape shows more or less the same range of variation as seen in LFN in Z sources: between strictly "red" and a broad peak (17, 36).

Both components have additional structure superimposed on their overall shapes; wiggles, such as those visible in Figure 7 (*bottom*), abound. Some have the character of QPO peaks (56, 74, 116). It has been argued (36) that they are not similar to QPO seen in Z sources. There is a tendency for gradual power spectral changes when a source moves from the island state to the left end of the banana, and from there to the upper right of the banana: VLFN becomes stronger and HFN weaker.

3.2 Understanding the Atoll Sources

No explanation has been proposed for atoll behavior. However, something can be said about the relation between atoll and Z sources (36). The division between atoll and Z sources does not coincide with earlier proposed groupings of bright LMXB (see 148, and references therein). Because, as a group, atoll sources are fainter than Z sources, they may have lower accretion rates. Interestingly, all four orbital periods measured in atoll systems are short (<5 hr), whereas both periods measured in Z sources are long (>10 hr). Thus the data are consistent with Z sources containing evolved companion stars, while atoll sources do not; this might account for different accretion rates (see Section 1.2; cf. 148). If the only difference between Z and atoll sources is accretion rate, then transitions between Z and atoll behavior are expected in LMXB that show a large range in $\dot{M}$. It is not clear, however, that a lower accretion rate alone explains all differences—in particular, the absence of HB-QPO in atoll sources. Possibly, the atoll neutron stars have lower magnetic fields, so that the beat-frequency mechanism does not work (36).

4. RELATED OBJECTS

The phenomena discussed in this section form in a sense the loose ends of the phenomenology of QPO and noise in X-ray binaries. Their relation with QPO and noise in the Z and atoll sources, of which they are in some cases very reminiscent, is in all cases uncertain.

4.1 Nonpersistently Bright LMXB

4.1.1 CIRCINUS X-1 Cir X-1 is a peculiar recurrent transient source showing large transitions in X-ray intensity related to a (probably orbital) period of 16.6 days (see 123, and references therein). During a bright X-ray outburst, 5–17 Hz QPO were seen for approximately 7 hr (123, 124). Their frequency was weakly correlated with X-ray intensity between 5 and 6 Hz and was strongly correlated between 7 and 17 Hz. This behavior is reminiscent of NB/FB-QPO, and spectral variations during this episode suggested the presence of two connected branches (125) in a hardness-intensity diagram; however, the exact relation to Z behavior is unclear. Excess power in the 30–300 Hz range detected simultaneously was described as a QPO broad peak (123); it may be an HFN component.

A very narrow (0.03–0.05 Hz) 1.4-Hz QPO peak was detected during another outburst (125). The QPO frequency did not significantly vary when the X-ray intensity changed by a factor of six. Very strong [40–80% (rms) down to 0.05 Hz] red noise was observed in this observation.

4.1.2 THE RAPID BURSTER   The Rapid Burster (MXB 1730−335; see 55) is unique in showing transient episodes during which the source exhibits Type II X-ray bursts, which are probably sudden increases in accretion rate lasting from seconds to minutes and recurring with intervals between 10 s and ∼1 hr. Long Type II bursts often have flat tops. Three different types of QPO were reported.

During long, flat-topped bursts, strong [up to 20% (rms)] QPO with frequencies in the 2–5 Hz range occur (109, 122). QPO peaks are narrow ($\Delta v/v$ down to 0.05; 109, 122), and red noise is weak (rms amplitude down to 10% of that of QPO; 109). L. Stella (see Note 2) argues that these QPO can be identified with NB-QPO in Z sources (see also 28). This is based on (a) similar QPO frequencies, (b) an anticorrelation between v_{QPO} and I_x (109) also seen in Sco X-1 in the NB, and (c) the shape of the X-ray color-color diagram.

Even stronger [up to 35% (rms)] 2–4.5 Hz QPO sometimes occur in the persistent (i.e. nonburst) flux, after short (1–2 min) Type II bursts. They often show a characteristic frequency evolution from ∼4 to ∼2 Hz within 3 to 6 min, correlated to a simultaneous X-ray spectral hardness evolution (109). In these QPO, time lags have been detected: For oscillation frequencies >3.6 Hz, the soft (1–5 keV) signal lags the hard (5–20 keV) one by ∼8 ms (108).

The third kind of QPO has frequencies between 0.4 and 1 Hz and also occurs in the persistent flux, preferentially before short Type II bursts. A strong second harmonic was observed in one case (109).

Recent observations with the *Ginga* satellite [F. Makino et al., reported by K. Mitsuda (see Note 2)] show QPO switching between frequencies of 5, 10, and 15 Hz during two long Type II bursts; the higher frequencies are consistent with their being harmonics of the lowest frequencies. The quality of the data is sufficient to see the oscillations directly in the X-ray light curve.

4.2 *Accreting Pulsars*

Two cases of a QPO peak in MXB accreting pulsars have recently been reported. In Cen X-3 (A. Tennant, see Note 2) the QPO frequency is between 35 and 68 mHz for a pulsar frequency of ~ 210 mHz. In EXO 2030+337 (L. Angelini, see Note 2) the QPO frequency is between 190 and 210 mHz for a ~ 24-mHz pulsar. The QPO frequency of the latter was reported to decrease by $\sim 10\%$ when the X-ray intensity dropped by $\sim 25\%$.

These observations are of great interest, as QPO caused by a beat-frequency mechanism (Section 2.4.2) in a source whose neutron star spin frequency is known allow us to measure the magnetospheric radius r_M. In both cases, the QPO frequency is consistent with a magnetospheric beat-frequency model with $r_M \approx 4000$ km, which indicates a magnetic field strength appropriate for an accreting pulsar (10^{12}–10^{13} G). In the case of EXO 2030+337, the slope of the correlated frequency-intensity change fits as well. However, in this case the data are also consistent with the QPO frequency being the Kepler frequency at the magnetospheric boundary. With a pulsar beam sweeping the surroundings, a beat-frequency model involving reflection or obscuration of X rays by circulating blobs in the disk is a possibility that should be considered in addition to an accretion-modulation mechanism.

Noise components have been reported from each of a sample of eight accreting pulsars by T. Belloni (see Note 2). The shape of the power spectrum of the noise is usually very reminiscent of HFN in atoll sources (it could also be LFN), including the power spectral wiggles. Cutoff frequencies are in the 0.1–10 Hz range. The strength of the noise can be large [up to 30% (rms)] and is quite variable.

4.3 *Black Hole Candidates*

Strong [$\gtrsim 20\%$ (rms)], rapid (>1 Hz) variability is no longer a good argument for black hole candidacy; Cir X-1 (Section 4.1.1) and V0332+53 were both suspected to be a black hole on these grounds (41, 121) and then later shown to be a burst source (126) and a pulsar (114), respectively. Many pulsars are now known to be strongly and rapidly variable (Section 4.2), and LMC X-3 [a strong black hole candidate on the basis of its companion star's orbital optical radial velocity curve (15)] does not show strong, rapid variability (129). The well-known rapid variability of another good black hole candidate, Cyg X-1 (see 60), has a noise power spectrum similar to that of pulsars (T. Belloni, see Note 2).

LMC X-1, a "tentative" black hole candidate on the basis of a relatively low-quality optical radial velocity curve (39; see also 67) and its so-called ultra-soft X-ray spectrum (157), shows slow (0.075 Hz) QPO (18). Slow QPO were earlier observed (77) in GX 339−4, a black hole candidate solely on the basis of its ultra-soft spectrum and its bimodal spectral behavior reminiscent of Cyg X-1 (see 157). It is unclear whether these two cases of slow QPO are related, as is their relation with the slow QPO mentioned in Section 1.3.

The full pattern, if there is one, has not yet emerged. Very recently, the surprising discovery of ∼6-Hz 3–4% (rms) QPO peaks with widths between 0.5 and 1.5 Hz was reported (63) from GX 339−4. Strong second harmonics and a broad noise component of strongly variable shape were observed as well. As noted by K. Mitsuda (see Note 2), if these QPO can be identified with NB-QPO in Z sources, then either the source is a black hole, and thus a neutron star surface and magnetic field are not required for NB-QPO, or it is a neutron star and then ultra-soft spectra are not a sign of black holes. However, the QPO in GX 339−4 are similar but not identical to NB-QPO: The frequency is the same, but harmonic content and red-noise strength are clearly larger.

5. CONCLUSIONS

Our understanding of the phenomenology of the bright LMXB has increased enormously over the last few years. Most progress was made in unraveling the properties of the very brightest among them, which are now called the Z sources (Section 2). These sources show three distinct X-ray spectral states, and two patterns of rapid temporal variability whose occurrence strongly correlates.

The "horizontal branch" quasi-periodic oscillations and associated low-frequency noise (HB-QPO/LFN) have properties that suggest a magnetospheric origin. The modulated-accretion magnetospheric beat-frequency model, involving a rapidly spinning neutron star with a surface field of roughly 10^{9-10} G has been successful in accounting for these variations. A prediction of this model is that at some (low) level, millisecond pulsations due to the magnetized neutron star's rotation are present.

The "normal-" and "flaring-branch" quasi-periodic oscillations (NB/FB-QPO) in Z sources are less easy to interpret, but the similar frequencies between sources in the NB state, the X-ray spectral variations to which the QPO correlate, and general considerations concerning the X-ray luminosity of the sources hint at Eddington-limited accretion flows as an important ingredient in their description.

Observations of subtle differences (time lags, amplitude variations) in the rapid variability between X-ray spectral bands provide links between X-ray spectroscopy and X-ray timing of Z sources. Compton-scattering models for this kind of data appear successful, and the data are of sufficient quality to test them at a detailed level.

The observational data on the atoll sources (Section 3) are as yet less constraining. However, the fact that there are two populations among bright LMXB is in itself important information. Present evidence is consistent with the two classes having different companion star types.

Broad noise components of very similar shape are observed in virtually all stellar-mass accreting compact objects out to relatively high frequencies (>1 Hz), and various QPO phenomena are seen in the more distant cousins of Z sources, such as pulsars and black hole candidates. Caution is indicated, however, in the interpretation of these results. As we have seen in the Z sources, apparently similar timing phenomena can turn out to be physically quite different. Although there are basic similarities, the detailed physical circumstances in these sources differ greatly. Even if a beat-frequency mechanism is at work, it might be physically quite distinct from that believed to operate in Z sources.

A few tasks that lie ahead can be clearly defined at this point. Further observational work on the apparently universal broad noise components is needed to clarify their relation to source type. The properties of accreting-pulsar and black-hole candidate QPO should be further investigated: Are they like those in Z sources, or are the observed similarities only superficial? The relation of atoll sources to Z sources needs to be clarified: Are they similar sources at a lower accretion rate, or are they in some respect qualitatively different? Observations of LMXB varying over a large range in X-ray luminosity (e.g. soft X-ray transients) could be crucial here.

It should be attempted, finally, to construct comprehensive models for the entire pattern of correlated temporal and spectral behavior of the Z sources. The observations have reached a level where it is possible to quantitatively test such models. This will in any case improve our understanding of poorly grasped physics, such as magnetospheric formation and near-Eddington accretion flows, and if successful, the models will allow us to estimate basic quantities of compact objects in LMXB, such as neutron star spin rates and magnetic field strengths.

ACKNOWLEDGMENTS

I gratefully acknowledge helpful discussions with Günther Hasinger and Luigi Stella during the writing of this review, and with most of the authors cited during the previous years. I thank Jan van Paradijs and Günther Hasinger for carefully reading the manuscript.

Literature Cited

1. Alpar, M. A. 1986. *MNRAS* 223: 469–78
2. Alpar, M. A. 1987. In *High Energy Phenomena Around Collapsed Stars, NATO ASI Ser. C*, ed. F. Pacini, 195: 359–66. Dordrecht: Reidel
3. Alpar, M. A., Shaham, J. 1985. *IAU Circ. No. 4046*
4. Alpar, M. A., Shaham, J. 1985. *Nature* 316: 239–41
5. Angel, J. R. P., Kestenbaum, H., Novick, R. 1971. *Ap. J. Lett.* 169: L57–61

6. Berman, N. M., Stollman, G. M. 1988. *MNRAS* 232: 487–95
7. Boyle, C. B., Fabian, A. C., Guilbert, P. W. 1986. *Nature* 319: 648–49
8. Bradt, H. V., Kelley, R. L., Petro, L. D. 1982. In *Galactic X-Ray Sources*, ed. P. W. Sandford, P. Laskarides, J. Salton, pp. 89–112. Chichester, Engl: Wiley
9. Bradt, H. V. D., McClintock, J. E. 1983. *Annu. Rev. Astron. Astrophys.* 21: 13–66
10. Brainerd, J., Lamb, F. K. 1987. *Ap. J. Lett.* 317: L33–38
11. Branduardi, G., Kylafis, N. D., Lamb, D. Q., Mason, K. O. 1980. *Ap. J. Lett.* 235: L153–57
12. Bussard, R. W., Weisskopf, M. C., Elsner, R. F., Shibazaki, N. 1988. *Ap. J.* 327: 284–93
13. Collmar, W., Kendziorra, E., Staubert, R. 1988. *Adv. Space Res.* 8(2): 411–14
14. Cooke, B. A., Stella, L., Ponman, T. 1985. *IAU Circ. No. 4116*
15. Cowley, A. P., Crampton, D., Hutchings, J. B., Remillard, R., Penfold, J. E. 1983. *Ap. J.* 272: 118–22
16. Cowley, A. P., Hutchings, J. B., Crampton, D. 1988. *Ap. J.* 333: 906–16
16a. Deeter, J. E. 1984. *Ap. J.* 281: 482–91
17. Dotani, T., Mitsuda, K. 1988. In *Physics of Neutron Stars and Black Holes*, ed. Y. Tanaka, pp. 143–48. Tokyo: Universal Acad. Press
18. Ebisawa, K., Mitsuda, K., Inoue, H., Dotani, T. 1988. In *Physics of Neutron Stars and Black Holes*, ed. Y. Tanaka, pp. 149–53. Tokyo: Universal Acad. Press
19. Eckmann, J.-P., Ruelle, D. 1985. *Rev. Mod. Phys.* 57: 617–56
20. Elsner, R. F., Shibazaki, N., Weisskopf, M. C. 1987. *Ap. J.* 320: 527–36
21. Elsner, R. F., Shibazaki, N., Weisskopf, M. C. 1988. *Ap. J.* 327: 742–49
22. Elsner, R. F., Weisskopf, M. C., Darbro, W., Ramsey, B. D., Williams, A. C., et al. 1986. *Ap. J.* 308: 655–60
23. Forman, W., Jones, C., Tananbaum, H. 1976. *Ap. J.* 208: 849–62
24. Giacconi, R., Gursky, H., Kellogg, E., Schreier, E., Tananbaum, H. 1971. *Ap. J. Lett.* 167: L67–73
25. Giacconi, R., Gursky, H., Paolini, F. R., Rossi, B. B. 1962. *Phys. Rev. Lett.* 9: 439–43
26. Hameury, J.-M., King, A. R., Lasota, J.-P. 1985. *Nature* 317: 597–99
27. Hasinger, G. 1986. In *The Evolution of Galactic X-Ray Binaries, NATO ASI Ser. C*, ed. J. Truemper, W. H. G. Lewin, W. Brinkmann, 167: 139–49. Dordrecht: Reidel
28. Hasinger, G. 1987. *Astron. Astrophys.* 186: 153–58
29. Hasinger, G. 1987. In *The Origin and Evolution of Neutron Stars, IAU Symp. No. 125*, ed. D. J. Helfand, J.-H. Huang, pp. 333–45. Dordrecht: Reidel
30. Hasinger, G. 1988. *Adv. Space Res.* 8(2): 377–81
31. Hasinger, G. 1988. In *Physics of Neutron Stars and Black Holes*, ed. Y. Tanaka, pp. 97–115. Tokyo: Universal Acad. Press
32. Hasinger, G., Langmeier, A., Sztajno, M., Pietsch, W., Gottwald, M. 1985. *IAU Circ. No. 4153*
33. Hasinger, G., Langmeier, A., Sztajno, M., Trümper, J., Lewin, W. H. G., et al. 1986. *Nature* 319: 469–71
34. Hasinger, G., Langmeier, A., Sztajno, M., White, N. 1985. *IAU Circ. No. 4070*
35. Hasinger, G., Priedhorsky, W. C., Middleditch, J. 1988. *Ap. J.* 337: 843–48
36. Hasinger, G., van der Klis, M. 1989. *Astron. Astrophys.* In press
37. Hertz, P. 1988. In *Physics of Neutron Stars and Black Holes*, ed. Y. Tanaka, pp. 211–14. Tokyo: Universal Acad. Press
38. Hjellming, R. M., Wade, C. M. 1971. *Ap. J. Lett.* 164: L1–7
39. Hutchings, J. B., Crampton, D., Cowley, A. P. 1983. *Ap. J. Lett.* 275: L43–47
40. Ilovaisky, S. A., Chevalier, C., White, N. E., Mason, K. O., Sanford, P. W., et al. 1980. *MNRAS* 191: 81–93
41. Jones, C., Giacconi, R., Forman, W., Tananbaum, H. 1974. *Ap. J. Lett.* 191: L71–74
42. Joss, P. C., Avni, Y., Rappaport, S. 1978. *Ap. J.* 221: 645–51
43. Kulkarni, S. R. 1988. *Adv. Space Res.* 8(2): 343–50
44. Kundt, W., ed. 1989. *Neutron Stars, AGN and Jets, NATO ASI Ser. C.* In press
45. Kylafis, N. D., Klimis, G. S. 1987. *Ap. J.* 323: 678–84
46. Lamb, F. K. 1984. In *High Energy Transients in Astrophysics, AIP Conf. Proc.*, ed. S. E. Woosley, 115: 179–214. New York: AIP
47. Lamb, F. K. 1986. In *The Evolution of Galactic X-Ray Binaries, NATO ASI Ser. C*, ed. J. Truemper, W. H. G. Lewin, W. Brinkmann, 167: 151–71. Dordrecht: Reidel
48. Lamb, F. K. 1988. *Adv. Space Res.* 8(2): 421–47
49. Lamb, F. K. 1988. *Ap. J.* In press

50. Lamb, F. K., Shibazaki, N., Alpar, M. A., Shaham, J. 1985. *Nature* 317: 681–87
51. Langmeier, A. 1988. PhD Thesis. Ludwig-Maximilians-Univ. München, Germ. (In German)
52. Langmeier, A., Hasinger, G., Sztajno, M., Trümper, J., Pietsch, W. 1985. *IAU Circ. No. 4147*
53. Lewin, W. H. G. 1986. In *The Physics of Accretion onto Compact Objects, Lect. Notes Phys.*, ed. K. O. Mason, M. G. Watson, N. E. White, 266: 177–93. Berlin: Springer-Verlag
54. Lewin, W. H. G., Clark, G. W., Smith, W. B. 1968. *Ap. J. Lett.* 152: L55–61
55. Lewin, W. H. G., Joss, P. C. 1983. In *Accretion-Driven Stellar X-Ray Sources*, ed. W. H. G. Lewin, E. P. J. van den Heuvel, pp. 41–115. Cambridge: University Press
56. Lewin, W. H. G., van Paradijs, J., Hasinger, G., Penninx, W. H., Langmeier, A., et al. 1987. *MNRAS* 226: 383–94
57. Lewin, W. H. G., van Paradijs, J., Jansen, F., van der Klis, M., Sztajno, M., et al. 1985. *IAU Circ. No. 4101*
58. Lewin, W. H. G., van Paradijs, J., van der Klis, M. 1988. *Space Sci. Rev.* 46: 273–377
59. Li, F. K., Joss, P. C., McClintock, J. E., Rappaport, S., Wright, E. L. 1980. *Ap. J.* 240: 628–35
60. Liang, E. P., Nolan, P. L. 1984. *Space Sci. Rev.* 38: 353–84
61. Makishima, K. 1986. *ISAS Res. Note 313*, ISAS, Tokyo. 29 pp.
62. Makishima, K., Mitsuda, K., Inoue, H., Koyama, K., Matsuoka, M., et al. 1983. *Ap. J.* 267: 310–14
63. Makishima, K., Miyamoto, S., and the Ginga Team. 1988. *IAU Circ. No. 4653*
64. Mason, K. O. 1986. In *The Physics of Accretion Onto Compact Objects, Lect Notes Phys.*, ed. K. O. Mason, M. G. Watson, N. E. White, 266: 29–57. Berlin: Springer-Verlag
65. Mason, K. O., Middleditch, J., Nelson, J. E., White, N. E. 1980. *Nature* 287: 516–18
66. Matsuoka, M. 1985. In *Catalysmic Variables and Low-Mass X-Ray Binaries*, ed. D. Q. Lamb, J. Patterson, pp. 139–42. Dordrecht: Reidel
67. McClintock, J. E. 1986. In *The Physics of Accretion Onto Compact Objects, Lect. Notes Phys.*, ed. K. O. Mason, M. G. Watson, N. E. White, 266: 211–28. Berlin: Springer-Verlag
68. McDermott, P. N., Taam, R. E. 1987. *Ap. J.* 318: 278–87
69. McHardy, I. M., Lawrence, A., Pye, J. P., Pounds, K. A. 1981. *MNRAS* 197: 893–919
69a. Meyer, F. 1986. In *Radiation Hydrodynamics in Stars and Compact Objects, Lect. Notes Phys.*, ed. D. Mihalas, K.-H. A. Winkler, 255: 249–67. Berlin: Springer-Verlag
70. Middleditch, J., Priedhorsky, W. 1985. *IAU Circ. No. 4060*
71. Middleditch, J., Priedhorsky, W. C. 1986. *Ap. J.* 306: 230–37
72. Mitsuda, K. 1984. *ISAS Res. Note 251*, ISAS, Tokyo
73. Mitsuda, K. 1988. In *Physics of Neutron Stars and Black Holes*, ed. Y. Tanaka, pp. 117–31. Tokyo: Universal Acad. Press
74. Mitsuda, K., Dotani, T., Yoshida, A. 1988. In *Physics of Neutron Stars and Black Holes*, ed. Y. Tanaka, pp. 133–42. Tokyo: Universal Acad. Press
75. Mitsuda, K., and the GINGA Lac Team. 1988. *Adv. Space Res.* 8(2): 391–95
76. Morfill, G. E., Truemper, J. 1986. In *The Evolution of Galactic X-Ray Binaries, NATO ASI Ser. C*, ed. J. Truemper, W. H. G. Lewin, W. Brinkmann, 167: 173–81. Dordrecht: Reidel
77. Motch, C., Ricketts, M. J., Page,C. G., Ilovaisky, S. A., Chevalier, C. 1983. *Astron. Astrophys.* 119: 171–76
78. Murakami, T., Inoue, H., Makishima, K., Hoshi, R. 1987. *Publ. Astron. Soc. Jpn.* 39: 879–86
79. Miyamoto, S., Matsuoka, M. 1977. *Space Sci. Rev.* 20: 687–755
80. Norris, J. P., Wood, K. S. 1985. *IAU Circ. No. 4087*
81. Norris, J. P., Wood, K. S. 1987. *Ap. J.* 312: 732–38
82. Oda, M., Gorenstein, P., Gursky, H., Kellogg, E., Schreier, E., et al. 1971. *Ap. J. Lett.* 166: L1–7
83. Ostriker, J. P. 1977. *Ann. N.Y. Acad. Sci.* 302: 229–43
84. Parmar, A. N., White, N. E. 1988. In *X-Ray Astronomy with EXOSAT, Mem. Soc. Astron. Ital.*, ed. R. Pallavicini, N. E. White, 59: 147–68
85. Parsignault, D. R., Grindlay, J. E. 1978. *Ap. J.* 225: 970–87
86. Patterson, J. 1979. *Ap. J.* 234: 978–92
87. Penninx, W., Hasinger, G., Lewin, W. H. G., van Paradijs, J., van der Klis, M. 1988. *MNRAS.* In press
87a. Penninx, W., Lewin, W. H. G., Mitsuda, K., van der Klis, M., van Paradijs, J., et al. 1989. *Astron. Astrophys.* In press
88. Penninx, W., Lewin, W. H. G., Zijlstra, A. A., Mitsuda, K., van Paradijs, J., et al. 1988. *Nature* 336: 146–48

89. Petro, L. D., Bradt, H. V., Kelley, R. L., Horne, K., Gomer, R. 1981. *Ap. J. Lett.* 251: L7–11
90. Ponman, T. 1982. *MNRAS* 201: 769–99
91. Ponman, T. J., Cooke, B. A., Stella, L. 1988. *MNRAS* 231: 999–1009
92. Priedhorsky, W. 1986. *Ap. J. Lett.* 306: L97–100
93. Priedhorsky, W., Hasinger, G., Lewin, W. H. G., Middleditch, J., Parmar, A., et al. 1986. *Ap. J. Lett.* 306: L91–95
94. Pringle, J. E. 1981. *Annu. Rev. Astron. Astrophys.* 19: 137–62
95. Rappaport, S. A., Joss, P. C. 1983. In *Accretion-Driven Stellar X-Ray Sources*, ed. W. H. G. Lewin, E. P. J. van den Heuvel, pp. 1–39. Cambridge: University Press
96. Sadeh, D., Byram, E. T., Chubb, T. A., Friedman, H., Hedler, R. L., et al. 1982. *Ap. J.* 257: 214–24
97. Schulz, N. S., Hasinger, G., Trümper, J. 1988. *Astron. Astrophys.* In press
98. Shaham, J. 1987. In *The Origin and Evolution of Neutron Stars, IAU Symp. No. 125*, ed. D. J. Helfand, J.-H. Huang, pp. 347–61. Dordrecht: Reidel
99. Shapiro, S. L., Lightman, A. P., Eardly, D. M. 1976. *Ap. J.* 204: 187–99
100. Shibazaki, N., Elsner, R. F., Bussard, R. W., Ebisuzaki, T., Weisskopf, M. C. 1988. *Ap. J.* 331: 247–60
101. Shibazaki, N., Elsner, R. F., Weisskopf, M. C. 1987. *Ap. J.* 322: 831–37
102. Shibazaki, N., Lamb, F. K. 1987. *Ap. J.* 318: 767–85
103. Shibazaki, N., Mitsuda, K. 1984. In *High Energy Transients in Astrophysics, AIP Conf. Proc.*, ed. S. E. Woosley, 115: 63–71. New York: AIP
104. Stella, L. 1986. In *Plasma Penetration into Magnetospheres*, ed. N. Kylafis, J. Papamastorakis, J. Ventura, pp. 199–214. Iraklion: Crete Univ. Press
105. Stella, L. 1988. *Adv. Space Res.* 8(2): 367–76
106. Stella, L. 1988. In *X-Ray Astronomy with EXOSAT, Mem. Soc. Astron. Ital.*, ed. R. Pallavicini, N. E. White, 59: 185–211
107. Stella, L., Chiappetti, L., Ciapi, A. L., Maraschi, L., Tanzi, E. G., Treves, A. 1986. *Proc. Marcel Grossman Meet., 4th*, ed. R. Ruffini, pp. 861–66. Amsterdam: North-Holland
108. Stella, L., Haberl, F., Lewin, W. H. G., Parmar, A. N., van der Klis, M., et al. 1988. *Ap. J. Lett.* 327: L13–16
109. Stella, L., Haberl, F., Lewin, W. H. G., Parmar, A. N., van Paradijs, J., et al. 1988. *Ap. J.* 324: 379–90
110. Stella, L., Kahn, S. M., Grindlay, J. E. 1984. *Ap. J.* 282: 713–18
111. Stella, L., Parmar, A. N., White, N. E. 1985. *IAU Circ. No. 4102*
112. Stella, L., Parmar, A. N., White, N. E. 1987. *Ap. J.* 321: 418–24
113. Stella, L., Parmar, A. N., White, N. E., Lewin, W. H. G., van Paradijs, J. 1985. *IAU Circ. No. 4110*
114. Stella, L., White, N. E., Davelaar, J., Parmar, A. N., Blissett, R. J., et al. 1985. *Ap. J. Lett.* 288: L45–49
115. Stella, L., White, N. E., Priedhorsky, W. 1985. *IAU Circ. No. 4117*
116. Stella, L., White, N. E., Priedhorsky, W. 1987. *Ap. J. Lett.* 315: L49–53
117. Stollman, G. M., Hasinger, G., Lewin, W. H. G., van der Klis, M., van Paradijs, J. 1987. *MNRAS* 227: 7P–12P
118. Stollman, G. M., Kuperus, M. 1988. *Astron. Astrophys.* 203: 104–10
119. Taam, R. E. 1983. *Ap. J.* 270: 694–99
120. Taam, R. E., van den Heuvel, E. P. J. 1986. *Ap. J.* 305: 235–45
121. Tanaka, Y., and the Tenma Team. 1983. *IAU Circ. No. 3891*
122. Tawara, Y., Hayakawa, S., Kunieda, H., Makino, F., Nagase, F. 1982. *Nature* 299: 38–41
123. Tennant, A. F. 1987. *MNRAS* 226: 971–78
124. Tennant, A. F. 1988. *Adv. Space Res.* 8(2): 397–404
125. Tennant, A. F. 1988. *MNRAS* 230: 403–14
126. Tennant, A. F., Fabian, A. C., Shafer, R. A. 1986. *MNRAS* 219: 871–81
127. Terrell, N. J. Jr. 1972. *Ap. J. Lett.* 174: L35–41
128. Terrell, J., Olsen, K. H. 1970. *Ap. J.* 161: 399–413
129. Treves, A., Belloni, T., Chiappetti, L., Maraschi, L., Stella, L., et al. 1988. *Ap. J.* 325: 119–27
130. van den Heuvel, E. P. J. 1983. In *Accretion-Driven Stellar X-Ray Sources*, ed. W. H. G. Lewin, E. P. J. van den Heuvel, pp. 303–41. Cambridge: University Press
131. van den Heuvel, E. P. J. 1988. *Adv. Space Res.* 8(2): 355–65
132. van der Klis, M. 1986. In *The Physics of Accretion Onto Compact Objects, Lect. Notes Phys.*, ed. K. O. Mason, M. G. Watson, N. E. White, 266: 157–75. Berlin: Springer-Verlag
133. van der Klis, M. 1987. In *The Origin and Evolution of Neutron Stars, IAU Symp. No. 125*, ed. D. J. Helfand, J.-H. Huang, pp. 321–31. Dordrecht: Reidel
134. van der Klis, M. 1987. In *Variability of Galactic and Extragalactic X-Ray Sources*, ed. A. Treves, pp. 185–92. Milano: Assoc. Avanz. Astron.

135. van der Klis, M. 1988. *Adv. Space Res.* 8(2): 383–89
136. van der Klis, M. 1989. In *Timing Neutron Stars, NATO ASI Ser. C*, ed. H. Ögelman, E. P. J. van den Heuvel, 262: 27–69. Dordrecht: Kluwer
137. van der Klis, M., Hasinger, G., Stella, L., Langmeier, A., van Paradijs, J., et al. 1987. *Ap. J. Lett.* 319: L13–18
138. van der Klis, M., Jansen, F. A. 1985. *Nature* 313: 768–71
139. van der Klis, M., Jansen, F. 1986. In *The Evolution of Galactic X-Ray Binaries, NATO ASI Ser. C*, ed. J. Truemper, W. H. G. Lewin, W. Brinkmann, 167: 129–37. Dordrecht: Reidel
140. van der Klis, M., Jansen, F., van Paradijs, J., Lewin, W. H. G., Trümper, J., et al. 1985. *IAU Circ. No. 4043*
141. van der Klis, M., Jansen, F., van Paradijs, J., Lewin, W. H. G., Trümper, J., et al. 1985. *IAU Circ. No. 4140*
142. van der Klis, M., Jansen, F., van Paradijs, J., Lewin, W. H. G., Sztajno, M., et al. 1987. *Ap. J. Lett.* 313: L19–23
143. van der Klis, M., Jansen, F., van Paradijs, J., Lewin, W. H. G., van den Heuvel, E. P. J., et al. 1985. *Nature* 316: 225–30
144. van der Klis, M., Jansen, F., White, N., Stella, L., Peacock, A. 1985. *IAU Circ. No. 4068*
145. van der Klis, M., Stella, L., White, N., Jansen, F., Parmar, A. N. 1987. *Ap. J.* 316: 411–26
146. van Paradijs, J. 1983. In *Accretion-Driven Stellar X-Ray Sources*, ed. W. H. G. Lewin, E. P. J. van den Heuvel, pp. 189–260. Cambridge: University Press
147. van Paradijs, J., Hasinger, G., Lewin, W. H. G., van der Klis, M., Sztajno, M., et al. 1988. *MNRAS* 231: 379–89
148. van Paradijs, J., Lewin, W. H. G. 1986. In *The Evolution of Galactic X-Ray Binaries, NATO ASI Ser. C*, ed. J. Truemper, W. H. G. Lewin, W. Brinkmann, 167: 187–93. Dordrecht: Reidel
149. Verbunt, F. 1988. In *Physics of Neutron Stars and Black Holes*, ed. Y. Tanaka, pp. 159–73. Tokyo: Universal Acad. Press
150. Verbunt, F., van Paradijs, J., Elson, R. 1984. *MNRAS* 210: 899–914
151. Wang, Y.-M., Schlickeiser, R. 1987. *Ap. J.* 313: 200–17
152. Warner, B. 1983. In *Cataclysmic Variables and Related Objects*, ed. M. Livio, G. Shaviv, pp. 155–72. Dordrecht: Reidel
153. Warwick, R. S., Marshall, N., Fraser, G. W., Watson, M. G., Lawrence, A., et al. 1981. *MNRAS* 197: 865–91
154. Warwick, R. S., Norton, A. J., Turner, M. J. L., Watson, M. G., Willingale, R. 1988. *MNRAS* 232: 551–64
155. Webbink, R. F., Rappaport, S., Savonije, G. J. 1983. *Ap. J.* 270: 678–93
156. White, N. E. 1986. In *The Physics of Accretion Onto Compact Objects, Lect. Notes Phys.*, ed. K. O. Mason, M. G. Watson, N. E. White, 266: 377–87. Berlin: Springer-Verlag
157. White, N. E., Marshall, F. E. 1984. *Ap. J.* 281: 354–59
158. White, N. E., Mason, K. O., Sanford, P. W., Ilovaisky, S. A., Chevalier, C. 1976. *MNRAS* 176: 91–102
159. White, N. E., Peacock, A., Taylor, B. G. 1985. *Ap. J.* 296: 475–80
160. White, N. E., Stella, L. 1988. *MNRAS* 231: 325–31
161. White, N. E., Stella, L., Parmar, A. N. 1988. *Ap. J.* 324: 363–78
162. Wijers, R. A. M. J., van Paradijs, J., Lewin, W. H. G. 1987. *MNRAS* 228: 17P–21P

KINEMATICS, CHEMISTRY, AND STRUCTURE OF THE GALAXY

Gerard Gilmore

Institute of Astronomy, University of Cambridge, Madingley Road, Cambridge CB3 0HA, England, and Canadian Institute for Theoretical Astrophysics, University of Toronto, 60 St. George Street, Toronto, Ontario M5S 1A1, Canada

Rosemary F. G. Wyse

Department of Physics and Astronomy, The Johns Hopkins University, Baltimore, Maryland 21218

Konrad Kuijken

Institute of Astronomy, University of Cambridge, Madingley Road, Cambridge CB3 0HA, England, and Canadian Institute for Theoretical Astrophysics, University of Toronto, 60 St. George Street, Toronto, Ontario M5S 1A1, Canada

INTRODUCTION

In principle, an understanding of the formation and early evolution of the Galaxy is a well-defined theoretical problem. All that one requires is a detailed knowledge of the spectrum of perturbations in the early Universe and their subsequent evolution; an understanding of the physics of star formation in a variety of environments, with particular emphasis on a prediction of the distribution of orbital elements of those intermediate-mass massive star binaries that will evolve to supernovae; a description of the hydrodynamics of a protogalaxy, particularly including the effects of a high supernova rate, the efficiency of mixing of the chemically enriched ejecta, and the incidence of thermal and gravitational instabilities; the growth and transport of angular momentum and their effect on the growth of a disk; and the effects of a time-dependent gravitational potential on the dynamics of any stars formed up to that time. In practice, there remain some limitations in our understanding of at least some of these physical processes. Hence, it is still useful on occasion to try to deduce the important physics involved in galaxy formation from observations of those old stars that were formed at the time of the formation of the Milky Way, and whose present properties contain some fossil record of the Galaxy's history.

The Galaxy offers a unique opportunity to set constraints on theories of galaxy formation and evolution, since only in the Milky Way can one obtain the true three-dimensional stellar spatial density distributions, stellar kinematics, and stellar chemical abundances. Knowledge of how stars move and how they are distributed in space constrains the Galactic potential, while knowledge of their kinematics, ages, and chemistry constrains the star formation history.

We review in Section 1 how the combination of density laws and data on chemical abundances, kinematics, and ages for stars near the Sun provides important information about the early evolution of the Galaxy. Section 2 summarizes some recent results regarding the shape of the stellar distribution in the Galactic spheroid, while Section 3 discusses the importance of observed relations between kinematics and chemistry. We review available data and analyses to show that the sum of all available information strongly suggests that the extreme Population II subdwarf system formed during a short-lived period of dissipative collapse of the proto-Galaxy. This subdwarf system now forms a flattened, pressure-supported distribution with axial ratio $\sim 2:1$. Section 4 is devoted to the nature and evolutionary status of the thick disk, which formed subsequent to the subdwarf system, with at least the metal-poor tail of the thick disk being comparable in age to the globular cluster system. The thick disk is probably chemically and kinematically discrete from the Galactic old disk, though the data remain inadequate for robust conclusions. The final section (Section 5) reviews the status of "missing" matter in the thin disk; new data and analyses lead to the conclusion that there is no statistically significant amount of nonluminous mass in the solar neighborhood and hence no evidence for dissipative dark matter.

1. THE FORMATION OF DISK GALAXIES

Current understanding of the formation and early evolution of disk galaxies allows a description of the important physical processes at various levels of complexity and generality. At one extreme, one simply considers the global evolution of a gas cloud and assumes that mean values of relevant parameters suffice for an adequate description of generic properties. Alternatively, one gives up general applicability, adopts instead specific numerical values for those parameters that quantify the important physics, and attempts a detailed confrontation of model predictions with observed stellar populations. The relation of any model prediction to detailed observations at a single radius in a specific galaxy clearly needs to be considered with some care. Mindful of this caveat, we outline here the most important time scales and physical processes that are likely to play a role in the determination of the observable properties of galaxies like the Milky Way.

1.1 *Dissipational Disk Galaxy Formation*

The existence of cold, thin, galactic disks has strong implications for galaxy formation. To see this, consider a standard picture whereby galaxies form from growing primordial density perturbations, which expand with the background universe until their self-gravity becomes dominant and they collapse upon themselves. Were there to be no loss of energy in the collapse, and if we neglect angular momentum, the transformation of potential energy into thermal (kinetic) energy would lead to an equilibrium system with final radius equal to half its size at maximum expansion, supported by random motions of the constituent particles. Thus an equilibrium, purely gaseous protogalaxy should have temperature

$$T \equiv T_{\text{virial}} \sim \frac{GMm_{\text{p}}}{kR}, \qquad \text{1a.}$$

and a stellar protogalaxy should equivalently have velocity dispersion

$$\sigma^2 \sim T_{\text{virial}} \frac{k}{m_{\text{p}}}, \qquad \text{1b.}$$

where k and m_{p} are the Boltzmann constant and mass of the proton, respectively. Numerically, we have $T_{\text{virial}} \sim 10^6 R_{50}^{-1} M_{12}$ K for gravitational (half-mass) radius R, in units of 50 kpc, and mass M, in units of $10^{12}\ M_\odot$. Since the disks of spiral galaxies are cold with $T \ll T_{\text{virial}}$, energy must have been lost. Since this lost energy was in random motions of individual particles, the only possible loss mechanism is through an inelastic collision, leading to the internal excitation of the particles and to subsequent energy loss through radiative deexcitation. Clearly, particles with small cross section per unit mass for collisions, such as stars, will not dissipate their random kinetic energy efficiently, and thus dissipation must occur prior to star formation, while the galaxy is still gaseous. The virial temperature of a typical galactic-sized potential well is $T_{galaxy} \sim 10^6$ K, with corresponding one-dimensional velocity dispersion of ~ 100 km s^{-1}.

The physical conditions of the Universe at the epoch of galaxy formation ($z \sim$ a few), as deduced from observations of quasar absorption lines (the Gunn-Peterson test for neutral hydrogen), are such that hydrogen is ionized and the temperature of the protogalactic gas is $\sim 10^4$ K, with a sound speed of only ~ 10 km s^{-1}. Thus, collapse of this gas in galactic potential wells will induce supersonic motions and lead to both thermalization of energy through radiative shocks and subsequent loss of energy by cooling. It is this conversion of potential energy—first to random kinetic energy as described by the virial theorem and then to radiation via atomic processes, the net result of which is an increase in binding energy of the system—that is termed dissipation.

The rate at which excited atoms can cool is obviously a fundamental limit on the amount and rate of dissipational energy loss and hence on the maximum rate at which a gas cloud can radiate its pressure support and collapse. A convenient measure of this time scale is the *cooling time* of a gas cloud, which is the time for radiative processes to remove the internal energy of the cloud. Defining the cooling rate per unit volume to be $n^2 \Lambda(T)$ (where n is the particle number density, and where the functional form of Λ is determined by the relative importances of free-free, bound-free, and bound-bound transitions and thus is an implicit function of the chemical abundance) gives

$$t_{cool} = \frac{3nkT}{n^2 \Lambda(T)} \propto \frac{T}{n\Lambda}. \qquad 2.$$

It is usually of most interest to compare this time scale with the global *gravitational free-fall collapse time* of a system, which is the time it would take for the system to collapse upon itself if there were no pressure support. This time scale depends upon only the mean density of the system and is given by

$$t_{ff} \sim 2 \times 10^7 n^{-1/2} \text{ yr}. \qquad 3.$$

The term *rapid* is often used to describe evolution that occurs on about a free-fall time.

An example of the role of atomic processes in allowing dissipation and increase of binding energy during galaxy formation, which though idealized is still of interest in comparisons with observation, was discussed in three contemporaneous papers—Rees & Ostriker (1977), Binney (1977), and Silk (1977), following the earlier work of Lynden-Bell (1967a). These authors investigated the nonlinear (collapse-phase) evolution of (baryonic) cosmological density perturbations in the density-temperature plane. [These ideas are straightforwardly adapted to allow for a significant non-baryonic component of galaxies (see e.g. White & Rees 1978, Fall & Efstathiou 1980, Blumenthal et al. 1984).] In the Rees & Ostriker model, the perturbation is hypothesized to be initially sufficiently lumpy and chaotic that collisions between local irregularities lead to efficient shock thermalization of the kinetic energy of collapse, resulting in a hot ($T \sim T_{\text{virial}} \sim 10^6$ K), pressure-supported system. The subsequent evolution will then depend on the efficiency with which the heated gas can radiate, and it can be calculated readily if one assumes for simplicity that the system is also of uniform density. (These simplifying assumptions, of course, remove the possibility of any useful discussion of star formation, which depends on *local* cooling and instability, in this model.) One may then define a curve in the density-temperature plane where the gas-cooling time equals the free-fall collapse time of the perturbation itself. Systems that formed with a *short* cooling time will occupy a locus inside this curve. The $t_{\text{cool}} = t_{\text{ff}}$ locus has an upper boundary corresponding to $M \sim 10^{12} M_\odot$, $R \sim 100$ kpc; the fact that these limits also correspond to the upper bound of masses and radii characteristic of observed galaxies is very suggestive. Indeed, when one translates observed surface brightnesses and velocity dispersions of galaxies to put them on the density-temperature plane, one finds that present-day galaxies of all Hubble types lie within the $t_{\text{cool}} = t_{\text{ff}}$ curve, whereas groups and clusters of galaxies lie outside (Silk 1983, Blumenthal et al. 1984). This can be interpreted to imply that the luminous parts of galaxies cooled and collapsed rapidly, at least for those galaxies of high enough central surface brightness to have been studied to date (see Disney 1976, Bothun et al. 1987). Indeed, Gunn (1982) finds that the "bulge" of the Milky Way individually also lies within the $t_{\text{cool}} \lesssim t_{\text{ff}}$ locus of the density-temperature plane, which suggests that it too dissipated on a time scale comparable to its free-fall collapse time. We defer discussion of the other observational constraints on the duration of the formation of the metal-poor spheroid of the Milky Way to Section 3 below.

The evolution of very massive ($M \gtrsim 10^{12} M_\odot$) protogalaxies is, however, only poorly determined by this theory. These galaxies may have had an early pressure-supported, quasi-static collapse phase, provided, of course, that the cooling time is less than the Hubble time; such density perturbations may plausibly evolve along a constant Jeans'-mass track (with density and temperature increasing and little or no star formation) in the density-temperature plane, until conditions are such that the cooling time is less than the dynamical time and rapid collapse is again expected to ensue. An alternative to this last conjecture was suggested by Fall & Rees (1985), who argued instead that conditions within a protogalaxy, once the global cooling and collapse times became comparable, may lead to thermal instability, with a background plasma of temperature $T \sim T_{\text{virial}} \sim 10^6$ K and embedded, dense condensations of temperature $T \sim 10^4$ K. The important mass scale of the condensations is still set by gravitational instability, however.[1] Regardless of the fine-tuning required in theories of the formation of globular clusters, the basic idea that thermal instability might cause a sufficiently massive protogalaxy to "hang up" with $t_{\text{cool}} \gtrsim t_{\text{ff}}$ in its early stages of evolution suggests that the spheroids of at least very massive galaxies may have collapsed less rapidly than on a free-fall time, with continuing star formation in thermally unstable condensations perhaps being possible over this longer time.

The discussion above is based on an extremely idealized model of a protogalaxy, in that only the *global* cooling and collapse time scales of a *uniform* gas cloud are considered. No analytic descriptions of more plausible models exist as yet. A first step has been made by White (1989a), who considers an idealized, spherically symmetric gas cloud that has an imposed initial density gradient. White's models are motivated by cosmologies dominated by cold dark matter (CDM) and assume that 90% of a galaxy is nonbaryonic, and that the remaining 10% is gas that is initially distributed in proportion to the total density, which has a profile consistent with observed flat rotation curves. The inner, more dense regions then cool on a shorter time scale than the outer regions, so that one can define a time-dependent "cooling radius" for each protogalaxy within which the gas is sufficiently dense to cool on a Hubble time. What happens to the gas within the cooling radius is as indeterminate in this model as in those discussed above. However, one can imagine earlier and slower star formation in the central regions than the previous models suggested.

The above discussion can say nothing about when or how local Jeans'-mass condensations actually form stars; the inherent assumption is that cooling is necessary and sufficient for efficient star formation, though the critical distinction between global and local time scales is rarely made explicit. However, it is clear that the existence of *gaseous* disks requires

---

[1] Fall & Rees suggest that this phase of galactic evolution represents an epoch of globular cluster formation, since the Jeans' mass under these conditions is $\sim 10^6 M_\odot$. (The *Jeans' mass* is that minimum mass at which gravity overwhelms pressure, so that density perturbations of mass $M \gtrsim M_J \sim 10^8 T_4^{3/2} n^{-1/2} M_\odot$ are unstable and collapse upon themselves, where the numerical factor is for temperature T, in units of 10^4 K, and number density n, in units of particles cm^{-3}.) However, the Jeans' mass will be continually reduced, presumably to stellar masses, unless further cooling by molecular hydrogen is suppressed in some way. Thus, continuing formation of globular cluster–sized objects requires some additional special conditions.

that the star-formation efficiency be low during the early stages of disk formation. A realistic discussion of galaxy formation must consider the hydrodynamics of the gas in a protogalaxy. Numerical computation of hydrodynamic models of galaxy formation can contain an explicit formulation of the rate of star formation, along with the other important time scales of gaseous dissipation, viscous transport of angular momentum, and free-fall collapse. Larson's (1976) models still offer the most detailed discussion of the effects of gas processes within the prescription of galaxy formation, despite the limitations imposed by his computational constraints. These models identify the major requirements for producing galaxies that contain *both* a high-central-surface-brightness, nonrotating stellar spheroid and an extended, cold, gas-rich disk—initially both the star formation rate and viscosity must be high to form the nonrotating but centrally concentrated stellar spheroid, but both these quantities must be suppressed later to form a thin, lower central surface density, centrifugally supported, gas-rich disk. Carlberg (1985) has pioneered the "sticky particle," modified N-body approach, and several groups have initiated studies of disk galaxy evolution using the smoothed-particle hydrodynamics (SPH) scheme. The general conclusion from available studies is that, while it is possible to build models that are somewhat like observations, it is necessary to specify the most sensitive parameters (viscosity and, in effect, the star formation rate) in an ad hoc way. Considerably more sophisticated numerical experiments are required to ensure a plausible treatment of the hydrodynamics of a multiphase interstellar gas in a system with a high supernova rate, even under the assumption that one understood how to parameterize viscosity and star formation and knew the initial conditions.

An important general feature of recent models of disk galaxy formation is exemplified by Gunn's (1982) continual-infall models. These models hinge on the existence of loosely bound material surrounding a density peak whose central regions are collapsing (rapidly) to form a galaxy; they are extensions of the cosmological secondary-infall paradigm of Gunn & Gott (1972), modified to include dissipation of the infalling gas. The free-fall collapse time scales for the outer regions can be of order a Hubble time, leading to a picture of disk galaxy formation whereby the central regions collapse rapidly to form the bulge, followed by accretion of protodisk material. In so far as the subsequent evolution of the gas may be modeled through the processes of shock heating, cooling, and star formation, the general features of continual-infall models are in good qualitative agreement with observed galactic disks. Thus one might reasonably expect that galactic *disks* formed on a longer time scale, though still without pressure support having played a major role, than did galactic spheroids, the slower collapse of disks being due simply to their lower density initial conditions implying a longer free-fall collapse time. The lack of disk-dominated systems in dense environments such as rich clusters of galaxies is consistent with this picture of unperturbed, continual accretion of disks (Larson et al. 1980, Frenk et al. 1985).

The angular momenta of galaxies that formed in environments of different density might also be expected to differ. The specific angular momentum distribution of the material surrounding density peaks has been investigated analytically by Ryden (1988) in the context of cold dark matter–dominated cosmological models, and by Barnes & Efstathiou (1987), using N-body techniques, for various assumed cosmological power spectra. The consensus is that the effect of tidal torques is to produce a system with specific angular momentum increasing with radius. Zurek et al. (1988) and Frenk et al. (1988) have shown that in CDM, or any scenario where chaotic aggregation of smaller systems is part of the formation of galaxy-sized systems, dynamical friction of dense clumps on the smoother background causes transport of both energy and angular momentum from the orbiting clumps to the smooth outer regions. Thus, during the buildup of structure, initially strongly bound particles lose both energy and angular momentum, whereas the weakly bound particles gain energy (become more weakly bound) and also gain angular momentum. There is overall alignment of the angular momentum vector of different shells in binding energy. These authors argue that slowly rotating stellar systems, such as giant elliptical galaxies or spheroids of disk galaxies, form in direct analogy to the dissipationless dark halos that they model (i.e. with lots of dynamical friction and merging of stellar clumps, the dark halo and outer stellar envelope taking up the angular momentum transported outward). Disks of spiral galaxies would then form without significant angular momentum transport, owing to the expectation that the baryons remain gaseous until the virialization of the dark halo, and shock heating (as described earlier) would homogenize the gas. The predictions of these models could be tested in detail if we knew the angular momentum distribution of the outer spheroid of our Galaxy; all we know at present is that the *kinematically selected* subdwarfs in the solar neighborhood have a lower specific angular momentum than do the disk stars, by roughly a factor of five, and that the metal-poor globular cluster system is consistent with zero net rotation to Galactocentric distances of ~ 30 kpc.

The angular momentum distribution of the material destined to form the disk controls both the range of galactocentric radii over which infall occurs at a given epoch and the duration of the infall at a given location. Thus, models of disk chemical and dynamical evolution that appeal to continual infall must also satisfy angular momentum constraints.

1.2 Specific Models of Milky Way Galaxy Formation

The most widely referenced model of the formation of our Galaxy is that of Eggen et al. (1962; henceforth ELS), which was developed primarily to understand their observations of the kinematics and chemical abundances of stars near the Sun. This model requires that the stellar spheroid formed during a period of rapid collapse of the entire proto-Galaxy, subsequent to which the remaining gas quickly dissipated into a metal-enriched cold disk, in which star formation has continued until the present. The ELS model was designed to provide conditions under which the oldest stars populated radially anisotropic orbits, whereas stars that formed later had increasingly circular orbits, in accord with their data, which implied that the most metal-poor stars (assumed to be the oldest stars) were on more eccentric, lower angular momentum orbits than were the more metal-rich stars. This model is based on two crucial assumptions. The first is that a pressure-supported, primarily gaseous galaxy (where $T = T_{\text{virial}}$) is stable against star formation. In this case, the *global* cooling time is the shortest time scale of interest, and thermal instabilities of the type invoked by Fall & Rees (1985) and discussed briefly above must be suppressed. The second assumption is that stellar orbits cannot be modified to become more radial after formation of the star.

If the first assumption is valid, the observed high-velocity stars must have formed from gas clouds that were not in equilibrium in a pressure-supported system. If the second assumption is valid, these clouds formed stars while on radial orbits at large distances from the Galactic center. Thus in this picture, these clouds must have turned around from the background universal expansion and be collapsing toward the center of the potential well. Hence, the oldest stars of the Galactic spheroid must have formed as the proto-Galaxy coalesced. To determine the *rate* of the collapse, ELS analyzed the evolution of the radial anisotropy of a stellar (or gas cloud) orbit as the Galactic potential changed and showed that it was approximately conserved during a slow collapse, but that it became more radially anisotropic in a fast collapse. They argued against a slow collapse on the grounds that such a collapse requires tangentially biased velocities (recall that pressure support has been excluded by assumption), and this tangential bias will be unaffected by the resulting slow changes of the gravitational potential. The observed radial anisotropy of the stellar orbits then implies an initially radially biased velocity ellipsoid, while the calculations of ELS show that such a velocity ellipsoid will have become more radially anisotropic during collapse. Hence, ELS deduced that the gas clouds were in free-fall radial orbits, and that the consequent collapse must have been rapid, "rapid" in this sense meaning that the time scale for collapse is comparable to an orbital or a dynamical time scale, which is a few $\times 10^8$ yr. It should be noted that Isobe (1974) came to the opposite conclusion from his analysis of the ELS data and favored a slow collapse, while Yoshii & Saio (1979) augmented the ELS sample and also concluded that the halo collapsed over many dynamical times. We return to the difficulty of inferring a time scale in Section 3 below.

Clearly, if either of ELS's assumptions were violated, there need be no correlation between the *time* of a star's formation—which they infer from a star's metallicity—and its *present* orbital properties. In a nonrotating, pressure-supported system, all stars formed would be on highly radial orbits, as the star has too small a surface area to be pressure supported by the gas. As we mentioned above, assumptions about star formation in pressure-supported systems must be treated as ad hoc until we understand better the physics of star formation, so that conclusions based on such assumptions are at best uncertain. If their second assumption were violated, then the stars that are now the high-velocity stars near the Sun could have originated from more circular orbits interior to the Sun and have present orbital properties that depend only on dynamical processes subsequent to their formation. The realization that a forming galaxy undergoes changes in its gravitational potential that are of order the potential itself (violent relaxation) means that stellar orbits can be modified considerably.

Recent N-body models [see, for example, May & van Albada (1984) and McGlynn (1984) for excellent descriptions of representative experiments] for systems in which dissipation does not play a major role show that the final state of the collapsed system depends on both the degree of homogeneity and on the temperature of the initial state. As seen in the cosmological N-body simulations discussed above, clumps cause angular momentum and energy transport. Violent relaxation never goes to completion, so that final and initial orbital binding energies and angular momenta are correlated, with the interior regions becoming more centrally concentrated and the outer regions being puffed up. The typical final steady-state velocity distribution is highly anisotropic exterior to (roughly) the half-light radius and more isotropic interior to that radius. If violent relaxation were completely efficient, all systems would reach the same final state with isotropic velocity distribution. In the Galaxy, the spheroidal half-light radius is ~ 3 kpc, well interior to the Sun's orbit. Thus the expected velocity distribution of old stars near the Sun after virialization of the spheroid is anisotropic, as observed by ELS, even though the dynamical evolution of the system is not as they envisage and a correlation between kinematics and age is no longer an inevitable conclusion. One might, for example, imagine a situation where later (rapid) collapse of either the disk or the dark halo, or the merger of a few large substructures, could lead to rapid dynamical evolution of a central spheroidal component that had previously formed on a longer time scale. Models of this type have yet to be studied in detail.

The implications of the ELS model and the current status of the relevant observational correlations are discussed more fully in Section 3.2 below.

1.3 The Time Scales of Galactic Chemical Evolution

In attempting to deduce the rate of star formation and dynamical evolution in a proto-galaxy, it is desirable to have available a clock whose rate can be calibrated independently of the naive discussions of global gas cloud properties noted above, and that runs sufficiently quickly to resolve the dynamical evolutionary time scales. Such a clock is provided by stellar evolution of high-mass stars, while the fossil record of the clock is observable in the chemical abundance enrichment patterns in long-lived low-mass stars. Fortunately, there exists a subset of common elements (most importantly oxygen) whose creation sites are restricted to very massive stars, and another subset (most importantly iron) that is created also during the evolution of lower mass stars (in merging-binary supernovae). Since the evolutionary time scales for high- and low-mass stars span the time-scale range of interest in galaxy formation, the differential enrichment of oxygen and iron provides an ideal clock to calibrate the rate of star formation in the proto-Galaxy.

Oxygen-to-iron element ratios have now been measured for a sufficient number of stars to define the systematic trends in the data. The observations are well reviewed by Wheeler et al. (1989) in this volume and are not discussed further here. The important result for present purposes is that a significant change of slope occurs in the relationship between the element ratio [O/Fe] and [Fe/H] (a similar relationship holds for [α/Fe], where the "alpha" elements are those synthesized by successive capture of alpha particles, such as Mg, Si, Ca, and Ti) close to metallicities where there also occurs a change in the stellar kinematics—that is, at [Fe/H] ~ -1([O/Fe], [α/Fe]) and ~ -0.4([α/Fe]), although the latter break is less well established. The [O/Fe] ratio is observed to be approximately constant, independent of [Fe/H] for the most metal-poor stars ($-2.5 \lesssim$ [Fe/H] $\lesssim -1$), while [O/Fe] declines for the more metal-rich stars ([O/Fe] $\sim -1/2$ [Fe/H]). Present data are summarized in Figure 1, in which all scatter is considered by the relevant observers to be consistent with observational error, i.e. there is *no* cosmic scatter (see also Wheeler et al. 1989). If we assume that [Fe/H] is a monotonically increasing function of time, the mean trend in the figure can be explained if the oxygen and iron in the more metal-poor stars have been produced in (massive) stars of the

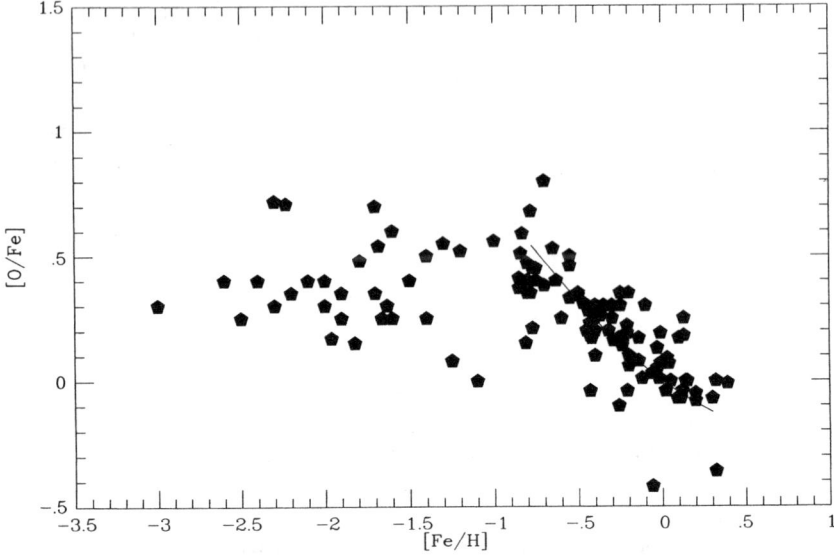

Figure 1 A compilation of oxygen to iron element ratio measurements from the literature and from B. Barbuy (private communication; her [O/Fe] values derived from the near-IR line have been offset by 0.2 dex for internal consistency). This figure is adapted from Wyse & Gilmore (1988). The smooth curve through the data for [Fe/H] $\gtrsim -1$ shows the prediction of a simple model with constant supernova rates in the ratio 1.5:1.0 for Type I: Type II, which results in twice as much oxygen as iron being produced per unit time (see Section 3.4).

same (short) lifetime, whereas for the more metal-rich stars, although the oxygen and iron continue to be produced together, an additional, longer time scale source now dominates the iron production. Such behavior is in good agreement with supernova nucleosynthesis calculations, which show that oxygen is produced only in Type II supernovae by massive stars ($M \gtrsim 20\ M_\odot$), while iron has a contribution from both massive and low-mass stars ($M \gtrsim 3\ M_\odot$, Type I supernovae) and thereby has an enhanced production once the much more numerous, lower mass stars contribute to its nucleosynthetic yield (Tinsley 1979, Matteucci & Greggio 1986). This results in the ratio [O/Fe] decreasing systematically with increasing metallicity [Fe/H]. The main-sequence lifetimes τ_{ms} of single stars of masses 0.08 $M_\odot$ to 100 $M_\odot$ can conveniently be estimated from the following (P. P. Eggleton, private communication):

$$\tau_{ms} = \frac{2.5 \times 10^3 + 6.7 \times 10^2 M^{2.5} + M^{4.5}}{3.3 \times 10^{-2} M^{1.5} + 3.5 \times 10^{-1} M^{4.5}}\ \text{Myr.} \qquad 4.$$

The main-sequence lifetime of massive stars approximates the time scale for oxygen enrichment, while the main-sequence lifetime of the stars responsible for Type I supernova explosions provides a lower limit to the time scale of "secondary" iron production.

In terms of models of the early chemical evolution of the Galaxy, one must explain why [O/Fe] is approximately constant, at three times the solar value, for [Fe/H] $\lesssim -1$, but decreases smoothly for [Fe/H] greater than this value, as well as why the mass of stars with [Fe/H] $\lesssim -1$ is only a few percent of the total stellar mass of the Galaxy, as discussed in detail in Section 4. Clearly, the approximate constancy of the [O/Fe] ratio independent of the [Fe/H] ratio at low metallicities requires that essentially *all* the stars with [Fe/H] $\lesssim -1$ formed on a time scale less than that on which a significant number of low-mass (Type I) supernovae exploded. This time scale is rather difficult to estimate precisely, owing to uncertainties in the mechanism of Type I supernovae and the fraction of all stars formed that are in binaries of the type that may be expected to be precursors (cf. Iben 1986); the lowest mass, and hence most numerous, progenitors of CO white dwarfs have main-sequence masses and lifetimes of $\sim 5\ M_\odot$ and $\sim 2.5 \times 10^8$ yr, respectively. Thus a reasonable estimate for the characteristic time after which one expects dominance of iron from Type I supernovae is $\lesssim 10^9$ yr (but bear in mind that some Type I systems will take a Hubble time to evolve). This general argument appears to be the strongest direct evidence for a rapid formation time scale for the extreme Population II stars in the Galaxy and is in agreement with the (currently more contentious) kinematic evidence discussed below (Section 3). It follows from this that a straightforward test of the formation time scale of the metal-rich central bulge would be to measure the [O/Fe] values for some metal-rich stars and compare them with those of corresponding field stars. For a constant IMF, if the oxygen is overabundant then the bulge must have formed on the same time scale as that of the metal-poor spheroid (i.e. rapidly and long ago), and vice versa if the oxygen is not overabundant.

This relatively short time scale for spheroid evolution to [Fe/H] ~ -1 may be compared with that inferred from the evolution of r- and s-process nucleosynthesis. Gilroy et al. (1988) find no evidence for the s-process from the heavy-element abundance in very metal-poor giants ([Fe/H] $\lesssim -2$), which implies that the time scale for this level of iron enrichment was shorter than that involved in nucleosynthesis via the s-process. Details of the s- and r-process sites and production mechanisms are not well understood, but it is generally thought that s-process isotopes are produced during helium shell flashes (thermal pulses) in intermediate-mass stars ($2 \lesssim M/M_\odot \lesssim 8$) on the asymptotic giant branch, with subsequent mixing (dredge-up) bringing these to the surface, where they are lost in a stellar wind (cf. Iben 1985). The r-process probably occurs during Type II supernovae. The lifetimes given by Equation 4 lead to the conclusion that the Galaxy had enriched to [Fe/H] ~ -2 in $\lesssim 10^8$ yr (cf. Gilroy et al. 1988). A further important finding of Gilroy et al. is that there *is* cosmic scatter in the relative abundances of r-process elements in these very metal-poor stars, in that stars of the same [Fe/H] show large variations in, for example, [Eu/Fe]. This contrasts with the apparent lack of cosmic scatter in other element ratios, such as oxygen. This is plausibly a manifestation of the inhomogeneity of the early spheroid; it should be remembered that the mass fraction of the metal-poor spheroid with [Fe/H] $\lesssim -2$ dex is $\lesssim 20\%$, adopting a mean metallicity of -1.5 dex and dispersion $\sigma_{[Fe/H]} \sim 0.5$ dex (Gilmore & Wyse 1985, Laird et al. 1988b). This corresponds to at most a few percent of the total stellar mass of the Galaxy, so that one expects local fluctuations in heavy-element enrichment due simply to Poisson noise in the very small number of supernovae that can have exploded at that stage of the Galaxy's evolution. This argument predicts that comparable inhomogeneities will be observed in all element ratios for stars with [Fe/H] $\lesssim -2$.

If the arguments above concerning the production time scales of different elements contained the whole story of chemical enrichment over the history of star formation in the Galaxy, then it would have to be mere coincidence that the change in the predominant production mechanism of iron occurred close to a metallicity, or epoch, at which the stellar kinematics change from those of a pressure-supported system (which formed stars rapidly) to those of an angular momentum–supported system. Rather, the coincidence of the value of [Fe/H] at which the Galaxy changed from a pressure-supported system to an angular momentum–supported system with the value of [Fe/H] at which the interstellar medium first became diluted by the products of long-lived stars provides a diagnostic of the relative star formation and dissipation rates in the proto-Galaxy.

One may limit the maximum allowed variations in the star formation rate by making the conservative assumption that the (2σ) scatter about the mean trend in [O/Fe] against [Fe/H] of a few tenths of a dex, or about a factor of two, reflects real cosmic scatter, rather than measurement uncertainties, even though all the observed scatter is consistent with measurement uncertainties. The number of Type I supernovae at a given time is determined by the star formation rate some 10^9–10^{10} yr previously, while that of Type II supernovae is determined by the star formation rate at that time. The scatter would then be caused by short-lived increases in the star formation rate leading to enhanced rates of Type II supernovae. Scatter of a factor of two implies that the Type II formation rate be increased by a factor of $\lesssim 4$ for a 2σ effect. Hence the mean star formation rate in the Galaxy cannot have increased by a factor of 4 since the Galaxy attained a metallicity of -1 dex. The average star formation rate during the formation of the low-mass, metal-poor spheroid was $\sim 10\ M_\odot\ \text{yr}^{-1}$, which is of the order of a typical present-day disk star formation rate. Thus disk galaxies, by far the most common nondwarf galaxies in the Universe, do not go through a sustained, highly luminous burst of star formation in their earliest stages and are not good candidates for detection in primeval galaxy searches.

When sufficient data and reliable massive-star evolutionary models all the way through the supernova explosion, with corresponding elemental yields, become available, it will be possible to quantify these arguments (subject to the assumption of a constant stellar initial-mass function) and provide a real time scale (in years) for the periods of protogalactic evolution that were dominated by collapse (possibly nondissipational) on a dynamical time scale, and for that period when angular momentum support became the dominant dynamical process.

One conclusion that is relatively independent of the details of elemental synthesis follows from the fact that features in both the stellar abundance and kinematics occur more or less together at $[\text{Fe/H}] \lesssim -1$. A discontinuity in kinematic properties implies that the ratio of the dissipation rate to the star formation rate changes rapidly. A possible explanation is that at metallicities $[\text{Fe/H}] \gtrsim -1.5$, the efficiency with which a gas cloud cools from $\sim 10^6$ K (a typical galactic virial temperature) increases markedly owing to a transition of the dominant cooling mechanism from free-free radiation (independent of metallicity) to line radiation (proportional to the number density of metals). Thus a rapid increase in the dissipation rate and collapse to a disklike angular momentum–supported structure is not implausible at a metallicity of ~ -1 dex. It is not crucial for these arguments that the breaks in kinematics and element ratios occur at *exactly* the same metallicity.

2. THE SPATIAL STRUCTURE OF THE MILKY WAY GALAXY

Counting stars is one of the few truly classical scientific techniques used to study high-latitude (and therefore low-obscuration) Galactic structure. The extensive data set and understanding available in 1965 are reviewed in many excellent articles in Volume 5 of the "Stars and Stellar Systems" series (Blaauw & Schmidt 1965). Relatively little further progress was achieved until the new deep high-quality data of I. R. King and collaborators at Berkeley became available in the late 1970s. The application of computer modeling to these data by van den Bergh (1979) led to a considerable resurgence of interest, continuing to the present. Recent relevant reviews include Bahcall (1986), Gilmore & Wyse (1987), Freeman (1987), and Buser (1988).

2.1 The Fundamentals of Star Count Analyses

The number of stars N countable in a given solid angle to a given magnitude limit m is given by a simple linear integral equation, often known as "the fundamental equation of stellar statistics." It is

$$N(m) = \int \Psi(M_v, \mathbf{x}) D(M_v, \mathbf{x}) d^3 x, \qquad 5.$$

where $\Psi(M_v, \mathbf{x})$ is the distribution function over absolute magnitude and position, $D(M_v, \mathbf{x})$ is the stellar space density distribution, and $d^3 x$ is a volume element. This (Fredholm) equation is rarely invertable, since it is ill conditioned. A detailed discussion of its use and approximate solution upon inversion is presented by Trumpler & Weaver (1953, Chap. 5.5). In general, the luminosity function is too broad to allow any solution for both $D(M_v, \mathbf{x})$ and $\Psi(M_v, \mathbf{x})$. The situation can be improved by restricting the data by color and/or spectral type, which is the technique usually followed. In this case, for an assumed form of the distribution function $\Psi(M_v, \mathbf{x})$, the density function $D(M_v, \mathbf{x})$ is recovered from $N(m, \text{color})$. This may be done by inverting the data (classical photometric parallax) or computer calculation of the integral with subsequent iterative comparison of data and model (cf. Bahcall 1986). These techniques are clearly entirely equivalent and should agree. They often do not.

The fundamental problem with use of the fundamental equation is that both the stellar luminosity function and the stellar density law are functions of many parameters. Few of these are sufficiently well known to be fixed. Consequently, a wide variety of combinations of Ψ and D are allowed mathematically. Other *astrophysical* constraints are necessary, whose choice has remained subjective until recently owing to the lack of adequate observational constraints.

2.2 The Choice of Astrophysical Constraints

The technique adopted by almost all workers to date is to fix the very large number of parameters by adopting empirically determined fitting functions for quantities such as the density laws and luminosity functions, and then fitting a set of these fitting functions to the observations [the exception being the analysis of Robin & Crézé (1986a,b), who derive all relevant relations from a model of Galactic evolution]. The empirical fitting functions are determined primarily from photometric observations of spiral galaxies thought to be similar to the Milky Way, from the Gliese catalogue of nearby stars, and from a small number of well-studied globular and open clusters. As most authors are forced to adopt the same few fitting functions, in the absence of any alternatives, it is not surprising that most conclusions are similar. Analyses of this type were pioneered by van den Bergh (1979). Later models have been published by Bahcall & Soneira (1980, 1981, 1984), Bahcall et al. (1985), Gilmore (1981, 1983, 1984b), Gilmore & Reid (1983), Gilmore et al. (1985), Brooks (1981), Yoshii (1982), Pritchet (1983), Buser & Kaeser (1985), Robin & Crézé (1986a,b), Friel & Cudworth (1986), Yoshii et al. (1987), Hartwick (1987), del Rio & Fenkart (1987), Sandage (1987a,b), and Fenkart (1989 et seq), and Rodgers & Harding (1989).

Some of the adopted constraints have recently been discovered to have been poorly justified. The most important example of this is the popular impression that the photometric properties of spiral galaxies are adequately described as a sum of a flat exponential (both radially and vertically) disk and a roughly round spheroid described by the $r^{1/4}$ law when seen in projection. Recent photometric analyses have shown that such a description is an unacceptably poor description of high-quality surface photometric data in almost all cases (Schombert & Bothun 1987, Shaw & Gilmore 1989). Rather, the luminosity profile of NGC 891, the galaxy often quoted to be most like the Milky Way, shows no evidence for a detectable $r^{1/4}$ component in its luminosity distribution (Shaw & Gilmore 1989). Thus, one should not necessarily expect a star count model based on an $r^{1/4}$ spheroid to be an adequate description of the Milky Way.

A specific point of interest in the recent star count literature has involved the parameters of the thick disk, with some conflicting claims as to its reality until a few years ago, when the evidence from several kinematic and spectroscopic surveys became overwhelming (cf. for example, Sandage & Fouts 1987, Sandage 1987a,b, Carney et al. 1989a). This apparent uncertainty in the star count modeling was due to the extreme sensitivity of computer models to the adopted stellar luminosity function and color-magnitude relations. Unless careful astrophysical constraints are imposed on these choices, a huge variety of models is possible that can reproduce the data. This is well illustrated by the "disproof" of the existence of a thick disk by Bahcall & Soneira (1984, Figure 19). They showed the complete disagreement of such a model with the faint stellar data of Kron in SA 57. In a later paper, however, Bahcall et al. (1985, Figure 19) showed the excellent fit of the same geometric model to the same data. The only difference in the latter case was the use of a luminosity function that is appropriate for an old stellar population, as required by the color data, expected for a spheroidal population, and earlier derived by Gilmore & Reid (1983).

The uncertainties above illustrate the fundamental limitation of the modeling of star count data in isolation, which is a severe restriction on its value—too few constraints are usually available to provide a unique model, and too few consistency arguments are usually applied during its use. One requires additional information other than star count data to constrain the appropriate color-magnitude relation and, if possible, kinematic data to allow segregation of different stellar populations. The color-magnitude relation of a stellar population is a function of chemical abundance and age, while the density law depends on both the stellar kinematics and the gravitational potential gradients. In general, none of these parameters is known adequately a priori.

Essential supplementary information is available from kinematics (particularly proper-motion and radial velocity surveys) and from spectroscopic surveys. The latter allow dwarf-giant discrimination, provide chemical abundance data, and hence allow the derivation of reliable distances. Both kinematic and spectroscopic data have become available in large quantities in the last few years, and are being included in star count modeling. Proper-motion data for bright stars ($V < 6$) have been included in a two-component disk/spheroid model by Ratnatunga et al. (1989). As a result of the restriction to bright, and hence nearby, stars, the results of such modeling are highly specific to the kinematic properties of young stars near the plane of the Milky Way—plausible mean values from the literature for the local kinematics were able to fit the data in their modeling only to within 25%. Extension of analyses of this type to the high-precision, high-statistical weight HIPPARCOS catalogue and to the Lick proper-motion survey (Klemola et al. 1987) will be of considerable interest. The results of the several spectroscopic surveys, which have provided conclusive evidence for the existence of a Galactic thick disk with descriptive parameters similar to those proposed by Gilmore & Reid (1983), are discussed in Section 4 below.

2.3 Analysis of Star Count Data

The most straightforward analysis technique for stellar number–magnitude–color data is photometric parallax. This involves use of the absolute magnitude–color relation for a galactic or globular cluster of appropriate chemical abundance. The absolute magnitude for a field star of measured color is read directly from this diagram and combined with the apparent magnitude to give a photometric distance. From a large set of distances, with appropriate Malmquist corrections, a density law can be derived directly. This technique has been extensively applied by the Basel group (R. P. Fenkart et al.) and by Gilmore & Reid (1983). The steep density profile from 2 kpc to 4 kpc was identified by these latter authors as a

Galactic thick disk, with exponential scale height ~ 1.3 kpc and local normalization $\sim 2\%$ of the old disk stars. They emphasized that a flattened $r^{1/4}$ law was an equally good fit to the data. A density profile of steep exponential form at distances of a few kiloparsecs from the Galactic plane was in fact very well established many years ago [cf. reviews by Elvius (1965, especially Figure 2) and Plaut (1965, especially Figure 7b)] though evidently forgotten. Similar results to those of Gilmore & Reid were earlier derived from the Basel surveys by Yoshii (1982), though not widely appreciated at that time. Yoshii showed that the Basel north Galactic pole star count data required the vertical density profile of the dominant stellar population more than ~ 1.5 kpc from the Galactic plane to follow an exponential density distribution with scale height a factor of ~ 6 larger than that of the old disk. The corresponding factor derived by Gilmore & Reid from their exponential fits to the south Galactic pole data was ~ 4. More recently Yoshii et al. (1987) have analyzed new data for the north Galactic pole and derived a scale height ratio of thick disk:old disk of ~ 3, again similar (though not identical) parameters to those of Gilmore & Reid.

The alternative analysis technique involves the direct calculation of the integral in the fundamental equation of stellar statistics. This is a straightforward computational exercise. Consequently, many attempts have been made recently to explore parameter space so that the uniqueness of the results from the direct analysis of star count data can be determined. In relevant form, this equation is

$$N(V, B-V) = w \int \Psi\left(M_v, \left[\frac{A}{H}\right], \tau, r, \ldots\right) D(r, M_v, \tau) r^2 \, dr, \qquad 6.$$

where τ is stellar age, $[A/H]$ is stellar elemental abundance, and w is observed solid angle. The luminosity function Ψ (stars mag^{-1} pc^{-3}) has been known for many years to be a function of distance from the Galactic plane (e.g. Bok & MacRae 1942). Similarly, the existence of age-velocity dispersion and age-metallicity relations for old thin disk stars is well known. This emphasizes the crucial and irreducible limitation of analyses of this type—both the luminosity function and the density law are functions of the other phase-space parameters. A unique solution of Equation 6 is therefore impossible. Instead, a large number of parameters must be fixed on external astrophysical grounds. Additionally, comparison with observational data requires adoption of an appropriate absolute magnitude–color relation, in exactly the way required in the more direct photometric parallax analysis technique described above. It is this nonuniqueness that must be overcome by the use of supplementary chemical abundance and stellar kinematic data. We emphasize that star count analyses *in isolation* are incapable of providing a unique description of the structure of the Galaxy. The results of any such analysis should be viewed as merely indicative of the type of combination of fitting functions that can be used to represent available data. They should not be accepted as a valid description of the stellar populations in the Galaxy until the chemical abundance, luminosity class, age, and kinematic assumptions are tested by spectroscopic observations.

2.4 Available Modern Star Count Data

The availability of high-efficiency, linear, two-dimensional detectors (CCDs) and fast, automated, photographic plate-scanning microdensitometers (PDS, COSMOS, APM) has revolutionized stellar statistics. Complete samples of stars can be measured to useful precision in several wavebands over sufficiently large areas of sky that random errors due to counting statistics are unimportant. The minimization of systematic errors still requires an enormous effort, however [cf. Gilmore (1984a) for a description of the requisite photometric techniques]. Table 1 of Gilmore & Wyse (1987) summarizes those recent high-Galactic-latitude studies in which the magnitude calibration was derived directly from photoelectric or CCD standards. With the conspicuous exception of work by the Basel group (R. P. Fenkart et al.; see below), no substantial new data sets have been published since that table was prepared.

The general features of the high-latitude sky are illustrated in Figure 2, which shows stellar photometric data derived from APM scans of high-

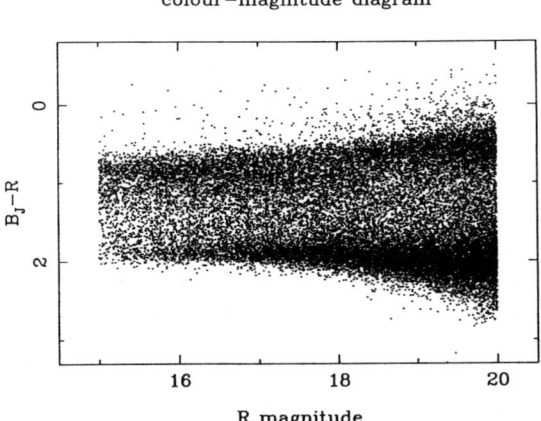

Figure 2 A representative color-magnitude diagram of the stellar distribution in the high-latitude sky. This is derived from APM scans of UK Schmidt Telescope plates.

Galactic-latitude plates from the UK Schmidt Telescope. The important aspects of these data are the following:

1. The sharp edge to the distribution near the equivalent of $B-V = 0.4$, with very few stars being seen significantly bluer than this limit. This corresponds to the main-sequence turnoff color of an old, metal-poor population. The absence of a younger turnoff shows that no substantial continuing star formation has taken place in the spheroid—though note the existence of some apparently young high-latitude metal-rich A stars, whose place of formation remains an extremely important mystery (Lance 1988), and some distant B stars that cannot have traveled from the thin disk in their main-sequence lifetimes (Keenan et al. 1986, Conlon et al. 1988).
2. The peak of the distribution near the equivalent of $B-V = 0.6$. This is similar to the main-sequence turnoff color of metal-rich globular clusters and shows the mean abundance of the field spheroid sampled here to be approximately -0.5 to -1 dex. The consequences of this for the age of these stars are discussed in Section 3.
3. The peak of the distribution near the equivalent of $B-V = 1.5$. This feature corresponds to the insensitivity of blue colors to effective temperature in cool main-sequence stars. It is therefore an artifact of the choice of photometric pass-bands rather than evidence of a structural property of the Galaxy. This is best illustrated by comparison with Figure 3b of Gilmore & Wyse (1987), which shows $V/V-I$ data for the same stars as are shown in Figure 2. The continuing temperature sensitivity of the $V-I$ color leads to the very different appearances of the two diagrams.
4. The absence of a large number of stars at the red edge of the distribution at faint apparent magnitudes. Such very red stars are very cool low-mass M dwarfs and have often been hypothesized as candidates for the local missing mass. Their absence in this diagram was the first direct evidence that low-mass luminous stars do not contribute significantly to the mass density in either the disk (Gilmore & Reid 1983) or the halo (Gilmore & Hewett 1983) of the Galaxy. [We also note that the observations and analyses that suggested the need for missing mass associated with the disk of the Galaxy have been superseded (see Section 5 below) by new data and analyses that suggest that there is no missing mass associated with the Galactic disk.]

The appearance of Figure 2 does not change significantly to $V \sim 22$ (Kron 1980). At fainter magnitudes it is expected that the blue edge will move to progressively redder colors as intrinsically fainter subdwarfs dominate the counts.

The most useful external estimates of the accuracy of the data come from observations of the Galactic poles, as only there do truly independent duplicating data exist. For the south Galactic pole, observations exist by Bok & Basinski (1962), Reid & Gilmore (1982), and Gilmore et al. (1985) in the Johnson BV system. For the north Galactic pole, similar data have been published by Weistrop (1972), Chiu (1980), and Stobie & Ishida (1987). The detailed counts from these sources are compared in Figures 5 and 6 of Stobie & Ishida (1987), showing that the various authors agree to within $\sim 5\%$ for apparent magnitudes brighter than about $V = 18$.

At apparent magnitudes fainter than $V = 18$ and in other directions, only one other such comparison is possible, and this is detailed in Gilmore et al. (1985). Similar precision is suggested to $V \lesssim 21$. At fainter magnitudes the precision of the star count data becomes dominated by the difficulties of reliable star-galaxy separation, leading to *systematic* errors whose amplitude is impossible to quantify from extant data (cf. Section 2.6). The best available estimate of the external accuracy of modern automated photographic data that is directly calibrated by a large number of photoelectric and/or CCD standards is about 10% in both apparent magnitude and color for $V \lesssim 21$. The internal precision is typically a factor of two better.

2.5 Star Counts and Stellar Components

Interpretation of the parameters of star count models—specifically the relative fractions of stars near the Sun assigned to discrete density distributions and characterized by a specific color-magnitude relation—in terms of stellar populations is an ambitious task. In principle, it is possible so long as supplementary chemical abundance, age, and kinematic data are available. (These supplementary data are discussed in Section 4 below.) Nevertheless, it is a worthwhile exercise to test the validity of the geometrical aspects of current star count models by comparing them with observations in directions other than those from which the model parameters have been derived.

2.5.1 STAR COUNTS AND THE THICK DISK By far the most extensive star count data set available is that obtained for the Basel Halo Program. This very large program has been in operation for 25 years, has provided three-color data in 15 separate fields, and is currently being summarized and analyzed in a series of four papers by R. Fenkart (1989 et seq; see also Buser 1988). Fenkart has emphasized the very good agreement between the thick disk density law derived by Gilmore & Wyse (1985), including an exponential profile for the thick disk and a color-magnitude relation like that of a metal-rich globular cluster, and that required by the Basel data in many fields at both high and low Galactic latitudes. This provides the first evidence for the global applicability of the thick disk parameters derived by Gilmore & Reid and by Gilmore & Wyse. Independent confirmatory evidence is provided by McNeil (1986), who derived the density profile of spectroscopically selected M giants toward the south Galactic pole and derived a density profile similar to that discussed above. [This sample has some overlap with the Hartkopf & Yoss (1982) sample of K giants, the kinematics of which is discussed in Section 4 below.]

When comparing observations with the predictions of the several star count models, however, it is essential to bear in mind the limitations of the available models. While the thick disk model is in good agreement with a wealth of star count, chemical abundance, spectroscopic luminosity-classification, and radial velocity data [see Freeman (1987) and Section 4 below], almost all high-precision data are restricted to the range from a few hundred to a few thousand parsecs from the Sun. The original set of parameters describing the thick disk were derived solely to reproduce the observed stellar distribution from ~ 1 kpc to ~ 4 kpc from the Galactic plane toward the south Galactic pole. While parameters similar to those originally derived have subsequently been found to provide a good description of the data in several other high-latitude fields [e.g. Yoshii et al. (1987) for the north Galactic pole; Fenkart (1989) et seq], extrapolation of the fitted exponential well outside the range over which it has been verified is certainly not guaranteed to be an adequate description of the Galaxy.

Determinations of the local thick disk normalization by extrapolation of the thick density law derived from star counts down to the Galactic plane—i.e. by extrapolation of the density from $z \sim 1$ kpc to $z \sim 0$ kpc—yield values from $\lesssim 1\%$ to a few percent of the local total stellar density, although Sandage (1987b) quotes 10%, with a lower value of the scale height than the other determinations. The *physical* local normalization can be determined from the distant data only if one knows the vertical Galactic force law as a function of z-distance from the plane, $K_z(z)$, and the velocity distribution function over the range of interest (Sandage 1987b). It is quite unrealistic that an extrapolated exponential will be a valid description of the space density of thick disk stars near the Sun, which complicates attempts to identify specific local candidate stars. The physical significance of the local normalization is further discussed by Evans (1987). The most recent determination of the vertical density profile of the Galaxy is derived from a sample of K dwarfs, with luminosity classification and abundance corrections to the absolute magnitude derived from high quality spectra, analyzed by Kuijken & Gilmore (1989b; cf. Section 5). Their density distribution is well described by the double-exponential fit

$$\frac{v_0(z)}{v_0(0)} = 0.959 e^{-z/249 \text{ pc}} + 0.041 e^{-z/1000 \text{ pc}}. \qquad 7.$$

This fit was made to the star count data over the z range 300–4000 pc, where the lower limit was chosen to avoid any residual giant contaminations of the sample. The individual exponential scale heights quoted are not to be interpreted as definitive. The numerical values are very highly anti-correlated, and other equally acceptable double-exponential fits exist. A density profile that does not include a few percent of the stars near the Sun in a component of the Galaxy with a characteristic scale height of ~ 1 kpc is, however, quite inconsistent with observations.

2.5.2 STAR COUNTS AND THE CENTRAL BULGE   One situation in which the current star count models appear not to provide an adequate description of the Galaxy is in lines of sight within $\sim 30°$ of the Galactic center. In these directions, available data (Rodgers et al. 1986, Rodgers & Harding 1989, Gilmore & Hewett 1989) show a substantial *excess* of stars above the predictions of the models. The available data are somewhat confusing, however, and are not consistent with a simple extra "bulge" population superimposed on the basic disk–thick disk–spheroid model of the Galaxy. It does seem that there is a contribution to the stellar distribution in the central regions of the Galaxy that has a scale length of ~ 1 kpc. (For comparison, the classical subdwarf system has a half-light scale length of ~ 3 kpc.) This population also seems to have a rather blue (F star) main sequence turnoff, suggestive of a young age and/or a low chemical abundance. The available photometry is, however, inadequate to produce definitive parameters, owing to the complexities of photometry in crowded fields, the effects of patchy reddening, and the need to survey many fields to describe the apparent distribution of stars adequately. Spectroscopic and photometric studies to clarify this uncertainty are underway.

One of the most important implications of data within a few kiloparsecs of the Galactic center is that one expects to see a flaring of the old disk and the thick disk in these regions. Lewis & Freeman (1989; see also Freeman 1987) have shown that the planar, radial velocity dispersion of the old disk rises exponentially toward the Galactic center, with a scale length twice that of the luminosity. This is the predicted behavior for a constant scale height, radially exponential disk, assuming that the ratio of the vertical to the radial velocity dispersions remains roughly constant (see Section 3.1 and van der Kruit & Freeman 1986). Thus there will exist a radius within which the old stars of the "disk" become of sufficiently high velocity dispersion—both radial and vertical—that their spatial configuration will no longer be specified predominantly by angular momentum support, but will also be determined by a significant contribution from pressure support. The stellar system will puff up, and the higher velocity dispersion thick disk will transform itself into a "bulge" population. This model (i.e. that the "bulge" is the inner regions of the thick disk) has a conspicuous advantage over explanations of the central bulge as being part of the inward extension of the local subdwarf system in that it naturally accounts for both the high chemical abundance and the high stellar rotation velocities observed in external galactic bulges. This point is discussed further by Jones & Wyse (1983), King (1986), and Wyse & Gilmore (1988).

The relation of the stars seen from ~ 15–$30°$ from the Galactic center to the highly concentrated (within $\lesssim 1$ kpc of the Galactic center) stellar distribution remains unclear. Frogel (1988) reviews the evidence concerning the age of the majority of stars in the central bulge and concludes that they are old, similar to the thick disk globular clusters (see also Terndrup 1988). However, there are contradictory indications. The late-type stars seen by the *IRAS* satellite (Habing 1987, Harmon & Gilmore 1988) are long-period variables with young ages, and both the *IRAS* sources and the late-spectral-type M giants form a stellar system with scale length an order of magnitude smaller than that of the subdwarf system (Blanco & Blanco 1986). The very uncertain direct age determinations for the K giants in Baade's Window (Terndrup 1988) allow a substantial age gap, of order 7 Gyr, between these stars and the oldest subdwarfs. Thus the relation of the dominant stellar population within the central few kiloparsecs of the Galaxy to the dominant population several kiloparsecs from the center and from the plane remains problematic (see also Section 1.3). The former, however, is metal rich, may be of intermediate age, and forms a high-velocity-dispersion yet predominantly rotationally supported system; the latter is metal poor, apparently exclusively old, and forms an entirely pressure-supported system. The assumption that the former is the inward extension of the latter is far from trivially consistent with observations.

2.6 *The Shape of the Metal-Poor Spheroid*

As an example of the type of analysis for which star counts are ideal, we discuss a recent determination (Wyse & Gilmore 1989) of the shape of the metal-poor spheroid, represented in the solar neighborhood by the high-velocity subdwarfs. The shape of the non–thin disk stars is important because of its implications for the early stages of galaxy collapse and star formation, the interpretation of the kinematics of high-velocity stars, and the shape of the underlying dark matter that generates the gravitational potential in which these stars move.

As discussed further in Section 3.1, the high-velocity, metal-poor field stars in the solar neighborhood have an anisotropic velocity dispersion tensor, with $\sigma_{rr}:\sigma_{\theta\theta}:\sigma_{zz} \sim 2:1:1$ derived from kinematically unbiased samples. Since the velocity dispersion tensor behaves as an anisotropic stress tensor in the equations governing stellar dynamics, one may expect this anisotropic "pressure" to result in an anisotropic shape, i.e. a flattened metal-poor spheroid (see Section 3.1 for the relevant equations). Binney & May (1986) investigated this idea in more detail for various Galactic potential–distribution function pairs and concluded that in the locally nonspherical potential felt by the subdwarfs, owing to the presence of the disk, the observed velocity dispersion anisotropy implies a substantially flattened spheroid, with shape $\sim$ E7 or axis ratio $\sim$ 0.25. A spherical potential would imply shape $\sim$ E3, with axis ratio $c/a \sim 0.7$. White (1989b) quantified the expectation of a flattened shape resulting from a flattened velocity ellipsoid using the tensor virial theorem for axisymmetric galaxies. This approach has the disadvantage of treating the two components of the velocity dispersions tangential to the vertical direction on an equal footing, which is not compatible with the observations, but the general results should be valid. Adjusting his results to reflect the *nonkinematically selected* velocity dispersions, one finds from the flattening of the velocity ellipsoid that in a spherical potential the subdwarfs should have axis ratio $c/a \sim 0.4$, while in a potential due to a mass distribution that is also flattened one obtains $c/a \sim 0.3$. The ratio of the vertical velocity dispersion to the circular velocity also constrains the flattening and predicts $c/a \sim 0.53$. The precise numerical value for this flattening is model dependent and hence uncertain, though the general conclusion that the subdwarf system must be appreciably flattened is robust.

The kinematic data of Ratnatunga & Freeman (1985, 1989) for distant metal-poor K giants can also most easily be explained by allowing these stars to form a flattened distribution. The most important feature of their data is the fact that the line-of-sight velocity dispersion in the south Galactic pole may not increase with distance, despite the increasing contribution of the radial (relative to the Galactic center) component of the velocity dispersion (σ_{rr}) to the observed line-of-sight stellar radial (relative to the Sun) speed. The assumption behind the expectation of a rising line-of-sight dispersion with distance is that the distant metal-poor K giants trace the same population as the local metal-poor K giants and the local subdwarfs, and hence they should have the same radially biased velocity-dispersion tensor. Adopting a global form of the distribution function in either a spherical (White 1985) or in an oblate (Levison & Richstone 1986) potential requires a flattened spatial distribution for the spheroid stars, again with axis ratio ~ 0.25. Alternatively, one can depress the observed velocity dispersion at large distances by allowing suitable discontinuities in the stellar distribution function—that is, by essentially assuming that all stars beyond a given Galactocentric radius are on circular orbits. This latter approach allows a fit that is consistent with a spherical spatial distribution for these stars, though at the expense of a somewhat contrived distribution function (Sommer-Larsen 1987, Dejonghe & de Zeeuw 1988, Sommer-Larsen & Christensen 1989), which as shown by White (1989b) cannot be globally applicable.

In the light of this kinematic evidence and the results of the more straightforward dynamical analyses, it is mildly puzzling that direct star count studies suggest that the subdwarf stellar system is approximately round [cf. Freeman (1987) for a review]. The most quoted evidence for a spherical distribution of field spheroid stars comes from the modeling by Bahcall & Soneira (1980, 1981, 1984; hereafter BS) of the faint star counts of Koo & Kron (1982) in two fields; BS conclude that the axis ratio of the spheroid stars is $c/a = 0.80^{+0.20}_{-0.05}$. Their technique is based on the fact that fields in the $l = 90°, 270°$ plane are at equal Galactocentric distances if at equal distances from the solar neighborhood, and hence a spherical distribution of stars will contribute equally to all fields in this plane. Thus, if one compares magnitude-limited samples in fields at high and low Galactic latitude, one should obtain equal numbers of spheroid stars in the two fields. A flattened distribution of stars will yield lower counts in the higher latitude field.

BS complicate their analysis somewhat by adopting different color magnitude relations for the two fields, and thus they do not predict equal numbers of stars for a spherical distribution—they are forced to do this to obtain an acceptable fit for their model in each of the two fields, owing to a combination of inadequacies in the model (such as lack of the thick disk component) and in the data (discussed below). There is no physical basis for such a variation of color-magnitude relations (metallicity gradients are not relevant, since the fields are supposed to be at the same Galactocentric distance), and it is a potential source of uncertainty. Adoption of a metal-poor color-magnitude relation has the effect of assigning a low intrinsic luminosity to main-sequence stars of a given color. Hence, in an apparent-magnitude-limited sample, one will be comparing lower luminosity, less distant stars in the "metal-poor" field with higher luminosity, more distant stars in the other field. The predictions of relative star counts are therefore sensitive to the shape of the subdwarf luminosity function, as well as to the shape and density profile of the stellar tracer population. It is then possible to produce predictions for the ratio of counts that can exceed unity in a spherical distribution, as BS derived.

A new study of this problem, utilizing a larger data set and using a more general model Galaxy program that requires internally consistent properties for a given stellar population in different fields, and that allows the inclusion of a thick disk, is described by Wyse & Gilmore (1989). Following BS, they counted stars blueward of a color limit ($B-V = 0.6$) chosen to minimize contamination of the tracer sample by nearby old disk stars, with the precise value of this limit not being critical. The two fields used were ($l = 0°$, $b = 90°$; area surveyed = 0.75 square degrees) and ($l = 272°$, $b = -44°$; area surveyed = 0.75 square degrees). The observed ratio of blue stars in the two fields was 0.59. This disagrees strongly with Koo & Kron's (1982) counts in two fields at similar Galactic latitudes but at much fainter magnitudes, which yield a ratio of 1.09 for blue stars with $20 \lesssim V \lesssim 22$; the more recent calibration of the Koo & Kron data by Koo et al. (1986) gives 1.3. (Note that these numbers are based on a somewhat uncertain color cut, but this should not matter provided one is blue enough to have isolated the metal-poor spheroid stars.) We suspect that this apparent disagreement reflects the uncertainty in the Koo & Kron counts due to the difficulty of reliable star-galaxy discrimination at faint magnitudes. This suspicion is based on the results shown in Table 1, which shows the Koo et al. (1986) "subdwarf" category counts for their north Galactic pole field, together with predictions from the Gilmore & Wyse (1985; henceforth GW) model and from the BS model, each model with an assumed spheroid axis ratio of 0.8. There is an obvious disagreement between the data and the models; the data fail to increase toward fainter magnitudes, contrary to both of the models and to intuition.

The BS model predictions (their Table 3) combined with the low value of the relative observed counts would imply that the spheroid had an axis ratio $c/a \lesssim 0.5$. When one considers the presence of the thick disk and also models the observed *total* counts as well as their ratio, the best estimate for the axis ratio of the metal-poor subdwarf stellar population within a few kiloparsecs of the Sun is $c/a \sim 0.6$. Note that this is 4σ below the quoted errors of BS, which illustrates the importance of systematic errors in these analyses.

One can also utilize direct counts of other spheroid tracers, such as RR Lyrae stars, to derive the density profile of spheroid light. Early work based on RR Lyrae stars in the Palomar-Groningen and Lick surveys, which were toward the Galactic center (Kinman et al. 1966, Oort & Plaut 1975), concluded that these stars were distributed in a nearly spherical system. These results have now been superseded by better photometric data (Wesselink et al. 1987); the more modern analysis finds, in contrast, that the RR Lyrae stars toward the Galactic center have a rather flattened distribution, with axis ratio $\lesssim 0.6$, in excellent agreement with the star count result above. A possible complication in this picture was introduced by the results of a kinematic analysis by Strugnell et al. (1986), who showed

Table 1 Star count constraints on the slope of the Galaxy

Koo et al. (1986) north Galactic pole data; relative star counts

J mag	Data	V mag	BS model	GW model
20.25	1.0	19.75	1.0	1.0
20.75	0.8	20.25	1.1	1.2
21.25	1.0	20.75	1.24	1.39
21.75	1.1	21.25	1.35	1.57
22.25	1.0	21.75	1.45	1.73
22.75	0.8	22.25	1.57	1.91

Constraints on the spheroid axis ratio

Model	$(l,b) = (0, -90)$		$(l,b) = (270, -45)$		Relative counts	
Axis ratio	Thick disk	Spheroid	Thick disk	Spheroid	Spheroid	Total
1.0	20	367	60	368	1.00	0.90
0.75	20	215	60	279	0.77	0.69
0.6	20	133	60	209	0.64	0.57
0.5	20	86	60	156	0.55	0.49

Observed number counts

$(l,b) = (0, -90)$	$(l,b) = (270, -45)$	Ratio
168	284	0.59

that the *c*-type (low amplitude of variability) and the small ΔS (metal-rich) RR Lyrae stars have significantly different kinematics from the more metal-poor RR Lyrae stars. Thus the RR Lyrae stars, like other field stars, may well form a two-component system, with the more metal-rich stars being part of the thick disk. The need for more data on these stars is clear (see Sections 3 and 4 below).

Yet another complication in this picture is due to the analysis of Hartwick (1987), who analyzed the available data for *metal-poor* RR Lyrae stars and concluded that the axis ratio of the metal-poor RR Lyrae system varies with Galactocentric radius, being flattened [axis ratio $c/a \sim 0.6$ and scale height ~ 1.5 kpc (i.e. rather similar to the parameters of the more metal-rich thick disk RR Lyrae stars)] interior to the solar circle. At very large Galactocentric distances the RR Lyrae data somewhat favor a more spherical distribution. Hartwick thus suggested that the *metal-poor* RR Lyrae stellar system is itself two component, in addition to the (third) metal-rich RR Lyrae system. As we discuss further below (Section 4.3), there are indications of a "metal-poor thick disk" in other samples of field stars.

Hartwick also finds a similar two-component structure for the metal-poor ([Fe/H] < −1) globular clusters from the distribution of their projected positions on the sky. Again, this two-component structure is additional to the well-established distinction between metal-rich and metal-poor clusters (Zinn 1985, Thomas 1989), where the metal-rich clusters form a clear disklike system. However, the small number of clusters involved gives Hartwick's multicomponent model small statistical weight. The kinematics of the metal-poor globular cluster system was found by Frenk & White (1980) to be describable by negligible net rotation and an isotropic velocity dispersion tensor. These properties suggest a spherical spatial distribution in a spherical potential and a flattened distribution in a flattened potential. However, Norris (1986) found there to be no statistically significant difference between the "isotropic" velocity dispersion tensor of the globular clusters and the markedly *anisotropic* velocity dispersions of the local subdwarfs, while Thomas (1989) has shown that one cannot in general draw *any* strong conclusions about the kinematics of the metal-poor globular cluster system, owing to the effects of distance errors.

Even if there were reliable evidence for a similarity between the kinematic parameters describing one or the other of the globular cluster systems and those describing the spheroid or thick disk field stars, this would not prove that the globular cluster and field star systems have a similar or common origin. Other parameters do appear to differ significantly. The most recent comparisons of the metallicity distributions of the globular clusters and the metal-poor field spheroid stars show statistically significant differences (Laird et al. 1988b), while Bell (1988a,b) and Schuster & Nissen (1989) provide suggestive evidence that the oldest field subdwarfs are older than the globular clusters. However, the latter conclusion rests on the uncertain need for an oxygen overenhancement in the stellar evolutionary calculations for globular cluster stars. The observational data regarding oxygen abundances in globular clusters are well reviewed by Kraft (1988), who shows the measured oxygen abundance to be quite different from star to star in a single cluster, with no clear conclusion regarding the appropriate abundance to be used for isochrone dating yet available. One should use considerable caution in deducing some property of the globular cluster systems from observations of the field star systems in the Galaxy, and vice versa.

In summary, the available evidence on the shape of the system of metal-poor field stars that make up the Galactic extreme Population II suggests that these stars form a distinctly nonspherical system, whose flattening may vary with radius but is $c/a \sim 0.5$ within a few kiloparsecs of the Sun, and within a few kiloparsecs of the Galactic center.

3. KINEMATICS AND CHEMISTRY OF OLD STARS

The kinematic properties of stars in the Galaxy are related, through the gravitational potential Φ, to their spatial distribution. The scale length of the spatial distribution is determined by the total energy (kinetic and potential) of the stellar orbits, as well as by the gradient of the potential (i.e. the force on the star). The shape of the spatial distribution depends on the relative populations of the orbits supported by the potential and on the relative amounts of angular momentum (rotational) and pressure (stellar velocity anisotropy) balance to the potential gradients. The total orbital energy and angular momentum of a star depend on the maximum distance from the center of the Galaxy that the gas from which it will form reached before falling out of the background expansion of the Universe, the angular momentum of the gas clouds' orbit at that time, the depth of the potential well through which it fell, the fraction of the total orbital energy that was lost (dissipated) before the gas formed into a star, and the subsequent dynamical evolution of the stellar orbit. That is, the present kinematic properties of old stars in the solar neighborhood are determined in part by initial conditions in the proto-Galaxy at the time of the first star formation and in part by physical processes during galaxy formation. Hence, local kinematic studies can help to determine both the detailed physics of galaxy formation and also some of the large scale structural properties of the Galaxy.

Stellar chemical abundance is determined by the fraction of the available interstellar medium (ISM) at the time and place of the star's formation that had been processed through the nuclear-burning regions of massive stars. It provides a valuable chronometer for the early evolution of the Galaxy. The chemical abundance of the ISM at any time depends both on the local history of formation and evolution of stars sufficiently massive to have created new chemical elements and on the mixing of local gas with more distant material. This more distant gas may or may not itself be enriched, so that the time-dependence of the chemical abundance of newly forming stars depends on both the local and the global star formation rates, the rate of infall of primordial gas, and the efficacy of mixing in the ISM. Thus, while the chemical abundance of newly formed stars is a valuable timepiece, this chronometer need not be a smooth or even a single-valued function of chronological time.

Clearly, however, the distribution function of stellar kinematics, chemistry, and age contains a wealth of information on the distribution of proto-Galactic gas, the dissipational and star formation history of that gas, the subsequent dynamical history of the resulting stars, and the Galactic gravitational potential.

3.1 Observable Stellar Dynamics

The dynamics of any collisionless system, such as a large number of stars, is governed by the Vlasov equation, which is more commonly referred to as the collisionless Boltzmann equation (CBE):

$$\frac{Df}{Dt} \equiv \frac{\partial f}{\partial t} + \frac{\partial \mathbf{x}}{\partial t} \cdot \frac{\partial f}{\partial \mathbf{x}} + \frac{\partial \mathbf{v}}{\partial t} \cdot \frac{\partial f}{\partial \mathbf{v}} = \frac{\partial f}{\partial t} + \mathbf{v} \cdot \frac{\partial f}{\partial \mathbf{x}} - \nabla\Phi \cdot \frac{\partial f}{\partial \mathbf{v}} = 0, \qquad 8.$$

where f is the phase space density at the point $(\mathbf{x}, \mathbf{v})$ in phase space [i.e. there are $f(\mathbf{x}, \mathbf{v})\,d\mathbf{x}\,d\mathbf{v}$ stars in a volume of size $d\mathbf{x}$ centered on spatial coordinate $\mathbf{x}$, and with velocity in the volume of size $d\mathbf{v}$ about velocity coordinate $\mathbf{v}$]. The collisionless nature of stellar interactions allows the substitution of the gradient of the smoothed gravitational potential Φ for the accelerations. Here f does not have to describe the entire Galaxy; one can concentrate on any subsample of stars and apply the CBE (Equation 8) to it. We refer to such subsamples as *tracer populations*, since one can use their kinematics to trace the gravitational potential of the Galaxy, irrespective of the nature of the mass distribution that generates this potential. For stellar populations whose mass *generates* the potential as well as traces it, it is necessary to consider joint solutions of both the Boltzmann equation and the Poisson equation. Such *self-consistent* solutions are discussed in Section 5 below, where the local thin disk is discussed.

Since this review is concerned mostly with the old stellar populations, and primarily with high-Galactic-latitude distributions, we may safely assume that we are concerned with a steady-state tracer population in an axisymmetric time-independent potential, so that time- and ϕ-derivatives are zero.

It is then convenient to write the collisionless Boltzmann equation in cylindrical polar coordinates (r, ϕ, z):

$$v_r \frac{\partial f}{\partial r} + v_z \frac{\partial f}{\partial z} + \left(K_r + \frac{v_\phi^2}{r}\right)\frac{\partial f}{\partial v_r} - \frac{v_r v_\phi}{r}\frac{\partial f}{\partial v_\phi} + K_z \frac{\partial f}{\partial v_z} = 0, \qquad 9.$$

where the accelerations $\dot{v}_r, \dot{v}_\phi, \dot{v}_z$ have been equated to the forces that cause them, ϕ-gradients in f and in the potential have been set to zero, and $K_r \equiv -\partial\Phi/\partial r$ and $K_z \equiv -\partial\Phi/\partial z$ are the components of the gravity force. Clearly, knowledge of $f(\mathbf{x}, \mathbf{v})$ allows the force components K_r and K_z to be derived. Note, though, that any general function f of two variables need not allow a solution for K_r and K_z. In view of the intractability of the general problem of solving the CBE, one proceeds in general by taking velocity moments. Multiplying through by v_z and by v_r and integrating over all of velocity space produces the Jeans' equations:

$$vK_z = \frac{\partial}{\partial z}(v\sigma_{zz}) + \frac{1}{r}\frac{\partial}{\partial r}(rv\sigma_{rz}), \qquad 10.$$

$$vK_r = \frac{1}{r}\frac{\partial}{\partial r}(rv\sigma_{rr}) + \frac{\partial}{\partial z}(v\sigma_{rz}) - \frac{v\sigma_{\phi\phi}}{r} - \frac{v}{r}\langle v_\phi \rangle^2, \qquad 11.$$

where $v(r,z)$ is the space density of the stars, $\sigma_{ij}(r,z)$ their velocity dispersion tensor, and the only mean streaming motion is rotation, $\langle v_\phi \rangle$.

Each of the two force components can, in principle, be derived from the measurements of the moments of the velocity distribution and of the spatial density distribution of a tracer stellar population. Such experiments have been carried out for the z-force and relate the stellar kinematics and space density to the potential of the Galactic disk. They are discussed further in Section 5. It is convenient for present purposes to rewrite the radial equation (Equation 11) in terms of observables in the Galactic plane ($z = 0$). We obtain

$$v_c^2 - \langle v_\phi \rangle^2 = \sigma_{\phi\phi} - \sigma_{rr} - \frac{r}{v}\frac{\partial(v\sigma_{rr})}{\partial r} - r\frac{\partial \sigma_{rz}}{\partial z}$$

$$= \sigma_{rr}\left\{\frac{\sigma_{\phi\phi}}{\sigma_{rr}} - 1 - \frac{\partial \ln(v\sigma_{rr})}{\partial \ln r} - \frac{r}{\sigma_{rr}}\frac{\partial \sigma_{rz}}{\partial z}\right\}. \qquad 12.$$

In this relation v_c is the circular velocity (i.e. $v_c^2 = r(\partial \Phi/\partial r) = -rK_r$, where we adopt a locally flat rotation curve with $v_c = 220$ km s^{-1} for this article), and $\langle v_\phi \rangle$ is the mean rotation velocity of the relevant sample of tracer stars, which has velocity dispersions $(\sigma_{rr})^{1/2}$, $(\sigma_{\phi\phi})^{1/2}$, and $(\sigma_{rz})^{1/2}$, and radial spatial density distribution $v(r)$ (recall that r is the *planar* radial coordinate). The quantity $v_c - \langle v_\phi \rangle \equiv v_a$ is often called the *asymmetric drift* of a stellar population.

Equation 12 relates measurable local moments of the stellar distribution function to global properties of the Galaxy. In order to illustrate its application, we discuss each term briefly.

♣ $\sigma_{\phi\phi}/\sigma_{rr}$: The velocity dispersions at $z = 0$ of old disk stars are probably best estimated from the nearby, spectroscopically selected K and M dwarfs with good parallax distances. These give $\sigma_{rr}:\sigma_{\phi\phi}:\sigma_{zz} = 39^2 : 23^2 : 20^2$ (Wielen 1974). For low-metallicity field stars ([Fe/H] ≤ -1) the weighted mean values from Carney & Latham (1986), Norris (1986), and Morrison et al. (1989) are $\sigma_{rr}:\sigma_{\phi\phi}:\sigma_{zz} = (131\pm7)^2 : (102\pm8)^2 : (89\pm5)^2$. (Note that these values are derived from *nonkinematically selected* samples.) The first term in Equation 12 then becomes $\sigma_{\phi\phi}/\sigma_{rr} = 0.35$ for the old disk and $\sigma_{\phi\phi}/\sigma_{rr} = 0.61$ for the low-abundance field stars.

♣ $\partial \ln(\nu\sigma_{rr})/\partial \ln r$: The thin exponential disks of spiral galaxies are apparently self-gravitating (see Section 5) and are observed to have a constant thickness with radius and to be approximately isothermal vertically (van der Kruit & Searle 1982). Thus, since the scale height of a self-gravitating disk varies as $h_z \propto \sigma_{zz}/\Sigma$, with $\Sigma \propto \exp(-r/h_r)$ being the disk surface mass density, one expects that the radial variation of the vertical velocity dispersion will be $\sigma_{zz} \propto \exp(-r/h_r)$. Assuming that the shape of the velocity ellipsoid is independent of position (or, more specifically, that $\sigma_{rr} \propto \sigma_{zz}$) leads to

$$\frac{\partial \ln(\nu\sigma_{rr})}{\partial \ln r} = 2 \times \frac{\partial \ln \nu}{\partial \ln r} = -2h_r^{-1}r. \qquad 13.$$

We had to assume a form for the velocity ellipsoid above to close the system of equations owing to the lack of an analogue of the equation of state. For simplicity, we restrict discussion of spheroidal distributions to an isothermal spheroid $\sigma_{rr} = $ constant, so that

$$\frac{\partial \ln(\nu\sigma_{rr})}{\partial \ln r} = \frac{\partial \ln \nu}{\partial \ln r}. \qquad 14.$$

♣ $(r/\sigma_{rr})(\partial\sigma_{rz}/\partial z)$: The term involving σ_{rz} describes the orientation of the velocity ellipsoid and has no general analytic solution. The velocity ellipsoid will be oriented along the coordinate system (if there is one) in which the Hamilton-Jacoby equation, which provides the equations of motion, is separable, though unfortunately there is no reason for there to be a simple relationship between this (Stäckel) coordinate system and the coordinate system in which observations are naturally available. We therefore consider two representative cases only. If the potential is that of an infinite, constant-surface-density sheet, the velocity dispersion tensor will be diagonal in cylindrical-polar coordinates and always point at the Galactic minor axis, so that $\sigma_{rz} \equiv 0$. This is the assumption most commonly adopted. An alternative idealization is to assume that the potential is dominated by a sufficiently centrally concentrated or spherical mass distribution that the local velocity ellipsoid points at the Galactic center. This assumption was made by Oort (1965), who found that $\sigma_{rz} \sim (z/r)(\sigma_{rr} - \sigma_{zz})$ for $z \ll r$, from which it follows that

$$\frac{r}{\sigma_{rr}} \frac{\partial \sigma_{rz}}{\partial z} \approx 1 - \frac{\sigma_{zz}}{\sigma_{rr}}. \qquad 15a.$$

An exact derivation for a disk in which the velocity ellipsoid has constant shape in spherical polar coordinates, with axis ratio $\sigma_{rr} = \alpha^2 \sigma_{zz}$ at $z = 0$, when spherical and cylindrical coordinates coincide (Kuijken & Gilmore 1989a) provides

$$\sigma_{rz} = \left[\frac{rz(\alpha^2 - 1)}{\alpha^2 z^2 + r^2}\right]\sigma_{zz},$$

so that

$$\frac{r}{\sigma_{rr}}\frac{\partial \sigma_{rz}}{\partial z} \approx (\alpha^2 - 1)\frac{\sigma_{zz}}{\sigma_{rr}}. \qquad 15b.$$

For the velocity dispersions quoted above, both Equation 15a and Equation 15b provide the same answer for the old disk, namely

$$\frac{r}{\sigma_{rr}}\frac{\partial \sigma_{rz}}{\partial z} \simeq 0.75, \qquad 15c.$$

while for the spheroidal field stars the numerical value from Equation 15a is 0.53. This range in values for the σ_{rz} term is one indication of the uncertainty in present applications of this equation. A second indication is that numerical orbit integrations in potentials derived from recent studies of the Galactic K_z force law (cf. Section 5 and Kuijken & Gilmore 1989a) show that the velocity ellipsoid tends to point somewhat above the direction to the Galactic center, so that these correction terms will slightly overestimate the amplitude of the term involving σ_{rz}. Neglecting this term (as is often done) is, however, clearly unjustified.

For an exponential disk of radial scale length h_r and a sample of stars observed in the solar neighborhood, Equation 12 therefore becomes

$$v_c^2 - \langle v_\phi \rangle^2 = \sigma_{rr}\left(2\frac{d_*}{h_r} - 1.4\right), \qquad 16a.$$

where d_* is the distance of the Sun from the Galactic center (~ 7.8 kpc; Feast 1987). Alternatively, for a spheroid with a power-law density distribution with exponent γ, i.e. $\nu(r) \propto r^{-\gamma}$, we have

$$v_c^2 - \langle v_\phi \rangle^2 = \sigma_{rr}(\gamma - 0.9). \qquad 16b.$$

Thus a stellar tracer population that belongs to an exponential distribution with scale length ~ 3.5 kpc (a plausible value for the Galactic old disk) will follow a similar asymmetric drift relation to that of a tracer population that is part of an isothermal distribution describing an r^{-4} spheroidal density distribution. Also, note that the largest allowed radial velocity dispersion, corresponding to zero net rotation, for such a stellar system with the observed anisotropic velocity dispersions is $\sim 1.2 v_c/\sqrt{4}$, or ~ 135 km s^{-1}. For a tracer population with any smaller radial velocity dispersion, Equation 12 describes the interplay between the pressure (velocity anisotropy) support and the angular momentum (rotation velocity) support to the spatial distribution. A higher velocity dispersion system than this has larger total energy and will form a more extended system. One might also assume it would have formed from less dissipated material, which is of course the clue to the *physical* significance of Equation 12 and the reason for this extensive discussion of it here.

3.1.1 THE ASYMMETRIC DRIFT The relevant observational data are shown in Figure 3, where the data points shown have been either collated or calculated from data available in the identified references. The data from the large surveys have been binned by metallicity. It is apparent that all data for tracer samples with a Galactic rotation velocity greater than about 50 km s^{-1} are consistent with a single density distribution, with the marginally significant exception of the metal-rich globular cluster system, whose radial velocity dispersion is rather low. This datum is, however, somewhat more uncertain than most of the other data shown owing to distance and reddening uncertainties. The range of uncertainty (see Figure 4a) shown covers the solutions found by Armandroff (1989); his favored solution, which has high rotation velocity, contains only nine clusters. The data of Norris & Ryan (1989a,b) have not been plotted, since although the data for their (kinematically selected) sample show the same trend as the Sandage & Fouts (1987) sample, they are offset in rotation velocity and obscure the important point of the figure. The origin of this systematic difference between the data of Sandage & Fouts and those of Norris & Ryan is not yet clear.

Observational selection effects can have a very substantial effect on the appearance of the diagram and have not been considered at all adequately. Obvious examples include explaining the apparent systematic difference in the deduced radial velocity dispersion between spectroscopically and kinematically selected samples (cf. Norris 1986), even for the highest velocity stars, which are far from the regions of phase space expected to be affected strongly by the selection criteria. In addition, proper-motion samples will not find metal-poor stars on circular orbits near the Sun if they exist (see Section 4) with the same low local normalization as the high-velocity subdwarfs. Local surveys will also miss stars on high-angular-momentum, high-energy orbits, since these will always lie beyond the solar circle.

The mean density law consistent with the majority of the data cor-

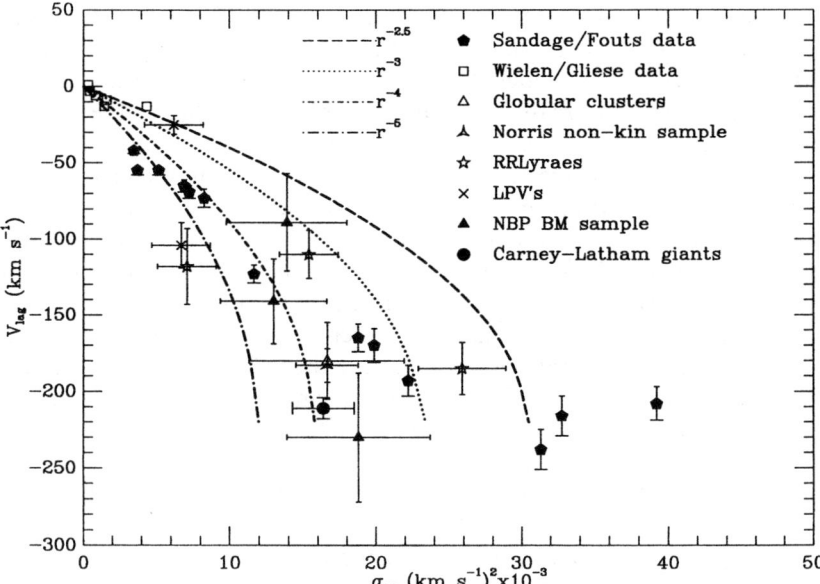

Figure 3 The relation between the radial velocity dispersion σ_{rr} and the asymmetric drift V_{rot} of samples of old stars in the Galaxy. The data for field stars are binned by metallicity. The key identifies data from the following sources: the proper-motion sample of Sandage & Fouts (1987) for stars with [Fe/H] $\gtrsim -2$ only, since photometric metallicities and hence distances are very uncertain for the more metal-poor stars; the Wielen (1974) analysis of the Gliese nearby star catalogue; the analyses by Norris (1986) of the globular cluster system and of kinematically unbiased low-abundance stars; the analysis by Strugnell et al. (1986) of field RR Lyrae stars; long-period variables (binned by period range) from the analysis by Osvalds & Risley (1961); spectroscopically selected low-abundance field stars from the analysis by Norris et al. (1985); and the spectroscopically selected local metal-poor giants from the study by Carney & Latham (1986). The model lines correspond to different solutions of Equation 12. The following density laws were used: $r^{-2.5}$ (long dashes), r^{-3} (dots), r^{-4} (short dash–dot), and r^{-5} (long dash–dot). The tendency for the data to cross the model lines at low V_{rot} shows that some star formation took place during dissipational collapse of the Galaxy.

responds to that of an isothermal spheroidal distribution with a power law having index ~ -4.5, or to an exponential disk with scale length ~ 3 kpc. The tendency for the tracer populations with the lowest mean rotational velocities to have larger radial velocity dispersions than would be consistent with this density profile is of considerable significance, if real. We note that systematic distance uncertainties move the data roughly parallel to the body of the data with smaller σ_{rr}, so they are unlikely to be relevant.

With the exception of the data for the metal-poor RR Lyrae stars, however, there are very few stars in the bins with the highest values of σ_{rr}. The stars with the highest radial velocity dispersions are also the most metal poor (see below) and hence are those that presumably formed first in the Galaxy. If they really do form a more extended spatial distribution than more metal-rich stars, as suggested by Figure 3, then one may conclude that these stars formed from less dissipated gas and hence preserve a fossil record of the star formation and dissipation history of the proto-Galaxy during its first condensation from the expanding background. That is, there is a shallow *radial* abundance gradient in the Galactic spheroid but no evidence for a *vertical* abundance gradient. The lack of any increase of rotation velocity for metal-poor stars during the collapse may reflect angular momentum transport during spheroid formation, as evident in the N-body simulations of Zurek et al. (1988). Alternatively, it may be that the gas that formed stars more metal poor than ~ -1.5 dex had simply not collapsed by a sufficiently large factor to have acquired a measurable mean rotation velocity.

Although the asymmetric drift arguments above provide strong evidence that star formation continued *during* a period of dissipational collapse, there is no information in this relation regarding the *rate* of this collapse. For this, one requires another clock, which is provided by the chemical abundances.

3.2 Correlations Between Kinematics and Chemistry

Stellar chemical abundance is a clock that measures age in units defined by the formation rate and lifetimes of those (mostly massive) stars that contribute to the enrichment of the ISM. The mean azimuthal streaming motion V_{rot} of a population of stars reflects both the amount by which the proto-Galaxy dissipated and collapsed and the amount of angular momentum transport that occurred prior to the formation of these stars. Thus, the existence of a correlation between kinematics, such as V_{rot}, and [Fe/H] allows one to relate the chemical evolutionary time scale to the dynamical and cooling time scales.

The existence or otherwise of an abundance gradient in the spheroid is often cited as an important diagnostic of the time scale of Galaxy formation, in the sense that the lack of a gradient is taken to mean a slow and/or chaotic collapse (cf. Searle & Zinn 1978, Carney et al. 1989b), and the presence of a gradient (manifest by a smooth correlation between kinematics and metallicity) to mean a rapid collapse (Sandage 1987a). In general, neither deduction need be correct, since although the presence of an abundance gradient means that dissipation was an important process during formation of the stellar component of the spheroid, this does not necessarily define time scales. In a dissipationless collapse there is no arrow of time in the kinematics, and so no correlation between kinematics and metallicity. In a dissipative collapse, however, a star-forming gas cloud will be continually transferred onto lower energy orbits with time and will be successively enriched with time owing to continuing star formation. This creates a correlation between chemical abundance and orbital energy, which is an abundance gradient. Only low-metallicity stars will be found in the outer regions of galaxies that formed dissipatively, and high-metallicity stars will be found only in the inner regions.

However, since the cooling time of a proto-galaxy may be expected to be less than the free-fall collapse time (cf. Section 1), dissipation need not slow a collapse significantly. In a rapid (free-fall) dissipationless collapse no abundance gradient will arise, although one present in the initial conditions may survive, since violent relaxation in practice never goes to completion, as noted above. A slow dissipationless formation of the spheroid could occur if the field stars originated in many independent "fragments" that were captured and disrupted over a Hubble time (cf. Searle & Zinn 1978), with each fragment having its own star formation history. This last model would also not lead to an abundance gradient, owing to the assumed randomness of the "fragments." However, consideration of the element ratios of spheroid stars, as discussed in Section 3.4, poses stringent constraints on such a model for formation of the field stars. It may remain viable for the formation of the outer parts of the globular cluster system, as originally motivated.

Correlations (or the lack thereof) between the angular momentum of a tracer population of stars and the stars' chemical abundances, however, remain one of the most powerful observationally feasible tests of the early star formation and dynamical history of the Galaxy.

The amplitude of the peculiar velocity of a star also may be expected to be correlated with that star's metallicity. Since this peculiar velocity is now manifest in the star's orbital eccentricity, the relationship between the eccentricity of stellar orbits, projected onto the plane of the Galaxy, and chemical abundance is also commonly discussed. The general arguments above may be applied in support of its use. In addition to the uncertainties arising from the potential effects of violent relaxation, an extra note of caution is required in its interpretation, however. Unless one is confident that the levels of substructure in the proto-Galaxy and the consequent interactions (more precisely, the hydrodynamics of the dissipational and heating processes during Galactic collapse, and the mixing of the hot stellar ejecta into the infalling gas) are clearly understood, then one should not be entirely confident of the evolution of the *random* motions of newly forming stars.

3.2.1 ROTATION VELOCITY VS. METALLICITY The direct correlation between stellar chemical abundance and Galactic rotational velocity was first convincingly established and discussed in detail by ELS. An extensive debate has arisen in the past few years as a result of several new surveys of the kinematics of metal-poor stars, with conflicting claims as to the reality of this correlation for stars more metal poor than ~ -1 dex (Norris et al. 1985, Sandage & Fouts 1987, Sandage 1987a, Norris 1987a, Carney et al. 1989b, Norris & Ryan 1989b).

The relevant recent data are collected in Figures 4a and 4b. The agreement between the various data sets is acceptable for present purposes for stars more metal poor than ~ -0.8 dex. (Higher abundance stars are discussed separately below.) The mean value of the rotation velocity decreases from ~ 160 km s^{-1} at [Fe/H] = -0.8 to near zero at [Fe/H] = -1.5 and shows no significant correlation at lower abundances.

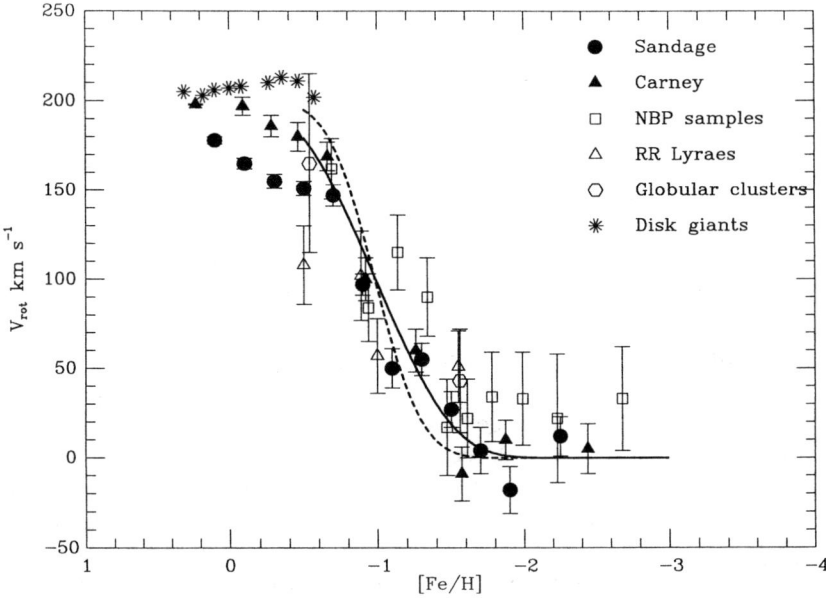

Figure 4a The relation between rotation velocity relative to the Galactic standard of rest (V_{rot}) and metallicity for those samples of field stars with good abundance data from Figure 3 and including the analysis of the disk globular cluster system by Armandroff (1989). The lines show alternative models; the solid line is a model involving a smooth correlation between V_{rot} and [Fe/H] over the range $-1.5 \lesssim$ [Fe/H] $\lesssim -0.5$. The dashed line shows a discontinuous relationship between V_{rot} and [Fe/H], with the discontinuity at [Fe/H] = -1.0. Both models have been convolved with a Gaussian of dispersion 0.25 dex in metallicity to represent measuring errors.

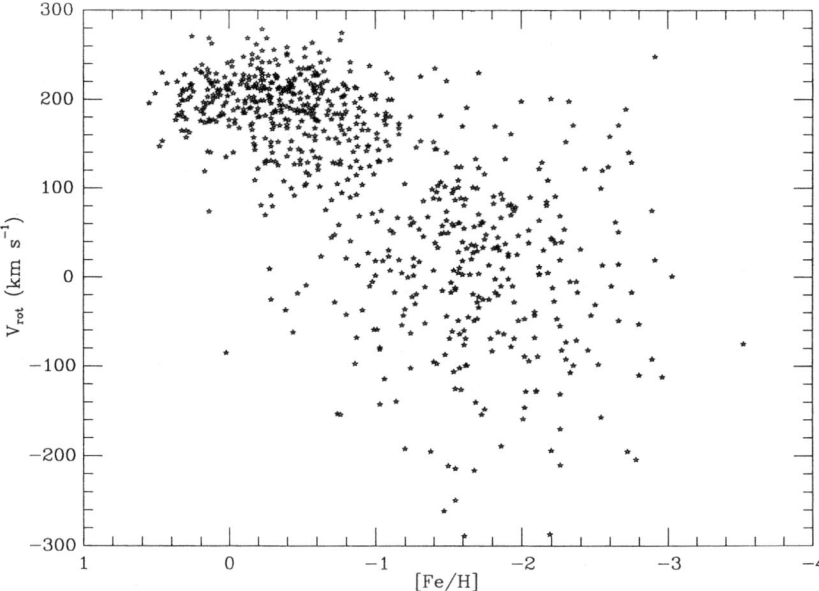

Figure 4b The relation between rotation velocity relative to the Galactic standard of rest, $V_{\rm rot}$, and metallicity for the sample of proper-motion stars studied by Laird et al. (1988a). Comparison of this figure with Figure 4a illustrates the difficulty in deducing the reality of a smooth correlation between kinematics and metallicity from data that are averaged into bins, particularly for metallicities near -1 dex.

Above a metallicity of ~ -1.5 dex, there is a transition to a mean rotational velocity of $\gtrsim 160$ km s^{-1} at ~ -0.8 dex. These latter values are evidently very poorly determined. Is there a systematic abundance-kinematic correlation in the interval $-1.5 \leq$ [Fe/H] ≤ -1.0? While this may seem such a narrow abundance interval as to be unimportant, it in fact contains about one half of the stellar mass of the spheroid. The answer to this question is confused by questions concerning the precision of the data and the validity of binning data with such a large and asymmetric scatter as a function of abundance (Figure 4b).

The two curves shown in Figure 4a are, in turn, a linear correlation between ($V_{\rm rot} = 0$ km s^{-1}, [Fe/H] $= -1.5$) and ($V_{\rm rot} = 200$ km s^{-1}, [Fe/H] $= -0.5$) (solid line) and a discontinuity between $V_{\rm rot} = 0$ km s^{-1} and $V_{\rm rot} = 200$ km s^{-1} at [Fe/H] $= -1$ (dashed line). Both curves have been convolved with a Gaussian of dispersion 0.25 dex in [Fe/H] to illustrate the effect of measuring errors. The greatest uncertainty in distinguishing between the models is now the systematic differences between the results from different data sets. The spectroscopically selected samples discussed by Norris et al. (1985) have a systematically higher rotation velocity than do the kinematically selected samples, for reasons that remain to be clarified. Nevertheless, the smoothed correlation is a slightly better description of the binned data than is the discontinuous model.

Sandage & Fouts (1987) interpreted their data as supporting the continuation of a smooth correlation between kinematics and abundance for *all* metallicities. It is clear from Figure 4a that there is no statistically significant difference between their sample and that of Norris et al. (NBP data points in the figure). Rather, the differences between the conclusions of the two groups result largely from different ways of binning data, which are clearly (Figure 4b) not adequately described by a mean and a standard deviation. One must discuss the distribution function of abundance and kinematic data to derive reliable conclusions. Fortunately, thanks to the very considerable efforts of Sandage & Fouts, Norris, and Carney & Latham, a sufficiently large and reliable data base is available for the first time to allow such a discussion.

One important conclusion from the discussion earlier in this section that is not widely appreciated is that the absence of a clear correlation between rotation velocity and metallicity at very low abundances does *not* mean that a rapid collapse model of the Galaxy is invalid. Thus, the conclusions reached by Sandage & Fouts (1987) are not incompatible with the data, even though their description of the data is not consistent with the appearance of Figure 4a. We emphasize that the presence or absence of such a correlation does *not* distinguish between fast and slow collapse models of the Galaxy. One may learn something about a complex combination of the importance of dissipation and the efficiency of violent relaxation, but time scales enter the interpretation of the relationship between rotation velocity and stellar metallicity only by assumption.

3.2.2 CORRELATIONS VS. DISCONTINUITIES A question of some interest is whether or not the appearance of Figure 4b is consistent with a continuous trend (indicative perhaps of significant star formation *during* the period when the proto-Galaxy was collapsing and spinning up) or instead represents a superposition of relatively discrete subsystems (indicative perhaps of the later merger of subsystems that retained a recognizable identity during the early stages of Galactic formation). Some suggestive but not conclusive evidence in favor of the picture of discrete substructure in phase space comes from the existence of several apparently intermediate groups of tracers that are identifiably discrete using astrophysical criteria. These include the metal-rich RR Lyrae stars ($\Delta S \lesssim 3$, $V_{\rm rot} \sim 110$ km s^{-1}; Strugnell et al. 1986), c-type RR Lyrae stars ($V_{\rm rot} \sim 100$ km s^{-1}; Strugnell et al. 1986), long-period variables with period between 150 and 200 days

($V_{rot} \sim 115$ km s^{-1}; Osvalds & Risley 1961), the metal-rich (G-type) globular clusters ([Fe/H] $\gtrsim -1$, $V_{rot} \sim 100$–200 km s^{-1}; Armandroff 1989), and the Arcturus moving group ($V_{rot} \sim 110$ km s^{-1}; Eggen 1987). The field type-II Cepheids (Harris 1981) are another closely related tracer sample, but they have less well-known kinematical properties at present.

It is remarkable that these different samples of objects all have almost exactly the same rotation velocity. Within the more considerable uncertainties, they also have similar abundances, $-1 \lesssim$ [Fe/H] $\lesssim -0.5$. [It was partially the similarity of this abundance range to that of the thick disk that motivated the suggestion by Wyse & Gilmore (1986) that the rotation velocity of the thick disk would also be ~ 100 km s^{-1}, lower than the angular momentum range compatible with recent data (see Section 4)]. Progress in understanding the amount of structure in the phase space distribution of old stars is most likely to follow further study of the Arcturus moving group, which, if a real and coeval feature in phase space, contains more new information about the dynamical evolution of the Galaxy than any other currently identified tracer population. The existence of such fine structure in phase space would require considerably more careful dynamical analyses of kinematic data (a "bowl of spaghetti" or "can of worms" model) but would also explain a variety of marginally significant observational phenomena that are otherwise inexplicable, such as the retrograde globular clusters (Rodgers & Paltoglou 1984) and field stars (Norris & Ryan 1989b) and the metal-poor moving groups (Eggen 1987, Sommer-Larsen & Christensen 1987). It is quite possible that a substantial fraction of the metal-poor stars now in the Galaxy were not formed there.

3.2.3 ORBITAL ECCENTRICITY VS. METALLICITY There is an alternative presentation of the relationship between stellar orbital properties and chemical abundance that is widely discussed. It involves the eccentricity of the stellar orbit in the Galaxy, projected onto the plane, and the star's metallicity. This presentation really depends on the ratio of the components of the star's velocity radially along and tangential to the line toward the Galactic center and on the stellar metallicity. The derivation of an orbital eccentricity requires the additional assumption of a Galactic potential, which adds further uncertainty. Nevertheless, the existence of a correlation between metallicity and the shape of the stellar orbit projected onto the plane of the Galaxy has been cited as the principal evidence leading ELS to support a rapid collapse model of the Galaxy (Sandage 1987a). Similarly, observational evidence that a tight correlation is violated by a significant fraction ($\sim 20\%$) of field metal-poor stars has been cited as evidence that a slow collapse is a preferred model (Yoshii & Saio 1979, Norris et al. 1985). This correlation is therefore worthy of explicit discussion.

The observational status of such a correlation is a little confused at present. The large kinematically selected samples studied recently by Sandage & Fouts (1987), Laird et al. (1988a), and Norris & Ryan (1989a) are naturally biased against stars with nearly circular orbits, as such stars will (owing to the trade-off between the various terms in Equation 12) tend to have smaller peculiar space velocities than stars on lower angular momentum orbits with the same total energy. Thus careful modeling of the available data will be required to test that any correlation between orbital planar eccentricity and stellar chemical abundance is not due to a selection bias. Low-abundance stars are intrinsically sufficiently rare, and those with both sufficiently large peculiar velocities to enter a kinematically selected sample and the correct angular momentum to reach the solar neighborhood occupy such a tiny fraction of the whole of phase space that one would not expect to find them readily in any case. Low-metallicity stars on nearly circular orbits may well exist in very large numbers nearer the Galactic center, irrespective of the rate of star formation during collapse of the Galaxy.

We emphasize that what is really at issue is the existence or otherwise of a significant number of stars with low metallicities, high angular momenta, and small peculiar velocities. These stars are further discussed in Section 4 under the guise of the "metal-poor thick disk." In the ELS rapid collapse picture, stars acquire large systematic rotation velocities (random velocities are ascribed to initial conditions) owing to the spin-up of the proto-Galaxy as it collapses, with conservation of angular momentum. Thus, mean rotational velocity is a direct measure of the collapse factor of the star-forming gas (if viscosity is unimportant) and is therefore a clock for the collapse, in the same way that stellar abundance is a clock for star formation. Since the natural time scale for an increase in rotational velocity due to Galactic collapse is a free-fall time, ELS deduced from their observed orbital shape–abundance correlation that star formation also occurred on a free-fall time scale. The basic rapid-collapse model of ELS assumes that star-formation, chemical enrichment, and collapse proceed at the same time, so that, in their model, one would not expect to find a substantial number of stars that are metal poor (i.e. among the first stars formed according to the chemical clock) but on roughly circular orbits (i.e. among the later stars formed according to the collapse clock). The ELS model would allow a substantial number of metal-poor stars with both high angular momentum and high orbital energy, as ELS have emphasized. Such stars would, however, still be on eccentric orbits (cf. ELS's discussion of their Figure 9 for this point).

Since there do exist some stars with low metallicity on nearly circular orbits (Norris et al. 1985, Norris 1987a), does this rule out the rapid collapse model, and does it support a slower collapse alternative? To answer these questions, one must rediscuss the important assumptions behind the argument (cf. Section 1.2). For present purposes, these are that the collapse was synchronized, so that metal enrichment follows the collapse factor locally (equivalently, the collapse clock and the enrichment clock are in phase everywhere, with a similar time scale for the two processes—this is necessary, since no correlation would be seen if the two time scales were very different); that the Galactic gravitational potential changed slowly at all times; and that the concept of a "gas cloud" retaining kinematic parameters during the collapse and enrichment process up to formation of a low-mass star is reasonable.

Since the correlation of orbital eccentricity and metallicity discussed by ELS is apparently violated by many stars, a variety of possibilities arise. Searle & Zinn (1978) developed a model in which the synchronization of the collapse and enrichment clocks was put out of phase by allowing the existence of substantial substructure, which survived as gravitationally bound entities for longer than a Galactic collapse time. In effect, their model is a sum of a large number of small ELS-like collapses, but with a different *rate* of enrichment (equivalently, star formation) in each element. [The most useful constraints on models of this type come from cooling-time arguments (cf. Section 1.1) and from study of the relative enrichment rates of different chemical elements, which provides a higher temporal resolution clock. Such constraints are discussed further below.] An alternative possibility has been suggested by Norris (1987a)—that the time scale for the Galactic collapse was much longer than a dynamical time scale, though it is not clear how this model can be consistent with the data either. The Larson (1976) model that Norris utilizes as an example predicts smooth, steep, vertical metallicity gradients. Merely changing the absolute time scale (in years) of both the collapse and the star formation rates will, of course, have no effect on the predictions of the ELS model if both rates are changed proportionately, since neither has been formulated in years. Although hydrodynamical support of the gas against collapse might well act to damp out random motions in the gas clouds during a slow collapse, this possibility on its own is not consistent either with the observations or with the cooling time scale arguments of Section 1.1. Clearly, changing the relative rates of the collapse and enrichment will change the *slope* of any resulting correlation, but it will not change the scatter. It is increased scatter that must be explained. Such scatter can be explained (under the assumptions that chemical abundance is a monotonic clock and that violent relaxation has not erased the fossil record) by inhomogeneity in either the chemical enrichment rate or the gaseous collapse factor (Searle & Zinn 1978), or by a wide diversity in the distribution of initial (precollapse) angular momenta among gas clouds, but not by merely changing the rapidity of the proto-Galactic collapse. The absence of a tight correlation between orbital eccentricity and metallicity does not support a slow collapse model of the Galaxy and in fact argues against a slow, dissipative collapse.

Perhaps the most important feature of the eccentricity-metallicity relation is that stars on highly radial orbits exist at all. A star can now be on a radial orbit either simply because it has dropped out of the quasi-static collapse of a pressure-supported gaseous Galaxy, or because it formed from gas that was on a highly radial orbit—and that would be undergoing rapid collapse owing to the cooling-time arguments—or because it formed on some other orbit that was subsequently made more radial by violent relaxation. The first alternative, which was not considered feasible in principle by ELS, is in fact ruled out by the lack of a steep abundance gradient in the metal-poor spheroid, as discussed above. The remaining two mechanisms involve rapid collapse of something—either the spheroid itself or the dominant mass of the Galaxy, which is not the spheroid. Thus, the combination of kinematic and chemical data implies that the existence of a significant number of subdwarf stars on radial orbits requires a rapid collapse of the spheroid and/or the dominant contribution to the mass of the central few kiloparsecs of the Galaxy.

Similarly, in a rapid collapse model one also expects any *tight* correlation that might have existed between stellar kinematic parameters and time-independent internal properties of the star (e.g. chemical abundance) to be smeared out by the violent relaxation. That is, contrary to the expectation of ELS [who it must be emphasized were working before the discovery of the concept of violent relaxation, by L (Lynden-Bell 1967b)], one does *not* expect a strong correlation of orbital eccentricity and metallicity in a rapid collapse model. The existence of a large fraction of spheroid stars on highly radial orbits at all is strong evidence for rapid collapse of most of the mass of the Galaxy, but it does not of course provide any evidence for determining whether or not the formation of most spheroid stars was completed before, or took place during, the collapse of the dominant contribution to the mass of the Galaxy.

3.3 *Correlations Between Abundances and Ages*

Calibration of the abundance enrichment rate onto a time scale that is calibrated independently of the collapse rate (i.e. in years) is necessary to provide direct evidence for the time scale of Galactic formation. In practice, only stars near the main-sequence turnoff have surface gravities that change sufficiently rapidly and monotonically that reliable comparison with evolutionary tracks is possible, although some useful information on a combination of age and chemical abundance can be derived from the color of field giant stars (e.g. Sandage 1987a). For *single* stars near the turnoff, the comparison of *uvbyβ* photometry with theoretical isochrones is by far the most reliable and precise age-dating technique available. If independent abundance estimates are available, then any photometric measure of the temperature of the hottest turnoff stars will measure the age of the *youngest* star in a tracer population. It is this method that is utilized to determine ages for globular clusters, where it also seems that all the member stars are coeval. A similar technique can be applied to field stars (cf., for example, Gilmore & Wyse 1987) and is illustrated in Figure 5. This figure shows the color-metallicity data for the high-proper-motion stars studied by Laird et al. (1988a), as well as the turnoff points of all those globular clusters with recent CCD photometry, and a representative isochrone for old metal-rich stars (VandenBerg & Bell 1985).

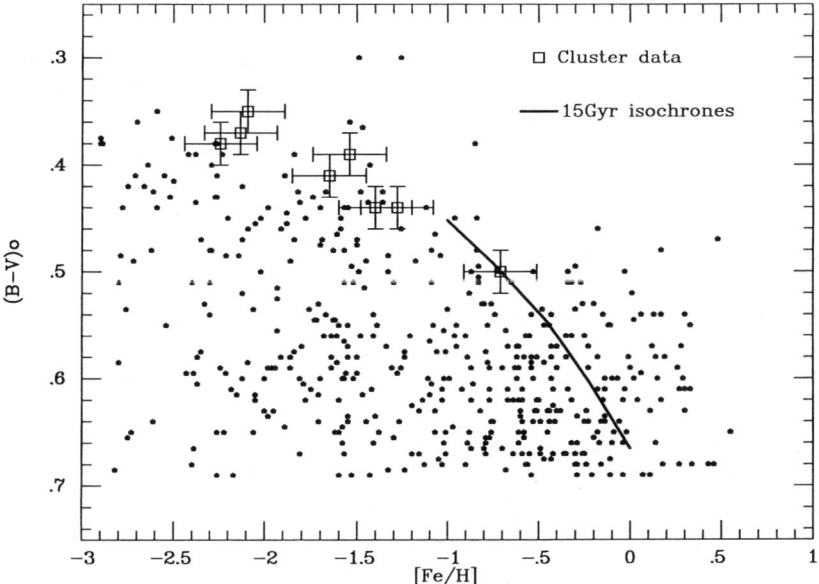

Figure 5 The $B-V$ vs. [Fe/H] relation for all stars observed by Laird et al. (1988a; points) and for those globular clusters with turnoff colors from recent CCD photometry (Stetson & Harris 1988; boxes). The photometric data are corrected for interstellar reddening. The solid line is a 15-Gyr isochrone calculated with oxygen-enhanced element ratios and scaled in $B-V$ to match the turnoff color of 47 Tuc. The blue edge of the stars with [Fe/H] $\lesssim -0.8$ is adequately defined by the isochrone and by the globular cluster data, which shows that effectively all stars more metal-poor than ~ -0.8 dex are as old as the globular clusters. At higher abundances, the trend for the data to move to the blue of the isochrone shows that at least some stars are younger than the globular clusters.

The important conclusion from Figure 5 is that essentially all stars with [Fe/H] $\lesssim -0.8$ are, insofar as is measurable, the same age as the globular cluster system. [Note, however, that Schuster & Nissen (1989) suggest that there is a real age spread of ~ 3 Gyr in the subdwarf system and even possibly an age-abundance relation in metal-poor stars. This time scale is much longer than those deduced in other ways and will radically alter our perception of Galactic evolution, if real. It deserves considerable attention.] Stars more metal rich than ~ -0.8 dex have a bluer turnoff, which implies that *at least some* of these stars are younger. The *distribution* of ages is, however, unmeasurable from a turnoff color. Some information on the age distribution for stars with [Fe/H] $\gtrsim -0.8$ is provided by studies of open clusters. These form a system with a very large scatter in the age-metallicity plane; for example, clusters exist near the Sun with solar abundance and an age of 12 Gyr (NGC 6791; Janes 1988), and with [Fe/H] ~ -0.5 but an age of only a few gigayears (e.g. Melotte 66). A similar scatter is evident in the age-metallicity relationship for F-dwarfs near the Sun (Twarog 1980, Carlberg et al. 1985, Knude et al. 1987), though unfortunately both qualitative and quantitative differences are found from author to author (even analyzing the same sample), which somewhat confuses the real situation. Thus, any attempt to deduce a representative age for a stellar population from the turnoff color of the *bluest* field stars with metallicity $\gtrsim -0.8$ dex (Norris & Green 1989) is fundamentally unreliable. This point is returned to in Section 4 below.

3.4 *Chemical Element Ratios and Galactic Collapse*

One of the most important constraints on the homogeneity of star formation in the spheroid, and hence on chemical evolution, is readily derivable from the observed small dispersion in the position of the breaks in element ratios as a function of [Fe/H]. As we show below, the inferred lack of cosmic scatter implies little spatial variation in the star formation history of the proto-Galaxy, which is obviously of considerable importance in determining the levels of independent substructures in the gas cloud that became the Galaxy. Models such as those of Searle (1977; Searle & Zinn 1978), whereby the spheroid is a result of the merging of many independent "fragments," each with its own chemical enrichment history, seek to explain the entire range of metallicity seen in halo stars as being due to the statistical effects of chemical inhomogeneities, rather than as a trend with age. The "fragments" are assumed to be disrupted by some process after some time, producing the field stars of the halo (cf. Fall & Rees 1977). Such models also predict a large spread in the element ratios observed from field star to field star if, for any one of the "fragments" (each of which is presumed to contain many globular cluster–sized objects), star formation continued for sufficiently long that Type I supernovae became a significant source of iron.

At metallicities above ~ -1 dex, the [O/Fe] ratio declines toward the solar value (see Figure 1). The slope of this relation is ~ -0.5 (thus the iron is mimicking the behavior of a "secondary" element), which implies that the amount of iron synthesized in association with oxygen is small compared with the amount synthesized independently of oxygen. This must hold even though the relevant [Fe/H] range encompasses the entire thick disk and most of the thin disk. We may utilize this constraint to determine the relative rates of Type I and of Type II supernovae during the epoch of formation of the disks, which in turn constrains the star formation rate during the formation of most of the stellar mass of the Galaxy. If we assume a constant initial-mass function (IMF), the star formation rates in the thick and thin disk formation stages must be low enough to prevent imprinting a feature through their associated massive stars. Assuming that single stars more massive than 8 $M_\odot$ explode as Type II supernovae, and that binary systems with a minimum primary mass of 5 $M_\odot$ eventually evolve into Type I supernovae, integration of the Miller & Scalo (1979) IMF predicts that the number of potential progenitors of Type I supernovae is a factor ~ 1.5 higher than the number of progenitors of Type II supernovae. The time for an individual star to become a supernova depends sensitively on the initial orbital properties and could easily be several Hubble times. The rate of Type I supernovae is therefore impossible to predict analytically. Observations of the present relative rates of Type I and Type II supernovae offer the most reliable constraint. Tammann (1982) finds that the Type I: Type II ratio is $\sim 1.5:1$ in a range of spiral types (Sab–Sd). Such galaxies have derived average star formation rates that vary little over the past $\sim 6 \times 10^9$ yr (Gallagher et al. 1984), despite differences in gas content and metallicity. Thus the present relative rates of Type I and Type II supernovae in disk galaxies may be assumed to be roughly constant over a large fraction of a Hubble time, and most potential candidates for Type I supernovae actually explode in less than a Hubble time.

Current models agree that each Type I event ejects ~ 0.6 $M_\odot$ of iron (e.g. Woosley & Weaver 1986), although there is less consensus about the yield from Type II supernovae, partly because of the uncertainties in how much of the progenitor envelope is actually ejected unchanged. The bare-core calculations of Arnett (1978) produced only ~ 0.25 $M_\odot$ of iron for stars of main-sequence mass $20 \lesssim (M/M_\odot) \lesssim 30$, assuming that 60% of the "Si" yield is ejected as iron and adopting his transformation between core and main-sequence masses. Woosley & Weaver (1986) find about a factor of two more iron than does Arnett for their standard 25-$M_\odot$ model.

Here we assume that each Type I event yields a factor of two more iron than every Type II event (0.6 $M_\odot$ and 0.3 $M_\odot$, respectively). The observation that current rates, averaged over a range of spiral galaxies with fairly constant star formation rates, are $\sim 1.5:1$ for Type I: Type II implies that roughly three fourths of the iron production averaged over the lifetime of the Galaxy is from Type I events. Thus, adopting an oxygen yield of 0.6 $M_\odot$ for each typical Type II event (Woosley & Weaver 1986) and the iron yields above implies that 1.2 $M_\odot$ of iron and 0.6 $M_\odot$ of oxygen, or twice as much iron as oxygen, are returned to the interstellar medium per unit time. The return of fixed amounts of iron and oxygen, rather than fixed increments, of course results in the prediction of a nonlinear slope to the [O/Fe]:[Fe/H] relation. As shown in Figure 1, the observations are well fit by this prediction, which suggests that the star formation rate has indeed been fairly constant over the lifetime of the disk and perhaps supports Tammann's *relative* supernovae rates. It must be noted, however, that the uncertainties in the supernova yields and in the observed supernova rates are so large that no unique model exists. The relative supernova rates of van den Bergh et al. (1987) (Type I: Type II $\sim 1:1.5$), where we have added together the Type Ia and Type Ib supernovae, would lead to approximately equal amounts of iron from each type of supernova, and equal amounts of iron and oxygen returned to the ISM per unit time, if we assume naively that both Type Ia and Type Ib have the same yield of iron. The enhanced rate of Type II supernovae leads to a predicted relation between [O/Fe] and [Fe/H] that has a less steep initial decay than that obtained using Tammann's (1982) relative rates, but that is also consistent with the observations owing to their large scatter.

The explanation, reiterated above, of the breaks in element ratio as a function of [Fe/H] as simply reflecting different element production sites depends on the existence of a simple correspondence between [Fe/H] and time. However, this mapping depends on the star formation rate and the stellar IMF. If we assume the IMF to be constant, "fragments" of higher star formation rate will have attained a higher [Fe/H] prior to the fixed time at which Type I supernovae dominate, so that the break in the [O/Fe]:[Fe/H] relation will occur at higher [Fe/H]. Similarly, "fragments" of substantially lower star formation rate would lead to a break at much lower [Fe/H]. Thus, samples of stars that originated in many different "fragments" that were subsequently disrupted would not show a well-defined break at one given metallicity, contrary to the observations. The small scatter seen then requires that all "fragments" had evolved to nearly the same gas fraction at the time when they were disrupted, which in all cases was $\lesssim 10^9$ yr after significant star formation began. Such an effect is implicit in the models of Hartwick (1976) and Gilmore & Wyse (1986), whereby the metallicity distribution of the extreme spheroid is a consequence of gas being lost from the star-forming process at a rate proportional to, and greater than, the star formation rate. The required rate of gas removal is a factor of 10 to 15 higher than the star formation rate, so it is possible for each "fragment" to evolve to completion (by losing all its gas) in a time shorter than that required by the break in the [O/Fe]:[Fe/H] relation (or $\lesssim 10^9$ yr).

"Fragment" models also require that star formation began over all of that part of the proto-Galaxy that became the extreme spheroid in the solar neighborhood at nearly the same time. This requirement arises because if some "fragments" lagged behind and began their chemical evolution substantially later than most, their oxygen-rich ejecta would have enriched the gas that was to form the younger disk populations, contrary to the observations. A further constraint on the spatial variations of the star formation process in the spheroid comes from the fact that metals may be transported at no more than the local sound speed, whereas the free-fall velocity may well be supersonic for an ISM at about 10^4 K (Fall 1987), which would lead to large local chemical inhomogeneities if the star formation rate varied rapidly from place to place. However, if the gas has been heated to the Galactic virial temperature, the sound speed will equal the free-fall velocity, by definition, so that efficient mixing is not constrained by the distance ejecta can be transported in a free-fall time. In general, it seems more plausible to identify any "fragments" with short-lived condensations that become the sites of star formation in a gaseous background, rather than as discrete structures merging to form the proto-Galaxy. The existence of structure in the element ratio relations is strong evidence that there was a continual increase with time of the heavy-element abundance of the Galaxy up to the time when the metallicity reached that at which the relations change slope, and it is difficult to reconcile with stochastic chemical evolution models.

3.5 *The Time Scale of Galaxy Formation*

In view of the complexity of the discussion above, we provide here a brief summary of the observational constraints on the time scale over which the extreme Population II stars in the Galaxy formed. The most important observational constraint is provided by the constancy of the oxygen-to-iron element abundance ratio as a function of iron abundance for metal-poor stars (Section 1.3). These data show the stars in the Galaxy with metallicities less than ~ -0.8 dex to have formed on a time scale of $\lesssim 1$ Gyr. Analysis of the asymmetric drift data (Section 3.1) shows that the most metal-poor stars formed during a period of dissipational collapse of unconstrained duration. Study of the correlations between stellar kinematics and stellar metallicities (Section 3.2) shows that the stars that formed during the dissipational collapse are the subset with metallicities $\lesssim -1.5$ dex of the same stars whose age range is set by the oxygen-to-iron element ratios. It is not yet clear if the apparent absence of evidence for dissipation during formation of the more metal-rich spheroidal stars [that is, the lack of a detectable abundance gradient (Section 3.2)] indicates that the dissipational collapse had effectively ceased, or at least had become dissipationless, during their formation. It may well be simply that a subsequent period of violent relaxation of the Galactic potential well has disturbed the fossil record of the state of collapse of the proto-Galaxy during the formation of the more metal-rich extreme Population II stars. Since a characteristic time scale for dynamical evolution is ~ 0.5 Gyr (Section 1.1), one may deduce that the period of star formation leading to the present extreme Population II stars occurred during a dissipational collapse that lasted for no more than a few dynamical times. Hence, the Galaxy formed its first generations of stars during a period of "rapid" (cf. Section 1.1) collapse. Further constraints on the homogeneity of the proto-Galaxy during this collapse are discussed in Section 3.4.

4. THE THICK DISK

The luminosity profile of the Milky Way, as given by star counts, provides evidence for a Galactic thick disk, as discussed in Section 2.5. Indeed, recent photometric and spectroscopic stellar surveys have emphasized the importance of the intermediate Population II stars, as introduced at the 1957 Vatican Conference (Oort 1958, O'Connell 1958). The modern characterization of this population assigns to this stellar component a vertical scale height of ~ 1–1.5 kpc, a vertical velocity dispersion of ~ 45 km s^{-1}, a typical stellar chemical abundance of $\sim 1/4$ of the solar metallicity, and a mean asymmetric drift of ~ 30–50 km s^{-1}. The detailed values of the descriptive parameters remain poorly determined, however, primarily because the offset in the mean values characterizing the thick disk distribution function over age, metallicity, and kinematics from those mean values characterizing the oldest thin disk stars is much less than the dispersions in these quantities. Reliable determination of the parameters of the distribution function is important, since it may allow a discrimination between the several currently viable models of the formation of the thick disk.

Possible formation mechanisms for the thick disk include the following:

1. A slow, pressure-supported collapse phase following formation of the

extreme Population II system, similar to the sequence of events in Larson's (1976) hydrodynamical models of disk galaxy formation.
2. Violent dynamical heating of the early thin disk by satellite accretion (cf. Hernquist & Quinn 1989) or by violent relaxation of the Galactic potential (Jones & Wyse 1983).
3. Accretion of the thick disk material directly—for example, by satellite accretion with a preferential population of suitable orbits (Statler 1988).
4. An extended period of enhanced kinematic diffusion of stars formed in the thin disk to high-energy orbits (Norris 1987a).
5. A rapid increase in the dissipation and star formation rates due to enhanced cooling once the metallicity is above ~ -1 dex (cf. Wyse & Gilmore 1988).

Discrimination among these several types of models is possible from appropriate age, metallicity, and kinematic data. The first type of model noted above will lead to an intermediate-age system, with an abundance gradient. The second will have a small internal age range but is unlikely to have an extant abundance gradient (*modulo* the details of the dynamical evolution). The third has a wide variety of allowed combinations of age and abundance, while the fourth will have a range of ages, a similar chemical abundance to the thin disk, but a kinematic discontinuity between the old disk and the thick disk. The last model predicts that the thick disk will be kinematically distinct from the metal-poor spheroid and will have metallicity $\gtrsim -1$ dex. In view of this possibility to determine the evolutionary history of the thick disk, an extensive debate is underway to describe reliably the kinematic, abundance, and age structure of the thick disk (cf. Sandage 1987a, Norris 1987a).

Here we summarize the data on the determination of the chemical abundance distribution and the age range of thick disk stars, and we also discuss the difficult questions of the relationships among the stellar populations near the Sun.

4.1 *Metallicity of the Thick Disk*

As discussed in Section 3, the chemical abundance of a stellar population contains much information about the population's early evolution, while detailed information about vertical metallicity gradients can test the reality of discrete stellar populations (Sandage 1981). The metallicity distribution of stars in situ above the thin disk plane has been the subject of several modern spectroscopic and photometric surveys. A population with a vertical velocity dispersion of ~ 45 km s^{-1} will dominate samples of stars presently at z-heights ~ 1 to a few kiloparsecs if it comprises of order 1% of the stars in the Galactic plane. Hence, distances of ~ 2 kpc from the plane are the most suitable environment to study the properties of the thick disk. Hartkopf & Yoss (1982) obtained DDO-photometric metallicity estimates for a spectroscopically selected sample of K giants at the Galactic poles; their data for stars with distances between 1 and 2 kpc are consistent with a Gaussian in log-metallicity, with mean $\langle[\text{Fe}/\text{H}]\rangle = -0.6$ and $\sigma_{[\text{Fe}/\text{H}]} = 0.3$ dex (Gilmore & Wyse 1985). This conclusion was confirmed by Yoss et al. (1987) for an augmented sample, though it should be noted that Norris & Green (1989) have reobserved many of the most metal-rich but distant Hartkopf & Yoss stars and conclude that both the distance and metallicity were overestimated by Hartkopf & Yoss. Norris & Green therefore suggest a revised, smaller dispersion for the thick disk metallicity distribution. The spectroscopically selected giant samples of Ratnatunga & Freeman (1985, 1989) and of Friel (1987, 1988) also contained relatively few metal-poor stars, in contradiction to the predictions of models that assume that the spheroid has metallicity similar to the metal-poor globular clusters. In particular, Friel's observations were in accord with the existence of a thick disk with scale height ~ 1 kpc and local normalization $\sim 3\%$, the stars of which have metallicity similar to the metal-rich globular cluster 47 Tuc (~ -0.7 dex). The kinematics of Ratnatunga & Freeman's stars is discussed further below (cf. also Freeman 1987).

The kinematically defined sample of Eggen (1979), when restricted to stars with vertical velocities $40 \lesssim |W| \lesssim 60$ km s^{-1}, has a metallicity distribution that is bimodal, with well-defined peaks at ~ -0.7 and -1.5 dex, in (remarkable) agreement with the globular cluster data of Zinn (1985). The metallicity distributions seen across other restricted velocity ranges are consistent with three discrete metallicity distributions for the metal-poor spheroid, the thick disk, and the thin disk, as are the proper-motion samples of Sandage & Fouts (1987) and Carney et al. (1989a). Distinct chemical abundance distributions for the thick and thin disks pose a problem for the fourth model of thick-disk formation briefly described above (scattering of extant thin-disk stars) and an embarrassment for the first model, where one expects smooth continuity.

This higher metallicity for the majority of spheroid stars than that implied by the solar neighborhood subdwarfs has important consequences for the chemical evolution of the solar neighborhood, and in particular for the "G-dwarf problem," which is the apparent lack of metal-poor, long-lived stars in the solar neighborhood compared with the predictions of the "simple closed-box" model of chemical evolution (van den Bergh 1958, 1962, Pagel & Patchett 1975). Whether or not the thick disk is a discrete entity is irrelevant here, since all that is of importance is that there exist G-dwarfs of metallicity ~ -0.6 dex that were not represented adequately in earlier surveys (cf. Yoshii 1984, Gilmore & Wyse 1986).

4.2 Is the Thick Disk Kinematically Discrete?

Here we address the kinematics of the thick disk—in particular, what one can infer about the evolutionary status of these stars from the similarities and differences of their kinematics from those of other populations in the Galaxy.

4.2.1 IS IT DISCRETE FROM THE SUBDWARF SYSTEM? The weighted mean average velocity dispersions for nonkinematically selected extreme Population II stars are determined to be $(\sigma_{rr})^{1/2} : (\sigma_{\phi\phi})^{1/2} : (\sigma_{zz})^{1/2} = 131 \pm 7 : 102 \pm 8 : 89 \pm 5$ (Carney & Latham 1986, Norris 1986, Morrison et al. 1989). The vertical velocity dispersion of the thick disk is ~ 45 km s^{-1} [cf. Ratnatunga & Freeman 1985, 1989, Gilmore & Wyse 1987 (Figure 12A), Sandage & Fouts 1987, Yoss et al. 1987, Carney et al. 1989a]. The lag behind solar rotation (the asymmetric drift) of the thick disk is apparently 30–50 km s^{-1} (Ratnatunga & Freeman 1985, 1989, Freeman 1987, Norris 1987c, Sandage & Fouts 1987, Morrison et al. 1989), while that of the subdwarf population is 180–220 km s^{-1} (Norris 1986; cf. Section 3.2). Thus the kinematics of the thick disk is dominated by rotational support, while that of the subdwarf system is dominated by "pressure" support from the anisotropic velocity dispersion tensor. One explanation for this dichotomy is simply that the rate of dissipation in the vertical direction was relatively high, compared with the star formation rate, as the protodisk collapsed.

The number of stars with abundances and kinematics such that they might plausibly be assigned either to the low-velocity tail of the extreme Population II or to the high-velocity tail of the thick disk (i.e. those stars with [Fe/H] ~ -1) is very small. Figure 4b shows the *distribution* over rotation velocity and metallicity for the sample of Laird et al. (1988a); the binned data are shown in Figure 4a. It is clear that there is a relative deficiency of stars with [Fe/H] ~ -1 and -200 km s$^{-1} \lesssim V \lesssim -100$ km s^{-1}. However, this overlap region is populated; in particular Norris et al. (1985) drew attention to metal-poor stars with thick disk kinematics, while Morrison et al. (1989), from their spectroscopically selected sample of G/K-giants in a field against Galactic rotation, suggest that stars with "disk kinematics" (large rotation velocity) and stars with "halo kinematics" exist in approximately equal numbers with metallicities ~ -1 dex. Thus Morrison et al. characterize the binned rotation velocity vs. metallicity plot of Figure 4a as two vertically offset, horizontal lines that overlap in metallicity, rather than either the step-function or the smooth correlation considered in Section 3. The relation of these stars to the flattened metal-poor component found by Hartwick in the metal-poor RR Lyrae stars and in the metal-poor globular clusters (discussed in Section 2.6) is not obvious, especially since Morrison et al. explicitly analyzed the kinematics of samples of RR Lyrae stars and of globular clusters and saw no behavior similar to that of their K-giants. The existence of a significant "metal-weak thick disk" obviously would have major ramifications for our understanding of the thick disk as a discrete entity; a few high-angular-momentum but metal-poor stars are predicted in many models of Galaxy formation, such as that of ELS, but these stars should be on highly eccentric orbits. Proper motions for the Morrison et al. sample would be highly desirable, as would the kinematics of similar samples of stars in fields that probe different components of the space motion. In this regard, Ratnatunga & Freeman (1985, 1989) interpret their K-giant south Galactic pole (SGP) data as consistent with only *one* metal-poor component, as do Yoss et al. (1987). Thus the situation remains confused, primarily as a result of the very small number of stars that might belong to such a metal-poor tail of the thick disk, and of the consequent difficulties in determining their properties and relationships to other groups of stars.

4.2.2 IS IT DISCRETE FROM THE THIN DISK? The relationship of the thick disk to the high-velocity tail of the old disk is equally problematic and has been discussed extensively by Sandage (1987a) and Norris (1987a). The main point at issue is whether there is a continuous relationship of vertical velocity dispersion with metallicity extending all the way to the ~ 45 km s^{-1} vertical velocity dispersion of the thick disk, or whether the old disk velocity dispersion becomes asymptotically constant at the value of ~ 22 km s^{-1} appropriate for spectroscopically selected samples of old dwarfs near the Sun (Fuchs & Wielen 1987, Sandage 1987b)? The deconvolution of *local* samples of proper-motion stars into components whose kinematics are similar (within a small multiplicative factor) to those of the old disk is fraught with difficulty. Preliminary attempts have been made (I. N. Reid & I. Lewis, private communication, 1988) to illustrate this uncertainty and to show that reliable results must await careful analysis of the several in situ surveys that are nearing completion.

The difficulty in deciding whether there is a continuous kinematic continuity from the thick disk to the old disk is illustrated by Figure 6a. This shows the vertical velocity distribution of those stars in the Gliese catalogue of nearby stars with photometric abundance parameter $\delta_{0.6} < 0.15$ ([Fe/H] $\gtrsim -0.9$; this abundance range was chosen so as to exclude high-velocity subdwarfs). The two models overlaying the histogram data are a two-component model, with discrete old thin disk and thick disk, and the four-component approximation to a continuous relation between the old disk and the thick disk fitted by Norris (1987a). The two models are clearly both excellent descriptions of the data, and they are equally clearly indistinguishable.

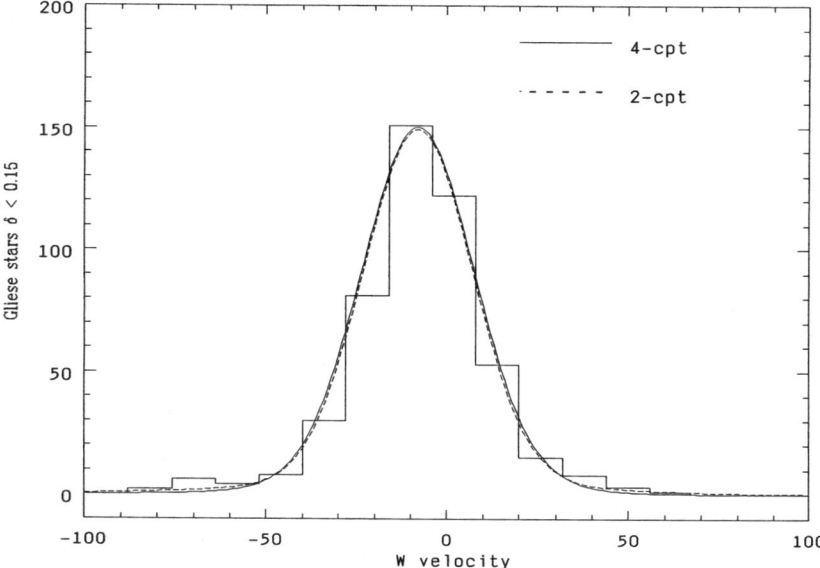

Figure 6a The distribution of vertical (W) velocities of stars in the Gliese catalogue, excluding stars with $\delta_{0.6} > 0.15$. The two lines illustrate models with a kinematically discrete thick disk (dashed lines) and with a continuous kinematic relationship between the old disk and the thick disk (solid line). Distinguishing between these models on the basis of these data is clearly not possible.

The data that motivated the four-component model are shown in Figure 6b, together with the data for the F-star sample of the Copenhagen group (Strömgren 1987). The important points to note here are the unsatisfactory disagreement between the two data sets, and the level of smoothness or otherwise of the observed trend.

It is evident from Figure 6 that available *local* data are incapable of determining if the old disk and the thick disk are kinematically discrete. Resolution of this uncertainty, with its important implications for the formation history of the Galaxy, must await completion of the several extant in situ surveys of the stellar distribution several kiloparsecs from the Galactic plane. The available data marginally favor a model in which the thick disk is a kinematically discrete component of the Galaxy, but the issue remains to be decided by observational test.

However, the theoretical predictions for how various secular scattering mechanisms act to increase the velocity dispersion of a population of stars as it ages are fairly clear. Scatterers that are confined to the plane of the Galaxy, such as giant molecular clouds (GMCs; Spitzer & Schwarzschild

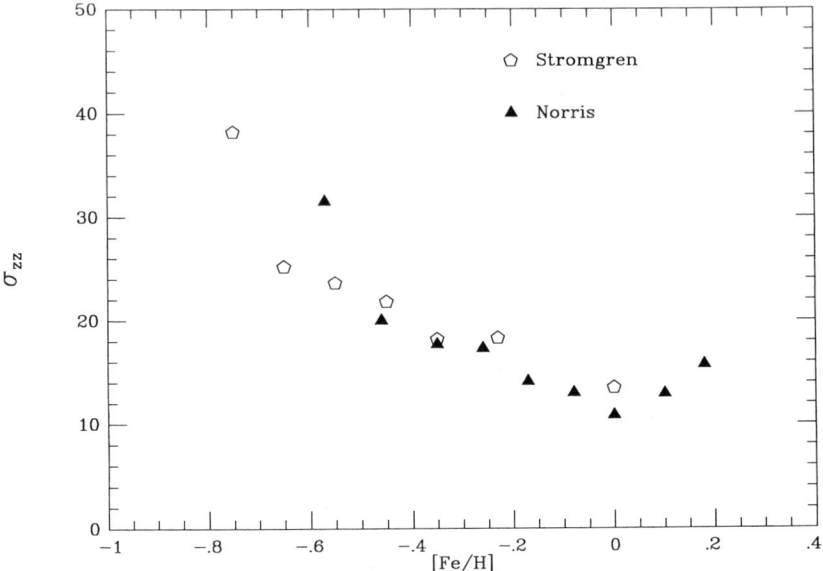

Figure 6b The relationship between [Fe/H] and vertical velocity dispersion for stars near the Sun, from data by Norris (1987c) and Strömgren (1987). These data provide the basis of the arguments that there exists a continuous kinematic relationship between the thick disk and the old disk. One should note particularly the degree of consistency between the data sets and the smoothness or otherwise of the trends at low metallicities, which is where the thick disk first contributes significantly to the data.

1953, Lacey 1984) or spiral density waves (Barbanis & Woltjer 1967, Carlberg & Sellwood 1985), cannot create a population of stars with a scale height such as that inferred for the thick disk, since this is at least an order of magnitude above that of the scatterers themselves, and the heating rate is proportional to the ratio of scale heights (Lacey 1984). A combination of GMCs and spiral density waves can plausibly account for the observed age–velocity dispersion relations for thin disk stars (Binney & Lacey 1988), and since GMCs are observed to occur preferentially in spiral arms, this may be a natural effect (though even this model cannot provide heating beyond $\sigma_z \sim 20$ km s^{-1}). Thus, although the data of Fuchs & Wielen (1987) may be described by known processes, one must appeal to more exotic, high-velocity scatterers to account for any further increase in velocity dispersion. Possibilities include a population of massive black holes (Lacey & Ostriker 1985) or "dark clusters" (Carr & Lacey 1987) in the Galactic halo, and these too have difficulties. The kinematics of thick and thin disks may be similar, but different mechanisms may well be responsible and the similarity misleading.

The rotation velocity of the thick disk is most probably close to that of the old thin disk, lagging behind solar rotation by 30–50 km s^{-1}, as noted above. The earlier determination of a lower rotational velocity, $V_{rot} \sim 100$ km s^{-1} (Wyse & Gilmore 1986), was derived from a proper-motion sample with inherent uncertainties (cf. Norris 1987c). However, as is seen in Figure 3, there do exist well-defined samples of stars with [Fe/H] ~ -0.7 and $V_{rot} \sim 100$ km s^{-1}—in particular, the metal-rich RR Lyrae stars (also of interest for age determinations; see below) and the long-period variables (periods of 150–200 days). Additionally, the spectroscopic survey of Gilmore & Wyse (1987, and in preparation) contains a significant number of stars with these kinematics. As discussed in Section 3.2.2, these stars may provide their own insight into the dynamical history of the Galaxy.

4.3 Age of the Thick Disk

An age determination for samples of thick disk stars is an extremely difficult observational problem. In part this is due to the usual difficulty in assigning a reliable age to anything in astronomy, but in this case the situation is complicated by the point noted above that there is no obvious a priori way to define a sample of purely "thick disk" stars. Any sample selected by abundance, kinematics, or chemistry will inevitably include old disk and/or extreme Population II stars in addition to the thick disk. Thus determination of the age of the *youngest* or the *oldest* star in a sample, while tractable, is not an obviously clever way to answer the question of interest. Some information may be derived from Figure 5. It is evident from this figure that the age of the oldest stars with [Fe/H] $\lesssim -0.7$ is comparable to that of the metal-rich globular cluster system, i.e. ~ 14 Gyr (Hesser et al. 1987). This abundance range is expected to be dominated by thick disk stars for $-1 \lesssim$ [Fe/H] $\lesssim -0.8$, which suggests that the most metal-poor thick disk stars are among the oldest in the Galaxy [*modulo* the age determinations of Bell (1988a,b) and Schuster & Nissen (1989) for field subdwarfs of 18 Gyr, which may be interpreted to imply that globular clusters are younger than the field]. More metal-rich thick disk stars may or may not be the same age. It is impossible to determine this from comparison with diagrams like Figure 5, as some old disk stars will contaminate the sample. This point is worth emphasizing, as it removes the rigorous basis for the conclusion of Norris & Green (1989) that the thick disk is several gigayears younger than the disk globular clusters; such a conclusion cannot be derived reliably from photometric data alone. This point is discussed in more detail by Sandage (1988); the interplay between age and metallicity in determining the photometric properties of stellar populations means that one can derive ages from turnoff colors only within the framework of an assumed age-metallicity relationship. The large intrinsic scatter seen in the age-metallicity relation for local F-dwarfs (Carlberg et al. 1985) and for open clusters (Geisler 1987) will inevitably complicate the determination of the ages of groups of stars from the color of the main-sequence turnoff. Of course, the amplitude of this scatter contains valuable information on the small-scale homogeneity of the interstellar medium in the Galaxy, the history of accretion of satellites (both gaseous and stellar), and the amplitude of large-scale abundance gradients in the Galactic disk.

The morphology of the horizontal branch also in principle contains important information on the age of the thick disk stars, *modulo* the metallicity and the well-known "second parameter" problem. Rose (1985) found a population of red horizontal branch (RHB) candidates with a thick disk configuration and kinematics, which would imply that the thick disk were old if these are truly RHB stars. The thick-disk nature of the kinematics of these stars was confirmed by the high-precision data of Stetson & Aikman (1987), who derived a vertical velocity dispersion of ~ 45 km s^{-1}. A complication is that Rose found such a large number of candidate RHB stars. In view of the intrinsic rarity of such stars, Rose's result would imply a very large local normalization of the thick disk near the Sun. However, Norris (1987b) and Norris & Green (1989) have argued that these stars are *not* bona fide RHB stars but rather core-helium-burning "clump" stars similar to those seen in open clusters. They argue that their interpretation would mean a higher metallicity for these stars than that of 47 Tuc and hence would lead to a younger age. They infer that the thick disk is at least 3-6 Gyr younger than the disk globular clusters and thus has an age of 8-11 Gyr. However, in addition to the complications with the number of these stars, Janes (1988) has derived an age of 12.5 Gyr for the metal-rich open cluster NGC 6791, which has a well-developed clump, showing explicitly that there is no universal age-metallicity relationship that can be applied.

The *youngest* dated population of RR Lyrae stars is that in the Small Magellanic Cloud cluster NGC 121, which has an age of ~ 12 Gyr [cf. Olszewski et al. (1987) for a detailed discussion of the age estimates for clump and horizontal branch stars]. The kinematics of the metal-rich RR Lyrae stars ($0 \lesssim \Delta S \lesssim 2$) is that of the thick disk (Sandage 1981, Strugnell et al. 1986), which suggests that 12 Gyr is a lower limit on the age of at least some of the thick disk. Further studies of these metal-rich RR Lyrae stars would be of considerable interest.

Similarly, while the relationship of the globular clusters to field stars is not obvious, *if* the disk globular cluster system studied so well by Zinn (1985) and collaborators [cf. Armandroff (1989) for the most recent analysis] is indeed part of the thick disk, then the antiquity of the thick disk is reliably established. The rotation velocity of the thick disk globular clusters found by Armandroff (1989) is closer to that of the field thick disk stars than the value derived earlier by Zinn (1985), which could be construed to support the identification of these two populations. However, samples of old disk open clusters *also* have similar kinematics to the field thick disk and hence to the disk globular clusters, which under the same logic would suggest an intimate connection between the open and globular clusters. An understanding of the kinematics of the extant sample of open clusters must take account of the destruction processes that operate preferentially to destroy clusters that are confined to the plane and hence lead to an artificially high scale height and to artificially high velocities for easily observed (i.e. high-latitude) open clusters. The same problem exists for globular clusters, but it will be worse for the more loosely bound open clusters.

All these arguments, however, leave open the possibility of a large age *range* in the thick disk. There is no reliable information yet available on this point; obviously, if one believed *all* of the ages inferred above, then an age spread of several gigayears results. However, this age scatter probably simply reflects the level of uncertainty associated with each of the available age estimates.

5. THE MASS DISTRIBUTION IN THE GALACTIC DISK

The distribution of mass in the Galactic disk is characterized by two numbers: its local *volume* density ρ_0 and its total *surface* density $\Sigma(\infty)$. These are fundamental parameters for many aspects of Galactic structure, such as chemical evolution (is there a significant population of white dwarf remnants from early episodes of massive star formation?), the physics of star formation (how many brown dwarfs are there?), disk galaxy stability (how important dynamically is the self-gravity of the disk?), the properties of dark matter (does the Galaxy contain *dissipational* dark matter, which may then be fundamentally different in nature from the dark matter assumed to provide flat rotation curves?), non-Newtonian gravity theories (where does a description of galaxies with Newtonian gravity and no dark matter fail?), and so on.

Although $\Sigma(\infty)$ and ρ_0 are different measures of the distribution of mass in the Galactic disk near the Sun, they are related. Of the two, the most widely used and commonly determined measure is the local *volume* mass density—i.e. the amount of mass per unit volume near the Sun, which for practical purposes is the same as the volume mass density at the Galactic plane. This quantity has units of $M_\odot \text{ pc}^{-3}$, and its local value is often called the "Oort limit" in honor of the early attempt at its measurement by Oort (1932). The contribution of identified material to the Oort limit may be determined by summing all local observed matter—an observationally difficult task, which leads to considerable uncertainties. These uncertainties arise in part from difficulties in detecting very low-luminosity stars, even very near the Sun (cf. Gilmore et al. 1985, Hawkins & Bessell 1988), in part from uncertainties both in the binary fraction among low-mass stars and in the stellar mass-luminosity relation, but mostly from uncertainties in determining the volume density of the ISM. This latter uncertainty is exacerbated by the fact that the physically important quantity (for dynamical purposes) is the mean volume density of the patchily distributed ISM at the solar Galactocentric distance. The best available determination of the local mass density in identified material is $\sim 0.1 \, M_\odot$ pc^{-3}.

The second measure of the distribution of mass in the solar vicinity is the integral surface mass density. This quantity has units of $M_\odot$ pc^{-2} and is the total amount of disk mass in a column perpendicular to the Galactic plane. It is this quantity that is required for the deconvolution of rotation curves into disk, bulge, and dark halo contributions to the large-scale distribution of mass in galaxies. The most recent determination of this surface mass density, prior to that discussed below, is by Bahcall (1984a), who derives values in the range 55–80 $M_\odot$ pc^{-2}. As an indication of the dynamical significance of this mass density, the contribution of a disk potential generated by this local mass density to the local circular velocity, if we assume an exponential disk with the Sun 2.5 radial scale lengths from the Galactic center, is

$$V_{\text{circ, disk}} \sim 150 \left(\frac{\Sigma_{\text{local}}}{60 \, M_\odot \, \text{pc}^{-2}} \right)^{1/2} \text{km s}^{-1}. \qquad 17.$$

The local circular velocity is ~ 220 km s^{-1}, and the contributions to this circular velocity from the various components generating the Galactic potential add in quadrature. Thus, the Galactic disk is far from dominating the local potential well.

Both these dynamical quantities are derived from a measurement of the vertical Galactic force field $K_z(z)$. If one knew both the local *volume* mass density and the integral *surface* mass density of the Galactic disk, one could immediately constrain the scale height of any contribution to the local volume mass density that was not identified. For example, one might suspect that some fraction of the local volume mass density was unidentified (i.e. a local "missing-mass" problem) but also determine a surface density that is effectively fully explained by observed mass. Then the unidentified contribution to the local volume density would have to have a small scale height in order that its contribution to the surface density be small. In view of the very small scale length on which it must be distributed, it would then be plausible to deduce that any local "missing" mass unidentified in the volume mass density near the Sun was not the "missing" mass that dominates the extended outer parts of galaxies.

Determination of the volume mass density and the integral surface mass density near the Sun requires similar observational data—namely, distances and velocities for a suitable sample of tracer stars—but rather different analyses.

5.1 *Measurement of the Galactic Potential*

All determinations of the mass distribution in the Galactic disk require a solution of the collisionless Boltzmann equation. In view of its intractability, in practice one utilizes its vertical moment, the vertical Jeans' equation (Equation 10), which we repeat here for convenience:

$$K_z = \frac{1}{v}\frac{\partial}{\partial z}(v\sigma_{zz}) + \frac{1}{rv}\frac{\partial}{\partial r}(rv\sigma_{rz}), \qquad 18.$$

where $v(r,z)$ is the space density of the stars, and $\sigma_{ij}(r,z)$ is their velocity dispersion tensor.

The first term on the right-hand side of this equation is dominant and contains a logarithmic derivative of the stellar space density $v(r,z)$ and a derivative of the vertical velocity dispersion, σ_{zz}. Since the stellar population in the solar neighborhood is, within a multiplicative factor of a few, tolerably well described by an isothermal stellar population, the term containing the derivative of the space density dominates the determination of $K_z(z)$ near the Sun. This point is not often appreciated adequately, but it means that one should determine stellar density profiles with even greater care than that required for the velocity dispersions.

The second term in the Jeans' equation describes the tilt of the stellar velocity ellipsoid away from the local cylindrical-polar coordinate system in which velocity dispersions are measured. One therefore needs the r-gradients of σ_{rz} and of v. There are no general analytical solutions for this term, but one may derive a realistic upper limit on its importance by considering velocity ellipsoids that are oriented toward the Galactic center. In this case, if the disk of the Galaxy is self-gravitating, radially exponential, and has a constant vertical scale height, as is seen in external disk galaxies, vertical balance implies (for disk surface density μ) that $\sigma_{zz} \propto \mu$, and hence that

$$\sigma_{zz} \propto \mu \propto v \propto e^{-r/h_r}. \qquad 19.$$

Thus we obtain

$$\frac{1}{rv}\frac{\partial}{\partial r}(rv\sigma_{rz}) = 2(\alpha^2-1)\sigma_{zz}\left[\frac{\alpha^2 z^3}{(\alpha^2 z^2+r^2)^2} - \frac{rz}{h_r(\alpha^2 z^2+r^2)}\right] \quad 20.$$

as the tilting term for a radially exponential population of constant vertical scale height with a velocity ellipsoid of constant axis ratio α that points at the Galactic center (Kuijken & Gilmore 1989a). Since this term is proportional to σ_{zz}, inserting it into the Jeans' equation (Equation 18) gives a linear equation in σ_{zz}, from which one can deduce K_z.

Given a measurement of the gravitational field $\mathbf{K}(r, z)$ in an axisymmetric galaxy, the total density ρ of gravitating matter follows from Poisson's equation:

$$\nabla \cdot \mathbf{K} = -4\pi G \rho. \quad 21.$$

In the case of a disk galaxy, we can express the r-gradient in $\nabla \cdot \mathbf{K}$ in terms of the observed circular velocity at the Sun, v_c, or in terms of the Oort constants of Galactic rotation, A and B (see, for example, Mihalas & Binney 1981):

$$\begin{aligned}\rho &= -\frac{1}{4\pi G}\left[\frac{\partial K_z}{\partial z} + \frac{1}{r}\frac{\partial}{\partial r}(rK_r)\right] \\ &= -\frac{1}{4\pi G}\left[\frac{\partial K_z}{\partial z} + \frac{1}{r}\frac{\partial(v_c^2)}{\partial r}\right] \\ &= -\frac{1}{4\pi G}\left[\frac{\partial K_z}{\partial z} + 2(A^2-B^2)\right]. \end{aligned} \quad 22.$$

For a disk galaxy with an approximately flat rotation curve the second term is small within a few kiloparsecs of the disk plane (for an exactly flat rotation curve, we have $A^2 - B^2 \equiv 0$ at $z = 0$; Kuijken & Gilmore 1989a), so we can integrate in z to obtain the total column density $\Sigma(z)$ between heights $-z$ and z relative to the disk plane $z = 0$:

$$\Sigma(z) = \int_{-|z|}^{|z|} \rho(z)\,dz = \frac{|K_z|}{2\pi G} - \frac{(A^2-B^2)}{\pi G}|z|. \quad 23.$$

It is evident from the equations above that determinations of the local volume mass density ρ_0 depend on the square of any distance scale errors in the tracer population, since they are derived from the second derivative of the stellar space density distribution, while determinations of the surface mass density are linearly proportional to the distance scale, being based on the first derivative.

Recently, Bahcall (1984a,b,c) has improved the theoretical methods with which to determine the local volume density of matter, ρ_0. He has reanalyzed the available F-dwarf and K-giant high-Galactic-latitude data with new, self-consistent Galaxy models (in the sense that the matter that generates the gravitational field itself responds to it via the collisionless Boltzmann equation and including a dark halo needed to support a flat rotation curve) to replace the simpler models that had been used up to that time. He found that (a) the gravitational field due to the 0.10 $M_\odot$ pc^{-3} of stars and gas that are identified in the solar neighborhood is inconsistent with the gravitational field derived from the data; and (b) depending on its scale height, a further 0.06–0.14 $M_\odot$ pc^{-3} of unidentified matter is required. This matter is not part of a spherical halo: The local volume density required in the dark halo to explain the rotation curve is only ~ 0.01 $M_\odot$ pc^{-3}, with this value being insensitive to the local disk mass. Hence, this result implies significant amounts of disklike, dissipational dark matter in the solar neighborhood.

The analytical techniques developed by Bahcall (1984a,b,c) represent a considerable improvement over those applied previously and for the first time allow a derivation of ρ_0 that is limited by the quality of the available observational data rather than by the approximate nature of the analysis. Bahcall's analysis is primarily appropriate for determination of ρ_0 and is less suitable for the determination of $\Sigma(\infty)$. Kuijken & Gilmore (1989a,b) therefore developed a new technique for the analysis of stellar kinematic data, which is more appropriate for determination of the integral surface mass density of the Galactic disk near the Sun. Their analysis involves maximum-likelihood comparison of observed and predicted *distribution functions* of stellar velocities as a function of distance from the plane. It thus removes the need to describe an array of distance-velocity data by moments, such as the rms velocity dispersion. It also provides the freedom to include important physical effects (the orientation of the stellar velocity ellipsoid far from the Galactic plane) and constraints (consistency with the Galactic rotation curve) in the modeling.

The study by Kuijken & Gilmore (1989a,b) utilized a new set of data for K-dwarf stars extending to 2 kpc from the disk plane. From these data, they measure a total disk surface mass density of $\Sigma(\infty) = 46 \pm 9$ $M_\odot$ pc^{-2}. The corresponding identified surface mass density, deduced by integrating local stellar data through their derived $K_z(z)$ law and adding the directly observed mass in the interstellar medium, is 48 ± 8 $M_\odot$ pc^{-2}. [The errors on this latter value arise from uncertainties in the mean value of the molecular gas density appropriate at the solar Galactocentric distance, and in the density of stars near the plane, where area-limited star count surveys are insensitive. The errors on the former value have been estimated as ± 12 $M_\odot$ pc^{-2} from analyses of simulated data by Gould (1989) and

Statler (1989).] Thus the 1σ *upper bound* on the surface density of any unidentified matter in the solar neighborhood is about $10\ M_\odot\ \mathrm{pc}^{-2}$, while the most likely value is no unidentified mass at all.

5.2 *Determination of the Local Volume Mass Density*

Determination of the local *volume* mass density near the Sun—the Oort limit—typically shows that perhaps 50% of the mass measured dynamically remains unidentified. Measurement of the *surface* mass density of the Galactic disk near the Sun shows no significant difference between the disk mass measured dynamically and the identified disk mass. Thus, either the unidentified mass in the Oort limit has a very small scale height, so that the *total* amount of dark mass remains insignificantly small—in effect, it must be distributed like the cold interstellar medium—or systematic errors remain in the determination of the Oort limit [cf. Kuijken & Gilmore (1989c) for details].

The specific limit derived to allow consistency between detection of significant unidentified mass in the local volume and no detection of significant unidentified mass in the local column, is that any local dark matter with volume density ρ_0 must be distributed with an effective scale height of $2H$ [such that $\Sigma = 2\rho_0 H$], where

$$\rho_\mathrm{unident} < \frac{10\ M_\odot\ \mathrm{pc}^{-2}}{2H}. \qquad 24.$$

Large amounts of missing matter in the volume density require a small scale height to be consistent with the measurement of $\Sigma(\infty)$. For example, if one really believed that $0.085\ M_\odot\ \mathrm{pc}^{-3}$ were unidentified in the local volume mass density, then this unidentified mass must be distributed with a scale height of $\lesssim 60$ pc.

In view of this rather severe scale height limitation, it is of interest to reexamine the uncertainties in the determination of the Oort limit. The sensitivity of determinations of the local volume mass density ρ_0 to uncertain data lies in the modeling of the stellar velocity distribution near the Galactic plane and in the determination of the stellar density distribution with distance from this plane. Both F-dwarf and K-giant tracer samples have been analyzed to determine ρ_0, with both producing a result of $\rho_0 \sim 0.20\ M_\odot\ \mathrm{pc}^{-3}$, where the identified mass provides $\rho_{0,\mathrm{obs}} = 0.10\ M_\odot\ \mathrm{pc}^{-3}$ (Bahcall 1984c).

The effect of *random* errors on determinations of the Oort limit has been discussed in detail by Gilden & Bahcall (1985), who conclude that these errors produce an unbiased uncertainty of $\sim 12\%$, and by Bienaymé et al. (1987) and Crézé et al. (1989), who conclude that random errors produce an uncertainty of ~50% and also produce a bias toward an erroneous detection of unidentified mass. The difference in these results is due to different techniques for handling observational errors in the simulations, which suggests that the appropriate uncertainty to apply to determinations of the Oort limit is substantially larger than that due to Poisson statistical noise. Potentially large *systematic* problems with the data remain and have been discussed by Kuijken & Gilmore (1989c).

The F-star sample analyzed is the sum of two subsamples (F5 and F8; Hill et al. 1979), with no evidence for a difference between their velocity distributions (Adamson et al. 1988). For steady-state stellar populations, two tracer populations with the same kinematics in the same gravitational potential must follow the same spatial density distribution. For the F5 and F8 samples this is not the case (Figure 7a). One or both of the data and the assumptions underlying the modeling of the F-star kinematics are thus clearly in error. The amplitude of the resulting uncertainty can be found by deducing ρ_0 from each of the three F-star samples (F5, F5+F8, and F8) by using the algorithm derived by Bahcall (1984a). The resulting values of ρ_0 are 0.29 $M_\odot$ pc^{-3}, 0.185 $M_\odot$ pc^{-3} [reproducing the result derived by

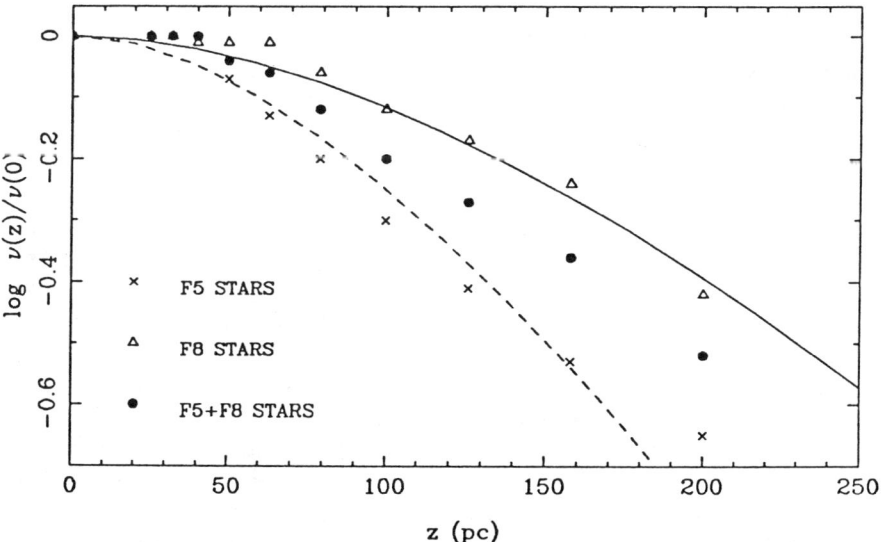

Figure 7a The Hill et al. (1979) F-star samples. The difference between the density profiles of the F5 and the F8 samples is evident. The curves show separate model fits calculated using the algorithm devised by Bahcall (1984a) to the F5 and the F8 subsets of the data defined by Hill et al. Only the averaged sample (solid dots) was analyzed by Bahcall (1984b). The models shown have local volume mass densities $\rho_0 = 0.11\ M_\odot$ pc^{-3} [i.e. with no missing mass (solid line)] and $\rho_0 = 0.29\ M_\odot$ pc^{-3} (dashed line).

Bahcall (1984b) exactly], and 0.11 $M_\odot$ pc^{-3}, respectively. Thus one may deduce that there is twice as much mass missing as observed in the local volume density, just as much mass missing as observed, or no missing mass at all, depending on which sample of stars one chooses to analyze. Clearly, the available F-star data are not capable of providing any evidence either for or against the concept of missing mass near the Sun.

The sample of K giants, which has been analyzed previously, has been shown to have a velocity distribution that is consistent with a single isothermal, with a velocity dispersion of ~ 20 km s^{-1} (Bahcall 1984c). Thus, unlike the F stars, in this model the K-giants consist entirely of old disk stars, with neither young disk nor thick disk star representatives. Since stars of a wide range of masses become K-giants, including the present F-dwarfs, this model is inherently implausible. Consistency with an isothermal is presumably an artifact of small number statistics near the Galactic plane, where lower velocity dispersion samples would be found. The K-giant density law is also uncertain, as the relevant color-magnitude relation is a strong function of age and metallicity, both of which appear to be correlated with distance from the Galactic plane. Higher precision data than those published to date are necessary to derive a reliable density profile.

A further complication follows from a feature of previous analyses, which assign high weight to the density profile near the plane (where the number of stars counted is smallest). Reanalysis of published data, including weighting of the density data by its Poisson noise and using a more detailed fit to the local velocity data from Hill (1960), leads to a value of $\rho_0 = 0.10$ $M_\odot$ pc^{-3}, that is, *no* missing mass (Figure 7b; cf. Kuijken & Gilmore 1989c). The previously derived value from the same data using the same analysis technique was $\rho_0 = 0.21$ $M_\odot$ pc^{-3}, that is, 50% missing mass (Bahcall 1984c).

We conclude that available determinations of the volume mass density near the Sun—the Oort limit—remain limited by systematic and random difficulties with the available data. One may deduce a local unexplained mass density that is up to a factor of two larger than the mass density that is identified with stars and the interstellar medium near the Sun from some samples of (young) F-stars. Other samples of (older) F-stars and of K-giants, when analyzed using velocity distributions consistent with the structure of the local Galactic disk, provide no evidence for any unexplained mass near the Sun. Determinations of the integral *surface* mass density of the Galactic disk near the Sun also show evidence for no missing mass in the Galactic disk. In brief, available data either are internally inconsistent or provide no robust evidence for the existence of any missing mass associated with the Galactic disk.

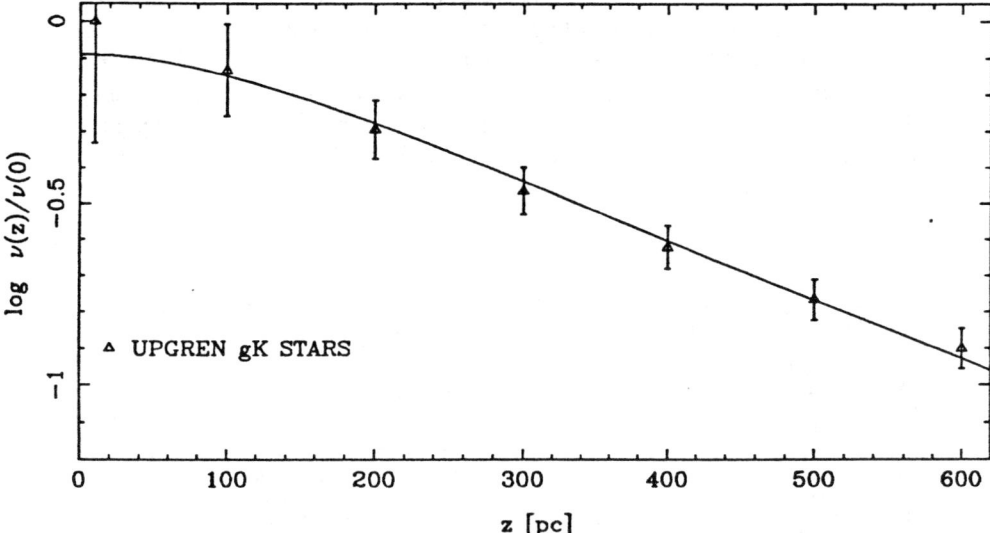

Figure 7b Weighted fit to the Upgren (1962) K-giant density distribution, using the velocity distribution measured by Hill (1960). The model shown contains no dark matter in the Galactic disk and has $\rho_0 = 0.10\ M_\odot\ pc^{-3}$.

This result has many important implications, some of which are discussed by Kuijken & Gilmore (1989b). It confirms that at least in our Galaxy, a "maximal disk" fit to explain extended flat rotation curves is not viable. It disproves available models of chemical (and luminosity) evolution that require preferential formation of high-mass stars (e.g. by a bimodal stellar initial mass function) in the early evolution of the Galactic disk. Extrapolations of the stellar initial mass function that result in significant mass in substellar mass "brown dwarfs" are ruled out. It implies that the disk is comfortably stable against axisymmetric local perturbations (the Toomre Q parameter for the stellar disk has a local value of ~ 2.1, where $Q > 1$ implies stability, though the destabilizing effect of the relatively large mass of cold gas should probably not be neglected) and just stable against global bar-mode instabilities. Severe constraints can be set on most available alternative theories to Newtonian gravity, with the local gravitational potential gradient deduced from recent K-dwarf data being inconsistent with the most recently suggested modifications to the theory of gravity.

Acknowledgments

GG and RFGW are grateful to the NATO Scientific Affairs Division for a travel grant to aid their collaboration. RFGW acknowledges partial support from the National Science Foundation (grant AST-88-07799) and thanks the UC Berkeley Astronomy Department and CITA for hospitality during the writing of some of this paper. We thank the many colleagues who facilitated the writing of this review by providing results and papers in advance of publication.

Literature Cited

Adamson, A. J., Hill, G., Fisher, W., Hilditch, R. W., Sinclair, C. D. 1988. *MNRAS* 230: 273
Armandroff, T. 1989. *Astron. J.* 97: 375
Arnett, W. D. 1978. *Ap. J.* 219: 1008
Bahcall, J. N. 1984a. *Ap. J.* 276: 156
Bahcall, J. N. 1984b. *Ap. J.* 276: 169
Bahcall, J. N. 1984c. *Ap. J.* 287: 926
Bahcall, J. N. 1986. *Annu. Rev. Astron. Astrophys.* 24: 577
Bahcall, J. N., Ratnatunga, K. U., Buser, R., Fenkart, R. P., Spaenhauer, A. 1985. *Ap. J.* 299: 616
Bahcall, J. N., Soneira, R. M. 1980. *Ap. J. Suppl.* 44: 73 (BS)
Bahcall, J. N., Soneira, R. M. 1981. *Ap. J. Suppl.* 47: 357
Bahcall, J. N., Soneira, R. M. 1984. *Ap. J. Suppl.* 55: 67 (BS)
Barbanis, B., Woltjer, L. 1967. *Ap. J.* 150: 461
Barnes, J., Efstathiou, G. 1987. *Ap. J.* 319: 575
Bell, R. A. 1988a. *Astron. J.* 95: 1484
Bell, R. A. 1988b. In *The Calibration of Stellar Ages*, ed. A. G. D. Philip, p. 163. Schenectady, NY: L. Davis Press
Bienaymé, O., Robin, A., Crézé, M. 1987. *Astron. Astrophys.* 180: 94
Binney, J. 1977. *Ap. J.* 215: 483
Binney, J., Lacey, C. G. 1988. *MNRAS* 230: 597
Binney, J., May, A. 1986. *MNRAS* 218: 743
Blanco, V. M., Blanco, B. M. 1986. *Astrophys. Space Sci.* 118: 365
Blaauw, A., Schmidt, M., eds. 1965. *Galactic Structure* (*Stars and Stellar Systems*, Vol. 5). Chicago: Univ. Chicago Press
Blumenthal, G. R., Faber, S. M., Primack, J. R., Rees, M. J. 1984. *Nature* 311: 517
Bok, B. J., Basinski, J. 1962. *Mem. Mt. Stromlo Obs.* 4: 1
Bok, B. J., MacRae, D. A. 1942. *Ann. NY Acad. Sci.* 42: 219
Bothun, G. D., Impey, C. D., Malin, D. F., Mould, J. R. 1987. *Astron. J.* 94: 23

Brooks, K. 1981. PhD thesis. Univ. Calif., Berkeley
Buser, R. 1988. In *Impacts des Surveys du Visible Sur Notre Connaisance de la Galaxie*. C. R. Journ. Strasbourg, p. 115
Buser, R., Kaeser, U. 1985. *Astron. Astrophys.* 145: 1
Carlberg, R. G. 1985. In *The Milky Way Galaxy, IAU Symp. No. 106*, ed. H. van Woerden, W. B. Burton, R. J. Allen, p. 615. Dordrecht: Reidel
Carlberg, R. G., Dawson, P., Hsu, T., Vanden Berg, D. A. 1985. *Ap. J.* 294: 674
Carlberg, R. G., Sellwood, J. A. 1985. *Ap. J.* 292: 79
Carney, B., Aguilar, L., Latham, D. W., Laird, J. B. 1989b. Submitted for publication
Carney, B., Latham, D. W. 1986. *Astron. J.* 92: 60
Carney, B., Latham, D. W., Laird, J. B. 1989a. *Astron. J.* 97: 423
Carr, B. J., Lacey, C. G. 1987. *Ap. J.* 316: 23
Chiu, L.-T. G. 1980. *Ap. J. Suppl.* 44: 31
Conlon, E. S., Brown, P. J. F., Dufton, P. L., Keenan, F. P. 1988. *Astron. Astrophys.* 200: 168
Crézé, M., Robin, A., Bienaymé, O. 1989. *Astron. Astrophys.* 211: 1
Dejonghe, H., de Zeeuw, P. T. 1988. *Ap. J.* 329: 720
del Rio, G., Fenkart, R. 1987. *Astron. Astrophys. Suppl.* 68: 397
Disney, M. 1976. *Nature* 263: 573
Eggen, O. J. 1979. *Ap. J.* 229: 158
Eggen, O. J. 1987. In *The Galaxy*, ed. G. Gilmore, B. Carswell, p. 211. Dordrecht: Reidel
Eggen, O. J., Lynden-Bell, D., Sandage, A. 1962. *Ap. J.* 136: 748 (ELS)
Elvius, T. 1965. See Blaauw & Schmidt 1965, Chap. 3
Evans, D. W. 1987. *MNRAS* 227: 13P
Fall, S. M. 1987. In *Towards Understanding*

Galaxies at High Redshift, ed. R. Kron, A. Renzini, p. 15. Dordrecht: Reidel
Fall, S. M., Efstathiou, G. 1980. MNRAS 193: 189
Fall, S. M., Rees, M. J. 1977. MNRAS 181: 37P
Fall, S. M., Rees, M. J. 1985. Ap. J. 298: 18
Feast, M. 1987. In The Galaxy, ed. G. Gilmore, B. Carswell, p. 1. Dordrecht: Reidel
Fenkart, R. P. 1989. Astron. Astrophys. Suppl. In press
Freeman, K. C. 1987. Annu. Rev. Astron. Astrophys. 25: 603
Frenk, C. S., White, S. D. M. 1980. MNRAS 193: 295
Frenk, C. S., White, S. D. M., Davis, M., Efstathiou, G. 1988. Ap. J. 327: 507
Frenk, C. S., White, S. D. M., Efstathiou, G., Davis, M. 1985. Nature 317: 595
Friel, E. D. 1987. Astron. J. 93: 1388
Friel, E. D. 1988. Astron. J. 95: 1727
Friel, E. D., Cudworth, K. M. 1986. Astron. J. 91: 293
Frogel, J. A. 1988. Annu. Rev. Astron. Astrophys. 26: 51
Fuchs, B., Wielen, R. 1987. In The Galaxy, ed. G. Gilmore, B. Carswell, p. 375. Dordrecht: Reidel
Gallagher, J. S., Hunter, D. A., Tutukov, A. V. 1984. Ap. J. 284: 544
Geisler, G. 1987. Astron. J. 94: 84
Gilden, D. L., Bahcall, J. N. 1985. Ap. J. 296: 240
Gilmore, G. 1981. MNRAS 195: 183
Gilmore, G. 1983. In Nearby Stars and the Stellar Luminosity Function, ed. A. G. D. Philip, A. R. Upgren, pp. 197, 221. Schenectady, NY: L. Davis Press
Gilmore, G. 1984a. In Astronomy With Schmidt-Type Telescopes, ed. M. Capaccioli, p. 77. Dordrecht: Reidel
Gilmore, G. 1984b. MNRAS 207: 223
Gilmore, G., Hewett, P. C. 1983. Nature 306: 669
Gilmore, G., Hewett, P. C. 1989. In preparation
Gilmore, G., Reid, I. N. 1983. MNRAS 202: 1025
Gilmore, G., Reid, I. N., Hewett, P. C. 1985. MNRAS 213: 257
Gilmore, G., Wyse, R. F. G. 1985. Astron. J. 90: 2015
Gilmore, G., Wyse, R. F. G. 1986. Nature 322: 806
Gilmore, G., Wyse, R. F. G. 1987. In The Galaxy, ed. G. Gilmore, B. Carswell, p. 247. Dordrecht: Reidel
Gilroy, K. K., Sneden, C., Pilachowski, C., Cowan, J. J. 1988. Ap. J. 327: 298
Gould, A. 1989. Ap. J. In press
Gunn, J. E. 1982. In Astrophysical Cosmology, ed. H. A. Brück, G. V. Coyne, M. S. Longair, p. 233. Vatican City: Pontif. Acad. Sci.
Gunn, J. E., Gott, J. R. 1972. Ap. J. 176: 1
Habing, H. J. 1987. In The Galaxy, ed. G. Gilmore, B. Carswell, p. 173. Dordrecht: Reidel
Harmon, R. T., Gilmore, G. 1988. MNRAS 235: 1025
Harris, W. 1981. Astron. J. 86: 719
Hartkopf, W. I., Yoss, K. M. 1982. Astron. J. 87: 1679
Hartwick, F. D. A. 1976. Ap. J. 209: 418
Hartwick, F. D. A. 1987. In The Galaxy, ed. G. Gilmore, B. Carswell, p. 281. Dordrecht: Reidel
Hawkins, M. R. S., Bessell, M. 1988. MNRAS 234: 177
Hernquist, L., Quinn, P. J. 1989. Submitted for publication
Hesser, J. E., Harris, W. E., Vanden Berg, D. A., Allwright, J. W. B., Shott, P., Stetson, P. B. 1987. Publ. Astron. Soc. Pac. 99: 739
Hill, E. R. 1960. Bull. Astron. Inst. Neth. 15: 1
Hill, G., Hilditch, R. W., Barnes, J. V. 1979. MNRAS 186: 813
Iben, I. 1985. In Nucleosynthesis, ed. W. D. Arnett, J. W. Truran, p. 272. Chicago: Univ. Chicago Press
Iben, I. 1986. In Cosmogonical Processes, ed. W. D. Arnett, C. J. Hansen, J. W. Truran, S. Tsuruta, p. 155. Utrecht: VNU Sci. Press
Isobe, S. 1974. Astron. Astrophys. 36: 333
Janes, K. 1988. In The Calibration of Stellar Ages, ed. A. G. D. Philip, p. 59. Schenectady, NY: L. Davis Press
Jones, B. J. T., Wyse, R. F. G. 1983. Astron. Astrophys. 120: 165
Keenan, F. P., Lennon, D. J., Brown, P. J. F., Dufton, P. L. 1986. Ap. J. 307: 694
King, I. R. 1986. In Stellar Populations, ed. C. Norman, A. Renzini, M. Tosi, p. 238. Cambridge: Univ. Press
Kinman, T. D., Wirtanen, C. A., Janes, K. A. 1966. Ap. J. Suppl. 13: 379
Klemola, A. R., Jones, B. F., Hanson, R. B. 1987. Astron. J. 94: 501
Knude, J., Schnedler Nielsen, H., Winther, M. 1987. Astron. Astrophys. 179: 115
Koo, D. C., Kron, R. G. 1982. Astron. Astrophys. 105: 107
Koo, D. C., Kron, R. G., Cudworth, K. 1986. Publ. Astron. Soc. Pac. 98: 285
Kraft, R. 1988. In New Ideas in Astronomy, ed. F. Bertola, J. W. Sulentic, B. F. Madore, p. 23. Cambridge: Univ. Press
Kron, R. G. 1980. Ap. J. Suppl. 43: 305
Kuijken, K., Gilmore, G. 1989a. MNRAS. In press (Pap. I)
Kuijken, K., Gilmore, G. 1989b. MNRAS. In press (Pap. II)

Kuijken, K., Gilmore, G. 1989c. *MNRAS*. In press (Pap. III)
Lacey, C. G. 1984. *MNRAS* 208: 687
Lacey, C. G., Ostriker, J. P. 1985. *Ap. J.* 299: 633
Laird, J. B., Carney, B. W., Latham, D. W. 1988a. *Astron. J.* 95: 1843
Laird, J. B., Rupen, M. P., Carney, B. W., Latham, D. W. 1988b. *Astron. J.* 96: 1908
Lance, C. M. 1988. *Ap. J.* 334: 927
Larson, R. B. 1976. *MNRAS* 176: 31
Larson, R. B., Tinsley, B. M., Caldwell, C. N. 1980. *Ap. J.* 237: 692
Levison, H. F., Richstone, D. O. 1986. *Ap. J.* 308: 627
Lewis, J., Freeman, K. C. 1989. *Astron. J.* 97: 139
Lynden-Bell, D. 1967a. *MNRAS* 136: 101
Lynden-Bell, D. 1967b. In *Radio Astronomy and the Galactic System, IAU Symp. No. 31*, ed. H. van Woerden, p. 257. London: Academic
Matteucci, F., Greggio, L. 1986. *Astron. Astrophys.* 154: 279
May, A., van Albada, T. J. 1984. *MNRAS* 209: 15
McGlynn, T. 1984. *Ap. J.* 281: 13
McNeil, R. C. 1986. *Astron. J.* 92: 335
Mihalas, D., Binney, J. 1981. *Galactic Astronomy*. San Francisco: Freeman
Miller, G. E., Scalo, J. M. 1979. *Ap. J. Suppl.* 41: 513
Morrison, H. L., Flynn, C., Freeman, K. C. 1989. Submitted for publication
Norris, J. 1986. *Ap. J. Suppl.* 61: 667
Norris, J. 1987a. In *The Galaxy*, ed. G. Gilmore, B. Carswell, p. 297. Dordrecht: Reidel
Norris, J. 1987b. *Astron. J.* 93: 616
Norris, J. 1987c. *Ap. J. Lett.* 314: L39
Norris, J., Bessell, M. S., Pickles, A. J. 1985. *Ap. J. Suppl.* 58: 463
Norris, J., Green, E. M. 1989. *Ap. J.* 337: 272
Norris, J., Ryan, S. G. 1989a. *Ap. J. Lett.* 336: L17
Norris, J., Ryan, S. G. 1989b. *Ap. J.* 340: 739
O'Connell, D. J. K., ed. 1958. *Stellar Populations*. Amsterdam: North-Holland
Olszewski, E. W., Schommer, R. A., Aaronson, M. 1987. *Astron. J.* 93: 565
Oort, J. H. 1932. *Bull. Astron. Inst. Neth.* 6: 249
Oort, J. H. 1958. See O'Connell 1958, p. 415
Oort, J. H. 1965. See Blaauw & Schmidt 1965, Chap. 21
Oort, J. H., Plaut, L. 1975. *Astron. Astrophys.* 41: 71
Osvalds, V., Risley, A. M. 1961. *Publ. Leander McCormick Obs.*, Vol. 11, Part 21
Pagel, B. E. J., Patchett, B. E. 1975. *MNRAS* 172: 13
Plaut, L. 1965. See Blaauw & Schmidt 1965, Chap. 13
Pritchet, C. 1983. *Astron. J.* 88: 1476
Ratnatunga, K. U., Bahcall, J. N., Casertano, S. 1989. *Ap. J.* 339: 106
Ratnatunga, K. U., Freeman, K. C. 1985. *Ap. J.* 291: 260
Ratnatunga, K. U., Freeman, K. C. 1989. *Ap. J.* 339: 126
Rees, M. J., Ostriker, J. P. 1977. *MNRAS* 179: 541
Reid, I. N., Gilmore, G. 1982. *MNRAS* 201: 73
Robin, A., Crézé, M. 1986a. *Astron. Astrophys. Suppl.* 64: 53
Robin, A., Crézé, M. 1986b. *Astron. Astrophys.* 157: 71
Rodgers, A., Harding, P. 1989. *Astron. J.* 97: 1036
Rodgers, A., Harding, P., Ryan, S. 1986. *Astron. J.* 92: 600
Rodgers, A. W., Paltoglou, G. 1984. *Ap. J. Lett.* 283: L5
Rose, J. 1985. *Astron. J.* 90: 803
Ryden, B. S. 1988. *Ap. J.* 329: 589
Sandage, A. 1981. *Astron. J.* 86: 1643
Sandage, A. 1987a. In *The Galaxy*, ed. G. Gilmore, B. Carswell, p. 321. Dordrecht: Reidel
Sandage, A. 1987b. *Astron. J.* 93: 610
Sandage, A. 1988. In *The Calibration of Stellar Ages*, ed. A. G. D. Philip, p. 43. Schenectady, NY: L. Davis Press
Sandage, A., Fouts, G. 1987. *Astron. J.* 92: 74
Schombert, J. M., Bothun, G. 1987. *Astron. J.* 93: 60
Schuster, W., Nissen, P. 1989. *Astron. Astrophys.* In press
Searle, L. 1977. In *The Evolution of Galaxies and Stellar Populations*, ed. B. M. Tinsley, R. B. Larson, p. 219. New Haven, Conn: Yale Univ. Press
Searle, L., Zinn, R. 1978. *Ap. J.* 225: 357
Shaw, M., Gilmore, G. 1989. *MNRAS* 237: 903
Silk, J. 1977. *Ap. J.* 211: 638
Silk, J. 1983. *Nature* 301: 574
Sommer-Larsen, J. 1987. *MNRAS* 227: 21P
Sommer-Larsen, J., Christensen, P. R. 1987. *MNRAS* 225: 499
Sommer-Larsen, J., Christensen, P. R. 1989. Preprint
Spitzer, L., Schwarzschild, M. 1953. *Ap. J.* 118: 106
Statler, T. S. 1988. *Ap. J.* 331: 71
Statler, T. S. 1989. *Ap. J.* In press
Stetson, P. B., Aikman, G. C. L. 1987. *Astron. J.* 93: 1439
Stetson, P. B., Harris, W. E. 1988. *Astron. J.* 96: 909

Stobie, R. S., Ishida, K. 1987. *Astron. J.* 93: 624
Strömgren, B. 1987. In *The Galaxy*, ed. G. Gilmore, B. Carswell, p. 229. Dordrecht: Reidel
Strugnell, P., Reid, I. N., Murray, C. A. 1986. *MNRAS* 220: 413
Tammann, G. 1982. In *Supernovae: A Survey of Current Research*, ed. M. J. Rees, R. J. Stoneham, p. 371. Cambridge: Univ. Press
Terndrup, D. M. 1988. *Astron. J.* 96: 884
Thomas, P. 1989. *Ap. J.* In press
Tinsley, B. M. 1979. *Ap. J.* 229: 1046
Trumpler, R. J., Weaver, H. F. 1953. *Statistical Astronomy*. Berkeley: Univ. Calif. Press
Twarog, B. A. 1980. *Ap. J.* 242: 242
Upgren, A. R. Jr. 1962. *Astron. J.* 67: 37
Vanden Berg, D., Bell, R. A. 1985. *Ap. J. Suppl.* 58: 711
van den Bergh, S. 1958. *Astron. J.* 63: 492
van den Bergh, S. 1962. *Astron. J.* 67: 486
van den Bergh, S. 1979. In *Scientific Research With the Space Telescope. NASA CP-2111*, ed. M. S. Longair, J. W. Warner, p. 151
van den Bergh, S., McClure, R. D., Evans, R. 1987. *Ap. J.* 323: 44
van der Kruit, P. C., Freeman, K. C. 1986. *Ap. J.* 303: 556
van der Kruit, P. C., Searle, L. 1982. *Astron. Astrophys.* 110: 61
Weistrop, D. 1972. *Astron. J.* 77: 849
Wesselink, Th., Le Poole, R. S., Lub, J. 1987. In *Stellar Evolution and Dynamics in the Outer Halo of the Galaxy*, ed. M. Azzopardi, F. Matteucci, p. 185. Garching: ESO
Wheeler, J. C., Sneden, C., Truran, J. W. Jr. 1989. *Annu. Rev. Astron. Astrophys.* 27: 279
White, S. D. M. 1985. *Ap. J. Lett.* 294: L99
White, S. D. M. 1989a. In *The Epoch of Galaxy Formation. NATO Adv. Study Inst., Durham, Engl. 1988*, ed. R. Ellis, C. Frenk, J. Peacock. Dordrecht: Reidel. In press
White, S. D. M. 1989b. *MNRAS* 237: 41P
White, S. D. M., Rees, M. J. 1978. *MNRAS* 183: 341
Wielen, R. 1974. In *Highlights of Astronomy*, ed. G. Contopoulos, 3: 395. Dordrecht: Reidel
Woosley, S. E., Weaver, T. A. 1986. *Annu. Rev. Astron. Astrophys.* 24: 205
Wyse, R. F. G., Gilmore, G. 1986. *Astron. J.* 91: 855
Wyse, R. F. G., Gilmore, G. 1988. *Astron. J.* 95: 1404
Wyse, R. F. G., Gilmore, G. 1989. *Comments Astrophys.* 13: 135
Yoshii, Y. 1982. *Publ. Astron. Soc. Jpn.* 34: 365
Yoshii, Y. 1984. *Astron. J.* 89: 1190
Yoshii, Y., Ishida, K., Stobie, R. S. 1987. *Astron. J.* 93: 323
Yoshii, Y., Saio, H. 1979. *Publ. Astron. Soc. Jpn.* 31: 339
Yoss, K. M., Neese, C. L., Hartkopf, W. I. 1987. *Astron. J.* 94: 1600
Zinn, R. 1985. *Ap. J.* 293: 424
Zurek, W. H., Quinn, P. J., Salmon, J. K. 1988. *Ap. J.* 330: 519

SUPERNOVA 1987A

W. David Arnett

Departments of Physics and Astronomy, University of Arizona, Tucson, Arizona 85721

John N. Bahcall

Institute for Advanced Study, Princeton, New Jersey 08540, and Space Telescope Science Institute, Baltimore, Maryland 21208

Robert P. Kirshner

Harvard-Smithsonian Center for Astrophysics, MS-19, 60 Garden Street, Cambridge, Massachusetts 02138

Stanford E. Woosley

Board of Studies in Astronomy and Astrophysics, University of California, Santa Cruz, California 95064, and General Studies Group, Lawrence Livermore National Laboratory, Livermore, California 94550

1. INTRODUCTION

On February 23.316 UT, 1987, light and neutrinos from the brightest supernova in 383 years arrived at Earth, shocking astrophysicists into a frenzied state of activity. Since that time Supernova 1987A (SN 1987A) has taken its place not only as a unique event in modern astronomy, but also as one of the most thoroughly studied objects outside the solar system. Detected by instruments on the ground, below the ground, in space, and from balloons, airplanes, and rockets, it has been observed from all continents, including Antarctica. Studied at all wavelengths from radio through gamma rays, SN 1987A is the only object besides the Sun to have been detected in neutrinos. From these extensive observations has emerged a picture, still evolving but the most complete yet, of the life and death of a massive star.

The circumstances of SN 1987A are nearly ideal for studying supernovae. The properties of the progenitor are known as a result of diligent work on the stellar populations of the Large Magellanic Cloud (LMC) carried out two decades earlier. The distance to the explosion is known to better precision than most Galactic distances, which allows the observed fluxes to be converted to luminosity at the explosion. Since the distance is not large and obscuration by dust is modest, SN 1987A is accessible to detectors of limited sensitivity. As a result, we have a relatively complete observational record. The Large Magellanic Cloud is fortuitously well placed in the sky, circumpolar for recently constructed southern observatories (which allows unbroken coverage of its evolution) and near the ecliptic pole (facilitating satellite observations). As a result of these practical advantages, a rich and varied set of observations can be used to test a well-developed, but hitherto poorly calibrated, theory for supernova explosions.

This review cannot explore every aspect of SN 1987A. We have tried to emphasize the main event: the test of stellar evolution represented by the observations. Some recent publications that give different coverage include the ESO Workshop on the SN 1987A (120), the George Mason Workshop "Supernova 1987A in the Large Magellanic Cloud" (192), IAU Colloquium No. 108 on "Atmospheric Diagnostics of Stellar Evolution" (254), and the White Conference on SN 1987A (116). Reviews of SN 1987A include (232, 321, 322, 359).

2. EARLY OPTICAL AND ULTRAVIOLET OBSERVATIONS

The supernova was discovered in Chile by Ian Shelton using the 10-inch astrograph at the Las Campanas Observatory on a plate taken February 24.23 UT, 1987 (*IAU Circular No. 4316*). A visual observation made at approximately February 24.2 UT by Oscar Duhalde was also reported from Las Campanas. Independently, the New Zealand amateur astronomer Albert Jones reported detecting the supernova at February 24.37 UT during his routine monitoring of variable stars in the LMC. This is of special interest because Jones made similar observations, but did not see the supernova on February 23.39. Another observation made on February 23.443 was reported by R. M. McNaught (234), who obtained a photograph of the LMC that showed the supernova at approximately mag = 6.5. This closely spaced pair of observations defines the time of brightening for SN 1987A. After the announcement, the *International Ultraviolet Explorer*

(*IUE*) satellite was employed to obtain ultraviolet spectra of SN 1987A. A series of photometric observations with the satellite's fine error sensor (FES) provided broadband flux monitoring that began February 24. These early optical observations are summarized in Figure 1, which also shows the timing of the neutrino events discussed in Section 5 and a theoretical model from Section 6. The neutrinos were seen first, and the brightening followed, as expected if the neutrinos signal the deposition of energy in the star from a collapsing core. The very rapid brightening of SN 1987A [by a factor of 100 in 3 hr, rather than days as seen in other Type II supernovae (248)] indicated that the atmosphere of this supernova was more compact than the red supergiants that give rise to normal Type II supernovae.

Spectra of the supernova taken in South Africa (*IAU Circular No. 4316*) and in Chile (*IAU Circular No. 4317*) on February 25 showed broad, shallow absorption features of the Balmer series extending up to $0.1c$ (30,000 km s^{-1}). The presence of hydrogen makes SN 1987A a Type II

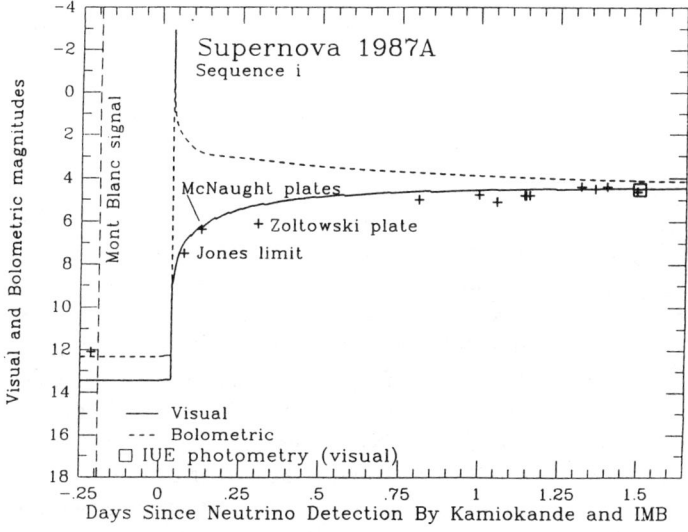

Figure 1 The *V*-magnitude light curve of SN 1987A during its first two days, with the time of core collapse defined by the neutrino signal from IMB and Kamiokande (February 23.316). The time of the Mont Blanc event is shown as a dashed vertical line. It is not consistent with the observed light curve. The upper bound at February 23.39 is due to Albert Jones, and other data points are taken from the *IAU Circulars*. Shown for comparison are two theoretical curves. The solid line is the *V* magnitude from (22–25); similar results were obtained in (296, 353–355). The bolometric luminosity predicted is shown as a dashed line. During the earliest times the bolometric correction is very large, and simple blackbody models, while qualitatively correct, do not provide an exact rendering of the observations.

supernova (SN II), but the very high velocity indicates that SN 1987A was distinct from most members of that class, for which the most rapidly approaching hydrogen absorption is typically at about 15,000 km s^{-1} (198). Other indications that SN 1987A was different included a very rapid decline in the UV flux, which dropped by a factor of 1000 in the first three days (62, 199, 200, 336), and similar, though less extreme changes in the optical colors. As pointed out by Blanco et al. (58), the range of colors was very similar to that observed in other SN II's, but the rate of change was much more rapid [cf., for example, SN 1969L (106)]. Spectrophotometry shows that the rapid color change was the result of a cooling continuum, which declined from a temperature near 14,000 K on February 24 to about 5500 K on the 20th day of observation.

Sampling is an important consideration in establishing typical values for astronomical objects. If the search for supernovae is flux limited (which is only approximately true), then the volume searched for ordinary SN II's ($M = -18$) is roughly 30 times larger than for intrinsically faint objects such as SN 1987A at $M = -15.5$. If SN 1987A's twin erupted in the Virgo cluster of galaxies, at a distance of 20 Mpc, it would be at apparent magnitude 16 and would be missed by most current searches. Events like SN 1987A may not be as rare in the Universe as in our samples of the Universe. A resemblance between SN 1987A and SN 1909A in M101 has been pointed out by Young & Branch (374).

We are in a unique and fortunate circumstance in the case of SN 1987A because the LMC has been intensively studied, and many individual stars, including some in the stellar association near SN 1987A [NGC 2204 (181, 213)], have magnitudes, colors, and spectra (278, 284). Accurate astrometry of the supernova by Girard et al. (164), Walborn et al. (333), West et al. (341), and White & Malin (342, 343) shows that the supernova is at R.A. = $5^h 35^m 49.992^s$, dec = $-69°17'50.08''$ (1950.0); this position coincides with that of a star from the Sanduleak catalog, #202 in his -69 declination band. For convenience we call it Sk -69 202, although it has a more ancient appellation of CPD -69 402.

Careful examination of the available data (for example, 58, 333, 334) indicates that the Sanduleak star was a B3 I supergiant, with $V = 12.4$, $B-V = +0.04$. The positional coincidence is important evidence that the Sanduleak star itself was the progenitor of the supernova; however, it had two close companions, located just 2.9'' and 1.6'' away (333, 341). Spatial analysis of *IUE* images made after the supernova had declined in the UV shows that the UV flux in March 1987 was due to two objects with the separation of these two stars, which have the UV flux of main sequence B stars (163, 300). There can be little doubt that the Sanduleak star has disappeared, and the direct evidence points to the blue supergiant as the progenitor of SN 1987A. While it is still not excluded that the 12th-mag Sanduleak star is buried within the supernova debris, the models for SN 1987A that assume that Sk −69 202 was not the progenitor (e.g. 191) provide little guidance to interpreting the observations of SN 1987A, whereas the models that take Sk −69 202 as their initial condition supply a detailed and accurate model for subsequent spectral and photometric evolution. The only previous case where a supernova progenitor has been seen is the very strange case of SN 1961V in NGC 1058 (127). The positive identification of the progenitor as a massive star is one of the most useful observations of SN 1987A: The fact that it was a blue supergiant, rather than a red supergiant, is one of the great surprises from the observations and accounts for most of the differences from other SN II's.

The Sk −69 202 star was unremarkable at the surface despite the approach to disaster in the core, with no sign of variation in the 100 years it has been observed (176), and no outstanding spectroscopic peculiarity noted (332, 334). The core evolution of SN 1987A is likely to resemble that of other SN II's, despite the differences observed at the surface. SN 1987A should not be regarded as a freak or a counterexample, but rather a variation on a familiar theme.

But SN 1987A did provide significant variations. The spectrum of SN 1987A showed much more rapid changes than are usually observed in SN II's. The spectrum, as observed in the UV, optical, and IR, showed rapid changes, illustrated in Figure 2. The changes, about 10 times as rapid as in a normal SN II, include a precipitous drop in the UV flux, a corresponding decline in the excitation of the atmosphere, and a decreasing velocity of expansion in the line-forming region.

The initial *IUE* observations of SN 1987A showed a strong UV continuum, with a color temperature of 14,000 K (89, 199, 200). The UV flux, however, rapidly decreased, and after only one week, the spectrum showed a very strong UV deficit. This has been interpreted as the effect of blanketing by a large number of resonance lines in the rapidly expanding envelope of the supernova (154, 214, 233). Ultraviolet photons are scattered, performing a random walk in space, and are redshifted in each scattering. Most far-UV photons can only emerge from the supernova atmosphere longward of 2700 Å.

Models for the optical spectrum have followed the notions developed for SN II's by Kirshner & Kwan (196) and by Branch et al. (65)—namely, that the lines are formed in the rapidly expanding atmosphere and may be treated with an appropriate version of the Sobolev method. In the case of SN 1987A, the excellent data have induced workers whose main interest is stellar atmospheres to apply a powerful array of computational techniques devised for other astrophysical settings. For example, Eastman & Kirshner

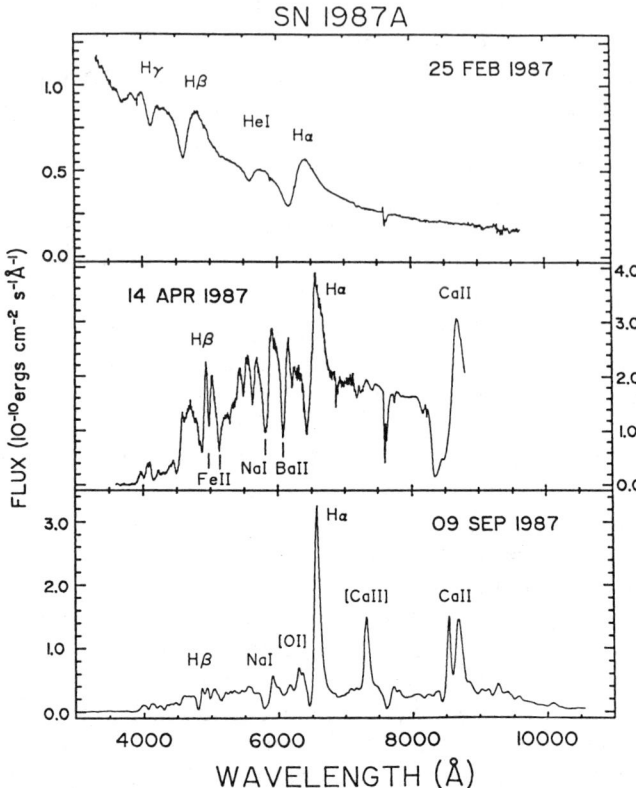

Figure 2 The optical spectrum of SN 1987A at three different epochs: (*a*) February 25, 1987, only 40 hr after core collapse. Note the broad profiles of the hydrogen and helium lines and the large blueshifts of the P Cygni absorption components. (*b*) April 14, 1987, 50 days after core collapse. The spectrum is now dominated by lines of low ionization elements. Note the strength of the barium line at 6142 Å. (*c*) September 9, 1987, more than 100 days after the maximum of the bolometric light curve. The spectrum has by this time taken on more of a nebular appearance, with strong emission lines of hydrogen, oxygen, calcium, and sodium dominating [observations from Cerro Tololo Inter-American Observatory; figure from (359)].

(139), Hoflich (182, 183), Lucy (214), and others have developed detailed models. A very satisfactory fit to the observations can be obtained in the case where the density drops off rapidly with velocity [$\rho \propto v^{-(9-11)}$]. As noted below, this is very similar to the results of hydrodynamic calculations for the disruption of an atmosphere like that of Sk −69 202 by a strong shock. The relative abundances of elements used in these calculations are roughly solar, often modified for the known chemical abundances of the

LMC (135), and the reasonable agreement with the data suggests that surface abundances for Sk −69 202 cannot have been very different. However, a few significant spectroscopic differences between SN 1987A and other SN II's have been pointed out, and these may provide clues to the evolution of Sk −69 202. The He I 5876-Å line is strong in the spectrum of SN 1987A for the first few days. Determining whether this implies a high helium abundance in the surface layers of the star requires a detailed non-LTE treatment of the atmosphere, but preliminary indications are that He may be enhanced. Similar He enhancements are reported for the atmospheres of other LMC blue supergiants by Kudritzki et al. (205), who also find enhanced N. The importance of these elements is that they are coupled to CNO burning, mixing, and mass loss in the star. As described below, there is very strong evidence from circumstellar UV emission lines that Sk −69 202 has mixed the products of CNO burning to the surface. More puzzling are strong lines of Ba, Sc, and Sr identified by Williams (345). These are s-process elements, whose implications for stellar evolution are more difficult to unravel. The absolute overabundance required to produce the observed lines is uncertain and requires a detailed treatment of the excitation and ionization of these lines in a realistic setting.

3. THE STAR THAT EXPLODED

The distance to the LMC is reasonably well established at 50 kpc $[(m-M) = 18.5$ (8, 9)]. This allows the observed properties of Sk −69 202 and SN 1987A to be converted to properties at the source. The best-estimate bolometric magnitude for Sk −69 202 was −7.8, with a probable range from −7.5 to −8.2. The value −7.9 was preferred by Humphreys & McElroy (185) prior to the explosion of the supernova. A bolometric magnitude of −7.8 translates into a luminosity of 4.0×10^{38} erg s^{-1}, with a probable range of $(3-6) \times 10^{38}$ erg s^{-1}. This range of luminosities corresponds to a range of main sequence masses. The critical quantity is the mass of the helium core, since at the time of the supernova, the hydrogen-burning shell contributes negligible energy generation. The helium core may be uniquely related to the main sequence mass and, for the stars we are considering, is insensitive to the amount of envelope mass that the star has lost. From examination of a variety of models for massive stars (13, 22–25, 255–257, 260, 326, 353, 362, 369), we conclude that the helium core mass of Sk −69 202 was in the range 5–7 $M_\odot$ and, prior to any mass loss, its main sequence mass was in the range 16–22 $M_\odot$.

Observations also constrain the radius of Sk −69 202. A B3 I supergiant has a surface temperature of ∼16,000 K (185). For the above range of luminosities and assuming a temperature in the range 15,000–18,000 K, one obtains a radius $3(\pm 1) \times 10^{12}$ cm. Observations of the presupernova star do not constrain the mass of its hydrogen envelope, but various analyses of the supernova itself discussed later in this paper suggest that the envelope mass was on the order of 10 $M_\odot$. In that case, Sk −69 202 was a 20-$M_\odot$ star, with a 6-$M_\odot$ He core, about 10 $M_\odot$ in the envelope, and perhaps a few solar masses having been lost during its evolution.

Some of the properties inferred from the observations of Sk −69 202 agree well with conventional stellar evolution theory. Sk −69 202 was a massive star and bright (about 10^5 $L_\odot$) even before it exploded. Theory predicts that only stars heavier than about 8 $M_\odot$ can ignite the advanced burning stages—carbon, neon, oxygen, and silicon burning—that ultimately lead to the formation of an unstable iron core. Lighter stars develop cores that are supported by the pressure of degenerate electrons before igniting carbon burning and thus experience a different fate.

The surprise, however, was that Sk −69 202 was not a red star when it exploded, but blue. The shock wave theory of SN II's (19, 30, 95, 145, 146, 166, 337, 338) showed that previously observed bright supernovae required massive progenitors with a large radius—that is, *red* supergiants. A large radius corresponds to a lower density and a faster diffusion time, so the thermal energy produced by the shock can escape to be seen before being converted into kinetic energy of the expanding matter. Other things being equal, dense stars produce a weaker luminous outburst. In retrospect, one may note that a number of papers published prior to 1987 had foreseen that core collapse might occur in a compact star and produce a supernova that, at least early on, was relatively faint (15, 16, 96, 297, 352, 366). Some of these also predicted a subsequent brightening of the supernova, as occurred with SN 1987A, owing to radioactivity (15, 16, 352, 366), and that such compact supernova progenitors (and faint supernovae) would be more abundant in regions of low metallicity (297, 366). A few stellar evolution calculations also yielded blue supergiant supernova progenitors (72, 73, 210). However, the paucity of observed supernovae corresponding to this description was generally taken to imply that core collapse generally occurred in the red supergiant stage, consistent with most previous studies of massive stellar evolution.

The specific reasons why Sk −69 202 was compact are still being debated, but from the earliest hours following its discovery two quite different explanations were offered—extensive mass loss and low metallicity.

3.1 *Mass Loss as a Cause for the Blue Color of Sk −69 202*

Well before SN 1987A, it had been recognized that a massive star that lost most of its hydrogen envelope would evolve back toward the blue in the

Hertzsprung-Russell (HR) diagram, exploding as a blue supergiant or, in the extreme limit of complete hydrogen evaporation, a Wolf-Rayet star (15, 105, 217, 218). A star with no hydrogen envelope is more compact; so, too, is one with just a little hydrogen. The last few solar masses of envelope contain a high helium concentration and hence have an average atomic weight much greater than material of solar composition. Such stars assume a small radius.

Immediately following the explosion of 1987A, mass loss was invoked by many theorists as the probable cause for the compact nature of the progenitor star (102, 103, 133, 144, 216, 296, 351, 364). In favor of this notion, and in addition to the need for a compact progenitor, one may note the existence of many Wolf-Rayet stars in and around 30 Doradus (241). Also, the spectrum of 1987A early on, the high velocities, and the rapid color evolution were quite different from what had been seen in other SN II's and could be modeled starting from a compact star. The slow-moving, nitrogen-rich circumstellar material in the ultraviolet spectrum of the supernova (Section 3.5; 89, 199, 200) has been taken by some to imply the need not just for mass loss, but for a lot of it (152, 216). It should be noted, however, that enhancements in the nitrogen-to-carbon ratio may exist naturally in *any* massive star that has passed through a red supergiant stage (e.g. 216, 219, 332, 339, 353) and do not necessarily indicate a large degree of mass loss, especially given uncertainties in the treatment of convection and of mixing.

There are several problems with mass loss as the sole explanation for the compact nature of Sk $-69\ 202$. First, 20 $M_\odot$ is rather light for a star to lose nearly all of its envelope. Humphreys (184) presents evidence that the bulk of the Wolf-Rayet stars originate from progenitors heavier than 40 $M_\odot$. Given the metal-deficient nature of the LMC, one might expect even less mass loss. However, we are dealing with a single event, so this argument is weak. Second, as (47, 282, 362) have emphasized, the radius and luminosity of Sk $-69\ 202$ allow two classes of solutions—stars that have lost little mass (a few solar masses at most), and stars that have lost almost their entire envelope. Intermediate values of envelope mass yield red solutions unless the envelope is extremely helium rich. This is consistent with the well-known tendency [cf. (72, 73)] of mass loss to drive the evolution of massive stars to the *red*, not to the blue. Thus Sk $-69\ 202$ had either a very low-mass hydrogen envelope or a high-mass one, not something in between.

Many convincing arguments now exist that SN 1987A exploded with a substantial hydrogen envelope in place. These reasons include the long, slow rise of the optical light curve to maximum (80 days); the late appearance of the slowest moving near-infrared lines of hydrogen (about 40 days); the slow speed of the slowest moving hydrogen observed in the infrared (~ 2000 km s^{-1}) (269); and the lack of early escape of gamma rays and X rays from radioactive decay. If the hydrogen envelope had been a few solar masses or less, a shock energy sufficient to reproduce the early light curve and also to prevent reimplosion of the radioactive elements that later powered the tail of the light curve would have violated these constraints (295, 353). The helium core of the supernova would have expanded more rapidly, which would have led to an early brightening and a brighter supernova [cf. (20)], somewhat like a Type Ib (143, 261, 364). The slowest moving hydrogen would have been uncovered after less than a week and would have had a substantially greater velocity than was observed. The X-ray and gamma-ray light curves would also have peaked much earlier and been at least a factor of 10 brighter. Numerical integrations of the density and velocity obtained from the spectra also yield values of hydrogen envelope mass more consistent with 10 $M_\odot$ than with 1 or a few $M_\odot$ (129–131, 193, 194). Finally, the fact that hydrogen was more abundant than helium in the spectrum of the progenitor star (334) implies that a substantial hydrogen envelope was still intact. Although Sk −69 202 surely had a significant episode of mass loss, substantial envelope mass remained at the time of explosion, and other causes for making Sk −69 202 a blue star need to be investigated.

3.2 Low Metallicity as a Cause of the Blue Star

The composition of the 30 Doradus region in the LMC is distinctly different from that of the Sun; heavy elements are less abundant, and the pattern of abundance is complex (128, 130–132, 135, 147, 148, 279, 280). Several analyses indicate, for example, an oxygen abundance compared with 10^{12} hydrogen atoms of log [O] = 8.3 ± 0.1, or about three times smaller than the abundance tabulated by Cameron (88) for the Sun. Nitrogen and carbon appear to be even more deficient than oxygen, and iron less so. These abundances affect the evolution of a massive star in two ways— first, by reducing the opacity of the hydrogen envelope (the deep interior is dominantly electron scattering), and second, by reducing the rate at which nuclear energy is released at a given temperature during hydrogen burning by the CNO cycle. The abundance of oxygen, which contains most of the mass residing in CNO nuclei, is the most important for the energy generation rate and, in most stellar cores, for the opacity as well.

Several groups calculated stellar models that, based upon an assumed metal-deficient composition similar to the LMC *and no mass loss*, gave blue supernova progenitors (26, 27, 282, 324, 325, 362). Details of the calculations varied in significant ways. The early models of Arnett (22–25) and Truran & Weiss (324, 325), for example, spent their entire lives as blue stars, never becoming red supergiants and probably never ejecting a low-velocity, nitrogen-rich wind as required by the (later) *IUE* observations (Section 3.5). The models of Woosley et al. (362) and Woosley (353), on the other hand, spent almost their entire helium-burning lifetimes as red supergiants, evolving back to the blue in the last 40,000 yr, just in time to explode. This fits the observed properties of Sk −69 202 nicely and would agree with the existence of many red supergiants in the LMC. It may fail, however, by itself to account for the large number of *blue* supergiants also seen in the LMC (185). The models of Saio et al. (282, 283) and Nomoto et al. (263, 264) move from the blue to the red during helium burning and spend less than one tenth of the helium-burning lifetime as a red supergiant. These models are consistent with the number ratios between the blue and red supergiants in the LMC.

The ways in which metallicity influences the evolution to yield a blue star are complex and still not understood. Altering the opacity or hydrogen energy generation rate in the presupernova star changes the entire evolution on the main sequence and during helium burning to yield, in the end, a different helium core structure. Figure 3 shows the evolution in the HR diagram of an 18-$M_\odot$ model (365) that fits many of the constraints on the progenitor. A composition typical of the LMC has been employed, as has a recent remeasurement of the reaction rate for $^{12}C(\alpha,\gamma)^{16}O$ (92) and a mass loss rate equal to three times that specified by de Jager et al. (125). Note that according to theory, the star Sk −69 202 *was* a red supergiant about 40,000 yr ago (the Kelvin-Helmholtz time for its 11-$M_\odot$ hydrogen envelope). During its red giant stage, which lasted from a few hundred thousand to one million years (depending upon parameters of the calculation) it lost several solar masses from its surface. The final mass of this 18-$M_\odot$ star was 16.0 $M_\odot$.

3.3 *Hybrid Schemes*

It now seems likely that a combination of factors caused the presupernova star to be blue, with low metallicity playing an important, but only partial, role. Woosley et al. (365), for example, have found that a blue presupernova star results in their calculations only if (*a*) the stellar abundances are those characteristic of the LMC (or lower in metallicity) and not of the Sun; (*b*) no more than a few solar masses are lost from the star before it explodes; (*c*) the mass of the star is not much more than 20 $M_\odot$ (not 25 $M_\odot$, for example); and (*d*) semiconvection occurs at a rate smaller than had been assumed in their previous calculations. (Specifically, the diffusion coefficient for compositional mixing in those zones satisfying the Schwarzschild criterion but not the Ledoux criterion should be about 0.1% that of radiation.)

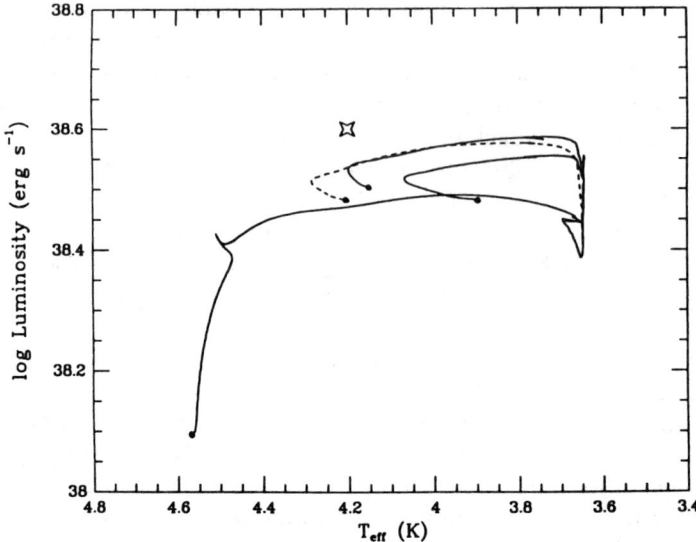

Figure 3 Hertzsprung-Russell (HR) diagram for three 18-$M_\odot$ stars evolved through hydrogen, helium, and carbon burning with two different mass loss rates. The leftmost of the two solid lines corresponds to a star having a mass loss rate three times that of de Jager et al. (125) and a final stellar mass of 16.2 $M_\odot$. The rightmost solid line is a model that uses three times the de Jager et al. rate on the main sequence (as before), but six times the de Jager et al. prescription when the star is a red supergiant. This star ended up with a mass of 14.7 $M_\odot$ and died in a state of thermal disequilibrium while making a transition back to the red during carbon burning. Any 18-$M_\odot$ star that has lost this much mass or more cannot be SN 1987A. The dashed line is a model that was artificially made to mix up more helium into the envelope. The four-pointed star is the presupernova star, Sk −69 202.

Saio et al. (282, 283) require another set of circumstances—moderate mass loss to drive their 21-$M_\odot$ star to the red [as demanded by observations of the nitrogen-rich circumstellar shell (see Section 3.5)], and mixing of helium out from the core during the late stages of evolution in order to make the star move back to the blue. They find it is easier to get a blue solution if the enhanced helium abundance is larger and the metallicity smaller.

Weaknesses remain in these schemes. The Woosley-Weaver scenario requires an artificial tuning of the semiconvective diffusion parameter in order to give both a blue Sk −69 202 and a reasonable ratio of blue to red supergiants in the LMC. The mixing invoked in the Saio et al. model is artificial both in the extent to which it occurs and in the time at which it is invoked—during the red supergiant phase but not earlier. Resolving the issue of the blue progenitor is of great importance because ultimately we would like to know the frequency of SN 1987A–like events. Because statistical samples of supernovae are strongly biased against faint supernovae, the answer can only come from a better understanding of what makes the presupernova star choose between blue and red solutions.

An important lesson from SN 1987A is that our evolutionary picture for the *early* stages of massive stars requires revision in ways yet to be determined.

3.4 Advanced Burning Stages

The advanced burning stages of massive stars have been reviewed in (17, 257, 352). The evolution of the surface of the star as shown in Figure 3 ceases after the star has completed helium burning in its center. Each of the subsequent burning stages (carbon, neon, oxygen, and silicon burning) occurs on a time scale of less than 1000 yr [17, and references therein; Table 1 (from 369)] that is very short compared with the 10^7 yr spent burning hydrogen and helium, and, most importantly, short compared with the Kelvin-Helmholtz time scale for the hydrogen-rich envelope. Evolution accelerates, because for temperatures above about 0.5×10^9 K (which are required to burn these fuels), neutrino losses transport more energy out of the core than does radiation. These losses, which arise from the annihilation of pairs produced by the thermal radiation, scale approximately as the ninth power of the temperature. Yet larger temperatures are required to burn each additional fuel because of the large Coulomb barrier of the reactants. Comparable changes in nuclear binding occur at each burning stage after helium burning, so the evolutionary time scale becomes quite short. The final conversion of silicon and sulfur to elements in the iron group, which is also accelerated by neutrino losses from electron capture, takes only a few days. Once iron has been formed, no more energy can be released by any rearrangement of the nucleons

Table 1 Burning stages in the evolution of a 20-$M_\odot$ star

Fuel	ρ_c (g cm^{-3})	T_c (10^9 K)	τ (yr)	L_{phot} (erg s^{-1})	L_ν (erg s^{-1})
Hydrogen	5.6(0)	0.040	1.0(7)	2.7(38)	—
Helium	9.4(2)	0.19	9.5(5)	5.3(38)	<1.0(36)
Carbon	2.7(5)	0.81	3.0(2)	4.3(38)	7.4(39)
Neon	4.0(6)	1.7	3.8(−1)	4.4(38)	1.2(43)
Oxygen	6.0(6)	2.1	5.0(−1)	4.4(38)	7.4(43)
Silicon	4.9(7)	3.7	2 days	4.4(38)	3.1(45)

to other nuclei, and thus the core must collapse. Calculations including Coulomb effects on the equation of state (15, 16, 26, 255–257, 362, 369) indicate that the mass of this iron core was close to 1.4 $M_\odot$. (A range of 1.3–1.6 $M_\odot$ seems reasonable.) Two instabilities beset the core: *photodisintegration* as the high-temperature radiation ($T \sim 10^{10}$ K) begins to rip iron nuclei apart into α-particles, absorbing energy that might otherwise have contributed to the pressure; and electron capture. Since the pressure in the core at this point is predominantly due to relativistic electrons, their removal makes it more difficult for the star to support its own weight. The core begins to collapse.

3.5 *Circumstellar Material and Observational Constraints on Presupernova Evolution*

Significant observational clues to the evolution of Sk −69 202 come from studies of the radio and ultraviolet emission from the supernova. Radio emission from SN 1987A was observed in the first month after maximum light (327). Other SN II's have been observed as radio emitters (340), with the emission generally appearing at high frequencies first, at a time approximately 1 yr after the explosion. A successful model for this emission from SN II's attributes the radio emission to a shock in the circumstellar medium, with the onset at high frequencies due to opacity effects (97). In this case, the flux appeared early, reaching a maximum at three days after discovery, and was roughly 10^3 lower than other SN II's. VLBI observations (189) showed that the radio source was larger than 1.25 milliarcsec (a linear extent of 8×10^{14} cm) five days after the neutrino burst. This implies that the radio source expanded at least as fast as 19,000 km s^{-1}, somewhat in excess of the photospheric velocity at that time. A reasonable explanation for this (98, 304) is that SN 1987A possessed a much lower density circumstellar shell. This is not surprising: In contrast to red supergiants, blue supergiant winds have high velocities and low densities.

Ultraviolet observations also provide information about circumstellar matter and the history of Sk −69 202. *IUE* spectra after 24 May 1987 show emission lines of He II, C III, N III, N IV, N V, and O III (152). These lines are narrow (<30 km s^{-1}) and could not arise within the opaque stellar envelope. The most likely origin is a photoionized, low-density circumstellar gas, lost by Sk −69 202 in an earlier stage of evolution when it was a red supergiant. The radial velocity of the UV emission lines is 284 ± 6 km s^{-1}, which coincides with the highest velocity interstellar absorption lines observed in the UV (56, 57, 124, 136).

The most likely excitation source is the EUV burst emitted by the stellar photosphere as the shock broke out. These lines grew in strength until April 1988 (299), which is consistent with a fluorescent light echo from a circumstellar shell. In this picture, larger segments of the circumstellar shell, at a distance of 5×10^{17} cm, are visible as time progresses. Optical observations at [O III] (335) show a width of 15 km s^{-1} and indicate a temperature of 50,000 K for this matter. Relative line strengths of the UV multiplet components of the C III], N III], and N IV] intercombination lines provide useful density diagnostics for this gas and are consistent with a density of $(1-3) \times 10^4$ cm^{-3}. A nebular analysis of the line strengths (152) reveals a large nitrogen overabundance, with N/C = 8 ± 4 and N/O = 1.6 ± 0.8. These are, respectively, factors of 37 and 12 higher than the solar values and seem to require that the gas seen has undergone substantial CNO processing. To have the products of hydrogen burning at the stellar surface probably requires a combination of mixing and mass loss for the progenitor of SN 1987A. This is consistent with some models in which Sk -69 202 was a red supergiant before it became a blue supergiant. In fact, similar conclusions about CNO processing were reached by Fransson et al. (152) based on UV observations of SN 1979C, which was a red supergiant when it exploded.

One interesting consequence of this circumstellar shell will be a violent interaction when it is hit by the fast-moving debris from the explosion. Since the debris has a velocity of approximately $0.1c$, and the shell is at a radius of about 1 lt-yr, the renaissance of the supernova can be expected in 10 yr. The collision will close this century with a high-temperature shock with copious X-ray emission (see Section 9.6), as calculated by (187, 222–224, 258).

4. THE EXPLOSION

4.1 *The Explosion Mechanism*

Much has been written about the Type II supernova explosion mechanism. The early speculations of Baade & Zwicky (35) have proved prophetic, but the details remain elusive. For discussion of current ideas, see (15, 16, 21, 53, 68, 80, 82, 178, 186, and references therein; 227, 243, 252, 348, 367, 368).

Once the unstable core begins to collapse, it continues until the central density in the star has risen by a factor of about 10^6. In 1 s a configuration the size of the Earth collapses to one with a radius of only 50 km. The velocity during the collapse reaches about 70,000 km s^{-1} in the outer portion of the iron core. However, because of the weaker gravity experienced by layers farther out and because the information that the core has collapsed propagates outward as a sound wave of finite speed, the neon, carbon, and helium shells (as well as the hydrogen envelope) do not participate in this collapse. Though pressure support has essentially disappeared in the center of the star, the outer layers hang suspended with inadequate time to respond.

The central density rises to several times that of the atomic nucleus (2.7×10^{14} g cm^{-3}) when the nuclear force, ordinarily attractive, changes sign and becomes repulsive. Once this occurs the resistance to further collapse is great. The nuclear pressure, the pressure due to the highly relativistic gas of electrons, and the extra pressure given as nuclei make the transition to a fluid of unbound nucleons cause the inner part of the core to halt and spring back. The inner region that rebounds as a unit consists of about 0.7 $M_\odot$ (that is, about one half of the collapsing iron core). Outside, matter is falling supersonically. As it runs abruptly into the "brick wall" of the rebounding inner core, a *shock wave* forms, a surface where matter meets matter at supersonic speed. For a time the expansion of the inner core, plus the energy that the infalling matter gets by bouncing off that core, pushes the shock out. If all goes well (unfortunately, it rarely does in the computer models of 20-$M_\odot$ stars), the shock continues moving out, finally exiting the collapsed core with enough energy (about 10^{51} erg) to eject the rest of the star with high velocity.

This is now called a "prompt hydrodynamical explosion" [see (110)]. In this picture the shock is out of the core and the explosion underway in only about 20 ms. The difficulty, however, is that the expanding shock wave loses a great deal of energy as it beats its way upstream against the infalling outer core. The problem is energy dissipation—neutrinos lost because of the high temperature interior to the shock and further *photodisintegration*. For every 0.1 $M_\odot$ that the shock disintegrates to neutrons and protons it loses 1.7×10^{51} erg, roughly equal to the final kinetic energy of a successful supernova explosion like SN 1987A. If the shock always starts at about the same place, then its success or failure depends sensitively upon the mass of the iron core. Larger iron cores experience more photodisintegration losses and are less likely to explode by this mechanism. It has proved difficult in practice to cause the explosion of iron cores larger than about 1.2 $M_\odot$ by the unaided prompt mechanism (81).

If the prompt mechanism fails, as it must for some critical mass of iron core, another means must be found to explain stellar explosion. Otherwise, the shock would halt its outward motion; the core would grow to several solar masses by accretion; and then, suddenly, the core would collapse inside its event horizon, to be followed within a few hours by the rest of the star. The optical event would be dim and definitely not a supernova. In recent years a second mechanism has been revived to avoid this dismal prospect. A return to earlier notions of the 1960s (12, 112, 346), this mechanism draws upon the enormous energy released by the collapsing core during its first second. At least 99% of the binding energy of the neutron star that forms (roughly $2–3 \times 10^{53}$ erg) comes out in neutrinos. The energy in these neutrinos is 100 times that needed for a shock wave to give a powerful supernova. The problem then is channeling some small fraction of the neutrino energy to the proper place and at the proper time to cause the explosion. Because of its longer time scale, this mechanism has now come to be known as the "delayed explosion mechanism."

Bahcall (39) discusses a third possibility: that most stellar collapses do not produce optically bright supernovae but instead emit essentially all of their energy as "neutrino bombs." This possibility implies a larger galactic stellar collapse rate than is inferred from optical surveys of supernovae, an implication that can be tested by monitoring with neutrino telescopes.

It was hoped that observations of SN 1987A would resolve the controversy over which mechanism dominates in the explosion of 20-$M_\odot$ stars—prompt or delayed. So far, the analysis has been inconclusive. Most of the observable properties of the supernova—velocities, spectra, light curve, and so on—are not sensitive to the core collapse and require only that about 10^{51} erg is deposited in the central regions of the star. Delayed explosions tend to be favored if the iron core mass exceeds 1.3 $M_\odot$. Model calculations (see above) and the energy of the neutrino signal (see below) suggest that the iron core in Sk -69 202 was probably heavier than this, perhaps as much as 1.6 $M_\odot$. On the other hand, the simulations of delayed explosions tend to have less kinetic energy ($\lesssim 10^{51}$ erg) than prompt ones ($\gtrsim 10^{51}$ erg). Current estimates of the explosion energy, based upon light curve and velocity (22–25, 295, 353), are in the range $0.6–1.5 \times 10^{51}$ erg with an uncertainty of perhaps 50%. Clearly this does not resolve the debate on the mechanism. The arguments relating explosion energy and iron core mass to mechanism are based upon theoretical estimates of modest precision. While SN 1987A did not resolve this issue, it has provided many new constraints. These are the neutrino burst spectrum, fluence, and time scale; an accurate measure of the kinetic energy of the ejecta; and a precise determination of the mass of ^{56}Ni produced by the shock wave just outside the core (see below). These constraints may ultimately define a unique solution to how SN 1987A exploded.

Stimulated in part by SN 1987A, a variety of calculations of iron core collapse have been carried out during the last year, but no prompt explosion has been achieved using a credible choice of input physics (48, 69–71, 228, 242). Various physical effects are responsible for the failure of some calculations that used to be more encouraging. These include neutrino losses from all three flavors of neutrinos, electron capture on free protons, and use of realistic nuclear symmetry energies. On the other hand, no other group has confirmed the Wilson-Mayle mechanism of delayed explosion by neutrino energy transport.

However it was born, a powerful shock *did* propagate through Sk −69 202, leading to its explosion. The very high emission temperature observed on the first day of the supernova was characteristic of a shock wave breaking through the surface of a star. We also know that the core collapsed to a neutron star because, for the first time, we saw the neutrino burst. The known radius of Sk −69 202 and the timing between the arrival of the neutrinos and the first optical observations are consistent with the supersonic propagation of a signal originating at the center of the star (22–25, 296). Finally, as we discuss later, the light curve of SN 1987A has, since the end of the first month, been powered by the decay of a radioactive nuclide, ^{56}Co. This short-lived species could only have been synthesized and ejected from a massive star by a strong shock.

4.2 Shock Propagation and Explosive Nucleosynthesis

The shock wave formed following the collapse of the iron core is initially strong enough to cause nuclear burning. During this stage a large fraction of the energy in the shock goes into internal energy, thus filling the mantle and envelope of the star with radiation. An approximate expression for the temperature behind the shock as a function of radius in the presupernova star, R, is given by setting the volume behind the shock, $4/3\pi R^3$, times the radiation energy density, aT^4, equal to E_0, the explosion energy (about 10^{51} erg). Thus, the peak temperature behind the shock is

$$T \approx \left(\frac{3E_0}{4\pi R^3 a}\right)^{1/4}. \qquad 1.$$

Any reasonable initial composition will burn to nuclear statistical equilibrium on a hydrodynamical time scale ($446/\rho^{1/2}$ s ~ 0.1–1 s) if it is heated to a temperature in excess of 5×10^9 K (356). Since all of the ejected material that was initially interior to a radius of ~ 3500 km is subjected to this temperature, it will be processed to the iron peak. With the possible exception of a small amount of neutron-rich material that might be blown away from the edge of the neutron star if the explosion is overly violent, most of this material will be in the form of ^{56}Ni.

This constraint in *radius* means that the amount of radioactive elements produced is very sensitive to the density gradient surrounding the iron core at the time it collapses. Lighter stars (8–12 $M_\odot$), for example, have partly degenerate cores with steep density gradients at their edges (251, 253). For this reason they produce little ^{56}Ni [e.g. 0.003 $M_\odot$ in an 8–10 $M_\odot$ supernova (347)]. But because Sk −69 202 had a helium core near 6 $M_\odot$, the density gradient was not so steep. A wide variety of parameters for core bounce and explosion yield values of ~ 0.08–0.4 $M_\odot$ of ^{56}Ni in simulated explosions of 6-$M_\odot$ cores (143). Thus, the fact that the light curve requires about 0.07 $M_\odot$ of ^{56}Ni to have been produced in the explosion (Section 7) is in reasonable agreement with the inferred mass for Sk −69 202, though small by about a factor of two compared with standard theory. The ^{56}Ni nucleosynthesis also strongly constrains the supernova mechanism. The boundary in mass between supernova ejecta and the neutron star should be situated 0.075 $M_\odot$ interior to roughly 3500 km in the presupernova model. Interestingly, the ejection of roughly 0.1 $M_\odot$ of ^{56}Ni is a consequence of the particular core-mantle structure predicted by modern theories of the weak interaction (17), which lead to a steep density gradient surrounding the iron core.

As the shock moves out, its temperature declines (Equation 1). When it falls below about 2×10^9 K, usually in the neon-oxygen layer, explosive nucleosynthesis ceases. Thus the elements heavier than magnesium are synthesized in the explosion, while lighter elements are produced during the burning stages before the explosion and simply ejected (Figure 4). The detailed isotopic composition of the ejecta has been calculated by (173, 261, 365) for a range of reasonable explosion energies, presupernova models, and an initial composition appropriate to the LMC. The results from the calculation of Woosley et al. are given in Table 2, which shows

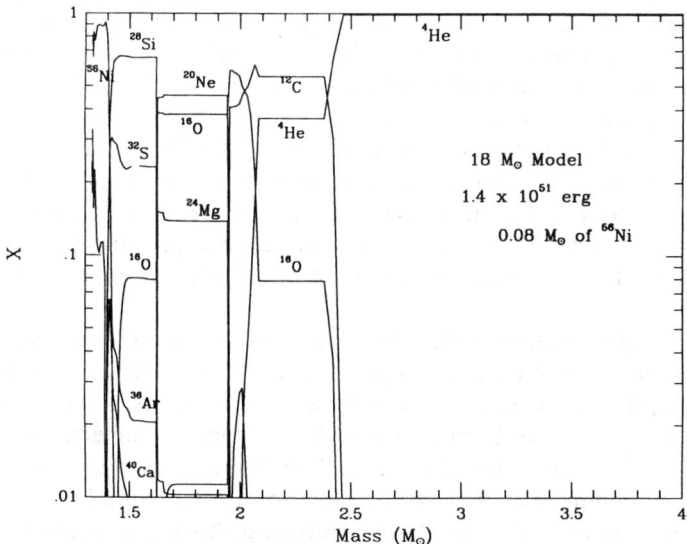

Figure 4 Final composition ejected by the 1.4×10^{51} erg explosion of an 18-$M_\odot$ model [(365); see also Table 2]. The neutron star and outer hydrogen-rich layers are not included in the plot.

Table 2 Explosive nucleosynthesis in 18-$M_\odot$ and 20-$M_\odot$ models

Element	$M_\odot{}^a$	$M_\odot{}^b$	Element	$M_\odot{}^a$	$M_\odot{}^b$
Fe Core	1.33	1.57	Cl^c	2.6(−4)	1.2(−4)
H	8.2(0)	9.0(0)	Ar	9.3(−3)	1.1(−2)
He	6.8(0)	7.0(0)	K^c	2.1(−4)	9.4(−5)
C	2.6(−1)	1.8(−1)	Ca	4.3(−3)	9.6(−3)
N	1.5(−2)	1.5(−2)	Sc^c	4.2(−7)	1.1(−6)
O	2.4(−1)	1.6(0)	Ti	1.4(−4)	3.3(−4)
F^c	1.0(−6)	1.1(−6)	V^c	8.6(−6)	1.2(−5)
Ne	1.5(−1)	1.8(−1)	Cr	8.8(−4)	2.2(−3)
Na	2.7(−3)	1.4(−3)	Mn^c	3.1(−4)	4.8(−4)
Mg	4.4(−2)	1.0(−1)	Fe	7.5(−1)	1.4(−1)
Al^c	2.3(−3)	7.8(−3)	Co^c	1.3(−4)	2.8(−4)
Si	1.4(−1)	1.1(−1)	Ni	4.8(−3)	5.9(−3)
P^c	3.1(−4)	6.0(−4)	Cu^c	3.8(−5)	1.0(−4)
S	5.9(−2)	5.3(−2)	Zn	3.1(−5)	1.0(−4)

^{a} 18-$M_\odot$ model.
^{b} 20-$M_\odot$ model.
^{c} Neglecting probable (v, v') contribution. See text.

a probable range of possible ejected compositions. [Note the especially large range for ^{16}O, which reflects different assumptions concerning nuclear reaction rates, especially for ^{12}C$(\alpha, \gamma)^{16}$O, and convective algorithm in the two models.] The ejecta of the 20-$M_\odot$ model contain too much ^{56}Fe (produced as ^{56}Ni) compared with the value known from the light curve. To correct for this, it is recommended that the masses in Table 2 for the 20-$M_\odot$ model be multiplied by a factor of 0.55 for all elements heavier than scandium. A substantial modification to the abundances of many of the rarer isotopes will also occur during the ejection process owing to inelastic scattering interactions with neutrinos escaping from the collapsed core (358). These interactions have not been included in preparing Table 2.

After approximately 1 min, the shock arrives at the outer edge of the helium core, whose radius is typically 5×10^{10} cm. Without a hydrogen envelope, the helium core would expand with velocities in the range 4000–25,000 km s^{-1}. This is one picture for a Type Ib supernova [see, for example, (143)]. However, the hydrodynamic interaction with the envelope slows the helium core, with the deceleration propagating into the core as a "reverse shock." Meanwhile, the outgoing shock moves ahead though the hydrogen just as though a strong point explosion had occurred in a hydrogen sphere. The radius of the shock as a function of time is given approximately by the Sedov solution:

$$R_s \approx \left(\frac{E_0}{\rho}\right)^{1/5} t^{2/5}. \qquad 2.$$

Using $\rho \sim 3M_{\text{env}}/4\pi R^3$ [with M_{env} the mass of the hydrogen envelope, and R the radius of the star (3.5×10^{12} cm)], we can estimate the time when the shock breaks through the surface of the envelope as (22–25, 296, 353, 362)

$$t_b \sim 2500 \left(\frac{M_{\text{env}}/M_\odot}{E_{51}}\right)^{1/2} \text{ s}, \qquad 3.$$

with E_{51} the explosion energy in units of 10^{51} erg.

5. THE NEUTRINO BURST

Most of the binding energy released when a neutron star forms is believed to be emitted as neutrinos. SN 1987A subjects this theory to an empirical test. The calculated binding energy for a neutron star is, in order of magnitude, about 10% of the gravitational mass of the star. The precise value depends upon the nuclear equation of state that is assumed and upon the gravitational mass of the star. Observations and theory both suggest that the mass of the residue star is between about 1 $M_\odot$ and 2 $M_\odot$ [see, e.g., (14, 38, 190, 293, 317)]. For this range of assumed gravitational masses, the total binding energy of the neutron star lies in the conservative range [see (29, 114)]

$$E_b = (2.5 \pm 1.5) \times 10^{53} \text{ erg}. \qquad 4.$$

Many different parameters are required to describe the neutrino emission. For each of the six types of neutrinos (three flavors of neutrinos and antineutrinos), there is a function that describes the shape of the energy spectrum as well as an absolute value that fixes the total intensity of that type: altogether six unknown functions and six absolute values. The shapes of the spectra are established by the complicated physics of the explosion and are affected by nonequilibrium as well as equilibrium processes, including nuclear, weak interaction, hydrodynamical, and radiative aspects of the production and transport of the neutrinos. Different authors calculate different spectrum shapes, depending upon the numerical scheme that is used and the physics that is included.

Detailed results of stellar collapse calculations of the type described in Section 4 may be summarized approximately in a few physically meaningful quantities [see (39), from which the following discussion is adapted]. For state-of-the-art calculations of neutrino spectra from a stellar collapse model made before SN 1987A, see (227, 229, 371). The total number of events that will be observed in a detector is determined by the fluences of each type of neutrino, the number per unit area received during the entire supernova pulse, as well as by their typical energies. The neutrinos that result from a stellar collapse may be divided crudely into two classes: the initial neutrinos produced dynamically in the collapse and early rebound phases (duration ~ 20 ms), and the much more numerous neutrinos produced in longer thermal cooling (of order seconds). The neutrinos from the dynamical phase contain only a small fraction ε of the total binding energy of the star. Most published models yield $\varepsilon \simeq 0.01$ (85). Nearly all of the remaining fraction of the binding energy $(1-\varepsilon)$ is emitted in neutrinos that thermally cool the star over a period of seconds. The shape of the energy spectrum may be represented approximately by a Boltzmann or a Fermi-Dirac distribution with a fixed temperature T. Either distribution is an adequate approximation to the somewhat uncertain overall spectrum shape computed from detailed numerical models of supernova collapse. The calculated departures from a thermal spectrum involve uncertain aspects of the physics, such as neutrino convection (22–25, 76–78, 227). While the star is cooling, the neutrino temperatures decrease somewhat, typically by order of 20%. To the accuracy that is required for approximate estimates of counting rates in experiments, the temperatures can be assumed to be constant. For fluences F of neutrinos that are produced in the dominant thermal phase, there is considerable agreement among the different groups that numerically calculate stellar collapses. The results can be parameterized as

$$F(v_j, T_j) = \frac{f_j(1-\varepsilon)E_b}{12\pi D^2 kT_j}, \quad j = 1, \ldots, 6, \qquad 5.$$

where D is the distance of the source, and f_j is the fraction of the (noninfall) binding energy that is carried off in the form of the six types of neutrinos (three flavors of neutrinos and antineutrinos). For the dynamical phase, $1-\varepsilon$ in the above equation is replaced by ε. The expected event rates in different detectors can be estimated by assuming that the neutrino temperatures in the dynamical phase are about the same as in the cooling phase.

The simplest approximation to the detailed numerical calculations suggests that each type of neutrino carries away from the thermal cooling phase approximately the same amount of energy (75). The temperatures of v_e and $\bar{v}_e$ are about the same, $T(v_e)$, and the temperatures of all the non-electron flavor neutrinos ($j = 3, \ldots, 6$) are about the same, $T(v_\mu)$. Thus,

$$f_{\text{cool},j} \equiv \tfrac{1}{6}, \quad T_{\text{cool}}(v_e) \equiv T_{\text{cool}}(\bar{v}_e), \quad \text{and}$$

$$T_{\text{cool},j} \equiv T_{\text{cool},\mu}, \quad j = v_\mu, \ldots, \bar{v}_\tau. \qquad 6.$$

Expressing the fluences in terms of characteristic values of the parameters for a Galactic supernova, we have

$$F(v_j, T_{\text{cool},j}) = 1.5 \times 10^{11} \text{ cm}^{-2} \left(\frac{10 \text{ kpc}}{D}\right)^2 \left(\frac{5 \text{ MeV}}{T_{\text{cool},j}}\right) (6f_{\text{cool},j})$$

$$\times \left[\frac{(1-\varepsilon)E_{\text{b}}}{2.5 \times 10^{53} \text{ erg}}\right]. \qquad 7.$$

The much smaller flux of neutrinos emitted during the dynamical phase is

$$F(v_j, T_{\text{dynamic},j}) = 1.5 \times 10^{9} \text{ cm}^{-2} \left(\frac{10 \text{ kpc}}{D}\right)^2 \left(\frac{5 \text{ MeV}}{T_{\text{dynamic},j}}\right)$$

$$\times (6f_{\text{dynamic},j}) \left[\frac{\varepsilon E_{\text{b}}}{3 \times 10^{51} \text{ erg}}\right]. \qquad 8.$$

There is a simple physical argument due to (121, 122, 292) that gives a useful numerical estimate for the temperature of antineutrinos of the electron type, which are most easily observed in the existing water detectors. The mean free path for neutrino absorption at the temperature and density at which the $\bar{v}_e$'s finally escape is defined to be some fraction α of the neutron star radius R, i.e.

$$\frac{1}{n_p \sigma_{\text{abs}}} \equiv \alpha R, \qquad 9.$$

where n_p is the proton number density. The main source of opacity for these $\bar{v}_e$'s is absorption by protons, for which the average cross section is $\sigma = 12\sigma_0 (T/10 \text{ MeV})^2$, where $\sigma_0 = 8.5 \times 10^{-42} \text{ cm}^2$. In the region of the star in which the $\bar{v}_e$'s escape to the outer world, the relation between the matter density and temperature is $\rho \simeq 10^{12} \text{ g cm}^{-3} (T/4.46 \text{ MeV})^3$. By definition, the proton number density is $Y_p \rho N_A$, where N_A is Avogadro's number. Thus, $n_p \sigma \propto T^5$. Inserting typical values for the different variables, one obtains

$$T \cong 4.8 \text{ MeV} \times \left[\left(\frac{0.1}{\alpha}\right)\left(\frac{20 \text{ km}}{R}\right)\left(\frac{0.3}{Y_p}\right)\right]^{1/5}. \qquad 10.$$

The main lesson to be learned from Equation 10 is that a simple physical argument gives approximately the same temperature for the $\bar{v}_e$'s as is obtained by detailed calculations. Equation 10 also shows that the expected temperature is insensitive to the input parameters because the temperature depends on the one-fifth power of the indicated numerical parameters. A factor of two change in one of the parameters only changes the inferred temperature by about 15%. A similar argument using the applicable cross sections for neutral current scattering by nucleons yields the temperature for the v_μ and v_τ neutrinos of ~ 6–7 MeV.

The order of magnitude of the time scale Δt over which the neutrinos are emitted can be understood on simple grounds. The opacity is sufficiently large [see (12, 37)] that the neutrinos diffuse slowly from the neutron star; the observed duration of the antineutrino pulse represents the time for escape from the stellar core [see, e.g., (16, 22–25, 83, 230, 286)]. The time scale is of the order of the total binding energy divided by the luminosity from all of the neutrinos. Most numerical models are not computed through this stage; see, however, (80, 86, 229), from which

$$\Delta t \cong \frac{E_b}{\sum_{i=1}^{6} L(v_i)} \cong 1\text{–}10 \text{ s}. \qquad 11.$$

5.1 SN 1987A

SN 1987A resulted from the only stellar collapse from which neutrinos have been observed [see the historic papers of (55, 179); cf. also (6)]. The observations are in satisfactory agreement with the conventional notions of a "standard stellar collapse" that are summarized in the previous section (80).

5.1.1 DETECTORS The first detection of extra–solar system neutrinos was registered by the Kamiokande II (179) and IMB (55) water Cherenkov detectors, with a possible supporting observation by the Baksan liquid scintillation telescope (6).

Both of these detectors use purified water as the target and as the detector. The detectors are sensitive to recoil electrons produced by neutrino electron scattering in the target,

$$v + e \to v' + e', \qquad 12.$$

and to recoil positrons produced by antineutrino absorption on free protons in the water molecules,

$$\bar{v}_e + p \to n + e^+. \qquad 13.$$

The electron-scattering reaction is dominant for solar neutrinos, while the absorption process, which has a cross section two orders of magnitude larger, is dominant for the antineutrinos from stellar collapses. At higher an oil-based liquid scintillator that is contained in 3156 standard units, each viewed by a single photomultiplier sensitive to Cherenkov light. The recoil electrons or positrons are contained within a single unit. Therefore, events in which only one unit is activated are relatively free from background. A total target mass of about 200 tons could be used to detect SN 1987A.

An Italian-Soviet collaboration has been operating since late 1984 a liquid scintillator detector (known as LSD) in the Mont Blanc Laboratory. The LSD neutrino telescope was designed to observe stellar collapses anywhere in the Galaxy and is located deep underground. The detector has a total active mass of 90 tons of liquid scintillator containing 8.4×10^{30} free protons. The scintillator is contained in 72 separate counters, each of which is observed by 3 photomultipliers. The facility has been described in detail in (1, 36).

5.1.2 DATA Table 3 gives the measured properties of the neutrino events from SN 1987A that were obtained with the water Cherenkov detectors, Kamiokande II and IMB, including the event time, electron recoil energy, and direction of the electron's momentum with respect to the vector connecting the Large Magellanic Cloud and the Earth. The indicated errors are the 1σ errors given in the original papers (55, 179) and include corrections made in later studies (66, 180). The first events detected in Kamiokande II and IMB were simultaneous to within the accuracy of the known zero point of time (about 1 min). The detection efficiencies, energy thresholds, and other experimental characteristics are described in the observational papers. For all these detectors in the Northern Hemisphere, the neutrinos from the LMC traversed the interior of the Earth before entering the detectors from below.

A detailed follow-up study by the Kamiokande II collaboration (180) showed that there were no other statistically significant neutrino bursts in a 10-hr period from 2:27 UT to 12:27 UT on February 23. Moreover, there was no evidence for a much larger signal at energies below the originally determined threshold. When the threshold was decreased to 5.6 MeV, the observed number of candidate events was 138 ± 12 compared with an expected value of 127 obtained from the average background trigger rate. One significant revision resulted from an intensive reanalysis of the data by the Kamiokande II collaboration—namely, the angle that the second event makes with the LMC was increased from 15° to 40°. The suggestion that v_e scattering caused some of the observed events is less compelling as a result of this revision of the Kamiokande II data.

The Baksan neutrino telescope reported a burst of six events that was originally believed [see (6)] to have occurred 25 s later than the first IMB temperatures ($T > 5$ MeV), v_e and $\bar{v}_e$ absorption on ^{16}O is comparable to neutrino-electron scattering (175).

Water is used in these detectors because it is cheap, transparent, and can be made radioactively pure. Also, detecting the recoil electrons or positrons by their Cherenkov radiation is efficient and allows the energies and directions of individual events to be measured.

The Kamiokande II experiment has been described in a number of informative articles. The technical and scientific details are available in (49, 179, 245, 246, 313, 320). The laboratory is located 1000 m underground (2700 m of water equivalent) in the Kamioka mine. The water detector is contained in a cylindrical tank, 15.6 m in diameter by 16 m in height. The entire volume of water weighs 3000 metric tons. Only the inner 680 tons is used for solar neutrino experiments because of the stringent background requirements. A larger volume (2.1 ktons) was used to observe neutrinos from SN 1987A. Large (20" diameter) photomultipliers constructed especially for Kamiokande constitute a key element of the experiment. Kamiokande II began taking data on solar neutrinos in December 1985, after a previous incarnation as a detector of proton decay.

The IMB detector (54, 171, 225) was designed, like the original Kamioka detector, to search for proton decay. The IMB detector is located in the Morton-Thiokol salt mine near Fairport, Ohio, at a depth of 1570 m of water-equivalent. A rectangular tank is filled with purified water, the six sides of which are instrumented with 2048 8-inch photomultiplier tubes arranged on an approximate 1-m grid. Just as for the Kamiokande II experiment, the water serves both as a target for incoming neutrinos and as a Cherenkov radiator for the charged products of such reactions. The timing, pulse height, and geometry of photomultiplier hits are used to reconstruct the vertex, direction, and energy of charged-particle tracks.

The main differences between the Kamiokande II and IMB detectors are in the fiducial volumes and in the thresholds for detecting neutrino events. The IMB detector has the larger fiducial volume, which is 6.8 ktons when studying a pulsed source like SN 1987A. However, the IMB detector is located at relatively modest depth, which makes the study of low-energy events difficult because of the high background rates. The detection threshold with IMB for pulsed neutrino events like those from SN 1987A is about 20 MeV (which also indicates why the IMB detector has not been applied to the study of solar neutrinos).

The Institute for Nuclear Research (INR) of the USSR Academy of Sciences has operated a neutrino telescope since June 1980 in search of stellar collapses in the Galaxy [see (6, 7)]. This telescope is located in the North Caucasus Mountains, under Mount Andyrchi, at a depth of 850 m of water-equivalent. The detector consists of approximately 330 tons of

Table 3 Measured properties of neutrino events observed in water Cherenkov detectors[a]

Event	Event time (s)	Electron energy (MeV)	Electron angle (degrees)
Kamiokande II:			
1	0.0	20.0 ± 2.9	18 ± 18
2	0.107	13.5 ± 3.2	40 ± 27
3	0.303	7.5 ± 2.0	108 ± 32
4	0.324	9.2 ± 2.7	70 ± 30
5	0.507	12.8 ± 2.9	135 ± 23
6	0.686	6.3 ± 1.7	68 ± 77
7	1.541	35.4 ± 8.0	32 ± 16
8	1.728	21.0 ± 4.2	30 ± 18
9	1.915	19.8 ± 3.2	38 ± 22
10	9.219	8.6 ± 2.7	122 ± 30
11	10.433	13.0 ± 2.6	49 ± 26
12	12.439	8.9 ± 1.9	91 ± 39
IMB:			
1	0.0	38 ± 7	80 ± 10
2	0.41	37 ± 7	44 ± 15
3	0.65	28 ± 6	56 ± 20
4	1.14	39 ± 7	65 ± 20
5	1.56	36 ± 9	33 ± 15
6	2.68	36 ± 6	52 ± 10
7	5.01	19 ± 5	42 ± 20
8	5.58	22 ± 5	104 ± 20

[a] The first events were detected on February 23, 1987, at about 7 hr 36 m UT. The angle in the last column is relative to the direction of the LMC. The errors are estimated 1σ uncertainties.

event, with an absolute uncertainty of 2 s in the Baksan time and much less than a second of uncertainty in the IMB time. An additional uncertainty of 54 s in the direction of reconciling the discrepancy was subsequently found in the time measurements for the Baksan detector [see (7)]. This detector had an active mass (200 tons) that is an order of magnitude smaller than the Kamiokande II and IMB detectors; the background rate in the Soviet detector was much higher than in the Kamiokande II or IMB experiments. The Baksan team estimates that approximately 1 event in the burst is caused by background and suggests, without convincing or detailed justification, that the first event they recorded (which had an energy of 17.5 MeV) was caused by background in the detector. The five events that the team regarded as real had arrival times (measured in seconds from the first pulse) and positron energies (in MeV) of, respectively, 0.0, 12±2.4; 0.45, 18±3.6; 1.73, 23.3±4.7; 7.75, 17±3.4; and 9.12, 20.1±4.0.

For the well-known ^{37}Cl detector of R. Davis, Jr., there were no events above the solar neutrino background that were attributable to SN 1987A. The lack of observation translates into a 90% confidence upper limit of about two ^{37}Ar atoms produced by the explosion (123).

The LSD scintillation detector with 90 tons of active mass, located in a tunnel underneath Mont Blanc, reported five events that they suggested were associated with the SN 1987A [see (2, 3)]. The events were detected at 2 hr 53 m (UT), about 4.7 hr before the simultaneous observations by the Kamiokande II and IMB experiments.

The burst of events observed in the Mont Blanc detector is unusual and occurred relatively close in time to the stellar collapse detected by Kamiokande II and IMB and before the observed stellar brightening, but we believe that the Mont Blanc events are not associated with SN 1987A. Our reasons for this belief are as follows: (*a*) No neutrino events (which were clearly different from background) were observed in the much larger Kamiokande II and IMB detectors at the earlier time reported by the scintillator detectors [(55, 179) and especially (180)]. The number of free protons in the Mont Blanc telescope (0.08×10^{32}) is more than an order of magnitude less than in the Kamiokande II detector (1.4×10^{32} protons) and the IMB detector (4.5×10^{32} protons). (*b*) The expected number of events in the Mont Blanc detector for a standard stellar collapse (see Table 3) is only ~1 event, assuming a 100% detection efficiency (40). The satisfactory agreement between the a priori model predictions and the observations made with the Kamiokande II and IMB detectors strengthens this argument. (*c*) The reported events have energies that are close to the threshold energy for the detection, which is between 5 and 7 MeV [depending upon which counters were excited; see (2)]. The measured energies are (in MeV) 7, 8, 11, 7, and 9. Theoretically, one expects a greater spread in energy, since the absorption cross section increases with the square of the neutrino energy for charged-current absorption, and the numerical models predict an average antineutrino energy of more than 10 MeV. (*d*) No plausible astrophysical scenario has been suggested for two distinct neutrino bursts [cf. (126)]. (*e*) It is difficult to obtain a satisfactory light curve for the visual supernova if the earlier time indicated by the scintillation experiments is adopted as the time at which the star collapsed [cf. (22–25, 353) and Figure 1].

5.1.3 PHENOMENOLOGICAL ANALYSES Many different papers have appeared analyzing the observed neutrino events from SN 1987A. A representative sample of this large set of papers is listed here: (10, 11, 42, 43, 76–78, 84, 157, 203, 209, 287, 288, 290, 302).

Many of these papers draw far-reaching conclusions based on the angular distribution and the time dependence of the Kamiokande II data. Since the first two electrons moved in the forward direction, the authors assumed that these events were caused by v_e–e scattering, not absorption of $\bar{v}_e$. There is no strong statistical basis for this assumption, especially since the direction of the second event was revised to be 40° instead of 15° [see (180)]. A number of the papers discuss the implications of assuming that there is a temporal "gap" between the last three events observed by Kamiokande II and all the other events (cf. Table 4). The number of observed events is too small to give any high significance to specific features of the time sequence of the registered pulses [see, e.g., (43, 274)].

For simplicity, consider first a constant-temperature analysis of all eight observed events detected by the IMB collaboration and of the first eight events seen by the Kamiokande II detector. The three subsequent Kamiokande II events, with somewhat lower average energy, may reflect a cooling of the neutrino photosphere with time and are discussed below.

Unfortunately, a "minimum" model (42) has proved adequate to describe the sparse data. In this oversimplified model, the (muon and tau) "scattering" neutrinos and antineutrinos have the same temperature (with no high-energy cutoff). The six types of neutrinos are separated into only two incident model fluxes. The electron antineutrinos $\bar{v}_e$'s have much the largest interaction cross sections in the water Cherenkov detectors. At the temperatures considered here, the v_e's as well as the v_μ's, v_τ's, and their antiparticles can only scatter off the electrons in the water targets. The scattering cross sections for v_e and $\bar{v}_e$ are about a factor of 10^2 less than the absorption cross section for $\bar{v}_e$ on protons; for muon and tau neutrinos and antineutrinos, the scattering cross sections are almost a factor of 10^3 less than the $\bar{v}_e$ absorption cross section. Since scattering is much less probable than $\bar{v}_e$ absorption, everything except $\bar{v}_e$ is lumped into one flux of "average" scatterers. In most detailed model calculations, $F_{v_e} \sim 0.5 \times F_{\text{scatt}}$. The best constant value for the temperature that is deduced with this simplified two-component model is (42)

$$T = 4.1 \text{ MeV}. \qquad 14.$$

The corresponding value of $F_{\bar{v}_e}$ is 0.5×10^{10} cm^{-2}. The single-temperature model represents a satisfactory fit to the observations. The statistical analysis shows that there is a well-defined range of anti-neutrino fluxes,

$$F_{\bar{v}_e} = (0.15\text{–}0.7) \times 10^{10} \text{ cm}^{-2} \qquad 15.$$

(95% confidence level), that is consistent with both the observed event rates. If we require that the count rates in IMB and Kamiokande II be compatible and assume that the detector efficiencies are accurately known, then the neutrino temperature must exceed 3.7 MeV.

For the temperature and flux ranges inferred from the Kamiokande II and IMB results, one would not expect a signal well above background in either the Mont Blanc or the Baksan scintillator detectors. The estimated total number of antineutrino absorption events in 100 tons of liquid scintillator is expected to be Neutrino Events = $0.8 \times (T/4.1 \text{ MeV})^2 \times (F_{\bar{\nu}_e}/0.5 \times 10^{10} \text{ cm}^{-2})$ per 0.1 kton. The most probable signal was less than of order 1 event in the Mont Blanc detector, and less than of order 2 events in the Baksan detector (42).

The flux of ν_{scatter} is poorly determined, since it is constrained experimentally only by the small number of events in the forward peak of the Kamiokande II and IMB detectors. Of course, one expects that the ν_{scatter} flux is harder to determine than the $\bar{\nu}_e$ flux because the scattering cross section is much smaller than the absorption cross section. The strongest experimental limit on F_{scatt} comes from the detection of 1 to 5 (forward-peaked) scattering events in Kamiokande II. This corresponds to

$$F_{\text{scatt}} = (0.1-5) \times 10^{10} \text{ cm}^{-2}. \qquad 16.$$

For the ^{37}Cl detector of R. Davis, Jr., the corresponding number of supernova-induced events is small. The calculated number of ^{37}Cl events varies from 0.02 to 2, depending upon how one manipulates the limits in the above equation. The a priori estimate (40) for the ^{37}Cl detector was 0.5 of an event for the standard model of a stellar collapse. According to these estimates, the background from solar neutrinos should have swamped Davis' supernova signal!

There is one unexpected feature of the LMC supernova neutrino data: the 7.3-s gap between the first nine and the last three events in the Kamiokande II data. The Kamiokande II detector observed nine events in the first 1.9 s, followed by a quiet period of 7.3 s, and then three events were detected within 3.2 s. However, similar gaps are found with appreciable frequency in Monte Carlo simulations of the sparse data with "events" that were produced by random sampling of a distribution that has a smooth time dependence [see (43)]. The simulated data contain remarkable, amusing, and completely spurious patterns that would be interpreted as a temperature that increases with time, huge deleptonization pulses, evidence for periodicity in the arrival times of the neutrino events, and gaps that correspond to a mass of the tau neutrino that just closes the Universe. The IMB detector observed six events in the first 2.7 s, followed by a quiet period of 2.3 s, and then two events were detected within 0.6 s. There are three IMB events in the Kamiokande II "time gap."

Table 3 shows that the average energy of an event declines with time and demonstrates the agreement of the data with a cooling blackbody model. A cooling hot neutron star model fits well all the observed data and provides an estimate of the radius of the hot neutron star [see (79, 209, 302)]. Spergel et al. (302) adopted a simplified model in which the neutrino source is a blackbody with an exponentially decaying temperature: $T = T_0 \exp(-t/4\tau)$. (The energy density at the surface is proportional to T^4; thus τ is the cooling time scale for the hot neutron star.) The joint (Kamiokande II and IMB) likelihood function is maximized at $T_0 = 4.2$ MeV, $\tau = 4.6$ s, and a total fluence $F = 1.3 \times 10^{10}$ $\bar{\nu}_e$ cm^{-2}. Using the multidimensional KS (Kolmogorov-Smirnov) test and Monte Carlo simulations to determine the significance of this solution, Spergel et al. (302) find that the observed KS measure is better than 55% of the cases obtained from the synthetic data. At 95% confidence level, they obtain

$$T_0 = 4.2^{+1.2}_{-0.8} \text{ MeV}; \qquad \tau = 4.5^{+1.7}_{-2.0} \text{ s};$$

$$F_0 T_0^2 = 4.0^{+2.4}_{-2.0} \times 10^{10} \text{ cm}^{-2} \text{ MeV}^2 \text{ s}^{-1}. \qquad 17.$$

There is a strong correlation between the inferred flux and temperature, which is why the fluence is given only in the combination $F_0 T_0^2$. The Kamiokande II data alone yield a peak temperature of $2.9^{+1.3}_{-0.5}$ MeV, and the IMB data alone yield a peak temperature of $4.9^{+4.2}_{-1.9}$ MeV. The rate at which absorption events occur in the Kamiokande II and IMB detectors is proportional to a higher power of the temperature. Since the event rate drops much faster than the temperature, the simplest constant-temperature model gives a reasonable fit to the data.

The total thermal energy emitted by the hot neutron star is

$$\int L_\nu \, dt = N_\nu F_0 (3.15 T_0)(4\pi D^2)\tau = 6.1^{+3.5}_{-3.6} \times 10^{52} N_{\text{all}} D_{50}^2 \text{ erg}, \qquad 18.$$

where N_{all} is the ratio of energy emitted in all neutrino species to the energy emitted in electron antineutrinos. The quantity N_{all} is expected to be of order six.

The results for the temperature, the cooling time scale, and the $\bar{\nu}_e$ flux are consistent with the standard picture of stellar collapse that is based upon detailed numerical models and on analytic arguments. The success of this simplified "standard" model suggests that it will be difficult to use the neutrino events observed from SN 1987A to establish more detailed models. The observations of SN 1987A have triumphantly confirmed the schematic picture of core collapse. The observational test of such a complex phenomenon is a great achievement. However, the data are not sufficient to discriminate between equations of state or to validate specific detailed models. There is no need to invoke new particle physics or complicated astrophysical scenarios. When a supernova is observed in the Galaxy, neutrino detectors should record many hundreds of events, and neutrino astronomy may then reveal surprises about stellar collapses and weak interaction physics.

5.1.4 NEUTRINO PROPERTIES

5.1.4.1 *Mass of v_e* The observations of neutrinos from SN 1987A place an interesting upper limit on the mass of the electron's neutrino, m_{v_e}. The basic idea was discussed first by Zatsepin (375), who pointed out that if neutrinos have a finite mass, then the higher energy neutrinos from a supernova explosion would arrive before the more slowly moving, lower energy neutrinos. The extra time Δt that a finite mass neutrino requires to reach the Earth compared with a zero-mass particle is

$$\Delta t_i = 2.57 \text{ s} \left(\frac{\text{Distance}}{50 \text{ kpc}}\right)\left(\frac{10 \text{ MeV}}{E_i}\right)^2 \left(\frac{m_{v_e}}{10 \text{ eV}}\right)^2, \qquad 19.$$

where E_i is the energy of the ith neutrino. A nonzero mass will cause particles of different energies to arrive at different times, even if they are emitted simultaneously. This dispersion with energy will typically stretch out a burst in time, with the lowest energy particles arriving last, unless there are unusual cancellations or special initial conditions.

Deducing a mass limit requires some assumptions. The dispersion relation for Δt_i determines only, for any assumed value of m_{v_e}, the emission time for each of the observed neutrinos. Unless we make a supporting statistical or physical assumption, we cannot infer a mass limit. The situation is similar to many laboratory experiments in which the final answer must be determined by Monte Carlo simulations.

Arnett & Rosner (32) and Bahcall & Glashow (41) derived an upper limit of about 11 eV for m_{v_e} by assuming that nature was not satanic, and therefore that the observed 2-s half-width of the observed neutrino pulse was not narrowed in transit by more than a factor of two. The argument is simple, but it does not provide a statistical confidence level for the inferred mass limit. If the delay time due to finite neutrino mass becomes comparable to the characteristic width of the pulse, the expected distribution of events should show a correlation of energy with time. This correlation is not present, which suggests in a model-independent way that the observed data do not imply a measurable mass for v_e. Bahcall & Glashow pointed out that more stringent limits could be inferred if the observed substructure of the neutrino burst reflects specific physical processes, and indeed many authors have obtained more stringent limits or specific mass values by identifying a substructure in the neutrino pulse (or by neglecting some possible way that nature could fool us). The interested reader can find different treatments of this problem in, for example, (10, 11, 79, 84, 201, 209, 287, 288, 314).

Spergel & Bahcall (301) provided a comprehensive statistical treatment of the mass limits that is based upon an extensive set of Monte Carlo simulations. They calculated confidence limits by performing Monte Carlo simulations that took account of the complexity of the actual experimental measurements, including the detector efficiencies and the uncertainties in the measured energies. This modeling also took account of the possible temperature variations that would emit low-energy neutrinos first. The model parameters were determined separately for each realization of a simulated observation and for each model by maximizing the joint likelihood function, which is a product over all events observed in the Kamiokande II and the IMB detectors. Spergel & Bahcall conclude that

$$m_{v_e} \leq 16 \text{ eV} \qquad\qquad 20.$$

at the 5% significance level. Burrows (80) independently obtained the same result.

The mass limit from SN 1987A shows that v_e's cannot supply enough missing matter to close the Universe.

5.1.4.2 *Charge of* v_e The electric charge Q of an electron neutrino can be limited by a similar argument (45). If Q is different from zero, then neutrinos of different energies will have different paths x_G in the Galactic magnetic field. Higher energy neutrinos would move along a straighter path and therefore arrive at Earth before lower energy neutrinos, just as is the case for neutrinos that are assumed massive. The upper limit (UL) on the charge Q (in units of the electron charge e) can be written in terms of the upper limit on the neutrino mass: $(Q/e)_{UL} = \sqrt{3(m_\nu c^2)}_{UL}/eBx_G$. Using the isomorphism between neutrino mass and charge, one finds [see (39)] that

$$\left(\frac{Q}{e}\right)_{UL} \leq 3 \times 10^{-17} \left[\left(\frac{1 \ \mu G}{B}\right)\left(\frac{1 \text{ kpc}}{x_G}\right)\right]. \qquad 21.$$

The above limit is expressed in terms of conservative estimates: A typical value for the ordered Galactic magnetic field in the solar vicinity may be of order 2 or 3 μG over an effective path length of order a kiloparsec or more [see (177)].

5.1.4.3 *Lifetime* Many authors noted that the detection of neutrinos from SN 1987A at a distance of 50 kpc implies that the lifetime of a 10-MeV $\bar{v}_e$ exceeds 10^5 yr. More formally, one can write

$$t_0 \gtrsim 5 \times 10^5 \text{ s} \left(\frac{m_v}{1 \text{ eV}}\right), \qquad 22.$$

where t_0 is the $\bar{v}_e$ lifetime at rest. There is a loophole in this argument. If, for example, $\bar{v}_e$ is a linear combination of two mass eigenstates, then the heavier state might decay and the lighter state could be stable and reach the Earth from SN 1987A. Frieman et al. (155) have discussed models in which fast decays of v_e are allowed when all the effects of MSW (Mikeyeev, Smirnov & Wolfenstine) mixing are included.

5.1.4.4 *Limiting velocity* The speed of light plays a dual role in the theory of special relativity. On the one hand, it is the limiting velocity for all objects, regardless of their nature. On the other hand, it is the velocity of a particular particle, the photon. Stodolsky (303) pointed out that the approximate equality of the arrival times of the neutrinos and the photons from SN 1987A allows an accurate check on the hypothesis of special relativity that the limiting velocity for all forms of radiation is the same. With a generous estimate of 10 hr for the uncertainty of the coincidence in arrival time between the first photons and the intense burst of neutrinos, Stodolsky noted that the speed of photons and of neutrinos cannot differ by more than 1 part in 10^8, which is close to the accuracy with which the speed of light is known.

5.1.4.5 *Geodesics* The observation of a neutrino burst within a few hours of the associated optical burst from SN 1987A provides a new test of the weak equivalence principle, demonstrating that neutrinos and photons follow the same trajectories in the gravitational field of the Galaxy (204, 212). A maximum possible violation can be calculated by assuming (implausibly) that the difference in arrival time for neutrinos and photons was caused solely by the passage through the Galactic gravitational field. The neutrinos arrived at February 23.32 UT (55, 179), and the first optical brightening was recorded at February 23.443 UT (234, 235), less than 3 hr later. Longo (212) and Krauss & Tremaine (204) considered a range of gravitational potentials for the Galaxy that bracket the possibilities allowed by measurements of the Galactic rotation curve. Their results show that photons and neutrinos move along the same trajectories to an accuracy of about 0.5% or better.

5.1.4.6 *Number of neutrino flavors* The supernova observations give, in principle, a limit on the number of different flavors of neutrinos that cooled the neutron star, since the total energy carried off by all the flavors cannot exceed the binding energy of a neutron star. One obtains

$$N_{\text{flavor}} \cong \frac{E_{\text{binding}}}{2E_{\text{observed}}(\bar{v}_e)} \lesssim \frac{(2.5 \pm 1.5) \times 10^1}{2(6.1^{+3.5}_{-3.6})D_{50}^2}, \qquad 23.$$

where N_{flavor} is the number of neutrino flavors (counting v and $\bar{v}$ as one flavor). Using the extreme limits in order to be conservative, one finds that there are solutions for all values of N_{flavor} between 1 and 8. This conclusion is more pessimistic than many discussions given in the literature and is less restrictive than the limits obtained from Big Bang nucleosynthesis and from laboratory measurements.

6. THE LIGHT CURVE

6.1 *Observations*

As SN 1987A evolved, observers measured the "bolometric" light curve through frequent brightness measurements obtained at optical and infrared wavelengths with ground-based telescopes, and (for the first few days) in the ultraviolet with *IUE*. The resulting energy emitted has been calculated by groups at Cerro Tololo Inter-American Observatory (172, 306, 307) in Chile and the South African Astronomical Observatory (90, 91, 344) and is given in Figure 5. There are small differences between the two curves due to different methods for integrating over wavelengths and different assumptions concerning the interstellar extinction (238, 319). Most of the energy emitted as photons from SN 1987A during the first year is in the optical region.

6.2 *Theory of the Light Curve and Implications for the Neutrino Burst Timing*

The light curve may be divided naturally into two phases: the early stage, which is dominated by the shock; and the late stage, for which radioactive decay is most important. The transition for SN 1987A occurs at about 4 weeks. For SN 1987A, both phases are well represented by theory, both numerically and analytically.

As the shock from deep inside the star breaks out through the surface of the hydrogen envelope, the electromagnetic display commences. Initially the temperature is so high ($2-5 \times 10^5$ K) that most of the radiation is in the ultraviolet (Figures 1, 6). This UV flash is thought to be the ionizing source for the circumstellar shell observed later in species as highly ionized as N V. The earliest *IUE* observations show a rapid decline in UV flux by a factor of 1000 in the first three days. This is presumably the tail of cooling in the days following shock breakout. Since the first UV observations were after the discovery, the initial UV burst was not directly observed. There is a possibility that the UV flash may be detectable as part of the "light

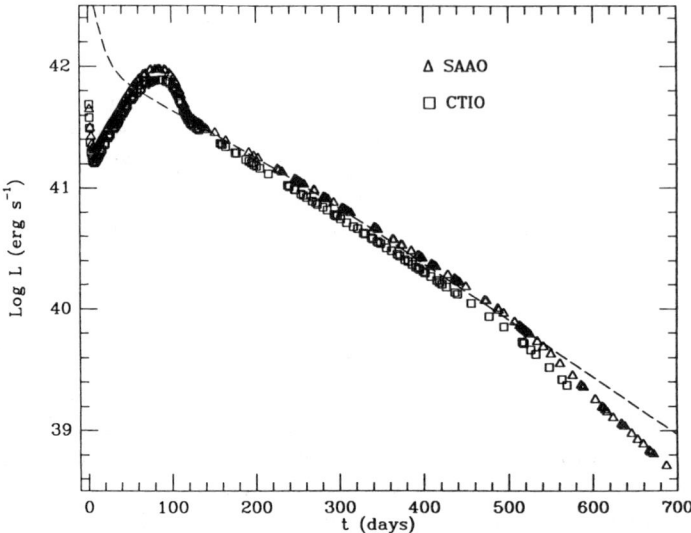

Figure 5 Bolometric light curve of Supernova 1987A. Points are inferred from data obtained at the Cerro Tololo Inter-American Observatory [CTIO (172, 306, 307)] and the South African Astronomical Observatory [SAAO (90, 91, 344)]. The dashed line would be the result of the 100% conversion of the decay energy of 0.075 $M_\odot$ of ^{56}Ni (and later ^{56}Co) to the forms of radiation detected by CTIO and SAAO (ultraviolet, optical, and infrared). The increasing difference in slope between CTIO and SAAO values has been identified with a difference in cutoff, which allows Ca II $\lambda 8542$ emission lines to be included in SAAO but not CTIO (238). In what follows we graph only the SAAO data for clarity. At late times, X rays and gamma rays escape from the supernova and the data fall below the dashed line.

echo" described in Section 7. The star expands by a factor of 10 to a radius of about $2-3 \times 10^{13}$ cm before a visual magnitude as bright as $V \sim 6.5$ can be observed. Thus, in order to accommodate McNaught's determination of $V = 6.4$ and $T = 0.128$ days (11,000 s), it is necessary and sufficient that appreciable matter travel at a velocity near 40,000 km s^{-1}. "Appreciable" here means sufficient to be optically thick at $2-3 \times 10^{13}$ cm, or about 10^{28} g. Subsequent spectra obtained a full day later indicate line absorption above 30,000 km s^{-1}, so this value for the initial photospheric expansion seems plausible.

Figure 1 showed the comparison between the first few days of optical data and the calculated visual magnitudes, normalized to an explosion at the time IMB and Kamiokande II saw the neutrino burst. If the time when energy was supplied to the star is shifted 3 hr earlier to the time of the Mont Blanc neutrino report, the model no longer fits the data.

This may be seen without a detailed model. As the shock breaks out, the kinetic and internal energies are comparable. The diffusion time for

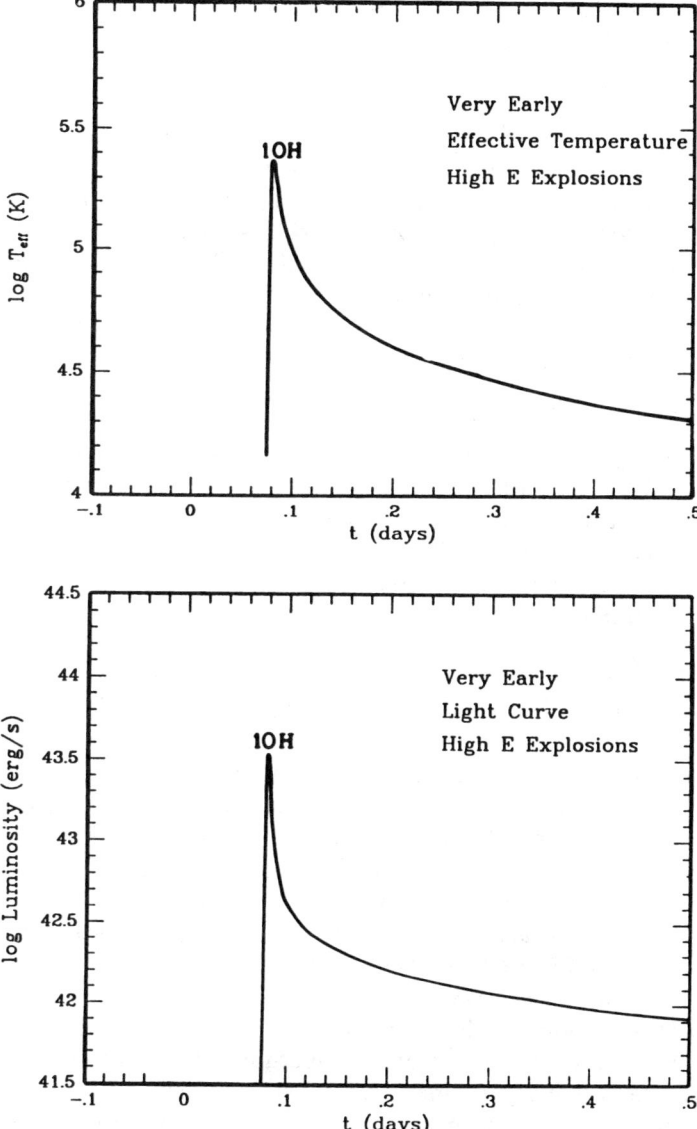

Figure 6 Bolometric luminosity and effective emission temperature during the first half-day of a theoretical model for SN 1987A [model 10H of (353)] that agreed with later observations of light curve, spectrum, and hard radiation. Approximately 10^{47} erg of hard UV radiation was emitted.

photons is $\tau \simeq R^2/\lambda c$, where $\lambda = 1/\rho\kappa$ is the radiation mean free path. The internal energy is $E \simeq \frac{4}{3}\pi R^3 \langle aT^4 \rangle \simeq \frac{1}{2}M\langle v^2 \rangle$.

Using $M = \frac{4}{3}\pi R^3 \rho$, we have the luminosity

$$L \simeq E/\tau \propto \langle v^2 \rangle R/\kappa. \qquad 24.$$

The time dependence of L is very weak at first, making the value of this initial luminosity proportional to the initial radius R (19). The time from core collapse to shock breakout is $t \simeq R/\langle v \rangle$ (see also Section 4.2). The opacity κ is dominated by electron scattering, so that these two equations may be solved for the observed t and L. The value of v may be inferred from Doppler-shifted line profiles as an independent test of the procedure. The results are consistent if we use the Kamiokande II–IMB time as the moment of energy deposition, but not when the Mont Blanc value is used.

6.3 Theory of the Light Curve Peak

The expanding star loses heat by adiabatic expansion until the decreasing density and temperature cause the photon diffusion time scale to decline, at which point radiation escapes. The energy lost to adiabatic expansion reappears as kinetic energy of the outflowing matter. Because even tenuous stars such as red supergiants initially have long diffusion times, less than a few percent of the explosion energy emerges as light. The relatively small initial radius of Sk $-69\ 202$ makes this effect even more pronounced for SN 1987A and accounts for its lower initial luminosity.

The internal specific energy at the time the shock breaks out is roughly constant throughout most of the hydrogen envelope. Given that the subsequent expansion of the envelope is homologous, this internal energy (and the temperature) will decline as t^{-1}, where t is the elapsed time. As the envelope expands, it flows through a recombination front that propagates into the mass of the star like a wave, releasing energy as it goes. During the plateau stage, when T_e is constant, the luminosity is proportional to the area ($L = 4\pi R^2 \sigma T_e^4 = 4\pi R^2 \rho v_p \varepsilon$), i.e. the mass flux through the photosphere times its internal energy is roughly constant. Early on, ε (which scales as t^{-1}) is very high, as is the density (which scales as t^{-3}), so the inward motion of the recombination surface is scarcely noticeable compared with the expansion of the supernova. The photosphere moves outward nearly linearly in time. Later, however, when $\rho\varepsilon$ declines and the recombination moves in to slower material, the photospheric velocity departs from that of the fast-moving outer layers, and the photosphere eventually recedes in space. The observations show that the photospheric temperature declined rapidly over the first 20 days, eventually reaching the temperature of the recombination zone at about 5500 K. The expansion of the photosphere was roughly linear, as inferred from observations of the flux and temperature [for example, (63, 64)]. The separation between $\dot{R}$ (the expansion of the photosphere) and the velocity of the *material* at the photosphere is observed for SN 1987A, as it has been for other SN II (197) in accord with this model.

Radioactive decay stores energy from adiabatic and diffusive loss for a time of order of the half-life $\tau_{1/2}$. If the decay occurs before radiative diffusion is effective, the energy is added to kinetic motion. For SN 1987A, this happens to ^{56}Ni ($\tau_{1/2} = 6.1$ days) but not to ^{56}Co ($\tau_{1/2} = 77.1$ days). With expansion, the diffusion time decreases, which reduces the trapping of radioactive energy. The luminosity increases until the escape matches the energy produced, after which the luminosity follows the decline of the radioactive source. Because the diffusion causes a time lag, a bump in luminosity occurs around the maximum. This is strictly analogous to Type I supernovae, where ^{56}Ni plays the important role because of the smaller mass ejected. The picture is not quantitatively complete without allowing for the increase in transparency due to recombination, which modifies the effective diffusion time (31, 146, 262–264, 290, 353).

The solid line in Figure 7 (31) shows the theoretical luminosity during the stage when the supernova is powered by radioactivity and is compared with the data from the South African Astronomical Observatory. After about 20 days, when the shock phase is over, the agreement is excellent. From 110 to about 300 days the luminosity follows the ^{56}Co decay, after which gamma escape causes the observed curve to sag. The Cerro Tololo Inter-American Observatory data are roughly similar but with a faster sag, perhaps due to a missed emission line of Ca II (238).

SN 1987A was no longer underluminous by the time the bolometric light curve had made the long, slow climb to maximum and then settled onto its radioactive powered tail. This observation tells us that the basic mechanisms of the explosion of Sk $-69\ 202$ (that is, the collapse of the iron core and the resulting outward propagation of the shock wave) were essentially identical to those that occur in other SN II's. The key difference was the fact that Sk $-69\ 202$ was a *blue* instead of a red supergiant and hence was considerably more compact initially. This means that insights from the remaining evolution of SN 1987A as the products of explosive nucleosynthesis and, perhaps, the neutron star are revealed should apply to other SN II's having similar mass progenitors.

6.4 *The Necessity of "Mixing"*

One of the surprises associated with SN 1987A has been observational evidence indicating that isotropic stratification of the composition has not been maintained in velocity space—that is, that some form of macroscopic "mixing" has gone on either during or following the explosion (26, 27, 31,

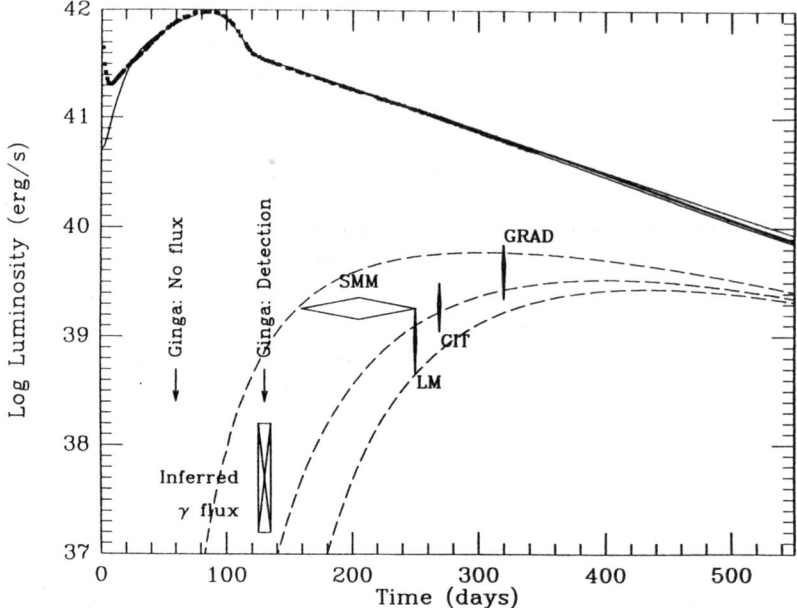

Figure 7 UVOIR (ultraviolet, optical, infrared) and gamma-line light curves for an ejected mass of 15 $M_\odot$, for different choices of mixing radius (31). The luminosity inferred from the SAAO data is shown as solid squares; see comments in Figure 5 legend and in text. The detections for the gamma-ray lines from the *Solar Maximum Mission* (*SMM*), Lockheed-Marshall (LM), Caltech (CIT), and Florida–Goddard Space Flight Center (GRAD) experiments are shown. A value inferred from the *Ginga* X-ray detection is shown as a crossed box. The theoretical luminosities for the gamma-ray lines are plotted as dashed curves. The higher dashed curve corresponds to mixing of ^{56}Co out to a fractional radius $b = 0.4$, while the middle and lower curves correspond to $b = 0.2$ and 0.1, respectively. Because of exponential attenuation, gamma-ray luminosity is sensitive to the amount of overlying matter. Gamma-ray escape causes the UVOIR curve to sag, with a corresponding increase in gamma luminosity.

140, 141, 169, 188, 206–208, 211, 260–264, 271, 272, 294, 295, 353–355, 362). Such an effect is required in order to understand the smoothness of the bolometric light curve, the early appearance of X rays and gamma rays from radioactive decay (Sections 9.3, 9.4), and the widths of spectral lines of heavy elements observed at infrared and gamma-ray wavelengths.

The analytic theory shown in Figure 7 uses a spatially averaged opacity and mixing of ^{56}Co. Apparently the mixing required is significant, though one need not go to the limit of full homogenization, which would be difficult to understand both in terms of hydrodynamics and the lack of strong X-ray and gamma-ray emission prior to August 1987. The effect of mixing on the light curve comes largely from smoothing out the opacity over regions that were initially composed predominantly of hydrogen and of helium. Moreover, a hydrogen recombination wave in the deep core plays a role in forming a "plateau"-like peak (262–264), which suggests that hydrogen is mixed down into the core. The recombination wave is not in perfect phase as it passes over the H-He-C interface.

Some mixing and clumping was expected (e.g. 104), given the precedent of chemical inhomogeneity in the Cas A supernova remnant (195) and even iron at the same velocity as oxygen in that remnant (193, 194). But how early did the mixing occur? Was the core explosion itself unstable when viewed in more than one dimension, or was the symmetry broken later?

Recent three-dimensional simulations using smoothed-particle hydrodynamics and a polytropic representation of the presupernova structure (244) show compositional mixing and density contrasts of approximately a factor of 8. These results were not confirmed in a three-dimensional calculation using a state-of-the-art finite difference method (E. Müller, private communication). Well-resolved one-dimensional calculations of realistic initial models (e.g. 26, 27, 353) show an entropy bump in the vicinity of the steep density gradient caused by the H-He interface, which may become Rayleigh-Taylor unstable. A relevant quantity is the velocity of the heavy elements, particularly the iron-group elements before the reverse shock has moved through and decelerated them. This speed, roughly 3000 km s^{-1} (26, 27, 353, 365), might be expected to provide a rough upper bound on the mixing that occurs during the explosion if the reverse shock is Rayleigh-Taylor unstable. Very recent, well-resolved two-dimensional calculations by W. D. Arnett, B. A. Fryxell & E. Müller show that for a *realistic* presupernova structure, such a Rayleigh-Taylor instability occurs.

Another source of mixing for the heavy elements is the "nickel bubble" (26, 27, 353–355, 362, 365). The decay of ^{56}Ni to ^{56}Co gives 3.0×10^{16} erg g^{-1} with a half-life of 6.1 days. The decay of ^{56}Co to iron gives 6.4×10^{16} erg g^{-1} with a half-life of 77.1 days. If all this energy were deposited in an homologously expanding sphere starting from rest (actually, in the supernova models one starts at a velocity around 1000 km s^{-1}) and radiation transport were neglected, the edge of the expanding sphere at late times would be moving at 5000 km s^{-1}. Taking the total energy available from the decay of 0.07 $M_\odot$ of ^{56}Ni to ^{56}Fe, Whitelock et al. (344) subtracted the fraction that came out in the bolometric light curve after day 40 (when the light no longer came from shock-deposited energy) and concluded that about 45% of the radioactive energy did work expanding the core and thus did not appear in the light curve. This energy argument leads to an upper bound to the velocity of the heavy elements synthesized in the core of about 3000–4000 km s^{-1}.

Of course the overlying material *does* lead to deceleration, but the core of the supernova is somewhat porous owing to clumping as the Rayleigh-Taylor instability develops into a nonlinear stage, with long tongues of material poking out through the overlying tamper. A value of several percent of the heavy elements mixed out to 4000 km s^{-1} seems to give good agreement with the gamma-ray light curve (31, 156, 188, 206–208, 262–264, 271, 272) and seems plausible.

6.5 *Later Time Behavior*

Since July 1987, the bolometric light curve of SN 1987A has followed a regular exponential decline (90, 91, 172, 306, 307, 344) having a characteristic decay rate very close to the laboratory value for the radioactive isotope ^{56}Co [half-life 77.12 days (221)]. This good agreement, together with the natural synthesis of approximately the correct amount of ^{56}Co [0.075 $M_\odot$; (26, 27, 259, 261, 295, 353–355, 362)] as another radioactive progenitor, ^{56}Ni, during the explosion, provides compelling circumstantial evidence that the decay of ^{56}Co is in fact powering the light curve at the present time. This inference receives even stronger support from the actual detection of the characteristic gamma-line spectrum (Section 9) resulting from the decay of ^{56}Co to ^{56}Fe and of infrared-line radiation from ^{56}Co itself (Section 7). It is interesting to note that a similar slope for the decay in Hα flux was observed in the SN II 1980K by Uomoto & Kirshner (329), who also inferred the presence of a mixed radioactive source. The light curve of SN Ib's also follows the slope of the ^{56}Co decay (291), which implies that these supernovae are more massive or expand slower than SN Ia's, which do not follow that slope (28).

If the light curve is powered solely by radioactivity, what should be its long-term history? The answer to this question is useful not only for predicting the future course of the supernova, but also for providing diagnostics of various theoretical models and, in particular, in aiding the recognition of deviations in the light curve that are *not* caused by the decreasing radioactivity and increasing transparency. For example, if a pulsar exists in the supernova, contributing to the luminosity, then the light curve should eventually cease its exponential decay and level off at a higher luminosity than a pure decay model would predict (26, 27, 169, 208, 240, 273, 363).

This is shown in Figure 8, where the two upper curves correspond to L_{pulsar} of 2×10^{39} and 10^{38} erg s^{-1}, respectively. The assumption that this luminosity was in the form of gamma-rays of energy 0.511 MeV (a value common in pulsar models) tended to minimize the deposition of this energy

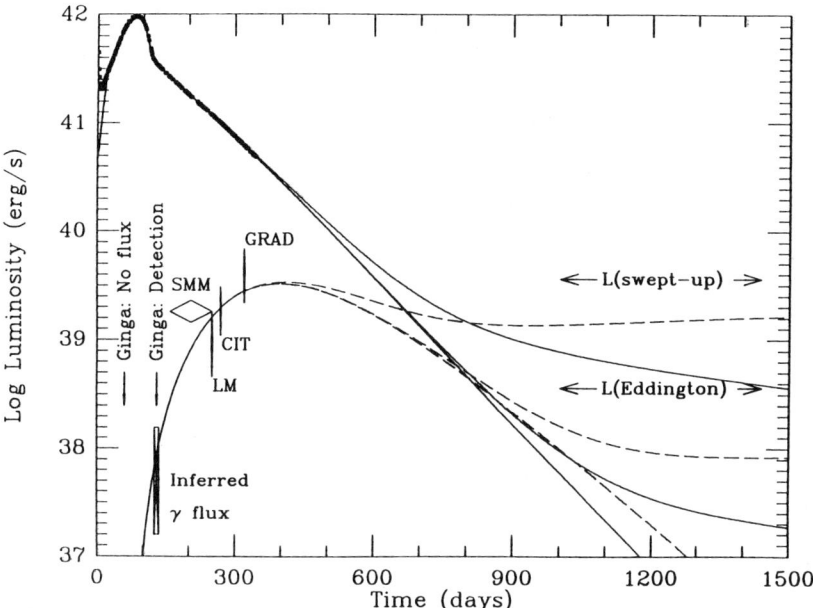

Figure 8 Light curves for an ejected mass of 15 $M_\odot$, for different choices of pulsar luminosity, under the assumption that luminosity is dominated by gamma radiation. The notation is the same as for Figure 7. The top curve represents the addition of a pulsar luminosity of $L_{\text{pulsar}} = 2 \times 10^{39}$ erg s^{-1}. The middle curve corresponds to $L_{\text{pulsar}} = 10^{38}$ erg s^{-1}. The lowest curve corresponds to $L_{\text{pulsar}} = 0$. Possible contributions to luminosity from circumstellar matter [L(swept-up)] and accretion [L(Eddington)] are shown (see text). Notice the dramatic effect of an embedded pulsar on late luminosity, even though the matter is becoming thin to gamma rays. The $L_{\text{pulsar}} = 10^{38}$ erg s^{-1} case mimics the effect of decay of ^{57}Co and ^{44}Ti by flattening the curve at $t \approx 1200$ days. Measuring and understanding this later behavior of the light curve is one of the most interesting challenges in the study of SN 1987A.

in the expanding nebula. One of the most surprising aspects of SN 1987A is the lack of evidence for energy input from a pulsar. There is no evidence for rotational effects in the neutrino burst either. The constraints from the exponential decline in luminosity are interesting, in that they limit the new pulsar to a luminosity comparable to or less than that of the Crab pulsar. The interpretation of the limits on the direct detection of a pulsar is complicated, since only a small, but unknown, fraction of the spin-down energy comes out in the form of pulsed electromagnetic radiation. Dust formation might cause a more rapid decline in the optical than given by the radioactive decay model (138, 202). A drop in electron temperature to the point that only low-energy fine-structure transitions can be excited might result in an "infrared catastrophe" (7, 33, 102, 109), i.e. the sudden shift of the majority of the supernova's luminosity to infrared wavelengths. This would produce a similar, more rapid decay in the optical luminosity, but the accompanying IR increase would be in the form of fine-structure line emission, not the continuum produced by dust. A light "echo" of supernova radiation from interstellar material along our line of sight (98–103, 119, 137, 289) could also cause a deviation from the regular radioactive decay curve, a possibility recognized by Zwicky (376). Clearly, the decay contribution to the light curve must itself be known before these other effects can be separated unambiguously.

The bolometric luminosity (Figure 5) measures the total ultraviolet, optical, and infrared (henceforth, UVOIR) emission of the supernova but not that in X and gamma rays. The bolometric light curve due to radioactive decay can be calculated with considerable accuracy because all the radiation that does not come out in gamma rays or hard X rays must eventually leave the supernova at UVOIR wavelengths. One is thus spared a detailed treatment of radiation transport involving myriads of lines in the strongly non-LTE supernova atmosphere. Although this work is needed for the detailed understanding of the optical spectra, only the absorption and scattering of high-energy photons need be computed. With each scattering, a gamma-ray photon loses approximately one half of its energy. Once a photon's energy has been degraded by scattering below about 10 keV, the enormous photoelectric opacity of the enriched supernova ejecta guarantees that the remaining energy will be deposited and thermalized. This energy is ultimately emitted in the UVOIR bands with energies on the order of kT, a few tenths of an electron volt. Moreover, the fact that the radiative diffusion time on the tail of the light curve is short compared with the expansion time (which is just the elapsed time for homologous expansion) means that the reprocessed energy can be assumed to escape immediately. The deviation of the UVOIR light curve from that given by assuming 100% deposition of the decay energy is then simply one minus the escape fraction of energy in X rays and gamma rays.

Figure 9a shows the continued evolution of a model (363) in which radioactive decay energy is the sole source of power and in which time-dependent deposition has been calculated. Gamma rays from the decay of ^{56}Co dominate the energy generation up until day 1200, though the contribution of ^{57}Co (given by the solar ratio of ^{57}Fe/^{56}Fe and the known production of mass 56 in SN 1987A) contributes at a steady increasing fraction before that. The half-life of ^{57}Co is 271 days. Since this species is unavoidably produced as ^{57}Ni in the same nuclear statistical equilibrium as ^{56}Ni, it is very likely present in a ratio that does not differ appreciably from solar. At still later times, the decay of ^{44}Ti (half-life uncertain but estimated at 54 yr) predominates (208, 363, 370). Ninety-five percent of

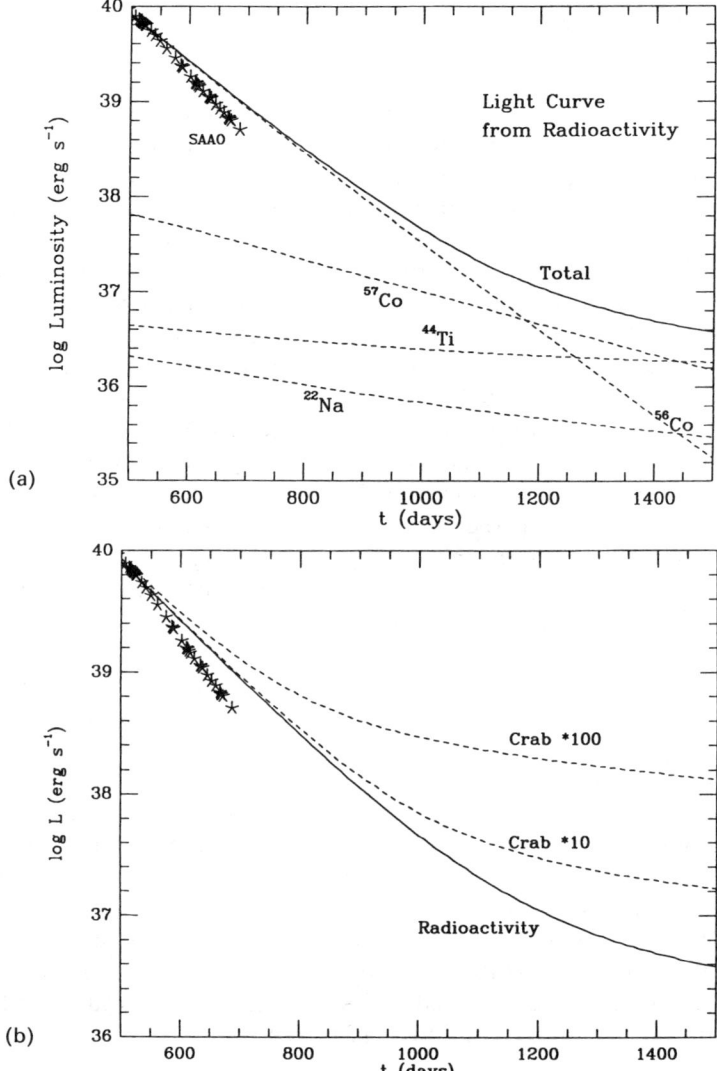

Figure 9 The projected future bolometric light curve (363) based upon a model (*a*) in which only radioactivity contributes or (*b*) in which an additional central energy source equal to 10 or 100 times that of the pulsed emission of the Crab pulsar (assumed to be 5×10^{36} erg s^{-1}) is present. An additional source having the total spin-down power of the Crab (about 100 L_{pulsed}) is already excluded. Data from the South African Astronomical Observatory up to day 700 shows a marked deviation below the calculated curve, which already includes time-dependent gamma-ray transparency. Perhaps clumping is indicated. As this paper goes to press, the late-time bolometric emission is becoming difficult to determine accurately owing to substantial emission in the *M*-band and longer wavelengths.

the time of the decay of ^{44}Sc to ^{44}Ca is accompanied not only by gamma-ray emission, but also by the creation of a positron having average kinetic energy of approximately 700 keV. Even a small, disordered magnetic field could trap the positrons in the ejecta long enough for them to deposit their kinetic energy (18, 33); thus the decay of ^{44}Ti should continue to power the supernova long after it becomes transparent to gamma rays. Fortunately (or unfortunately, depending upon one's viewpoint) ^{44}Ti nucleosynthesis is very uncertain and model dependent. It comes from the deepest zones to be ejected by the supernova and is therefore most sensitive to matter falling back during and after the explosion. Thus we would learn a lot from its detection at any level. Both ^{57}Co and ^{44}Ti might also eventually be recognized by their gamma-ray line spectrum (170, 208, 363). Interestingly, if the suggested amount (173, 208) of ^{44}Ti has been synthesized, the luminosity of SN 1987A will not decrease below about 10^{36} erg s^{-1} for many decades to come. Any additional power source below that value may be difficult to disentangle. Without identification of their gamma-ray line spectra, the effects of these radioactivities on the light curve resemble those of an embedded pulsar (see Figure 8).

Figure 9b depicts in more detail the possibility that a pulsar similar to the Crab has been created in SN 1987A and contributes at an appreciable level to the bolometric light curve. Just what fraction of the Crab's appreciable spin-down power (about 5×10^{38} erg s^{-1}) is converted into bolometric radiation (as opposed to driving mass motion or producing hard gamma rays) is very uncertain. Here only the measured *pulsed* component (about 1% of the spin-down energy) has been included, multiplied by factors of 10 and 100. Energy-dependent transmission of the pulsar radiation has been included in the same model as Figure 9a. A third alternative (169, 170, 239, 273, 363) is that accretion onto the neutron star has suffocated the radio pulsar mechanism and has turned the central object into an X-ray pulsar like Her X-1. If so, a similar contribution to that depicted in Figure 9b, but scaled to the accretion luminosity of the central object, would result. Ultimately, pulsed emission (best sought around 30 keV) might distinguish these possibilities (170, 363).

7. SPECTRAL EVALUATION OF SN 1987A

In the first weeks past maximum light, the optical spectrum of SN 1987A was dominated by a continuum with broad P-Cygni absorption and emission lines of hydrogen and other elements of the stellar photosphere. This early stage of the supernova spectrum is of special interest because the expanding photosphere's angular size can be inferred from the observed flux and temperature, and the linear expansion rate can be inferred from the line shapes. Combining these estimates provides an extragalactic distance estimate that is independent of the many steps on the extragalactic ladder (63, 64, 196). This has been applied by Branch (63, 64) using very simple (blackbody) assumptions, and he finds a distance consistent with conventional estimates for the LMC near 55 kpc. More refined models for the photosphere are in progress; for example, Wagoner et al. (331) use non-LTE calculations to derive a somewhat smaller distance to the LMC. Very accurate spectrophotometry available from a variety of sources makes continued theoretical work on this problem valuable. The results may prove widely applicable to SN II's.

The ultraviolet behavior of SN 1987A has been summarized by Kirshner (193, 194) and Panagia (266). The rapid initial decline was followed by a very slow variation in the spectrum, with low-ionization species becoming increasingly important. Because of the very large UV opacity, very little of the energy leaks out shortward of 3000 Å, and the ultraviolet behavior does not reflect the overall energetics. In fact, although the ultraviolet flux increased from a minimum reached just a few days after the outburst to a new maximum at the time of the bolometric maximum near 20 May 1987, the ultraviolet did not share in the exponential decline observed at other wavelengths. As a result, the relative contribution of the UV flux has increased, but it has remained a small fraction of the bolometric luminosity.

In contrast, the optical flux has not only declined at very nearly the bolometric rate [see (266)], but also the nature of the spectrum has changed (Figure 2) from a continuum with superposed lines to an emission-line spectrum as the supernova changes into a "supernebula." The principal optical lines are Hα, [Ca II] λ7290, and [O I] λ6300, but many others are present. The late-time spectra resemble those seen for other SN II's as reported by (149, 198, 329). The simplicity of a nebular analysis of optically thin lines is extremely attractive, since one goal of this work is to determine the chemical abundances of the stellar interior in order to compare these values with the results of stellar evolution calculations. Models for the late-time spectra (150, 151, 153) point the way to analysis of the spectra in terms of abundance, but no firm conclusions on the composition of the Sk -69 202 interior are warranted yet, since the opacity of the material remains significant.

One interesting observational development associated with SN 1987A has been deep views into the core of the supernova using infrared observations. X and gamma rays arise from core material, but their energy is mostly thermalized before escape. Optical and infrared photons can scatter nonresonantly only off free electrons; in a low-ionization medium such as the supernova, the scattering optical depth is much lower. While optical photons suffer little from this *continuous* opacity, the large number of lines present in the heavy-element-rich supernova coupled with the large velocity gradient produce a *line-scattering* opacity that is still large. Most optical photons diffuse through several transitions of increasing rest wavelength on their way out from the core.

In the infrared (from 1 or 2 μm to ~ 20 μm), there are fewer overlapping lines, so the stellar core is transparent to the infrared-line emission resulting from collisional excitation of fine-structure transitions (33, 34). Beyond about 20 μm, the free-free opacity becomes large, and one is again prevented from seeing deep into the interior. Flights of the *Kuiper Airborne Observatory* during fall 1987 and spring 1988 have produced particularly revealing results showing strong emission lines of Ni I, Ni II, Ar II, Co II, Fe I and II, CO, SiO, and a host of long-wavelength hydrogen lines (275, 276, 350). Some of these lines are also visible from the ground (4). The Co II feature is particularly interesting because, using recent atomic data (265), the observed strength of that line agrees with the abundance of radioactive ^{56}Co expected to be present in the supernova [cf. (4, 5), based upon the initial creation of 0.075 $M_\odot$ of ^{56}Ni required to explain the light curve and gamma-line observations]. Apparently, the line comes from ions of the *radioactive* cobalt isotope.

Figure 10 shows a study at high resolution made in April 1988 of two lines of Ni II and Ar II (350). The FWHM of the roughly symmetric line cores (~ 3000 km s^{-1}) shows that heavy elements are moving at least twice as fast as calculated in the one-dimensional models that ignore the Rayleigh-Taylor instability ($\lesssim 1000$ km s^{-1}), and it also indicates that significant amounts of nickel and argon must be moving at velocities in excess of 2000 km s^{-1}. Since the slowest moving hydrogen has also been measured at 2100 km s^{-1}, this is a strong indication that macroscopic mixing has occurred. Figure 10 shows red wings on both emission lines and a redshift of the line cores' centroid. Witteborn et al. (350) have interpreted these red wings as due to Compton scattering in the expanding supernova ejecta above the emitting heavy elements (153).

8. OF MYSTERIES AND ECHOES

SN 1987A provided the first opportunity to apply many techniques of limited sensitivity to supernovae. Among the most intriguing was the attempt to resolve the supernova image by speckle interferometry. Observations at the Cerro Tololo Inter-American Observatory 4-m telescope by the Center for Astrophysics (CfA) group (249) in March and April 1987 indicated the presence of a second bright source 60 milliarcseconds from the supernova having a brightness within a factor of 12 of the supernova itself. Observations at the Anglo-Australian Observatory (AAT) in April 1987 (236, 237) using speckle techniques also produced evidence for the

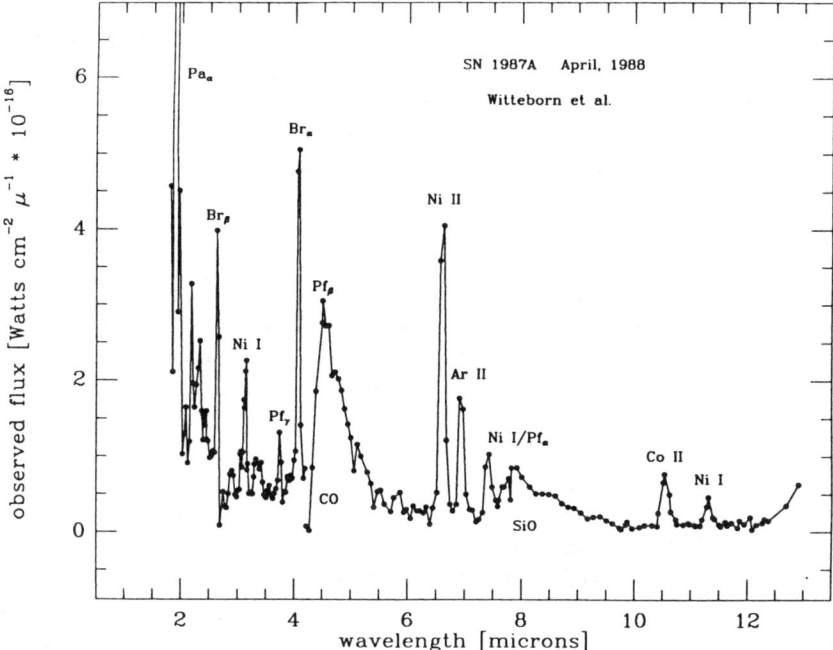

Figure 10 The infrared spectrum of SN 1987A, taken in mid-April 1988 (410 days) from the *Kuiper Airborne Observatory*, showing the presence of many heavy elements created in the explosion (350).

second source. The great luminosity and large separation of this "mystery spot" have proved difficult to understand [for a summary, see (270)], and the novelty of speckle methods makes it difficult to assess the observations. The CfA group has continued its work and is now measuring the angular diameter of the expanding debris (250). The rate at which the supernova angular diameter has changed size at Hα is consistent with mean expansion of 2850 km s^{-1} over the first 410 days. Similar results are reported from the AAT.

While the radio and ultraviolet results briefly summarized in Section 3.5 indicate the presence of circumstellar matter, SN 1987A has also been a novel probe of interstellar matter. During the early days of strong UV flux, *IUE* was employed at high dispersion to measure interstellar lines both in the LMC and in the Galaxy (56, 57, 136). Similarly, ground-based observations provided a unique opportunity for high-resolution and high-sensitivity observations (330). Perhaps the most striking feature was the detection of lines from Fe X by Pettini et al. (268). The overall pattern of absorption-line strengths includes a very large range of ionization, with the high-ionization lines especially strong at the LMC velocities. One especially interesting absorption component is observed at $+286$ km s^{-1}, which is identical (within the errors) with the emission-line velocity observed a year later from the *circumstellar* matter. It is tempting to identify this component with the circumstellar matter, but the conclusive evidence—rapid changes due to ionization from the UV flash [see (215)]—was not observed because of the supernova's rapid decline in the UV continuum.

Another phenomenon associated with the interstellar matter is the light echo (Figure 11) detected in March 1988 (118, 165, 308). Two semicircular arcs of scattered supernova light were imaged at distances about 33 and 55 arcsec from the supernova. They are light echoes of the supernova near its maximum of May 1987 that have accumulated 10 light-months of additional travel time in their path from the supernova to the scattering dust clouds and then to us. The geometry was worked out in detail by Couderc (117) and for SN 1987A by Chevalier (98) and Emmering & Chevalier (142). The observed echoes are consistent with the dust inferred from the reddening to Sk $-69\ 202$ at distances 115 and 305 pc in front of SN 1987A. One amusing aspect of the dust echoes is that they have large proper motions, corresponding to a transverse velocity 16 times the speed of light. This is due to the changing locus of constant time delay, which corresponds to an ellipse with the supernova at one focus and the Earth at the other. Another is that the dust allows observations of the supernova's *past*. It may allow UV observations of the initial UV outburst, even from light emitted before the first measurement of SN 1987A in February 1987.

9. HIGH-ENERGY EMISSIONS—X RAYS AND GAMMA RAYS

Past observations of X rays and gamma rays in supernovae have been dominated by the emission of either a central radio pulsar or the blast wave interacting with circumstellar material. While these emissions may also exist at some level in SN 1987A (Section 9.6), observations thus far show that the dominant source of both X rays and gamma rays is something never directly observed in a supernova before—*radioactivity*.

9.1 *A Brief History of Radioactivity in Supernovae*

The idea that radioactivity might provide an important source of energy in stellar environments is an old one. Originally suggested (281, 349) as a possible mechanism for powering the Sun, before nuclear fusion was understood, the idea later took hold as a source of energy for powering

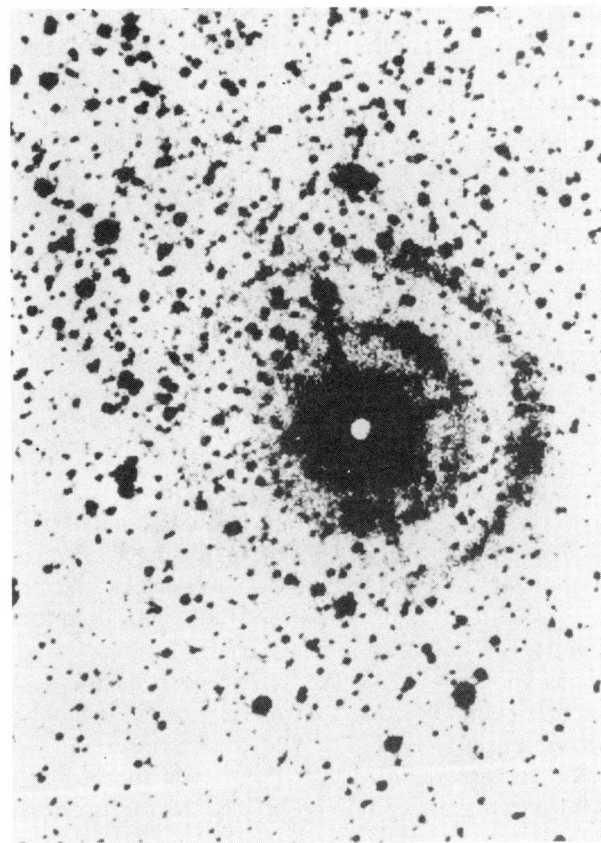

Figure 11 Optical light echoes from SN 1987A, showing two concentric rings at radii 33 and 55 arcsec (as observed in March 1988). The light has the spectrum observed at optical maximum in May 1987 and is likely to be scattered from two sheets of interstellar dust at ~ 122 and 316 pc from the supernova (figure courtesy of the European Southern Observatory).

the light curves of supernovae. Various nuclei [^{7}Be (61), ^{254}Cf (74), and ^{56}Ni (111, 267)] have been suggested as the isotope of interest. In 1956 Suess & Urey (305) noted the double magic shell structure ($Z = N = 28$) of the isotope ^{56}Ni that made it seem a likely progenitor for the abundant nucleus ^{56}Fe. Because of the short half-life of ^{56}Ni, they suggested that its synthesis would occur under conditions of high temperature and brief duration. This inference presaged, though probably unintentionally, the development more than a decade later of the notion of *explosive* nucleosynthesis (nuclear activity occurring in a transitory high-temperature state behind the shock wave). Influenced by Suess & Urey's remarks, Titus Pankey suggested in 1962, in an unpublished appendix to his PhD thesis on meteorite abundances, that the decay of ^{56}Ni and its daughter ^{56}Co were responsible for powering the light output of Type I supernovae.

Colgate & McKee (111), independently and without knowledge of Pankey's previous work, reinvented his argument that ^{56}Ni and ^{56}Co decay powers Type I supernova light curves. Influenced by the success of the light curve model, but even more so by the rapidly developing theory of explosive stellar nucleosynthesis (59, 323), Clayton et al. (108) predicted that those gamma rays that escape without interacting with the supernova ejecta might be detectable. Thus the identity of the specific gamma rays observed from SN 1987A were correctly predicted almost 20 years ago.

However, Clayton et al. (108) were initially concerned only with the gamma-ray signal expected from a *Type I* supernova (SN I), roughly 1000 times brighter than what was observed in SN 1987A. Subsequent work (67, 357) noted that the signal from a Type II supernova was likely to be much fainter owing to the large column depth associated with the larger mass and slower moving ejecta. In 1981 a Monte Carlo calculation of gamma-ray transport in a model 25-$M_\odot$ supernova (357) predicted line fluxes and widths within a factor of two of those measured for SN 1987A.

The observation of "tails" on the light curves of several SN II's showed a slope that was about right (46, 328, 337, 338) to be due to the decay of ^{56}Co. This led to expectations that the synthesis of ^{56}Ni might be a common property of SN II's as well as SN I's. Weaver & Woosley (337, 338) showed that a shock wave with sufficient energy to eject the helium mantle and envelope of a massive star would synthesize ^{56}Ni for any reasonable pre-supernova model star.

9.2 Theory of Gamma-Ray Emergence

For about three years after a SN II explosion, the radioactive contribution to the hard electromagnetic spectrum is dominated by the gamma rays from the decay of ^{56}Co. During each decay of ^{56}Co, a number of gamma-ray lines are produced, having (on average) a total energy of 3.59 MeV (including 1.02 MeV for positron annihilation). The two most prominent lines at 847 keV and 1238 keV are produced in 100% and 68% of decays, respectively. In addition, 19% of the time ^{56}Co decay will release a positron of average kinetic energy 660 keV, which brings the total average decay energy to 3.72 MeV. Positron annihilation also produces a characteristic line at 511 keV. Early successful attempts at gamma-ray detection centered on the 847 and 1238 keV lines.

The unscattered flux of gamma rays from a mass M_{56} of ^{56}Co (in solar masses) located in the LMC (50 kpc) is (353)

$$F = 0.81 \left(\frac{M_{56}}{0.1 M_\odot}\right) \exp[-t/111.3 \text{ days} - \kappa_\gamma \phi_0 (t_0/t)^2] \text{ cm}^{-2} \text{ s}^{-1}, \qquad 25.$$

where t is the elapsed time since the explosion, t_0 is some fiducial time at which the column depth to the edge of the ^{56}Co layer, ϕ_0, is to be determined, and κ_γ is the opacity to 1-MeV gamma rays. Here F is the flux of some line, such as 847 keV, through which all decays proceed and homologous expansion has been assumed. An appropriate value of κ_γ is 0.06 cm^2 g^{-1}, and a reasonable time to evaluate the column depth is $t_0 = 10^6$ s. This flux will have a maximum at time

$$t_{max} = (2\tau_{Co}\kappa_\gamma \phi_0 t_0^2)^{1/3} = 263(\phi_0/10^4)^{1/3} \text{ days}. \qquad 26.$$

The maximum flux is easily obtained by evaluating Equation 25 at time t_{max}, i.e.

$$F_{max} = 0.602 \left(\frac{M_{56}}{0.10 M_\odot}\right) \exp\left[-3\left(\frac{\kappa_\gamma t_0^2 \phi_0}{4\tau_{Co}^2}\right)^{1/3}\right]$$

$$= 0.602 \left(\frac{M_{56}}{0.10 M_\odot}\right) \exp(-0.161 \phi_0^{1/3}) \text{ cm}^{-2} \text{ s}^{-1}. \qquad 27.$$

This result is extremely sensitive to the column depth at t_0, i.e. to the expansion rate. Models appropriate to SN 1987A have ϕ_0 in the range 5–7 × 10^4 g cm^{-2} at 10^6 s [e.g. models 5L and 10H of (353)] *provided that the ^{56}Co remains concentrated at the center of the supernova*. Since we know from the light curve that the supernova produced 0.075 $M_\odot$ of ^{56}Ni, these same models would have peak fluxes in the range (0.6–1.1) × 10^{-3} photon cm^{-2} s^{-1} roughly 450 to 500 days after the explosion. The gamma-ray optical depth at maximum emission is $1.4(\phi_0/10^4 \text{ g cm}^{-2})^{1/3}$, which is in the range 2–3 for any reasonable model. Thus there is a pronounced continuum even when the gamma rays are at their peak luminosity.

The widths and energy profiles of the gamma-ray lines reflect the velocity distribution of the ^{56}Co as modified by optical depth effects. At late times, one sees a FWHM roughly twice the average velocity of the ^{56}Co, or $\sim 6v_8$ keV for the 847-keV line, where v_8 is the velocity in thousands of kilometers per second. Unmixed models suggest that $v_8 \sim 1$–2 on average. Note that $v_8 \simeq 2$ is the speed of the slowest moving hydrogen seen in the infrared (269). At early times, the cobalt on the far side of the supernova will be optically thick and one will see narrower, blueshifted lines. The gamma-ray telescope technology of 1987 was just barely adequate to detect the supernova, and some of the other interesting diagnostics that come from line shapes at early times may not have been obtained.

A number of groups have calculated the expected time histories of the 847- and 1238-keV lines of ^{56}Co decay (93, 156, 162, 167–170, 188, 211, 272, 294, 298, 363, 372). Generally the results agree with analytic estimates of peak fluxes and timing. However, the detailed models also give the time evolution of the hard X-ray flux resulting from Comptonization of the gamma rays and accommodate arbitrary distributions of ^{56}Co source and various density structures and compositions for the overlying material. The simple model in which the ^{56}Co emitters are all concentrated in the center of the supernova proved inadequate to account for the early appearance of the gamma-ray lines, and thus some mixing must have occurred (see Figure 7).

As was discussed in Section 6.5, the signal from ^{56}Co will decay in the future, and the species ^{57}Co, ^{44}Ti, and ^{22}Na will present challenging targets of great interest (93, 94, 107, 170, 272, 360, 363, 364). Because of its longer lifetime (e-folding lifetime 391.91 days), ^{57}Co emissions at 122.06 keV (81%) and 136.47 keV (11%) persist as the supernova becomes optically thin. While the mass of ^{56}Co in SN 1987A is known from the bolometric luminosity of the tail of the light curve, the mass of ^{57}Co must be inferred from theory (which means that its detection will provide unique and important information). If we assume that ^{57}Co exists in that proportion to mass 56 required to give a final solar abundance ratio to ^{56}Fe/^{57}Fe (0.0243), there should be 1.7×10^{-3} $M_\odot$ of ^{57}Co produced in SN 1987A (173, 363, 365). Because this species is produced by the same process as ^{56}Co, the line profile should in theory be like that of the 847-keV line. Early on, the 122-keV line would be obscured by Comptonized photons from ^{56}Co decay; thus, prospects for detection would be best at late times, so long as a sufficiently sensitive detector is available. If the *Gamma Ray Observatory* is launched close to its scheduled time, it might recover the supernova in lines of ^{57}Co. The Soviet space mission *GRANAT*, to be launched late in 1989, will very probably see evidence for ^{57}Co in the hard X-ray spectrum.

Supernova 1987A is also likely to produce ^{44}Ti, with a mass estimated to be around 10^{-4} $M_\odot$ (173, 263, 365). The mean lifetime of ^{44}Ti is 78 yr, and its decay to ^{44}Sc produces two gamma rays of energy 78.4 and 67.85 keV. ^{44}Sc decays immediately to ^{44}Ca, almost always producing a gamma ray of 1.157 MeV and a positron. The flux of any of these lines (at sufficiently late times that the supernova is optically thin) should be given by $\sim 2 \times 10^{-6} e^{-t/78\,\mathrm{yr}}$ cm^{-2} s^{-1}; detecting these lines will be a goal perhaps for the graduate students of our graduate students.

9.3 *X Rays*

As long as the gamma rays interact with the supernova ejecta, downscattering to hard X rays will occur. At very late times this scattering leads to a Compton tail on the red side of the lines. At early times, when the supernova is thick to gamma radiation, the scattered X rays are the only escapees. The column depth for models in good agreement with the light curve is about 7×10^4 g cm^{-2} at 10^6 s, which leads to an optical depth for 1-MeV gamma rays of about $5t_{yr}^{-2}$. Thus the supernova becomes optically thin to 1-MeV gamma rays after about 2 yr. Since the number of scatterings is τ^2 and the energy change per scattering is about $\Delta\varepsilon = -\varepsilon^2/m_ec^2$, the photon energy after n scatterings is about m_ec^2/n (18, 231, 233, 298). Once the energy of the photon has declined substantially below the electron rest mass, one should use an (X-ray) Compton depth about three times larger; thus a typical gamma-ray photon will be degraded to an X-ray photon of about $2t_{yr}^4$ keV. Actually, all photons scattered below the photoelectric thresholds for heavy elements in the core will be totally absorbed and thermalized. This is the mechanism for the radioactive contribution to the bolometric light curve (Section 6). The photoelectric optical depth for X rays is about $10^3 t_{yr}^{-2}(\varepsilon/20 \text{ keV})^{-3}$ (231). Thus, for example, the supernova becomes optically thin at 10 keV after about 12 yr.

The optical depth to X rays in the hydrogen envelope is much less owing to the much smaller abundance of heavy elements. McCray & Li (232) estimate a photoelectric optical depth of $2t_{yr}^{-2}(\varepsilon/20 \text{ keV})^{-3}$ for the envelope. At early times the hard X-ray emission will be dominated by those relatively few gamma rays that scatter in the envelope because the X rays produced there will have a greater chance of escaping without degradation by the photoelectric effect. Based upon the photoelectric optical depth of the hydrogen envelope alone, one can estimate that the lowest energy X rays escaping the supernova will have an energy of roughly (25 keV) $t_{yr}^{-2/3}$. At early times this energy will characterize the most abundant X-ray photons to escape the supernova. Later, however, the average energy will be greater, since the gamma rays experience fewer scatterings in the envelope. Setting $2t_{yr}^4$ equal to $(25 \text{ keV}) t_{yr}^{-2/3}$ and solving gives a time when the X rays peak (about 1.7 yr). At later times, the gamma rays will experience fewer scatterings and the spectrum becomes harder.

While these qualitative arguments are useful in understanding the X-ray emission process and in estimating roughly the spectrum and light curve of the hard radiation, accurate calculations require a detailed supernova model and radiation transport including the full angle and energy dependence of the cross sections. Such Monte Carlo simulations have been carried out by many groups (87, 93, 156, 162, 167–170, 188, 206–208, 271, 272, 294, 298, 311, 312, 360, 372). It is interesting to compare calculations prior to fall 1988, when the X rays were first detected by *Ginga* (134) and by *Mir* (311, 312), to what later developed. The smooth rise of the optical light curve (Section 6.2) indicated that mixing might have occurred in the explosion, but anticipated X-ray fluxes calculated during late summer 1987, based upon 0.075 $M_\odot$ of ^{56}Co, were estimated using chemically stratified (unmixed) models. These predicted that the X rays would rise to a detectable level about 250 days after the explosion. Instead, X rays were discovered about 170 days after the explosion (Figure 7). Their early appearance showed that a larger fraction of the gamma rays were depositing in the supernova envelope than predicted by unmixed models. A remedy immediately tried by many was to stir a small fraction of the ^{56}Co, which had been situated at very low velocity (about 1000 km s^{-1}) in the center of the supernova, out to greater distances (i.e. greater velocities), where the column depth of the core (where most gamma rays deposit) was less. The mechanisms have been discussed in Section 6.4.

In one successful model [model 10HM (271, 272)] ^{56}Co was mixed out to velocities as great as 3000 km s^{-1}, but the mass fraction at the edge of the helium core (1700 km s^{-1}), for example, was only 2% of the central value. In models of this sort, an increasing fraction of ^{56}Co is revealed to compensate for the exponential decay in its overall abundance. These parameters are similar to the preferred model (26, 27, 156, 207, 208). The hard X-ray flux, once it peaks at about 200 days, remains relatively constant for about 200 days thereafter (Figure 5 of 272).

9.4 Gamma-Ray Observations

Because gamma rays are totally absorbed by the Earth's atmosphere (though see Section 9.6), the high-energy radiation from SN 1987A could only be studied using satellites in space [in particular, the *Solar Maximum Mission (SMM)*] and detectors carried aloft by high-altitude (~ 20–25 mi) balloon flights in the Southern Hemisphere. With one exception, all the balloons flew from Alice Springs, Australia. Owing to the long exposure time necessary to make a gamma-ray observation using balloon-borne detectors (typically several hours) and the difficulty in tracking and retrieving instruments that have been blown a great distance by the wind, measurements were best carried out during a few weeks in the spring and fall when the wind subsided and "turned around." One successful flight was also launched in Antarctica (277), and one long-duration flight was carried out from Alice Springs across Africa to an eventual landing near Rio de Janeiro, Brazil.

Since it is in space, *SMM* can carry out repeated long-duration measurements of the supernova. It carried the first gamma-ray instrument to detect the supernova (226). A typical data accumulation period was at least several weeks. Unfortunately, in order to keep the Sun in view at all times and avoid the possibility of losing orientation in space, *SMM* could not turn and point at the supernova. Thus the gamma rays that were detected came through the side of the instrument, with a consequent reduction in sensitivity to just under 10^{-3} photon $cm^{-2} s^{-1}$, a value that (to within a factor of two) also characterized most of the balloon flights. The gamma-ray detectors on *SMM* are sodium iodide and thus could offer only limited broadband spectroscopic resolution. However, *SMM* can continue to integrate gamma-ray emission over a very long period and may ultimately prove the most sensitive instrument to study SN 1987A.

The balloon flights carried both high-resolution germanium and sodium iodide detectors. While observations and data analysis are continuing as this paper goes to press, the experimental situation has been recently summarized in various papers (113, 161, 220, 285, 318) and is given in Table 4 and Figures 12 and 13 as extracted from (161). These data are shown in Figure 7, along with predicted gamma-ray luminosities (dashed lines).

9.5 X-Ray Observations

The first announcement of the detection of X rays from SN 1987A came late in the summer of 1987. The X rays were first seen by a set of large area proportional counters on board the Japanese satellite *Ginga* (134) and several instruments on board the Soviet space station *Mir* (311, 312). Since that time, the supernova has been monitored by both groups at frequent intervals in the energy range 2–400 keV (Figure 14). Observations of SN 1987A from August 1987 through February 1988 with the HEXE and Pulsar X-1 instruments on the Kvant module of the Soviet space station showed a noticeable growth in the intensity of hard radiation in the 20–400 keV range (310), but by April 1988 the flux in this energy range had peaked and was beginning an apparent decline. During the August–February interval, the flux in the 45–105 keV band was given approximately by $F = I(1+At)$, where I is the intensity (0.392 ± 0.018 count s^{-1}), $A = (2.8 \pm 0.5) \times 10^{-3}$, and t the time in days since August 1, 1987 (310). The magnitude of the hard flux and its time history are in good agreement with expectations based upon the radioactive model (31, 140, 141, 156, 167, 168, 188, 261, 271, 272, 353), provided that sufficient mixing is invoked to give a hard X-ray flux that is initially detectable in August.

Owing to a fortuitous launch date, the satellite *Ginga* began to monitor the X-ray emission of SN 1987A only two days after the supernova explosion. SN 1987A was first clearly detected in two energy bands of 6–16 keV and 16–28 keV during July 1987 (130 days after the explosion) and was not visible at the same sensitivity during April 1987. Though background subtraction generates a source of systematic error, these errors have been accounted for, and the flux from SN 1987A and its time variation have been determined. The flux in the 16–28 keV band is relatively constant and consistent, both with observations by *Mir* and with theoretical expectations. The soft component is time variable and still not well understood.

Table 4 Gamma-ray line observations of SN 1987A [from [161]]

Instrument[a]	E_0 (keV)	Date	Day #	Line width (keV)	Flux (10^{-4} photon cm^{-2} s^{-1})	Comments
SMM	847	2/23/87–4/2/87	0–38	—	0.8 ± 4.2	—
SMM	1238	2/23/87–4/2/87	0–38	—	1.3 ± 3.5	—
SMM	847	4/18/87–5/21/87	54–87	—	0.6 ± 4.4	—
SMM	1238	4/18/87–5/21/87	54–87	—	0.4 ± 3.8	—
Caltech	847	5/20/87	86	—	<15.0	3-sigma limit
Caltech	1238	5/20/87	86	—	<23.0	3-sigma limit
Caltech	2599	5/20/87	86	—	<14.0	3-sigma limit
Lockheed/MSFC	847	5/29/87–5/30/87	95–96	3 (assumed)	<17.0	3-sigma limit
Lockheed/MSFC	1238	5/29/87–5/30/87	95–96	3 (assumed)	<13.0	3-sigma limit
SMM	847	6/10/87–7/16/87	107–143	—	1.6 ± 4.6	—
SMM	1238	6/10/87–7/16/87	107–143	—	0.6 ± 4.0	—
SMM	847	8/1/87–9/7/87	159–196	—	8.6 ± 4.0	—
SMM	1238	8/1/87–9/7/87	159–196	—	5.3 ± 3.5	—
SMM	847	9/27/87–10/30/87	216–249	—	6.2 ± 4.1	—
SMM	1238	9/27/87–10/30/87	216–249	12	11.0 ± 3.5	—
Lockheed/MSFC	847	10/29/87–10/31/87	248–250	18 (assumed)	10.0 ± 2.0	—
Lockheed/MSFC	1238	10/29/87–10/31/87	248–250	81 (assumed)	<8.5	3-sigma limit
Caltech	847	11/19/87	269	81 (assumed)	$\leq 11.0 \pm 5.0$	Includes continuum
Caltech	1238	11/19/87	269	93 (assumed)	$\leq 11.0 \pm 6.0$	Includes continuum
JPL	847	12/7/87	287	8 (assumed)	5.0 ± 7.0	—
JPL	1238	12/7/87	287	8.2 ± 3.4	21.0 ± 7.0	Line centroid at 1240.8 ± 1.7 keV
SMM	847	11/16/87–12/22/87	267–303	—	5.8 ± 3.4	—
SMM	1238	11/16/87–12/22/87	267–303	—	8.8 ± 2.9	30% systematics, all three lines
Florida/GSFC	847	1/8/88–1/11/88	319–322	7.5	11.0 ± 4.0	Peaks at 1226 & 1240 keV
Florida/GSFC	1238	1/8/88–1/11/88	319–322	~14	19.0 ± 4.0	Peaks at 2582 & 2603 keV
Florida/GSFC	2599	1/8/88–1/11/88	319–322	~21	8.6 ± 3.2	—
SMM	847	1/8/88–2/14/88	319–356	—	3.9 ± 4.2	—
SMM	1238	1/8/88–2/14/88	319–356	—	2.2 ± 3.6	—
SMM	847	3/1/88–4/6/88	372–408	—	8.2 ± 3.7	—
SMM	1238	3/1/88–4/6/88	372–408	—	3.6 ± 3.1	—
Caltech	847	4/12/88–4/13/88	414–415	94 (assumed)	$\leq 8.0 \pm 4.0$	Includes continuum
Caltech	1238	4/12/88–4/13/88	414–415	93 (assumed)	<14.0	3-sigma limit. Includes continuum
GSFC/Bell/Sandia	847	5/2/88	434	15–25	—	847 keV not yet reported
GSFC/Bell/Sandia	1238	5/2/88	434	—	8.1 ± 1.7	—

[a] Acronyms: SMM, *Solar Maximum Mission*; MSFC, Marshall Space Flight Center; JPL, Jet Propulsion Laboratory; GSFC, Goddard Space Flight Center.

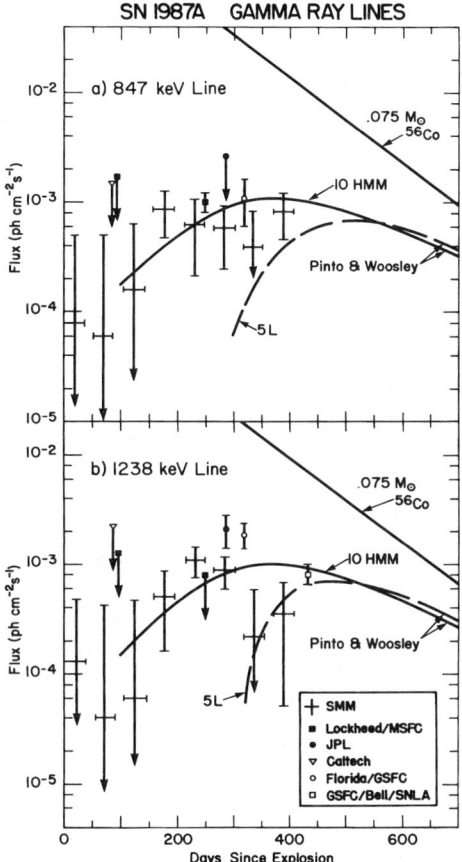

Figure 12 Measurements of the 847- and 1238-keV line fluxes from ^{56}Co decay. Also shown is the line for 0.075 $M_\odot$ of unobscured ^{56}Co and the curves for models 10HMM and 5L of (271, 272) and (294). Figure is from (161).

Early data reported by the TTM module on *Mir* gave a flux a factor of two lower in the soft energy range studied by *Ginga*. Unfortunately, the TTM module became inoperational a short time later. More recently, the TTM module has been repaired and should shortly provide a useful comparison to the *Ginga* soft X-ray observations (309).

9.6 *Nonradioactive Sources of X-Ray and Gamma-Ray Emission*

Radioactivity is certainly not the only source capable of contributing to the X-ray emission of SN 1987A: The interaction of the blast wave

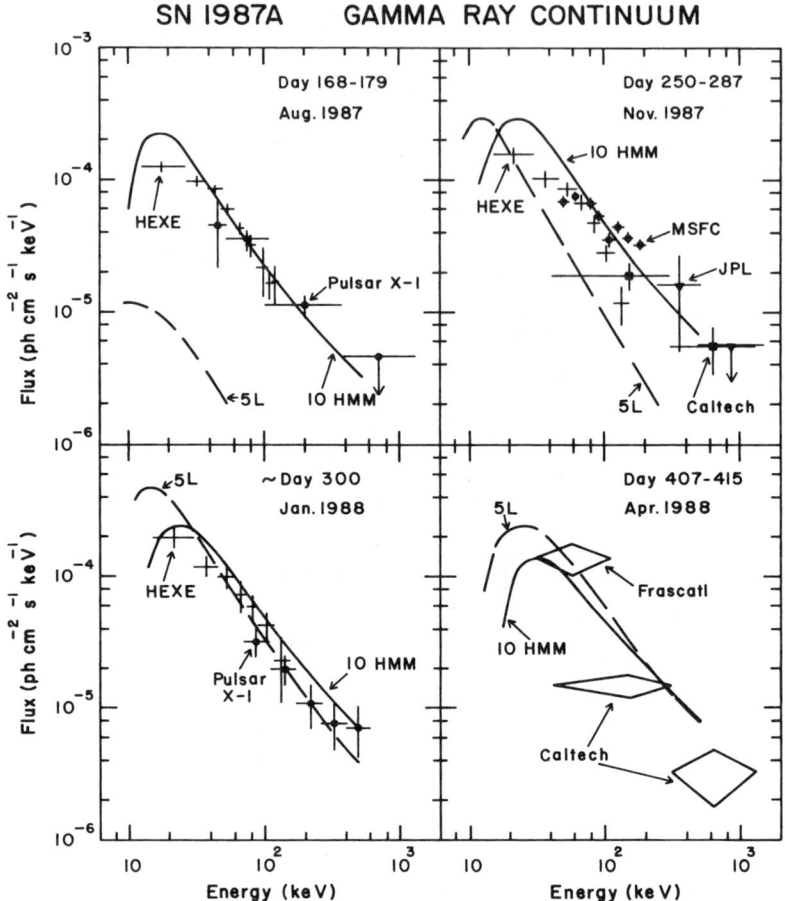

Figure 13 Measurements of the gamma-ray continuum compared with the same models as Figure 12. Figure is from (161).

with circumstellar material and the emissions of a young compact object in the center of the supernova can also contribute to the high-energy phenomena.

From the first detection of SN 1987A by *Ginga*, the measured spectrum, while consistent with Comptonized gamma rays of ^{56}Co decay at high energy, has been inconsistent with the radioactive model at energies below about 15 keV, where total absorption of hard radiation was expected owing to an enormous photoelectric opacity. During January 1988, the X-ray flux in this soft (6–16 keV) band, as measured by *Ginga*, underwent a dramatic and temporary increase by about a factor of almost four com-

SUPERNOVA 1987A   689

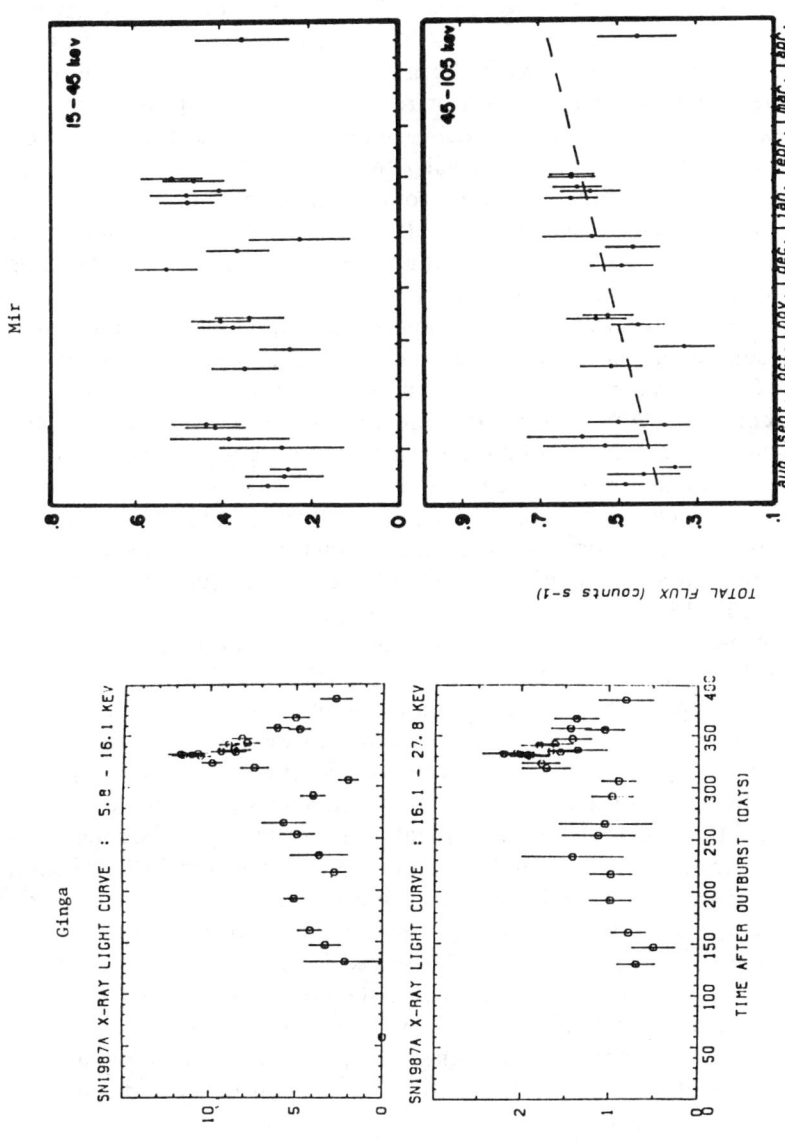

Figure 14 History of X-ray emission monitored by *Ginga* (316) and *Mir* (309).

pared with its nearly constant level during September–December 1987 (315, 316). The rise and fall took place over a period of several weeks and has not repeated in the following year. This soft component is clearly not due to radioactivity. Bandiera et al. (44) have attributed it to X rays leaking out from a synchrotron source in the middle of the supernova. Either well-evacuated holes in the supernova or relativistic plasma sprayed into the outer extremity are required in order to avoid the photoelectric absorption problem. Other researchers, especially Itoh et al. (187) and Masai et al. (222–224), have attributed the softer component to interaction of the blast wave with a circumstellar medium. In order to obtain the rapid time variation, they posit a clumpy medium resulting from inhomogeneous mass loss from the presupernova star. A variation on this model has been suggested by Sunyaev et al. (310), who point out that electrons accelerated by the supernova shock might pump up optical and IR photons by the inverse Compton effect. A still more ambitious claim (174) is that the soft X-ray component may originate from the decay of massive neutrinos during flight. The future evolution of the supernova and its X-ray emission may help to distinguish between these models.

At very high energies ($> 10^{12}$ eV), a potentially detectable flux of gamma rays may originate from the interaction of charged particles accelerated by a young pulsar in the middle of SN 1987A with the supernova interior (50, 51, 158, 159, 247, 373). The energetic particles are decelerated in a shock where gamma rays, especially from the decay of neutral pions, are produced. Owing to their high energy, these photons should escape with much greater ease than the gamma-line photons of ^{56}Co decay, and the supernova should have become transparent to them early in 1988. The possibility that emissions of this type are occurring is increasingly constrained by the exact tracking of the radioactive model by the bolometric light curve (Section 6.5). Any additional source of energy in the supernova that became thermalized would contribute to the optical and IR emission. [An unknown fraction could go into driving mass motion; see, for example (115).] The contribution to the light curve over and beyond that from radioactivity cannot (conservatively) exceed 10% of the present (August 1988) luminosity of the supernova (i.e. the contribution cannot exceed 5×10^{38} erg s^{-1}). The curve for pulsar luminosity of 2×10^{39} erg s^{-1} in Figure 8 falls well above the data; these curves were calculated for gamma-ray energies of 1 MeV, which give the loosest constraint and might be expected from an e^-e^+ cascade.

One brief burst of very high-energy (TeV) gamma rays has been inferred from an air shower reported by the JANZOS Collaboration (60) on January 14–15, 1988. Because such high-energy photons have a large cross section for interacting with low-energy photons, the source would need to be outside the supernova photosphere (52). It is also difficult to understand why the emission would have such a short duration [though see (52)].

10. THE FUTURE

In the immediate future the attention of observers will be focused upon obtaining further information on the composition and distribution of matter in the exploding debris. What elements were made in the explosion, and how much of each? Is the material homogeneous, or has it begun to clump? Is the supernova spherically symmetric or deformed? Are there jets? Has there been extensive mixing of material created in the deep interior with material farther out, or has the spherical distribution been approximately retained? Further study of the gamma rays from ^{56}Co and ^{57}Co decay will aid in answering some of these questions. The intensity of the gamma rays as a function of time will tell how much material lies between us and the decaying atoms. How the bolometric light curve deviates from the strict exponential decay of its radioactive power source will give similar information. The shapes of the gamma-ray lines may give a handle on the extent to which the supernova has mixed. So, too, will the shapes of emission lines observed in the infrared and optical from heavy elements like iron, oxygen, silicon, sulfur, and calcium.

But what of the collapsed object that lies at the center? Theory and observation both tell us that either a neutron star or black hole has been born. There *was* a neutrino signal, and the energy can only be explained by gravitational collapse of the stellar core to one of these two objects, most likely a neutron star. Theory predicts that the gravitational mass of the collapsed remnant is near 1.4 $M_\odot$, a value consistent with the properties of the neutrino signal (80), in which case it is very likely a neutron star. Is it a pulsar? That depends upon the magnetic field strength and rotation rate of the neutron star. It also critically depends upon the density of the material surrounding the neutron star. Even a little bit of matter falling back from the expanding debris could choke the pulsar mechanism.

To actually see pulsed emission from a central source the expanding supernova must become transparent. So long as a typical light ray coming from the neutron star is absorbed or scatters along the way, we will observe at Earth a hodgepodge of signals that have come along paths of varying length. This will wash out any regular pulsation at the source. The optical depth to electron scattering remains large for hard X rays ($\gtrsim 30$ keV) for about three years, after which time one might begin to see some modulated signal leaking out (170, 363). In and between some of the infrared lines, the supernova is already nearly thin (lower energy photons do not scatter off bound electrons, and the supernova is only partially ionized), but the infrared continuum (mostly free-free) is quite faint and so, too, is the pulsed infrared emission expected from any pulsar. In the UV and optical there is, and will remain for years to come, a large depth for scattering into and out of myriads of Doppler-broadened lines. So optical pulsations may be difficult to see, and, again, pulsars are faint emitters of pulsed optical radiation. Optical pulsations from SN 1987A have been reported (204a) but not confirmed. If there is a pulsar, it might make its presence known by its contribution to the bolometric luminosity of the remnant long before pulsations are seen. On the other hand, the radio pulsar mechanism may be shorted out by accretion or be masked by the decay of long-lived isotopes ^{57}Co and ^{44}Ti. If the neutron star is an accreting X-ray emitter, it might make its presence known when the supernova has declined in luminosity to that of the brighter of these sources (about 10^{38} erg s^{-1} which is the current limit as this paper goes to press) [also note L(Eddington) in Figure 8].

So much for the collapsed remnant. Farther out, and still moving at about one tenth the speed of light, the shock wave is bound for interstellar space. By now it has gone about one-half trillion miles. In other supernovae in the past this shock wave, by interacting with gas around the supernova, has generated intense radio and X-ray emission. The present supernova is somewhat anomalous in having a lower density in its vicinity, a property attributed to the fact that it originated from a blue supergiant rather than a red one. Blue supergiants have low-density stellar winds. But observations and at least some of the theoretical models suggest that the progenitor star, Sk -69 202 was at some point in its life (perhaps as recently as 20,000 yr ago) a red supergiant. Observations from *IUE*, for example, show spectroscopic evidence for low-velocity, nitrogen-rich material surrounding the supernova. As time passes, from one year to several decades, the blast wave should impact this circumstellar shell and give rise to strong radio and X-ray emission [i.e. "L(swept-up)" in Figure 8].

Whatever occurs from this point on will be new and exciting. The great beauty of this supernova is that, because it is near, we will be able to observe it at all wavelengths for a long time to come. Direct measurements of the radioactive decay of radioactive elements, the birth of a pulsar, and the evolution of a young supernova remnant are all likely spectacles over the next few years.

ACKNOWLEDGMENTS

R. Kirshner would like to acknowledge support from NSF grant AST-8516537, and from NASA NAG5-645 and NAG5-841. This work was also supported in part by NSF grant AST 8802533 and NASA NAS5-29225 to J. N. Bahcall, and by NSF grants AST 8813649, AST 8418185, and

NASA NAGW1273 to S. Woosley. D. Arnett wishes to acknowledge support from the University of Arizona.

While personally responsible for any omissions or inaccuracies contained in this review, the authors are grateful to a number of people for carefully reading earlier drafts and suggesting improvements. These people include Adam Burrows, Neil Gehrels, Alice Harding, Dick McCray, Ken Nomoto, and Nolan Walborn. The authors are also particularly indebted to Ms. Maggie Best, who carried out the daunting task of assembling a complex, lengthy paper written by four distant authors over electronic mail. Without her tireless efforts you would not be reading this paper, at least not this year.

Literature Cited

1. Aglietta, M., Badino, G., Bologna, G. F., Castagnoli, C., Fulgione, W., et al. 1986. *Nuovo Cimento* 9C: 185
2. Aglietta, M., Badino, G., Bologna, G., Castagnoli, C., Castellina, A., et al. 1987. *Europhys. Lett.* 3: 1315
3. Aglietta, M., Badino, G., Bologna, G., Castagnoli, C., Castellina, A., et al. 1987. *Europhys. Lett.* 3: 1321
4. Aitken, D. K. 1988. See Ref. 116. In press
5. Aitken, D. K., Smith, C. H., James, S. D., Roche, P. F., Hyland, A. R., McGregor, P. J. 1988. *MNRAS* 231: 7P
6. Alekseev, E. N., Alekseeva, L. N., Volchenko, V. I., Krivosheina, I. V. 1987. *JETP Lett.* 45: 589
7. Alekseev, E. N., Alekseeva, L. N., Krivosheina, I. V., Volchenko, V. I. 1988. *Phys. Lett. B* 205: 209
8. Andreani, P., Ferlet, R., Vidal-Madjar, A. 1987. *Nature* 326: 770
9. Andreani, P., Ferlet, R., Vidal-Madjar, A. 1987. See Ref. 120, p. 697
10. Arafune, J., Fukugita, M. 1987. *Phys. Rev. Lett.* 59: 367
11. Arafune, J., Fukugita, M., Yanagida, T., Yoshimura, M. 1987. *Phys. Rev. Lett.* 59: 1864
12. Arnett, W. D. 1967. *Ap. J.* 153: 341
13. Arnett, W. D. 1972. *Ap. J.* 173: 393
14. Arnett, W. D. 1975. *Ap. J.* 192: 727
15. Arnett, W. D. 1977. *Ann. NY Acad. Sci.* 302: 90
16. Arnett, W. D. 1977. *Ap. J.* 218: 815
17. Arnett, W. D. 1978. In *Physics and Astrophysics of Neutron Stars and Black Holes*, ed. R. Giacconi, R. Ruffini, p. 356. Amsterdam: North-Holland
18. Arnett, W. D. 1979. *Ap. J. Lett.* 230: L37
19. Arnett, W. D. 1980. *Ap. J.* 237: 541
20. Arnett, W. D. 1982. *Ap. J.* 253: 785
21. Arnett, W. D. 1983. *Ap. J. Lett.* 263: L55
22. Arnett, W. D. 1987. *Ap. J.* 319: 136
23. Arnett, W. D. 1987. *Bull. Am. Astron. Soc.* 19: 1072
24. Arnett, W. D. 1987. See Ref. 120, p. 373
25. Arnett, W. D. 1987. In *The Origin and Evolution of Neutron Stars, IAU Symp. No. 125*, ed. D. J. Helfand, J.-H. Huang, p. 273. Dordrecht: Reidel
26. Arnett, W. D. 1988. *Ap. J.* 331: 377
27. Arnett, W. D. 1988. See Ref. 192, p. 301
28. Arnett, W. D. 1989. *Stars, Nuclei and Supernovae*. Princeton: Princeton Univ. Press. In preparation
29. Arnett, W. D., Bowers, R. L. 1977. *Ap. J. Suppl.* 33: 415
30. Arnett, W. D., Falk, S. W. 1976. *Ap. J.* 210: 733
31. Arnett, W. D., Fu, A. 1989. *Ap. J.* In press
32. Arnett, W. D., Rosner, J. L. 1987. *Phys. Rev. Lett.* 58: 1906
33. Axelrod, T. S. 1980. PhD thesis. Univ. Calif., Santa Cruz (available from Lawrence Livermore Natl. Lab. as *UCRL 52994*)
34. Axelrod, T. S. 1988. See Ref. 254. In press
35. Baade, W., Zwicky, F. 1934. *Proc. Natl. Acad. Sci. USA* 20: 254
36. Badino, G., Bologna, G., Castagnoli, C., Fulgione, W., Galeotti, O., et al. 1984. *Nuovo Cimento* 7C: 573
37. Bahcall, J. N. 1964. *Phys. Rev.* 136: 1164
38. Bahcall, J. N. 1978. *Annu. Rev. Astron. Astrophys.* 16: 241

39. Bahcall, J. N. 1989. *Neutrino Astrophysics*. Cambridge: Univ. Press. In press
40. Bahcall, J. N., Dar, A., Piran, T. 1987. *Nature* 326: 135
41. Bahcall, J. N., Glashow, S. L. 1987. *Nature* 326: 476
42. Bahcall, J. N., Piran, T., Press, W. H., Spergel, D. N. 1987. *Nature* 327: 682
43. Bahcall, J. N., Spergel, D. N., Press, W. H. 1988. See Ref. 192, p. 172
44. Bandiera, R., Pacini, F., Salvati, M. 1988. *Nature* 332: 418
45. Barbiellini, G., Cocconi, G. 1987. *Nature* 329: 21
46. Barbon, R., Capellaro, E., Turatto, M. 1984. *Astron. Astrophys.* 135: 27
47. Barkat, Z., Wheeler, J. C. 1988. *Ap. J.* 332: 247
48. Baron, E. 1988. Private communication
49. Beier, E. W. 1986. *Proc. Workshop Grand Unification, 7th, ICOBAN'86, Toyama, Jpn.*, ed. J. Arafune, p. 79. Singapore: World Sci.
50. Berezinsky, V. S., Ginzburg, V. L. 1987. *Pisma Astron. Zh.* 13: 931
51. Berezinsky, V. S., Ginzburg, V. L. 1987. *Nature* 329: 807
52. Berezinsky, V. S., Stanev, T. 1988. Bartol Res. Found. Preprint No. BA-88-49
53. Bethe, H. A., Brown, G. E. 1985. *Sci. Am.* 252: 60
54. Bionta, R. M., Blewitt, G., Bratton, C. B., Cortez, B. G., Errede, S., et al. 1983. *Phys. Rev. Lett.* 51: 27
55. Bionta, R. M., Blewitt, G., Bratton, C. B., Casper, D. 1987. *Phys. Rev. Lett.* 58: 1494
56. Blades, J. C., Wheatley, J. M., Panagia, N., Grewing, M., Pettini, M., et al. 1988. *Ap. J.* 334: 308
57. Blades, J. C., Wheatley, J. M., Panagia, N., Grewing, M., Pettini, M., et al. 1988. See Ref. 192, p. 261
58. Blanco, V. M., Gregory, B., Hamuy, M., Heathcote, S. R., Phillips, M. M., et al. 1987. *Ap. J.* 320: 589
59. Bodansky, D., Clayton, D. D., Fowler, W. A. 1968. *Ap. J. Suppl.* 16: 299
60. Bond, I. A., Budding, E., Conway, M. J., Fenton, K. B., Fujii, H., et al. 1988. *Phys. Rev. Lett.* 60: 1110
61. Borst, L. B. 1950. *Phys. Rev.* 78: 807
62. Boyarchuk, A. A., Gershberg, R. E., Zvereva, A. M., Petrov, P. P., Severnyi, A. B., et al. 1987. *Sov. Astron. Lett.* 13: 311
63. Branch, D. 1987. *Ap. J. Lett.* 320: L23
64. Branch, D. 1987. *Ap. J. Lett.* 320: L121
65. Branch, D., Falk, S. W., McCall, M. L., Rybski, P., Uomoto, A. K., Wills, B. J. 1981. *Ap. J.* 244: 780
66. Bratton, C. B., Casper, D., Ciocio, A., Claus, R., Crouch, M., et al. 1988. *Phys. Rev. D* 37: 3361
67. Brown, R. T., 1973. *Ap. J.* 179: 607
68. Bruenn, S. W. 1987. *Phys. Rev. Lett.* 59: 938
69. Bruenn, S. W. 1988. *Astrophys. Space Sci.* 143: 15
70. Bruenn, S. W. 1988. *Ap. J.* In press
71. Bruenn, S. W. 1988. Fla. Atl. Univ. Preprint
72. Brunish, W. M., Truran, J. W. 1982. *Ap. J.* 256: 247
73. Brunish, W. M., Truran, J. W. 1982. *Ap. J. Suppl.* 49: 447
74. Burbidge, G. R., Hoyle, F., Burbidge, E. M., Christy, R. F., Fowler, W. A. 1956. *Phys. Rev.* 103: 1145
75. Burrows, A. 1984. *Ap. J.* 283: 848
76. Burrows, A. 1987. *Ap. J. Lett.* 318: L57
77. Burrows, A. 1987. See Ref. 120, p. 315
78. Burrows, A. 1987. In *Neutrino Masses and Neutrino Astrophysics*, ed. B. Barger, F. Halzen, M. Marshak, K. Olive, p. 28. Singapore: World Sci.
79. Burrows, A. 1987. *Ap. J. Lett.* 328: L51
80. Burrows, A. 1988. *Ap. J.* 334: 891
81. Burrows, A., Lattimer, J. 1983. *Ap. J.* 270: 735
82. Burrows, A., Lattimer, J. 1985. *Ap. J. Lett.* 299: L15
83. Burrows, A., Lattimer, J. 1986. *Ap. J.* 304: 179
84. Burrows, A., Lattimer, J. 1987. *Ap. J. Lett.* 318: L63
85. Burrows, A., Mazurek, T. J. 1983. *Nature* 301: 315
86. Burrows, A., Mazurek, T. J., Lattimer, J. M. 1981. *Ap. J.* 251: 325
87. Bussard, R. W., Burrows, A., The, L. S. 1988. Preprint
88. Cameron, A. G. W. 1982. In *Essays in Nuclear Astrophysics*, ed. C. A. Barnes, D. D. Clayton, D. N. Schramm, p. 23. Cambridge: Univ. Press
89. Cassatella, A., Fransson, C., van Santvoort, J., Gry, C., Talavera, A., et al. 1987. *Astron. Astrophys.* 177: L29
90. Catchpole, R. M., Whitelock, P. A., Feast, M. W., Menzies, J. W. 1988. *MNRAS* 231: 75P
91. Catchpole, R. M., Whitelock, P. A., Feast, M. W., Menzies, J. W., Monk, A. S., et al. 1987. *MNRAS* 229: 15P
92. Caughlan, G. R., Fowler, W. A. 1988. *At. Data Nucl. Data Tables.* In press
93. Chan, K. W., Lingenfelter, R. E. 1987. *Ap. J. Lett.* 318: L51
94. Chan, K. W., Lingenfelter, R. E. 1988. In *Nuclear Spectroscopy of Astrophysical Sources, AIP Conf. Proc. 170*, ed. N. Gehrels, G. Share, p. 110. New York: AIP

95. Chevalier, R. A. 1976. *Ap. J.* 207: 872
96. Chevalier, R. A. 1976. *Ap. J.* 208: 826
97. Chevalier, R. A. 1982. *Ap. J.* 259: 302
98. Chevalier, R. A. 1987. See Ref. 120, p. 481
99. Chevalier, R. A. 1988. *Proc. Astron. Soc. Aust.* In press
100. Chevalier, R. A. 1988. *Nature* 332: 514
101. Chevalier, R. A. 1988. In *The Early Evolution of Supernova Remnants, IAU Colloq. No. 10*, ed. R. S. Roger, T. L. Landecker, p. 31. Cambridge, Mass: Cambridge Univ. Press
102. Chevalier, R. A., Fransson, C. 1987. *Nature* 328: 44
103. Chevalier, R. A., Fransson, C. 1987. *Ap. J. Lett.* 322: L15
104. Chevalier, R. A., Klein, R. I. 1978. *Ap. J.* 219: 994
105. Chiosi, A., Maeder, A. 1986. *Annu. Rev. Astron. Astrophys.* 24: 329
106. Ciatti, F., Rosino, L., Bertola, F. 1971. *Mem. Soc. Astron. Ital.* 42: 163
107. Clayton, D. D. 1974. *Ap. J.* 188: 155
108. Clayton, D. D., Colgate, S. A., Fishman, G. J. 1969. *Ap. J.* 155: 75
109. Colgan, S. W. J., Hollenbach, D. J. 1988. *Ap. J. Lett.* 329: L25
110. Colgate, S. A., Johnson, H. L. 1960. *Phys. Rev. Lett.* 51: 235
111. Colgate, S. A., McKee, C. 1969. *Ap. J.* 157: 623
112. Colgate, S. A., White, R. H. 1966. *Ap. J.* 143: 626
113. Cook, W. R., Palmer, D. M., Prince, T. A., Schindler, S. M., Starr, C. H., et al. 1988. *Ap. J. Lett.* 334: L87
114. Cooperstein, J. 1988. *Phys. Rev. C* 37: 786
115. Coroniti, F. V., Kennel, C. F. 1985. In *The Crab Nebula and Related Supernova Remnants*, ed. M. Kafatos, R. B. C. Henry, p. 25. Cambridge: Univ. Press
116. Couch, W., ed. 1988. *Elizabeth and Frederick White Symposium on Supernova 1987A. Proc. Astron. Soc. Aust.* In press
117. Couderc, P. 1939. *Ann. Astrophys.* 2: 271
118. Crotts, A. 1988. *IAU Circ. No. 4561*
119. Crotts, A. 1988. *Ap. J. Lett.* 333: L51
120. Danziger, I. J., ed. 1987. *Workshop on the SN 1987A. ESO Workshop Proc. No. 26.* Garching bei München: ESO
121. Dar, A. 1988. *Proc. Int. Conf. Extrasol. Neutrino Astron., Los Angeles, 1987*, ed. D. Cline. In press
122. Dar, A. 1988b. *Proc. Rencontre Phys. Valle d'Aosta, 3rd, La Thuile, Italy*, ed. G. Belletini, M. Greco, P. Galeotti. In press
123. Davis, R. Jr. 1988. Private communication
124. de Boer, K. S., Grewing, M., Richtler, T., Wamsteker, W., Gry, C., Panagia, N. 1987. *Astron. Astrophys.* 177: L37
125. de Jager, O. C., Nieuwenhuijzen, H., van der Hucht, K. A. 1987. Preprint. See also Ref. 105, p. 336
126. de Rujula, A. 1987. *Phys. Lett. B* 193: 514
127. Doggett, J., Branch, D. A. 1985. *Astron. J.* 90: 2303
128. Dopita, M. A. 1986. In *Star Forming Regions, IAU Symp. No. 115*, ed. M. Peimbert, J. Jugaku, p. 501. Dordrecht: Reidel
129. Dopita, M. A. 1988. *Nature* 331: 506
130. Dopita, M. A. 1988. See Ref. 254. In press
131. Dopita, M. A. 1988. *Space Sci. Rev.* 46: 225
132. Dopita, M. A. 1988. Private communication
133. Dopita, M. A., Achilles, J. A., Dawe, J. A., Flynn, C., Meatheringham, S. J., McNaught, R. D. 1987. *Proc. Astron. Soc. Aust.* 7: 141
134. Dotani, T., Hayashida, K., Inoue, H., Itoh, M., Koyama, K., et al. 1987. *Nature* 330: 230
135. Dufour, R. 1984. In *Structure and Evolution of the Magellanic Clouds, IAU Symp. No. 108*, ed. S. van den Bergh, K. S. de Boer, p. 353. Dordrecht: Reidel
136. Dupree, A. K., Kirshner, R. P., Nassiopoulos, G. E., Raymond, J. C., Sonneborn, G. 1987. *Ap. J.* 320: 597
137. Dwek, E. 1983. *Ap. J.* 274: 175
138. Dwek, E. 1988. *Ap. J.* 329: 814
139. Eastman, R. G., Kirshner, R. P. 1987. *Bull. Am. Astron. Soc.* 19: 951
140. Ebisuzaki, T., Shibazaki, N. 1988. *Ap. J.* 328: 699
141. Ebisuzaki, T., Shibazaki, N. 1988. *Ap. J. Lett.* 327: L5
142. Emmering, R. T., Chevalier, R. A. 1988. *Ap. J. Lett.* 331: L105
143. Ensman, L., Woosley, S. E. 1988. *Ap. J.* In press
144. Fabian, A. C., Rees, M. J., van den Heuvel, E. P. J., van Paradijs, J. 1987. *Nature* 328: 323
145. Falk, S. W., Arnett, W. D. 1973. *Ap. J. Lett.* 180: L65
146. Falk, S. W., Arnett, W. D. 1977. *Ap. J. Suppl.* 33: 515
147. Feast, M. W. 1988. See Ref. 192, p. 51
148. Feast, M. W. 1988. *Proc. Workshop Monde des Galaxies, Paris.* In press
149. Filippenko, A. V. 1988. *Astron. J.* 96: 1941
150. Fransson, C. 1987. See Ref. 120, p. 467
151. Fransson, C. 1988. See Ref. 254, p. 383
152. Fransson, C., Cassatella, A., Gilmozzi,

R., Kirshner, R. P., Panagia, N., et al. 1989. *Ap. J.* 336: 429
153. Fransson, C., Chevalier, R. A. 1989. *Ap. J. Lett.* In press
154. Fransson, C., Grewing, M., Cassatella, A., Panagia, N., Wamsteker, W. 1987. *Astron. Astrophys.* 177: L33
155. Frieman, J. A., Haber, H. E., Freese, K. 1988. *Phys. Lett. B* 200: 115
156. Fu, A., Arnett, W. D. 1989. *Ap. J.* In press
157. Gaisser, T. K., Stanev, T. 1987. *Phys. Rev. Lett.* 58: 1695 (Erratum, 59: 844)
158. Gaisser, T. K., Stanev, T. 1988. *Ap. J.* 332: 314
159. Gaisser, T. K., Harding, A. K., Stanev, T. 1987. *Nature* 329: 314
160. Deleted in proof
161. Gehrels, N., Leventhal, M., MacCallum, C. J. 1988. In *Nuclear Spectroscopy of Astrophysical Sources, AIP Conf. Proc. No. 170*, ed. N. Gehrels, G. Share, p. 87. New York: AIP
162. Gehrels, N., MacCallum, C. J., Leventhal, M. 1987. *Ap. J. Lett.* 320: L19
163. Gilmozzi, R., Cassatella, A., Clavel, J., Fransson, C., Gonzalez, R., et al. 1987. *Nature* 328: 318
164. Girard, T., van Altena, W. F., Lopez, C. E. 1988. *Astron. J.* 95: 58
165. Gouiffes, C., Rosa, M., Melnick, J., Danziger, I. J., Remy, M., et al. 1988. *Astron. Astrophys.* 198: L9
166. Grasberg, E. K., Imshennic, V. S., Madyoshin, D. K. 1971. *Astrophys. Space Sci.* 10: 28
167. Grebenev, S. A., Sunyaev, R. A. 1987. *Pisma Astron. Zh.* 13: 1042
168. Grebenev, S. A., Sunyaev, R. A. 1987. *Pisma Astron. Zh.* 13: 945
169. Grebenev, S. A., Sunyaev, R. A. 1988. *Pisma Astron. Zh.* 14: 675
170. Grebenev, S. A., Sunyaev, R. A. 1988. *Pisma Astron. Zh.* 14: 1066
171. Haines, T. J., Bionta, R. M., Blewitt, G., Bratton, C. B., Casper, R., et al. 1986. *Phys. Rev. Lett.* 57: 1986
172. Hamuy, M., Suntzeff, N. B., Gonzales, R., Martin, G. 1988. *Astron. J.* 95: 63
173. Hashimoto, M., Nomoto, K., Shigeyama, T. 1988. *Astron. Astrophys.* In press
174. Hatsuda, T., Lim, C. S., Yoshimura, M. 1988. *Phys. Lett. B* 203: 462
175. Haxton, W. C. 1987. *Phys. Rev. D* 36: 2283
176. Hazen, M. 1987. *IAU Circ. Nos. 4365, 4367*
177. Heiles, C. 1987. In *Interstellar Processes*, ed. D. J. Hollenbach, H. A. Thronson, Jr., p. 171. Dordrecht: Reidel
178. Hillebrandt, W. 1985. In *High-Energy Phenomena Around Collapsed Stars*, p. 73. Dordrecht: Reidel
179. Hirata, K., Kajita, T., Koshiba, M., Nakahata, M., Oyama, Y., et al. 1987. *Phys. Rev. Lett.* 58: 1490
180. Hirata, K. S., Kajita, T., Koshiba, M., Nakahata, M., Oyama, Y., et al. 1988. *Phys. Rev. D* 38: 448
181. Hodge, P., Wright, F. 1967. *The Large Magellanic Cloud*. Cambridge, Mass: Smithson. Inst. Press
182. Hoflich, P. 1987. *Lect. Notes Phys.* 287: 307
183. Hoflich, P. 1987. *Mitt. Astron. Ges.* 70: 192
184. Humphreys, R. M. 1984. In *Observational Tests of the Stellar Evolution Theory, IAU Symp. No. 105*, ed. A. Maeder, A. Renzini, p. 279. Dordrecht: Reidel
185. Humphreys, R. M., McElroy, D. B. 1984. *Ap. J.* 284: 565
186. Imshennik, V. S., Nadezhin, D. K. 1983. *Astrophys. Space Sci. Rev. (Sov. Sci. Rev., Sect. E)* 2: 75
187. Itoh, H., Hayakawa, S., Masai, K., Nomoto, K. 1987. *Publ. Astron. Soc. Jpn.* 39: 529
188. Itoh, M., Kumagai, S., Shigeyama, T., Nomoto, K., Nishimura, J. 1987. *Nature* 330: 233
189. Jauncey, D. L., Kemball, A., Bartel, N., Whitney, A. R., Rogers, A. E. E., et al. 1988. *Nature* 334: 412
190. Joss, P. C., Rappaport, S. A. 1984. *Annu. Rev. Astron. Astrophys.* 22: 537
191. Joss, P. C., Podsiadlowski, Ph., Hsu, J. J. L., Rappaport, S. A. 1987. *Nature* 331: 237
192. Kafatos, M., Michalitsianos, A., eds. 1988. *Supernova 1987A in the Large Magellanic Cloud*. Cambridge: Univ. Press. 487 pp.
193. Kirshner, R. P. 1988. In *Supernova Remnants and Their Supernovae, IAU Colloq. No. 101*, ed. R. S. Roger, T. L. Landecker, p. 1. Cambridge: Univ. Press
194. Kirshner, R. P. 1988. See Ref. 192, p. 87
195. Kirshner, R. P., Chevalier, R. A. 1977. *Ap. J.* 218: 142
196. Kirshner, R. P., Kwan, J. 1974. *Ap. J.* 193: 27
197. Kirshner, R. P., Kwan, J. 1975. *Ap. J.* 197: 415
198. Kirshner, R. P., Oke, J. B., Penston, M. V., Searle, L. 1973. *Ap. J.* 185: 303
199. Kirshner, R. P., Sonneborn, G., Casatella, A., Gilmozzi, R., Wamsteker, W. 1987. *IAU Circ. No. 4435*
200. Kirshner, R. P., Sonneborn, G., Cren-

shaw, D. M., Nassiopoulos, G. E. 1987. *Ap. J.* 320: 602
201. Kolb, E. W., Stebbins, A. J., Turner, M. S. 1987. *Phys. Rev. D* 35: 3598 (Addendum and erratum, 36: 3820)
202. Kozasa, T., Hasegawa, H., Nomoto, K. 1989. *Ap. J.* In press
203. Krauss, L. M. 1987. *Nature* 329: 689
204. Krauss, L. M., Tremaine, S. 1988. *Phys. Rev. Lett.* 60: 176
204a. Kristian, J. A., Pennypacker, C. R., Middleditch, J., Hamuy, M. A., Imamura, J. N., et al. 1989. *Nature* 338: 234
205. Kudritzki, R., Groth, H. G., Butler, K., Husfeld, D., Becker, S., et al. 1987. See Ref. 120, p. 39
206. Kumagai, S., Itoh, M., Shigeyama, T., Nomoto, K., Nishimura, J. 1988. See Ref. 192, p. 414
207. Kumagai, S., Itoh, M., Shigeyama, T., Nomoto, K., Nishimura, J. 1988. *Astron. Astrophys.* 197: L7
208. Kumagai, S., Shigeyama, T., Nomoto, K., Itoh, M., Nishimura, J., Tsuruta, S. 1988. Submitted for publication
209. Lamb, D. Q., Melia, F., Loredo, T. J. 1988. See Ref. 192, p. 204
210. Lamb, S. A., Iben, I. Jr., Howard, W. M. 1976. *Ap. J.* 207: 209
211. Leising, M. D. 1988. *Nature* 332: 516
212. Longo, M. J. 1988. *Phys. Rev. Lett.* 60: 173
213. Lucke, P. B., Hodge, P. W. 1970. *Astron. J.* 75: 171
214. Lucy, L. B. 1987. *Astron. Astrophys.* 182: L31
215. Lundqvist, P., Fransson, C. 1988. *Astron. Astrophys.* 192: 221
216. Maeder, A. 1987. See Ref. 120, p. 251
217. Maeder, A. 1987. *Astron. Astrophys.* 173: 247
218. Maeder, A., Lequeux, J. 1982. *Astron. Astrophys.* 114: 409
219. Maeder, A., Meynet, G. 1987. *Astron. Astrophys.* 182: 243
220. Mahoney, W. R., Varnell, L. S., Jacobson, A. S., Ling, J.C., Radocinski, R. G., Wheaton, W. A. 1988. *Ap. J. Lett.* 334: L81
221. Martin, M. J. 1987. *Nucl. Data Sheets* 51: 67
222. Masai, K., Hayakawa, S., Inoue, H., Itoh, H., Nomoto, K., Shigeyama, T. 1988. See Ref. 254, p. 450
223. Masai, K., Hayakawa, S., Inoue, H., Itoh, H., Nomoto, K. 1988. *Nature* 335: 804
224. Masai, K., Hayakawa, S., Itoh, H., Nomoto, K. 1987. *Nature* 330: 235
225. Matthews, J. 1988. See Ref. 192, p. 151
226. Matz, S. M., Share, G. H., Leising, M. D., Chupp, E. L., Vestrand, W. T., et al. 1988. *Nature* 331: 416
227. Mayle, R. W. 1986. PhD thesis. Univ. Calif., Berkeley
228. Mayle, R., Wilson, J. R. 1988. Private communication
229. Mayle, R. W., Wilson, J. R., Schramm, D. N. 1987. *Ap. J.* 318: 288
230. Mazurek, T. J. 1974. *Nature* 252: 287
231. McCray, R. 1989. *Proc. Yellow Mountain Summer Sch. Structure and Evolution of Galaxies, 1987*, ed. L.-Z. Fang. Singapore: World Sci. In press
232. McCray, R., Li, H. W. 1989. *Proc. Yellow Mountain Summer Sch. Structure and Evolution of Galaxies, 1987*, ed. L.-Z. Fang. Singapore: World Sci. In press
233. McCray, R., Shull, J. M., Sutherland, P. 1987. *Ap. J. Lett.* 317: L73
234. McNaught, R. M. 1987. *IAU Circ. No. 4316*
235. McNaught, R. M. 1987. *Astronomer* 23(Mar.): 174
236. Meikle, W. P. S., Matcher, S. J., Morgan, B. L. 1987. *Nature* 329: 608
237. Meikle, W. P. S., Matcher, S. J., Morgan, B. L. 1987. *Bull. Am. Astron. Soc.* 19: 950
238. Menzies, J. W. 1989. *MNRAS.* In press
239. Michel, F. C. 1988. *Nature* 333: 644
240. Michel, F. C., Kennel, C. F., Fowler, W. A. 1987. *Science* 238: 938
241. Moffat, A. F. J., Niemela, V. S., Phillips, M. M., Chu, Y. H., Seggewiss, W. 1987. *Ap. J.* 312: 612
242. Myra, E. S., Bludman, S. A. 1988. Preprint
243. Myra, E. S., Bludman, S. A., Hoffman, Y., Lichtenstadt, J., Sack, N., Van Riper, K. A. 1987. *Ap. J.* 318: 744
244. Nagasawa, M., Nakamura, T., Miyama, S. M. 1988. *Publ. Astron. Soc. Jpn.* In press
245. Nakahata, M. 1988. PhD thesis. Univ. Tokyo, Jpn. (Preprint UT-ICEPP-88-01)
246. Nakahata, M., Arisaka, K., Kajita, T., Koshiba, M., Oyama, Y., et al. 1986. *J. Phys. Soc. Jpn.* 55: 3786
247. Nakamura, T., Yamada, Y., Sato, H. 1987. *Prog. Theor. Phys.* 78: 1065
248. Niemela, V., Ruiz, M. T., Phillips, M. M. 1985. *Ap. J.* 289: 52
249. Nisenson, P., Karovska, M., Noyes, R., Papaliolios, C. 1987. *Bull. Am. Astron. Soc.* 19: 950
250. Nisenson, P., Papaliolios, C., Karovska, M., Noyes, R. 1988. *Ap. J. Lett.* 320: L15 (Erratum, 324: L35)
251. Nomoto, K. 1984. *Ap. J.* 277: 791

252. Nomoto, K. 1986. *Ann. NY Acad. Sci.* 470: 294
253. Nomoto, K. 1987. *Ap. J.* 322: 206
254. Nomoto, K., ed. 1988. *Atmospheric Diagnostics of Stellar Evolution, Proc. IAU Colloq. No. 108.* Berlin: Springer-Verlag. In press
255. Nomoto, K., Hashimoto, M. 1986. *Prog. Part. Nucl. Phys.* 17: 267
256. Nomoto, K., Hashimoto, M. 1987. In *Chemical Evolution of Galaxies With Active Star Formation,* ed. K. Takakubo, p. 93. Tokyo: Universal Acad. Press
257. Nomoto, K., Hashimoto, M. 1988. *Phys. Rep.* 163: 13
258. Nomoto, K., Hayakawa, S., Itoh, H., Masai, K., Shigeyama, T. 1987. See Ref. 120, p. 503
259. Nomoto, K., Shigeyama, T. 1988. See Ref. 192, p. 273
260. Nomoto, K., Shigeyama, T., Hashimoto, M. 1987. See Ref. 120, p. 325
261. Nomoto, K., Shigeyama, T., Hashimoto, S. 1988. See Ref. 254, p. 319
262. Nomoto, K., Shigeyama, T., Kumugai, S., Itoh, M., Nishimura, J., et al. 1988. In *Physics of Neutron Stars and Black Holes,* ed. Y. Tanaka, p. 441. Tokyo: Universal Acad. Press
263. Nomoto, K., Shigeyama, T., Kumagai, S., Hashimoto, M. 1988. See Ref. 116. In press
264. Nomoto, K., Hashimoto, M., Shigeyama, T., Kumagai, S., Yamaoka, H., Saio, H. 1988. In *Big Bang, Active Galactic Nuclei and Supernovae,* ed. S. Hayakawa, K. Sato. Tokyo: Universal Acad. Press. In press
265. Nussbaumer, H., Storey, P. 1988. *Astron. Astrophys.* 200: L25
266. Panagia, N. 1988. See Ref. 192, p. 96
267. Pankey, T. 1962. PhD thesis. Howard Univ., Washington, DC (Diss. Abstr. 23: 4)
268. Pettini, M., Strathakis, R., D'Odorico, S., Molaro, P., Vladilo, G. 1989. *Ap. J.* In press
269. Phillips, M. M. 1988. See Ref. 192, p. 16
270. Phinney, E. S. 1988. *Nature* 331: 566
271. Pinto, P. A., Woosley, S. E. 1988. *Ap. J.* 329: 820
272. Pinto, P. A., Woosley, S. E. 1988. *Nature* 333: 534
273. Pinto, P. A., Woosley, S. E., Ensman, L. M. 1988. *Ap. J. Lett.* 331: L101
274. Piran, T., Spergel, D. 1988. *Proc. Moriond Astrophys. Meet. Dark Matter,* ed. J. Audouze, J. Tran Thanh Van. In press
275. Rank, D. M., Bregman, J., Witteborn, F. C., Cohen, M., Lynch, D. K., Russell, R. W. 1988. *Ap. J. Lett.* 325: L1
276. Rank, D. M., Pinto, P. A., Woosley, S. E., Bregman, J. D., Witteborn, F. C., et al. 1988. *Nature* 331: 505
277. Rester, G., et al. 1988. *IAU Circ. No. 4535.* Also, preprint submitted for publication
278. Rousseau, J., Martin, N., Prévot, L., Rebeirot, E., Robin, A., Brunet, J. P. 1978. *Astron. Astrophys. Suppl.* 31: 243
279. Russell, S. C., Bessell, M. S., Dopita, M. A. 1988. *Proc. Montreal Star Formation Workshop.* In press
280. Russell, S. C., Bessell, M. S. 1988. Submitted for publication
281. Rutherford, E. 1904. *Radioactivity,* p. 342 ff. Cambridge: Univ. Press
282. Saio, H., Kato, M., Nomoto, K. 1988. *Ap. J.* 331: 388
283. Saio, H., Nomoto, K., Kato, M. 1988. *Nature* 334: 508
284. Sanduleak, N. 1969. *CTIO Contrib. No. 89,* Cerro Tololo Inter-Am. Obs., La Serena, Chile
285. Sandie, W. G., Nakano, G. H., Chase, L. F. Jr., Fishman, G. J., Meegan, C. A., et al. 1988. *Ap. J. Lett.* 334: L91
286. Sato, K. 1975. *Prog. Theor. Phys.* 53: 595
287. Sato, K., Suzuki, H. 1987. *Phys. Rev. Lett.* 58: 2722
288. Sato, K., Suzuki, H. 1987. *Phys. Lett. B* 196: 267
289. Schaefer, B. E. 1987. *Ap. J. Lett.* 323: L47
290. Schaeffer, R., Declais, Y., Jullian, S. 1987. *Nature* 330: 142
291. Schlegel, E. M., Kirshner, R. P. 1989. Submitted for publication
292. Schramm, D. N. 1987. *Comments Nucl. Part. Phys.* 17: 239
293. Shapiro, S. L., Teukolsky, S. A. 1983. *Black Holes, White Dwarfs, and Neutron Stars.* New York: Wiley
294. Shibazaki, N., Ebisuzaki, T. 1988. *Ap. J. Lett.* 327: L9
295. Shigeyama, T., Nomoto, K., Hashimoto, M. 1988. *Astron. Astrophys.* 196: 141
296. Shigeyama, T., Nomoto, K., Hashimoto, M., Sugimoto, D. 1987. *Nature* 328: 320
297. Shklovskii, I. S. 1987. *Sov. Astron. Lett.* 10(5): 302
298. Shull, J. M., Xu, Y. 1988. See Ref. 192, p. 371
299. Sonneborn, G., et al. 1988. *IAU Circ. No. 4685*
300. Sonneborn, G., Altner, B., Kirshner, R. P. 1987. *Ap. J. Lett.* 323: L35
301. Spergel, D. N., Bahcall, J. N. 1988. *Phys. Lett. B* 200: 366
302. Spergel, D. N., Piran, T., Loeb, A.,

Goodman, J., Bahcall, J. N. 1987. *Science* 237: 1471
303. Stodolsky, L. 1988. *Phys. Lett.* B 201: 353
304. Storey, M. C., Manchester, R. N. 1987. *Nature* 329: 421
305. Suess, H. E., Urey, H. C. 1956. *Rev. Mod. Phys.* 28: 53 (Note also private communication from O. Haxel, 1946)
306. Suntzeff, N. B., Heathcote, S., Weller, W. G., Caldwell, N., Huchra, J. P., et al. 1988. *Nature* 334: 135
307. Suntzeff, N. B., Hamuy, M., Martin, G., Gomez, A., Gonzalez, R. 1988. *Astron. J.* 96: 1864
308. Suntzeff, N. B., Heathcote, S., Weller, W. G., Caldwell, N., Huchra, J. P., et al. 1988. *Nature* 334: 135
309. Sunyaev, R. A. 1989. Private communication
310. Sunyaev, R. A., Efremov, V. V., Kaniovskii, A. S., Stepanov, D. K., Yunin, S. N., et al. 1988. Preprint
311. Sunyaev, R. A., Kaniovsky, A., Efremov, V., Gilfanov, M., Churazou, E., et al. 1987. *Pisma Astron. Zh.* 13: 1027
312. Sunyaev, R. A., Kaniovsky, A., Efremov, V., Gilfanov, M., Churazov, E., et al. 1987. *Nature* 330: 227
313. Suzuki, H. 1986. *Proc. Int. Conf. Neutrino Phys. and Astrophys., 12th*, Sendai, Jpn., p. 306
314. Takahara, M., Sato, K. 1987. *Mod. Phys. Lett.* A 2: 293
315. Tanaka, Y. 1988. See Ref. 254, p. 399
316. Tanaka, Y. 1988. In *Physics of Neutron Stars and Black Holes*, ed. Y. Tanaka, p. 431. Tokyo: Universal Acad. Press
317. Taylor, J. H., Weisberg, J. M. 1982. *Ap. J.* 253: 908
318. Teegarden, B. J., Barthelmy, S. D., Gehrels, N., Tueller, J., Leventhal, M., MacCallum, C. J. 1988. Submitted for publication
319. Terndrup, D. M., Elias, J. H., Gregory, B., Heathcote, S. R., Phillips, M. M., et al. 1988. See Ref. 116. In press
320. Totsuka, Y. 1987. *Proc. Workshop in Grand Unification, 7th, ICOBAN'86*, Toyama, Jpn., ed. J. Arafune, p. 118. Singapore: World Sci.
321. Trimble, V. 1988. *Rev. Mod. Phys.* 60: 859
322. Trimble, V. 1988. *Comments Astrophys.* 12: 203
323. Truran, J. W., Arnett, W. D., Cameron, A. G. W. 1967. *Can. J. Phys.* 45: 2315
324. Truran, J. W., Weiss, A. 1987. *Proc. Workshop Nucl. Astrophys.* In press
325. Truran, J. W., Weiss, A. 1987. *Lect. Notes Phys.* 287: 293
326. Truran, J. W., Weiss, A. 1988. See Ref. 192, p. 331
327. Turtle, A. J., Campbell-Wilson, D., Bunton, J. D., Jauncey, D. L., Kesteven, M. J., et al. 1987. *Nature* 327: 38
328. Uomoto, A., Kirshner, R. P. 1985. *Astron. Astrophys.* 149: L7
329. Uomoto, A., Kirshner, R. P. 1986. *Ap. J.* 308: 685
330. Vidal-Madjar, A. 1987. *Astron. Astrophys.* 177: L17
331. Wagoner, R. V., Linder, E. V., Hershkowitz, S. 1987. *Bull. Am. Astron. Soc.* 19: 740
332. Walborn, N. R. 1988. See Ref. 254, p. 70
333. Walborn, N. R., Lasker, B. M., Laidler, V. G., Chu, Y.-H. 1987. *Ap. J. Lett.* 321: L41
334. Walborn, N. R., Prévot, M. L., Prévot, L., Wamsteker, W., Gonzalez, R., et al. 1989. Submitted for publication
335. Wampler, E. J., Richichi, A. 1988. *ESO Messenger No. 52*, p. 14
336. Wamsteker, W., Panagia, N., Barylak, M., Cassatella, A., Clavel, J., et al. 1987. *Astron. Astrophys.* 177: L21
337. Weaver, T. A., Woosley, S. E. 1980. *Ann. NY Acad. Sci.* 336: 335
338. Weaver, T. A., Woosley, S. E. 1980. In *Supernova Spectra, AIP Conf. Proc. No. 63*, ed. R. Meyerott, p. 15. New York: AIP
339. Weaver, T. A., Zimmerman, G. B., Woosley, S. E. 1978. *Ap. J.* 225: 1021
340. Weiler, K. W., Sramek, R. A. 1988. *Annu. Rev. Astron. Astrophys.* 26: 295
341. West, R. M., Lauberts, A., Jorgensen, H. E., Schuster, H.-E. 1987. *Astron. Astrophys.* 177: L1
342. White, G. L., Malin, D. 1987. See Ref. 120, p. 11
343. White, G. L., Malin, D. F. 1987. *Nature* 327: 36
344. Whitelock, P. A., Catchpole, R. M., Menzies, J. W., Feast, M. W., Winkler, H., et al. 1988. *MNRAS* 234: 5P
345. Williams, R. E. 1987. *Ap. J. Lett.* 320: L117
346. Wilson, J. R. 1971. *Ap. J.* 163: 209
347. Wilson, J. R., Mayle, R. 1987. Submitted for publication
348. Wilson, J. R., Mayle, R., Woosley, S. E., Weaver, T. A. 1986. *Ann. NY Acad. Sci.* 470: 267
349. Wilson, W. E. 1903. *Nature*, July 9 letter
350. Witteborn, F. C., Bregman, J. D., Woode, D. H., Pinto, P. A., Rank, D. M., et al. 1989. *Ap. J. Lett.* 338: L9
351. Wood, P. R., Faulkner, D. J. 1987. *Proc. Astron. Soc. Aust.* 7: 75
352. Woosley, S. E. 1986. In *Nucleosynthesis*

and Chemical Evolution, Saas-Fee Lect. Notes, ed. B. Hauck, A. Maeder, G. Meynet, p. 1. Geneva: Geneva Obs.
353. Woosley, S. E. 1988. *Ap. J.* 330: 218
354. Woosley, S. E. 1988. See Ref. 192, p. 289
355. Woosley, S. E. 1988. See Ref. 254, p. 361
356. Woosley, S. E., Arnett, W. D., Clayton, D. D. 1973. *Ap. J. Suppl.* 26: 231
357. Woosley, S. E., Axelrod, T. S., Weaver, T. A. 1981. *Comments Nucl. Part. Phys.* 9: 185
358. Woosley, S. E., Haxton, W. C. 1988. *Nature* 334: 45
359. Woosley, S. E., Phillips, M. M. 1988. *Science* 240: 750
360. Woosley, S. E., Pinto, P. A. 1988. In *Nuclear Spectroscopy of Astrophysical Sources, AIP Conf. Proc. 170*, ed. N. Gehrels, G. Share. New York: AIP
361. Deleted in proof
362. Woosley, S. E., Pinto, P. A., Ensman, L. 1988. *Ap. J.* 324: 466
363. Woosley, S. E., Pinto, P. A., Hartmann, D. 1989. *Ap. J.* In press
364. Woosley, S. E., Pinto, P. A., Martin, P., Weaver, T. A. 1987. *Ap. J.* 318: 664
365. Woosley, S. E., Pinto, P. A., Weaver, T. A. 1988. See Ref. 116, p. 355
366. Woosley, S. E., Weaver, T. A. 1983. In *Collapse and Numerical Relativity*, ed. D. Bancel, M. Signore, p. 27. Dordrecht: Reidel
367. Woosley, S. E., Weaver, T. A. 1986. *Annu. Rev. Astron. Astrophys.* 24: 205
368. Woosley, S. E., Weaver, T. A. 1986. In *Radiation Hydrodynamics in Stars and Compact Objects*, ed. D. Mihalas, K.-H. A. Winkler, p. 91. Berlin: Springer-Verlag
369. Woosley, S. E., Weaver, T. A. 1988. *Phys. Rep.* 163: 79
370. Woosley, S. E., Weaver, T. A., Taam, R. E. 1980. In *Type I Supernovae*, ed. J. C. Wheeler, p. 96. Austin: Univ. Tex.
371. Woosley, S. E., Wilson, J. R., Mayle, R. W. 1986. *Ap. J.* 302: 19
372. Xu, Y., Sutherland, P., McCray, R., Ross, R. R. 1988. *Ap. J.* 327: 197
373. Yamada, Y., Nakamura, T., Kasahara, K., Sato, H. 1988. Preprint KUNS 898
374. Young, T. R., Branch, D. 1988. *Nature* 333: 305
375. Zatsepin, G. I. 1968. *JETP Lett.* 8: 205
376. Zwicky, F. 1940. *Rev. Mod. Phys.* 12: 66

CHEMICAL ANALYSES OF COOL STARS

Bengt Gustafsson

Astronomiska Observatoriet, Box 515, S-75120 Uppsala, Sweden

Scepticism is the beginning of faith

Oscar Wilde: *The Picture of Dorian Gray*

1. INTRODUCTION

Out of 747 stars with [Fe/H] determinations in the catalogue of Cayrel de Strobel et al. (40), there are 203 classified as G stars and 181 as K stars but only 12 as M stars. There are no stars of types N or S in the catalogue. Several reasons for this absence of cool stars may be thought of:

1. These stars are not among the apparently brightest stars in the blue-violet spectral region, where traditional photographic plates were sensitive.
2. The late-type stellar spectra are so crowded with lines from atoms and molecules that there is little hope of seeing any continuum even at high spectral resolution, at least in the visual spectral region. Therefore, it is difficult to measure reliable equivalent widths.
3. Heavy blanketing seriously affects the structure of the stellar atmospheres and must be considered when constructing model atmospheres of late-type stars.
4. The cool giants and supergiants are extended objects with small atmospheric pressures, and they are almost always variables—thus it is not evident that standard plane-parallel model atmospheres in hydrostatic equilibrium and LTE may be used to represent them.
5. The derivation of fundamental parameters for the adequate models, such as effective temperatures and surface gravities, is nontrivial—in particular, the surface gravities for cool giants and supergiants are highly uncertain.
6. Many molecular data needed when a model atmosphere for a cool star

is to be constructed, or when molecular lines are to be used in the abundance analysis, are still not accurately known.

However, in recent years a number of important new developments have led to a very significant increase in the number of chemical analyses of cool stars. Among these developments are the following:

(a) the construction of multielement detectors such as Reticons and CCDs that are sensitive in the infrared and their use in efficient spectrometers, as well as the systematic use of Fourier Transform Spectrometers (FTSs) in the infrared beyond 1 μm [in particular, the FTS at the 4-m Mayall telescope at Kitt Peak National Observatory (90)];

(b) the systematic use of synthetic spectrum calculations to analyze crowded spectra with many blends and with continuum definition problems;

(c) the development of adequate methods for treating blanketing in model-atmosphere calculations, and the enormous increase in computing power that makes it possible to calculate extensive grids of models and detailed synthetic spectra;

(d) the promising development of empirical and theoretical methods for investigating spherical extension effects, convective energy transport, velocity fields, and departures from LTE in cool stars;

(e) the establishment of temperature scales for cool giants and supergiants from angular diameter measurements—in particular by lunar occultation techniques (197, and references therein) or by comparison of observed fluxes with those of model atmospheres (230, 231);

(f) the production of many new and reliable measurements and ab initio calculations of wave numbers, transition probabilities, dissociation energies, etc., for numerous molecular bands and states, though certainly not all that are needed.

In fact, these current improvements in the arsenal of basic methods and data for detailed chemical analysis of cool stars are so impressive that a great number of new important results have recently been obtained and others are to be foreseen in the immediate future. Thus, while a review of the results already achieved may hopefully serve as a first starting point for new investigations, it will certainly run the risk of becoming out of date quite soon.

This review is confined to the chemical analysis of the photospheres of M stars, S stars, and N stars (including the intermediate MS and SC stars). From an astrophysical point of view, this subject is rather artificially delineated—it would have been reasonable to include barium stars, R stars, CH stars, and possibly also RCrB stars; nor is the distinction between K stars and M stars very natural here. Furthermore, the chemical analysis of circumstellar shells would have been a reasonable inclusion. However, our focus lies more on the methodological aspects than on the astrophysical interpretation of the abundance results—from this point of view the restriction to M, S, and N photospheres is reasonable, since most of the difficulties mentioned above are much more severe for these stars than for the somewhat hotter ones, and these difficulties are different from those of analyses of circumstellar shells. A recent review on Ba star and CH star abundances was given by Lambert (133), and the analysis of R stars was discussed in detail by Dominy (49). Recent results for RCrB stars were summarized by Lambert (134). The chemical analysis of late-type stars in general was reviewed by Gustafsson (87), and of late Population II stars by Gustafsson (89) and Spite & Spite (213), while Lambert (132) discussed the chemical composition of red giants in the first dredge-up phase. Theories of mixing in red giants and related observations of abundances were reviewed by Scalo (194), and Baschek (12) summarized the chemical analyses of evolved stars. For recent discussions of the chemistry in circumstellar shells, see, e.g., (85, 171, 249).

The subject of the present review partially overlaps the following scientific fields summarized in a number of reviews from the last two decades:

The monograph by Fujita (69) covered the knowledge then available (1970) on the chemical analysis of cool stars—a very great fraction of these pioneering efforts were in fact contributed by Fujita himself and his Japanese colleagues. The knowledge about carbon stars was reviewed by Wallerstein (245) and by Alksne et al. (1) [cf. also Alksne et al. (2)]. Basic properties of M, S, and C stars were recently reviewed by F. Querci (181), and spectroscopy and nonthermal processes of these stars by M. Querci (182). Spectroscopic problems for carbon stars were discussed by Fujita (70). Infrared spectroscopy, so vital for the recent progress in the present field, was summarized by Merrill & Ridgway (152). Model atmospheres of late-type stars were discussed by Carbon (33), Gustafsson (88) and Johnson (114, 115). The testing of red giant models against observations was recently reviewed by Tsuji (238). Basic molecular data used in the chemical analysis of cool stars were reviewed by Nicholls (165) and Lambert (131; see also 141, 142). Lambert (135) summarized the use of molecules in such analyses, and he also reviewed the identification of molecular lines in carbon star spectra (138). Various aspects of molecules in cool stars were also discussed by Tsuji (236). The results of abundance analyses of red giant stars in late stages were recently reviewed and discussed by Lambert (137) and Smith (205), and a more general review was given by Wallerstein (247).

In the present review some general methodological aspects are first discussed—especially the calculation and use of model atmospheres and synthetic spectra (Section 2). In Section 3, abundance determinations of (mainly) metals from atomic lines will be reviewed. The discussion of lithium abundances is deferred to Section 4. In Section 5 the derivation of abundances of carbon, nitrogen, and oxygen is discussed, and in Section 6 the estimates of isotopic ratios are considered. The question of whether the hydrogen abundances, relative to helium, are nonsolar in evolved cool stars is not discussed here—we refer the interested reader to the review given by Lambert (134, and references therein) for a discussion of this topic.

Recent progress in the study of abundances in red giants has led to a fairly consistent qualitative picture of the nuclear and mixing processes that are of importance in different types of such stars, as well as of their evolutionary status. Here, however, the emphasis will be on methodological aspects—the reader mainly interested in the astrophysical implications of the abundance results is invited to consult the original papers or the reviews by Lambert (137), Smith (205), or Wallerstein (247).

Our discussion will be confined to methods and results based on high-resolution spectroscopy, in spite of the fact that many future abundance data will probably be derived from low-resolution observations, calibrated on stellar samples for which high-resolution methods have been applied, or calibrated directly by synthetic spectroscopy [for recent examples of the latter method used for analyses of M- and C-type stars, see (29) and (256), respectively]. This limitation is not entirely due to lack of space. Although synthetic spectroscopy is vital in this field of research, high-resolution analysis will be the dominant primary method for abundance determination for these stars for years to come. Low-resolution analyses based on synthetic spectra will have to be checked repeatedly against results from high-resolution analyses before they can be trusted for accurate abundance work for M, S, and N stars.

2. GENERAL METHODS

2.1 Model Atmospheres

When a stellar spectrum is to be used for estimating chemical abundances it must be confronted with some models of stellar atmospheres. Even in the seemingly least theoretical approach—the use of empirical curves of growth—important though implicit assumptions concerning the structure of the stellar atmosphere are being made. The most important of these assumptions is the adoption of only one temperature and pressure as characterizing the atmosphere (i.e. the neglect or very schematic consideration of so-called stratification effects). Such effects are often quite important in cool stars—because of the low temperatures involved, the Boltzmann factors for excited states [$\exp(-\chi/kT)$] tend to be quite small and vary rapidly through the atmosphere; moreover, the changes due to the dissociation of molecules and the ionization of atoms may well be very drastic, leading to significantly different depths of formation for spectral lines of equal strengths. The systematic errors that this leads to in the curve-of-growth (hereafter abbreviated as COG) analyses may be partly compensated for by a differential approach whereby lines of roughly equal strength and, if possible, equal excitation are intercompared. However, it is possible to check for the systematic errors only after comparisons with results from a detailed model atmosphere analysis. Nevertheless, COG analysis and single-layer approximation was not merely the basic method for most of the pioneering work in this field—it has remained in use, for example, in the analysis of spectra of long-period variables, where the classical model atmospheres in hydrostatic equilibrium seem highly inadequate and lead to very unrealistic abundances (51). In such applications, the "excitation temperature" tends to lose its physical meaning and is instead reduced to a fitting parameter. The method merely turns into an inter-/extrapolation scheme, the validity of which must remain in serious doubt until more realistic models have been constructed.

Progress in the theoretical modeling of cool stellar atmospheres has been considerable during the last few decades. In particular, numerical methods have been developed and put into practice for calculating models, permitting detailed consideration of the enormous number of spectral lines affecting the structure of the atmospheres. Similarly, the molecular data are currently being increased and improved. This development has already been summarized in comprehensive reviews (33, 88, 114, 115), so that in the following, only a very schematic outline is presented, with some comments on the most recent developments.

The grid models published hitherto are classical in the sense that basic assumptions have been made of plane-parallel stratification, hydrostatic equilibrium, convective energy transport according to the at-best approximate mixing-length recipe, and LTE. Within this framework, the frequency and depth dependence of the spectral line absorption has been considered in a reasonably realistic way. Model grids for M giants and supergiants have been published by Tsuji (225, 228) and by Johnson et al. (116). A small grid of models for M dwarfs was computed by Mould (156). A grid of models for S-type stars has been published by Johnson (113); this grid also includes models for cool carbon stars. Carbon star models have also been produced by Querci et al. (178) and by Querci & Querci (179). An extensive carbon star grid, as yet unpublished, has been calculated by the Uppsala group (64). These grids of models include a realistic consideration of most important opacity contributors, with certain exceptions. The data for the H_2O absorption are still preliminary. The polyatomic opacities of HCN and C_2H_2 in the carbon star models, recently found to be of vital importance for the structure of models with $T_{eff} < 3000$ K (63, 117, 118), are included in the Uppsala grid. The importance of these opacities has been found to be less by Tsuji (232), but this seems to be because his models contain fewer molecular bands for the polyatomic molecules. The calculated intensities of the weak HCN overtone bands, which are of great significance for the structure of the Uppsala models, have recently been verified in laboratory experiments for some bands (204). The polyatomic opacity also seems necessary for these stars to evolve to the observed low effective temperatures, an aspect consistent with the hypothesis that dust formation plays a role for their mass loss mechanism (145). Also, other polyatomic molecules are of importance for the structures of these stars [e.g. Jørgensen et al. (118) found the contribution of C_3 to be of significance]. In addition, grain opacities may severely affect the atmospheric structures and the spectra for the cooler stars.

The numerical methods for treating blanketing are different for the grids of Johnson et al. (113, 116)—models that we hereafter call the Indiana models—and for the grid calculated at Uppsala. The different methods are discussed in (33) and (61). In spite of these differences, the Indiana and the Uppsala models tend to agree well when the same input physics is used.

The models of Mould (156) and Tsuji (225, 228) have a less detailed blanketing treatment.

An important property of cool star atmospheres that is demonstrated by these grids is the great sensitivity of the atmospheric structures to the chemical compositions of the stars (cf. Figure 1). This mainly reflects the dominant importance of the line blanketing, e.g. of CNO molecules in the carbon stars or the oxides (including H_2O) in the M stars. For the cooler stars, the abundances of rather rare electron donors (alkali metals) are also of key importance for the H^- opacity. This abundance sensitivity of the model atmospheres must be considered in the chemical analysis, such that a model consistent with the abundances derived for the important elements is ultimately used. It should be noted that it is not clear that this procedure leads to a unique solution: Model atmospheres with different chemical compositions may, in certain cases, show very similar spectra [see, e.g., (78)]. Models with identical fundamental parameters may also show different spectra. It has been suggested that the thermal inhomogeneity or bifurcation traced in the upper atmospheres of the Sun and Arcturus is related to the surface cooling of CO (8–10). Kneer (128) suggested that the atmospheres of cool stars may be affected by a radiative instability due to the high temperature sensitivity of the formation of CO and other cooling molecules. This instability may be reflected in double solutions to the classical problem (128, 161, 162, 167). A similar effect, due to SiO, may exist for cooler stars (163). Polyatomic opacities in the upper

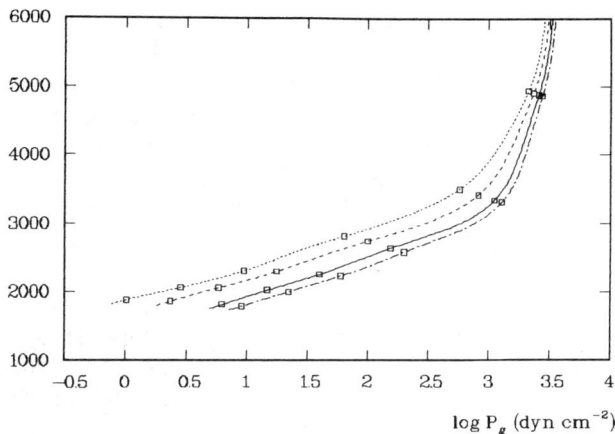

Figure 1 The temperature-pressure structures of four different carbon-star model atmospheres from the Uppsala grid (64), all with $T_{\rm eff} = 2800$ K and $\log g = 0.0$ (cgs units). The chemical abundances are solar with the following exceptions: solid line—$\log$ (C/O) = 0.01, dashed line—$\log$ (C/O) = 0.04, dotted line—$\log$ (C/O) = 0.13 (all three with solar oxygen and enhanced carbon), dashed-dotted line—[O/H] = -0.3 and $\log$ (C/O) = 0.01. The squares along each curve, going from left to right, mark logarithmic Rosseland optical depths of $-4, -3, -2, -1, 0.0$, and 1, respectively. The figure illustrates the rather great dependence of the atmospheric structure on the chemical composition.

layers of M-star or carbon-star models may, under certain conditions, lead to drastically different double solutions to the model atmosphere problem (62, 200). These results should be borne in mind for any classical model atmosphere analysis for cool stars. Further studies of the conditions for such multiple solutions should be performed, and physically more realistic dynamical simulations of these instabilities should be developed.

An important question is how reliable are these classical models, i.e. how large are the errors that the errors in opacities and in the basic assumptions of the plane-parallel stratification, etc., generate when the models are used for deriving the chemical abundances?

The study of model errors in abundance analyses of cool stars tends to show that with the high S/N ratios obtainable today in observed spectra, the model uncertainties (including errors in the calculated spectra) may be the dominant source of errors in the resulting abundances. There are different ways of studying and reducing these errors. One may try to construct more self-consistent models and use them, at least for checks; or one may look for differences between observations and model predictions and study whether such differences carry information on the shortcomings of the models.

The first approach demands construction of (a) models that are spherically symmetric; or (b) models that are out of hydrostatic equilibrium with systematic flows or that have improved treatments of convection and thermal inhomogeneities; or (c) models in statistical equilibrium, where LTE is no longer assumed. In view of the theoretical problems involved, supporting observational evidence is of key importance in such efforts. In fact, as was nicely illustrated by Tsuji (237), the theoretical and empirical approaches toward better modeling of cool stellar atmospheres are intimately linked. Some recent developments are mentioned here.

Spherically symmetric models for M giants were calculated by Watanabe & Kodaira (252, 253), Schmid-Burgk et al. (199), Kipper (125), and Scholz (200). A positive feedback was found—the cooling of the surface layers changed the molecular equilibria and the opacity such that the sphericity effects increased. The effects are serious where the atmospheric extensions are greater than 5% of the stellar radius, which occurs when the luminosity of the star is greater than about 10^3 $L_\odot$ (199). These effects are sensitive to the chemical composition (200, 254), which may in principle be used in the future to derive radii and masses of stars spectroscopically (203). This possibility was recently investigated further by Bessell et al. (20) on the basis of more detailed models and synthetic spectra, with combinations of narrowband colors being suggested for this purpose. For carbon stars, the sphericity effects were found to be less by Scholz & Tsuji (202); this study, however, was based on straight mean opacities for diatomic and some polyatomic molecules and should be repeated with more realistic absorption.

Measurements of the wavelength dependence of the angular diameters of cool giants and supergiants (including the phase dependence for Miras) is a promising tool of photospheric diagnostics. When reducing these measures, one should note, however, that the center-to-limb variation may show a complex behavior in these extended stars (14, 201). Interferometric techniques may be used to trace some of this behavior, e.g. as regards the extended atmospheres around nearby supergiants [cf. (237) for further references], and to give valuable information on departures from spherical symmetry and hydrostatic equilibrium [see, e.g., (36)].

The hydrogen ionization convection zone of cool stars starts deep in the atmospheres and does not, in the local mixing-length approximation, affect the thermal structure of the layers of the atmosphere where the visible spectrum is formed. The H_2 dissociation convection zone, however, is located in the upper atmospheric layers. Even though the convective fluxes are not great, they may well affect the thermal structures considerably, at least in dwarfs and subgiants, since the derivative of the convective flux relative to the depth, $dF_{conv}/d\tau$, may be substantial.

Very little is actually known about the importance of convective overshooting in the cool giants and supergiants. One should stress the point made by Nordlund (167) for the Sun—namely, that the heat balance in the upper standard model atmospheres is qualitatively different from that in the actual Sun. In the model atmospheres, radiation at different frequencies heats and cools the gas, whereas in the star, the adiabatic expansion in the upward-moving gas cools and radiation heats. Self-consistent simulations of convection in solar-type stars, with radiative transfer in the inhomogeneous atmospheres taken into account as appropriate, have been carried out by Dravins and Nordlund [Dravins et al. (55–59)] and show a very satisfactory agreement with observations of line shifts and asymmetries (55–59). These simulations are, however, not too easily extended to cool giants and supergiants, since empirical information as regards the relevant scales of convection is missing. CO line asymmetries and line shifts in cool giants have been observed by Ridgway & Friel (185) and Nadeau & Maillard (164).

An interesting possibility for empirically studying atmospheric inhomogeneities in cool stars is offered by spectropolarimetry. The enhanced polarization of Ca I $\lambda 4226$, found in several Miras, is thought to have a photospheric origin and may be a suitable diagnostic in such studies (28).

Our empirical knowledge about systematic motions in the photospheres of cool giants has grown enormously as a result of the Fourier Transform Spectrometer—from Maillard's discovery of double CO lines in Miras (146) to studies by Hinkle et al. of shocks in long-period variables (105, 106). Knowledge about less dramatic dynamic phenomena has also increased considerably—from the observations of low-amplitude mass motions in the M-supergiant α Ori (30) and in the irregularly variable carbon star 19 Psc (174) to the probable discovery of quasi-static molecular envelopes between the classical photospheres and the wind regions (236) or of cool quasi-static turbulent condensations in the chromospheres (236, 238). No self-consistent modeling with these phenomena taken into consideration has been possible as yet. The details of the physical processes are complicated; the interaction between macroscopic phenomena (e.g. shocks, radiation, and magnetic fields) and microscopic processes (formation of molecules and dust) is only tentatively understood. However, important steps have been taken in this direction [see, e.g., (77, 83)]. Models for the outer layers of long-period variables were recently published by Bowen (27). Periodic shocks are obtained, and the models are enormously extended by them (cf. Figure 2). The gas compression in the shocks greatly enhances the dust formation, and the mass loss increases. Although these

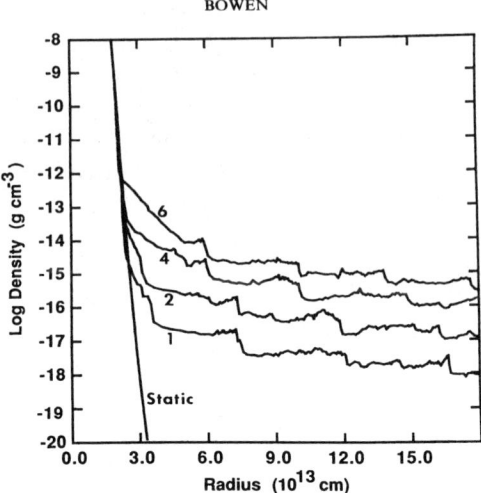

Figure 2 Density vs. radius at phase 0.0 for Mira star models ($P_0 = 350$ days, $T_{\rm eff} = 3000$ K, $M = 1.2\ M_\odot$), calculated by Bowen (27). The structures are denoted by the different assumed velocity amplitudes (in kilometers per second) for a driving piston at the lower boundary. The enormous extension of the models, relative to the static model, is obvious. Figure taken from Bowen (27).

models are still schematic, they very clearly illustrate the need for a realistic treatment of the dynamic processes in the atmospheres of the variables. Bessell et al. (21) have recently constructed Mira model photospheres, based on pulsational models and measured CO line profiles. The observed properties, such as the wavelength dependence of the diameter and the IR low-resolution spectra, are reproduced quite well for shorter period Miras. For the cooler stars, the agreement between observed and calculated fluxes is not so satisfactory. The Mt. Stromlo–Heidelberg group (21) has convincingly demonstrated the great sensitivity of the observable line profiles and fluxes to the fundamental parameters of the Miras.

A first theoretical study of departures from LTE in cool model atmospheres was that of Auman & Woodrow (7), who obtained considerable overionization effects; however, these effects were overestimated, since the ultraviolet blocking by spectral lines was not considered. Subsequently, empirical studies (168, 169, 183, 184) tended to support these results, even though the first reports gave highly exaggerated effects due to underestimated stellar temperatures. The existence of the overionization effects was more firmly established by Ruland et al. (189)—the effects found for Fe I in β Gem (K0 III) are also qualitatively reproduced by statistical equilibrium calculations (216).

Molecular lines within the electronic ground state were found by Thompson (218) and Hinkle & Lambert (102) to be formed in LTE—as regards the electronic systems, however, there may be serious depatures [see, e.g., (11, 35) for CO V-R bands and (160) for the violet CN system]. The electronic transitions may in fact be strongly decoupled from the local gas temperature and would then probably affect the thermal structure of the surface layers less than the LTE models imply [cf., e.g., the great surface heating of the absorption from the TiO electronic systems found in M-giant LTE models (129)].

Methods are now available for treating complex non-LTE problems with model atoms having a considerable number of states and transitions in static or moving atmospheres (4, 41, 196). In principle, such relatively detailed statistical-equilibrium calculations for the appropriate atoms and molecules can be coupled into the computation of model atmospheres. A major problem, however, is the continuing absence of relevant collision cross sections for inelastic collisions of atoms and molecules with electrons, hydrogen atoms, and hydrogen molecules.

2.2 Fluxes and Temperatures

The possibility of testing the model atmospheres by comparing them with observed flux distributions and spectra may contribute further evidence concerning the failure of the LTE approximation and of other basic assumptions. Here, only a few comments are made on fluxes and colors, with emphasis on the determination of effective temperatures [for more complete discussions, see (115, 237)].

A systematic comparison between observed and calculated fluxes and colors for cool stars has yet to be made. However, at least for the early M-star models, the present situation looks satisfactory. Tsuji (227) was able to rely on calculated colors sufficiently to establish a theoretical calibration of $R-I$ vs. T_{eff} for M giants, in satisfactory agreement with angular-diameter measurements. Later, applying the so-called IR flux method by Blackwell & Shallis (23), he used the L-band flux, relative to the integrated total flux, to establish a new temperature calibration (231) that was found to agree with the angular-diameter-based T_{eff} scale of Ridgway et al. (187). A great advantage of the IR flux method is its rather small model sensitivity due to the small temperature sensitivity of the flux in a wavelength band far to the red of the flux maximum.

Piccirillo et al. (176) calculated the near-infrared Wing colors, using the Indiana M-star models, and found good agreement with observations. In particular, they demonstrated the importance of the veil of weak TiO lines, which significantly decreases the "continuum band" around 7540 Å even in α Tau (K5 III). Steiman-Cameron & Johnson (217) reported a satisfactory agreement between calculated and observed red and IR colors for early-type M giants. However, problems seem to occur for $T_{\text{eff}} < 3500$ K, in that the calculated TiO strengths seem too weak. This is interpreted as a consequence of the sphericity of the stellar atmospheres (neglected in the models), a conclusion that seems to be supported by the improved agreement reached by Bessell et al. (20) for their spherically extended models. However, Bessell et al.'s TiO band strengths tend to be exaggerated as a result of their piecewise straight-mean opacities.

For the cooler M dwarfs the temperature calibration is still uncertain. The recent spectrophotometry beyond 2.2 μm by Berriman & Reid (19) suggests unexpectedly low effective temperatures as compared with those of main sequence models. This may indicate that these stars are contracting brown dwarfs instead of hydrogen-burning stars. Detailed modeling of their fluxes seems well motivated.

Spectrophotometric scans of S stars were compared with Indiana models by Augason et al. (6) with satisfactory, though not perfect, agreement, and a temperature scale could be attempted.

For the carbon stars, Tsuji (230) established a temperature scale, again using the IR flux method with the L band as the reference band. He corrected empirically for the 3-μm polyatomic absorption, which was not considered in the models [from (178, 179)]. For stars with lunar occultation measurements by Ridgway et al. (187), a fair agreement was found. However, the models with HCN and C_2H_2 absorption that have been calculated by Eriksson et al. (64) seem to produce polyatomic bands that are too strong [cf. (142, 255)]. This discrepancy is currently being investigated. The reason for it seems nontrivial, and some doubts as regards the temperature scale for carbon stars are in order as long as the discrepancy has not been satisfactorily explained. The well-known problem with the unknown violet opacity source in the N stars [cf. (237) for references] is less important for the temperature scale in view of the small fraction of the flux in that wavelength region.

The empirical lunar occultation temperature scales, as well as the semiempirical scales derived by Tsuji (230, 231) by means of the IR flux method, may be used to calibrate other temperature criteria [cf., e.g., Lambert et al. (142), who calibrated several criteria for N stars].

The present temperature scales seem to admit effective temperature determinations for M giants and N stars to an accuracy of about 200 K. For S stars the scale is somewhat more uncertain, partly owing to the lack of relevant photometric data [cf. (52)]. One should note, however, that possible systematic errors in temperatures (e.g. as a function of chemical composition) may well occur and are hard to trace, since the number of lunar occultation diameter measurements is too small to reveal such effects.

Instead, reliable model fluxes must be obtained and used in systematic comparisons with observations.

2.3 Surface Gravities

It is important to obtain a good estimate of the gravity of a cool star for a reliable determination of its abundances. For instance, an error of 0.3 dex in the surface gravity of M giants may lead to errors in the Y and Zr abundances of typically 0.2 dex (206), and a similar error in log g of cool carbon stars may cause errors of about 0.1 dex in the O and C abundances (142).

The problem of determining the surface gravities of red giant stars is, however, nontrivial [cf. the illuminating discussions for α Boo (K2 III) (222) and for the dwarf carbon star G77−61 (78)].

The spectroscopic methods available for estimating surface gravities of G and K giants—based on the study of ionization equilibria, dissociation equilibria, and damping wings of strong lines—have not yet been used much, nor even tested, for the cool giants. The ionization equilibria are difficult to use owing to the shortage of sufficiently strong lines from ions and the low degree of ionization for abundant elements. A way to circumvent this is to use lines of neutral atoms with different ionization potentials (e.g. Ca and Fe). In typical cool stars, Fe is mostly neutral while Ca is partly ionized, and thus if a solar Fe/Ca ratio is assumed, one may derive the surface gravity from the strengths of Ca I and Fe I lines. This method was used, for example, by Kilston (124) for N-type stars and tested by Lambert et al. (141) for α Ori (M1–2 Ia–Iab), but it seems dangerous, in that the results may be systematically affected by overionization.

The use of molecular equilibria for obtaining log g is not frequent, not even for earlier giants [cf. (18) for an attempt to derive log g for α Boo from MgH lines]. The temperature sensitivity of many pressure-sensitive molecules is problematic; however, the richness of the molecular spectra should be a great advantage in attempts to avoid blends and find suitable lines. Studies of the possibilities that this method offers, even at low resolution [cf. (20)], for cool giants should be rewarding.

One may also try to find pressure-sensitive anomalies in molecular band strengths and exploit them for gravity determinations. One example is the abrupt termination of the CuH $A-X$ (0, 0) band at low J'' values. This band was identified in the spectrum of 19 Psc (N0, C6, 2) by Wojslaw & Peery (258) and used for estimating the surface gravity.

The method of Blackwell & Willis (22), in which the damping wings of strong lines are measured as pressure criteria, and the corresponding abundances are derived from weak lines with the same lower excitation energy, has the advantage of being insensitive to the thermal structure of the model atmosphere [cf. (18)]. This method, as far as I know, has not yet been used for cool giants. Attempts to do so should be made, initially for S stars, where the blending problems should be at a minimum.

The method that is totally dominant today for deriving the surface gravity is to adopt an absolute magnitude of the star and a stellar mass (e.g. from evolutionary tracks) and then derive the surface gravity from these quantities, an appropriate bolometric correction, and the effective temperature. This method leads to surface gravities accurate to about 0.5 dex, which may be sufficient for many purposes. A drawback, however, is that possible variations of, for example, mass with chemical composition are not taken into consideration, which may lead to systematic errors and erroneous tendencies in the relative abundances derived for a set of stars. Examples and discussions of this method are provided in, for example, (206) for M and MS stars, (54) for SC stars, and (142) for C stars.

2.4 Line Broadening, Synthetic Spectra

In the determination of chemical abundances, every effort should be made to measure weak lines [log $W(\lambda)/\lambda < -5$], whose equivalent widths are most dependent on the chemical composition and not very sensitive to the physically badly understood photospheric velocity field. However, the blending problems and continuum location problems may prohibit the use of weak lines or at least enforce additional use of a number of intermediate-strength lines. Even if weak lines are the primary criteria, it often is necessary to correct them and the continuum points for the blending effects of stronger lines by means of synthetic spectra.

In the absence of unblended weak lines, a microturbulence parameter ξ_t may be estimated from comparisons of observed spectra with synthetic ones with different values of ξ_t. The great number of free parameters in such fittings (assuming that the chemical abundances are not known beforehand) and the uncertainties of the upper layers of the model atmospheres (which affect the strong lines) should, however, enforce a rigorous care in such a procedure and a simultaneous variation of all free parameters in the estimation of errors. This is still more the case if there are reasons to distrust the continuum owing to interference from lines not included in the synthetic spectra. One should also note that the macroturbulence parameter is important in many of these fittings [cf. (234)].

Olson & Richer (170) have used an interesting cross-correlation technique in the comparison between observed and synthetic spectra for carbon stars. A minimum of judgment is needed in the application of the procedure: No continuum has to be drawn, and no spectral features of particular interest have to be selected. However, the weights given by the method to different features are not under full control—therefore, uncertain features in the calculated spectra may disturb the final results severely without this being obvious. Systematic errors that would be revealed by a traditional spectrum comparison may thus remain undiscovered. A detailed inspection of the spectra and the final fits seems to be necessary as a complement to this method.

It is important to realize that a good fit of an observed crowded spectral region does not prove that the abundances of the corresponding models are those of the star. Even if synthetic spectra with different abundances are presented and seem to be rather different, one might still fit an observed spectrum with astonishingly different sets of abundances and microturbulence parameters, owing to the strong correlation between spectral effects of different parameter changes. In particular, we must warn against COG effects—even if calculated spectra seem to imply that a particular abundance cannot be reduced, it may well be possible to increase it significantly without any noticeable spectral effects.

It has been found difficult to reproduce the really strong molecular lines, such as the $\Delta v = 1$ and $\Delta v = 2$ CO V-R lines and the strong V-R lines of OH for cool giants, with a microturbulence parameter derived from somewhat weaker lines [cf., e.g., (141, 142, 236, 238)]. Whether this may indicate velocity gradients, failures of the models to reproduce the upper layers of the atmospheres, thermal inhomogeneities, circumstellar absorption, or (possibly) departures from LTE is not clear—one should, however, be careful not to ascribe weight to the inner cores of the strong spectral lines in the abundance determinations. In the dwarfs, the pressure-broadened wings of strong lines may also cause serious problems, owing to missing data for the damping constants.

Another problem in the determination of abundances in the cool stars is the hyperfine splitting (hfs) and isotope shifts of the atomic spectral lines. If these effects are neglected, severe errors may occur in the abundances when derived from saturated lines. [Thus, for example, for the Tc $\lambda 4262.24$ line in MS and S stars, the corresponding abundance correction is 0.5–0.7 dex (207).] If these effects are taken into consideration, it is often done in an approximate way, by adding a nonthermal Gaussian broadening to the line profiles, the amount of which is determined from, for example, solar spectra. A detailed synthetic spectrum calculation is to be preferred, provided that hfs and isotopic splitting data exist—which is, however, often not the case.

3. METALS

3.1 *Criteria and Methods*

The overwhelming richness of molecular lines in spectra of cool stars makes any chemical analysis from a set of scattered atomic lines highly risky. These lines should be chosen to be free of blends, or the effects of these blends must be estimated. Also, the continuum level must be possible to trace in spite of the mat of molecular lines. Obviously, the importance of high spectral resolution and of high signal/noise ratios cannot be overemphasized in this situation. If the continuum cannot be traced, the only possibility left is synthetic spectroscopy. This method is, however, not very reliable in the extreme cases of heavy line blocking, since it requires a detailed knowledge of all the important opacity contributors, including the disturbing molecular lines.

The difficulties may make even a differential approach dangerous or sometimes impossible. For example, the analysis of s-element abundance changes along the spectral sequence M–MS–S–SC–C in practice requires that the abundance criteria are changed, not only because the abundances tend to increase such that the lines gradually saturate, but also because the intervening molecular bands change drastically. In particular, the difficulties in measuring weak atomic lines are often very great. These difficulties are less severe for abundance determinations from sets of rotational molecular lines, since each molecular band produces a fair number of lines with a predictable variation of line strength. This can be used for selecting lines as free of blends as possible. Also, lines from several different bands, different systems, or even different molecules may be used for cross-checking in an abundance determination. It is reasonable to assume that future abundance determinations for metals in cool stars will exploit the molecular bands further—at present, they have mainly been used for estimating CNO abundances, for studies of isotopic ratios, and as low-resolution abundance criteria (calibrated by synthetic spectra or by high-resolution studies of atomic lines).

Another problem with the atomic lines is that our need for accurate gf values for rather weak lines of neutral atoms is often not met by existing laboratory measurements, nor by accurate calculations. Also, in order to avoid blending, one often prefers lines in the infrared, where laboratory data are still more scanty. The possibilities of using the Sun as a standard source and deriving "astrophysical gf values" may be restricted, owing to the weakness of these lines in the solar spectrum. Instead, a late-type star may be used as a standard, provided that its fundamental parameters are known well enough. For instance, α Tau (K5 III) has been used for this purpose [cf., e.g., (207)].

Beyond the fact that the lines for abundance analysis ought to be weak and not strongly affected by blends, what recommendations may be given for selecting them? If possible, one should choose lines with such excitation of the lower level that they are not particularly temperature sensitive, when ionization and molecular formation are taken into consideration, in order to avoid effects of errors in the effective temperature and in the temperature structure of the model atmosphere. Since the gravities of late-type giants and supergiants are notoriously uncertain, one should also try to avoid lines that are very gravity sensitive.

If possible, one should try to select lines from the dominant species of the relevant element. This is particularly tricky for elements like Ti, which varies rapidly with decreasing effective temperature from mainly Ti II to mainly TiO [cf., e.g., (156, Figure 4)]. This recommendation is of importance, not only for compensating the temperature uncertainty, but also because the ionization and the dissociation equilibria may be severely affected by non-LTE effects.

The detailed study by Ruland et al. (189) of β Gem showed systematically low Fe I and Ti I abundances from low-excitation lines as compared with those from high-excitation lines. This may be understood as a result of overionization by nonlocal hot radiation in the outer layers of the stellar atmosphere, where the low-excitation lines are predominantly formed.

Several other suspected non-LTE effects on abundance determinations have been reported for G and K giants, although these are not fully understood [17, 31; cf. also (122, 123)].

One should try to minimize the non-LTE effects by selecting high-excitation and weak spectral lines, formed at large atmospheric depths, where the collisions may be more active in establishing the excitation, ionization, and dissociation equilibria, and the radiation fields should be more local and isotropic.

Obviously, the combination of all the recommendations listed above may be unrealistic—in practice, one may have little choice in selecting the appropriate criteria. One example is the Tc abundance determination in S stars, for which just a few lines in heavily crowded regions are available— a still more extreme example is the Li $\lambda 6708$ line, heavily blocked by molecular absorption in cool stars. For such features, critical studies with detailed synthetic spectroscopy are necessary. In general one should explore further spectral regions (in particular, in the infrared beyond 1 μm) and try molecular lines. The need for such efforts may become clear from the subsequent survey of metal abundance determinations.

3.2 M Giants and Supergiants

The first pioneering COG analyses of β Peg (M2 III) and of other M giants by Yamashita (262, 263) and Yamashita & Utsumi (264) led to astonishingly low metal abundances. The situation was further investigated in an important study of Huggins (109), who used photographic spectra at 17 Å mm^{-1} in the wavelength range 8000–8350 Å, where the TiO absorption was assumed to be small. Scaled solar model atmospheres were employed in a differential analysis relative to α Boo for eight M giants of different temperatures. He found that the titanium abundance, which was judged to be the most well-determined abundance, decreased successively with decreasing effective temperature. Also, the microturbulence parameter derived from the Ti I lines deviated systematically from that derived from Fe I lines. These tendencies were ascribed to an error in the model atmospheres—the inability of scaled solar models to properly account for the molecular blanketing effects, which grow with decreasing effective temperature. Other possible explanations, such as errors in the effective temperature scale, in the ionization equilibrium of Ti, or in the continuum drawing, were not found to be very probable. Huggins illustrated the importance of calculating refined model atmospheres for cool stars, and he thus stimulated model-maker interest in contributing new grids of models.

In a study of supergiants and the Galactic abundance gradient, Luck & Bond (143) included chemical analyses of six M supergiants earlier than M3. This study was based on Reticon spectra at a resolution of 0.3 Å in the wavelength region 6260–6800 Å. The observations were analyzed relative to the Sun, and Indiana models were used for the M stars. The effective temperatures were estimated from excitation equilibria and the gravities from adopted absolute magnitudes and masses. The [Fe/H] values were in general estimated to be accurate to 0.2 dex. Abundances for 16 different chemical elements were derived. Luck & Bond did not find any evidence for s-processing or other anomalous element abundances in their stars. The overall metal abundances of the M stars, however, seem low (86)—for 5 out of the 6 stars, the [Fe/H] range was $-0.40 \leq$ [Fe/H] ≤ -0.11. These values are significantly lower than those of most G and K supergiants analyzed in the same study. It seems tempting to ascribe this not very probable result to effects of weak absorption lines across their spectra, in particular from the γ' system of TiO [cf. (144)]. Other explanations (departures from LTE or from plane-parallel stratification) could also be possible.

In his recent thesis, Vieira (244) analyzed α Ori (M1–2 Ia–Iab), for the first time using FTS spectra in the wavelength region 11,100–13,500 Å. He succeeded in identifying a great number of mostly weak lines from 12 different neutral atoms. Uppsala models (with no TiO blanketing) were used in the analysis, but they were shown to lead to results close to those when an appropriate model from the Indiana grid was used. An effective temperature of 3900 K [following Tsuji (229)] and a logarithmic gravity of 0.0 were adopted. Vieira found a depletion of iron relative to its solar abundance by -0.6 dex, while most of the other elements were less depleted. This result is dependent on the effective temperature adopted—

a value of $T_{eff} = 3600$ K would, however, only increase the iron abundance by about 0.2 dex, while the abundances of the other elements would be affected more. One could fear absorption across Vieira's wavelength region by TiO or other molecules; however, if present, this absorption must be evenly distributed in wavelength in view of the numerous continuumlike regions in the spectrum. Differential studies of M giants relative to α Boo or to α Tau in this wavelength region would be of interest.

Smith & Lambert (206, 207) have recently derived CNO abundances and metal abundances for 11 M giants, 8 MS stars, and 2 S stars. The metal lines were observed in Reticon spectra at 0.2-Å resolution, covering the interval 7400–7580 Å [redward of the (0, 0) band of the TiO γ system] and a region between 9980 and 10,100 Å, which was "quite free from blanketing." About 30 atomic lines (from Ti I, Fe I, Ni I, Sr II, Y II, Zr I, Ba I, and Nd II) were used. The analysis was performed relative to α Tau using Indiana models. Effective temperatures were essentially derived from $V-K$ colors, with the calibration of Ridgway et al. (187). Although Fe II lines could be identified in the earliest M stars, ionization equilibria were not used to establish the gravities, since an attempt to use this method was considered a failure for α Tau. (It led to $\log g = 0.8$, corresponding to a mass of about 0.3 $M_\odot$. This result was ascribed to overionization.) Instead, $M_V(K)$ and masses from evolutionary tracks yielded gravity estimates with an estimated error of about 0.3 dex. The microturbulence parameters (with consistent results from Fe, Ti, Ni, and Co) range in the interval 1.8–3.1 km s^{-1}, with a scatter that seems partly physical.

The metal abundances derived by Smith & Lambert (206, 207) constitute a major improvement in the analysis of M and MS stars. Spectral lines on the linear part of the COG were reached for many elements. The errors in the abundances of several elements may be estimated to be about 0.2 dex, and on the basis of consistency the authors conclude that "the differential analysis relative to α Tau has reduced systematic errors to an insignificant level." The lack of correlation between [Fe/H] and T_{eff} suggests that no serious non-LTE effects or unconsidered TiO blocking affects the results. The authors, however, note a tendency for [Ni/H] and [Ti/H] to decrease with T_{eff}, reaching a value around -0.3 dex near $T_{eff} = 3200$ K. They ascribe this tendency to departures from LTE, qualitatively similar to those traced for Fe and Ti by Ruland et al. (189) but quantitatively greater for the M stars than for the K stars. The Ni and Ti lines have lower excitation energies and are formed closer to the surface, where overionization may occur, than the typical Fe lines.

Another possible non-LTE problem is posed by the strength of the strong Sr II λ10,036 line, which implies enhanced LTE abundances (by up to 0.5–0.8 dex), even for stars that show no signs of enrichments of other s-elements, such as Y. This effect is strongly luminosity dependent and may be caused by model inadequacies in the outer layers or by overionization, similar to the effects traced for δ Oph (M0.5 III) by Oinas (169).

A major result of Smith & Lambert, expected but satisfactory in view of previous studies, is that the [Fe/H] values for the M and MS stars all lie in the interval -0.2 to $+0.2$. A more unexpected result is that four of the stars classified as MS do not show any measurable enhancements ($[s/\text{Fe}] \leq 0.2$) of the s-process elements. The additional results of Smith & Lambert for S stars, and for CNO abundances, are commented on below.

3.3 M Dwarfs

The analysis of metal contents in M dwarfs is complicated by several circumstances. The high pressures, which grow with decreasing effective temperature and decreasing metal abundance to 10^6–10^7 dyn cm^{-2} in the line-forming region, may broaden the strong lines such that they show a small or even reversed metal abundance sensitivity (157, 172). Furthermore, the depletion of neutral atoms by formation of molecules, such as TiO, may be more important in metal-rich stars and thus lead to a decrease in the Ti I line strength with increasing Ti metal abundance (157). This occurs, in particular, for strong lines of low excitation formed in the surface layers, where the formation of molecules with moderate dissociation energies is most important. Obviously, weak lines of high excitation should be preferred in abundance studies—however, they are difficult to use owing to the uncertain gf values, to the blending from molecular lines and from wings of strong pressure-broadened lines, and to the general faintness of the stars, which may lead to noisy spectra and/or unsatisfactory spectral resolution.

Pioneering work in the quantitative analysis of M dwarfs was done by Mould (157, 158). In his first paper he presented an analysis of Kapteyn's star (HD 33793, M0 V) and Yale 5817 (M4 V), based on image-tube spectra at 10 Å mm^{-1} in the wavelength region 8380–8590 Å. About 15 lines from Ti I and Fe I were used, mostly located on the shoulder of the curve of growth. However, the equivalent widths were hardly small or accurate enough for a confident separation between abundance and microturbulence effects. A risk that the $\Delta v = -2$ sequence of the TiO γ bands depresses the continuum for cooler M dwarfs in this spectral region was noted. Mould's second analysis (158) included four M stars—for a fifth (Barnard's star), the temperature is so low ($T_{\text{eff}} \simeq 3000$ K) that the spectrum could not be "guaranteed to be uncontaminated by H$_2$O absorption." This latter work was based on Kitt Peak National Observatory (KPNO) FTS spectra at 0.3 cm^{-1} resolution, covering the 4000–6600 cm^{-1} region. About 20 lines from Na I, Ca I, Al I, Mg I, Ti I, and Fe I were used. Most of the lines were fairly strong, but for Ti I a number of weak lines were also measured. However, the analysis of these lines is hazardous for the reasons mentioned above; errors of ± 0.3 dex are estimated for the Ti abundances.

Both of these analyses were based on Mould's (156) grid of models for M-dwarf atmospheres. The effective temperatures were obtained from the photometric gradient around 1 μm (159). Gravities were chosen in the interval $4.5 \leq \log g \leq 4.9$; although there is a considerable uncertainty in this parameter, it does not seem very important for most abundances. The microturbulence parameters adopted were 2.5 and 2.0 km s^{-1}, respectively—the need for a higher microturbulence parameter for, e.g., Kapteyn's star than for the Sun seems to be significant and might possibly reflect the convective instability due to H_2 formation at shallow optical depths in these atmospheres. One should note, however, that the convective velocities for these stars, as estimated from the mixing-length recipe, are quite small (0.5 km s^{-1}). The gf values were estimated from the solar spectrum with certain modifications: In the case of the weak Ti I lines, they were derived from a cooler star [61 Cyg B (K7 V)], since the lines were not visible in the Sun.

Mould derived mean metal abundances around [Fe/H] = -0.3 for three of his five M stars. The two remaining stars are found to be more metal rich, one of them (Gliese 205) quite so ([M/H] $\simeq 0.5$, with the uncertainty claimed to be about 0.15 dex). This latter star seems to be a member of the HR 1614 group, which was found to be metal rich by Eggen (60).

No significant nonsolar abundance ratios (X/Fe) are apparent in Mould's results, with the exception of oxygen, which is discussed below.

Another early study of M dwarfs and late K dwarfs was the work by Hartmann & Anderson (100) on six stars, three of which are emission-line flare stars. Echelle spectra in the interval 5900–6600 Å with a resolution of 0.10–0.15 Å were used. Note that the central part of this region may be depressed by the $\Delta v = 0$ bands of the TiO γ' system, and that an adequate continuum location is particularly difficult in echelle spectra when such broad spectral features are of importance [cf. (82)]. Hartmann & Anderson located lines from a considerable number of metal atoms, and using model atmospheres extrapolated from the grid of Carbon & Gingerich (34) (which, as they noted, are underblanketed owing to the lack of molecular blanketing), they derived roughly solar abundances with estimated errors on the order of 0.2 dex.

In spite of the considerable difficulties that meet those who attempt accurate analyses of M-dwarf spectra, it should be possible with existing techniques to substantially enrich our knowledge about the chemical composition of these stars. One major motivation for such studies (as was stressed by Mould) is the lack of knowledge about the physical spread in [Fe/H] and other elemental abundances for M dwarfs—with further bearings on the understanding of star formation and Galactic evolution. An accurate calibration of broadband colors of M dwarfs [e.g. their location in the $(J-H)$–$(H-K)$ diagram in terms of metal abundance] would be particularly valuable for many applications [see, e.g., (215)].

3.4 S Stars

Since the early recognition by Merrill (153, and references therein) that the lines of Zr and Y and their oxides are enhanced in S-type stars and his remarkable subsequent discovery of Tc in S stars (154), much effort has been spent in attempts to trace these enhancements for other heavy elements as well and to estimate the magnitude of the corresponding chemical overabundances.

The first attempts at quantitative abundance analysis of an S star were by Fujita (68) for χ Cyg; later, Tsuji (223) tried to analyze HD 216672 and HD 22649 using near-infrared spectra. The results were certainly not definitive. Boesgaard (24) estimated the Zr/Ti abundance ratios in a differential COG study of late-type giants, including 15 M or S stars. This was based on high-dispersion spectra in the yellow-red spectral region and produced abundance ratios that tend to agree within one order of magnitude or better with results of more recent work, to be discussed below. Another important early study was that of Tsuji (224) on the composition of two S stars.

In an empirical COG analysis, Smith & Wallerstein (209) determined Tc abundances in four S and SC stars, together with one Ba star. The Tc I $\lambda 5924.57$ intercombination line was measured in coudé spectra at 5–7 Å mm^{-1}; in addition to this, about 30 fairly strong lines of Ti, V, Zr, Nb, Mo, and Ru were measured and corresponding abundances estimated. The analysis is regarded as preliminary by the authors—in a more definitive analysis, synthetic spectra should be employed, not least for the Tc I $\lambda 5924$ blend, and weaker lines should (if possible) be used. Hyperfine splitting was not considered for Tc, which has led to somewhat overestimated Tc abundances. All of Smith & Wallerstein's stars show a tendency for (V/Ti) to be greater than the solar ratio; the reality of this is, however, disputable. The Zr, Mo, and Ru abundances, relative to that of Ti, are found to be significantly increased (by one or two orders of magnitude) and follow rather well the s-process predictions as regards relative enrichments. Tc is found to contribute to the equivalent widths of the $\lambda 5924.57$ feature in R CMi (SC4–6.5/10e) and CY Cyg (SC2/7.5). The Tc/Mo ratios indicate that s-elements were produced in these stars and mixed to the surface about 2–4×10^5 yr ago, while the Nb/Zr ratios suggest that such processes also occurred prior to this event a few million years ago. [The only stable Nb isotope is bypassed in the s-process and formed from the decay of ^{93}Zr, with a half-life of 1.5×10^6 yr (147).]

In a more recent study, Dominy & Wallerstein (51) measured many V, Zr, Nb, and Tc lines in the violet spectrum in high-resolution (2.4 Å mm^{-1}) photographic spectra of χ Cyg (S6–9/1–2e), Mira (o Ceti, M6e) and R Lyr (M5 III). Here as well, an empirical COG method was used and motivated by the fact that the line-forming regions in Miras are not adequately represented by classical hydrostatic equilibrium (cf. Section 2.1, above). The temperatures of the single layers are obtained from the Zr excitation equilibrium, while the electron pressures are fortunately not vital for the relative abundance ratios, since the elements studied are mostly neutral and do not associate much in molecules. Hyperfine splitting is taken into consideration in the equivalent Doppler-width approximation. Important error sources are the continuum drawing difficulties, the blending problems (including blending by weak emission lines), the uncertainties in gf values and in the velocity parameter, as well as the errors due to the stratification; nevertheless, the authors estimate the maximum errors in the abundance ratios relative to vanadium to be on the order of 0.3 dex, which may be optimistic. Important results are that Zr and Tc are found to be enriched in χ Cyg by 0.4 and 1.7 dex, respectively, whereas Nb is solar in χ Cyg and R Lyr. Although Tc is present in Mira, remarkably Nb seems significantly underabundant in this star by about 1 dex. Smith & Wallerstein suggest as an explanation that a very small pulse of neutrons has converted some ^{98}Mo into ^{99}Tc and most of the ^{93}Nb into ^{94}Mo.

Later, Wallerstein & Dominy (248) extended this study to three S stars, one MS, and one M giant. They found ranges for the Nb/Tc ratios and Nb abundances of 1.5 dex and 2.1 dex, respectively, indicating great differences in the degree of s-process enhancement and the time since the most recent shell flash occurred. The authors ascribe a Tc abundance greater by 0.4 dex than that derived by Smith & Lambert (207) for a common star to their own lower spectral resolution. In view of the uncertainties inherent in the analysis, these results must still be considered preliminary, in particular for the long-period variables.

The enrichments of Y, Zr, Ba, and Nd obtained by Smith & Lambert (207) for their S stars lie in the interval 0.8–0.6 dex for HR 1105 and 0.45–0.20 for HR 8714. These authors also estimated Tc abundances for six stars from the λ4262.24 resonance line by making Reticon observations at 0.2-Å resolution and then applying a synthetic spectrum technique and allowing for hfs (which was found to be very essential). Comparison between observed and calculated flux ratios $[F(\lambda = 4250 \text{ Å})/F(\lambda = 10{,}000 \text{ Å})]$ showed that the Indiana models are severely underblanketed in this spectral region—moreover, the line strengths calculated for an appropriate model with solar abundances are far too strong to match the M-giant prototype β Peg (M2 III). This shortcoming of the theory was ascribed to a veil of densely packed lines that is missing in the models. In an attempt to circumvent this, Smith & Lambert reduced all atomic gf values by a factor of around 3 until a good match to the spectra of β Peg was obtained. The same reduction factor was applied for all program stars and for the Tc line. Also, Zr I lines were observed in the same wavelength region, and the Tc/Zr abundance ratio was first determined. By means of the Zr abundances derived from lines in the infrared, this ratio could then be converted to a Tc abundance.

The Tc abundances found are more or less reduced relative to the neighboring elements and the predicted s-process production, probably as a result of β-decay of Tc. The Tc/Zr ratios range from no measurable underabundances to a reduction of almost a factor of 10, which indicates that some of the stars have spent on the order of 10^6 yr in the asymptotic giant branch (AGB) phase. This is consistent with expected lifetimes on the AGB. [A considerable fraction of the MS and S stars show, however, no visible Tc lines at all; cf. the survey by Smith & Lambert (208), who ascribe this fact to the existence of evolved barium stars, presumably produced at earlier stages by mass transfer from an AGB companion.]

Kipper & Kipper (127) also analyzed the violet Tc I lines, using model atmospheres and synthetic spectra, but they obtained significantly higher Tc abundances. Smith & Lambert (207) suggest that this discrepancy is due to the fact that Kipper & Kipper neglect hfs and use the heavily blended $\lambda 4297$ line; this may give Tc abundances almost one order of magnitude greater than the $\lambda 4262$ line.

Smith et al. (210) recently extended the study of s-elements to a survey of Y, La, and Nd abundances (relative to Fe and Cr) in 34 MS and S stars, with 14 M giants as a control group. One line from each element was measured in the wavelength region 7430–7530 Å, and a calibration of the line strength ratios was established with model atmospheres and synthetic spectra. A range of neutron exposures was derived and found compatible with the hypothesis that the s-elements in the Galaxy were produced in such stars.

Finally, we mention the analysis of the SC star UY Cen (S6/8-) by Catchpole (37), based on spectra at 13 Å mm^{-1}, covering the wavelength region 5630–8550 Å. A great number of mostly strong lines from many chemical elements were measured and analyzed in an empirical COG analysis. Alternatively, model atmospheres from the early grid of Johnson (112) and, as a third alternative, from the grid of Gingerich et al. (84) were used in the analysis. These different approaches were found to give drastically different results in absolute abundances; for example, Ti was found to be underabundant by a factor of 15, overabundant by a factor of 21, or solar for the three approaches! The choices of effective temperature and gravity were, however, different for the different models selected. The problems with continuum drawing in this spectral region, as well as the problems of blends, of hfs and uncertain gf values, and of the fact that most lines are very strong and thus sensitive to the temperature structure in the uppermost layers of the models, suggest that the error estimates (< 0.4–0.5 dex relative to Ti) may be too low for several elements.

The s-element enrichments relative to Ti, found by Catchpole (37) for UY Cen, are typically 1 dex or more.

3.5 N Stars

The strong lines from s-process elements in cool carbon stars were noted early (119, 192, 198), but the first pioneering abundance analyses were not attempted until the middle of the 1960s by Fujita & Tsuji (72) for Y CVn and by Hirai (107) for WZ Cas.

In an attempt to avoid the heavy molecular line blocking in the red and near-infrared, Utsumi (240, 241) studied the N-star spectra between 4400 and 5000 Å and found two regions (4400–4500 Å and 4750–4900 Å) that seemed relatively free of strong CN, C_2, and Merrill-Sanford (SiC_2) bands. He used spectra of 22 N stars and an absolute empirical COG analysis to derive abundance ratios relative to iron for almost 20 metals. This study has been modified and extended recently (242, 243), and results have now been presented for 24 N stars and 6 cool R stars. About 170 lines were measured from photographic spectra in the wavelength regions mentioned above at a resolution of 13 Å mm^{-1}. The lines are fairly strong and are located on the flat part of the COG. Again, an absolute empirical curve of growth method was used, with gf values mainly adopted from the list of Kurucz & Peytremann (130). Microturbulence parameters ranging from 6 to 12 km s^{-1} were obtained. These parameters are much greater than those derived by Lambert et al. (142; see below) for the same stars. Abundances relative to iron were derived for 13 metals. The accuracy in these values was estimated to be about 0.4 dex, except for a few elements in some stars. Thus, the few lines from Sc or La might be seriously blended by ^{13}CN in the J (^{13}C-rich) stars.

The derived abundances must be regarded as quite preliminary in view of the great problems involved in blends and continuum drawing [cf. the spectral tracings published by Utsumi (240)], the considerable uncertainties in fitting parameters [θ(exc), θ(ion), ξ_t] and gf values, and the neglect of hfs and stratification effects. This is also true for the errors estimated for these abundance values. Note that modifications implemented into the analysis between the years 1970 and 1985 caused reductions of the abundance ratios of the s-elements relative to Ti by a factor of 10–100, due to the new choices of gf values, of ionization temperatures, and some further attempts to consider blending in the ^{13}C-rich (J-type) stars. The main result is that in the "normal" N-type stars, the heavy metals Y, Zr, and Ba are enriched relative to Ti by a factor of 10–100, whereas the rare earth enrichment is about a factor of 10. On the contrary, in the J-type stars the abundances of the s-process elements are found to be nearly solar.

This interesting tendency is supported by the survey of line strengths in carbon stars made by Dominy (50), although a couple of counterexamples are brought up, such as DY Per (C4, 5) which displays weak Sr I, Zr I, and Ba II lines but is not a J-type star. Dominy tentatively explained the anticorrelation of ^{13}C and s-process enhancements as the result of ^{13}C destruction by the ^{13}C$(\alpha, n)^{16}$O reaction, which provides the neutrons for the formation of neutron-rich isotopes.

Kilston (124) performed a study of eight representative N-type stars, using 8 Å mm^{-1} spectra in the wavelength region 5600–6800 Å. Lines of 10 metals were identified, in addition to C_2 and CN bands—the discussion of the latter two bands is deferred to Section 5. The analysis was performed by using a constant-temperature Minnaert model (based on the Milne-Eddington model) to calculate synthetic spectra. The gf values were based on laboratory measurements or were derived from the solar spectrum. The temperature of the model was determined from the excitation equilibrium of Fe I, supported by CN excitation, while the electron pressure was derived from the ionization equilibrium of Ca I as estimated from the Ca I/Fe I ratio [assuming the abundance ratio $\varepsilon(Ca)/\varepsilon(Fe)$ to be solar]. Hyperfine splitting was not taken into account explicitly. A microturbulence parameter of 6 km s^{-1} was derived.

Kilston found the Zr/Ti ratios to be a factor of 2–10 greater than the solar value, while Ba was enriched only in some high-Zr stars. There is only moderate agreement between Kilston's results and those of Utsumi (242, 243); in particular, Kilston found the J-star Y CVn (N3, C5, 4) to be overabundant in Y, Zr, and La by about a factor of 10 or more. A modern model atmosphere study with detailed synthetic spectroscopy of optimal spectral criteria will be needed before more definitive statements can be made about the s-element abundances in cool carbon stars of various types.

One illustration of the difficulties is the recent attempt by Kipper (126) to derive the Tc abundances of two carbon stars from the $\lambda5924.47$ line. Kipper's spectrum synthesis shows that this line is severely blended by CN lines, and only an upper limit of the abundance could be derived.

In a study of CNO abundances in 30 cool carbon stars by Lambert et al. (142; to be commented on below), based on FTS spectra in the 1.5–2.5 µm spectral region at 0.07 cm^{-1} resolution, a search for useful atomic lines in the 2-µm region was performed. Out of the 20 candidate lines, 5 were selected that were sufficiently weak [log $W(\lambda)/\lambda < -4.8$, which is, however, still not on the linear part of the curve of growth in several cases], not too seriously blended, and measurable in the solar photospheric spectrum or in that of α Tau. The list contains one (tentative) Na line, one Ca line, and three Fe lines. The [Fe/H] values derived from the model atmosphere analysis of these lines lie in the interval $-0/3 <$ [Fe/H] < 0.3, while the means of [Na/H] and [Ca/H] come out somewhat smaller (about -0.3 dex). This may suggest an overionization effect. It seems safe to conclude that N stars are neither significantly metal poor nor metal rich, relative to hydrogen, as compared with disk population and Population I stars. Since the continuous opacity is provided by H$^-$ free-free absorption, this result also seems to exclude a drastic hydrogen deficiency among most normal N stars.

An exceptional star is the dwarf carbon star G77−61, which as judged from the very weak or absent atomic lines (e.g. Na D lines) and the absent metal hydride bands, has been estimated to be very metal poor (79). The discovery of periodic radial velocity variations by Dearborn et al. (47) suggests that this very extreme Pop II dwarf has been polluted by carbon-enriched material from a more massive companion star.

4. LITHIUM ABUNDANCES

The determination of Li abundances in cool stars is of considerable interest. Lithium is destroyed in the interior of stars and is further diluted in red giant atmospheres by the deep convection zone, in which Li-poor matter is dredged up from the interior, but it may be produced again (e.g. in the AGB stage), probably by the "beryllium transport mechanisms" (32). The amount of Li visible on the surface is sensitive to mixing processes that are not very well known. Therefore, Li abundances are interesting diagnostics for studies of these processes. The Li abundance results discussed below are solely based on the strength of the Li I $\lambda 6708$ resonance doublet. The surrounding wavelength region is severely blended by molecular lines in cool stars.

4.1 M Stars

The first quantitative survey of lithium in M giants and supergiants was made by Merchant (Boesgaard) (26, 151) on the basis of photographic high-dispersion spectra. Abundances were estimated with a standard COG technique for 30 stars. Lithium abundances ranging from $\log \varepsilon(\text{Li})\simeq 1.5$ (two times the solar value) to less than 1/200 thereof were found.

Luck & Lambert (144) published abundances of Li (and Al, as derived from the neighboring doublet) for 31 M giants and supergiants. These results were based on high-resolution (0.1-Å) high-S/N Reticon spectra and a synthetic spectrum analysis. A tendency for the Al abundances to decrease with decreasing effective temperatures was found and explained as the result of a veil of weak TiO lines (e.g. from the γ' system) not included in the line list of the synthetic spectrum program. In the coolest stars this depression may exceed 20% of the continuum intensity. The abundance results of Luck & Lambert agree reasonably well with those of Merchant (Boesgaard). This is, however, fortuitous and results from the fact that the low-temperature scale used by Merchant (Boesgaard) compensates her neglect of blends with atomic and TiO lines. The mean Li abundance found by Luck & Lambert is about $\log \varepsilon(\text{Li}) \simeq -0.2$, with several stars showing $\log \varepsilon(\text{Li}) < -1.0$. Two supergiants did not show any detectable Li line.

Hänni (93) derived Li abundances in nine M giants on the basis of photographic spectra at 4.6 Å mm^{-1} and synthetic spectra. For the three stars in common with the Luck & Lambert study, the abundances found by Hänni are systematically smaller by 0.08–0.24 dex, which seems to be the result of her neglect of the type of veil absorption assumed by Luck & Lambert across the Li doublet wavelength region.

4.2 S Stars and N Stars

The fact that the Li I $\lambda 6708$ feature is enhanced in S-star spectra was noted by Keenan & Teske (120). Boesgaard (25) also made the pioneering COG analyses of Li in 13 MS and S stars and found that these stars could be grouped into Li-poor stars [$\log \varepsilon(\text{Li}) \sim 0.5$], Li-rich stars ($\sim 2$), and the super-Li-rich star T Sgr (S5/6-e) (~ 4.3). Since then, several quite Li-rich Se, SC, and CS stars have been reported and analyzed (38, 39, 251). [In passing, one should note the interesting fact that super-Li-rich stars, apart from having a carbon/oxygen abundance ratio close to 1.0, also seem to show abnormally strong CO absorption at 2.3 μm (257).]

The first reported super-Li-rich star was, however, WZ Cas (N1p, C9, 1) [McKellar (149)], with an equivalent width of $\lambda 6708$ of 8 Å. A first attempt to estimate the Li/Na abundance ratio for this star was made by Spitzer (214). Subsequently, more N stars were added to this group (65, 193). In their COG analyses, Fujita & Tsuji (72) and Hirai (107) found values of the Li/Na ratio for Li-rich N stars more than 10 times greater than the primordial solar system value [$\log \varepsilon(\text{Li}) = 3.3$]. Kilston (124; cf. Section 3.5) derived Li abundances for eight N stars with his synthetic spectrum technique and found log ε(Li) values ranging from 3.5 to about 0.3.

The strong Li I resonance lines are formed in the upper atmospheres. The possibility that the super-Li-rich phenomenon may be caused by cooled outer atmospheric layers was discussed by Cohen (46), who compared the Li line strength with the K I λ7699 resonance line (which is of about equal strength in these stars, but not significantly strengthened as compared with more Li-poor stars). From this, she concluded that the Li enhancement is real and not due to surface cooling. Another possibility would be that the formation of λ6708 is affected by departures from LTE (e.g. due to the hot radiation field from the chromosphere). De la Reza & Querci (48) performed statistical equilibrium calculations for Li in carbon-rich model atmospheres and found substantial effects that seem to mainly weaken the lines as a result of overionization. These results are, however, given for models hotter than the temperature of, for example, WZ Cas (142), in the atmosphere of which Li should be more neutral. Systematic surveys of Li abundances for S and N stars with contemporary techniques would be of great interest, but a deeper physical study of these lines and their formation is also required. For instance, it would be of interest to monitor them at high S/N and high spectral resolution in various cool stars.

5. CNO ABUNDANCES

5.1 *Criteria and Methods*

The determination of the abundances of carbon, nitrogen, and oxygen in cool giants is of particular interest, since we may expect the atmospheres of these stars to be affected by interior H burning in the CNO cycles and by He burning and subsequent mixing. The CNO analyses are in practice founded on lines of molecules and radicals. Atomic lines that could be tried would be the forbidden C I lines λ8727 (186), λ9824, and λ9850, or the [O I] lines λ6300 and λ6363. The first-mentioned line has been identified in carbon stars (71), but its equivalent width is difficult to measure, since it is blended with CN, as are the rest of these carbon lines (136). The [O I] lines are severely affected by overlying absorption from TiO in the M stars (141) and from CN in the N stars.

The molecular equilibrium calculations of Russell (190) demonstrated the drastic effects on the molecular equilibria that occur in cool stellar atmospheres when the C/O ratio varies from C/O < 1 to C/O > 1. [In fact, Russell referred to C. D. Shane (198) and to R. H. Curtiss—"years ago," according to Russell—as regards the origin of this explanation for the carbon-star spectra.] This branching of the late-type spectral sequence was further discussed in terms of chemical abundances by Fujita (66, 67). Of key importance for this phenomenon is the fact that the CO, SiO, and CS molecules are so strongly bound that almost all available atoms are locked into these molecules, to the extent that this is possible.

This dichotomy makes the methods for measuring CNO abundances different depending on which side of the C/O = 1.0 point the star is (i.e. whether it is an M, MS, or S star, or a C star). For the M, MS, and S stars, spectral CO features are primarily carbon abundance criteria, since almost all the carbon is bound in CO molecules. For the carbon stars, the CO abundance instead reflects the oxygen abundance. The electronic systems of CO are located in the vacuum ultraviolet. Of the V-R bands, the lines of the fundamental ($\Delta v = 1$) at 4.6 μm are in practice too strong in cool stars to be suitable abundance criteria. The lines of the first overtone ($\Delta v = 2$) at 2.3 μm are also saturated to a very great extent, although a smaller number of weak high-excitation lines may be found and used. The most suitable CO lines are in many cases the second-overtone ($\Delta v = 3$) bands around 1.6 μm. This spectral region is, however, heavily affected by CN and H_2O absorption in the M stars and by CN and C_2 in the carbon stars, so that great care must be exercised in picking out CO lines free of blends and in drawing the continuum. One should also note that the continuous opacity is at a minimum at 1.6 μm, which strengthens the spectral lines and makes the continuous flux dependent on the temperature structures of the deeper layers of the atmospheres. In certain cases this may introduce further uncertainties into the analysis, as compared with the use of lines at longer or shorter wavelengths.

For M and MS stars, other oxides may be used as oxygen abundance criteria. In particular, the V-R lines of the OH radical ($\Delta v = 1$ around 3–4 μm or $\Delta v = 2$ around 1.7 μm) have been used with consistent results {with the exception of abundances from the strongest and lower excitation bands, the lines of which form in the outer atmospheres, where the structure is uncertain [cf. (207)]}.

Other possible oxygen abundance criteria for M stars are bands from H_2O or TiO. When using metal oxides, one must make independent, accurate determinations of the abundance of the metal atoms and, for certain temperature intervals, very accurate determinations of the temperature and gravity.

The carbon abundances of M and S stars may also be derived from C_2 lines and CH lines, or even CN lines provided that the nitrogen abundance is measured by other means. In order to find spectral features of these molecules free of blends and weak enough for an accurate abundance estimate in M stars, one would probably nevertheless have to proceed out into the infrared beyond the J window, where the CO V-R lines can be measured, anyhow.

The carbon abundances in N stars have been estimated from the strengths of C_2 lines (Swan bands, Ballik-Ramsay, or Phillips systems) and from the CH V-R lines, first identified by Ridgway et al. (188) and, independently, by Lambert et al. (142).

The nitrogen abundances are usually derived from lines of the red CN system; the NH V-R lines in the 3–4 μm region have also been used for M stars (141) and N stars (142). The abundances of CN and NH molecules in a stellar atmosphere are controlled by the formation of the relatively stable N_2 molecules. This reduces the sensitivity of NH and CN lines to nitrogen abundance variations: The equivalent widths of such weak lines are proportional to the square root of the nitrogen abundance.

The determination of abundances from molecular lines relies on accurate data for dissociation energies, partition functions, and oscillator strengths. The data for the CNO molecules are discussed in (142) and (141) [see also (236)]. As an example, we note the ~0.1-eV uncertainty in the dissociation energy of the CN radical, which leads to errors in nitrogen abundances for M stars and C stars of about 0.4 dex.

5.2 M Stars and S Stars

A first attempt to determine CNO abundances in M (and S) stars from infrared molecular bands was made by Spinrad & Vardya (211). As a result of crude model atmospheres, misidentifications, and saturated lines, this and other early attempts (e.g. 15, 81) have mostly historical interest. A more recent model atmosphere analysis of α Ori (M1–2 Ia–Iab) by Tsuji (229) was based on V-R lines of CO ($\Delta v = 2$) and OH ($\Delta v = 1$) and CN lines from the Red system. It yielded solar O, enhanced N, and severely depleted C abundances (by a factor of 20 relative to that of the Sun). Tsuji adopted a microturbulence parameter of 9 km s^{-1}, and, as was pointed out by Lambert et al. (141), this choice is a major reason for the low C abundance—the CO $\Delta v = 2$ lines are saturated and thus more sensitive to microturbulence than to CO abundance.

Lambert et al. (141) performed a detailed and fundamental study of α Ori using high-resolution FTS spectra through the atmospheric windows from 1.4 to 5.3 μm. At least the region between 3 and 4 μm was found to be clear after removal of the telluric absorption by ratioing to a line-free reference spectrum, while the 1.6-μm region is more crowded. Here, a relation between line depth and equivalent width was first established for unblended spectral lines, and then line depths of the remaining lines were converted to equivalent widths by using this relation. Lambert et al. used lines from the CO $\Delta v = 3$ and the OH V-R $\Delta v = 1$ bands to derive the C and O abundances, respectively. The NH V-R $\Delta v = 1$ lines were used for obtaining the N abundance. Indiana models were used with $T_{\text{eff}} = 3800 \pm$

100 K and $\log g = 0.0 \pm 0.25$. The former value was derived from IR fluxes and was found to be consistent with the excitation of the CO lines. The gravity was estimated from the absolute magnitude and the location of the star in the HR diagram as compared with evolutionary tracks and checked with angular diameter measurements. A number of lines around 1 μm of neutral metal atoms with different ionization energies gave a gravity estimate of $\log g = -1$, provided that solar relative abundances were assumed. This lower value was tentatively explained as the result of departures from LTE (overionization of the easily ionized elements), but it may also simply reflect an underestimate of T_{eff} by 100 K. A microturbulence parameter of 4 km s^{-1} [i.e. a value considerably lower than that adopted by Tsuji (229)] was derived from the CO $\Delta v = 3$ lines of different strengths.

A marginal inconsistency appears in the analysis of Lambert et al. (141) for the N abundance, which is 0.25 dex lower as obtained from the CN lines than from the NH lines. The reason for this may be thermal inhomogeneities—the temperature sensitivity of the CN lines is systematically different from that of the other lines—or departures from LTE for the CN Red electronic system. An overestimated dissociation energy of CN is not a very probable explanation, since the choice of $D_0^0 = 7.52$ eV is close to the low end of the wide range suggested by recent determinations [cf. (13)]. Another inconsistency is the deviating oxygen abundance (by about 0.4 dex) from the weak 5-4 OH lines, as compared with that of the 4-3 and 3-2 bands; the reason for this discrepancy is probably erroneous gf values. Also, the strong OH and CO lines are not successfully predicted by the theory. This might reflect a systematic velocity gradient, departures from LTE, or the existence of a molecular shell around the star (236). However, this should not be important for the success of the abundance analysis, since it is based on much weaker, and in many cases unsaturated, lines. The star is found to be somewhat carbon poor relative to the Sun (by 0.4 dex) and to have an enhanced nitrogen abundance (+0.6 dex) and a somewhat reduced oxygen abundance (-0.2 dex). These results are consistent with theoretical predictions of a first dredge-up of CNO-processed material. The oxygen depletion may be primordial or reflect the operation of the ON cycles, but it is not significant. The errors in the abundances are at least ± 0.15 dex and may well be considerably greater. Yet, this work denotes a major step forward in the analysis of CNO abundances in M stars, since weak lines in spectra where the continuum is clearly seen are used.

A similar approach was used by Smith & Lambert (206, 207) in their studies of 21 giants of type M, MS, and S (discussed above), with some slight methodological modifications relative to the α Ori investigation. It should be noted that the nitrogen abundances derived from the NH lines and the CN lines are in excellent agreement in (207), which suggests that the dissociation energy (here chosen to be 7.60 eV) is not very much in error. The problem with the strong OH lines found for α Ori also shows up for the other giants. The OH lines, and thus the oxygen abundances, are also found to be more sensitive to the temperature structure of the models than the CO, CN, and NH lines. The C/O abundance ratio increases along the sequence M–MS–S, as expected, but this result is dependent on the consideration of the structural changes of the increasing C/O ratio in the model atmospheres.

The resultant CNO abundances of these studies are consistent with the dredge-up of material enriched by hydrogen burning in the CNO cycles (corresponding to carbon converted to nitrogen) for the M stars, while the MS and, in particular, the S stars show signs of an addition of fresh ^{12}C. One should note, however, that the C/O ratios obtained for the S-type stars span a rather wide range, and that high Zr abundances relative to Ti may be more of a reason for an S-type classification for many stars [cf. (137)] than is a carbon abundance close to that of oxygen.

Tsuji (234) has analyzed FTS spectra of the 2.2–2.5 μm region for 18 M giants. He determined microturbulence parameters and CO abundances from the $\Delta v = 2$ CO lines, using his own model atmospheres (228) and effective temperature calibration (231). The gravities were determined from the giant locus in the HR diagram, with a mass of 3 $M_\odot$ adopted for all stars. Temperatures and gravities agree well with those of Smith & Lambert (206, 207) for the four stars in common. However, the microturbulence parameters of Tsuji are systematically higher by typically 1.0 km s^{-1}, and the [C/H] abundances are considerably smaller, by typically 0.5 dex. If the smaller microturbulences of Smith & Lambert were adopted, it seems that the carbon abundances would have increased to values close to those derived by Smith & Lambert [cf. Tsuji (234), Figures 1b,c]. Tsuji has found a considerably smaller number of reasonably weak spectral lines in the $\Delta v = 2$ bands than Smith & Lambert measured in the generally weaker $\Delta v = 3$ bands, and it may be tempting to ascribe the difference between the results of the two studies to the small number of reliable weak lines in Tsuji's study. However, Tsuji's (237) further analysis indicates that the inconsistency in ξ_t and C/H derived from the $\Delta v = 2$ and $\Delta v = 3$ lines, respectively, is real. It may imply serious difficulties in the classical models for the line-forming layers (presumably for the stronger $\Delta v = 2$ lines) or systematic differences in line blanketing, continuum location, or in unrecognized opacity sources.

In a special study of α Her (M5 Ib–II), Tsuji (233) also used high-resolution FTS observations in the 1.6-μm region. He derived C, N, and O abundances from lines in CO $\Delta v = 3$ bands, in the $\Delta v = -1$ sequence of CN, and in the 4–0 band of OH, respectively, using a model atmosphere and spectrum synthesis. The microturbulence parameter ($\xi_t = 3$ km s^{-1}) and isotopic ratios were derived from CO $\Delta v = 2$ lines. The carbon abundance was found to be reduced by 0.6 dex relative to the solar value (with an inconsistency between $\Delta v = 2$ and $\Delta v = 3$ lines); N is correspondingly increased by 0.9 dex, while oxygen is slightly deficient (-0.1 dex). Errors of 0.3–0.4 dex were ascribed to these values. The low carbon abundance may be at least partly the result of the neglect of sphericity: Höflich et al. (108) found a solar carbon abundance from analyzing the first overtone CO lines with spherically extended models. However, this study used rather arbitrarily chosen parameters.

The CNO abundances of a few M dwarfs were discussed by Mould (157, 158; see above). By combining data on Ti I and TiO for Kapteyn's star, Mould found an O/C ratio close to the solar value, $-0.3 < [O/H] < 0.3$. In his second study, FTS observations at 0.3 cm^{-1} resolution of the CO $\Delta v = 2$ and OH lines were used to estimate the C and O abundances of six M dwarfs. The CO line strengths were found to be consistent with solar C/Fe ratios, while the old disk subdwarfs Gliese 15A and 411 (with $[M/H] \leq -0.5$) were found to have greater O/C ratios than the M dwarfs with higher metallicities. A value of about $+0.15$ seemed to be required for [O/C] for the subdwarfs. Such a tendency would be consistent with more recent, and more accurate, oxygen abundance determinations for F-type dwarfs of different metallicities [see, e.g., (166)].

5.3 SC Stars and Carbon Stars

Bands of oxides and carbon molecules are weak or absent in SC star spectra except for CO bands. This is interpreted as the result of a near-balance between carbon and oxygen and the high binding energy of the CO molecule. Molecular equilibrium calculations (e.g. 110, 175) suggest that the balance is very even; the abundance ratio C/O can be estimated to an accuracy of a few percent or better for these stars by a mere inspection of their spectra [cf. (121)].

Assuming equal carbon and oxygen abundances, Dominy et al. (54) analyzed three SC stars on the basis of FTS spectra in the 2.3-μm region. The CO $\Delta v = 2$ lines were used for estimating the C (and O) abundance, while the $\Delta v = -2$ sequence of the Red CN system was used for the abundance of N. For a fourth star, the Mira variable RZ Peg (SC6–9/9e), which shows stronger C_2 bands and thus clear signs of being more carbon rich, the authors used lines from the C_2 Phillips system (0, 2) band for estimating the carbon abundance. The effective temperatures were obtained from IR photometry, which was compared with calculated model colors and supplemented with temperature estimates from spectra. As usual, surface gravities were derived from estimates of absolute magnitudes and masses. The microturbulence parameters were obtained from unblended ^{12}CN lines.

The spectra were analyzed with synthetic spectra and Indiana models. Typical errors in the abundance results due to uncertainties in fundamental parameters are ± 0.1 dex, while the uncertain C/O ratio (with a possible range from 1.00 to 1.01, except for RZ Peg) causes a range for the N abundance of about 0.2 dex. The uncertainty in $D_0^0(CN)$ ($= 7.52$ eV) leads to an additional N uncertainty of about 0.3 dex. The resulting C and O abundances are below solar values by typically 0.4 dex, while the N abundances are greater than the solar value, though not by enough to be explained only in terms of CNO burning (i.e. a plain conversion of C and O to N). Dominy et al. (54) invoked CNO burning at high temperatures ($T \sim 25 \times 10^6$ K) to reduce the oxygen abundance, succeeded by α-captures on ^{14}N and ^{18}O up to ^{22}Ne. [The ^{18}O abundance is found to be low (see below).] Adopting these CNO abundances, one must assume a rather complicated nuclear history, with extensive mixing, for these stars. The CNO abundances of RZ Peg are about 1 dex lower than the solar values. The authors explain this as a consequence of the inadequacy of using model atmospheres in hydrostatic equilibrium for the line-forming region in this Mira variable.

Fujita & Tsuji (72) attempted the first determination of the N/C ratio for a carbon star (Y CVn), using CN and C_2 bands and a COG method. Other pioneering (but now obsolete) studies of carbon abundances in N stars (177, 221) led to high N/C ratios and quite low O abundances—similar to those expected from CNO burning at equilibrium. The reasons for these failures were errors in the oscillator strengths and inadequate model atmospheres or COG analyses.

Kilston's (124) work questioned these results. In his synthetic spectrum analysis with primitive model atmospheres of eight N-type stars, he used C_2 Swan lines ($\Delta v = -3$) and CN Red system lines ($\Delta v = 4, 5$). The C_2 lines were used for deriving $(C-O)/H$, and the CN lines for obtaining N/H. The nitrogen abundances were found to be essentially the same as the solar value for all the stars. Since such a low N abundance did not support the idea of CNO-cycle equilibrium, Kilston argued that an enrichment by fresh ^{12}C from helium burning was a more probable explanation for the carbon-star phenomenon.

Querci & Querci (180) made a new study of UU Aur and Y CVn, based on FTS spectra from 1 to 2.5 μm at 0.3 cm^{-1} resolution and using their model atmospheres and synthetic spectra with CO $\Delta v = 2$ and $\Delta v = 3$ lines, as well as the Phillips and Ballik-Ramsay bands of C_2 and the Red CN lines. The observed spectra could be fitted with synthetic ones, provided that low nitrogen abundances were adopted, which gave further support to Kilston's conclusions. The effective temperatures needed to fit the spectra were, however, found to be significantly higher than those obtained by other methods.

Thompson (220) analyzed the CO $\Delta v = 3$ bands in two long-period variables with comparatively weak C_2 bands (i.e. with $C/O \simeq 1.0$), using model atmospheres and synthetic spectra. He showed that it was not possible to reach the observed strength of the CO bands if one assumes that carbon and oxygen are significantly depleted below solar abundances.

In a recent study by Lambert et al. (142), the CNO abundances and $^{12}C/^{13}C$ ratios of 30 Galactic cool carbon stars were derived from FTS spectra in the 1.6- and 2.2-μm regions. Some of the most important results from this and other studies by Lambert and collaborators are summarized in Figure 3. Lines from the C_2 Phillips system $\Delta v = -2$ sequence were used as a carbon abundance criterion (or, rather, a criterion for the abundance difference between carbon and oxygen), with the C_2 Ballik-Ramsay system (0, 0) lines as checks; unfortunately, the number of weak C_2 lines was not great enough to ensure a carbon abundance determination independent of the microturbulence parameter. The CO $\Delta v = 3$ lines were used for determining the oxygen abundance, and the $\Delta v = -2$ lines from the CN

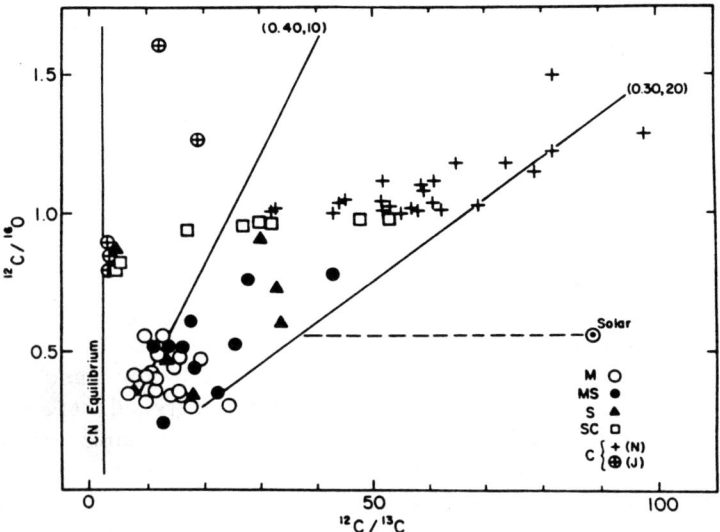

Figure 3 $^{12}C/^{16}O$ and $^{12}C/^{13}C$ ratios for the cooler giant stars of different types. Solid lines represent addition of pure ^{12}C to M giants, with $^{12}C/^{16}O$ and $^{12}C/^{13}C$ ratios indicated within the parentheses. This figure, taken from Lambert (137), is one representation of the significant progress achieved in the chemical analysis of cool stars in recent years.

Red system were used for the nitrogen abundance. In addition to this, vibration-rotation lines of CH and NH and the $\Delta v = -3$ sequence of the CN Red system were measured in the 3.6–4.2 μm region for some stars; moreover, the 3-μm HCN lines were studied in 19 Psc. These additional measurements were used as consistency checks for the analysis. Note that Morris & Wyller (155) early on stressed the importance of including NH and HCN lines in the analysis.

The analysis was carried out with Uppsala models, with polyatomic opacities from HCN and C_2H_2 taken into account. The measured strengths of the H_2 quadrupole 1–0 $S(0)$ transition, which are sensitive to the pressure in the outer photospheric layers, were found to be reproduced reasonably well by these models. Effective temperature estimates were based on the scales of Ridgway et al. (187) and Tsuji (230), on which further photometric and spectrophotometric criteria were calibrated. Some confirmation of these temperatures was obtained from the excitation equilibrium of CO. The logarithmic gravities were estimated from available data on absolute magnitudes and masses and were found to lie in the interval $-1.1 < \log g < 0.3$; a typical value of $\log g = 0.0$ was adopted for all the stars, as well as a solar overall abundance of metals (i.e. essentially $[M/H] = 0.0$ for the important electron donors). This latter assumption is supported by the simultaneous analysis of a small number of metal lines, discussed above (Section 3.5). The microturbulence parameters were derived from ^{12}CO $\Delta v = 3$ and ^{12}CN $\Delta v = -2$ lines, and these different determinations were found to agree quite well. A mean value of $\xi_t = 2.2$ km s^{-1} was established, with a (probably real) scatter of about 0.5 km s^{-1}.

A detailed error analysis was carried out with the following results: For the warmer stars with lower C/O ratios, the "probable maximum errors" were estimated to be 0.3 dex in [O/H], 0.5 dex in [N/H], and 0.01 dex in the logarithmic abundance ratio $\log(C/O)$; for the cooler stars, the corresponding numbers were 0.3, 0.7, and 0.04. Dominating the (known) sources of error were errors in continuum location, in polyatomic opacities, and effects of possible dust emission, as well as errors in the adopted surface gravity and metal abundance. For the N abundance, the errors in the atmospheric structure and in the uncertain dissociation energy of CN (a value of $D_0^0 = 7.60$ eV was preferred) and of C_2 ($D_0^0 = 6.11$ eV) are also important.

The consistency checks show a very good agreement between calculated and observed 4-μm CN and NH lines, which lends support to the low nitrogen abundances derived. The single 3-μm HCN lines of 19 Psc also agree fairly well, in contrast to the low-resolution observations, which give HCN band strengths considerably weaker than the calculated ones (cf. Section 2.2 above). The observed CH V-R lines are much weaker than the calculated ones—the difference corresponds to a factor of 5–50 in the CH abundance, depending on the effective temperature. The reason for this discrepancy is as yet unknown, but it is not trivial. Different possibilities discussed by Lambert et al. are great systematic errors in the effective temperature scale, and the effects of departures from LTE in the dissociation and excitation equilibrium of CH. Neither of these explanations seem particularly probable [see also (91)]. An explanation in terms of low hydrogen abundances can be discarded on the basis of the agreement achieved for H_2 and NH. The CH problem deserves further study.

The results of the Lambert et al. study show that O is moderately underabundant (by typically -0.25 dex), which agrees with results obtained from other AGB stars and from planetary nebulae. The C/O ratios range from 1.00 to 2, with a preference for small ratios [half of the stars have $\varepsilon(C)/\varepsilon(O) < 1.10$]. This result seems to be consistent with predictions from simulated populations of AGB stars. A more astonishing result is the low nitrogen abundances (solar or less, with a median value around -0.20 for [N/H]). This result is, however, dependent on the dissociation energies adopted for CN and C_2—not unrealistic revisions of these values would raise the [N/H] values by 0.5–0.6 dex. An interesting result is the lack of correlation between the nitrogen abundances and the $^{12}C/^{13}C$ ratios of the stars; this seems to exclude the possibility of explaining the high ^{13}C abundances of J-type stars as the consequence of normal CNO burning.

6. ISOTOPIC RATIOS

6.1 *Methods*

The small isotopic shifts of atomic lines may make the determination of isotopic ratios from such lines "an exercise in futility" (245), one that was nevertheless attempted by Wallerstein (246), who found the $^6Li/^7Li$ ratio of Y CVn to lie in the interval 0.0–0.2 from an accurate measurement of the wavelength of the $\lambda 6708$ blend of resonance lines. The molecular isotopic shifts are, however, clearly measurable, often even at rather low spectral resolution.

One might expect that it would be almost trivial to determine, for example, the $^{12}C/^{13}C$ ratio for carbon stars from $^{12}C^{14}N$ and $^{13}C^{14}N$ lines in a differential procedure. However, this is not the case—in fact, the $^{12}C/^{13}C$ ratio of carbon stars has been an issue of rather intensive debate for two decades, with widely different results obtained in different studies. A basic reason for these difficulties is that the isotopic ratios may deviate from unity by several orders of magnitude. This means that reasonably equivalent lines of molecules with different isotopes cannot be used in the comparison, since the lines of the more abundant molecule may be saturated, or those of the rare molecule too faint to be measured. Also, different line strengths will correspond to different depths of formation and thus will be dependent on the errors in the model atmospheres. The unequal strengths may be avoided by choosing lines with higher excitation for the most abundant molecule, but this will again lead to differences in the depths of formation. Another possibility is to intercompare lines belonging to different systems or sequences with different transition probabilities; this possibility will, however, be dependent on the accuracy with which the differences in continuous opacity between the two bands are known. Fujita & Tsuji (76) proposed that one could use the weak satellite lines, for example, of the Red $^{12}C^{14}N$ system. The low oscillator strengths of these lines may lead to equivalent widths of the same order of magnitude as those of $^{13}C^{14}N$ lines of roughly equal excitation. This possibility was exploited in studies of $^{12}C/^{13}C$ ratios by Fujita & Tsuji (73), and later by Lambert et al. (142) for N stars and by Dominy et al (54) for SC stars.

6.2 $^{12}C/^{13}C$

Since the discovery by Johnson & Mendez (111) that ^{13}CO V-R band heads were apparent in various late-type spectra, many estimates of $^{12}C/^{13}C$ ratios for M stars have been made. Only for the early M stars has it been possible to use CN lines as alternatives—for example, Red ^{13}CN lines from the $\Delta v = 2$ sequence were used in studies of α Ori (139) and α Sco (104).

Pioneering studies based on the V-R CO lines for the determination of isotopic ratios were made by Spinrad et al. (212) and by Maillard (146) ($\Delta v = 2$), Geballe (80) ($\Delta v = 1$), and Lambert et al. (139) ($\Delta v = 3$). These first results were partly criticized [see, e.g., (104, 219)]. Hinkle et al. (104) carried out a study on α Ori, α Sco, α Her (M4 II–III), β Peg (M2 II–III), and the Mira-type variable and mild S-star χ Cyg, mainly based on $\Delta v = 2$ and $\Delta v = 3$ lines of CO. The resolution was high enough (about 0.1 cm^{-1}) for weak lines to be measured and a reasonable continuum level to be established. The observed line strengths were analyzed in a careful differential COG analysis. The resulting $^{12}C/^{13}C$ ratios in the interval from 7 (α Ori, β Peg) to 25 (χ Cyg) were estimated to have uncertainties ranging from 20% to 40%.

Smith & Lambert (206, 207) derived $^{12}C/^{13}C$ ratios for a number of M giants, MS stars, and two S stars. They used spectrum synthesis of the 2–0 head of $^{13}C^{16}O$ for the hotter stars ($T_{\text{eff}} > 3400$ K); for the cooler stars this feature became too saturated, and individual weaker $^{13}C^{16}O$ lines were used instead. The ^{12}C abundance was derived from $^{12}C^{16}O$ $\Delta v = 3$ lines and from higher excitation $\Delta v = 2$ lines. The typical difference in excitation energy between the $^{12}C^{16}O$ and $^{13}C^{16}O$ lines used is on the order of 1 eV.

The $^{12}C/^{13}C$ ratios derived range from typically 13 for the M stars to 30 for the MS stars and even more for the S stars. The results for the M stars match those for G and K giants, and the tendency for $^{12}C/^{13}C$ and [C/H] is consistent with the assumption that ^{12}C from He burning is added to the M stars to form MS and S stars. However, Tsuji (234) reported still lower $^{12}C/^{13}C$ ratios—from FTS spectra of the CO $\Delta v = 2$ lines he finds that all stars in a sample of 12 M giants have isotope ratios below 10. More details of this study are needed before a judgment about the reason for the discrepancy with the Smith & Lambert results can be made. Note, however, that the value reported by Tsuji (234) for α Her is in agreement with that derived by Hinkle et al. (104).

Hinkle & Scharlach (103) report $^{12}C/^{13}C$ ratios, determined from CO second-overtone lines, for 34 oxygen-rich Miras or SRa variables. They find that the shorter period stars have an isotopic ratio about three times smaller than the longer period stars. They ascribe this to a more significant mixing of 3α-processed material to the surfaces of the latter stars. Details of this study are also eagerly awaited.

The ^{13}CN lines of the Red system are measurable in S, SC, and N stars and have been extensively used to derive $^{12}C/^{13}C$ ratios for such stars.

In Catchpole's (37) study of the SC star UY Cen, a ratio of 40 was obtained, but this value was found to be very dependent on the C/O value adopted. This dependency reflects the different depths of formation for the $^{12}C^{14}N$ and $^{13}C^{14}N$ lines.

Dominy et al. (54; cf. Section 3.4) derived $^{12}C/^{13}C$ ratios for four SC stars from the $\Delta v = -2$ sequence of the Red CN system (2.1–2.4 μm). They investigated the sensitivity to structural uncertainties of the atmospheres caused by the uncertainty in the C/O ratio. For the $^{12}C/^{13}C$ ratio of GP Ori (which can be assumed to show the greatest sensitivity to C/O), they found that the value varies from 20 to 37 when the model C/O varies through the (exaggerated) interval 0.6–2.0. Thus, the effects for this method and star are smaller than those found by Catchpole (37), though not entirely negligible.

Using FTS spectra in the 2.2-μm region, Dominy & Wallerstein (52) determined $^{12}C/^{13}C$ ratios for four additional SC stars, including WZ Cas (SC7/10e) and LX Cyg (SC7/9-e). For the latter star, which is a Mira variable, as well as for the S-star and Mira R Cyg (S5–8/6e), a COG method was used instead of hydrostatic model atmospheres. Again, $^{12}C^{14}N$ and $^{13}C^{14}N$ lines from the $\Delta v = -2$ sequence were used.

The $^{12}C/^{13}C$ ratios derived by Dominy et al. (54) and by Dominy & Wallerstein (52) for SC stars range from 27 to 53 with two exceptions—the Mira variable CY Cyg (for which the $^{12}C/^{13}C$ ratio of 5.6 is obtained), and the well-known "lithium star" WZ Cas (with an isotopic ratio of 4.7).

These two ratios are in good agreement with studies based on other criteria by Dominy, Hinkle, and Lambert [cited in (54)] and by Lambert et al. (142), respectively. The $^{12}C/^{13}C$ ratios of the main group of SC stars nicely continue the progression mentioned above from M, MS, and S stars; in fact, it is further continued for the N-star $^{12}C/^{13}C$ ratios for which Lambert et al. (142) find values around 60 (J-type stars excluded). The explanation in terms of a gradual (or random) enrichment of ^{12}C is tempting and supported by the CNO abundance results.

Although the cool carbon stars provided the first discovery of ^{13}C in any extraterrestrial object [Sanford (191)], there has been a long-standing controversy concerning the $^{12}C/^{13}C$ ratios in these stars (as was mentioned above) since the first estimates by Menzel (150)—an account of the early history of this field of research is given by McKellar (149). Important contributions were later made by Climenhaga (44), Wyller (260, 261), Utsumi (239), and Fujita et al. (74, 75). These and many later studies were based on C_2 and CN lines in the visual and near-infrared wavelength regions. Sometimes, quite ambitious synthetic spectrum fits were used but still rather disparate results were obtained; for example, Climenhaga et al. (45) derived quite low $^{12}C/^{13}C$ ratios, whereas Olson & Richer (170) obtained isotopic ratios that were considerably greater, though not as great as those derived by Lambert et al. (142; see below). It should be noted that Climenhaga et al. found microturbulence parameters significantly greater than those of Olson & Richer, which in turn are greater than those of Lambert et al. Since the $^{12}C^{14}N$ lines are often more saturated than the $^{13}C^{14}N$ lines in these comparisons, one may expect a correlation between $^{12}C/^{13}C$ and ξ_t of this character. The low $^{12}C/^{13}C$ ratios of Climenhaga et al. were also discussed by Dominy et al. (53) in their study of V460 Cyg (N1; C6,3), and they point out the key importance of the assignment of the continuum level when the near-infrared wavelength region is used.

The detailed study by Fujita & Tsuji (73) was based on lines from the $\Delta v = 3$ and $\Delta v = 2$ sequences of the CN Red system between 7700 and 8300 Å. The problems of blends, of defining the continuum, and of comparing lines of different (and not well-known) degrees of saturation were partly circumvented by these authors by a critical and highly restrictive choice of criteria and a strictly differential approach. Fujita & Tsuji found that the "normal" N-type carbon stars show $^{12}C/^{13}C$ ratios in the interval 20–100, whereas the J-type stars have $^{12}C/^{13}C < 10$. This conclusion is confirmed in the recent study by Lambert et al. (142).

Lambert et al. (142) used lines in the 2.2-μm region from the $\Delta v = -2$ bands of the CN Red system, CO $\Delta v = 2$ lines, and CO $\Delta v = 3$ lines from the 1.6-μm region. These wavelength regions have the great advantage, relative to the near-infrared, of offering seemingly acceptable continuum points, at least in most carbon-star spectra. For the CN line comparison, the presence of weak satellite $^{12}C^{14}N$ lines made a comparison possible between weak $^{12}C^{14}N$ and $^{13}C^{14}N$ lines of similar equivalent widths and lower excitation potentials. For the CO $\Delta v = 3$ lines, enough lines of both ^{12}CO and ^{13}CO were available but the ^{13}CO lines were systematically weaker and thus formed at greater atmospheric depths, which made that estimate of $^{12}C/^{13}C$ sensitive to model atmosphere errors. A third comparison was also made, that between the ^{13}CO $\Delta v = 2$ lines and the ^{12}CO $\Delta v = 3$ lines. With very few exceptions the three measurements of $^{12}C/^{13}C$ agreed well. The estimates from CN are assumed to be least sensitive to systematic errors; the errors should be 0.1 dex or less in the isotopic ratios.

The four J-type stars RY Dra (N4; C4, 4), T Lyr (Np; C5, 5), WZ Cas, and Y CVn were found to have $^{12}C/^{13}C$ ratios between 3.2 and 4.5, close to the CNO-burning equilibrium value, while the majority (24 stars) show values in the interval 30–100. In the sample of Lambert et al. (142) there are two stars, VX And (N7) and R Scl (Np; C6, 5), with intermediate $^{12}C/^{13}C$ ratios.

6.3 $^{12}C/^{14}C$

The relative role of the two neutron sources in AGB stars, $^{13}C(\alpha, n)^{16}O$ and $^{22}Ne(\alpha, n)^{25}Mg$, is still not clear. An interesting consequence of the first-mentioned process is the possible production of significant amounts of ^{14}C through the reaction $^{14}N(n, p)^{14}C$.

Olson & Richer (170) first looked for ^{14}CN lines in N-type spectra and derived a lower limit $^{12}C/^{14}C > 450$ for Y CVn. Recently, Harris & Lambert (96) have derived considerably more restrictive limits ($\geq 10^4$) for 26 MS, S, and N stars. The first overtone bands of CO (and in particular the 3–1 band head at 4132.2 cm^{-1}) were used; as is shown by Harris & Lambert, even higher limits are obtainable from good spectra of the fundamental bands. Improvements in wavelengths for the ^{14}CO lines and in S/N would help too. Observations of extended samples of stars would also increase the chances of catching stars that have undergone shell flashes so recently that the decaying ^{14}C is still observable.

6.4 $^{14}N/^{15}N$

A first attempt to derive stellar $^{14}N/^{15}N$ abundance ratios was made by Querci & Querci (177), who tentatively identified in the spectrum of the N-star UU Aur (N3; C5, 3) six lines from $^{12}C^{15}N$ in the $\Delta v = 2$ sequence of the CN Red system. These authors estimated a $^{14}N/^{15}N$ ratio of about $(3.6–12.3) \times 10^3$.

More recently, Olson & Richer (170) attempted a derivation of $^{14}N/^{15}N$ in the near-infrared spectra of five carbon stars. Using their cross-correlation technique, they found a significant maximum in the coherency measure for $^{14}N/^{15}N = 150$ for the J-star Y CVn, whereas the other stars gave no significant ^{15}N signal. An examination of the spectrum of Y CVn reveals a number of possible $^{12}C^{15}N$ features—one at 8037.4 Å is particularly noteworthy. Contrary to this, Lambert (136) was not able to trace $^{12}C^{15}N$ lines in the 2.2-μm region of the spectrum of Y CVn, which should have been possible if the identification of Olson & Richer had been correct. In view of the problems involved in understanding the nature of the J-type stars, further attempts to measure the nitrogen isotopic ratios of these stars could be important. We note that $HC^{15}N$ lines have been identified in two dense carbon-rich circumstellar envelopes (5, 250) at millimeter wavelengths, suggesting $^{14}N/^{15}N$ ratios of about 3000.

6.5 $^{16}O/^{17}O$ and $^{16}O/^{18}O$

A pioneering attempt to derive the relative abundances of oxygen isotopes was made by Maillard (146), who used CO $\Delta v = 2$ lines to estimate the $^{16}O/^{17}O$ ratio in α Her as ~ 450. Geballe et al. (81) used Fabry-Perot scans in the 5-μm window of the $\Delta v = 1$ band of CO in an attempt to determine $^{16}O/^{17}O$, $^{16}O/^{18}O$, and $^{12}C/^{13}C$ in three M supergiants (α Ori, α Her, α Sco). From synthetic spectrum fits with parametrized model atmospheres, they obtained similar oxygen isotopic ratios with estimated errors of a factor of two to three. These errors are largely due to the saturation of the $\Delta v = 1$ line; in practice, the strengths of the stronger lines are more dependent on velocity fields and the uncertain temperature structure of the outer atmospheres than on the abundances. Also, the few observed lines of $^{12}C^{17}O$ and $^{12}C^{18}O$ are at least partially blended with stronger $^{12}C^{16}O$ transitions.

In a series of papers Harris, Lambert, and collaborators have studied the oxygen isotopic abundances in different types of red giants. Their first study (94) dealt with α Ori and α Sco and was again based on the spectra of the CO $\Delta v = 1$ bands, now observed with the KPNO FTS, which were analyzed with synthetic spectra from Indiana models. The problem of locating the true continuum relative to the "local continuum" within the short intervals between the strong telluric lines was attacked by means of an interesting method: A $^{13}C^{16}O$ abundance was derived from the weak $^{12}C^{16}O$ $\Delta v = 3$ lines at 1.6 μm and a $^{12}C/^{13}C$ ratio, obtained from suitable CO and CN lines. The $^{13}C^{16}O$ abundance derived was used to predict the $^{13}C^{16}O$ $\Delta v = 1$ lines, and to fit these predictions to the observed lines, by adjusting the continuum. Next, the $^{12}C^{17}O$ and $^{12}C^{18}O$ $\Delta v = 1$ lines provided the $^{12}C^{17}O$ and $^{12}C^{18}O$ abundances on the same scale—which led to $^{16}O/^{17}O$ and $^{16}O/^{18}O$ ratios. Each narrow window between regions of telluric absorption was treated independently, but a smoothly varying true continuum resulted.

The resulting isotopic ratios are on the order of 500–850 for both supergiants and for both $^{16}O/^{17}O$ and $^{16}O/^{18}O$, and they agree well within the errors with the values given by Geballe et al. (81). Relative errors estimated by Harris & Lambert were about $+50\%$ to -30% in these ratios.

Harris & Lambert (95) presented new determinations of oxygen isotopic ratios for seven red stars, four of which were M giants. Again FTS observations of the CO $\Delta v = 1$ bands were used, but they were supplemented with measurements of the considerably weaker isotopic $\Delta v = 2$ bands.

In cool star spectra, suitable $^{12}C^{17}O$ (2, 0) lines are available in the 4269–4294 cm^{-1} region (97, 146). Although a number of them are unaffected by the strong $^{12}C^{16}O$ (3, 1) lines that also reside in that interval, blending by CN and terrestrial lines is of importance and has to be taken into account properly. It is far more difficult to find suitable $^{12}C^{18}O$ lines—they fall in regions rich in $^{12,13}C^{16}O$ lines and CN lines, and blends from these and from telluric lines make the $^{16}O/^{18}O$ ratios less well determined.

For the stars where the isotopic $\Delta v = 2$ lines were strong enough to be measurable, Harris & Lambert (95) found a convincing agreement with the results from the $\Delta v = 1$ lines. The $^{16}O/^{17}O$ ratios ranged from 160 to 1000 [the solar system value is 2630 (3)], while the $^{16}O/^{18}O$ ratios were all around 500 (which is consistent with the solar system value of 490). The reason for the wide range in $^{16}O/^{17}O$ is not known, but Harris & Lambert suggested that the stars with low ratios may have undergone the second dredge-up and are making their second ascent up the giant branch, which, however, implies that their masses are greater than 4.5 $M_\odot$. The $^{16}O/^{18}O$ ratios, and the absence of a correlation between the $^{12}C/^{13}C$ and $^{16}O/^{18}O$ ratios, give interesting bounds for the rate of the $^{18}O(p, \alpha)^{15}N$ reaction.

Tsuji (235) reported analysis results for FTS spectra of the 2.3-μm region for M giants. He found $^{16}O/^{17}O$ ratios not very different from the solar system value for the stars less luminous than $M_{bol} = -5$; for the more brilliant ones, however, there is a sudden decrease to values in the range 200–1000. For the few stars in common, Tsuji's isotopic ratios seem to agree well with those of Harris & Lambert (95). Details of Tsuji's study are awaited with interest.

In a subsequent paper Harris et al. (98) proceeded to the analysis of oxygen isotopes in eight MS and S stars. They now used isotopic lines of the CO $\Delta v = 2$ bands and compared their strengths with $^{12}C^{16}O$ lines of approximately the same strengths (but with different excitation energies) of the $\Delta v = 3$ band region. Again, synthetic spectra calculated from Indiana models (with solar composition) were used. The most important uncertainties in the isotopic ratios are caused by the errors in the ^{12}C abundances, which are due to errors in the microturbulence parameters, effective temperatures, and model atmospheres; the latter sources of error reflect the different excitation and continuous opacities for the $^{12}C^{16}O$ and $^{12}C^{17,18}O$ lines. Resulting errors are estimated to be on the order of ± 0.15 dex. The isotopic ratios found are astonishingly great, even exceeding the solar $^{16}O/^{18}O$ ratio and significantly greater than what is expected from stellar evolution theory. Both ratios correlate with the ^{12}C abundance and with the neutron exposure needed to explain the s-element abundances. No simple modifications of current models of AGB stars seem to be capable of explaining these results.

Using similar methods, Harris et al. (99) derived oxygen isotopic ratios for 26 cool carbon stars. The synthetic spectra were calculated from Uppsala models with chemical compositions consistent with the stellar abundances derived by Lambert et al. (142) (cf. Figure 4). Important sources of error are the uncertainties in the ^{16}O abundances and blending effects (in particular, for $^{16}O/^{18}O$). For this reason, only lower limits for $^{16}O/^{18}O$ were obtained for seven stars. Since even the isotopic $\Delta v = 2$ CO lines are somewhat saturated, the results are also sensitive to the microturbulence parameter and the placement of the continuum (e.g. a hypothetic "veil" of lines from polyatomic molecules, depressing the continuum by 5%, would cause a true ratio $^{16}O/^{17}O \sim 700$ to be estimated to be around 1500). The ratios derived resemble those obtained for MS and S stars, namely $550 \leq {}^{16}O/^{17}O \leq 4100$ and $700 \leq {}^{16}O/^{18}O \leq 2400$. Again, these ratios are much higher than expected and are extremely difficult to explain in terms of present theories of stellar evolution.

In their study of four SC stars, Dominy et al (54; cf. Section 3.4) derived $^{16}O/^{17}O$ ratios from a synthesis of the 4269–4294 cm^{-1} region, while the ^{16}O abundances were obtained from $\Delta v = 2$ $^{12}C^{16}O$ lines in a nearby spectral region. The resulting $^{16}O/^{17}O$ ratios, which are estimated to have relative errors of 50% or less, are within the range $160 \leq {}^{16}O/^{17}O \leq 400$; that is, they are significantly less than the values derived for MS, S, and C stars by Harris and collaborators. This interesting fact, which might suggest that SC stars are not evolutionarily related to the M–MS–S–C sequence, stimulated Dominy & Wallerstein (52) to further investigate $^{16}O/^{17}O$ ratios for seven stars of type MS, S, and SC, three of which were previously studied by Harris et al. There is a significant difference between the procedures of these authors: Dominy & Wallerstein compared the $^{12}C^{17}O$ $\Delta v = 2$ lines with high-excitation $^{12}C^{16}O$ lines from the same bands (the difference in χ being about 1.5 eV), whereas Harris et al. relied more on the $^{12}C^{16}O$ lines of the $\Delta v = 3$ bands (which are close in excitation energy). For two of the stars in common between the studies, there is good agreement; for the third, Dominy & Wallerstein derive a significantly lower

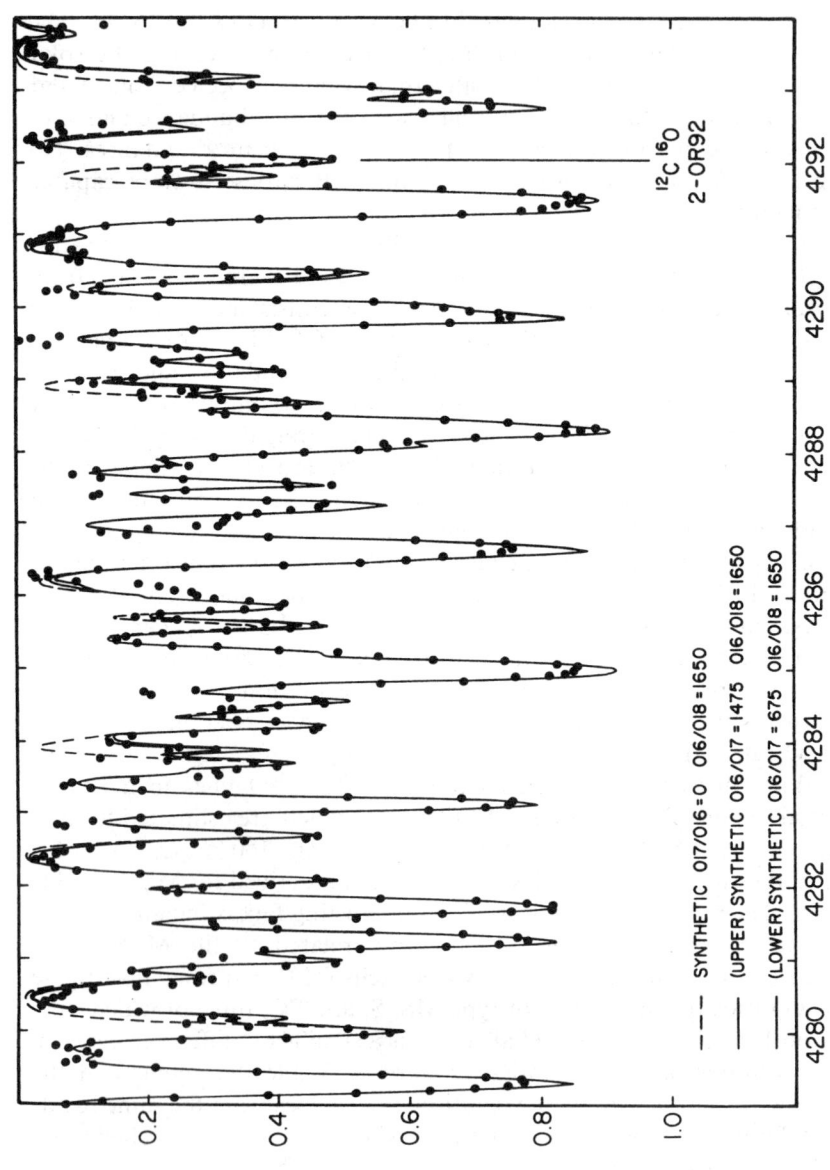

Figure 4 Comparison of observed (dots) and synthetic spectra for BL Ori (NO), used for estimating the $^{16}O/^{17}O$ ratio by Harris et al. (99) from CO VR $\Delta v = 2$ lines. A value of 1500 ± 400 for $^{16}O/^{17}O$ was obtained, which is higher than expected. Figure taken from (99).

value, although possibly still within the error bars of the Harris et al. value. Lambert (137) has reported on a reanalysis of oxygen isotopic ratios in MS and S stars, using higher quality data. The ratios found span narrower ranges than previously reported, and a fair consistency is found with previous independent studies [cf. (137, Table 1)]. It is important to make further detailed comparisons between the results from different oxygen isotopic criteria, since the ratios derived for AGB stars are too large to be easily understood in terms of stellar evolution theory. The normal oxygen abundances show that such high ratios should be explained in terms of ^{17}O and ^{18}O destruction rather than ^{16}O production. Possible hypotheses discussed by Harris et al. include shell helium burning combined with extreme mass loss, CNO-cycle reactions in low-mass main sequence stars, and explosive nucleosynthesis in the envelope after core or shell flashes. Further detailed studies, based on high-quality spectra with due consideration of possible systematic errors, are important before the observed estimates can be taken as firm evidence for any of the hypotheses.

6.6 Isotopes of Mg, Si, Cl, Ti, and Zr

Attempts to use molecular bands of metal compounds (e.g. metal hydrides or oxides) for deriving isotopic ratios in late-type stars have not been frequent. In a recent analysis, Smith & Lambert (207) attempted determinations of Mg isotopic ratios for two M giants and three S stars. The $\Delta v = 0$ region of the $A^2\Pi-X^2\Sigma$ system of MgH near 5140 Å was observed at 0.05-Å resolution. (The isotopic shift amounts of 0.15 Å.) The line absorption across the region is heavy and not well reproduced by the synthetic spectra; the authors tried to compensate for this by scaling the gf values with a fudge factor for each star to match the central line depths. ^{24}Mg : ^{25}Mg : ^{26}Mg proportions not far from the solar ones of 79 : 10 : 11 (3) were found for all stars, including those where the s-elements suggest that a heavy neutron exposure has occurred. Clegg et al. (43) also found similar isotopic ratios for β Peg (M2.5 II–III) and HR 1105 (S3.5/2). The absence of great proportions of ^{25}Mg and ^{26}Mg indicates that the reaction ^{22}Ne$(\alpha, n)^{25}$Mg is not responsible for the s-element in MS and S stars, which in turn suggests that the temperature of the s-process site is below 3×10^6 K, and thus that the MS/S stars have masses below 2 $M_\odot$.

The first-overtone SiO bands in the 4-μm region have been found in many M-star spectra. Beer et al. (16) attempted estimates of the isotopic ratios ^{28}Si : ^{29}Si : ^{30}Si from an FTS spectrum at 0.1-cm^{-1} resolution of α Ori. These authors found no significant departures from the solar-terrestrial isotopic ratios. Merrill & Ridgway (152) reported similar results for several M giants. Recently, Lambert et al. (140) have estimated Si isotopic ratios for four red giants, determined from FTS spectra at about 0.05 cm^{-1}

resolution. They found terrestrial ^{28}Si : ^{29}Si ratios in β Peg and HR 1105 but a probable underabundance of ^{29}Si in 10 Dra (M3.5 III) and o^1 Ori (S3.5/1 −). Moreover, the heavier isotope ^{30}Si seems deficient, relative to its terrestrial abundance, in all four stars. This sample of four stars also suggests that the abundances of heavier Mg and Si isotopes are anticorrelated. The results are not well understood in terms of stellar nucleosynthesis, and further studies of Si isotopes are needed.

Strong H^{35}Cl and H^{37}Cl lines were identified by Ridgway et al. (186) in the spectrum of R And (S5/4.5e). From them, a value of ^{35}Cl/^{37}Cl close to the terrestrial value of 3 was found (152, 188). A similar value was derived for the N-type star 19 Psc (136).

Different Ti isotopes were first detected in stellar atmospheres by Herbig (101), who found that each of the isotopes ^{46}Ti, ^{47}Ti, ^{49}Ti, and ^{50}Ti was less abundant than 30% of the dominant isotope ^{48}Ti. The terrestrial ratios are ^{46}Ti : ^{47}Ti : ^{48}Ti : ^{49}Ti : ^{50}Ti = 7.9 : 7.7 : 73.4 : 5.5 : 5.3. Differences between terrestrial and stellar isotopic ratios are conceivable, since different nuclear processes contribute to the different isotopes. An attempt to measure the Ti isotopic ratios in Mira (o Ceti) was made by Wyckoff & Wehinger (259), using photographic spectrograms of the γ_3 (0, 0) band. A more recent study based on Reticon spectra at 0.07-Å resolution of the same feature was performed by Clegg et al. (43). These authors also included portions of the γ (0, 1) bands and the δ (0, 0) bands. Seventeen stars were studied, 15 of which were M giants, MS stars, or S stars. Indiana models were used for the calculation of synthetic spectra. The continuum level could be well defined shortward of the γ_3 (0, 0) $R(3)$ band head at 7054 Å—from this point, the model atmospheres were used to extrapolate the true continuum level into the key region near 7080 Å; this method was vital for the analysis of the γ (0, 0) lines. The Ti isotopic ratios were found to be terrestrial within a relative error of 50% or less for all stars including Mira, a result that is in some conflict with that obtained by Wyckoff & Wehinger (259). The normal abundance of the magic nucleus ^{50}Ti relative to ^{48}Ti in the MS and S stars, together with the observed Mg isotopic ratios and the ^{37}Cl/^{35}Cl ratios, again indicates that the ^{22}Ne(α, n)^{25}Mg reaction is not the neutron source in these S stars.

The attempt by Hänni (92) to derive Ti isotopic ratios from the γ (1, 0) band met with difficulties, probably owing to the overlying bands for the TiO γ' system [cf. also (43)].

An early attempt to derive Zr isotopic ratios for the S-star HR 1105 was made by Schadee & Davis (195), but with somewhat ambiguous results. Peery & Beebe (173) tried an analysis of Zr isotopes in the long-period variable S-star R Cyg, the spectrum of which is much more free from contaminating TiO lines than HR 1105. They fitted the spectrum of the γ_2

(0, 0) band head of ZrO around 6400 Å with calculated spectra using the Minnaert method and single-slab LTE model atmospheres. The resulting isotopic abundance proportions were ^{90}Zr : ^{91}Zr : ^{92}Zr : ^{93}Zr : ^{94}Zr : ^{96}Zr = 50 : 5 : 10 : 5 : 30 : 0, to be compared with the terrestrial proportions 51 : 11 : 17 : 0 : 17 : 3. Note the suggested presence in the star of the unstable ^{93}Zr (with a half-life of 1.5×10^6 yr) and the absence of the r-nucleus ^{96}Zr. The authors claim the errors in the proportions to be less than 5 units.

In a model-atmosphere analysis of the ZrO $^1\Pi$–$^1\Sigma$ (0, 1) band head around 6930 Å, Zook (265) obtained Zr isotopic ratios for R Cyg and V Cnc (S0–4/6e). The proportions derived are similar for the two stars—for R Cyg, Zook found the array 47 : 10 : 17 : 6 : 20 : 0, which supersedes the result of Peery & Beebe (173). Zook estimates the errors to be on the order of 10 units in these proportions. An extensive survey of ZrO in about 30 S stars (mostly long-period variables) reported by Lambert (137) gave, however, no evidence for an appreciable abundance of ^{96}Zr.

The possibilities for determining isotopic abundance ratios for a great number of elements present in molecular form and visible as molecular lines in cool stars should be further investigated systematically. As illustrated in, for example, the discussion by Clegg & Lambert (42) about such possibilities for CeO in S stars and FeH in M stars, this will not always be easy, owing to the crowded spectra, the weak isotopic lines, the frequently small isotopic shifts, and the absence of adequate molecular data.

7. CONCLUDING REMARKS

It should be clear from this review that progress in the chemical analysis of cool stars has been impressive during the last decade. Systematic efforts by several groups and individuals to develop the methods of this art have brought it into the field of exact research where not only many results, but even estimated errors in these results, may be trusted.

However, this progress has not been easily achieved. As is illustrated by, for example, the long series of different results during the 1960s and 1970s as regards the ^{12}C/^{13}C ratios of carbon stars, ambitious attempts may fail owing to the difficulties with blends, blocking, saturated spectral lines, unsatisfactory models for the atmospheric structures, errors in the fundamental parameters, lack of adequate atomic and molecular data, etc. Undoubtedly, the pioneers in these fields have been aware of the preliminary character of their results. For many types of analyses, one may fear that some of the difficulties mentioned still cause severe systematic errors.

It is worrisome that several of the more ambitious studies discussed above, where different alternative abundance criteria are being used or tested in the same analysis, show internally inconsistent results. Another disturbing circumstance, still prevailing in some recent investigations, is the fact that unexpectedly low abundances not seldom are derived for the red giants and supergiants. Reasons for this may be the existence of unconsidered veils of molecular lines, of dust emission, of thermal inhomogeneities or spherical extension, or even of the depletion of molecules by shocks or hot radiation fields.

Common to most of the studies discussed above is that arguments based on the intrinsic consistency and intercomparisons between different studies suggest that the typical systematic errors in any abundance may well be about 0.3 dex or greater. Only some of the most recent studies may provide more accurate results. Further improvements will have to be based on the use of more spectral features in reasonably clear spectral regions, on very detailed synthetic spectroscopy in more crowded regions in combination with observations at high spectral resolution and high S/N, on a better basic physical understanding of stellar atmospheres, and on improved atomic and molecular data for the model and spectrum calculation.

What qualitative improvements can be hoped for in the near future? First, on the observational side the development of new efficient IR spectrometers, such as cryogenic echelles with multielement InSb arrays, will certainly open up dramatic new opportunities for observing wide samples of cool stars, including faint M dwarfs, at high resolution. The study of stellar spectra at wavelengths inaccessible from the ground will also give new important information, in particular as regards the V-R bands of several molecules. Investigations of cool stellar disks with interferometry and spectrum polarimetry will certainly provide much more empirical understanding of the structure and dynamics of these atmospheres. This is certainly necessary—one cannot expect the theorists of stellar atmospheres to contribute considerably improved atmospheric models without the challenges, and tests, that only observations can provide.

More realistic modeling of the atmospheric structures of cool stars is of key importance for improving the abundance analyses. The line opacity data for diatomic and polyatomic molecules will hopefully gradually improve as a result of further systematic ab initio calculations and laboratory investigations. Self-consistent dust opacities will require a better theory of dust condensation than is presently available. A better understanding of convection, velocity fields, inhomogeneities, mechanical fluxes, and magnetic fields in the line-forming regions is expected as a result of large-scale computer simulations and new observations. Also, our understanding of departures from LTE will grow gradually as systematic empirical studies of these effects, and the measurements or calculations of the relevant collisional cross sections, are carried out.

One would expect that interest in the as yet not well understood mixing processes and detailed evolution of red giants and supergiants will continue to stimulate work on the chemical analysis of cool stars. In addition, the importance of applying cool star analyses to problems in Galactic evolution and extragalactic astronomy will increase considerably as the result of improvements in telescope and spectrometer technology, which will make it possible to study faint Galactic stars and individual stars in external galaxies at sufficient spectral resolution. Hopefully, the methodological developments will by then have turned much of the presently remaining scepticism as regards the validity of the analysis results into well-motivated faith.

ACKNOWLEDGMENTS

All those who have sent reprints and preprints for this review are thanked. J. Brett, U. Ekberg, K. Eriksson, Y. Fujita, G. Hammarbäck, U. G. Jørgensen, D. L. Lambert, D. Minugh, O. Morell, H. Olofsson, M. Scholz, and T. Tsuji contributed valuable suggestions and comments on the manuscript.

Literature Cited

1. Alksne, Z. K., Ikaunieks, Ya. Ya., Baumert, J. H. 1981. *Carbon Stars*. Tucson: Pachart
2. Alksne, Z. K., Alksnis, A., Dzervitis, U. 1983. *Properties of Carbon Stars of the Galaxy*. Riga: "Zinatne"
3. Anders, E., Ebihara, M. 1982. *Geochim. Cosmochim. Acta* 46: 2363
4. Anderson, L. S. 1985. *Ap. J.* 298: 848
5. Andersson, B.-G., Wannier, P. G., Olofsson, H. 1987. *Bull. Am. Astron. Soc.* 19: 645
6. Augason, G. C., Johnson, H. R., Bregman, J. D., Witteborn, F. C. 1983. *Bull. Am. Astron. Soc.* 15: 948
7. Auman, J. R., Woodrow, J. E. J. 1975. *Ap. J.* 197: 163
8. Ayres, T. R. 1986. *Ap. J.* 308: 246
9. Ayres, T. R., Testerman, L. 1981. *Ap. J.* 245: 1124
10. Ayres, T. R., Testerman, L., Brault, J. W. 1986. *Ap. J.* 304: 542
11. Ayres, T. R., Wiedemann, G. R. 1989. *Ap. J.* 338: 1033
12. Baschek, B. 1979. In *Les Elements et leurs Isotopes dans l'Universe*, p. 327. Liège: Inst. Astrophys., Univ. Liège
13. Bauschlicher, C. W. Jr., Langhoff, S. R., Taylor, P. R. 1988. *Ap. J.* 332: 531
14. Beach, T. E., Willson, L. A., Bowen, G. H. 1988. *Ap. J.* 329: 241
15. Beer, R., Hutchinson, R. B., Norton, R. H., Lambert, D. L. 1972. *Ap. J.* 172: 89
16. Beer, R., Lambert, D. L., Sneden, C. 1974. *Publ. Astron. Soc. Pac.* 86: 806
17. Begley, M. J., Cottrell, P. L. 1987. *MNRAS* 224: 633
18. Bell, R. A., Edvardsson, B., Gustafsson, B. 1985. *MNRAS* 212: 497
19. Berriman, G., Reid, N. 1987. *MNRAS* 227: 315
20. Bessell, M. S., Brett, J. M., Scholz, M., Wood, P. R. 1989. *Astron. Astrophys. Suppl.* 77: 1
21. Bessell, M. S., Brett, J. M., Scholz, M., Wood, P. R. 1989. *Astron. Astrophys.* 213: 209
22. Blackwell, D. E., Willis, R. B. 1977. *MNRAS* 180: 169
23. Blackwell, D. E., Shallis, M. J. 1977. *MNRAS* 180: 177
24. Boesgaard, A. M. 1970. *Ap. J.* 161: 163
25. Boesgaard, A. M. 1970. *Ap. J.* 161: 1003
26. Boesgaard, A. M. 1970. *Astrophys. Lett.* 5: 145
27. Bowen, G. H. 1988. *Ap. J.* 329: 299

28. Boyle, R. P., Aspin, C., Coyne, G. V., McLean, I. S. 1986. *Astron. Astrophys.* 164: 310
29. Brett, J. M. 1989. *MNRAS*. In press
30. Brooke, A. L., Lambert, D. L., Barnes, T. G. III. 1974. *Publ. Astron. Soc. Pac.* 86: 419
31. Brown, J. A., Tomkin, J., Lambert, D. L. 1983. *Ap. J. Lett.* 265: L93
32. Cameron, A. G. W., Fowler, W. A. 1971. *Ap. J.* 164: 111
33. Carbon, D. F. 1979. *Annu. Rev. Astron. Astrophys.* 17: 513
34. Carbon, D. F., Gingerich, O. 1969. *Proc. Harvard-Smithson. Conf. Stellar Atmos.*, 3rd, ed. O. Gingerich, p. 377. Cambridge, Mass: MIT Press
35. Carbon, D. F., Milkey, R. W., Heasley, J. N. 1976. *Ap. J.* 207: 253
36. Carlsson, M. 1986. *Uppsala Astron. Obs. Rep. 33*
37. Catchpole, R. M. 1982. *MNRAS* 199: 1
38. Catchpole, R. M., Feast, M. W. 1971. *MNRAS* 154: 197
39. Catchpole, R. M., Feast, M. W. 1976. *MNRAS* 175: 501
40. Cayrel de Strobel, G., Bentolila, C., Hauck, B., Duguennoy, A. 1985. *Astron. Astrophys. Suppl.* 59: 145
41. Christou, J. C., Hebden, J. C., Hege, E. K. 1988. *Ap. J.* 327: 894
42. Clegg, R. E. S., Lambert, D. L. 1978. *Ap. J.* 226: 931
43. Clegg, R. E. S., Lambert, D. L., Bell, R. A. 1979. *Ap. J.* 234: 188
44. Climenhaga, J. L. 1960. *Publ. Dominion. Astrophys. Obs.* 11: 307
45. Climenhaga, J. L., Harris, B. L., Holts, J. T., Smolinski, J. 1977. *Ap. J.* 215: 836
46. Cohen, J. G. 1974. *Publ. Astron. Soc. Pac.* 86: 31
47. Dearborn, D. S. P., Liebert, J., Aaronson, M., Dahn, C., Harrington, R., et al. 1986. *Ap. J.* 300: 314
48. de la Reza, R., Querci, F. 1978. *Astron. Astrophys.* 67: 7
49. Dominy, J. F. 1984. *Ap. J. Suppl.* 55: 27
50. Dominy, J. F. 1985. *Publ. Astron. Soc. Pac.* 97: 1104
51. Dominy, J. F., Wallerstein, G. 1986. *Ap. J.* 310: 371
52. Dominy, J. F., Wallerstein, G. 1987. *Ap. J.* 317: 810
53. Dominy, J. F., Hinkle, K. H., Lambert, D. L., Hall, D. N. B., Ridgway, S. T. 1978. *Ap. J.* 223: 949
54. Dominy, J. F., Wallerstein, G., Suntzeff, N. B. 1986. *Ap. J.* 300: 325
55. Dravins, D. 1982. *Annu. Rev. Astron. Astrophys.* 20: 61
56. Dravins, D. 1987. In *The Impact of Very High S/N Spectroscopy on Stellar Physics, IAU Symp. No. 132*, ed. G. Cayrel de Strobel, M. Spite, p. 239. Dordrecht: Kluwer
57. Dravins, D., Nordlund, Å. 1989. *Astron. Astrophys.* In press
58. Dravins, D., Larsson, B., Nordlund, Å. 1986. *Astron. Astrophys.* 158: 83
59. Dravins, D., Lindegren, L., Nordlund, Å. 1981. *Astron. Astrophys.* 96: 345
60. Eggen, O. J. 1978. *Ap. J.* 222: 203
61. Ekberg, U., Eriksson, K., Gustafsson, B. 1986. *Astron. Astrophys.* 167: 304
62. Ekberg, U., Eriksson, K., Gustafsson, B., Jørgensen, U. G. 1989. In preparation
63. Eriksson, K., Gustafsson, B., Jørgensen, U. G., Nordlund, Å. 1984. *Astron. Astrophys.* 132: 37
64. Eriksson, K., Gustafsson, B., Jørgensen, U. G. 1986. Unpublished research
65. Feast, M. W. 1954. *Mem. Soc. R. Sci. Liège* 15: 413
66. Fujita, Y. 1939. *Jpn. J. Astron. Geophys.* 17: 17
67. Fujita, Y. 1940. *Jpn. J. Astron. Geophys.* 18: 45, 177
68. Fujita, Y. 1952. *Trans. IAU* 8: 828
69. Fujita, Y. 1970. *Interpretation of Spectra and Atmospheric Structure in Cool Stars*. Tokyo: Univ. Tokyo Press
70. Fujita, Y. 1980. *Space Sci. Rev.* 25: 89
71. Fujita, Y. 1985. In *Cool Stars With Excesses of Heavy Elements*, ed. M. Jaschek, P. C. Keenan, p. 31. Dordrecht: Reidel
72. Fujita, Y., Tsuji, T. 1965. *Publ. Dominion Astrophys. Obs.* 12: 339
73. Fujita, Y., Tsuji, T. 1977. *Publ. Astron. Soc. Jpn.* 29: 711
74. Fujita, Y., Tsuji, T., Maehara, H. 1966. In *Colloquium on Late-Type Stars*, ed. M. Hack, p. 75. Trieste: Oss. Astron. Trieste
75. Fujita, Y., Tsuji, T., Maehara, H. 1966. *Proc. Jpn. Acad.* 42: 765
76. Fujita, Y., Tsuji, T. 1976. *Proc. Jpn. Acad.* 52: 226
77. Gail, H. P., Sedlmayr, E. 1987. *Astron. Astrophys.* 171: 197
78. Gass, H. 1988. *Astron. Astrophys.* 193: 185
79. Gass, H., Liebert, J., Wehrse, R. 1987. *Astron. Astrophys.* 189: 194
80. Geballe, T. R. 1974. PhD thesis. Univ. Calif., Berkeley
81. Geballe, T. R., Wollman, E. R., Lacy, J. H., Rank, D. M. 1977. *Publ. Astron. Soc. Pac.* 89: 840
82. Geisler, D. 1986. *Ap. J. Lett.* 304: L41

83. Gillet, D., Lafon, J.-P. J. 1983. *Astron. Astrophys.* 128: 53
84. Gingerich, O., Latham, D. W., Linsky, J., Kumar, S. S. 1966. In *Colloquium on Late-Type Stars*, ed. M. Hack, p. 291. Trieste: Oss. Astron. Trieste
85. Glassgold, A. E., Huggins, P. J. 1986. *Ap. J.* 306: 605
86. Grenon, M. 1987. *J. Astrophys. Astron.* 8: 123
87. Gustafsson, B. 1980. In *ESO Workshop on Methods of Abundance Determination for Stars*, ed. P. E. Nissen, K. Kjär, p. 31. Munich: ESO
88. Gustafsson, B. 1981. In *Physical Processes in Red Giants*, ed. I. Iben Jr., A. Renzini, p. 25. Dordrecht: Reidel
89. Gustafsson, B. 1983. *Publ. Astron. Soc. Pac.* 95: 101
90. Hall, D. N. B., Ridgway, S. T., Bell, E. A., Yarborough, J. M. 1979. *Proc. Soc. Photo-Opt. Instrum. Eng.* 172: 121
91. Hammarbäck, G. 1986. Preprint
92. Hänni, L. 1981. *Tartu Astrophys. Obs. Publ.* 49: 138
93. Hänni, L. 1983. Acad. Sci. Estonian SSR Preprint No. A-2
94. Harris, M. J., Lambert, D. L. 1984. *Ap. J.* 281: 739
95. Harris, M. J., Lambert, D. L. 1984. *Ap. J.* 285: 674
96. Harris, M. J., Lambert, D. L. 1987. *Ap. J.* 318: 868
97. Harris, M. J., Lambert, D. L., Smith, V. V. 1985. *Ap. J.* 292: 620
98. Harris, M. J., Lambert, D. L., Smith, V. V. 1985. *Ap. J.* 299: 375
99. Harris, M. J., Lambert, D. L., Hinkle, K. H., Gustafsson, B., Eriksson, K. 1987. *Ap. J.* 316: 294
100. Hartmann, L., Anderson, C. M. 1976. Washburn Obs. Preprint No. 40
101. Herbig, G. H. 1949. *Publ. Astron. Soc. Pac.* 60: 378
102. Hinkle, K. H., Lambert, D. L. 1975. *MNRAS* 170: 447
103. Hinkle, K. H., Scharlach, W. W. G. 1985. In *Cool Stars With Excesses of Heavy Elements*, ed. M. Jaschek, P. C. Keenan, p. 255. Dordrecht: Reidel
104. Hinkle, K. H., Lambert, D. L., Snell, R. L. 1976. *Ap. J.* 210: 684
105. Hinkle, K. H., Hall, D. N. B., Ridgway, S. T. 1982. *Ap. J.* 252: 697
106. Hinkle, K. H., Scharlach, W. W. G., Hall, D. N. B. 1984. *Ap. J. Suppl.* 56: 1
107. Hirai, M. 1969. *Publ. Astron. Soc. Jpn.* 21: 91
108. Höflich, R., Lowe, R. P., Moorhead, J., Scholz, M., Wehlau, W., Wehrse, R. 1986. *MNRAS* 220: 377
109. Huggins, P. J. 1973. *Astron. Astrophys.* 28: 217
110. Irgens-Jensen, S. 1976. *Astron. Astrophys.* 51: 107
111. Johnson, H. L., Mendez, M. E. 1970. *Astron. J.* 75: 785
112. Johnson, H. R. 1974. *NCAR Tech. Note NCAR-TN/STR 95*
113. Johnson, H. R. 1982. *Ap. J.* 260: 254
114. Johnson, H. R. 1985. In *Cool Stars With Excesses of Heavy Elements*, ed. M. Jaschek, P. C. Keenan, p. 271. Dordrecht: Reidel
115. Johnson, H. R. 1986. In *Nonthermal Phenomena in Stellar Atmospheres, CNRS/NASA Monogr. Ser., The M-Type Stars*, ed. H. R. Johnson, F. Querci, p. 323
116. Johnson, H. R., Bernat, A. P., Krupp, B. M. 1980. *Ap. J. Suppl.* 42: 501
117. Jørgensen, U. G., Almlöf, J., Gustafsson, B., Larsson, M., Siegbahn, P. 1985. *J. Chem. Phys.* 83: 3034
118. Jørgensen, U. G., Almlöf, J., Siegbahn, P. 1989. *Ap. J.* In press
119. Keenan, P. C. 1957. *Publ. Astron. Soc. Pac.* 69: 5
120. Keenan, P. C., Teske, R. G. 1956. *Ap. J.* 124: 449
121. Keenan, P. C., Boeshaar, P. C. 1980. *Ap. J. Suppl.* 43: 379
122. Kelch, W. L. 1975. *Ap. J.* 195: 679
123. Kelch, W. L., Milkey, R. W. 1976. *Ap. J.* 208: 428
124. Kilston, S. 1975. *Publ. Astron. Soc. Pac.* 87: 189
125. Kipper, T. 1982. *Tartu Astrophys. Obs. Teated* 66: 3
126. Kipper, T. 1987. *Sov. Astron. Lett.* 13: 429
127. Kipper, T., Kipper, M. A. 1984. *Sov. Astron. Lett.* 10: 363
128. Kneer, F. 1983. *Astron. Astrophys.* 128: 311
129. Krupp, B. M., Collins, J. G., Johnson, H. R. 1978. *Ap. J.* 219: 963
130. Kurucz, R. L., Peytremann, E. 1975. *Smithson. Obs. Spec. Rep. No. 362*
131. Lambert, D. L. 1977. In *Les Spectres des Molecules Simples au Laboratoire et en Astrophysique*, p. 173. Liège: Inst. Astrophys., Univ. Liège
132. Lambert, D. L. 1981. In *Physical Processes in Red Giants*, ed. I. Iben Jr., A. Renzini, p. 115. Dordrecht: Reidel
133. Lambert, D. L. 1985. In *Cool Stars With Excesses of Heavy Elements*, ed. M. Jaschek, P. C. Keenan, p. 191. Dordrecht: Reidel
134. Lambert, D. L. 1986. In *Hydrogen-Deficient Stars and Related Objects*, ed. K. Hunger, D. Schönberner, N. K. Rao, p. 127. Dordrecht: Reidel
135. Lambert, D. L. 1987. In *Astro-*

chemistry, *IAU Symp. No. 120*, ed. M. S. Vardya, S. P. Tarafdar, p. 583. Dordrecht: Reidel
136. Lambert, D. L. 1987. Private communication
137. Lambert, D. L. 1988. In *Evolution of Peculiar Red Giant Stars, IAU Colloq. No. 106*, ed. H. R. Johnson, B. M. Zuckerman. Cambridge: Univ. Press. In press
138. Lambert, D. L. 1988. Preprint
139. Lambert, D. L., Dearborn, D. S. P., Sneden, C. 1974. *Ap. J.* 193: 621
140. Lambert, D. L., McWilliam, A., Smith, V. V. 1987. *Astrophys. Space Sci.* 133: 369
141. Lambert, D. L., Brown, J. A., Hinkle, K. H., Johnson, H. R. 1984. *Ap. J.* 284: 223
142. Lambert, D. L., Gustafsson, B., Eriksson, K., Hinkle, K. H. 1986. *Ap. J. Suppl.* 62: 373
143. Luck, R. E., Bond, H. E. 1980. *Ap. J.* 241: 218
144. Luck, R. E., Lambert, D. L. 1982. *Ap. J.* 256: 189
145. Lucy, L. B., Robertson, J. A., Sharp, C. N. 1986. *Astron. Astrophys.* 154: 267
146. Maillard, J. P. 1974. In *Highlights of Astronomy*, ed. G. Contopoulos, 3: 269
147. Mathews, G. J., Takahashi, K., Ward, R. A., Howard, W. M. 1986. *Ap. J.* 302: 410
149. McKellar, A. 1940. *Publ. Astron. Soc. Pac.* 54: 407
149. McKellar, A. 1948. *Publ. Dominion Astrophys. Obs.* 7: 395
150. Menzel, D. H. 1930. *Publ. Astron. Soc. Pac.* 42: 34
151. Merchant, A. E. 1967. *Ap. J.* 147: 587
152. Merrill, K. M., Ridgway, S. T. 1979. *Annu. Rev. Astron. Astrophys.* 17: 9
153. Merrill, P. W. 1947. *Ap. J.* 105: 360
154. Merrill, P. W. 1952. *Ap. J.* 116: 21
155. Morris, S., Wyller, A. A. 1967. *Ap. J.* 150: 877
156. Mould, J. R. 1976. *Astron. Astrophys.* 48: 443
157. Mould, J. R. 1976. *Ap. J.* 210: 402
158. Mould, J. R. 1978. *Ap. J.* 226: 923
159. Mould, J. R., Hyland, A. R. 1976. *Ap. J.* 208: 399
160. Mount, G. H., Ayres, T. R., Linsky, J. L. 1975. *Ap. J.* 200: 383
161. Muchmore, D. O. 1986. *Astron. Astrophys.* 155: 172
162. Muchmore, D. O., Ulmschneider, P. 1985. *Astron. Astrophys.* 142: 393
163. Muchmore, D. O., Nuth, J. A. III, Stencel, R. E. 1987. *Ap. J. Lett.* 315: L141
164. Nadeau, D., Maillard, J. P. 1987. *Ap. J.* 327: 321
165. Nicholls, R. W. 1977. *Annu. Rev. Astron. Astrophys.* 15: 197
166. Nissen, P. E., Edvardsson, B., Gustafsson, B. 1985. In *ESO Workshop on Production and Distribution of C, N, O Elements*, ed. I. J. Danziger, F. Matteuchi, K. Kjär, p. 131. Garching: ESO
167. Nordlund, Å. 1985. In *Theoretical Problems in High Resolution Solar Physics, Max-Planck-Inst. No. 212*, ed. H. U. Schmidt, p. 1. Munich: Max-Planck-Inst. Phys. Astrophys.
168. Oinas, V. 1974. *Ap. J. Suppl.* 27: 405
169. Oinas, V. 1977. *Astron. Astrophys.* 61: 17
170. Olson, B. I., Richer, H. B. 1979. *Ap. J.* 227: 534
171. Omont, A. 1987. In *Astrochemistry, IAU Symp. No. 120*, ed. M. S. Vardya, S. P. Tarafdar, p. 357. Dordrecht: Reidel
172. Pagel, B. E. J. 1962. *R. Obs. Bull.* 55
173. Peery, B. F. Jr., Beebe, R. F. 1970. *Ap. J.* 160: 619
174. Peery, B. F. Jr., Wojslaw, R. S. 1977. *Bull. Am. Astron. Soc.* 9: 365
175. Piccirillo, J. 1980. *MNRAS* 190: 441
176. Piccirillo, J., Bernat, A. P., Johnson, H. R. 1981. *Ap. J.* 246: 246
177. Querci, M., Querci, F. 1970. *Astron. Astrophys.* 9: 1
178. Querci, F., Querci, M., Tsuji, T. 1974. *Astron. Astrophys.* 31: 265
179. Querci, F., Querci, M. 1975. *Astron. Astrophys.* 39: 113
180. Querci, M., Querci, F. 1976. *Astron. Astrophys.* 49: 443
181. Querci, F. 1986. In *Nonthermal Phenomena in Stellar Atmospheres, CNRS/NASA Monogr. Ser., The M-Type Stars*, ed. H. R. Johnson, F. Querci, p. 1
182. Querci, M. 1986. In *Nonthermal Phenomena in Stellar Atmospheres, CNRS/NASA Monogr. Ser., The M-Type Stars*, ed. H. R. Johnson, F. Querci, p. 113
183. Ramsey, L. W. 1977. *Ap. J.* 215: 827
184. Ramsey, L. W. 1981. *Ap. J.* 245: 984
185. Ridgway, S. T., Friel, E. D. 1981. In *Effects of Mass Loss on Stellar Evolution, IAU Colloq. No. 59*, ed. C. Chiosi, R. Stalio, p. 119. Dordrecht: Reidel
186. Ridgway, S. T., Hall, D. N. B., Carbon, D. F. 1977. *Bull. Am. Astron. Soc.* 9: 636
187. Ridgway, S. T., Joyce, R. R., White, N. M., Wing, R. F. 1980. *Ap. J.* 235: 126
188. Ridgway, S. T., Carbon, D. F., Hall, D. N. B., Jewell, J. 1984. *Ap. J. Suppl.* 54: 177
189. Ruland, F., Holweger, H., Griffin, R.,

Biehl, D., Griffin, R. 1980. *Astron. Astrophys.* 92: 70
190. Russell, H. N. 1934. *Ap. J.* 79: 317
191. Sanford, R. F. 1929. *Publ. Astron. Soc. Pac.* 41: 271
192. Sanford, R. F. 1947. *Publ. Astron. Soc. Pac.* 59: 333
193. Sanford, R. F. 1950. *Ap. J.* 111: 262
194. Scalo, J. M. 1981. In *Physical Processes in Red Giants*, ed. I. Iben Jr., A. Renzini, p. 77. Dordrecht: Reidel
195. Schadee, A., Davis, D. N. 1968. *Ap. J.* 152: 169
196. Scharmer, G., Carlsson, M. 1985. *J. Comput. Phys.* 59: 56
197. Schmidtke, P. C., Africano, J. L., Jacoby, G. H., Joyce, R. R., Ridgway, S. T. 1986. *Astron. J.* 91: 961
198. Shane, C. D. 1919. *Lick Obs. Bull.* 10: 79
199. Schmid-Burgk, J., Scholz, M., Wehrse, R. 1981. *MNRAS* 194: 383
200. Scholz, M. 1985. *Astron. Astrophys.* 145: 245
201. Scholz, M., Takeda, Y. 1987. *Astron. Astrophys.* 186: 200
202. Scholz, M., Tsuji, T. 1984. *Astron. Astrophys.* 130: 11
203. Scholz, M., Wehrse, R. 1982. *MNRAS* 200: 41
204. Smith, A. M., Jørgensen, U. G., Lehmann, K. K. 1987. *J. Chem. Phys.* 87: 5649
205. Smith, V. V. 1988. In *Origin and Distribution of the Elements*, ed. G. J. Mathews, p. 535. Singapore: World Sci.
206. Smith, V. V., Lambert, D. L. 1985. *Ap. J.* 294: 326
207. Smith, V. V., Lambert, D. L. 1986. *Ap. J.* 311: 843
208. Smith, V. V., Lambert, D. L. 1988. *Ap. J.* 333: 219
209. Smith, V. V., Wallerstein, G. 1983. *Ap. J.* 273: 742
210. Smith, V. V., Lambert, D. L., McWilliams, A. 1987. *Ap. J.* 320: 862
211. Spinrad, H., Vardya, M. S. 1966. *Ap. J.* 146: 399
212. Spinrad, H., Kaplan, L. D., Connes, P., Connes, J., Kunde, V. G., Maillard, J. P. 1970. *Proc. Conf. Late-Type Stars*, ed. G. W. Lockwood, H. M. Dyck, p. 59. Tucson, Ariz: Kitt Peak Natl. Obs.
213. Spite, M., Spite, F. 1985. *Annu. Phys. Astron. Astrophys.* 23: 225
214. Spitzer, L. Jr. 1949. *Ap. J.* 109: 548
215. Stauffer, J. R., Hartmann, L. W. 1986. *Ap. J. Suppl.* 61: 531
216. Steenbock, W. 1985. In *Cool Stars With Excesses of Heavy Elements*, ed. M. Jaschek, P. C. Keenan, p. 231. Dordrecht: Reidel
217. Steiman-Cameron, T. Y., Johnson, H. R. 1986. *Ap. J.* 301: 868
218. Thompson, R. I. 1973. *Ap. J.* 181: 1039
219. Thompson, R. I. 1973. *Ap. J.* 184: 187
220. Thompson, R. I. 1977. *Ap. J.* 212: 754
221. Thompson, R. I., Schnopper, H. W., Rose, W. K. 1971. *Ap. J.* 163: 533
222. Trimble, V., Bell, R. A. 1981. *Q. J. R. Astron. Soc.* 22: 361
223. Tsuji, T. 1962. *Publ. Astron. Soc. Jpn.* 14: 222
224. Tsuji, T. 1971. *Publ. Astron. Soc. Jpn.* 23: 275
225. Tsuji, T. 1976. *Publ. Astron. Soc. Jpn.* 28: 543
226. Deleted in proof
227. Tsuji, T. 1978. *Astron. Astrophys.* 62: 29
228. Deleted in proof
229. Tsuji, T. 1979. *N. Z. J. Sci.* 22: 415
230. Tsuji, T. 1981. *J. Astrophys. Astron.* 2: 95
231. Tsuji, T. 1981. *Astron. Astrophys.* 99: 48
232. Tsuji, T. 1984. *Astron. Astrophys.* 134: 24
233. Tsuji, T. 1985. In *Cool Stars With Excesses of Heavy Elements*, ed. M. Jaschek, P. C. Keenan, p. 295. Dordrecht: Reidel
234. Tsuji, T. 1986. *Astron. Astrophys.* 156: 8
235. Tsuji, T. 1986. *Astrophys. Space Sci.* 118: 227
236. Tsuji, T. 1986. *Annu. Rev. Astron. Astrophys.* 24: 89
237. Tsuji, T. 1988. In *Evolution of Peculiar Red Giant Stars, IAU Colloq. No. 106*, ed. H. R. Johnson, B. M. Zuckerman. Cambridge: Univ. Press. In press
238. Tsuji, T. 1988. *Astron. Astrophys.* 197: 185
239. Utsumi, K. 1963. *Publ. Astron. Soc. Jpn.* 15: 482
240. Utsumi, K. 1967. *Publ. Astron. Soc. Jpn.* 19: 342
241. Utsumi, K. 1970. *Publ. Astron. Soc. Jpn.* 22: 93
242. Utsumi, K. 1985. In *Cool Stars With Excesses of Heavy Elements*, ed. M. Jaschek, P. C. Keenan, p. 243. Dordrecht: Reidel
243. Utsumi, K. 1985. *Proc. Jpn. Acad.* 61: 193
244. Vieira, T. 1986. *Uppsala Astron. Obs. Rep. No. 32*
245. Wallerstein, G. 1973. *Annu. Rev. Astron. Astrophys.* 11: 115
246. Wallerstein, G. 1977. *Publ. Astron. Soc. Pac.* 89: 35
247. Wallerstein, G. 1988. *Science* 240: 1743
248. Wallerstein, G., Dominy, J. F. 1988. *Ap. J.* 330: 937

249. Wannier, P. G. 1985. In *ESO Workshop on Production and Distribution of C,N,O Elements*, ed. I. J. Danziger, F. Matteucci, K. Kjär, p. 233. Munich: ESO
250. Wannier, P. G., Linke, R. A., Penzias, A. A. 1981. *Ap. J.* 247: 522
251. Warner, B., Dean, C. A. 1970. *Publ. Astron. Soc. Pac.* 82: 904
252. Watanabe, T., Kodaira, K. 1978. *Publ. Astron. Soc. Jpn.* 30: 21
253. Watanabe, T., Kodaira, K. 1979. *Publ. Astron. Soc. Jpn.* 31: 61
254. Wehrse, R. 1981. *MNRAS* 195: 553
255. Wennfors, B. 1986. PhD thesis. Stockholm Univ., Swed.
256. Westerlund, B. 1987. In *ESO Workshop on Stellar Evolution and Dynamics in the Outer Halo of the Galaxy*, ed. M. Azzopardi, F. Matteucci, p. 207. Garching: ESO
257. Whitelock, P. A., Catchpole, R. M. 1985. *MNRAS* 212: 873
258. Wojslaw, R. S., Peery, B. F. Jr. 1976. *Ap. J. Suppl.* 31: 75
259. Wyckoff, S., Wehinger, P. 1972. *Ap. J.* 178: 481
260. Wyller, A. A. 1960. *Astrophys. Norv.* 7: 13
261. Wyller, A. A. 1966. *Ap. J.* 143: 828
262. Yamashita, Y. 1965. *Publ. Astron. Soc. Jpn.* 17: 27
263. Yamashita, Y. 1965. *Publ. Astron. Soc. Jpn.* 17: 55
264. Yamashita, Y., Utsumi, K. 1962. *Publ. Astron. Soc. Jpn.* 14: 128
265. Zook, A. C. 1985. *Ap. J.* 289: 356

SUBJECT INDEX

A

A stars, high-latitude metal-rich, 575
Abell 1367, 116
Abundance ratios, as function of metallicity, 279–315
Abundances
 carbon, see Carbon, abundance of
 in carbon stars, 725–38
 chromium, 298
 cobalt, 301, 319, 676
 in cool stars, see Abundances in carbon stars; in M stars; in N stars; in S stars; in SC stars; and Cool stars, abundances in
 copper, 301–2, 311
 in globular cluster stars, 292, 307–13, 336
 in H II regions of Magellanic Clouds, 314–15
 importance of, in calculation of model atmospheres, 706–7
 iron, see Iron, abundance of
 of iron-group elements, 297–303
 light-metal, 293–97, 310–11
 in Local Group galaxies, 140–56
 in M dwarfs, 720–22, 734
 in M giants and supergiants, 717–20, 727–28, 730–34
 in Magellanic Clouds, 141–42, 149–50, 313–15, 638–39
 in N stars, 725–29, 731
 nitrogen, see Nitrogen, abundance of
 in Orion molecular cloud, 65–69
 oxygen, see Oxygen, abundance of
 in S stars, 717, 722–25, 728–34
 in SC stars, 734–35
 in Sk -69 202, 634–35, 639–40
 solar, 284–85
 technetium, 321, 717, 722–24, 726
 of very heavy elements, 303–7, 311
Accretion disks, see also Circumstellar disks
 in active galactic nuclei, 413, 415–16
 in binary stars, 402
 in low-mass X-ray binary stars, 521, 524–25, 539–42
 in T Tauri stars, 376–90
Active galactic nuclei, *IUE* observations of, 412–16
Age-metallicity relation, 140–56, 325–35, 600–2, 613–14
Alpha Bootis, 291–92
Alpha Herculis, 733–34, 739, 743
Alpha Orionis
 abundances in, 709, 718–19, 731–32
 isotopic ratios in, 739, 743, 747
 IUE observations of, 401
Alpha Scorpii, 739, 743
Arcturus moving group, 597
Aromatic infrared (AIR) bands, 166–67
Arp 220, 102

B

B stars
 formation and evolution of, 4–5, 575
 supergiant, 632–33, 635–40
Barnard's loop, 51
BD $-18°5550$, 291
Becklin-Neugebauer-Kleinmann-Low nebula region, 46–48, 55
 infrared emission of, 60–65
 luminosities of sources in, 60–61
 masers in, 49
 polarization in, 61–63
Becklin-Neugebauer object
 central star in, 73–74
 circumstellar region of, 73–75
 infrared emission of, 60–65
Beta Geminorum, 710, 717
Beta Lyrae, 402
Beta Pegasi, 717, 739, 747–48
Betelgeuse, see Alpha Orionis
Binary star evolution, 401–5
Binary stars
 accretion in, 402
 chromospheres of, 401–2
 hot companions in, 403
 IUE observations of, 401–5
 mass flow in, 401–5
 among T Tauri stars, 364
 X-ray emission from close accreting, 88–96, 100–4
Binary stars, Algol-type, 402–4
Binary stars, low-mass X-ray, 517–53
 accretion disks in, 521, 524–25, 539–42
 atoll-source, 529, 533, 542–45, 548–49
 black hole candidates among, 523, 529, 533, 547–48
 high-frequency noise (HFN) in, 531–33, 544–45
 low-frequency noise (LFN) in, 525, 528, 531–35, 538, 541–42
 luminosities of, 522–23
 neutron stars in, 522–25, 528–29, 539–42
 noise components in, 518–20, 524–28, 548–49
 in normal galaxies, 96, 100–1
 quasi-periodic oscillations in, see Quasi-periodic oscillations
 very low-frequency noise (VLFN) in, 528, 531–33, 544–45
 Z-source
  flaring branch (FB) in, 530, 536–38
  horizontal branch (HB) in, 530–38
  models for, 538–42
  noise in, 531–33
  quasi-periodic oscillations in, 534–38
Binary stars, massive X-ray, 100–1, 521–22, 524
Binary stars, W Ursae-Majoris-type, 402
Black hole candidates, 92, 523, 529, 533, 547–48
Black holes, 238–39
BP Tauri, 366, 379–80

C

Carbon
 abundance of
  in cool stars, 729–47
  in dwarf stars, 283–89
  in Magellanic Clouds, 314–15
  in red giant stars, 291–93
 isotopic ratios of, 739–42
Carbon stars
 abundances in, 725–38
 isotopic ratios in, 738, 741–45
 model atmospheres of, 705–8
 temperature scale for, 712

757

SUBJECT INDEX

Carina dwarf galaxy, 156
CD $-38°245$, 298–300
Centaurus A, 112, 122, 129, 244
Centaurus X-3, 524, 546–47
Chi Cygni, 722–23, 739
Chi Persei, 185
Chlorine, isotopic ratios of, 748
Chromium, abundance of, 298
Circinus X-1, 545–47
Circumstellar disk, around HL Tauri, 373–75, 382–83
Circumstellar disks, 373–90
 accretion in, 376–90
 around FU Orionis objects, 377, 384
 and stellar evolution, 383–85
 and stellar winds, 385–90
 in T Tauri stars, 373–90
 models for, 375–81
Cobalt
 abundance of, 301, 319, 676
 radioactive decay of, in supernovae, 322, 646, 667–74, 680–91
Cold dark matter, 560, 562
Comet Halley, 416
Comet *IRAS*-Araki-Alcock, 416
Comets, *IUE* observations of, 416
Cool stars, 701–56. See also Carbon stars; M stars; N stars; S stars; and SC stars
 chromospheres of, 401
 CNO abundances in, 729–38
 isotopic abundance ratios in, 738–49
 lithium abundances in, 727–29
 metal abundances in, 715–27
 model atmospheres of, 704–11
 surface gravities of, 713–14
 synthetic spectra of, 714–15
Copernicus satellite, 12–13
 observations with, 7–8, 13, 400, 408–10
Copper, abundance of, 301–3, 311
Coronal mass ejections (CMEs), 422, 424, 447
COS-B satellite observations, 469–70, 476–88, 495–98, 506, 508–11
Cosmic-ray anomalous component, 207–8, 224–25
Cosmic-ray flares, 423
Cosmic-ray halo, 503–4, 507
Cosmic-ray production, 499–500, 503
 X-ray emission and, 101
Cosmic-ray spectrum
 electron, 500–1, 507

variations in, 506–8
Cosmic rays
 abundances in, 498
 distribution in Galaxy of, 477–83, 499–506
 and gamma-ray production, 472–76
 in heliosphere, 223–25
 and interstellar medium, 205–6, 469–76, 483, 498–508
 in molecular clouds, 493, 497
 in solar neighborhood, 500–2
CY Cygni, 722, 740
Cygnus X-1, 524, 547
Cygnus X-2, 519, 523, 526–32, 534–38, 541
Cygnus X-3, 101, 525

D

Diffuse interstellar bands (DIBs), 191–92
DF Tauri, 356, 358–59, 366–67
DN Tauri, 356, 358–59
DR Tauri, 356–59, 369
Draco dwarf galaxy, 156
DY Persei, 726

E

Einstein Observatory satellite observations
 of galaxies, 87–88, 99, 104–12, 120–24, 130
 of T Tauri stars, 353, 360
EXO 2030+337, 547
EXOSAT observations, 525–27

F

F stars, Galactic distribution of, 619–622
Fairall 9, 415
Fifth Orbiting Geophysical Observatory, 425
Flare stars, 367–69
Fornax A, 129, 244–45
Fornax dwarf elliptical galaxy, 142, 153–54
FU Orionis objects, 357, 377, 384

G

Galactic center, gamma-ray emission from, 497–98
Galactic disk
 age of, 326–27
 gamma rays from, 476–86
 mass distribution in, 615–23
 surface density of, 615–22
 volume density of, 615–22

Galaxies
 evolution of spheroids of, 567–69
 formation of disk, 556–69
  angular momentum distribution in, 562
  continual-infall models of, 561–62
  dissipation in, 557–62
  models of, 559–62
  star formation in, 560–61, 565–69
 formation and evolution of very massive disk, 559–60
 gaseous halos around, 108–29
  cooling flows from 117–19, 124
  radio sources and, 128–29
  temperature of 114–17
 masses of early-type, 120–27
 mergers of, 239–40, 242–43, 265–69
 nuclei of, 10–11
  X-ray emission from, 103–8
 radio emission in, 100–1, 128–29
 stellar populations of, 139–59
 X rays from normal, 87–138. See also Galaxies, elliptical and Galaxies, spiral
Galaxies, BCM, 266–69
Galaxies, cD, 126, 265–69
Galaxies, dwarf elliptical, 142, 153–56
Galaxies, dwarf spheroidal, 238–39, 251–53, 263
Galaxies, elliptical, 108–29, 235–77
 accretion in, 239–40, 242, 246–48
 brightness profiles of, 248–50
 color gradients in, 261–65
 cooling flows in, 117–19
 cores of, 236–37, 239–40
 dust in, 240–46
 formation and evolution of, 236–69
 fundamental plane of, 254–58
 luminosities of, 253–58, 260–65
 mergers of, 239–40, 242–43, 265–69
 metallicity in, 262–65
 nuclei of, 237–39
 shapes of, 243–48
  isophote, 258–61
 shells and ripples in, 246–48
 stellar population gradients in, 263
 structure of, 258–61

SUBJECT INDEX 759

tidal effects in, 250–51
X-ray emission from, 108–29
Galaxies, S0
  classification of, 246, 258–61
  X-ray emission from, 108–29
Galaxies, spiral
  interstellar gas in, 4–5, 94, 98–99
  X-ray emission from, 88–102
    correlation with optical, 99–100
    correlation with radio continuum, 100–1
Galaxy
  asymmetric drift of stellar population of, 587, 589–92
  chemical evolution of, 279–49, 565–69, 585, 592–606
  cosmic-ray distribution in, 477–83, 499–506
  density distribution of stars in, 587–92
  formation of, 559
  models of, 563–65
  gamma-ray emission in, 469–516
  high-latitude stellar distribution of, 574–76
  infrared emission of, 163–66
  interstellar extinction curve of, 183–85, 190
  interstellar matter in, 5–6, 161–98
  metal-poor spheroid of, 563–65, 567–68, 579–85
  molecular gas content of, 476–83, 496–97
  radio spectrum of, 507–8
  star counts in, 569–85
  star formation in, 563–69, 585–606
  stellar population of, 139–46
  thick disk of, 571–73, 606–15
    age of, 613–15
    discreteness of, 609–13
    formation of, 606–7
    metallicity of stars in, 607–8
  star counts in, 576–78
  velocity dispersions of stars in, 587–92, 609–13
  X-ray sources in, 92–93, 103
Gamma-ray emission
  from Galactic center, 497–98
  from Galactic disk, 476–86
  in local interstellar medium, 486–92
  from supernova remnants, 477, 491, 508, 680–82
  from Supernova 1987A, 672, 684–90

Gamma-ray emission, Galactic, 469–516
  at medium latitudes, 488–92
  as molecular gas tracer, 492–98
Gamma-ray halo, 487–92
Gamma-ray line and gamma-ray proton solar flares, see Solar flares, gamma-ray line and gamma-ray proton
Gamma-ray production processes, Galactic
  bremsstrahlung, 472–76
  cosmic-ray role in, 469–76
  inverse Compton scattering, 474–76, 491
  nuclear interactions in, 472–76
Gamma-ray sources, searches for, 508–11
Gamma-ray spectrum, variations of, in Milky Way, 483–86, 506–7
Gamma-rays, Galactic, intensities of, 475–76, 482, 496–97
Gamma-rays, solar nuclear, 426–35. See also Solar flares, gamma-ray proton
Ginga satellite observations
  of galaxies, 87
  of low-mass X-ray binaries, 529, 546
  of Supernova 1987A, 683, 685–88
Gleise 15A, 734
Gleise 205, 721
Gleise 411, 734
Globular clusters
  abundances in, 292, 307–13, 336, 584
  age determinations of, 326–27, 338, 601–2, 614–15
  distribution of, 584
  evolution of, 9–11
  in Local Group galaxies, 146–47, 149
  metallicity in, 308–9
  X-ray sources in, 92–93
Gould's belt, 44, 487
Ground-level events (GLEs), 423
GX 3+1, 523
GX 5+1, 519, 525–28, 534–36
GX 9+9, 523
GX 17+2, 526, 530–31, 535, 537
GX 339−4, 523, 547–48
GX 340+0, 531, 536, 538
GX 349+2, 525–26, 536
G 77−61, 288–89, 727

H

H I regions, 5–7
H II regions
  abundances in, in Magellanic Clouds, 314–15
  in Orion molecular cloud, 42, 50, 55–60, 73–75
  temperatures of, 5–6
HD 122563, 284, 289, 303–4
HD 140283, 291
HD 207739, 403
Heliopause, 211, 216, 229–31
Heliosphere, 199–234
  cosmic rays in, 223–25
  interstellar neutral gas in, 217–23
  magnetic field of, 215–217, 229–31
  structure of, 226–31
  terminal shock in, 226–29
Hinotori satellite observations, 422, 435–36, 445
HL Tauri, 373–75, 382
HR 1105, 723, 747–48
HR 8714, 723
HS 1700+6416, 416
Hubble Space Telescope, 13–14

I

IC 342, 103–4
IC 1459, 240
IC 1613, 152
ICE (International Cometary Explorer), 416
Infrared emission
  from circumstellar disks, 375–76
  of interstellar grains, 162–89, 194–96
  from normal galaxies, 100–4
  in Orion molecular cloud, 44–46, 60–77
Infrared observations
  of Supernova 1987A, 675–77
  of T Tauri stars, 375–76, 381–84
Interplanetary protons, 423, 431–35, 441–48
Interstellar dust, 206, 225–26. See also Interstellar grains
Interstellar extinction curve, 182–85, 190
Interstellar gas
  heating of the, 192–93
  neutral, in heliosphere, 203–4, 217–23
  in Orion molecular cloud, 41–47
    high-velocity, 51
    kinematics and energetics of, 47–48

SUBJECT INDEX

Interstellar grains, 161–98. See also Interstellar dust
  infrared emission of, 162–78, 185–87, 194–96
  size spectrum of, 178–80
  ultraviolet radiation of, 162, 183–88
Interstellar magnetic field, 7, 213–17, 226–31
Interstellar matter, 4–8. See also Interstellar grains; Very small grains; Polycyclic aromatic hydrocarbons
  Supernova 1987A and, 677–78
Interstellar medium
  gamma rays and, 470–92
  at high Galactic latitudes, 411
  interaction of cosmic rays with, 205–6, 469–76, 483, 498–508
  IUE observations of, 408–11
  solar wind and, 199–234
  as source of X-ray emission, 98–99, 108–29. See also Galaxies, gaseous halos around
Interstellar medium, local, see Local interstellar medium
Interstellar medium, very local, see Very local interstellar medium
Interstellar pickup ions, 207, 221, 225
Interstellar plasma, 210–15
IRAS bands, colors in the, 163–66, 173–76, 183–89, 194–96
IRAS observations
  of diffuse Galactic radiation, 164, 182, 184–85
  of galaxies, 242
  of Orion molecular cloud, 42
  of T Tauri stars, 375, 383
IRc2, 61–65, 68–71, 73
  circumstellar region of, 75–78
IRc3–7, 61, 63–64
IRc9, 64
Iron
  abundance of, 297–98
  as measure of metallicity, 280–81, 284–85, 290, 327–28
  production of, in star formation, 333–34, 340–45, 565–69, 603–6
Isotopic ratios, in spectra of cool stars, 738–49
IUE (International Ultraviolet Explorer), 397–420
  instrumentation, 397–99

observations
  of active galactic nuclei, 412–16
  of close binary stars, 401–5
  of comets, 416
  of galaxies, 150
  of the interstellar medium, 408–11
  of planets, 416
  of stellar chromospheres, 400
  of stellar winds, 405–7
  of Supernova 1987A, 411–12, 630–33, 642–43, 663–64, 677–78, 692
  of T Tauri stars, 360
operation of, 397–99

K

K stars
  dwarf, 403, 577–78, 619–20, 623
  giant, 580, 608, 610, 619, 622
Kapteyn's star, 720–21, 734
Kitt Peak National Observatory, predicted light pollution at, 25–26
Kleinmann-Low nebula, see also Becklin-Neugebauer-Kleinmann-Low nebula
  abundances in, 65–69
  core of
    chemistry of, 65–69
    outflows from, 69–71
    shocked gas in, 71–73
  infrared emission of, 60–65
  polarization in, 49, 61–63

L

Large Magellanic Cloud, 314–15
  abundances in, 638–39
  30 Doradus region in, 638
  stellar population in, 145–46, 149
  Supernova 1987A in, 630, 632, 635–39. See also Supernova 1987A
  X-ray sources in, 89–92, 98
Leo I dwarf elliptical galaxy, 154–55
Leo II dwarf elliptical galaxy, 155
Lick Observatory, predicted light pollution at, 24–25
Light pollution, 19–40
  model of, 20–22
  predicted for future observatory sites, 32–35

predicted for present observatory sites, 23–32
Light sources, effect on sky brightness of, 36
Lithium, abundance of, 282–83, 289, 361, 717, 727–29
LMC X-1, 92, 525, 547
LMC X-3, 92, 547
Local Bubble, 202–3
Local Fluff, 202–5
Local Group of galaxies
  stellar populations of, 139–59
  X-ray emission from, 88–94
Local interstellar medium, 202
  IUE observations of, 408–10
  gamma-ray emission from, 486–87
Long-period variable stars, 709–10
Low-mass X-ray binary stars, see Binary stars, low-mass X-ray
Lowell Observatory, predicted light pollution at, 30–31
Low-pressure sodium street lamps, 36
Luminaires, 36–37
LX Cygni, 740
L1630, 42
L1640, 42
L1641, 42, 48, 52
L1647, 42

M

M stars
  dwarf, abundances in, 720–22, 734
  dwarf emission, 367–68
  flux distribution in, 711–12
  giant and supergiant, abundances in, 717–20, 727–28, 730–34
  isotopic ratios in spectra of, 739–40, 743–44, 747–48
  model atmospheres for, 705–11
  temperature scale for, 712
Magellanic Clouds, see also Large Magellanic Cloud; Small Magellanic Cloud
  abundances in, 141, 149–50, 313–15, 638–39
  globular cluster population of, 146
  interstellar medium in, 181, 411
  mass-loss rates in, 406
  metallicities in, 315
  wind velocities of stars in, 406

SUBJECT INDEX 761

Magnesium, isotopic ratios of, 747–48
Magnetic field
  interstellar, 7, 213–17, 226–31
  in Orion molecular cloud, 48–50
  solar, 447–48, 450–60
Magnetic fields in T Tauri stars, 365–66
Manganese, abundance of, 301, 319
Masers, in Orion molecular cloud, 47, 49, 67–68, 70–71, 75–76
Mauna Kea Observatory, predicted light pollution at, 27–28
McDonald Observatory, predicted light pollution at, 26–27
Metallicity
  abundance ratios as a function of, 279–315
  correlation of age with, 140–56, 325–34, 600–2, 613–14
  correlation of color with, in galaxies, 263
  correlation of orbital eccentricity with, 597–600
  correlation of rotation velocity with, 593–97, 609–10
Milky Way
  gamma-ray observations of, 470–71
  gamma-ray spectral variations along, 483–86
Milky Way galaxy, see Galaxy
Mir satellite, 683, 685–87
Mira, see Omicron Ceti
Mira variable stars, 709–10, 723, 734–35, 739–40, 748
Missing mass, 575, 616–17, 620–23
Molecular clouds, 10. See also Orion molecular cloud
  cosmic-ray penetration of, 493
  gamma-ray emission in, 492–97, 504
  infrared emission in, 166–67, 185–87
  T Tauri stars associated with, 352–53
  UV radiation in, 186–87
Molecular gas
  content in Galaxy of, 496–97
  gamma-ray radiation as tracer of, 492–98

Monoceros R2 molecular cloud, 42–44
Mount Hopkins, predicted light pollution at, 28–29
Mount Lemmon, predicted light pollution at, 30
Mount Wilson Observatory, predicted light pollution at, 24
Mrk 335, 413–14
MXB 1730−335, see Rapid Burster
M13, 312
M15, 310
M17, 164
M31
  nucleus of, 237
  stellar population of, 147–48
  stellar winds in, 406
  X-ray sources in, 89–96, 112
M32
  stellar population of, 151
  tidal effects in, 250, 252
M33
  nucleus of, 239
  stellar population of, 148–49
  stellar winds in, 406
  X-ray emission in, 94, 98, 107
  X-ray sources in, 89–92
M33 X-7, 91
M51
  X-ray emission in, 97, 103, 107
  X-ray sources in, 95–96
M55, 311
M81
  nucleus of, 237–38
  X-ray emission in, 95, 97, 107
M82
  infrared emission in, 166
  X-ray emision in, 95, 103–6, 130
M83, 95–97, 103–4
M86, 108, 125
M87, 108, 120–24, 126
M92, 310, 312
M100, 95
M101, 95, 98–99

N

N stars
  abundances in, 725–31
  isotopic ratios in, 741–42, 745, 748
  temperature scale for, 712
Neutrino burst, 645–46, 649–63
Neutrinos
  detectors of, 652–56
  properties of, 660–63

in supernova explosions, 644–46, 649–52
  from Supernova 1987A, 629, 631, 645–46, 652–60
Neutron star; possible, in V Sagittae system, 403
Neutron stars
  in low-mass X-ray binary stars, 522–24, 528–29
  magnetic field decay in, 522–24
  magnetospheres of, 539–42
  in supernovae, 646, 649–52, 659, 691–92
NGC 121, 614
NGC 147, 152–53, 252
NGC 185, 152, 252
NGC 205, 150–51, 252
NGC 253, 95, 97, 103–6
NGC 720, 125
NGC 891, 571
NGC 1275, 121
NGC 1316, see Fornax A
NGC 1395, 125
NGC 1399, 124, 129
NGC 1600, 237
NGC 1961, 99
NGC 1977, 42, 46
NGC 2023, 42, 177
NGC 2024, 42, 47, 49
NGC 2149, 42
NGC 3115, 238
NGC 3628, 104, 106
NGC 4038/9, 102
NGC 4151, 415
NGC 4406, 119, 240
NGC 4438, 99
NGC 4472, 108–10, 122–26
NGC 4486B, 250
NGC 4546, 244
NGC 4589, 244
NGC 4594, 238
NGC 4631, 95, 99
NGC 4636, 125
NGC 4649, 125
NGC 4696, 121
NGC 5053, 309
NGC 5128, see Centaurus A
NGC 5266, 245
NGC 5322, 240
NGC 5363, 244
NGC 5813, 239–40
NGC 5846A, 251
NGC 6822, 151–52
NGC 6946, 95–96, 103
NGC 7714, 103
Nickel
  abundance of, 298–300
  decay of, in supernovae, 322, 669–70
Night sky, brightness of, see Light pollution

SUBJECT INDEX

Nitrogen
  abundance of
    in cool giant stars, 729–38
    in giant stars, 291–93
    in Magellanic Clouds, 314
    in metal-poor dwarfs, 286–89
  isotopic ratios of, 742–43
Nova Centauri 1986, 189
Nucleosynthesis, 316–17
  in intermediate-mass stars, 320–21, 323
  in massive stars, 317–20, 323
  in supernova explosions, 316, 566–69, 646–48, 679–80
  in Type Ia supernovae, 321–23
  in very massive stars, 317, 323

O

O stars, 4–7
OB association, in Orion, 50–51
OB stars
  IUE observations of, 408
  in Orion A H II region, 55–60
Observatory sites
  predicted light pollution at present, 19, 23–32
  predicted light pollution at prospective, 19, 32–35
Omega Centauri, 308
Omicron Ceti, 723, 748
Oort limit, 616, 620–22
OMC 1, 44–47, 52–54
OMC 2, 44, 46
Orion molecular cloud, 41–85
  Becklin-Neugebauer object in, see Becklin-Neugebauer object
  distribution of gas in, 41–42
  H II region interface with, 55–60
  H II regions in, 42–46, 50, 73–76
  infrared sources in, 44–47, 60–77
  Kleinmann-Low nebula in, see Kleinmann-Low nebula
  low-mass-star distribution in, 51–52
  low-mass-star formation in, 52–55
  magnetic fields in, 48–50
  masses in, 42, 44–47, 65, 67
  molecular gas clumps in, 44–47, 65
  OB star interaction with, 55–60

  origin of, 42–44
  outflows from stars in, 47–48
  photo-dissociation region in, 56–60
  polarization in, 48–50, 61–63
  star formation in, 44–55, 60–77
  supernova explosions in, 50–51
  Trapezium in, see Trapezium Cluster
  velocities of gas in, 47–48, 51, 67, 71–73
Oxygen
  abundance of
    in cool stars, 729–38
    in giant stars, 291–93
    in globular cluster stars, 310, 312–13
    in Magellanic Clouds, 314
    as measure of relative abundances, 327–28, 334–35
    in metal-poor dwarf stars, 285–86, 289–91
  isotopic ratios of, 743–47
  production of, in stellar evolution, 323–24, 340–45
Oxygen/iron ratio, stellar, 565–69, 603–6

P

Palomar Observatory, predicted light pollution at, 23–24
Photo-dissociation regions, 56–60
Pioneer 10, 208
19 Pisces, 709, 713, 737, 748
Planetary nebulae
  PAH particles in, 166, 187, 189, 196
  winds in nuclei of, 406–7
Plasma physics, 14–16
Polarization, 8
  in Orion molecular cloud, 48–50, 61–63
Polycyclic aromatic hydrocarbon molecules (PAHs), 166–96
  dehydrogenation of, 177–78
  destruction and formation of, 187–90
  as heating mechanisms for interstellar gas, 192–93
  infrared spectroscopy of, 169–78
  in Orion molecular cloud, 57
  in planetary nebulae, 166, 187, 189, 196
  temperature fluctuations of, 168–71

Population growth, effect on night sky brightness of, 21–22
Population I stars,
  in Local Group galaxies, 139–40, 143–44
  as X-ray sources, 89, 92, 100–1
Population II stars
  abundances in, 291–93
  extreme, formation of, 567, 605–6, 609
  intermediate, evolution of, 606–7
  in Local Group galaxies, 139–40, 143
Population III stars, 335–36
Project Stratoscope, 12
Proton flares, 423–24
Pulsar, in Supernova 1987A, 670–71, 674, 691–92
Pulsars
  accreting, 529, 533, 546–47
  millisecond radio, 521–23, 526

Q

Quasars, 416
Quasi-periodic oscillations
  in accreting pulsars, 546–47
  in black hole candidates, 547–48
  in Circinus X-1, 545–46
  in low-mass X-ray binary stars
    characteristics of, 518–10
    discovery of, 525–29
    flaring-branch, 536–38, 548
    horizontal-branch, 534–38, 548
    normal-branch, 536–38, 548
    in Z-source, 529–31, 537–42
  in Rapid Burster, 546

R

R Andromedae, 748
R Canis Minoris, 722
R Cygni, 740, 748–49
R Lyrae, 723
R Sculptoris, 742
Radio bursts, solar-flare, 423–25, 427, 429, 432, 440–43, 458, 460
Radio emission
  in normal galaxies, correlation with X-ray, 97–103, 128–29
  from Supernova 1987A, 642

SUBJECT INDEX 763

Radio observations of solar wind flow, 206–7
Radio spectrum of Galaxy, variation in, 507–8
Radioactivity, supernova, 667, 678–91
Rapid Burster, 525–26, 546
Rho Ophiuchi dark cloud
  gamma-ray sources in, 509
  interstellar extinction of, 408
  T Tauri stars in, 352–53, 368
RR Lyrae stars
  ages of, 614
  Galactic distribution of, 582–84, 596–97
  in Local Group galaxies, 146–47, 149
RY Draconis, 742
RY Lupi, 369
RY Tauri, 357
RZ Pegasi, 734–35

S

S stars
  abundances in, 717, 722–25, 728–34
  isotopic ratios in, 739–40, 744, 747–49
  model atmospheres for, 705
  temperature scale for, 712
Sacramento Peak Observatory, predicted light pollution at, 28–29
SAS-2 satellite observations, 469–70, 476–78, 482–83, 486–87, 506, 508
SC stars
  abundances in, 724–25, 734–38
  isotopic ratios in, 740–41, 745
Scandium, abundance of, 300–1, 319
Scorpius X-1, 101, 519, 523–24, 526–28, 536–37, 541
Sculptor dwarf elliptical galaxy, 142, 154
Seyfert galaxies, 415
Sigma Scorpii, 185
Silicon, isotopic ratios of, 747–48
SK 21−65, 314–15
SK 41−68, 314–15
Sk −69 202, 632–43. See also Supernova 1987A
  abundances in, 634–35, 639–40
  circumstellar shell of, 642–43
  evolution of, 636–43
  mass loss in, 636–40
  metallicity of, 638–39

Skylab observations, 424–25
Small Magellanic Cloud
  stellar evolution in, 141–42
  stellar population of, 150
  X-ray sources in, 89–92
SN 1980k, 95
SN 1987A, see Supernova 1987A
Solar abundances, 284–85
Solar atmospheres, model, 284–85
Solar cycle, influence on night sky brightness of, 22
Solar flares. See also Proton flares; Radio bursts; Cosmic-ray flares
  classification of, 421–67
  from *Hinotori* observations, 435–36
  from *Skylab* observations, 424–25
  energy release processes in, 450–57
  filament activity and, 446–48, 453–61
  first phase of, 422, 425–29
  frequency of different classes of, 448–50
  gamma-ray line, 427–35
  gamma ray/proton, gradual, 431–37, 441–45, 461
    filament activity in, 447, 453
    frequency of, 448–49
  gamma ray/proton, impulsive, 431–34, 436, 440–41, 460–61
    filament activity in, 447, 453
    frequency of, 448–49
  nonthermal hard X-ray, 436, 438–40, 450, 460
  nuclear gamma rays in, 426–29, 431–32
  phases of, 458–60
  quiescent filament-eruption, 437, 445–46, 449–50, 461
  second phase of, 425–26
  acceleration in, 431–35
  thermal hard X-ray, 436–38, 460
  X-ray emission from, 424–60
Solar Maximum Mission (SMM) observations
  of solar flares, 422, 426–35
  of Supernova 1987A, 684–85
Solar wind
  effect of neutral gas flow on, 220–23
  flow speed of, 200–2, 209–11, 215–17, 221–23

interstellar medium and, 199–234
proton flux density in 200–2
ram pressure in, 200–2, 208–11
shock front in, 206–8
terminal shock in, 215, 226–29, 231
Speckle interferometry, of Supernova 1987A, 676–77
SS 433, 101
Star-count analysis, 570–74
Star counts, 569–85
  and central bulge of Galaxy, 578–79
  density profile from, 572–73, 576–78
  Galactic thick disk and, 576–78
  metal-poor spheroid and, 579–85
Star formation, 4–8
  in disk galaxies, 560–61
  in Galactic spheroid, 563–65, 568–69, 602–5
  in Galaxy, 338–45, 585–86, 605–6
  in Local Group galaxies, 141–45, 148–56
  low-mass, 52–55
  in Orion molecular cloud, 44–55, 60–77
Starburst galaxies
  low-activity nuclei of, 106–8
  nuclei of, 103–6
  X-ray emission from, 102–8
Stars
  A-type, see A stars
  abundances in, see Abundances
  age determinations of 601–2, 613–15
  asymptotic giant branch, 320–21
  B-type, see B stars
  binary, see Binary stars; Binary stars, Algol-type; Binary stars, low-mass X-ray
  carbon, see Carbon stars
  chromospheres of, 399–402
  cool, see Cool stars
  distribution of, in Galaxy, 569–85
  F-type, see F stars
  giant, abundances in, 291–96
  high-velocity, 563–64
    metallicity of, 333–34
    subdwarf, 579–82
  intermediate-mass, 320–21
  low-mass, 51–55
  K-type, see K stars

764 SUBJECT INDEX

long-period variable, see
  Long-period variable
  stars
low-mass, 565–67
M-type, see M stars
mass loss in, 406–7
massive
  evolution of, 565–67, 638–42
  mass function of, 323–25
  mass loss in, 636–37
  nucleosynthesis in, 317–20, 323
  oxygen abundance in, 323–25
  as progenitors of supernovae, 629–30, 636–42
metal-poor, 565–68, 594, 598–99, 608
  abundances in, 283–307
  dwarf, 328–30
metal-rich, 565–66
metallicity in, 279–349
pre-main-sequence, 352–55
in Orion molecular cloud, 51–55
S-type, see S stars
SC-type, see SC stars
subdwarf, 562
  high-velocity, 579–82
  winds in, 406–7
supergiant, 636–41
very massive, nucleosynthesis in, 317, 335–36
Stellar dynamics, 9–11, 586–92
Stellar evolution, see also Nucleosynthesis
  binary-star, 401–5
  in Local Group galaxies, 139–56
  metallicity as measure of, 279–315
  pre-main-sequence, 353–55, 376, 383–91
  stellar winds and, 407
Stellar populations, 139–59
Stellar winds
  IUE observations of, 400, 405–7
  in hot subdwarfs, 406
  in planetary nebulae, 406–7
  stellar evolution and, 407
  in T Tauri stars, 385–90
Sun
  current interruption in magnetic fields of, 452–53, 459
  filaments in, 446–48, 453–61
  flares in, see Solar flares

magnetic field of
  heliospheric winds and, 230–31
  flares and, 447, 450–56, 459
  particle acceleration in, 422–45
  tearing-mode process in, 451–52
Supernova events
  core collapse in, 643–46
  gamma-ray emission in, 680–82
  X-ray emission in, 95, 682–84
Supernova nucleosynthesis, 316–20, 328, 646–49, 679–80
Supernova remnants
  and cosmic-ray production, 499
  as gamma-ray sources, 477, 491, 503, 508, 680–82
  X-ray emission of, 88–89, 94, 102–3, 115
Supernova 1987A, 629–700
  core collapse in, 645–46, 659
  discovery of, 630–31
  distance to, 630, 675
  early observations of, 630–35
  gamma-ray emission of, 672, 684–90
  infrared spectrum of, 675–77
  interstellar matter and, 677–79
  light-curve of, 663–74
    early-stage, 663–670
    late-stage, 670–74
  radioactive decay influence on, 670–74, 680
  light echoes from, 664, 672, 678–79
  "mixing" in early stage of, 667–70
  neutrinos from, 629, 631, 645–46, 652, 654–60
  neutron star in, 646, 649–52, 691–92
  progenitor of, see Sk −69 202
  pulsar in, 670–71, 674, 691–92
  radio emission from, 642
  radioactivity in, 667–74, 678–91
  speckle interferometry of, 676–77
  spectrum of, 631, 633–34, 674–78, 692
  ultraviolet spectrum of, 630–33, 642–43, 663–64, 675, 677–78, 692

X-ray emission of, 682–90
Supernovae
  neutrinos produced in explosions of, 644–46, 649–52
  in Orion molecular cloud, 50–51
  relative rates of Types I and II, 603–5
  shock propagation in explosions of, 646–49
  Type I, 566–68, 603–4, 680
  Type II, 566–68, 603–4, 631–32, 642, 667, 680
  explosion mechanism of, 643–46
SX Cassiopeiae, 402

T

T Lyrae, 742
T Tauri stars, 351–95
  accretion disks in, 376–90
  accretion rates in, 383–85
  chromospheres in, 366–67
  circumstellar region of, 373–90
  classical (CTTS), 355–91
  disk-star boundary layer in, 378–81
  flares in, 367–70
  infrared excess in, 389
  infrared spectra of, 375–85
  jets in, 388
  lithium abundance in, 361
  magnetic fields in, 365–73
  mass loss in, 385–90
  and molecular clouds, 52, 352–53
  photospheres of, 360–73
  photospheric spots in, 365–66
  rotation-activity correlation in, 370–73
  rotational velocity of, 361–63
  spectra of, 355–61
  spectroscopic binaries among, 364, 385
  veiling in spectra of, 360–61
  weak-line (WTTS), 355–65, 372, 390–91
  winds in, 385–90
  X-ray emission of, 370–73
TAP 57, 356, 358–59
Technetium, abundance of, 321, 717, 722–24, 726
Terminal shock, 206–31
Titanium, isotopic ratios of, 748
Trapezium cluster, 50, 52–55
47 Tucanae, 327

SUBJECT INDEX 765

U

U Cephei, 403–4
Ultraviolet backscatter radiation, 204, 208, 231–32
Ultraviolet observations, see *IUE* observations
Ultraviolet extinction, 162, 183–91, 195
US Naval Observatory, Flagstaff, predicted light pollution at, 31–32
Ursa Minor elliptical galaxy, 155
UU Aurigae, 735–36, 742
UV Ceti stars, 367–68
UY Centauri, 724–25, 740

V

V Cancri, 749
V Sagittae, 403
Vanadium, abundance of, 301, 319
Very local interstellar medium (VLISM), 199–234
 components of, 203–5
 cosmic rays in, 205–6
 densities and temperatures of, 204–5
 dust in, 206, 225
 magnetic field of, 205
 neutral atomic component of, 203–4
 solar wind interaction with, 206–8
  models of, 208–31
 velocity of, 204

Very small grains (VSGs), 161–65, 180–90
 chemical composition of, 180–81
 destruction and formation of, 187–90
Virgo cluster, X-ray emission in, 108–10, 120
VLA observations
 of Orion molecular cloud, 63, 65
 of T Tauri stars, 353, 368, 373, 387
Voyager 1 and *2*, 206
22 Vulpeculae, 403
VX Andromedae, 742
V410 Tauri, 365, 368
V471 Tauri, 403
V1057 Cygni, 357, 377
V0332+53, 547

W

W Serpentis, 402
W Ursae Majoris-type binary stars, 402
White dwarfs, 408–10
Wolf-Rayet stars, 407
WZ Cassiopeiae, 725, 728, 740, 742

XYZ

X-ray background emission
 extragalactic, 129–31
 Galactic, 94
X-ray binary stars, see Binary stars, low-mass X-ray; Binary stars, massive X-ray

X-ray emission
 from normal galaxies, 87–138
 correlation with cosmic-ray production of, 101
 correlation with infrared emission of, 100–4
 correlation with optical emission of, 99–100, 116–17, 126–27, 130–31
 correlation with radio emission of, 97–101, 125–29
 and starburst activity, 102–8
 from solar flares, 424–50
 from spiral galaxies, 88–102
 from supernova events, 95, 682–84
 from supernova remnants, 88–89, 94, 102–3, 115
 from Supernova 1987A, 682–90
 from T Tauri stars, 370–73
XZ Tauri, 373–75
Y Canum Venaticorum, 725–26, 735–36, 738, 742–43
YY Orionis stars, 357, 380–81
Zinc, abundance of, 297, 300
Zirconium
 abundances of, 723–24, 726
 isotopic ratios of, 748–49

MISCELLANEOUS

3C 273, 509
4U 1626−67, 525
4U 1820−30, 525–26

CUMULATIVE INDEXES

CONTRIBUTING AUTHORS, VOLUMES 17–27

A

Abbott, D. C., 25:113–50
Abt, H. A., 21:343–72
Adams, F. C., 25:23–81
Akasofu, S.-I., 20:117–38
Ambartsumian, V. A., 18:1–13
Angel, J. R. P., 18:321–61
Arnett, W. D., 27:629–700
Athanassoula, E., 23:147–68

B

Backer, D. C., 24:537–75
Bahcall, J. N., 24:577–611; 27:629–700
Bahcall, N. A., 26:631–86
Bai, T., 27:421–67
Balick, B., 20:431–68
Baliunas, S. L., 23:379–412
Baym, G., 17:415–43
Beckwith, S., 20:163–90
Beichman, C. A., 25:521–63
Bertout, C., 27:351–95
Bignami, G. F., 21:67–108
Binggeli, B., 26:509–60
Binney, J., 20:399–429
Bloemen, H., 27:469–516
Bodenheimer, P., 26:145–97
Boesgaard, A. M., 23:319–78
Boggess, A., 27:397–420
Böhm-Vitense, E., 19:295–318
Borra, E. F., 20:191–220
Bosma, A., 23:147–68
Bracewell, R. N., 17:113–34
Bradt, H. V. D., 21:13–66
Brault, J. W., 22:291–317
Bridle, A. H., 22:319–58
Brown, R. L., 22:223–65

C

Cameron, A. G. W., 26:441–72
Carbon, D. F., 17:515–49
Carswell, R. F., 19:41–76
Cassinelli, J. P., 17:275–308
Caughlan, G. R., 21:165–76
Cesarsky, C. J., 18:289–319
Chapman, G. A., 25:633–67
Chincarini, G. L., 22:445–70
Chiosi, C., 24:329–75
Chupp, E. L., 22:359–87

Conti, P. S., 25:113–50
Coulman, C. E., 23:19–57
Cowie, L. L., 24:499–535
Cowling, T. G., 19:115–35; 23:1–18
Cox, A. N., 18:15–41
Cox, D. P., 25:303–44

D

Davidson, K., 23:119–46
Davis, M., 21:109–30
Deubner, F.-L., 22:593–619
Djorgovski, S., 27:235–77
Dravins, D., 20:61–89
Dressler, A., 22:185–222
Dulk, G. A., 23:169–224
Dupree, A. K., 24:377–420

E

Edmunds, M. G., 19:77–113
Elliot, J. L., 17:445–75
Ellis, G. F. R., 22:157–84
Elson, R., 25:565–601

F

Fabbiano, G., 27:87–138
Faber, S. M., 17:135–87
Feast, M. W., 25:345–75
Fesen, R. A., 23:119–46
Ford, W. K. Jr., 17:189–212
Forman, W., 20:547–85
Fowler, W. A., 21:165–76
Freeman, K. C., 19:319–56; 25:603–32
Frogel, J. A., 26:51–92
Fujimoto, M., 24:459–97
Fusi Pecci, F., 26:199–244

G

Gallagher, J. S., 17:135–87; 22:37–74
Garstang, R. H., 27:19–40
Gehrz, R. D., 26:377–412
Genzel, R., 25:377–423; 27:41–85
Gillett, F. C., 19:411–56
Gilmore, G., 27:555–627
Giovanelli, R., 22:445–70

Goldreich, P., 20:249–83
Golub, L., 23:413–52
Gough, D., 22:593–619
Greenstein, J. L., 22:1–35
Gustafsson, B., 27:701–56

H

Habing, H. J., 17:345–85
Harris, M. J., 21:165–76
Harris, W. E., 17:241–74
Hartmann, L. W., 25:271–301
Haynes, M. P., 22:445–70
Heckman, T. M., 20:431–68
Hellings, R. W., 24:537–75
Hermsen, W., 21:67–108
Hillas, A. M., 22:425–44
Ho, P. T. P., 21:239–70
Hoag, A. A., 17:43–71
Hodge, P. W., 19:357–72; 27:139–59
Hollenbach, D. J., 18:219–62
Holt, S. S., 20:323–65
Holzer, T. E., 27:199–234
Houck, J. R., 25:187–230
Howard, R., 22:131–55
Hoyle, F., 20:1–35
Hudson, H. S., 26:473–507
Hunter, D. A., 22:37–74
Hurford, G. J., 20:497–516
Hut, P., 25:565–601

I

Iben, I. Jr., 21:271–342
Inagaki, S., 25:565–601
Ionson, J. A., 19:7–40
Israel, F. P., 17:345–85

J

Jones, C., 20:547–85
Joss, P. C., 22:537–92
Joyce, R. R., 19:411–56

K

Kaler, J. B., 23:89–117
Kellermann, K. I., 19:373–410
Kirshner, R. P., 27:629–700
Kleinmann, S. G., 19:411–56
Kondo, Y., 27:397–420

CONTRIBUTING AUTHORS 767

Kormendy, J., 27:235–77
Kuijken, K., 27:555–627
Kuperus, M., 19:7–40

L

Lada, C. J., 23:267–317
Landstreet, J. D., 20:191–220
Larson, H. P., 18:43–75
Lebofsky, M. J., 17:477–511
Léger, A., 27:161–98
Leovy, C. B., 17:387–413
Liebert, J., 18:363–98; 25:473–519
Linsky, J. L., 18:439–88
Liszt, H. S., 22:223–65
Lizano, S., 25:23–81
Lubow, S. H., 19:227–93

M

Mackay, C. D., 24:255–83
Maeder, A., 24:329–75
Maran, S. P., 27:397–420
Margon, B., 22:507–36
Mariska, J. T., 24:23–48
Marsh, K. A., 20:497–516
Mathews, W. G., 24:171–203
Mathis, J. S., 17:73–111
McAlister, H. A., 23:59–87
McClintock, J. E., 21:13–66
McCray, R., 17:213–40; 20:323–65
McCrea, W. H., 25:1–22
McKee, C. F., 18:219–62
Mendis, D. A., 26:11–49
Merrill, K. M., 17:9–41
Mestel, L., 20:191–220
Miley, G., 18:165–218
Monet, D. G., 26:413–40
Moore, R., 23:239–66
Moran, J. M., 19:231–76
Morgan, W. W., 26:1–9
Morris, M., 20:517–45
Morrison, D., 20:469–95
Mould, J. R., 20:91–115

N

Narayan, R., 24:127–70
Ness, N. F., 20:139–61
Neugebauer, G., 25:187–230
Newkirk, G. Jr., 21:429–67
Nityananda, R., 24:127–70
Norris, J., 19:319–56
Noyes, R. W., 25:271–301

O

Oort, J. H., 19:1–5; 21:373–428
Osterbrock, D. E., 24:171–203

P

Pagel, B. E. J., 19:77–113
Pauliny-Toth, I. I. K., 19:373–410
Pearson, T. J., 22:97–130
Peebles, P. J. E., 21:109–30
Perley, R. A., 22:319–58
Pethick, C., 17:415–43
Phillips, T. G., 20:285–321
Pollack, J. B., 22:389–424
Popper, D. M., 18:115–64
Pringle, J. E., 19:137–62
Probst, R. G., 25:473–519
Puget, J. L., 27:161–98

R

Rabin, D., 23:239–66
Racine, R., 17:241–74
Rappaport, S. A., 22:537–92
Raymond, J. C., 22:75–95
Readhead, A. C. S., 22:97–130
Rees, M. J., 22:471–506
Reid, M. J., 19:231–76
Renzini, A., 21:271–342; 26:199–244
Reynolds, R. J., 25:303–44
Rickard, L. J, 20:517–45
Ridgway, S. T., 17:9–41; 22:291–317
Rieke, G. H., 17:477–511
Rood, H. J., 26:245–94
Rosner, R., 23:413–52

S

Saikia, D. J., 26:93–144
Salter, C. J., 26:93–144
Sandage, A., 24:421–58; 26:509–60, 561–630
Savage, B. D., 17:73–111
Schwartz, R. D., 21:209–37
Sellwood, J. A., 25:151–86
Shu, F., 19:277–93; 25:23–81
Shull, J. M., 20:163–90
Smith, A. G., 17:43–71
Smith, M. G., 19:41–76
Sneden, C., 27:279–349
Snow, T. P. Jr., 17:213–40
Sofue, Y., 24:459–97
Soifer, B. T., 21:177–207; 25:187–230
Songaila, A., 24:499–535
Spicer, D. S., 19:7–40
Spinrad, H., 25:231–69
Spite, F., 23:225–38
Spite, M., 23:225–38
Spitzer, L. Jr., 27:1–17
Sramek, R. A., 26:295–341
Steigman, G., 23:319–78
Stein, W. A., 21:177–207
Stern, D. P., 20:139–61

Stinebring, D. R., 24:285–327
Stockman, H. S., 18:321–61
Strömgren, B., 21:1–11
Sturrock, P. A., 27:421–67
Stutzki, J., 27:41–85
Sunyaev, R. A., 18:537–60
Swings, P., 17:1–7
Syrovatskii, S. I., 19:163–229

T

Tammann, G. A., 26:509–60
Taylor, J. H., 24:285–327
Telesco, C. M., 26:343–76
Tenorio-Tagle, G., 26:145–97
Townes, C. H., 21:239–70; 25:377–423
Tremaine, S., 20:249–83
Trimble, V., 25:425–72
Truran, J. W. Jr., 27:279–349
Tsuji, T., 24:89–125

V

Vaiana, G. S., 23:413–52
van Altena, W. F., 21:131–64
van der Klis, M., 27:517–53
Vauclair, G., 20:37–60
Vauclair, S., 20:37–60
Vaughan, A. H., 23:379–412

W

Wagner, W. J., 22:267–89
Walker, A. W., 25:345–75
Wannier, P. G., 18:399–437
Weaver, T. A., 24:205–53
Weiler, K. W., 26:295–341
Weiss, R., 18:489–537
Wetherill, G. W., 18:77–113
Weymann, R. J., 19:41–76
Wheeler, J. C., 27:279–349
Whitford, A. E., 24:1–22
Wielebinski, R., 24:459–97
Woody, D. P., 20:285–321
Woolf, N. J., 20:367–98
Woosley, S. E., 24:205–53; 27:629–700
Wynn-Williams, C. G., 20:587–618
Wyse, R. F. G., 27:555–627

Y

York, D. G., 20:221–48
Yorke, H. W., 24:49–87

Z

Zel'dovich, Ya. B., 18:537–60
Zimmerman, B. A., 21:165–76
Zuckerman, B., 18:263–88
Zwaan, C., 25:83–111

CHAPTER TITLES, VOLUMES 17–27

PREFATORY CHAPTER

A Few Notes on My Career as an Astrophysicist	P. Swings	17:1–7
On Some Trends in the Development of Astrophysics	V. A. Ambartsumian	18:1–13
Some Notes on My Life as an Astronomer	J. H. Oort	19:1–5
The Universe: Past and Present Reflections	F. Hoyle	20:1–35
Scientists I Have Known and Some Astronomical Problems I Have Met	B. Strömgren	21:1–11
An Astronomical Life	J. L. Greenstein	22:1–35
Astronomer by Accident	T. G. Cowling	23:1–18
A Half-Century of Astronomy	A. E. Whitford	24:1–22
Clustering of Astronomers	W. H. McCrea	25:1–22
A Morphological Life	W. W. Morgan	26:1–9
Dreams, Stars, and Electrons	L. Spitzer, Jr.	27:1–17

SOLAR SYSTEM ASTROPHYSICS

Martian Meteorology	C. B. Leovy	17:387–413
Stellar Occultation Studies of the Solar System	J. L. Elliot	17:445–75
Infrared Spectroscopic Observations of the Outer Planets, Their Satellites, and the Asteroids	H. P. Larson	18:43–75
Formation of Terrestrial Planets	G. W. Wetherill	18:77–113
Planetary Magnetospheres	D. P. Stern, N. F. Ness	20:139–61
The Dynamics of Planetary Rings	P. Goldreich, S. Tremaine	20:249–83
The Satellites of Jupiter and Saturn	D. Morrison	20:469–95
Origin and History of the Outer Planets: Theoretical Models and Observational Constraints	J. B. Pollack	22:389–424
Comets and Their Composition	H. Spinrad	25:231–69
A Postencounter View of Comets	D. A. Mendis	26:11–49
Origin of the Solar System	A. G. W. Cameron	26:441–72

SOLAR PHYSICS

On the Theory of Coronal Heating	M. Kuperus, J. A. Ionson, D. S. Spicer	19:7–40
High Spatial Resolution Solar Microwave Observations	K. A. Marsh, G. J. Hurford	20:497–516
Variations in Solar Luminosity	G. Newkirk, Jr.	21:429–67
Solar Rotation	R. Howard	22:131–55
Coronal Mass Ejections	W. J. Wagner	22:267–89
High-Energy Neutral Radiations From the Sun	E. L. Chupp	22:359–87
Helioseismology: Oscillations as a Diagnostic of the Solar Interior	F.-L. Deubner, D. Gough	22:593–619
Radio Emission From the Sun and Stars	G. A. Dulk	23:169–224
Sunspots	R. Moore, D. Rabin	23:239–66
The Quiet Solar Transition Region	J. T. Mariska	24:23–48
Elements and Patterns in the Solar Magnetic Field	C. Zwaan	25:83–111
Variations of Solar Irradiance due to Magnetic Activity	G. A. Chapman	25:633–67
Observed Variability of the Solar Luminosity	H. S. Hudson	26:473–507
Interaction Between the Solar Wind and the Interstellar Medium	T. E. Holzer	27:199–234
Classification of Solar Flares	T. Bai, P. A. Sturrock	27:421–67

768

STELLAR PHYSICS

Title	Author(s)	Reference
Infrared Spectroscopy of Stars	K. M. Merrill, S. T. Ridgway	17:9–41
Stellar Winds	J. P. Cassinelli	17:275–308
On the Nonhomogeneity of Metal Abundances in Stars of Globular Clusters and Satellite Subsystems of the Galaxy	R. P. Kraft	17:309–43
Physics of Neutron Stars	G. Baym, C. Pethick	17:415–43
Model Atmospheres for Intermediate- and Late-Type Stars	D. F. Carbon	17:513–49
The Masses of Cepheids	A. N. Cox	18:15–41
Stellar Masses	D. M. Popper	18:115–64
Envelopes Around Late-Type Giant Stars	B. Zuckerman	18:263–88
White Dwarf Stars	J. Liebert	18:363–98
Stellar Chromospheres	J. L. Linsky	18:439–88
Mass, Angular Momentum, and Energy Transfer in Close Binary Stars	F. H. Shu, S. H. Lubow	19:277–93
The Effective Temperature Scale	E. Böhm-Vitense	19:295–318
Element Segregation in Stellar Outer Layers	S. Vauclair, G. Vauclair	20:37–60
Photospheric Spectrum Line Asymmetries and Wavelength Shifts	D. Dravins	20:61–89
Magnetic Stars	E. F. Borra, J. D. Landstreet, L. Mestel	20:191–220
The Search for Infrared Protostars	C. G. Wynn-Williams	20:587–618
The Optical Counterparts of Compact Galactic X-Ray Sources	H. V. D. Bradt, J. E. McClintock	21:13–66
Galactic Gamma-Ray Sources	G. F. Bignami, W. Hermsen	21:67–108
Herbig-Haro Objects	R. D. Schwartz	21:209–37
Asymptotic Giant Branch Evolution and Beyond	I. Iben, Jr., A. Renzini	21:271–342
Normal and Abnormal Binary Frequencies	H. A. Abt	21:343–72
Observations of Supernova Remnants	J. C. Raymond	22:75–95
High Angular Resolution Measurements of Stellar Properties	H. A. McAlister	23:59–87
Planetary Nebulae and Their Central Stars	J. B. Kaler	23:89–117
Radio Emission From the Sun and Stars	G. A. Dulk	23:169–224
The Composition of Field Halo Stars and the Chemical Evolution of the Halo	M. Spite, F. Spite	23:225–38
Stellar Activity Cycles	S. L. Baliunas, A. H. Vaughan	23:379–412
On Stellar X-Ray Emission	R. Rosner, L. Golub, G. S. Vaiana	23:413–52
Molecules in Stars	T. Tsuji	24:89–125
The Physics of Supernova Explosions	S. E. Woosley, T. A. Weaver	24:205–53
Recent Progress in the Understanding of Pulsars	J. H. Taylor, D. R. Stinebring	24:285–327
The Evolution of Massive Stars With Mass Loss	C. Chiosi, A. Maeder	24:329–75
Mass Loss From Cool Stars	A. K. Dupree	24:377–420
The Population Concept, Globular Clusters, Subdwarfs, Ages, and the Collapse of the Galaxy	A. Sandage	24:421–58
Pulsar Timing and General Relativity	D. C. Backer, R. W. Hellings	24:537–75
Star Formation in Molecular Clouds: Observation and Theory	F. H. Shu, F. C. Adams, S. Lizano	25:23–81
Wolf-Rayet Stars	D. C. Abbott, P. S. Conti	25:113–50
Rotation and Magnetic Activity in Main-Sequence Stars	L. W. Hartmann, R. W. Noyes	25:271–301
Very Low Mass Stars	J. Liebert, R. G. Probst	25:473–519
Tests of Evolutionary Sequences Using Color-Magnitude Diagrams of Globular Clusters	A. Renzini, F. Fusi Pecci	26:199–244
Supernovae and Supernova Remnants	K. W. Weiler, R. A. Sramek	26:295–341
The Infrared Temporal Development of Classsical Novae	R. D. Gehrz	26:377–412

Abundance Ratios as a Function of Metallicity	J. C. Wheeler, C. Sneden, J. W. Truran, Jr.	27:279–349
T Tauri Stars: Wild as Dust	C. Bertout	27:351–95
Quasi-Periodic Oscillations and Noise in Low-Mass X-Ray Binaries	M. van der Klis	27:517–53
Supernova 1987A	W. D. Arnett, J. N. Bahcall, R. P. Kirshner, S. E. Woosley	27:629–700
Chemical Analyses of Cool Stars	B. Gustafsson	27:701–56

DYNAMICAL ASTRONOMY

Astrometry	W. F. van Altena	21:131–64
Dynamical Evolution of Globular Clusters	R. Elson, P. Hut, S. Inagaki	25:565–601
The Galactic Spheroid and Old Disk	K. C. Freeman	25:603–32
Recent Advances in Optical Astrometry	D. G. Monet	26:413–40

INTERSTELLAR MEDIUM

Observed Properties of Interstellar Dust	B. D. Savage, J. S. Mathis	17:73–111
The Violent Interstellar Medium	R. McCray, T. P. Snow, Jr.	17:213–40
Compact H II Regions and OB Star Formation	H. J. Habing, F. P. Israel	17:345–85
Interstellar Shock Waves	C. F. McKee, D. J. Hollenbach	18:219–62
Cosmic-Ray Confinement in the Galaxy	C. J. Cesarsky	18:289–319
Nuclear Abundances and Evolution of the Interstellar Medium	P. G. Wannier	18:399–437
Interstellar Molecular Hydrogen	J. M. Shull, S. Beckwith	20:163–90
Herbig-Haro Objects	R. D. Schwartz	21:209–37
Interstellar Ammonia	P. T. P. Ho, C. H. Townes	21:239–70
Observations of Supernova Remnants	J. C. Raymond	22:75–95
The Influence of Environment on the H I Content of Galaxies	M. P. Haynes, R. Giovanelli, G. L. Chincarini	22:445–70
Planetary Nebulae and Their Central Stars	J. B. Kaler	23:89–117
Cold Outflows, Energetic Winds, and Enigmatic Jets Around Young Stellar Objects	C. J. Lada	23:267–317
The Dynamical Evolution of H II Regions—Recent Theoretical Developments	H. W. Yorke	24:49–87
High-Resolution Optical and Ultraviolet Absorption-Line Studies of Interstellar Gas	L. L. Cowie, A. Songaila	24:499–535
Star Formation in Molecular Clouds: Observation and Theory	F. H. Shu, F. C. Adams, S. Lizano	25:23–81
The Local Interstellar Medium	D. P. Cox, R. J. Reynolds	25:303–44
Large-Scale Expanding Superstructures in Galaxies	G. Tenorio-Tagle, P. Bodenheimer	26:145–97
Supernovae and Supernova Remnants	K. W. Weiler, R. A. Sramek	26:295–341
The Orion Molecular Cloud and Star-Forming Region	R. Genzel, J. Stutzki	27:41–85
A New Component of the Interstellar Matter: Small Grains and Large Aromatic Molecules	J. L. Puget, A. Léger	27:161–98
Interaction Between the Solar Wind and the Interstellar Medium	T. E. Holzer	27:199–234
Diffuse Galactic Gamma-Ray Emission	H. Bloemen	27:469–516

SMALL STELLAR SYSTEMS

Mass, Angular Momentum, and Energy Transfer in Close Binary Stars	F. H. Shu, S. H. Lubow	19:277–93
The Chemical Composition, Structure, and Dynamics of Globular Clusters	K. C. Freeman, J. Norris	19:319–56
Normal and Abnormal Binary Frequencies	H. A. Abt	21:343–72
Dynamical Evolution of Globular Clusters	R. Elson, P. Hut, S. Inagaki	25:565–601

The Galactic Nuclear Bulge and the Stellar Content of Spheroidal Systems	J. A. Frogel	26:51–92
Tests of Evolutionary Sequences Using Color-Magnitude Diagrams of Globular Clusters	A. Renzini, F. Fusi Pecci	26:199–244
Quasi-Periodic Oscillations and Noise in Low-Mass X-Ray Binaries	M. van der Klis	27:517–53

THE GALAXY

Cosmic-Ray Confinement in the Galaxy	C. J. Cesarsky	18:289–319
The Chemical Composition, Structure, and Dynamics of Globular Clusters	K. C. Freeman, J. Norris	19:319–56
Stellar Populations in the Galaxy	J. R. Mould	20:91–115
Gas in the Galactic Halo	D. G. York	20:221–48
The Optical Counterparts of Compact Galactic X-Ray Sources	H. V. D. Bradt, J. E. McClintock	21:13–66
Galactic Gamma-Ray Sources	G. F. Bignami, W. Hermsen	21:67–108
Sagittarius A and Its Environment	R. L. Brown, H. S. Liszt	22:223–65
Neutron Stars in Interacting Binary Systems	P. C. Joss, S. A. Rappaport	22:537–92
Star Counts and Galactic Structure	J. N. Bahcall	24:577–611
Physical Conditions, Dynamics, and Mass Distribution in the Center of the Galaxy	R. Genzel, C. H. Townes	25:377–423
The IRAS View of the Galaxy and the Solar System	C. A. Beichman	25:521–63
The Galactic Spheroid and Old Disk	K. C. Freeman	25:603–32
The Galactic Nuclear Bulge and the Stellar Content of Spheroidal Systems	J. A. Frogel	26:51–92
Large-Scale Expanding Superstructures in Galaxies	G. Tenorio-Tagle, P. Bodenheimer	26:145–97
Diffuse Galactic Gamma-Ray Emission	H. Bloemen	27:469–516
Kinematics, Chemistry, and Structure of the Galaxy	G. Gilmore, R. F. G. Wyse, K. Kuijken	27:555–627

EXTRAGALACTIC ASTRONOMY

Masses and Mass-to-Light Ratios of Galaxies	S. M. Faber, J. S. Gallagher	17:135–87
Globular Clusters in Galaxies	W. E. Harris, R. Racine	17:241–74
Infrared Emission of Extragalactic Sources	G. H. Rieke, M. J. Lebofsky	17:477–511
The Structure of Extended Extragalactic Radio Sources	G. Miley	18:165–218
Optical and Infrared Polarization of Active Extragalactic Objects	J. R. P. Angel, H. S. Stockman	18:321–61
Absorption Lines in the Spectra of Quasistellar Objects	R. J. Weymann, R. F. Carswell, M. G. Smith	19:41–76
Abundances in Stellar Populations and the Interstellar Medium in Galaxies	B. E. J. Pagel, M. G. Edmunds	19:77–113
Compact Radio Sources	K. I. Kellermann, I. I. K. Pauliny-Toth	19:373–410
Dynamics of Elliptical Galaxies and Other Spheroidal Components	J. Binney	20:399–429
Extranuclear Clues to the Origin and Evolution of Activity in Galaxies	B. Balick, T. M. Heckman	20:431–68
Molecular Clouds in Galaxies	M. Morris, L. J Rickard	20:517–45
X-Ray-Imaging Observations of Clusters of Galaxies	W. Forman, C. Jones	20:547–85
Dust in Galaxies	W. A. Stein, B. T. Soifer	21:177–207
Superclusters	J. H. Oort	21:373–428
Structure and Evolution of Irregular Galaxies	J. S. Gallagher, III, D. A. Hunter	22:37–74
The Evolution of Galaxies in Clusters	A. Dressler	22:185–222
Extragalactic Radio Jets	A. H. Bridle, R. A. Perley	22:319–58
Black Hole Models for Active Galactic Nuclei	M. J. Rees	22:471–506

772 CHAPTER TITLES

Shells and Rings Around Galaxies | E. Athanassoula, A. Bosma | 23:147–68
Emission-Line Regions of Active Galaxies and QSOs | D. E. Osterbrock, W. G. Mathews | 24:171–203
Global Structure of Magnetic Fields in Spiral Galaxies | Y. Sofue, M. Fujimoto, R. Wielebinski | 24:459–97
The IRAS View of the Extragalactic Sky | B. T. Soifer, J. R. Houck, G. Neugebauer | 25:187–230
Cepheids as Distance Indicators | M. W. Feast, A. R. Walker | 25:345–75
Existence and Nature of Dark Matter in the Universe | V. Trimble | 25:425–72
Polarization Properties of Extragalactic Radio Sources | D. J. Saikia, C. J. Salter | 26:93–144
Voids | H. J. Rood | 26:245–94
Enhanced Star Formation and Infrared Emission in the Centers of Galaxies | C. M. Telesco | 26:343–76
The Luminosity Function of Galaxies | B. Binggeli, A. Sandage, G. A. Tammann | 26:509–60
Observational Tests of World Models | A. Sandage | 26:561–630
Large-Scale Structure in the Universe Indicated by Galaxy Clusters | N. A. Bahcall | 26:631–86
X Rays From Normal Galaxies | G. Fabbiano | 27:87–138
Populations in Local Group Galaxies | P. Hodge | 27:139–59
Surface Photometry and the Structure of Elliptical Galaxies | J. Kormendy, S. Djorgovski | 27:235–77

OBSERVATIONAL PHENOMENA

Infrared Spectroscopic Observations of the Outer Planets, Their Satellites, and the Asteroids | H. P. Larson | 18:43–75
Optical and Infrared Polarization of Active Extragalactic Objects | J. R. P. Angel, H. S. Stockman | 18:321–61
Measurements of the Cosmic Background Radiation | R. Weiss | 18:489–537
Preliminary Results of the Air Force Infrared Sky Survey | S. G. Kleinmann, F. C. Gillett, R. R. Joyce | 19:411–56
Spectra of Cosmic X-Ray Sources | S. S. Holt, R. McCray | 20:323–65
Galactic Gamma-Ray Sources | G. F. Bignami, W. Hermsen | 21:67–108
The Evolution of Galaxies in Clusters | A. Dressler | 22:185–222
Observations of SS 433 | B. Margon | 22:507–36
Recent Developments Concerning the Crab Nebula | K. Davidson, R. A. Fesen | 23:119–46
The IRAS View of the Extragalactic Sky | B. T. Soifer, J. R. Houck, G. Neugebauer | 25:187–230
Existence and Nature of Dark Matter in the Universe | V. Trimble | 25:425–72
Polarization Properties of Extragalactic Radio Sources | D. J. Saikia, C. J. Salter | 26:93–144
X Rays From Normal Galaxies | G. Fabbiano | 27:87–138
Populations in Local Group Galaxies | P. Hodge | 27:139–59
Astrophysical Contributions of the International Ultraviolet Explorer | Y. Kondo, A. Boggess, S. P. Maran | 27:397–420

GENERAL RELATIVITY AND COSMOLOGY

Measurements of the Cosmic Background Radiation | R. Weiss | 18:489–537
Microwave Background Radiation as a Probe of the Contemporary Structure and History of the Universe | R. A. Sunyaev, Ya. B. Zel'dovich | 18:537–60
The Extragalactic Distance Scale | P. W. Hodge | 19:357–72

Evidence for Local Anisotropy of the Hubble Flow	M. Davis, P. J. E. Peebles	21:109–30
Alternatives to the Big Bang	G. F. R. Ellis	22:157–84
Big Bang Nucleosynthesis: Theories and Observations	A. M. Boesgaard, G. Steigman	23:319–78
Pulsar Timing and General Relativity	D. C. Backer, R. W. Hellings	24:537–75
Existence and Nature of Dark Matter in the Universe	V. Trimble	25:425–72
Observational Tests of World Models	A. Sandage	26:561–630
Large-Scale Structure in the Universe Indicated by Galaxy Clusters	N. A. Bahcall	26:631–86

INSTRUMENTATION AND TECHNIQUES

Advances in Astronomical Photography at Low Light Levels	A. G. Smith, A. A. Hoag	17:43–71
Computer Image Processing	R. N. Bracewell	17:113–34
Digital Imaging Techniques	W. K. Ford, Jr.	17:189–212
Millimeter- and Submillimeter-Wave Receivers	T. G. Phillips, D. P. Woody	20:285–321
High Resolution Imaging from the Ground	N. J. Woolf	20:367–98
Astrometry	W. F. van Altena	21:131–64
Image Formation by Self-Calibration in Radio Astronomy	T. J. Pearson, A. C. S. Readhead	22:97–130
Astronomical Fourier Transform Spectroscopy Revisited	S. T. Ridgway, J. W. Brault	22:291–317
Fundamental and Applied Aspects of Astronomical "Seeing"	C. E. Coulman	23:19–57
High Angular Resolution Measurements of Stellar Properties	H. A. McAlister	23:59–87
Maximum Entropy Image Restoration in Astronomy	R. Narayan, R. Nityananda	24:127–70
Charge-Coupled Devices in Astronomy	C. D. Mackay	24:255–83
The Art of N-Body Building	J. A. Sellwood	25:151–86
Recent Advances in Optical Astrometry	D. G. Monet	26:413–40
The Status and Prospects for Ground-Based Observatory Sites	R. H. Garstang	27:19–40
Astrophysical Contributions of the International Ultraviolet Explorer	Y. Kondo, A. Boggess, S. P. Maran	27:397–420

PHYSICAL PROCESSES

Interstellar Shock Waves	C. F. McKee, D. J. Hollenbach	18:219–62
Nuclear Abundances and Evolution of the Interstellar Medium	P. G. Wannier	18:399–437
The Present Status of Dynamo Theory	T. G. Cowling	19:115–35
Accretion Discs in Astrophysics	J. E. Pringle	19:137–62
Pinch Sheets and Reconnection in Astrophysics	S. I. Syrovatskii	19:163–229
Masers	M. J. Reid, J. M. Moran	19:231–76
Interaction Between a Magnetized Plasma Flow and a Strongly Magnetized Celestial Body With an Ionized Atmosphere: Energetics of the Magnetosphere	S.-I. Akasofu	20:117–38
Interstellar Molecular Hydrogen	J. M. Shull, S. Beckwith	20:163–90
The Optical Counterparts of Compact Galactic X-Ray Sources	H. V. D. Bradt, J. E. McClintock	21:13–66
Thermonuclear Reaction Rates, III	M. J. Harris, W. A. Fowler, G. R. Caughlan, B. A. Zimmerman	21:165–76
The Origin of Ultra-High-Energy Cosmic Rays	A. M. Hillas	22:425–44
Observations of SS 433	B. Margon	22:507–36
The Physics of Supernova Explosions	S. E. Woosley, T. A. Weaver	24:205–53
Quasi-Periodic Oscillations and Noise in Low-Mass X-Ray Binaries	M. van der Klis	27:517–53

Annual Reviews Inc.
A NONPROFIT SCIENTIFIC PUBLISHER
4139 El Camino Way
P.O. Box 10139
Palo Alto, CA 94303-0897 • USA

ORDER FORM

ORDER TOLL FREE
1-800-523-8635
(except California)

Telex: 910-290-0275

Annual Reviews Inc. publications may be ordered directly from our office by mail, Telex, or use our Toll Free Telephone line (for orders paid by credit card or purchase order*, and customer service calls only); through booksellers and subscription agents, worldwide; and through participating professional societies. Prices subject to change without notice. ARI Federal I.D. #94-1156476

- **Individuals:** Prepayment required on new accounts by check or money order (in U.S. dollars, check drawn on U.S. bank) or charge to credit card—American Express, VISA, MasterCard.
- **Institutional buyers:** Please include purchase order number.
- **Students:** $10.00 discount from retail price, per volume. Prepayment required. Proof of student status must be provided (photocopy of student I.D. or signature of department secretary is acceptable). Students must send orders direct to Annual Reviews. Orders received through bookstores and institutions requesting student rates will be returned. You may order at the Student Rate for a maximum of 3 years.
- **Professional Society Members:** Members of professional societies that have a contractual arrangement with Annual Reviews may order books through their society at a reduced rate. Check with your society for information.
- **Toll Free Telephone orders:** Call 1-800-523-8635 (except from California) for orders paid by credit card or purchase order and customer service calls only. California customers and all other business calls use 415-493-4400 (not toll free). Hours: 8:00 AM to 4:00 PM, Monday-Friday, Pacific Time. ***Written confirmation** is required on purchase orders from universities before shipment.
- **Telex: 910-290-0275**

Regular orders: Please list the volumes you wish to order by volume number.
Standing orders: New volume in the series will be sent to you automatically each year upon publication. Cancellation may be made at any time. Please indicate volume number to begin standing order.
Prepublication orders: Volumes not yet published will be shipped in month and year indicated.
California orders: Add applicable sales tax.
Postage paid (4th class bookrate/surface mail) **by Annual Reviews Inc.** Airmail postage or UPS, extra.

ANNUAL REVIEWS SERIES		Prices Postpaid per volume USA & Canada/elsewhere	Regular Order Please send: Vol. number	Standing Order Begin with: Vol. number
Annual Review of **ANTHROPOLOGY**				
Vols. 1-14	(1972-1985)	$27.00/$30.00		
Vols. 15-16	(1986-1987)	$31.00/$34.00		
Vol. 17	(1988)	$35.00/$39.00		
Vol. 18	(avail. Oct. 1989)	$35.00/$39.00	Vol(s). _____	Vol. _____
Annual Review of **ASTRONOMY AND ASTROPHYSICS**				
Vols. 1, 4-14, 16-20	(1963, 1966-1976, 1978-1982)	$27.00/$30.00		
Vols. 21-25	(1983-1987)	$44.00/$47.00		
Vol. 26	(1988)	$47.00/$51.00		
Vol. 27	(avail. Sept. 1989)	$47.00/$51.00	Vol(s). _____	Vol. _____
Annual Review of **BIOCHEMISTRY**				
Vols. 30-34, 36-54	(1961-1965, 1967-1985)	$29.00/$32.00		
Vols. 55-56	(1986-1987)	$33.00/$36.00		
Vol. 57	(1988)	$35.00/$39.00		
Vol. 58	(avail. July 1989)	$35.00/$39.00	Vol(s). _____	Vol. _____
Annual Review of **BIOPHYSICS AND BIOPHYSICAL CHEMISTRY**				
Vols. 1-11	(1972-1982)	$27.00/$30.00		
Vols. 12-16	(1983-1987)	$47.00/$50.00		
Vol. 17	(1988)	$49.00/$53.00		
Vol. 18	(avail. June 1989)	$49.00/$53.00	Vol(s). _____	Vol. _____
Annual Review of **CELL BIOLOGY**				
Vol. 1	(1985)	$27.00/$30.00		
Vols. 2-3	(1986-1987)	$31.00/$34.00		
Vol. 4	(1988)	$35.00/$39.00		
Vol. 5	(avail. Nov. 1989)	$35.00/$39.00	Vol(s). _____	Vol. _____

| ANNUAL REVIEWS SERIES | Prices Postpaid per volume USA & Canada/elsewhere | Regular Order Please send: Vol. number | Standing Order Begin with: Vol. number |

Annual Review of **COMPUTER SCIENCE**
    Vols. 1-2   (1986-1987)................$39.00/$42.00
    Vol. 3   (1988) ....$45.00/$49.00
    Vol. 4   (avail. Nov. 1989)............$45.00/$49.00   Vol(s). _____   Vol. _____

Annual Review of **EARTH AND PLANETARY SCIENCES**
    Vols. 1-10   (1973-1982)................$27.00/$30.00
    Vols. 11-15   (1983-1987)................$44.00/$47.00
    Vol. 16   (1988) ....$49.00/$53.00
    Vol. 17   (avail. May 1989)............$49.00/$53.00   Vol(s). _____   Vol. _____

Annual Review of **ECOLOGY AND SYSTEMATICS**
    Vols. 2-16   (1971-1985)................$27.00/$30.00
    Vols. 17-18   (1986-1987)................$31.00/$34.00
    Vol. 19   (1988) ....$34.00/$38.00
    Vol. 20   (avail. Nov. 1989)............$34.00/$38.00   Vol(s). _____   Vol. _____

Annual Review of **ENERGY**
    Vols. 1-7   (1976-1982)................$27.00/$30.00
    Vols. 8-12   (1983-1987)................$56.00/$59.00
    Vol. 13   (1988) ....$58.00/$62.00
    Vol. 14   (avail. Oct. 1989)............$58.00/$62.00   Vol(s). _____   Vol. _____

Annual Review of **ENTOMOLOGY**
    Vols. 10-16, 18   (1965-1971, 1973)
        20-30   (1975-1985)................$27.00/$30.00
    Vols. 31-32   (1986-1987)................$31.00/$34.00
    Vol. 33   (1988) ....$34.00/$38.00
    Vol. 34   (avail. Jan. 1989)............$34.00/$38.00   Vol(s). _____   Vol. _____

Annual Review of **FLUID MECHANICS**
    Vols. 1-4, 7-17   (1969-1972, 1975-1985).......$28.00/$31.00
    Vols. 18-19   (1986-1987)................$32.00/$35.00
    Vol. 20   (1988) ....$34.00/$38.00
    Vol. 21   (avail. Jan. 1989)............$34.00/$38.00   Vol(s). _____   Vol. _____

Annual Review of **GENETICS**
    Vols. 1-19   (1967-1985)................$27.00/$30.00
    Vols. 20-21   (1986-1987)................$31.00/$34.00
    Vol. 22   (1988) ....$34.00/$38.00
    Vol. 23   (avail. Dec. 1989)............$34.00/$38.00   Vol(s). _____   Vol. _____

Annual Review of **IMMUNOLOGY**
    Vols. 1-3   (1983-1985)................$27.00/$30.00
    Vols. 4-5   (1986-1987)................$31.00/$34.00
    Vol. 6   (1988) ....$34.00/$38.00
    Vol. 7   (avail. April 1989)............$34.00/$38.00   Vol(s). _____   Vol. _____

Annual Review of **MATERIALS SCIENCE**
    Vols. 1, 3-12   (1971, 1973-1982)............$27.00/$30.00
    Vols. 13-17   (1983-1987)................$64.00/$67.00
    Vol. 18   (1988) ....$66.00/$70.00
    Vol. 19   (avail. Aug. 1989)............$66.00/$70.00   Vol(s). _____   Vol. _____

Annual Review of **MEDICINE**
    Vols. 9, 11-15   (1958, 1960-1964)
        17-36   (1966-1985)................$27.00/$30.00
    Vols. 37-38   (1986-1987)................$31.00/$34.00
    Vol. 39   (1988) ....$34.00/$38.00
    Vol. 40   (avail. April 1989)............$34.00/$38.00   Vol(s). _____   Vol. _____